PROBLEMS AND THEIR SOLUTION IN ORGANIC CHEMISTRY

I. L. Finar, Ph.D., D.Sc., A.R.I.C.

Principal Lecturer in Organic Chemistry,

The Polytechnic of North London, Holloway.

Longman

LONGMAN GROUP LIMITED
London
Associated companies, branches and representatives throughout the world

First published 1973
ISBN 0582 44223 0

PRINTED BY Unwin Brothers Limited
THE GRESHAM PRESS OLD WOKING SURREY ENGLAND

Produced by 'Uneoprint'

A member of the Staples Printing Group (G4407)

Contents

(The figures following each heading refer to the problem number)

Preface

The study of chemistry requires at least three processes: *learning*, i.e., acquiring knowledge (factual and theoretical); *understanding* of this knowledge; *application* of learning and understanding. A very good way of achieving these objectives is through the solving of problems.

The first part of this book is the collection of the problems set at the end of each chapter in the author's Textbook of Organic Chemistry, Vol. 1, Longman, 6th edn., 1972. There are 415 problems (which include about 1200 items), and they cover a wide range of topics described in the *Text*.

The second part of this book deals with possible solutions to these problems. Obviously, it is highly desirable that students, particularly those who work very much on their own, should have a means of checking their answers. It is, however, just as important that students should also have the opportunity of seeing the reasoning used to obtain a solution to a problem. I have tried to help the student in this respect, and this has meant that answers I have given are not necessarily in the form in which the student should present an answer. I have often introduced a section on the relevant facts before offering a solution. This serves the purpose of drawing attention to the more important points that are required to solve the problem under consideration, and will also help the student in his revision. Thus, I have discussed many points covered in the *Text*, often in more detail and, in a number of cases, in an argumentative fashion. This should make the student aware that opinions may differ about what the best solution may be. Too often the student thinks that there is only *one* solution to a particular problem. It is far better for him to offer several possibilities if he can and then to assess their relative merits. In this way he will increase his knowledge, understanding, application, and his power to discuss problems based on reasoned argument.

I have made no attempt to arrange the problems in order of difficulty. What may be easy to one student may be difficult to another, and vice versa. Nothing discourages a student more than that he should be told that he ought to be able to answer a given question because it is easy, and he finds he cannot give the answer.

The following points may be helpful. The student should carefully read the question and make sure he understands what is required. A point in this connection is well worth noting. The answer may appear so obvious that the student does not stop to think about it. It may well be that the problem is really very simple, but it may also be that there is something odd about it, i.e., the expected answer is the wrong one. The student should always think before he writes. If, after about five minutes, he cannot offer a solution, he should read the appropriate section in the *Text*, wait a few days, and then try again to answer the question. Only after this should he look up the solution. It is far better to have tried and failed than never to have tried at all. The student should also write out in full some of his answers to get practice in presenting his arguments in a clear manner.

In this book, reference to the *Text* is to Vol. 1 (6th edn.). If the reference is simply to the *Text*, then the relevant material is dealt with in the chapter corresponding to the problem being considered; otherwise, the chapter reference is given. Many cross-references are given in solutions to other solutions; these are indicated by specifying the chapter and the number of the problem.

Some students may prefer not to work through the problems chapter by chapter, but would rather solve problems of one type, e.g., problems involving the use of spectroscopy, synthetic problems, mechanistic problems. To help the student in this approach, I have given a list of Contents in which all types of problems have been classified under the following headings which are given in alphabetical order for each chapter: Acids and bases; Calculations; Energy profiles; Equations; Mechanisms; Nomenclature; Physical methods (i.r., u.v.,

NMR, mass spectrometry, dipole moments); Stereochemistry; Structure determination; Syntheses; Unclassified.

In some cases, problems appear under more than one heading. In each set of problems there is at least one which requires completion of equations. These will help the student to revise a large amount of subject matter in that chapter.

I. L. FINAR
1971

PROBLEMS

1 Determination of structure

(Solutions on page 65)

1. Calculate the percentage composition of the elements of $C_{14}H_{14}N_2O_4S$; C_8H_9NO; $C_8H_8ClNO_2S$; $C_{15}H_9BrO_2$.

2. Calculate the empirical formulae of the following compounds from the given elemental analyses:

	C%	H%	Br%	Cl%	N%	S%
(i)	40·1	6·7				
(ii)	42·6	3·5			20	
(iii)	73·8	5·3			16·4	
(iv)	60·6	4·8			11·8	9·0
(v)	51·2	4·0	21·35		14·95	
(vi)	24·3	4·0		71·7		

3. The sulphur content of cystine is 26·7 per cent. Given that cystine contains *two* sulphur atoms, what is the molecular weight of cystine?

4. How many proton signals would be expected in the NMR spectra of:
(i) $ClCH_2CH_2Cl$; (ii) CH_3OCH_3; (iii) $CH_3OC_2H_5$; (iv) $CH_3CHClC_2H_5$; (v) CH_3COCH_3; (vi) CH_3CH_2CHO; (vii) $CH_3CO_2CH_3$; (viii) $(CH_3)_2CHCH_2Br$; (ix) $CH_3CH(OCH_3)CH_2CH_2OH$?

5. Give, for each of the following compounds: (*a*) the number of signals and their relative intensities, (*b*) the multiplicity of each signal and the relative intensities of the components, (*c*) the relative field positions of the signals with respect to TMS:

(i) CH_3OCH_2Cl; (ii) CH_3CHO; (iii) $BrCH_2CHBr_2$; (iv) $CHCl_2CHBr_2$; (v) $CD_3CH_2CH_3$.

6. How many signals would be given by

(i) $CH_3OCH_2CH_2CH_2Br$;

(ii) $CH_3CHOHCH_3$ in the presence of a trace of acid? Using the simple rules of splitting, calculate the multiplicity of each signal. If overlapping of components of a multiplet can occur, in which signals will this happen? What now would be the number of signals, the multiplicities, and the relative intensities of the components in each signal? What ratios would be indicated by the integration trace?

7. Suggest molecular formulae, in terms of ^1H and ^{12}C, for the following fragment ions obtained from ethanol: *m*/*e* 25, 26, 28, 42.

8. The ion with the highest mass obtained from $CBrCl_3$ is 167. Discuss.

9. Calculate the double bond equivalents of the following compounds:

(i) $C_7H_6O_2$; (ii) C_3H_7N; (iii) $C_{10}H_7Cl$; (iv) $C_{12}H_{16}N_2O$; (v) $C_{13}H_9BrS$.

2 Properties of molecules

(Solutions on page 68)

1. Draw the electronic configurations (of the valency electrons) in:

(i) the carbon atom; (ii) CH_4; (iii) $CH_2{=}CH_2$; (iv) $CH{\equiv}CH$; (v) oxygen atom; (vi) H_2O; (vii) nitrogen atom; (viii) NH_3; (ix) NH_4^+.

Where bonds are involved, name their types, and describe the geometry of the molecules.

2. Draw the V.B. and M.O. structures of:

(i) CH_3COCH_3; (ii) CH_3CONH_2; (iii) $CH_2{=}CHCOCH_3$; (iv) $CH_3CO_2C_2H_5$; (v) CH_3NO_2.

3. Which of the following compounds have a dipole moment? Comment. Br_2, HI, CH_2Cl_2, $CHCl_3$, CCl_4, CH_3OH, CH_3OCH_3, $(CH_3)_2NH$, H_2O, CH_4, CO_2, BrCl, BF_3, CH_3COCH_3.

4. Draw the energy profiles for a three-step reaction in which (i) the first step is slowest and the last step fastest; (ii) the first step fastest and the last step slowest; (iii) second step slowest and last step fastest.

5. The order of bond lengths in the halogen acids is: HI > HBr > HCl > HF; the order of their dipole moments is: HF > HCl > HBr > HI. Offer an explanation.

6. Arrange the four halide ions in order of decreasing basicity. Explain your order.

7. $CH_4(g) + 2O_2(g) \longrightarrow CO_2(g) + 2H_2O(g)$

$\Delta H^{\ominus} = -802{\cdot}5\ \text{kJ mol}^{-1}\ (-191{\cdot}8\ \text{kcal mol}^{-1})$;

$\Delta G^{\ominus} = -800{\cdot}8\ \text{kJ mol}^{-1}\ (-191{\cdot}4\ \text{kcal mol}^{-1})$.

What is $\Delta S^{\ominus}$? Comment on its value.

8. Consider the reaction:

$A \rightleftharpoons B$	$\Delta H_f^{\ominus}$ kJ	$S^{\ominus}$ JK^{-1}
A	$-24{\cdot}27$	$246{\cdot}4$
B	$-25{\cdot}44$	$252{\cdot}3$

What is the composition of the equilibrium mixture at 25°C?

9. Arrange the acids with the following pK_a values in order of increasing strength: 6·3, 4·7, 9·5, −2·1, 15·8. Do the same for bases with pK_a values: 8·1, 4·2, 16, 10, 25; and for bases with pK_b values: 7, 11·4, 9·2, 4·7, 12·8.

3 Alkanes

(Solution on page 72)

1. $\Delta G_f^{\ominus}$ for the formation of n-butane is $-15{\cdot}69$ kJ mol^{-1}, and for isobutane is $-17{\cdot}99$ kJ mol^{-1}. Starting with 1 mole of n-butane, calculate the moles of n- and isobutane in the mixture after heating (in the presence of a catalyst) until equilibrium is obtained.

2. Write out all the possible structures for the formula C_6H_{14}. State how many primary, secondary, tertiary and quaternary carbon atoms there are in each isomer, and name the isomers by the I.U.P.A.C. system.

3. Complete the following equations:

(i) $Me_2CO \xrightarrow{?} Me_2CH_2$; (ii) $PrBr \xrightarrow{?} PrH$; (iii) $Me_3CH \xrightarrow{?} Me_3COH$;

(iv) $MeCH{=}CHMe \xrightarrow{?} MeCH_2CH_2Me$; (v.) $MeI + EtI \xrightarrow{Na} ?$; (vi) $EtMgI \xrightarrow{H_2O} ?$;

(vii) $EtCO_2^- + e \longrightarrow ?$; (viii) $EtCO_2^-Na^+ \xrightarrow{heat} ?$;

(ix) $EtCH{=}CH_2 \xrightarrow{B_2H_6} ? \xrightarrow[NaOH]{AgNO_3} ?$

4. An alkane has a M.W. of 72, and monochlorination produces one product only. What is the compound?

5. Draw Newman projection formulae for the eclipsed and staggered forms of: (*a*) ethanol; (*b*) 1-bromo-2-chloroethane; (*c*) propane; (*d*) 1, 2-dibromopropane.

6. The table lists a number of fragment ions and relative abundances for A, B and C, which are isomers of molecular formula C_5H_{12}. What are A, B, and C? Give reasons for your answer.

m/e	15	27	29	39	41	42	43	44	56	57	58	72
A(%)	7	41	46	21	68	87	100	33	17	55	24	6
B(%)	—	34	25	14	41	59	100	34	—	13	6	9
C(%)	5·3	15	—	13	42	—	—	—	—	100	44	—

7. How many signals can be expected in the NMR spectrum of $MeCHBrCH_2Br$? Give reasons for your answer.

N.B. The reader should test himself on the general methods of preparation and general properties of any homologous series.

4 Alkenes and alkynes

(Solutions on page 76)

1. Write out the structures (ignoring stereochemistry) of the isomeric pentenes, and name them by the IUPAC system. Give the structures of the products formed from each on ozonolysis.

2. Draw the energy profiles for the reactions:

(i) $MeCH{=}CH_2 + HX \longrightarrow MeCH_2CH_2^+ + X^-$

(ii) $MeCH{=}CH_2 + HX \longrightarrow Me\overset{+}{C}HMe + X^-$

Give reasons for your answers.

3. Using the Boord synthesis, how would you prepare:

(i) but-2-ene; (ii) hex-3-ene?

4. Draw only the geometrical isomeric forms of: (i) n-pentenes; (ii) isopentenes; (iii) monobromopropenes.

5. Can you compare the stabilities of pent-1-ene, *cis*- and *trans*-pent-2-enes and 3-methyl-but-1-ene by measuring their heats of hydrogenation? If not, why not? What method could you use?

6. Arrange the following carbonium ions in order of increasing stability, and give your reasons.

$Me\overset{+}{C}HMe$ $\quad CH_2{=}CHCH_2^+$ $\quad MeCH_2CH_2^+$

7. What alkenes could you expect from the following compounds? Give the experimental conditions:

(i) Me_3CCl; (ii) $Me_2C(OH)CMe_3$; (iii) Me_2CHCH_2OAC; (iv) $MeCH_2CH(OAc)Me$.

8. Addition of HCl to 3,3-dimethylbut-1-ene gives two isomeric alkyl chlorides. Explain.

9. Using bond energies (Table 2.2, p. 36), estimate $\Delta H^\ominus$ for the following reactions:

(i) $MeCH{=}CH_2 + HBr \longrightarrow MeCH_2CH_2Br$

(ii) $MeCH{=}CH_2 + HBr \longrightarrow MeCHBrMe$

What would you predict from the results? Are these predictions correct? If not, why not? Under what experimental conditions are the *equations* correct?

10. Arrange the following alkenes in order of increasing stability, and give reasons for your answer.

$CH_2{=}CH_2$ $\quad CH_2{=}CMe_2$ $\quad MeCH{=}CHMe$

$MeCH{=}CH_2$ $\quad Me_2C{=}CHMe$ $\quad Me_2C{=}CMe_2$

11. Draw the energy profiles for: (i) the hydrogenation of ethylene with and without a catalyst; (ii) the catalytic hydrogenation of *cis*- and *trans*-but-2-enes.

12. Isobutene, in the presence of H_2SO_4, forms a mixture of two isomeric alkenes C_8H_{16}. Formulate the reaction. How could you distinguish between the isomers chemically?

13. Formulate the reactions between but-1-ene, in the presence of a small amount of benzoyl peroxide, and (i) CCl_4; (ii) $CBrCl_3$. Give your reasons.

14. Draw the energy profile for the addition of HI to propene. Explain your diagram.

15. Compound A (C, 88·9%; H, 11·1%), M.W. 54, on catalytic hydrogenation, gives B (C, 82·8%; H, 17·2%). The i.r. spectrum of A shows a band in the region 2240 cm^{-1}. What is A?

16. Discuss the mechanisms of the following reactions:

(i) $MeCH{=}CH_2 + HOBr \longrightarrow MeCHOHCH_2Br$

(ii) $CCl_3CH{=}CH_2 + HOBr \longrightarrow CCl_3\check{C}HBrCH_2OH$

17. $\Delta G_f^{\ominus}$ for *cis*-but-2-ene is 65·86 kJ mol^{-1} and that for *trans*-but-2-ene is 62·97 kJ mol^{-1}. What is the value of K at 25°C for the equilibrium mixture? Does either isomer form this equilibrium mixture at 25°C? Discuss.

18. Explain the course of the following reaction:

$$CH_2{=}CHCHOHCH{=}CHMe \xrightarrow[H_2SO_4]{MeOH} CH_2{=}CHCH{=}CHCH(OMe)Me$$

19. Compound A (C, 87·1%; H, 12·8%), on catalytic reduction, gives B (C, 84·1%; H, 15·9%). Ozonolysis of A gives acetic acid, acetone, and pyruvic acid ($MeCOCO_2H$). What are A and B?

20. Write out the structures of the isomeric hexynes, and name them by the I.U.P.A.C. system.

21. Show how you could distinguish between ethane, ethylene, and acetylene by (i) chemical methods; (ii) i.r. spectroscopy; (iii) NMR spectroscopy.

22. Complete the following equations:

$$CH_3COCH_3 \xrightarrow{NaNH_2} ? \xrightarrow{C_2H_2} ? \xrightarrow{H^+} (CH_3)_2C(OH)C{\equiv}CH \xrightarrow[Pd]{1H_2} ? \xrightarrow[400°C]{Al_2O_3} ?$$

23. What are the geometries of:

(i) but-1-yne; (ii) but-2-yne; (iii) hept-2-en-5-yne?

24. Complete the following equations, and comment where necessary:

(i) $MeCH_2CH_2CMe{=}CHMe \xrightarrow{SeO_2} ?$

(ii) $MeCH_2C{\equiv}CBr + CH{\equiv}CMe \xrightarrow{Cu^+} ?$

(iii) $Me_2CHCH{=}CH_2 + HBr \longrightarrow ?$

(iv) $MeCH_2C{\equiv}CH \xrightarrow{Na/NH_3} ? \xrightarrow{EtBr} ?$

(v) $BrCH_2CH_2Br \xrightarrow{?} C_2H_2$

(vi) $MeCH_2CH_2OH \xrightarrow{?} MeCH{=}CH_2$

(vii) $MeCHBrCH_2Br \xrightarrow[MeOH]{Zn} ?$

(viii) 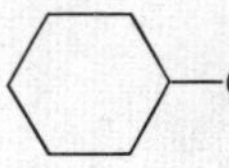 $-C{\equiv}CH \xrightarrow[H_2SO_4]{Hg^{2+}} ?$

25. Some of the mass spectra peaks given by the three methylpent-1-enes, A, B, and C are listed in the table. What are A, B, and C? Give reasons for your answer.

m/e	27	28	29	41	42	43	55	56	57	69	84
A(%)	48	7	40	79	21	7·5	100	39	—	71	30
B(%)	35	8	28	80	29	8	43	100	44	36	31
C(%)	33	4	6	72	33	100	8	44	20	13	11

26. How many NMR signals would you expect for the following compounds? If a signal is other than a singlet indicate this by (m):

(i) 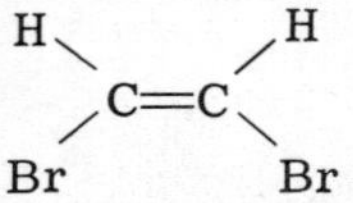

(ii) 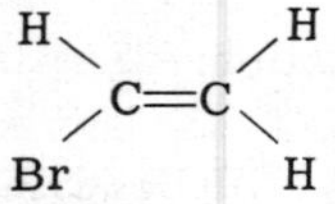

(iii) $CH{\equiv}CCH_2Br$

(iv) $CH_2{=}C{=}CH_2$ (v) $CH_3C{\equiv}CH$ (vi) 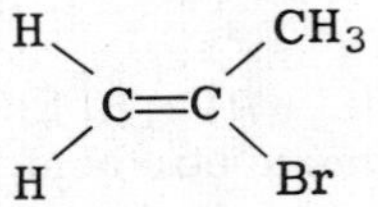

5 Halogen derivatives of the alkanes

(Solutions on page 85)

1. Write out the structures and IUPAC names of all the dichloro-derivatives of:

(i) butane; (ii) isobutane.

2. Distinguish between ethylene dichloride and ethylidene dichloride:

(i) chemically; (ii) by one physical method.

3. Draw all possible stereoisomers for:

(i) $MeCHBrCH{=}CH_2$; (ii) $MeCHOHCH{=}CHMe$; (iii) $MeCHClCO_2H$; (iv) $BrCH_2CH_2CO_2H$; (v) $Me_2C(OH)CO_2H$.

4. Explain the following reactions:

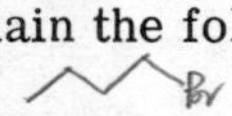
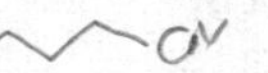

$$\text{(i) n-BuBr + KCN} \xrightarrow{EtOH-H_2O} \text{n-BuCN}$$

$$\text{(ii) t-BuBr + KCN} \xrightarrow{EtOH-H_2O} Me_2C{=}CH_2$$

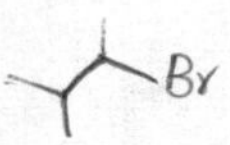

5. Explain the following reactions:

(i) $EtOH + HI \longrightarrow EtI$ (ii) $EtOH + HCN \longrightarrow$ no reaction

6. Calculate $\Delta H^{\ominus}$ for the reaction:

$$C_2H_6 + Br_2 \longrightarrow C_2H_5Br + HBr$$

If the reaction thermodynamically favourable? If so, will it proceed in the dark at room temperature? If not, why not?

7. Treatment of $Me_3CCH{=}CH_2$ and $Me_3CCHOHMe$ with concentrated hydrochloric acid gives the same two isomeric alkyl chlorides. What are these two products? Explain.

8. Complete the following equations:

(i) $ClCH_2CH_2CH_2Br + KCN \xrightarrow[\text{heat}]{EtOH-H_2O} ?$

(ii) $PhCHO \xrightarrow{SF_4} ?$

(iii) $BrCH_2CH{=}CHCO_2Me \xrightarrow{AgF} ?$

(iv) $\text{isoPrBr} \xrightarrow{?} PrH$

(v) $Me_2C{=}CH_2 + HCl \xrightarrow{\text{peroxide}} ?$

9. $EtCH_2CHBrMe \xrightarrow[(-HBr)]{EtOK} \underset{A}{\text{1-ene}} + \underset{B}{\text{2-ene}} + \underset{C}{\text{2-ene}}$

What are A, B, and C, and which one predominates? (Assume the mechanism is E2.)

10. Discuss the reaction: $ClCH_2CHCl_2 \xrightarrow{OH^-} CH_2{=}CCl_2$

11. The hydrolysis of n-BuCl in aqueous ethanol is accelerated in the presence of NaI. Explain, and draw the energy profiles for the uncatalysed and catalysed reactions.

12. What is the simplest alkane that is optically active?

13. The NMR spectrum of the compound $C_3H_5Cl_3$ showed two signals, one a doublet and the other a quintet. What is the structure of the compound?

14. State the number of signals and their multiplicity for the NMR spectra of:

(i) CH_2F_2; (ii) Me_3CCH_2Cl; (iii) Me_3CCl; (iv) $ClCH_2CHClCCl_3$; (v) $ClCH_2CCl_2CH_3$.

15. The table gives some of the peaks in the mass spectra of two isomeric straight-chain alkyl halides, A and B. One is a primary halide. What is this one? Give your reasons.

m/e	41	43	57	85	135	164	166
A(%)	38	100	22	49	50	2·3	2·3
B(%)	38	100	12	57	1	1	1

6 Monohydric alcohols

(Solutions on page 91)

1. Write out the structures and I.U.P.A.C. names of the isomeric secondary heptanols containing one methyl side-chain.

2. The dehydration of butan-2-ol with acid gives a mixture of 1- and 2-butenes. Write out the equations, state which predominates, and give your reasons. Draw the energy profiles of the reactions.

3. Convert ethanol into (i) n-propanol; (ii) n-butanol.

4. Arrange EtOH, CF_3CH_2OH, and CCl_3CH_2OH in order of increasing strength as acids and give your reasons.

5. The dehydration of n-BuOH with acid gives 1- and 2-butene. Explain this, state which product predominates, and why.

6. Convert propene into (i) n-PrOH; (ii) isoPrOH.

7. How could you show the existence of the following equilibrium?

$$ROH + H^+ \xrightleftharpoons{H_2SO_4 - H_2O} R^+ + H_2O$$

8. What are the disadvantages of preparing alkenes by:

(i) RX + base; (ii) ROH + acid ?

9. 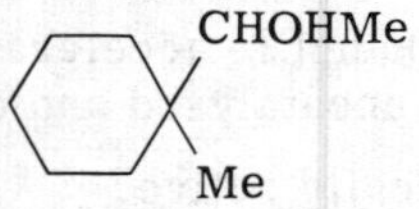

$\xrightarrow{H^+}$ $A(C_9H_{16})$

A, on ozonolysis, gives nonane-2, 8-dione. What is A and how is it formed?

10. Complete the following equation and comment:

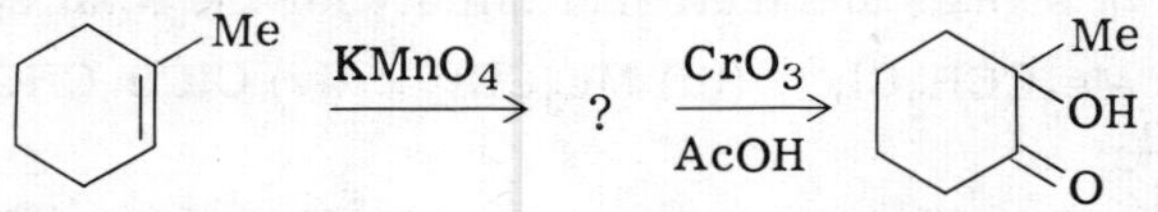

11. Convert (i) n-C_4H_9OH into MeC≡CMe; (ii) n-PrOH into *cis*-hex-2-ene.

12. A concentrated aqueous solution of HBr reacts with EtOH to give EtBr, but a concentrated aqueous solution of NaBr does not. Explain.

13. What are A to E in the following reactions?

$$(A)\ C_4H_{10}O \xrightarrow{SOCl_2} C_4H_9Cl\ (B)$$

$$(A)\ C_4H_{10}O \xrightarrow{H_2SO_4} C_4H_8(C)$$

$$(D)\ C_3H_6O_2 \xleftarrow[(ii)\ H_2O_2]{(i)\ O_3} C_4H_8(C) \xrightarrow{HCl} C_4H_9Cl\ (E)$$

14. Complete the following equations:

(i) $n\text{-}C_3H_7CO_2H \xrightarrow{?} n\text{-}C_4H_9OH$

(ii) $Me_2CO + EtMgI \longrightarrow ? \xrightarrow{H^+} ?$

(iii) $EtCO_2Et + 2MeMgI \longrightarrow ? \xrightarrow{H_2O} ?$

(iv) 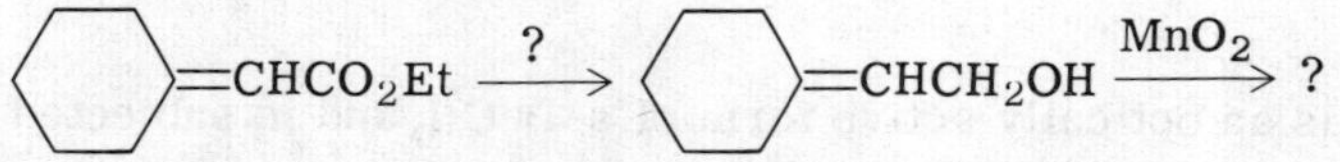

(v) $CH_3CH_2CH_2OH \xleftarrow{?} BrCH_2CH_2COCl \xrightarrow{?} BrCH_2CH_2CH_2OH$

15. How many NMR signals would be given by the following alcohols in the presence of a trace of acid? Indicate all signals other than singlets by (m).

(i) $CH_3CH_2CH_2OH$; (ii) Me_2CHCH_2OH; (iii) Me_3COH; (iv) Me_3CCH_2OH; (v) $MeCH_2CMe_2OH$.

16. The table lists some peaks in the mass spectra of the four isomeric butyl alcohols, A, B, C, D. Assign each alcohol to a letter, and give your reasons.

m/e	18	29	31	41	43	45	56	59	74
A(%)	—	34	100	63	60	7	86	—	1
B(%)	4	24	67	55	100	52	3	5	9
C(%)	—	15	22	11	10	100	1	18	0·2
D(%)	5	13	34	20	15	—	1·4	100	0·1

17. What reagents could you use for the following conversions?

(i) $MeCO(CH_2)_2CO_2Et \longrightarrow MeCHOH(CH_2)_2CO_2Et$

(ii) $HO_2C(CH_2)_4COCl \longrightarrow HO_2C(CH_2)_4CH_2OH$

(iii) $O_2N(CH_2)_2CN \longrightarrow O_2N(CH_2)_2CH_2NH_2$

(iv) $O_2N(CH_2)_2CH{=}CH_2 \longrightarrow H_2N(CH_2)_2CH{=}CH_2$

(v) $O_2N(CH_2)_2CH{=}CH_2 \longrightarrow O_2N(CH_2)_3CH_3$

(vi) $Me_2CHCOCl \longrightarrow Me_2CHCHO$

(vii) $O_2N(CH_2)_3CHO \longrightarrow O_2N(CH_2)_3CH_2OH$

7 Ethers

(Solutions on page 98)

1. The starting material is an optically active form of s-BuOH, and is subjected to the following reactions:

$$\text{s-BuOH} \xrightarrow{\text{Na}} ? \xrightarrow{\text{MeI}} ?$$

Complete the equations and state the configuration of the s-Bu group in each product. Explain your answer.

2. Which of the two reactions would you use to prepare Me_3COMe, and why?

(i) $Me_3CBr + MeO^-K^+ \longrightarrow$; (ii) $Me_3CO^-K^+ + MeBr \longrightarrow$.

3. Complete the following equations and comment:

(i) $\text{MeOEt} \xrightarrow{\text{HI}} ?$ (ii) $Et_2O \xrightarrow{\text{Na}} ?$

(iii) $Me_2C{=}CH_2 \xrightarrow[\text{pressure}]{H_2SO_4-H_2O} ? \xrightarrow[\text{pressure}]{Me_3COH} ?$

4. The compound $C_6H_{14}O$ showed two NMR signals. What is the compound?

5. The reaction between HI and C_2H_4 in EtOH gives predominantly EtI, whereas the reaction with HCl under the same conditions gives predominantly Et_2O. Explain.

6. Show the fragmentation paths whereby the ether, Et—O—CHMeEt, could produce ions *m/e*: 102, 87, 59, 57, 45, 31, 29.

8 Aldehydes and ketones

(Solutions on page 100)

1. Show, by equations, how acetaldehyde may be converted into:

(i) EtOH; (ii) CH_3COCl; (iii) CHI_3; (iv) MeCOEt; (v) CH_3CO_2Et; (vi) $CH_3CHOHCO_2H$; (vii) $EtNH_2$; (viii) $CH_3CHBrCHBrCO_2H$.

2. Convert $ClCH_2CH_2CHO$ into $HOCH_2CHOHCHO$.

3. Draw the energy profile for the following, and give your reasons.

$$CH_3COCH_3 + NaOH + Br_2 \longrightarrow CH_3COCH_2Br + H_2O + NaBr$$

4. Acetone, in dilute aqueous solution is 100 per cent unhydrated. When acetone is dissolved in water enriched with ^{18}O, recovered acetone contains ^{18}O. Explain.

5. Convert: (i) n-butanol into 2-ethylhexan-1-ol; (ii) bromocyclohexane into cyclopentanecarboxylic acid.

6. Calculate ΔH for each reaction (use Table 2·2, p. 36):

$$CH_3CHO + OH^- \rightleftharpoons \bar{C}H_2CHO + H_2O$$

$$CH_3COCH_3 + \bar{O}H \rightleftharpoons \bar{C}H_2COCH_3 + H_2O$$

Are these reactions equally thermodynamically favourable? Are the experimental results in agreement with your calculations? If not, why not?

7. Compound $C_5H_{10}O$ had a band at 1715 cm^{-1} in its i.r. spectrum, and showed two signals, a triplet and a quartet, in its N.M.R. spectrum. What is the compound?

8. One of the i.r. bands of compound **A**, C_3H_6O, shifts to a higher frequency when its solution in CCl_4 is progressively diluted. Write out possible acyclic structures [4] for the formula C_3H_6O, and suggest which is most likely for **A**. Using i.r. and N.M.R. spectroscopy, how would you distinguish between the four structures?

9. The bromination of acetone is catalysed by acids, and the rate law is zero order with respect to bromine. Discuss this statement.

10. Suggest what compounds the letters stand for in the following equations:

(i) $MeCOMe + CH{\equiv}CH \xrightarrow[\text{(ii) A}]{\text{(i) NaNH}_2} B$

(ii) $PhCOMe + ClCH_2CO_2Et \xrightarrow{NaNH_2} C$

(iii) 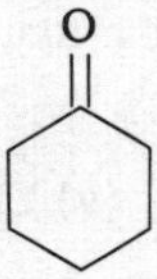 $+ ClCH_2CO_2Et \xrightarrow[\text{t-BuOH}]{\text{t-BuOK}} D$

(iv) $CH_3COC_2H_5 + EtONO \xrightarrow[\text{gas}]{HCl} E$

(v) $Et_2CO + NH_4Cl + NaCN \xrightarrow{NH_4OH} F$

11. Complete the following equations and give reasons for your answer.

(i) $Me_2C(OH)CH_2OH \xrightarrow{H^+} ?$ (ii) $PhMeC(OH)C(OH)MePh \xrightarrow{H^+} ?$

12. Describe paths that could lead to the formation of the following fragment ions:

(i)

m/e	29	43	44	71	72
n-Butyraldehyde (%)	54	79	100	5	73
Isobutyraldehyde (%)	23	100	—	1	53

(ii)

m/e	29	43	57	58	71	86
Pentan-2-one (%)	16	100	26	9	9	25
Pentan-3-one (%)	99	—	100	—	1	17

9 Saturated monocarboxylic acids

(Solutions on page 111)

1. Convert n-$C_5H_{11}CO_2H$ into n-$C_4H_9CO_2H$ by three methods.

2. Which of the following would be most and least readily hydrolysed with NaOH, and why?

(i) $MeCO_2Me$; (ii) Me_2CHCO_2Me; (iii) $MeCO_2Bu^t$

3. Using MeI, $\overset{*}{M}eI$, NaCN, $Na\overset{*}{C}N$ ($\overset{*}{C}={}^{14}C$) and any other chemicals, suggest a synthesis for each of the following:

(i) $\overset{*}{C}H_3CO_2H$; (ii) $CH_3\overset{*}{C}O_2H$; (iii) $\overset{*}{C}H_3\overset{*}{C}O_2H$; (iv) $\overset{*}{C}H_3COCH_3$; (v) $\overset{*}{C}H_3\overset{*}{C}HO$.

4. Complete the equations and comment:

(i) $CHCl_2COCl \xrightarrow{LAH} ?$ (ii) $HOCH_2CH_2CN \xrightarrow{HBr-H_2O} ?$

(iii) $B_{AC}2$ mech. $Me_3CCH_2OAc + OH^- \longrightarrow ?$

(iv) $B_{AL}1$ mech. $Me_3CCH_2OAc + OH^- \longrightarrow ?$ (v) $ClCH_2CO_2H + NaI \xrightarrow{acetone} ?$

(vi) $Me_2CBrCONH_2 \xrightarrow[NaOH]{Br_2} ?$

5. Calculate $\Delta G^{\ominus}$ for the equilibrium:

$$AcOH + EtOH \rightleftharpoons AcOEt + H_2O$$

moles $\frac{1}{3}$ $\frac{1}{3}$ $\frac{2}{3}$ $\frac{2}{3}$

What information does the value of $\Delta G^{\ominus}$ give on the time required for this equilibrium to be reached?

6. Draw the energy profile for the $A_{AC}2$ esterification mechanism and comment.

7. A solution of Ph_3CCO_2H in conc. H_2SO_4 gives $MeOCPh_3$ when poured into methanol. Discuss.

8. Convert: (i) n-BuOH into n-$PrNH_2$; (ii) Et_2CHCO_2H into Et_2CHCHO.

9. Can $MeCH_2COCl$, $MeCHClCO_2H$ and $ClCH_2CH_2CO_2H$ be differentiated by N.M.R. spectroscopy? Give reasons for your answer.

10. Draw the diagrammatic N.M.R. spectra of: (i) $MeCO_2Et$; (ii) $EtCO_2Me$.

11. Give the fragmentation paths that could produce the following ions:

(i)

m/e	31	43	59	71	74	102
Me n-butyrate (%)	6	100	25	55	69	2

(ii)

m/e	16	43	44	59	71	87
n-Butanamide (%)	3	40	94	100	6	3

10 Polycarbonyl compounds

(Solutions on page 118)

1. Draw the energy profiles for tautomerism between A and B and for resonance between A and B.

2. Explain the following equation:

$$MeCOCH_2CO_2Et + MeCOCHR^1CO_2Et \xrightarrow[\text{(ii) } 1R^2X]{\text{(i) } 1EtONa}$$

$$MeCOCHR^2CO_2Et + MeCOCHR^1CO_2Et$$

3. How could you differentiate, by i.r. and N.M.R. spectroscopy, between A, $MeCOCH_2CH_2COMe$ and B, $MeCOCH_2COCH_2CH_3$?

4. Starting with E.A.A. or D.E.M., give *one* synthesis (using equations) for: (i) 4-methylpentan-2-one; (ii) 2, 3-dimethylbutanoic acid; (iii) methylsuccinic acid; (iv) cyclopentane-1, 2-dicarboxylic acid; (v) $PhCOCH_2CO_2Et$; (vi) 3-carboxy-2-methylpentanedioic acid.

5. The β-keto-acid shown is decarboxylated with difficulty. Suggest an explanation.

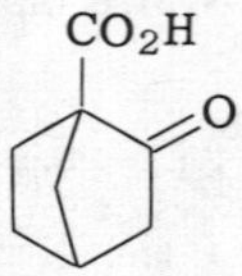

6. What *two* compounds will condense to give each product shown (after acidification)?

(i) $A + B \xrightarrow{EtONa} MeCOCH_2COCO_2Et$

(ii) $C + D \xrightarrow{EtONa} PhCH(CN)COCO_2Et$

(iii) $E + F \xrightarrow{EtONa} PhCH(CN)COMe$

7. Starting with E.A.A., synthesise 3-methylpentan-2-ol.

8. Complete the equations:

(i) $CH_2(CO_2Me)_2 + ? \xrightarrow[\text{(ii) AcOH}]{\text{(i) Na}} CH(CO_2Me)_3$

(ii) $MeCH(CO_2Et)COCO_2Et \xrightarrow{\text{heat}} ?$

(iii) $MeCH_2CO_2Et + ? \xrightarrow[\text{(ii) AcOH}]{\text{(i) EtONa}} MeCH(CHO)CO_2Et$

(iv) $MeCH_2CO_2Et + (CO_2Et)_2 \xrightarrow[\text{(ii) AcOH}]{\text{(i) EtONa}} ? \xrightarrow{\text{heat}} ?$

9. Suggest a synthesis for ethyl 4-keto-3-methylpentanoate.

10. A compound $C_4H_6O_2$, shows a very strong i.r. band at 1721 cm^{-1} and its N.M.R. spectrum shows only one singlet signal. What is the compound?

11 Polyhydric alcohols

(Solutions on page 124)

1. Write out the structures of all the possible isomeric diols derived from the butanes, and indicate which can be oxidised with periodic acid.

2. Arrange the following in order of increasing b.p., and give your reasons. $(CH_2OH)_2$, $(CH_2OMe)_2$, $HOCH_2CH_2OMe$.

3. Complete the following equations:

(i) $HOCH_2CH_2OH + ? \xrightarrow{\text{pyridine}}$ 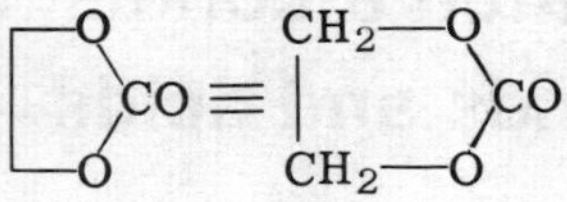

(ii) $Me_2CO + ? \xrightarrow{H^+}$ 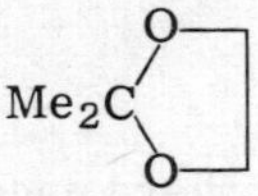

(iii) $MeCHO + HOCH_2CH_2SH \xrightarrow{H^+} ?$

(iv) $R_2C(OH)CH_2R \xleftarrow{?}$ 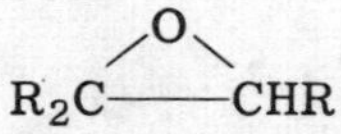 $\xrightarrow{?} R_2CHCHOHR$

(v) 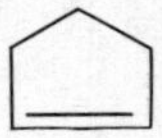 $\xrightarrow[\text{(ii) } H^+]{\text{(i) } HCO_3H} ?$

(vi) $CH_2{=}CH(CH_2)_2CH{=}CH_2 \xrightarrow[\text{(ii) } H_2O_2;\ OH^-]{\text{(i) } B_2H_6} ?$

(vii) $(CH_2)_8(CO_2Et)_2 \xrightarrow{?} (CH_2)_8(CH_2OH)_2$

4. Write out the order of rates of oxidation with periodic acid of the following diols, and give your reasons: $HOCH_2CH_2OH$, MeCHOHCHOHMe, Me_2COHCH_2OH, $Me_2C(OH)C(OH)Me_2$. $Me_2COHCHOHMe$.

5. Convert : 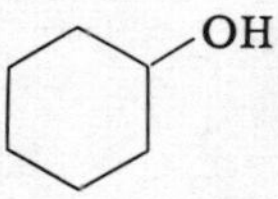 into 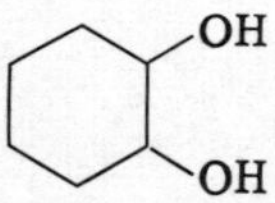 (*cis*) and 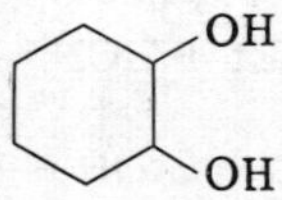 (*trans*)

6. Suggest a mechanism for the conversion of ethylene glycol into dioxan by conc. H_2SO_4.

7. How many N.M.R. signals can be expected for:

(i) 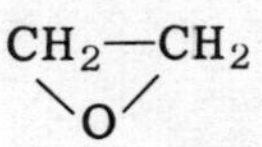

(ii) 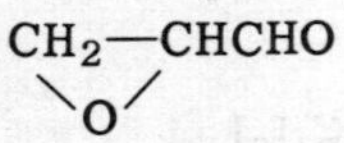

(iii) 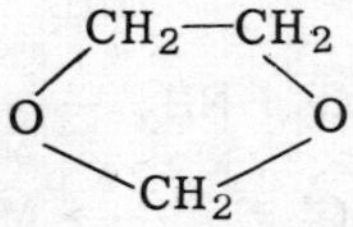 ?

8. Tetramethylethylene, on oxidation with acid dichromate, gives the compound $C_6H_{12}O$. This compound showed an i.r. band at about 1700 cm^{-1}, and undergoes the haloform reaction. Suggest a structure for this compound, and give your reasons.

12 Unsaturated alcohols, ethers, carbonyl compounds and acids

(Solutions on page 128)

1. Complete the following equations and comment where necessary:

(i) $Me(C{\equiv}C)_3COCH{=}CH_2 \xrightarrow{NaBH_4} ? \xrightarrow{H^+} ?$

(ii) $\overbrace{CH_2{-}CH_2}^{O} + HCN \longrightarrow ? \xrightarrow[300^\circ C]{Al_2O_3} ?$

(iii) $CH_2{=}CHCl + Cl_2 \longrightarrow ? \xrightarrow[Ca(OH)_2]{aq.} ?$

(iv) $CH_2{=}CHCN + EtOH \xrightarrow{H_2SO_4-H_2O} ?$

(v) $CH_2{=}CHCOCH_3 \xrightarrow{NaOCl} ?$

(vi) $MeCH{=}CHCOMe \xrightarrow[H_2(1\ atm)]{Pd-C} ?$

(vii) $MeCH{=}CHCOEt \xrightarrow[press.]{Ni;\ H_2} ?$

(viii) $MeCH{=}CHCO_2Et + HCN \longrightarrow ?$

(ix) $CH_2{=}CHCO_2Et + EtOH \xrightarrow{EtONa} ?$

(x) $CH_2{=}CHCOMe \xrightarrow{H_2O_2 : OH^-} ?$

(xi) $CH{\equiv}CNa + Cl(CH_2)_3I \xrightarrow[NH_3]{liq.} ? \xrightarrow[(ii)\ H^+]{(i)\ NaCN} ?$

(xii) $MeCH{=}CHCOMe + NH_3 \longrightarrow ?$

(xiii) $MeCH{=}CHCH_2Cl + ? \xrightarrow{?} MeCH{=}CHCH_2I$

(xiv) $MeCH{=}CHCO_2Et + CH_3NO_2 \xrightarrow{EtONa} ?$

(xv) $MeCH{=}CHCHO + HCN \xrightarrow{OH^-} ?$

(xvi) $MeCH{=}CHCHO \xrightarrow{LAH} ?$

(xvii) $PhCHO + ? \xrightarrow{pyridine} PhCH{=}CHCO_2H$

(xviii) $EtCHO + 2CH_2(CO_2Et)_2 \xrightarrow{piperidine} ?$

(xix) $EtCHO + CH_2(CO_2H)_2 \xrightarrow{pyridine} ?$

(xx) $EtCH{=}CHCH_2CONH_2 \xrightarrow{NaOCl/NaOH} ? \xrightarrow{HCl} ?$

(xxi) 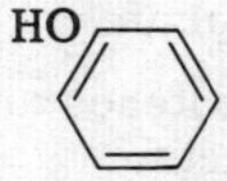 $NH_2 + CH_2{=}CO \longrightarrow$?

2. A compound has structure $MeCH{=}CHCH_2CHO$ (A) or $MeCH_2CH{=}CHCHO$ (B). Distinguish between the two by (i) a chemical method; (ii) i.r.; (iii) NMR

3. Explain the following:
(i) μ of acraldehyde > μ of propionaldehyde; (ii) μ of ethyl bromide > μ of vinyl bromide.

4. Complete the following equation for:
(i) $B_{AC}2$; (ii) $B_{AL}1$; (iii) $A_{AC}1$ mechanism.

$$CH_2{=}CHCHMeOOCCH_3 \xrightarrow{OH^- \text{ or } H^+} ?$$

5. How could you demonstrate the S_N1' mechanism in the reaction

$$CH_2{=}CHCH_2OH + HCl \longrightarrow CH_2{=}CHCH_2Cl?$$

6. Calculate $\Delta H^\ominus$ for the 1, 2- and 1, 4-additions of HZ to C=C—C=O. Which addition is favoured?

7. Draw the M.O. for the group C=C—C=O, and suggest a reason for the 1, 4-addition properties.

8. Convert $MeCH{=}CHCO_2H$ into $MeCH(CH_2CO_2Et)_2$.

9. Synthesise $EtO_2CCH{=}CHCH_2COCO_2Et$ starting from compounds not containing more than *four* carbon atoms.

10. Aldol may undergo dehydration in two possible ways:

$$CH_2{=}CHCH_2CHO \xleftarrow{-H_2O} MeCHOHCH_2CHO \xrightarrow{-H_2O} MeCH{=}CHCHO$$

One of these products is obtained almost exclusively. Which one and why?

11. 2, 2-Dimethylbut-3-enoic acid, on warming, gives trimethylethylene. Suggest a mechanism.

12. Suggest a synthesis for:

(i) $\begin{matrix} Me_2CCH_2CO_2H \\ | \\ Me_2CCH_2CO_2H \end{matrix}$; (ii) $Me_2C(CH_2CO_2H)_2$; (iii) $CH_3CH{=}CHCH_2CH_2OH$.

13 Nitrogen compounds

(Solutions on page 134)

1. Write out the structures and names of the isomeric amines of formula $C_4H_{11}N$.

2. Of a group of isomeric amines, the t-amine has the lowest b.p. Explain.

3. Deamination of n-$BuNH_2$ with $NaNO_2/HCl$ gives two butanols, two butenes and two butyl chlorides. Give a possible mechanism for the formation of these products.

4. What products would you expect to get by the application of the Hofmann exhaustive methylation to:

(i) $Me_2CHCH_2CH_2NH_2$; (ii) Me_2CHCH_2NHMe; (iii) $Me_2CHCH_2NHCH_2CH_2Me$; (iv) $EtNHCH_2CH_2Cl$?

5. RCONHMe does not undergo the Hofmann amine preparation reaction. Offer an explanation.

6. Complete the following equations:

(i) $CH_2{=}CHCN + PhCH_2OH \xrightarrow{H_2SO_4} ?$

(ii) $PhCH_2COCHN_2 + HCl(g) \xrightarrow{Et_2O} ?$

(iii) $NH_2CONH_2 + n\text{-}BuOH \xrightarrow{heat} ?$

(iv) $MeNH_2 + ? \longrightarrow MeNHCO_2Et$

(v) $HOCH_2CH_2CN \xrightarrow[heat]{aq.\ HBr} ?$

(vi) 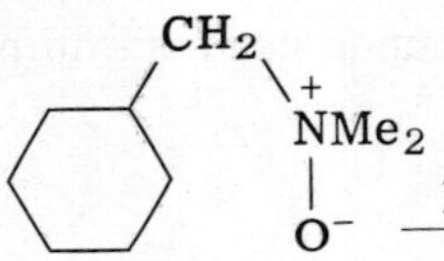

$\xrightarrow{Heat} ?$

(vii) $PhCH(CN)CH(CN)CO_2Et \xrightarrow[heat]{aq.\ HCl} ?$

(viii) 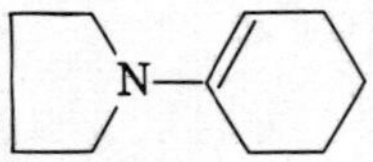

$+ ClCO_2Et \xrightarrow{Et_3N} ? \xrightarrow{H^+} ?$

(ix) 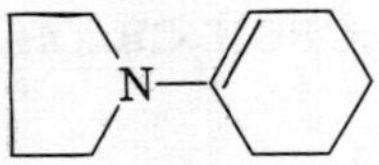

$+ CH_2{=}CHCN \longrightarrow ? \xrightarrow[warm]{H^+} ?$

(x) $Me_2CHNO_2 + HCHO + EtNH_2 \xrightarrow{HCl} ?$

7. Discuss the reactions:

(i) $R_2C(OH)CH_2CR_2NH_2 \xrightarrow{HNO_2} R_2CO + R_2C{=}CH_2$

(ii) $PhCH{=}NR \xrightarrow{MeI} ? \xrightarrow{H^+}$ aldehyde + s-amine

8. Devise an experiment to show whether the Me group in n-$PrNH_2$ undergoes migration during deamination.

9. The NMR spectrum of Me_2NNO shows two peaks at room temperature and one peak at ~200°C. Offer an explanation.

10. t-$BuNH_2$ and neoPenNH_2 cannot be prepared by the action of NH_3 on the corresponding alkyl bromide. Why not? Each can be prepared from a carboxylic acid. How?

11. Complete the following equations:

(i) $Me_2C(CN)NHNHC(CN)Me_2 \xrightarrow{HOCl} ?$

(ii) $MeNHNHMe \xrightarrow{HgO} ?$

(iii) $Me(CH_2)_2Me \xrightarrow[\sim 500°C]{HNO_3} ?$

(iv) $RCH{=}CH_2 \xrightarrow{B_2H_6} ? \xrightarrow{ClNH_2} ?$

(v) Me_3N + D-EtMeCHCl $\longrightarrow ?$

(vi) $PhNCO \xrightarrow{H_2O} ? \xrightarrow{-CO_2} ? \xrightarrow{PhNCO} ?$

(vii) $CH_2(CO_2Et)_2 \xrightarrow[\text{(ii) BrCN}]{\text{(i) EtONa}} ?$

(viii) $EtNH_2 + KCN + Br_2 \xrightarrow{KOH} ?$

12. $MeCOCH_2Cl + CH_2N_2 \longrightarrow$ (A) C_4H_7ClO + (B) C_4H_7ClO
(A) is a ketone and (B) is an epoxide. Suggest structures for (A) and (B) and mechanisms for their formation.

13. Complete and comment ($\overset{*}{C}={}^{13}C$):

$RMgBr + \overset{*}{C}O_2 \longrightarrow ? \xrightarrow{H^+} ? \xrightarrow[\text{(ii) } CH_2N_2]{\text{(i) } SOCl_2} ? \xrightarrow{Ag_2O} ? \xrightarrow{H_2O} ?$

14. Amine (A), $C_6H_{15}N$, on treatment with MeI and then KOH, gives (B), $C_8H_{20}N^+OH^-$. This, on heating, produces isobutene and amine (C), $C_4H_{11}N$. What is (A)?

15. Complete the following equations:

(i) $R_2CO + Ph_3P{=}NH \longrightarrow ?$

(ii) $Ph_3CCl + RNH_2 \longrightarrow ?$

(iii) $EtNO_2 + 2CH_2O \xrightarrow{KHCO_3} ?$

(iv) $RCH_2NH_2 + 2CH_2O + 2HCO_2H \longrightarrow ?$

(v) $CH_2{=}CO + HCN \longrightarrow ? \xrightarrow{CH_2=CO} ?$

(vi) $Me_2C{=}CHCOMe + RCH_2NH_2 \longrightarrow ? \xrightarrow{HNO_2} ? \xrightarrow{EtONa} ?$

(vii) $EtCHO + CH_2N_2 \longrightarrow ?$

(viii) $MeCOEt + CH_2N_2 \longrightarrow ?$

(ix) $? \xleftarrow{MeBr} Et_3\overset{+}{N} - CH_2^- \xrightarrow{Me_2CO} ?$

(x) 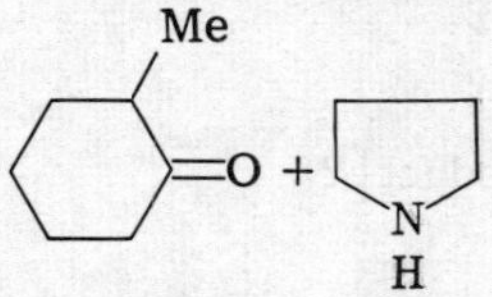 $\xrightarrow{TsOH} ?$

(xi) $CH_2{=}CHCOMe + CH_2N_2 \longrightarrow ?$

16. Discuss the NMR spectrum of $MeCH_2CH_2NO_2$ and give the relative positions of the signals.

17. Convert: (i) Me_2CHCO_2H into $Me_2CH(CH_2)_6CO_2H$; (ii) $(CH_2)_3Br_2$ into $Et_2N(CH_2)_3NH_2$; (iii) $O_2NCH_2CH_2CN$ into $(CH_2)_3(NH_2)_2$; (iv) $O_2NCH_2CH_2CN$ into $O_2NCH_2CH_2CH_2NH_2$.

18. How many NMR signals would you expect for:

(i) $MeCONHCH_2CO_2H$; (ii) $MeCH{=}CHCN$; (iii) $MeCH_2CH(NH_2)Me$; (iv) $(MeCO_2)_2NMe$?

19. Show the fragmentation paths that would account for the ions m/e:

(i) 30 and 58 from ethyl-n-propylamine; (ii) 30, 58, and 86 from triethylamine; (iii) 41 from n-butyl cyanide.

14 Aliphatic compounds of sulphur, phosphorus, silicon and boron

(Solutions on page 144)

1. The reaction $RX + RS^-Na^+ \longrightarrow R_2S + NaX$ is faster than the reaction $RX + RO^-Na^+ \longrightarrow R_2O + NaX$. Explain.

2. Complete the following equations:

(i) $CH_2{=}CHCMe{=}CH_2 + SO_2 \longrightarrow ?$

(ii) $MeNH_2 + CS_2 + NaOH \xrightarrow[\text{heat}]{H_2O} ?$

(iii) $MeCONH_2 + P_2S_5 \xrightarrow{C_6H_6} ?$

(iv) $CH_2{=}CHCN + HOCH_2CH_2SH \xrightarrow{\text{heat}} ?$

(v) $\underset{\diagdown O \diagup}{CH_2{-}CH_2} + H_2S \longrightarrow ?$

(vi) $MeCH{=}CHCH_2SCN \xrightarrow{\text{heat}} -CH(NCS)-$[part of product] ?

(vii) $CH_2{=}CH_2 + (SCN)_2 \longrightarrow ?$

(viii) $Me_2C(SEt)_2 \xrightarrow{KMnO_4} ?$

(ix) $PhSO_2Me \xrightarrow{EtONa} ? \xrightarrow{PhCO_2Me} ? \xrightarrow{-OMe^-} ?$

(x) $MeCOCl + EtSH \longrightarrow ?$

(xi) $(RS)_2Hg \xrightarrow{heat} ?$

(xii) $Et_2S + H_2SO_4 \longrightarrow ? \xrightarrow{H_2O} ?$

(xiii) $Et_2S + BrCN \longrightarrow ?$

3. Name the following compounds:
(i) $PhPO(OH)_2$; (ii) $HOP(OEt)_2$; (iii) $EtPCl_2$; (iv) $MePOCl_2$; (v) $H_2PCH_2CH_2Cl$; (vi) $H_3P{=}NH$; (vii) $(HO)POCl_2$; (viii) $POCl_2(NH_2)$; (ix) $HPOCl(OH)$; (x) H_2POCl; (xi) $MeHPO(NH_2)$; (xii) Et_2PCl.

4. Suggest a mechanism for the removal of oxygen from epoxides by Ph_3P to form alkenes (use ethylene oxide as your example).

5. Suggest a preparation for each of the following:
(i) Ph_3P; (ii) MeEtPH; (iii) Et_3PO; (iv) diethylphosphinous chloride; (v) diethyl methylphosphonate; (vi) dimethylphosphonic chloride; (vii) ethylphosphonic diamide.

6. Complete the following equations:

(i) $PhCH{=}CHCH_2PPh_3{}^+Cl^- + PhCHO \xrightarrow{EtOLi}$

(ii) 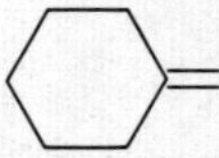 $=O + Ph_3P{=}CH_2 \longrightarrow ?$

(iii) $Ph_3P{=}CCl_2 + p\text{-}Me_2NC_6H_4CHO \longrightarrow ?$

(iv) $Ph_3P + CHCl_3 \xrightarrow{t\text{-}BuOK} ?$

(v) $Ph_3P + ClCH_2COMe \longrightarrow ? \xrightarrow{Na_2CO_3} ? \xrightarrow{PhCHO} ?$

(vi) $PhCO(CH_2)_4Br \xrightarrow{Ph_3P} ? \xrightarrow[EtOH]{EtONa} PhCO(CH_2)_3CH{=}PPh_3 \longrightarrow ?$

(vii) $PhNO + Ph_3P{=}CHPh \longrightarrow ?$

(viii) $Et_2PCl + Et_2PNa \longrightarrow ?$

(ix) $Ph_3P + Et_3\overset{+}{N}{-}\overset{-}{O} \longrightarrow ?$

(x) $R^1_3P + R^2_2SO \longrightarrow ?$

(xi) $Ph_3P{=}CH_2 + MeCH{=}CHCOMe \longrightarrow ?$

(xii) $(EtO)_3P + Cl_2 \longrightarrow ?$

(xiii) $(EtO)_3P + Me_3\overset{+}{N}{-}\overset{-}{O} \longrightarrow ?$

(xiv) $Me_2SO + Ph_3P \longrightarrow ?$

(xv) $Ph_4P^+I^- + PhLi \longrightarrow ?$

7. Name the following compounds:
(i) $H_3Si(SiH_2)_2SiH_3$; (ii) $MeHSi(NH_2)_2$; (iii) $EtSi(OH)_3$; (iv) Me_2SiCl_2; (v) $Et_2SiHOOCMe$; (vi) $EtOSiH_2(OSiH_2)_3SiH_2NH_2$; (vii) $SiH_3NMeSiH(OEt)NHSiH_3$.

8. Suggest a synthesis for each of the following:
(i) n-Pr_4Si; (ii) $MeSiH_3$; (iii) $Et_3SiSiEt_3$; (iv) Me_3SiCH_2Cl; (v) Me_3SiNEt_2; (vi) $Me_3SiNMeSiMe_3$; (vii) Me_3SiEt.

9. Complete the following equations:

(i) $MeCH{=}CHMe + Me_2SiHCl \xrightarrow{\text{cat.}} ?$

(ii) $2CH_2{=}CHCH_2Cl + Si \xrightarrow{Cu} ?$

(iii) $Me_3SiCl + BrCl \longrightarrow ?$

(iv) $Me_3SiCl \xrightarrow{H_2O} ?$

(v) $Et_3SiCl + Me_3SiOH \longrightarrow ?$

(vi) $Me_3SiH + NaOH \xrightarrow{H_2O} ?$

(vii) $Me_3SiCH_2CHClCH_3 + OH^- \longrightarrow ?$

10. Write out the structures of:

(i) *B*-trichloroborazine; (ii) dimethoxyborane; (iii) methyl dimethylborinate; (iv) diethyl methylboronate; (v) trimethyl borate; (vi) ethoxy(methoxy)borane; (vii) acetoxyborane.

11. Write out the structures of the *five* possible methyldiboranes (mono, di, etc.). Comment.

12. Complete the following equations:

(i) $B(OMe)_3 + 3EtMgBr \longrightarrow ?$

(ii) $3PhMgBr + BF_3 \longrightarrow ?$

(iii) $Et_3B + Br_2 \longrightarrow ?$

(iv) $2(MeO)_3B + NaH \longrightarrow ?$

(v) $Me_3B + HCl \longrightarrow ?$

(vi) $EtBCl_2 + 2MeOH \longrightarrow ?$

(vii) $B(OMe)_3 + 2EtMgBr \longrightarrow ? \xrightarrow{H_2O} ?$

(viii) $MeB(OH)_2 \xrightarrow{\text{heat}} ?$

13. Suggest mechanisms for the following rearrangements:

(i) $EtSCHMeCH_2OH \xrightarrow{HCl} EtSCH_2CHClMe$

(ii) $Ph_3P + ROOR \longrightarrow Ph_3PO + R_2O$ [*Not* free radical]

14. Convert:

(i) 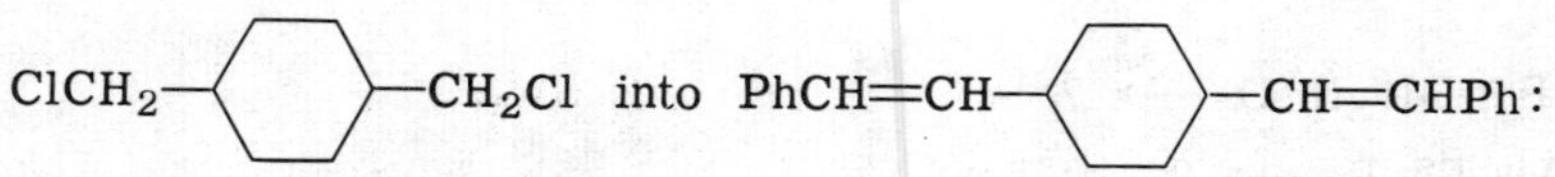

(ii) EtMeCHBr into EtMeC=$CHPr^n$.

15. Complete the following equations:

(i) $Ph_2CO + Ph_3P{=}CHOMe \longrightarrow ? \xrightarrow{H^+} ?$

(ii) $Ph_2CN_2 + SO_2 \longrightarrow\ ? \xrightarrow{ROH}\ ?$

(iii) $MeSC{\equiv}CH + HCl \longrightarrow\ ?$

(iv) $(R^1O)_3P + R^2COCl \longrightarrow\ ?$

(v) $Et_3B + Et_3N \longrightarrow\ ?$

(vi) $Et_3B + O_2(H_2O) \longrightarrow\ ?$

(vii) $B(OMe)_3 + 2EtMgBr \longrightarrow\ ? \xrightarrow{H_2O}\ ?$

(viii) $EtB(OH)_2 \xrightarrow{heat}\ ?$

(ix) $?\ \xleftarrow{EtOH} \underset{\diagdown S \diagup}{CH_2{-}CH_2} \xrightarrow{Me_2NH}\ ?$

(x) $(CH_2SH)_2 + R_2CO \xrightarrow[BF_3.Et_2O]{AcOH}\ ? \xrightarrow[H_2]{Ni}\ ?$

(xi) $MeP(OEt)_2 + EtI \longrightarrow\ ?$

(xii) $Et_2PNEt_2 + MeI \longrightarrow\ ?$

(xiii) $R^1SSR^2 \xrightarrow[Ni]{Raney}\ ?$

(xiv) $3BCl_3 + 3EtNH_2 \xrightarrow{heat}\ ?$

16. Each of the following compounds showed only one singlet line in its N.M.R. spectrum: (i) $C_2H_3F_3O_2S$; (ii) $C_2H_4O_4S$. Suggest a structure for each compound.

15 Organometallic compounds

(Solutions on page 153)

1. Comment on the following equations:

$$HOCH_2CH_2COCH_2CH_2CO_2Et \xrightarrow{1RMgX}\ ? \xrightarrow{1RMgX}\ ? \xrightarrow{2RMgX}\ ?$$

2. Complete the following equations and comment where necessary:

(i) $n\text{-}C_4H_9MgBr +\ ? \xrightarrow[(ii)H^+]{(i)\,Et_2O} n\text{-}C_7H_{15}OH$

(ii) $3EtMgBr + (EtO)_2CO \xrightarrow[\text{(ii)aq. } NH_4Cl]{\text{(i)}Et_2O} ?$

(iii) $PhMgBr + MeOCH_2CN \xrightarrow[\text{(ii)}H^+]{\text{(i)}Et_2O} ?$

(iv) $C_6H_{11}MgBr + BrCH_2CBr{=}CH_2 \xrightarrow{Et_2O} ?$

(v) $CH_2{=}CO + EtMgBr \xrightarrow[\text{(ii)}H^+]{\text{(i)}Et_2O} ?$

(vi) $\underset{\text{(epoxide)}}{MeCH{-}CH_2} + Me_2CHMgBr \xrightarrow[\text{(ii) } H^+]{\text{(i) } Et_2O} ?$

(vii) $EtCH{=}CHCOEt \xrightarrow[\text{(ii)}H^+]{\text{(i)}EtMgBr + CuBr} ?$

(viii) $PrMgBr + MeC(OMe)_3 \xrightarrow[\text{(ii)}H^+]{\text{(i)}C_6H_6} ?$

(ix) $ClCH_2CO_2Et + 2RMgX \xrightarrow[\text{(ii)}H^+]{\text{(i)}Et_2O} ?$

(x) $\underset{\text{(epoxide)}}{MeCH{-}CH_2} + EtMgBr \xrightarrow[\text{(ii) } H^+]{\text{(i) } Et_2O} ?$

(xi) $(CH_2)_5Br_2 \xrightarrow[Et_2O]{2Mg} ? \xrightarrow[\text{(ii)}H^+]{\text{(i)}2MeONH_2} ?$

(xii) $MeCOCH_2CO_2Et + Me_2Cd \xrightarrow[\text{(ii)}H^+]{\text{(i)}Et_2O} ?$

(xiii) $Et_2Cd + CCl_3CHO \xrightarrow[\text{(ii)}H^+]{\text{(i)}Et_2O} ?$

3. Starting with EtOH and n-PrOH, prepare pent-2-ene via a Grignard reagent. Is there more than one way? If so, which is the best and why?

4. Convert n-$C_5H_{11}CO_2H$ into n-$C_4H_9CO_2H$ by *two* methods.

5. $R^1CH{=}CHCOR^2 + R^3MgX \longrightarrow$ 1, 2- + 1, 4-products
How would the ratio of the products change if R^2 = Et, isoPr, t-Bu? Explain.

6. Convert CHD_2I into CHD_2CH_2OH. What alkenes would you expect to get by the acid-catalysed dehydration of the alcohol? Explain.

7. Carry out the following conversions via Grignard reagents, and use any other compounds you like: (i) $MeCO_2H \longrightarrow MeCOEt$; (ii) $EtBr \longrightarrow EtCOEt$; (iii) $Me_2CHOH \longrightarrow Me_2CHNH_2$; (iv) $MeCOEt \longrightarrow EtCH(Me)CHBrMe$.

8. Complete the following equations:

(i) $4PhLi + (CH_2{=}CH)_4Sn \longrightarrow ?$

(ii) $2EtNa + Hg \longrightarrow ?$

(iii) $BuLi + MeONH_2 \longrightarrow ? \xrightarrow{H_2O} ?$

(iv) $Me_2CHLi + C_2H_4 \longrightarrow ? \xrightarrow[(ii) H_2O]{(i) CO_2} ?$

(v) $Me_3CCl + Me_2Zn \longrightarrow ?$

(vi) $2EtI + Na_2/Hg \longrightarrow ?$

(vii) $Et_2Hg + 2K \longrightarrow ?$

(viii) $4RMgCl + SnCl_4 \longrightarrow ?$

(ix) $Me_2Hg + Zn \longrightarrow ?$

(x) $2CH_2(CN)COCl + Et_2Cd \longrightarrow ?$

(xi) $3Ph_2Hg + 2Al \xrightarrow{heat} ?$

(xii) $CH_2{=}CHCOMe \xrightarrow{Me_2SnH_2} ?$

(xiii) $R_2CO + BrCH_2CH{=}CHCO_2Et \xrightarrow[(ii) H^+]{(i) Zn} ?$

(xiv) $EtMgX \xrightarrow{PbCl_2} ?$

(xv) $(PhCH_2)_4Sn \xrightarrow{heat} ?$

(xvi) $2MeI + Zn/Cu \longrightarrow ?$

(xvii) $EtLi + Br_2 \longrightarrow ?$

9. $Me_2CHNO_2 \xrightarrow[Br_2]{NaOH} A \xrightarrow{Me_2Zn} B$

What are A and B? Formulate the reaction. Could you use MeMgI instead of Me_2Zn? Explain.

10. Give the structures of A-E in the reactions:

$$EtMeCHCOPh \xrightarrow[(ii)\ aq.\ NH_4Cl]{(i) PhMeBr} A \xrightarrow{H^+} B \xrightarrow{O_3} C + D$$

$$C \xleftarrow{E} EtCHO$$

11. Suggest a mechanism for the reaction:

$R_2CO + Et_3Al \longrightarrow R_2CHOAlEt_2 + C_2H_4$

12. Using any compounds you like, convert:

(i) $Me_2CO \longrightarrow Me_2C(OH)CH_2CO_2H$; (ii) $EtOH \longrightarrow Me_2CHCHO$;
(iii) $Me_3COH \longrightarrow Me_3CCH_2CH{=}CH_2$; (iv) $MeCHO \longrightarrow EtMeCHCO_2H$;
(v) $HOCH_2(CH_2)_2CH_2OH \longrightarrow Me_2C(OH)(CH_2)_4C(OH)Me_2$.

16 Saturated dicarboxylic acids

(Solutions on page 162)

1. Write out all the possible isomeric dicarboxylic acids of molecular formula $C_6H_{10}O_4$ and name them by the I.U.P.A.C. system.

2. Complete the following equations and comment where necessary:

(i) 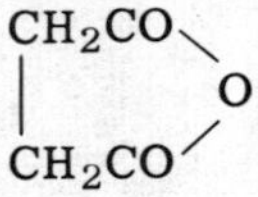 $+ \text{MeOH} \xrightarrow{\text{heat}} ?$

(ii) $CH_2(CO_2H)_2 + 2Me_2C{=}CH_2 \xrightarrow[Et_2O]{H_2SO_4} ?$

(iii) $HO_2C(CH_2)_4CO_2Me \xrightarrow{AgNO_3} ? \xrightarrow[CCl_4]{Br_2} ?$

(iv) 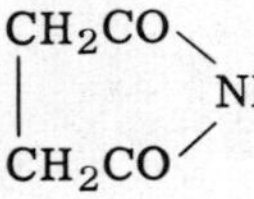 $\xrightarrow[KOH]{Br^2} ?$

(v) $PhCH(CO_2H)(CH_2)_2CO_2H \xrightarrow[\text{heat}]{Ac_2O} ?$

(vi) $CH_2{=}CH(CH_2)_nCO_2R^1 \xrightarrow[R^2OH]{O_3} ?$

(vii) 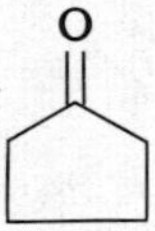 $\xrightarrow{HNO_3} ?$

(viii) $Bu^tOH + NH_2CONH_2 \xrightarrow{H_2SO_4} ?$

(ix) $KNCS + NH_2NH_2 \xrightarrow[H_2O]{H_2SO_4} ?$

(x) $CH_3CH{=}CHCH_2CH_3 \xrightarrow{NBS} ?$

(xi) $HOCH_2(CH_2)_nCH_2OH \xleftarrow{?} HO_2C(CH_2)_nCO_2Et \xrightarrow{?} HO_2C(CH_2)_nCH_2OH$

3. Synthesise from readily accessible materials:
(i) $EtCH(CN)CO_2Et$; (ii) $EtN(CO_2Et)_2$; (iii) $EtCH(CO_2H)CH_2CO_2H$.

4. Suggest a mechanism for the reaction:

$$NH_2CONH_2{\cdot}HNO_3 \xrightarrow{H_2SO_4} NH_2CONHNO_2 + H_2O$$

5. Convert succinic acid into cyclopentanone.

6. Acetic anhydride, when refluxed with potassium permanganate, gives compound A, $C_4H_4O_3$. The i.r. spectrum of A shows bands at 1860cm^{-1} and 1785 cm^{-1}. What is A? Suggest a mechanism for its formation.

7. The electrolysis of the sodium salt of β,β-dimethylglutaric acid gives A, C_5H_{10}, which, on ozonolysis, gives B, C_4H_8O, as one of the products. The i.r. spectrum of B showed a band at 1720cm^{-1} and the NMR spectrum showed three signals, τ 9·1, 7·8, 7·6. What are A and B? Give your reasons.

8. The N.M.R. spectrum of the compound $C_8H_{14}O_4$ shows three signals: triplet, τ 8·74; quartet, τ 5·82, singlet, τ 7·4. Suggest a structure for the compound.

17 Hydroxyacids, stereochemistry, unsaturated dicarboxylic acids

(Solutions on page 167)

1. Draw all possible stereoisomers for the following and state which are optically active, inactive, and which are *meso* forms:

(i) MeCHOHCHOHMe; (ii) MeCHOHCHOHCHO; (iii) MeCHOHCOCHOHMe; (iv) MeCHClCO$_2$H; (v) HO$_2$CCHBrCH$_2$CO$_2$Et; (vi) CH$_2$=CHMe; (vii) MeCH=CHEt; (viii) MeCHBrCH=CH$_2$; (ix) MeCHOHCH=CHMe; (x) CMe$_2$=C=CH$_2$; (xi) MeEtPrN; (xii) MeEtPrPr^{i}N$^+$Br$^-$; (xiii) PhMeEtN$^+$—O$^-$; (xiv) PhMeSO; (xv) CMeCl=C=CHMe.

2. Complete the following equations and state the relations between the configurations of the reactant and product. Give your reasons.

(i) D-MeCHOHEt + Ac$_2$O $\longrightarrow$?

(ii) L-MeCH(OAc)Et + NaOH $\longrightarrow$?

(iii) L-MeCHOHEt + HBr $\longrightarrow$?

(iv) D-MeCHClCO$_2$H + 2NaOH $\longrightarrow$?

(v) L-MeEtPrCOMe + HI $\longrightarrow$?

3. Complete the equations of the reactions of maleic and fumaric acids with the following reagents and comment:

(i) bromine; (ii) dilute alkaline permanganate solution; (iii) perbenzoic acid; (iv) catalytic hydrogenation.

4. Label the following compounds with *R* or *S* and give your reasons:

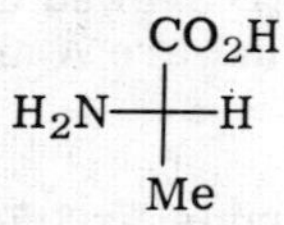
(i)

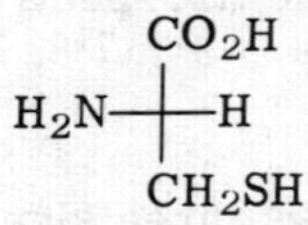
(ii)

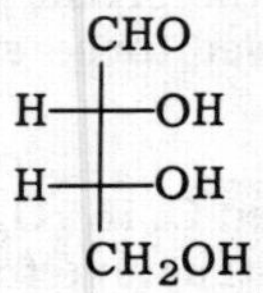
(iii)

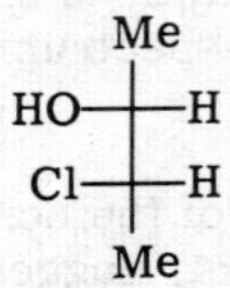
(iv)

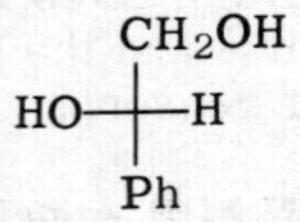
(v)

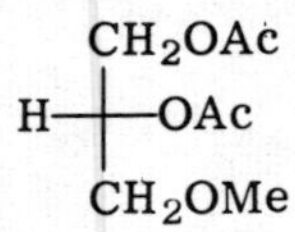
(vi)

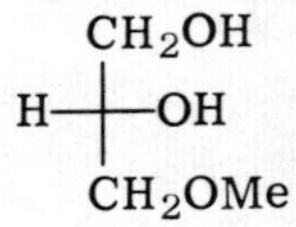
(vii)

5. Draw the Fischer projection formulae for:

(i) (*R*)-3-methylpentan-1-ol; (ii) (*S*)-2, 3-dimethylhexane; (iii) (*R*)-3-methoxy-2-methylpropan-1-ol; (iv) (2*S*, 3*R*)-2, 3-dichloropentane.

6. Write out the structures and I.U.P.A.C. names of the isomeric hydroxyacids with the molecular formula $C_4H_8O_3$. How would each acid behave towards:

(i) heat; (ii) dilute sulphuric acid on warming?

7. Complete the equations, and assuming all but one of the steps are S_N2, label each product with D or L. Comment.

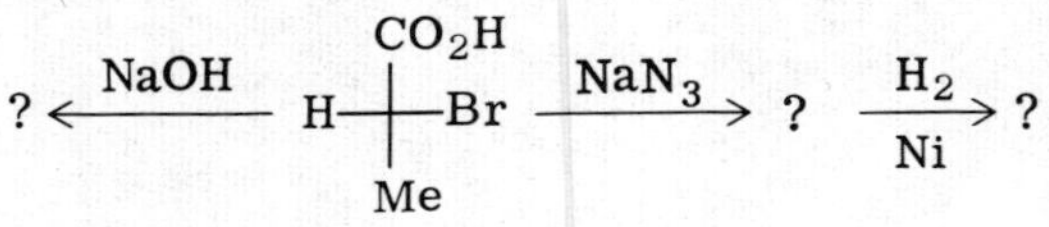

D

8. Complete and comment:

(i) 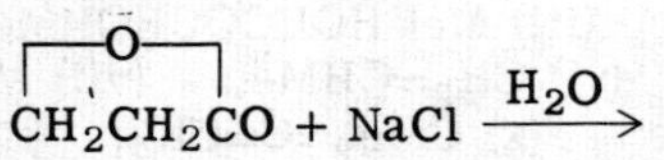

(ii) $(HO_2CCH_2)_2C(OH)CO_2H \xrightarrow[H_2SO_4]{\text{fuming}} ? \xrightarrow[HCl]{EtOH}$

(iii) $EtO_2CC{\equiv}CCO_2Et + CH_2(CO_2Et)_2 \xrightarrow[EtOH]{EtONa} ?$

(iv) 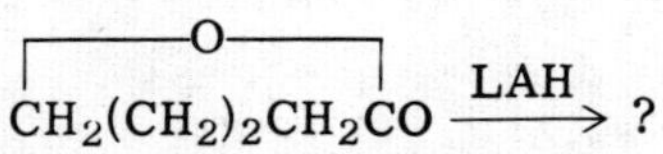

9. Discuss the following data:

	ΔH_f^θ(kJ)	S^θ (JK^{-1})
Maleic acid	−787·85	159·4
Fumaric acid	−810·9	166·1

10. At which steps do inversions occur? Explain.

$$PhCH_2CHMeOH \xrightarrow[(i)]{TsCl} PhCH_2CHMeOTs \xrightarrow[(ii)]{AcONa}$$

$$PhCH_2CHMeOAc \xrightarrow[\text{(iii)}]{OH^-} PhCH_2CHMeOH$$

11. *meso*-2, 3-Dibromobutane reacts with iodide ion more rapidly than does the (±)-compound. What are the respective alkenes formed? Explain.

12. Complete the equations:

(i) $n\text{-}C_5H_{11}\overline{CH(CH_2)_2CO} + 2MeMgBr \xrightarrow[\text{(ii) aq. } NH_4Cl]{\text{(i) } Et_2O} ?$ (ring closed through O)

(ii) $\overline{CH_2(CH_2)_2CO} + EtOH + HBr(g) \longrightarrow ?$ (ring closed through O)

(iii) $CH_3CHOHCO_2H \xrightarrow{Na} ?$

13. Mustard gas, $(ClCH_2CH_2{-})_2S$, is hydrolysed by water to $ClCH_2CH_2SCH_2CH_2OH$ *much faster* than expected for a primary alkyl halide. Offer an explanation.

14. An aliphatic ketone, M.W. 86, when converted into its oxime which is then reduced, gives an amine that can be resolved. What is the ketone? If more than one is possible, how could they be distinguished by a physical method?

15. What are A-D in the following reactions?

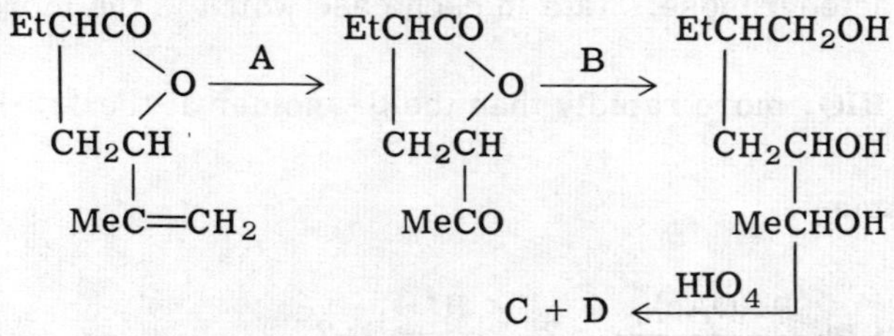

18 Carbohydrates

(Solutions on page 181)

1. Outline the evidence to justify the structure $HOCH_2(CHOH)_3CHO$ for arabinose. Is this a fair question? Comment.

2. Formulate the equations for the action of periodic acid on:
(i) $CH_3CHOHCH_2OH$; (ii) $HOCH_2COCH_2OH$; (iii) $HOCH_2CHOHCHOHCH_2OH$;
(iv) $HOCH_2CH_2CHOHCO_2H$; (v) $HO_2CCHOHCHOHCHO$;
(vi) $OCHCH_2CHOMeCHOHCH_2OH$; (vii) $OCHCHOMeCHOHCH_2OEt$.

3. Convert:
(i) D-erythrose to the next higher aldose; (ii) D-mannose to the next lower aldose;
(iii) D-arabinose to the next higher ketose; (iv) D-fructose to the next lower aldose.

4. Designate, using *R* and *S* symbols, D(+)-glucose

5. Complete and comment:

(i) D-glucose $\xrightarrow[HCl]{EtSH}$? $\xrightarrow[NaOH]{Me_2SO_4}$? $\xrightarrow[CdCO_3]{HgCl_2}$?

(ii) D-glyceraldehyde $\xrightarrow[HCl]{Me_2CO}$? $\xrightarrow{CH_2{=}CHMgCl}$? $\xrightarrow{O_3}$?

(iii) D-glucose $\xrightarrow[EtONa]{MeNO_2}$? $\xrightarrow{H_2SO_4}$? $\xrightarrow[Ni]{H_2}$?

6. Draw the conformations of the α- and β-forms of:

(i) D-arabinopyranose; (ii) D-galactopyranose. State in each case which is the more stable form.

7. α-Glucopyranose is oxidised by HIO_4 more rapidly than the β-anomer at the 1, 2-bond. Suggest a reason.

8. Complete the equations and comment.

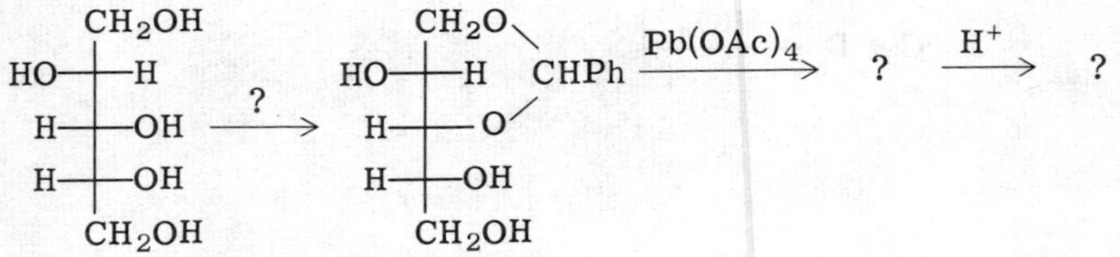

9. Ruthenium tetroxide is known to oxidise a *single* CHOH group to ketone. How could you convert $HOCH_2(CHOH)_4CHO$ into $HOCH_2(CHOH)_2'COCHOHCHO$?

10. What sugars are obtained by epimerising, respectively, C-2, C-3, C-4 of D-glucose?

11. Glucose, mannose and fructose give identical osazones. Explain.

12. Compound A, $C_5H_{10}O_5$, gives a tetra-acetate with Ac_2O, and oxidation of A with $Br_2{-}H_2O$ gives an acid, $C_5H_{10}O_6$. Reduction of A with HI gives isopentane. What are possible structures of A and how could you distinguish between the possibilities?

13. Compound A, $C_5H_{10}O_4$, is oxidised by $Br_2{-}H_2O$ to the acid, $C_5H_{10}O_5$, which readily forms a lactone. A forms a triacetate with Ac_2O and an osazone with $PhNHNH_2$. A is oxidised by HIO_4, only one molecule of which is consumed. What is the structure of A?

14. Compound A, $C_5H_{10}O_4$, on oxidation with $Br_2{-}H_2O$, gives acid B, $C_5H_{10}O_5$. A forms a triacetate (Ac_2O) and is reduced by HI to n-pentane. Oxidation of A with HIO_4 gives, among other products, 1 molecule of CH_2O and 1 molecule of HCO_2H. What are the possible structures of A and how could you distinguish between them?

19 Alicyclic compounds

(Solutions on page 194)

1. Convert cyclohexanone into cyclopentanone.

2. Would you expect the N.M.R. spectra of *cis*- and *trans*-1, 2-dibromocyclopropane to be identical? Explain.

3. Formulate the following reaction:

$$CH_3CO_2Et + (CH_2)_5(MgBr)_2 \xrightarrow[\text{(ii)}H^+]{\text{(i)}Et_2O} \text{(A) } C_7H_{14}O \xrightarrow{H^+} \text{(B) } C_7H_{12} \xrightarrow[\text{reagent}]{\text{Lemieux}} \text{(C) } C_7H_{12}O_3$$

4. Complete the following equations:

(i) 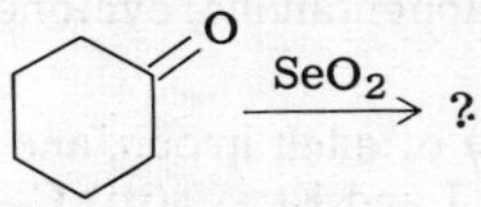

(ii) 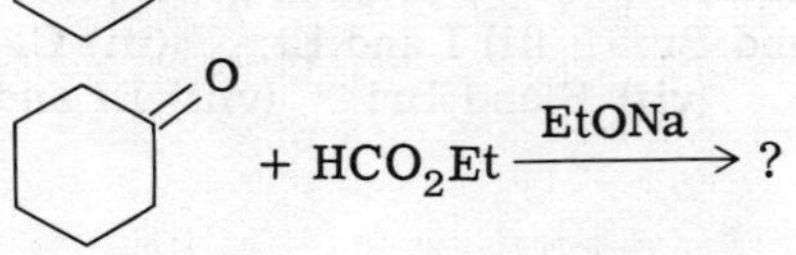

(iii) $ClCH_2CH_2CH_2CN \xrightarrow{NaOH}$ (cyclic compound)?

(iv) $PhCHBrCH_2CH_2Br \xrightarrow{Zn-Cu}$?

(v) 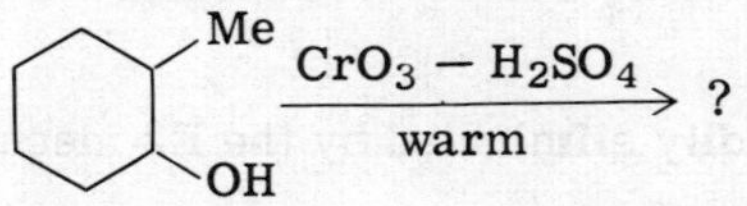

(vi) $C_3H_7CHClCOCH_2Me \xrightarrow{EtONa}$?

(vii) $MeCH(CH_2CO_2Et)CH_2CH_2CO_2Et \xrightarrow{EtONa}$?

(viii) $MeCO(CH_2)_3CO_2Et \xrightarrow{EtONa}$?

(ix) 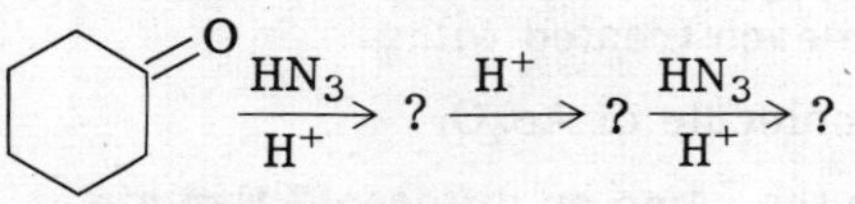

(x) 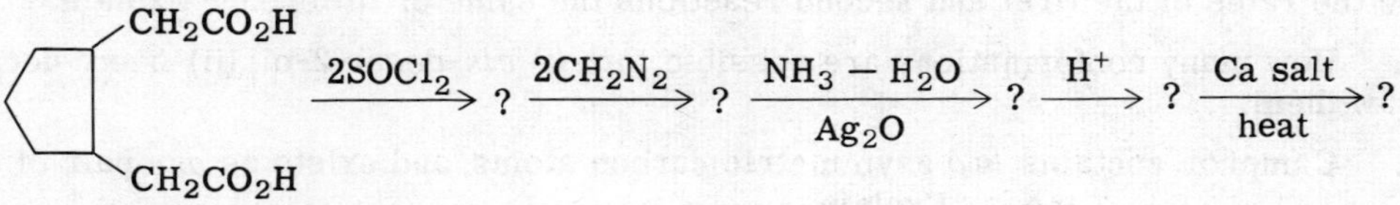

(xi) 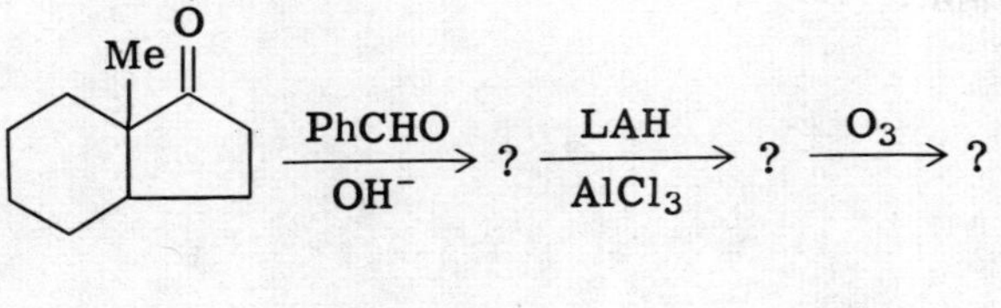

5. Synthesise 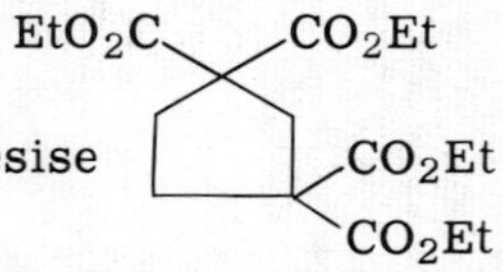

6. Treatment of $Me_2C{=}CHCH_2CH_2Cl$ with $CaCO_3$–H_2O gives cyclopropyldimethylmethanol. Suggest a mechanism.

7. Discuss:

(i) 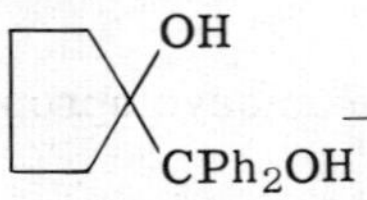

$\xrightarrow{H^+}$ a cylcohexanone derivative

(ii) 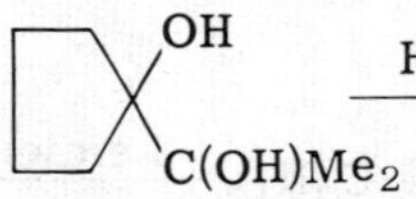

$\xrightarrow{H^+}$ a cyclopentane deriv. + a cyclohexanone deriv.

(iii) 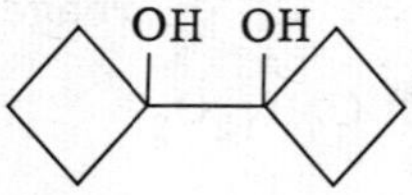

$\xrightarrow{H^+}$ a spiro-octanone

8. Using i.r. spectroscopy, differentiate between 2-methylcyclopentanone, cyclohexanone, cyclohex-2-enone, and cyclohex-3-enone.

9. Given the conformation shown, state: (*a*) the nature (*e* or *a*) of each group, and (*b*) the geometrical relationship (*cis* or *trans*) of: (i) Cl and Br; (ii) I and Et; (iii) Cl and Et; (iv) Et and Br; (v) Me and F; (vi) Cl and Me; (vii) F and Br; (viii) Et and Me; (ix) Cl and I; (x) I and Me.

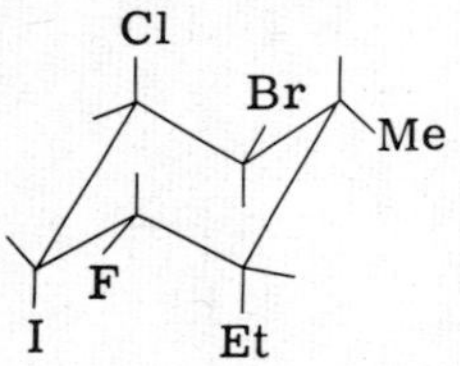

10. Which halogen atom in Q. 9 would be most readily eliminated by the E2 mechanism to form the ene compound and why?

11. Draw all possible conformations for the disubstituted cyclohexanes:

(i) 1-Br-2-Cl; (ii) 1, 3-Br_2; (iii) 1-Cl-4-I; (iv) 1, 3-$(OH)_2$.

Which conformation is the most stable one in each set? Which are optically active compounds?

12. *cis*- and *trans*-Cyclohexane-1, 2-diol are each treated with:

(i) 1 molecule of Ac_2O; (ii) a second molecule of Ac_2O.

Are the rates of the first and second reactions the same or different? Explain.

13. How many conformations are possible for: (i) *cis*-decal-2-ol; (ii) *trans*-decal-2-ol? Draw them.

14. Camphor contains *two* asymmetric carbon atoms, and exists as *one* pair of enantiomers. Explain.

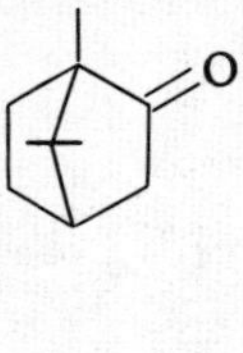

15. Draw the equation (conformational) for the pyrolysis of cyclohexyl acetate and explain.

16. Complete the equations:

(i) 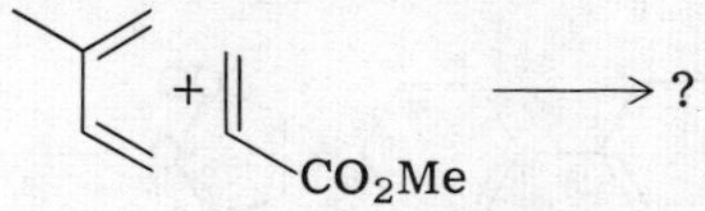

(ii) 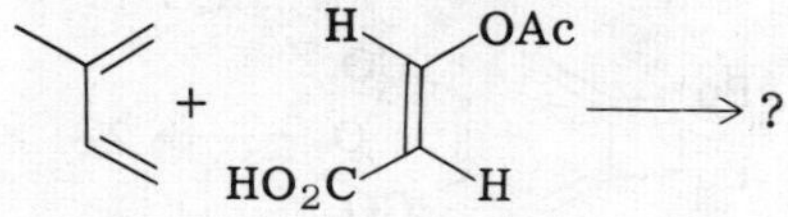

(iii) 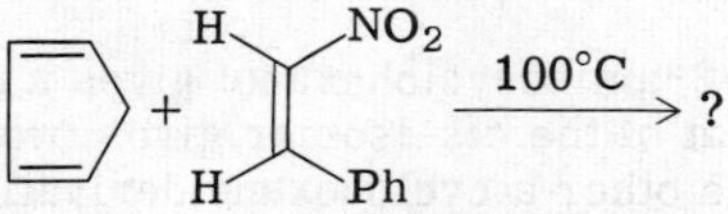

17. *One* isomer of hexachlorocyclohexane undergoes dehydrochlorination with great difficulty. Suggest its conformation. Explain. How many N.M.R. signals would this isomer show?

18. 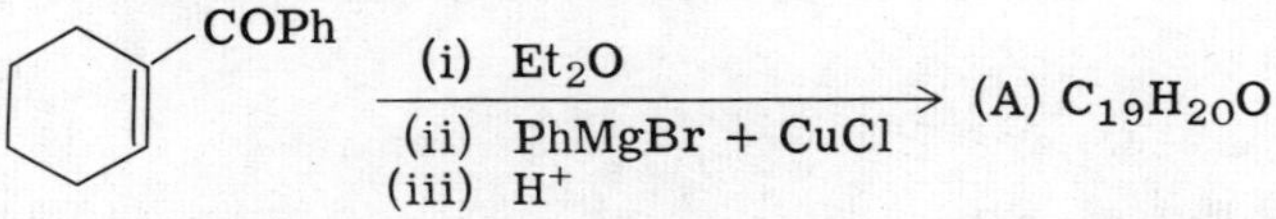

(A) showed an i.r. band at about 1680cm^{-1}. Formulate the course of the reaction and discuss the geometry of the product.

19. Menthyl and neomenthyl chlorides are stereoisomers of 1-methyl-3-chloro-4-isopropylcyclohexane and in both the alkyl groups are *trans*. On treatment with EtONa, menthyl chloride gives menth-2-ene (100%) and neomenthyl chloride gives menth-2-ene (25%) and menth-3-ene (75%). Deduce the conformation of the chlorine atom in each chloride, and formulate the reactions.

20. Discuss the data of the heats of formation (kJ mol^{-1}) of the following methylcyclohexanols: *cis*-2-Me-, –396·6; *trans*-2-Me-, –422·2; *cis*-3-Me-, –422·6; *trans*-3-Me-, –400·8.

21. *cis*-Cyclohexane-1, 2-diol is oxidised by periodic acid more rapidly than the corresponding *trans*-isomer. Discuss.

22. (A) 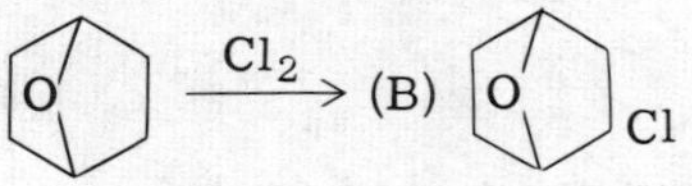

Two forms of (B) are produced, one with μ = 1·07D and the other 3·06D. Discuss.

23. Starting from D.E.M. and any other compounds you like, synthesise cyclohexane-1, 4-dicarboxylic acid.

24. Convert 1, 2, 4-trihydroxycyclohexane into 3, 4-dihydroxycyclohexanone.

25. Suggest a method for the conversion:

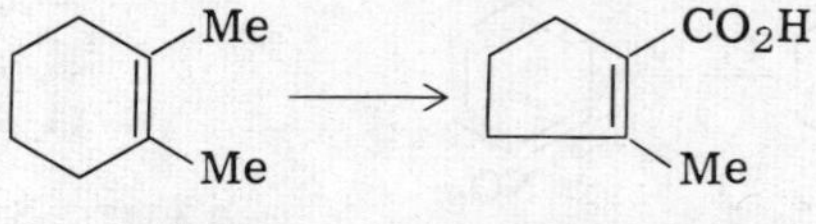

26. Expoxidation of 4-methylcyclopentene gives two isomeric epoxides, with one predominating. Discuss.

27. Convert cyclohexanol into *cis*- and *trans*-cyclohexane-1, 2-diol.

28. Discuss: *cis*-2-chlorocyclohexanol with aqueous alkali gives cyclohexanone, whereas the *trans* isomer gives cyclohexene oxide.

29. Complete the equations:

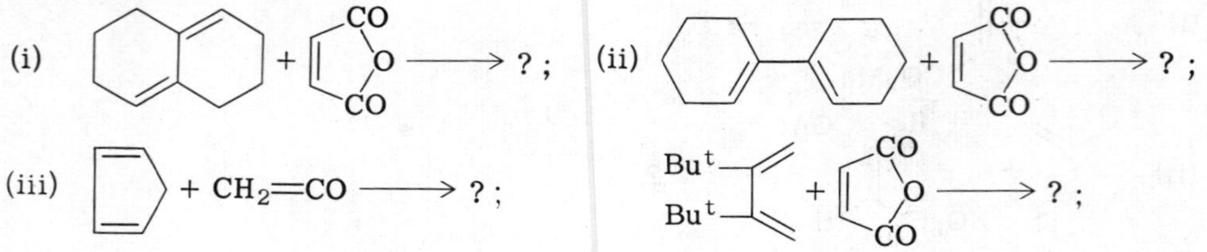

30. The deamination (with nitrous acid) of *trans*-2-aminocyclohexanol gives a cyclopentane derivative, $C_6H_{10}O$, as the only product and that of the *cis*-isomer gives two products, one the same as that from the *trans*-isomer and the other a cyclohexane derivative, $C_6H_{10}O$. Formulate the reactions and explain them.

20 Monocyclic aromatic hydrocarbons

(Solutions on page 215)

1. Write out the structures of the benzene isomers with the formula C_9H_{12}.

2. Assign orientations to:

(i) the three chlorotoluenes with dipole moments 1·3, 1·9, and 1·78D; (ii) the three chloronitrobenzenes with dipole moments 3·4, 2·5, and 4·3D.

3. Suggest oxidizing agents for the conversions:

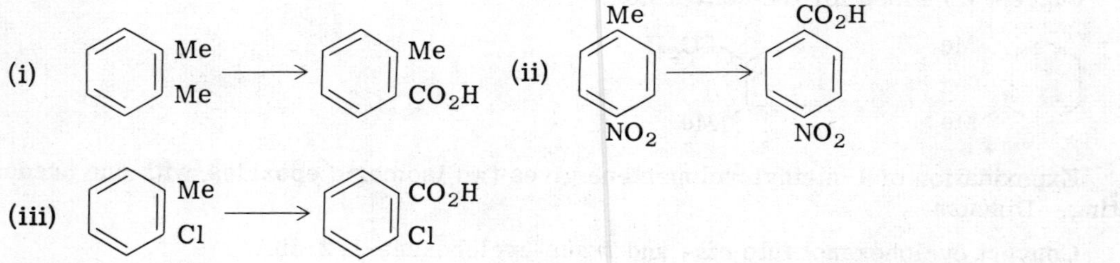

4. Indicate the positions of monosubstitution and, where possible, the predominant product:

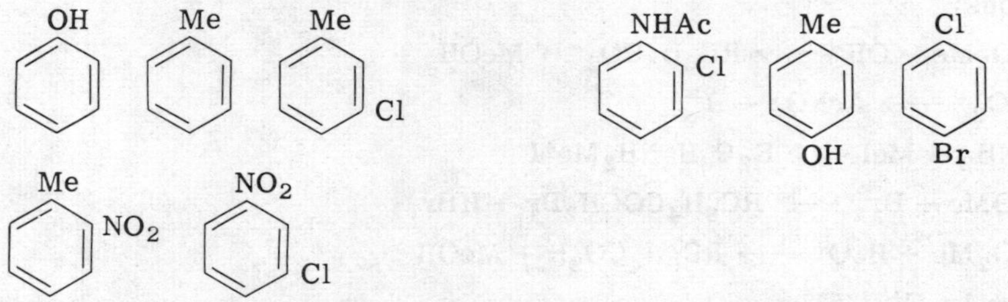

5. Suggest a reason for the catalytic hydrogenation of benzene giving cyclohexane and the inability to isolate intermediates.

6. What are the signs of the I and R effects for the following groups attached to the benzene ring: NO_2, NH_2, Br, COMe, NMe_3^+, SO_3H, OMe? When in opposition, which effect is greater?

7. Benzene, toluene, xylenes (*o*, *m*, *p*), and mesitylene dissolve in HBF_4 to form salts. Explain the order of basicity: mesitylene > *m*-xylene > *o*- and *p*-xylene > toluene > benzene.

8. Complete the equations and comment:

(i) $PhBr + Ac_2O \xrightarrow{AlCl_3} ?$

(ii) $PhCH_2CN \xrightarrow{HNO_3} ?$

(iii) 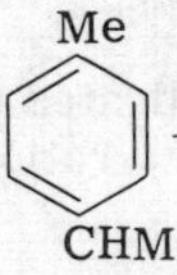 $\xrightarrow{M.A.} ?$

(iv) 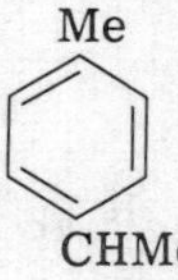 $+ AcCl \xrightarrow{AlCl_3} ?$

(v) $Ph(CH_2)_3Me \xrightarrow{KMnO_4} ?$

(vi) $PhNO_2 + EtCl \xrightarrow{AlCl_3} ?$

(vii) $PhH + Me_3CCH_2Cl \xrightarrow{AlCl_3} ?$

9. Suggest a method for each of the following conversions:
(i) PhEt ⟶ PhH; (ii) PhH ⟶ *p*-$Et_2C_6H_4$; (iii) PhH ⟶ *p*-MeC_6H_4Pr;
(iv) PhH ⟶ $PhCH_2CO_2H$; (v) PhH ⟶ PhLi;
(vi) 2-OH-3-Me-$C_6H_3CO_2H$ ⟶ 2-OH-1,3-$C_6H_3(CO_2H)_2$.

10. Complete the equation and give your reasons:

PhCH=CHMe + HI ⟶ ?

11. Predict the sign of the Hammet σ constant for the following substituents and give your reasons: *m*-NO_2, *p*-MeO, *p*-Me_2S^+, *m*-Et, *p*-CF_3, *m*-OH, *p*-OH, *m*-NMe_3^+, *p*-CO_2H, *m*-CN, *p*-NMe_2, *p*-SO_2NH_2, *m*-IO_2.

12. Predict the sign of ρ for the following reactions (R is *m* or *p* to the functional group) and give your reasons:

(i) $RC_6H_4CO_2Me + OH^- \longrightarrow RC_6H_4CO_2^- + MeOH$

(ii) $ArH + NO_2^+ \longrightarrow ArNO_2 + H^+$

(iii) $R_2C_6H_3NH_2 + MeI \longrightarrow R_2C_6H_3NH_2Me^+I^-$

(iv) $RC_6H_4COMe + Br_2 \longrightarrow RC_6H_4COCH_2Br + HBr$

(v) $RC_6H_4CO_2Me + H_3O^+ \longrightarrow RC_6H_4CO_2H + MeOH$

(vi) $Ar_3CCl \rightleftharpoons Ar_3C^+ + Cl^-$

(vii) $ArH + Br_2 \longrightarrow ArBr + HBr$

(viii) $ArCOCl + EtOH \longrightarrow ArCO_2Et + HCl$

13. The acid-catalysed hydrolysis of ethyl benzoates has a ρ value of +0.144. What will be the effect of −I groups on the rate? Explain.

14. Methyldiaryl chlorides (ArCHClAr) undergo solvolysis in EtOH by the S_N1 mechanism. This reaction has $\rho = -5{\cdot}090$. What would be the effects of electron-withdrawing and releasing groups on the rate of reaction?

15. Given the following ionisation constants at 25°C (in H_2O): $PhCO_2H$, $6{\cdot}3 \times 10^{-5}$; *m*-$CH_3COC_6H_4CO_2H$, $1{\cdot}48 \times 10^{-4}$; *p*-$CH_3COC_6H_4CO_2H$, $2{\cdot}0 \times 10^{-4}$, calculate σ_m and σ_p of the CH_3CO group.

16. The partial rate factor for the *p*-postion for electrophilic substitution of fluorobenzene is greater than 1. Discuss.

17. Discuss the statement: In a monosubstituted benzene, the *m*/*p* ratio indicates the polar influences of the substituent at these two positions, whereas the *o*/*m* and *o*/*p* ratios indicate both polar and steric effects.

18. Discuss the reaction:

$$PhCH_2CH{=}CH_2 \xrightarrow[\text{heat}]{\text{KOH}} PhCH{=}CHMe$$

19. Which of the following ring compounds obey Hückel's rule: $C_{10}H_{10}$, $C_{12}H_{12}$, $C_{12}H_{12}^{2-}$, $C_{12}H_{12}^{2+}$, $C_{20}H_{20}$, $C_{20}H_{20}^-$, $C_{20}H_{20}^{2-}$, $C_{20}H_{20}^+$, $C_{20}H_{20}^{2+}$?

20. Arrange the following in order of increasing rate of nitration and give your reasons: PhH, PhMe, C_6D_6, $PhNO_2$, PhCl.

21. How many N.M.R. signals and their multiplicity (assume the benzene ring is a singlet) would be expected from:

(i) $PhCMe_3$; (ii) $PhCH_2CHMe_2$; (iii) $PhCH_2CClMe_2$; (iv) $PhCH_2OMe$;
(v) PhCHBrMe; (vi) $PhCHBrCH_2Br$?

22. The following fragment ions are derived from n- and s-butylbenzene. Which is which? Give your reasons.

m/*e*	51	65	77	91	92	105	134	135
A (%)	5	10	6	100	60	9	30	3
B (%)	7	3	10	14	1	100	18	2

21 Aromatic halogen compounds

(Solutions on page 228)

1. Write out the possible isomers of the aromatic compounds A, C_7H_7Cl. How could you distinguish between them? What would each give on oxidation?

2. Convert:

(i) 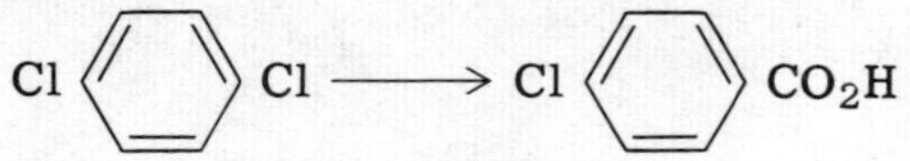

(ii) 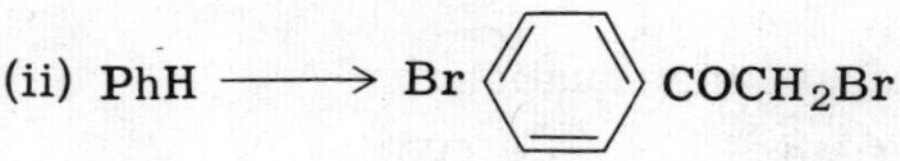

(iii) PhH ⟶ *o*-bromotoluene.

3. Starting from benzene and any other compounds you like, prepare *p*-bromostyrene.

4. Complete the following equations:

(i) 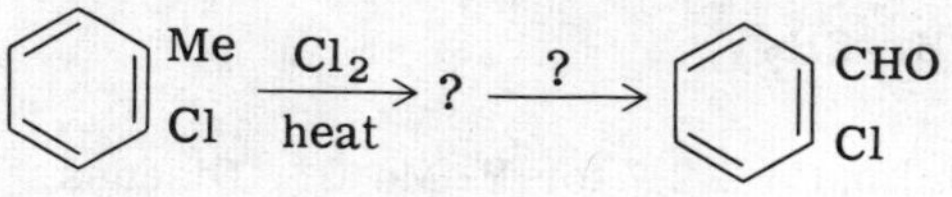

(ii) $2Ph_2CCl_2 \xrightarrow[PhH]{Cu} ?$

(iii) 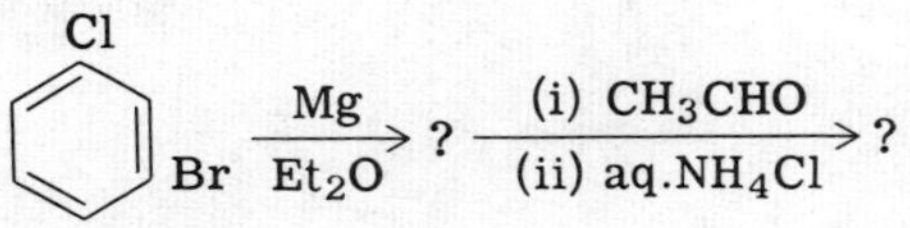

(iv) 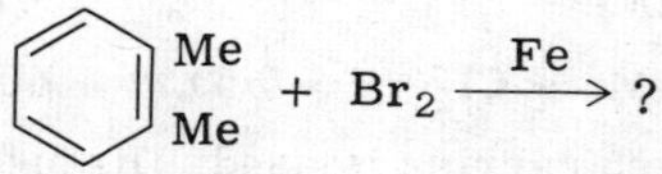

(v) 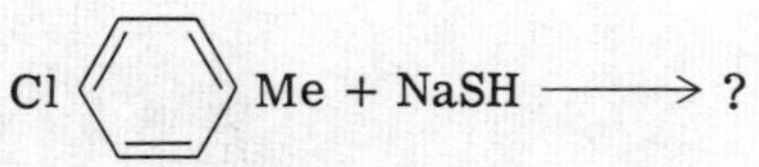

(vi) $PhCH_3 + CBrCl_3 \xrightarrow{hv} ?$
(not ring substitution)

(vii) 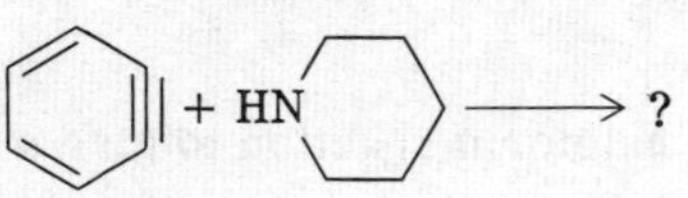

(viii) $PhBr \xrightarrow[liq.\,NH_3]{NaNH_2} ? \xrightarrow{[EtCOCHMe]^- Na^+} ?$

(ix) 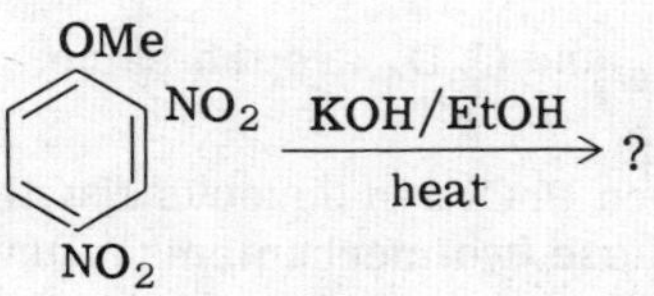

(x) 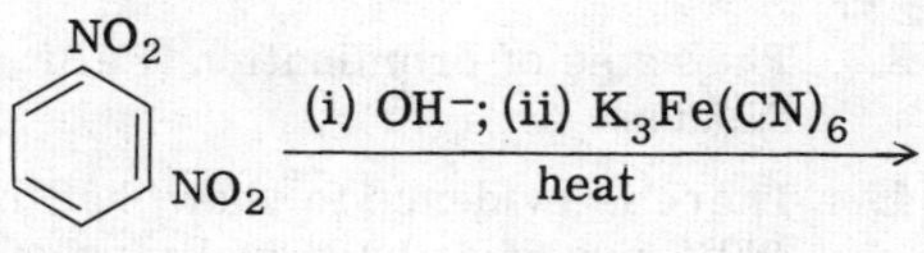

(xi) 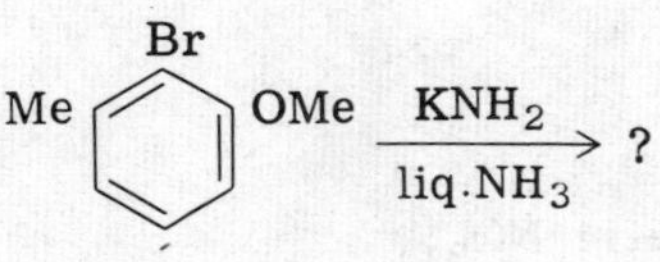

(xii) 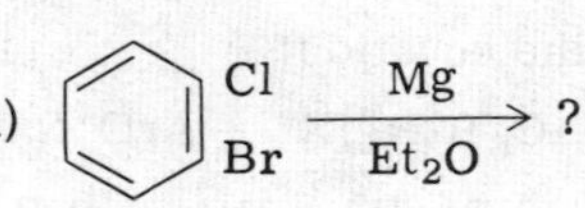

(xiii) 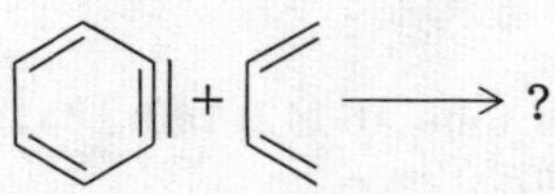

5. Bromine adds to styrene to give styrene dibromide only; no nuclear substitution occurs. Discuss.

6. Discuss the following reaction:

$$RC_6H_4X \xrightarrow[\text{liq. } NH_3]{NaNH_2} RC_6H_4NH_2$$

for: (i) $R = o\text{-OMe}, X=Br$; (ii) $R = m\text{-}CF_3, X=Cl$; (iii) $R = m\text{-Me}, X=Br$.

7. Complete and comment:

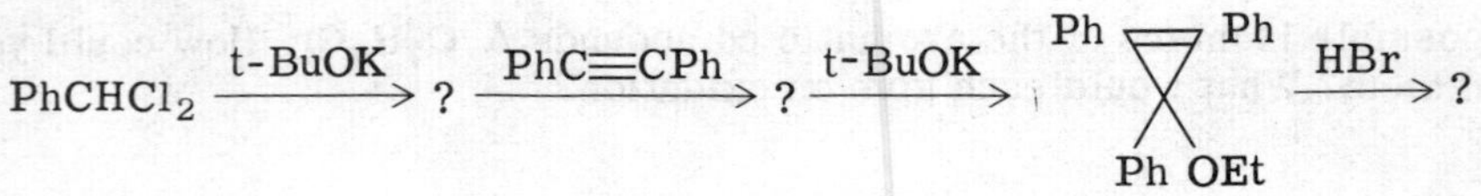

8. Discuss the equation:

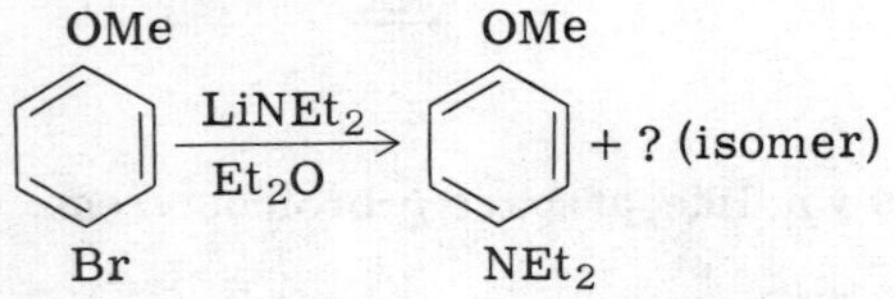

9.

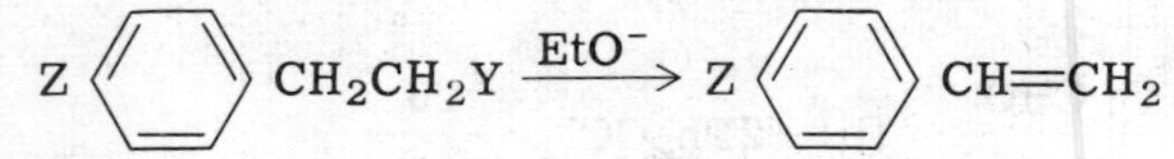

This is an E2 reaction, and $\rho = 2{\cdot}1$ for $Y=Br$, and $\rho = 2{\cdot}7$ for $Y=Me_2S^+$. Discuss.

10. $PhMe + Br_2 \longrightarrow 0{\cdot}3\%\, m + 67\%\, p$

$PhMe + Cl_2 \longrightarrow 0{\cdot}5\%\, m + 40\%\, p$

No halogen carrier is used. Discuss.

11. Discuss the reaction:

$$Me_3CCHOHPh \xrightarrow[S_N1]{HCl} Me_3CCHClPh$$

12. $PhCH_2MgCl + MeCHO \longrightarrow PhCH_2CHOHMe + A$

A is an isomer of the other product and undergoes the haloform reaction to give *o*-toluic acid. What is A? Suggest a mechanism for its formation.

13. The rates of bromination ($FeBr_3$ catalyst) of C_6H_6 and C_6D_6 are the same. Discuss.

14. There is evidence to show that the reaction between PhCl and aqueous NaOH at 300°C proceeds by a dual mechanism. What are these two mechanisms? Devise an experiment to demonstrate them.

15. Complete the equation:

$$R^1C_6H_4\text{—}\overset{+}{I}\text{—}C_6H_4R^2\}X^- + ArO^-Na^+ \longrightarrow ?$$

where: (i) $R^1 = p\text{-}NO_2, R^2 = H$; (ii) $R^1 = p\text{-}NO_2, R^2 = m\text{-}NO_2$;
(iii) $R^1 = m\text{-MeO}, R^2 = p\text{-MeO}$.

16. The mass spectrum of 1-chloro-4-ethylbenzene shows ions at 142, 140, 125, and 105. What are they? Suggest paths for their formation.

22 Aromatic nitro-compounds

(Solutions on page 240)

1. Arrange in order of relative rates the *o*-nitration of PhMe, 2-D-toluene, and 2,6-D_2-toluene. Give your reasons.

2. Draw the energy profiles for (and discuss):
(i) $PhH + NO_2^+ \longrightarrow PhNO_2$; (ii) $C_6D_6 + NO_2^+ \longrightarrow C_6D_5NO_2$.

3. Suggest a method for each of the following conversions (do not describe separations of isomers):
(i) benzene ⟶ *p*-bromonitrobenzene; (ii) benzene ⟶ *m*-chloronitrobenzene;
(iii) benzene ⟶ 1,2-dibromo-4-nitrobenzene;
(iv) benzene ⟶ *m*-nitrobenzenesulphonic acid; (v) benzene ⟶ $PhCH_2NO_2$;
(vi) benzene ⟶ 1-Br-4-Cl-2-NO_2-benzene;
(vii) PhMe ⟶ 2,4-dinitrobenzoyl chloride; (viii) PhH ⟶ *o*-nitrophenol;
(ix) PhH ⟶ 1-Cl-2,4-$(NO_2)_2$-benzene; (x) PhH ⟶ 2,4,6-Br_3-1-NO_2-benzene;
(xi) PhH ⟶ $PhN{=}C(CO_2Et)_2$.

4. Discuss the following data:

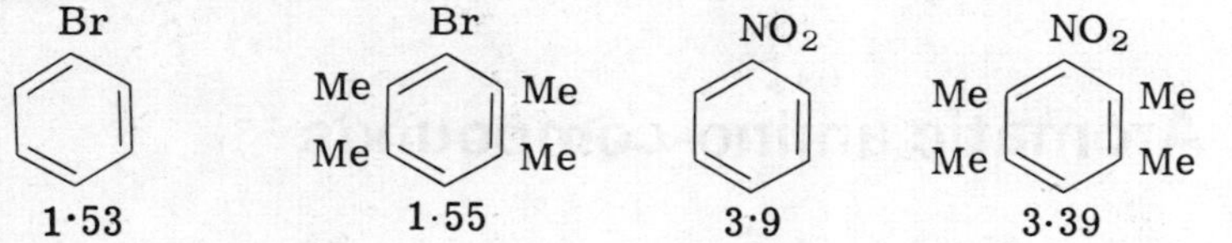

μ(D) 1·53 1·55 3·9 3·39

5. The nitration of phenylboronic acid, $PhB(OH)_2$, with M.A. at −20°C gives predominantly the *m*-derivative. With HNO_3–Ac_2O, the predominant product is the *o*-derivative. Suggest an explanation.

6. Complete the equations:

(i) 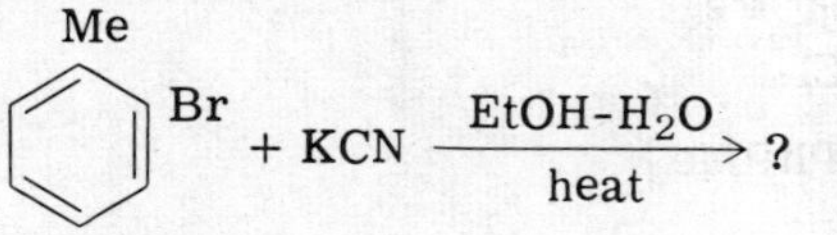

(ii) 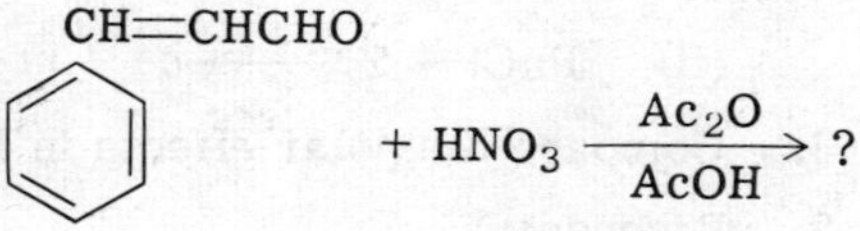

(iii) 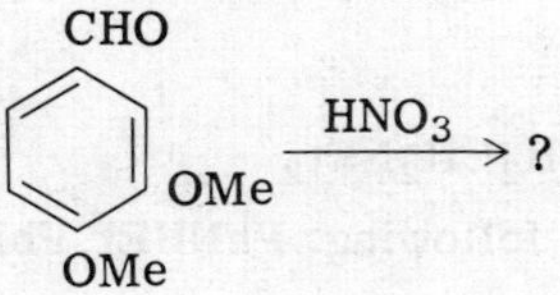

(iv) 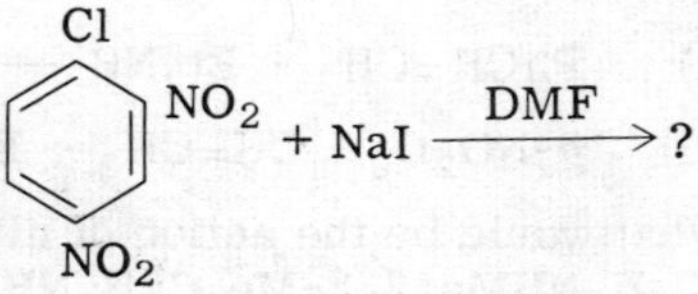

(v) 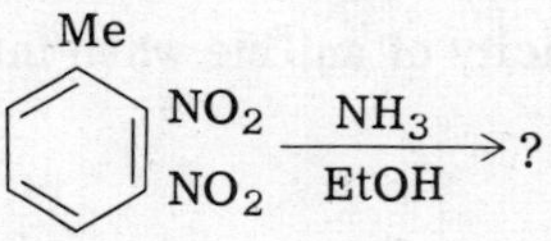

(vi) 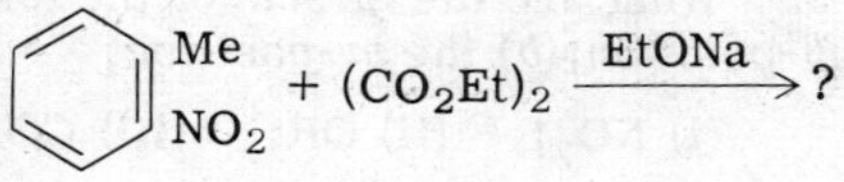

(vii) 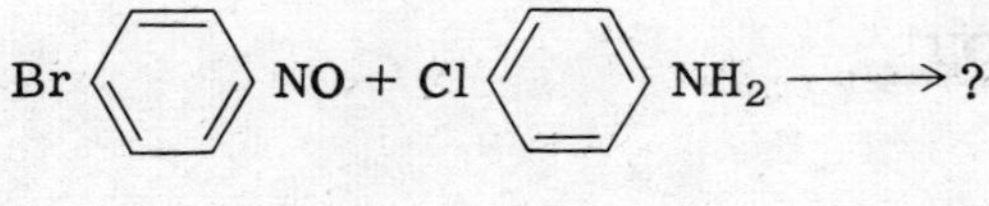

(viii) 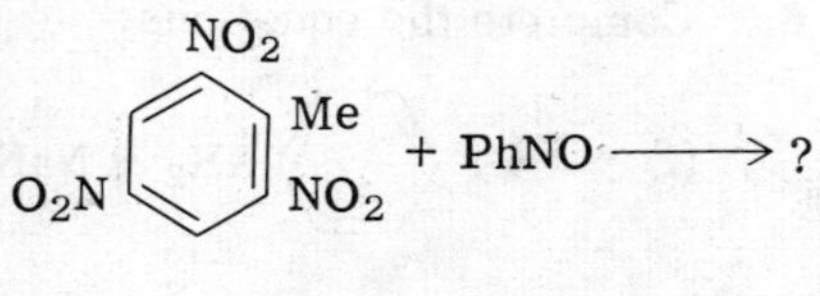

7. The aromatic compound A, C_8H_9Br, on treatment with DEM/EtONa, followed by reflux with H_2SO_4, gave B, $C_{10}H_{12}O_2$. B, on vigorous oxidation, gave C, $C_8H_6O_4$, which, on mononitration, gave only one mononitro derivative, D. What are A–D?

8. Offer an explanation for the following data of the mononitration of acetanilide (PhNHAc): M.A.: *o*-, 19%, *p*-, 79%: HNO_3–Ac_2O: *o*-, 68%: *p*-, 30%.

9. The rate of nitration of toluene in nitromethane with excess of HNO_3 is constant and the reaction is zero order with respect to toluene. If the rate is temperature-dependent, what conclusion can you draw?

10. Describe the fragmentation paths which could give the following ions (abundance ratio in brackets) from 1-chloro-3-nitrobenzene: 76(8), 87(2), 85(5·5), 113(32), 111(100), 129(2), 127(5), 157(73), 159(23). At least one ion that might have been expected is missing. Which is it?

23 Aromatic amino-compounds

(Solutions on page 255)

1. Draw the energy profiles for the reactions:

(i) $PhNH_2 + Z^+ \longrightarrow$ *o*- + *m*- + *p*-$ZC_6H_4NH_2$

(ii) $PhCl + Z^+ \longrightarrow$ *o*- + *m*- + *p*-ZC_6H_4Cl

2. Compare the polar effects in $Ph\overset{+}{N}Me_3$ and $Ph\overset{+}{P}Me_3$.

3. Discuss:

(i) $PhCH{=}CH_2 + Et_2NH \longrightarrow$ no reaction

(ii) *p*-$NO_2C_6H_4CH{=}CH_2 + Et_2NH \longrightarrow$ *p*-$NO_2C_6H_4CH_2CH_2NEt_2$

4. What would be the action of nitrous acid on each of the following: PhNHEt, $PhNMe_2$, *p*-MeC_6H_4NHMe, 2, 5-$Me_2C_6H_3NH_2$?

5. What are the effects of the following groups on the basicity of aniline when in: (*a*) the *p*-position; (*b*) the *m*-position;

(i) NO_2; (ii) OR; (iii) CO_2R; (iv) NR_2?

6. Complete the equations:

(i) Br 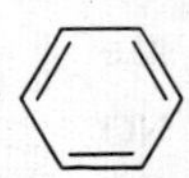 NH_2 + NaNCO $\xrightarrow{AcOH}$?

(ii) $PhNH_2 + ? \xrightarrow{Et_2O} PhNHCOCH{=}CHCO_2H$

(iii) 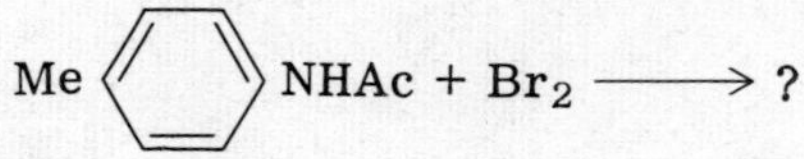

(iv) 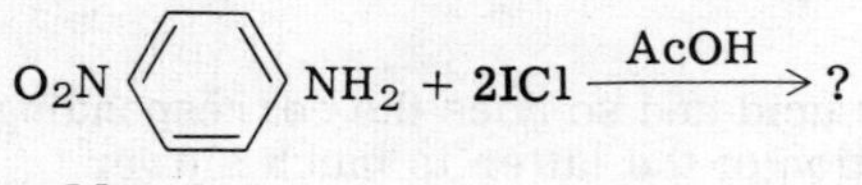

(v) 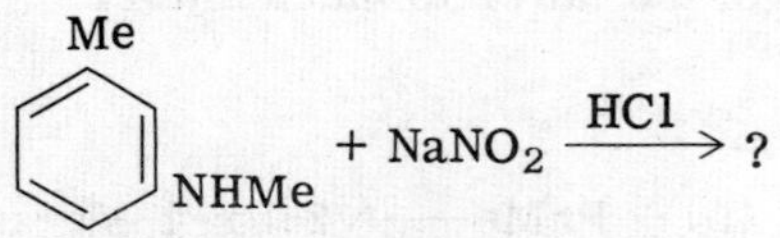

(vi) $PhCH_2NH_2 + ? \longrightarrow PhCH_2NHCO_2Et$

(vii) 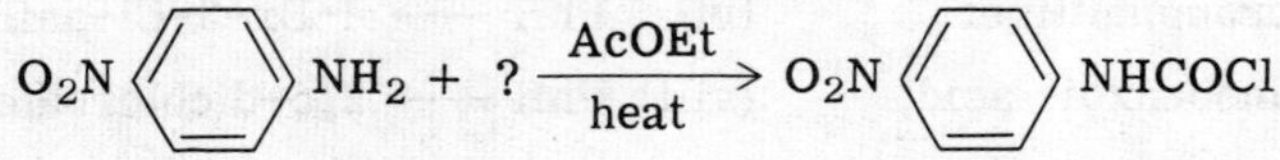

(viii) 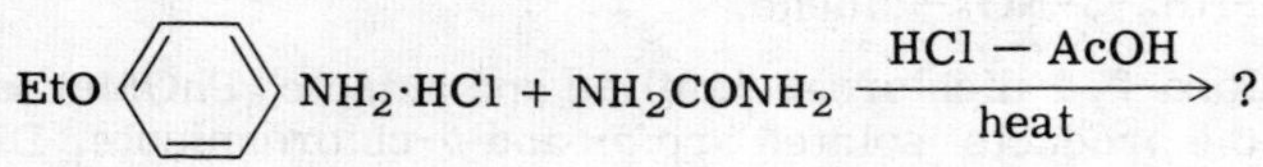

(ix) 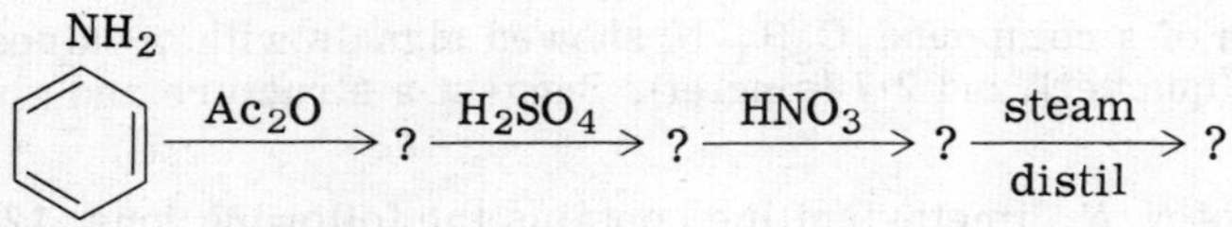

(x) $Ph_2NH \xrightarrow{Li} ? \xrightarrow[CuI]{PhI} ?$

(xi) 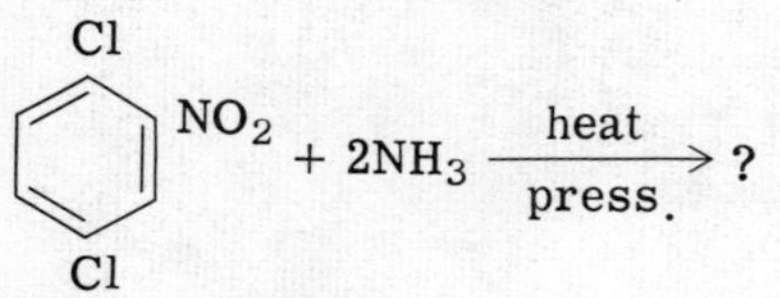

(xii) $Ph_3N + MeI \longrightarrow ?$

(xiii) 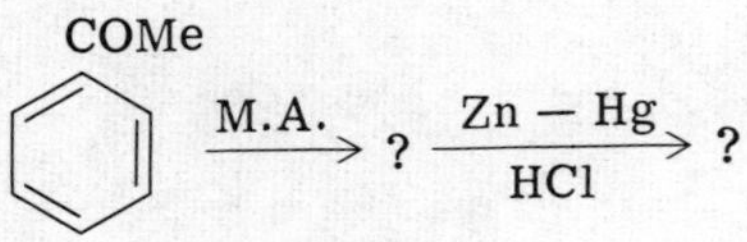

(xiv) 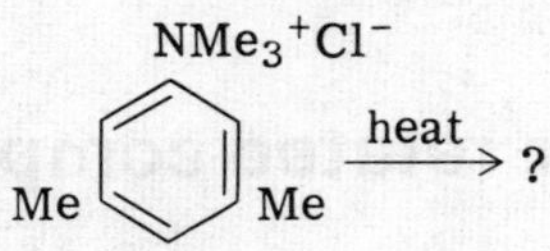

(xv) 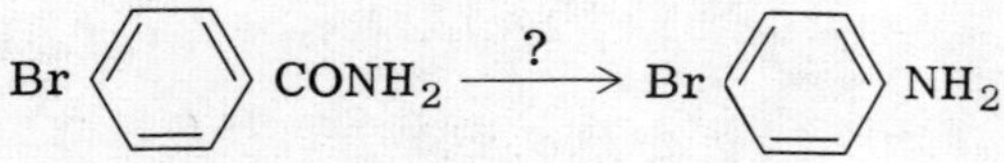

(xvi) 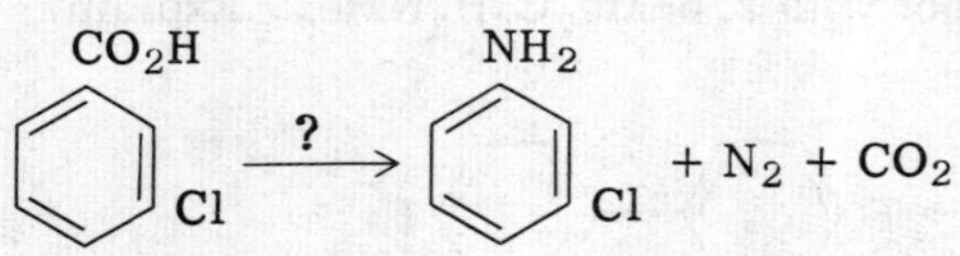

(xvii) $PhNH_2 + PhSO_2Cl \longrightarrow ? \xrightarrow[NaOH]{Me_2SO_4} ? \xrightarrow{H_3O^+} ?$

(xviii) $PhNH_2CH_2CHMe_2{}^+Br^- \xrightarrow{heat} ?$

(xix) $? \xleftarrow{LAH} PhNO_2 \xrightarrow[Pd-C]{NaBH_4} ?$

7. *N,N*-Dimethylaniline reacts with nitrous acid and so does the corresponding 2,6-dimethyl derivative, but the rate of reaction of the latter is much slower. Discuss.

8. Convert:

(i) PhH ⟶ *o*-ethylaniline; (ii) PhMe ⟶ 2-Br-4-Me-aniline;

(iii) PhH ⟶ *m*-chloroaniline; (iv) PhH ⟶ *p*-iodoaniline;

(v) PhH ⟶ 2,4-dibromoaniline; (vi) PhH ⟶ 4-Br-3-Cl-aniline;

(vii) PhMe ⟶ *p*-aminobenzoic acid; (viii) PhH ⟶ 2,5-dichloroaniline;

(ix) *p*-toluidine ⟶ 4-NH_2-3-NO_2-toluene.

9. When a mixture of *N*-chloro-2,4-dichloroacetanilide and anisole (PhOMe) is dissolved in hydrochloric acid, among the products isolated are *o*- and *p*-chloroanisole. Discuss.

10. The N.M.R. spectrum of a compound, $C_8H_{11}N$, showed signals with τ-values at 8·6 (doublet), 8·4 (singlet), 5·9 (quartet), and 2·7 (singlet). Suggest a structure and give your reasons.

11. The mass spectrum of *N,N*-dimethylaniline contains the following ions: 121 (70%), 120 (100%), 105 (13%), 77 (25%), 51 (16%). Suggest paths by which they may be formed.

24 Diazonium salts and their related compounds

(Solutions on page 268)

1. $PhN_2{}^+Cl^-$ couples with $PhNMe_2$ but not with 2,6-$Me_2C_6H_3NMe_2$. Explain.

2. Convert:

(i) 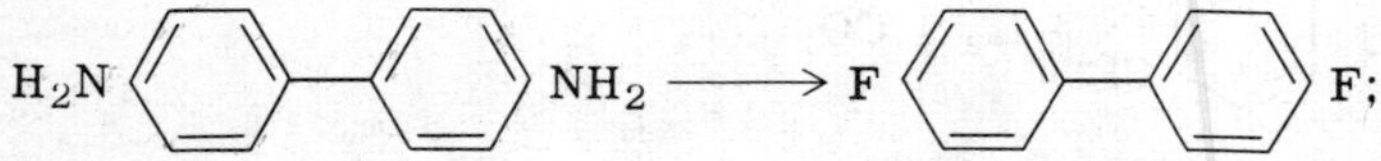

(ii) *m*-dinitrobenzene ⟶ *m*-chloronitrobenzene;
(iii) benzaldehyde ⟶ *m*-chlorobenzaldehyde;
(iv) *o*-chloronitrobenzene ⟶ *o*-bromochlorobenzene;
(v) *p*-bromoaniline ⟶ *p*-bromoiodobenzene;
(vi) 2-Br-4-Me-aniline ⟶ *m*-bromotoluene;
(vii) *m*-aminobenzaldehyde ⟶ *m*-hydroxybenzaldehyde;
(viii) *o*-toluidine ⟶ *o*-methylphenyl cyanide;
(ix) *p*-nitroaniline ⟶ *p*-dinitrobenzene; (x) *o*-nitrotoluene ⟶ *o*-cresol;
(xi) nitrobenzene ⟶ pentabromobenzene.

3. *p*-Toluidine reacts with benzenediazonium chloride to form a compound which, on boiling with H_2SO_4, gives *four* products (excluding nitrogen). Discuss.

4. Convert:
(i) PhH ⟶ *m*-dibromobenzene; (ii) PhMe ⟶ 3-Br-4-NO_2-toluene;
(iii) PhH ⟶ *m*-difluorobenzene; (iv) PhMe ⟶ 3, 4-dinitrotoluene;
(v) $PhNH_2$ ⟶ 1, 3, 5-tribomobenzene.

5. Discuss, including the use of thermochemical data, the reaction:

$$PhN_2^+Cl^- + CH_2(CN)CO_2Et \xrightarrow{AcONa} PhNHN{=}C(CN)CO_2Et$$

6. Complete the following equations:

(i) 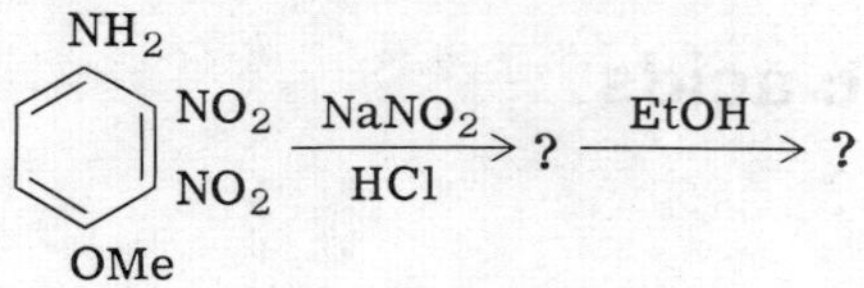

(ii) $PhN{=}NNHPh + PhOH \xrightarrow{HCl} ?$

(iii) 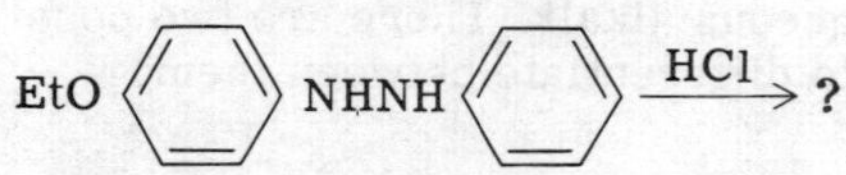

(iv) 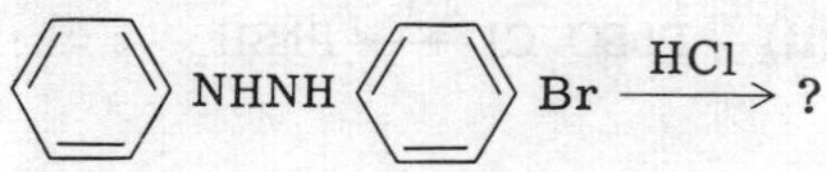

(v) 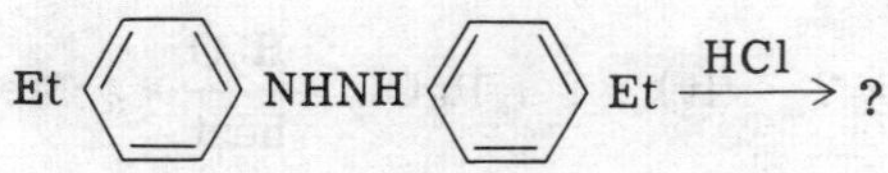

(vi) 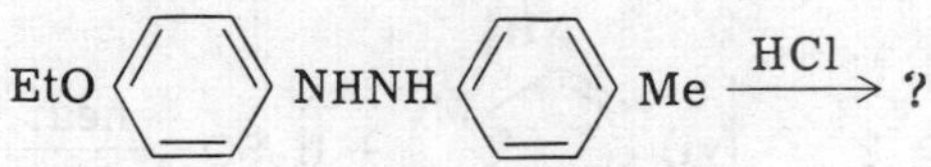

(vii) 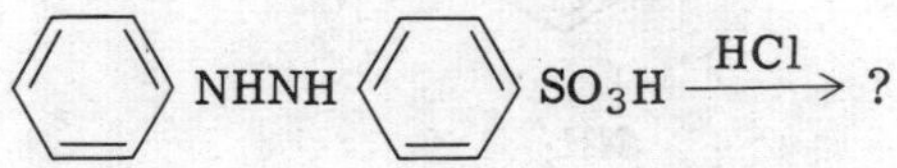

(viii) 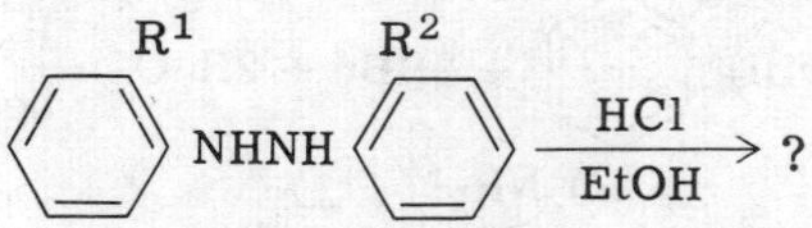

(I) + (II), where (I): $R^1 = R^2 = OMe$; (II): $R^1 = R^2 = OEt$

(ix)

(x) $PhN_2{}^+BF_4{}^- + PhNO_2 \xrightarrow{\text{heat}} ?$

(xi) $p\text{-}R^1C_6H_4N{=}NC_6H_4R^2\text{-}p \xrightarrow{MeCO_3H} ?$

(xii) 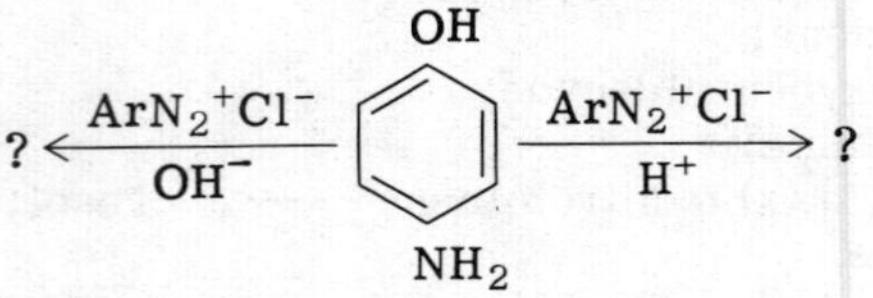

(xiii) $p\text{-}ClC_6H_4N_2{}^+Cl^- + CH_2{=}CHCHO \xrightarrow{Cu^{2+}} ?$

25 Aromatic sulphonic acids

(Solutions on page 276)

1. Arenesulphonic esters are hydrolysed by aqueous alkali. There are two possible mechanisms. What are they? Devise an experiment to differentiate between them.

2. Complete the equations:

(i) $p\text{-}MeC_6H_4SO_2Cl \xrightarrow{?} p\text{-}MeC_6H_4SO_2H$

(ii) $PhSO_2Cl \xrightarrow{?} PhSH$

(iii) 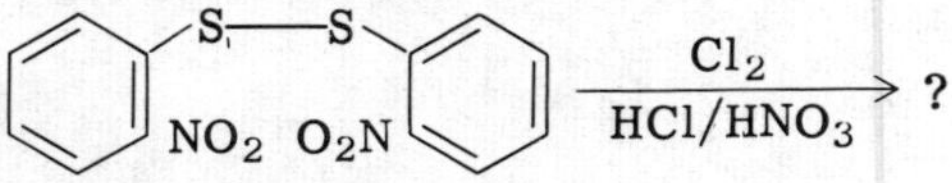

(iv) $? + HgCl_2 \xrightarrow[\text{heat}]{H_2O} p\text{-}MeC_6H_4HgCl$

(v) 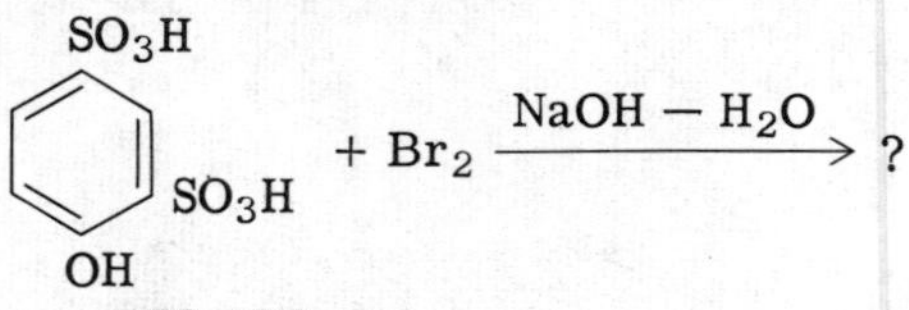

(vi) 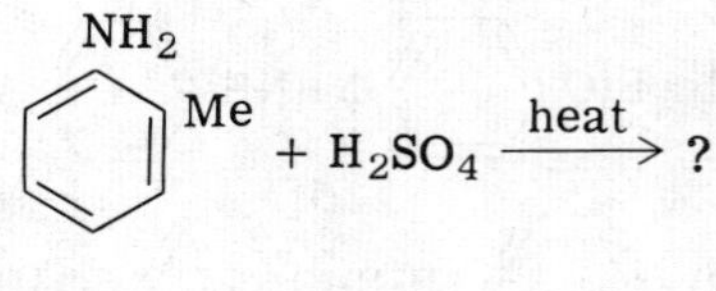

(vii) 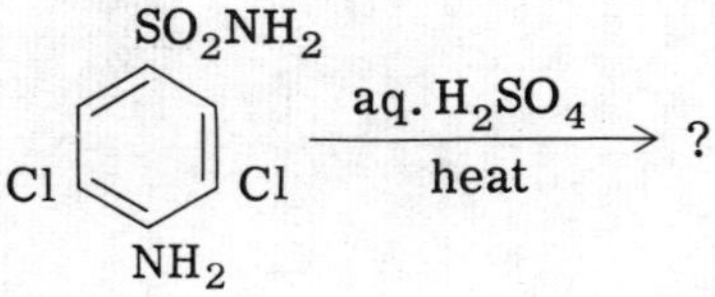

(viii) 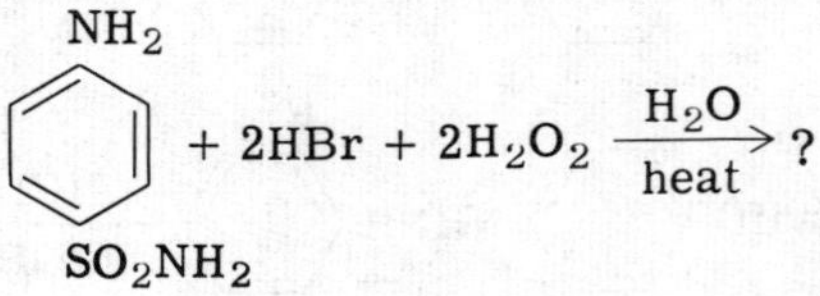

(ix) 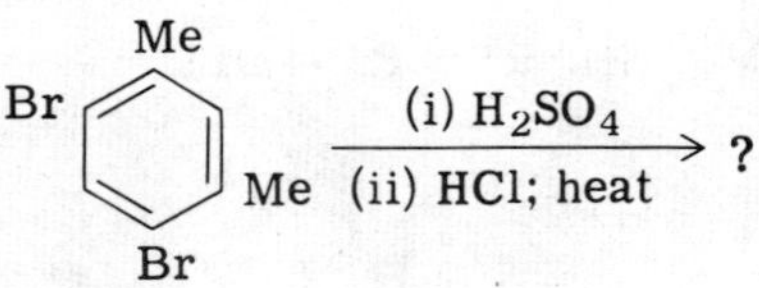

(x) $PhSO_3{}^-Na^+ + PCl_5 \longrightarrow ?$

(xi) $p\text{-MeC}_6H_4SO_2Cl + PhH \xrightarrow{AlCl_3}$? (xii) $PhSO_3Et + 2MeNH_2 \longrightarrow$?

(xiii) 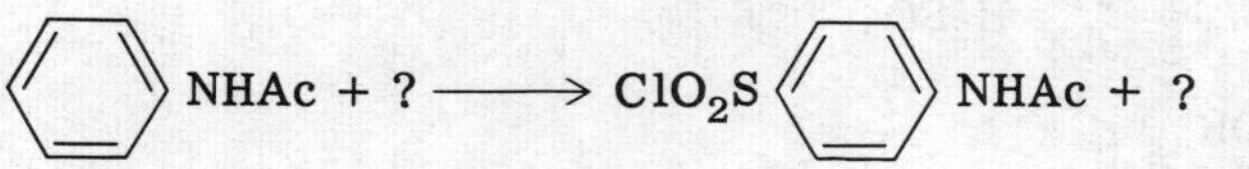

(xiv) 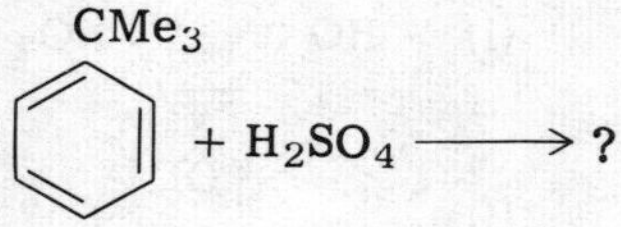

(xv) $PhSO_3Et + EtMgI \xrightarrow{Et_2O}$?

3. The sodium salt of sulphanilic acid can be readily acetylated with Ac_2O, but not the free acid. Explain.

4. Suggest a mechanism for the reaction

$$ArH + SO_2Cl_2 \xrightarrow{AlCl_3} ArSO_2Cl + HCl$$

5. Starting from *p*-chloronitrobenzene, prepare bis(4-aminophenyl)sulphone.

26 Phenols and quinones

(Solutions on page 280)

1. How would the i.r. spectra of *o*- and *p*-hydroxybenzaldehyde differ?

2. How could you separate chemically a mixture of benzene, aniline, phenol, and benzoic acid?

3. Alcohols are weaker acids than phenols but are stronger nucleophiles. Explain.

4. Which in each pair is the more stable and why?

(i) $Ph\overset{+}{C}HMe$ and $PhCH_2CH_2^+$; (ii) PhO^- and $p\text{-NO}_2C_6H_4O^-$

5. *o*-Nitrophenol has a lower b.p. and lower water-solubility than the *p*-isomer. Explain.

6. *m*-t-Butylphenol reacts with chlorine to give a trichloro-derivative, with bromine to give a dibromo-derivative, and with iodine to give a mono-iodo-derivative. What are these products? Explain

7. 2,6-Di-t-butylphenol is a much weaker acid than phenol. Explain.

8. Discuss the following data:

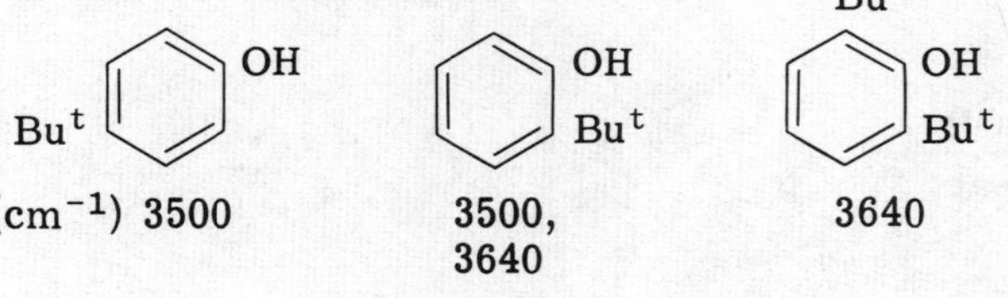

$OH(cm^{-1})$ 3500   3500, 3640   3640

9. Complete the equations:

(i) 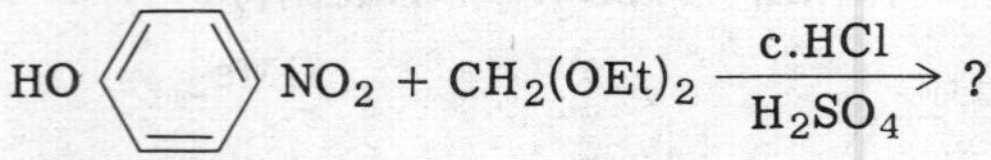

(ii) 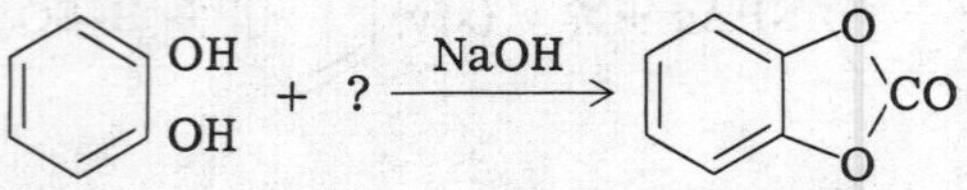

(iii) $PhOH + (AcO)_2Hg \xrightarrow{\text{heat}} ? \xrightarrow[H_2O]{NaCl} ?$

(iv) 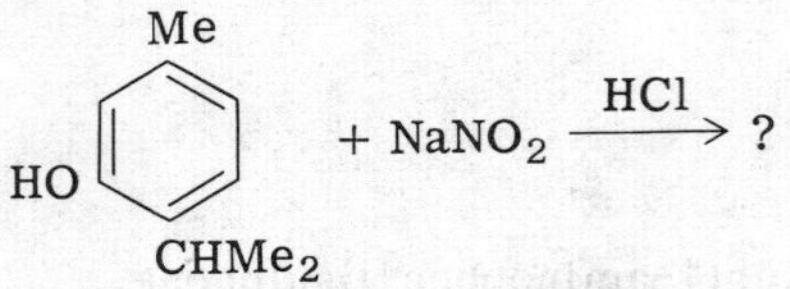

(v) $PhOH + ClCH_2CO_2H \xrightarrow[\text{heat}]{NaOH} ?$

(vi) 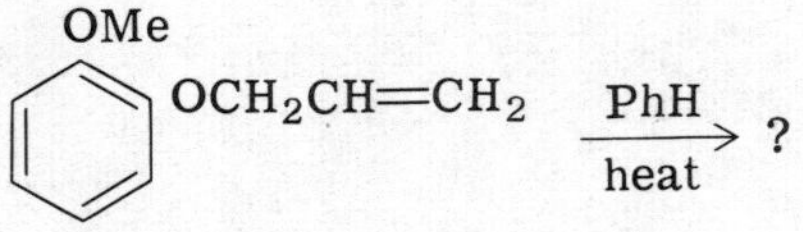

(vii) 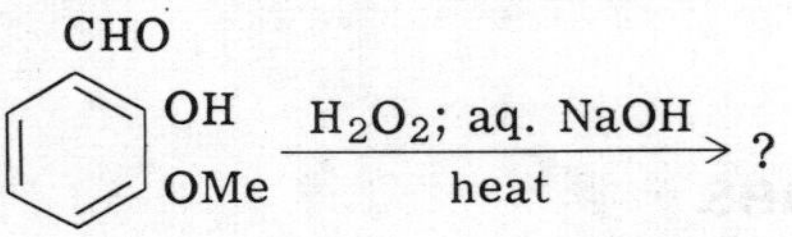

(viii) 2, 4, 6-tri-t-butylphenol $\xrightarrow[C_6H_6]{PbO_2} ? \xrightarrow{Br_2} ?$

(ix) $PhOH + CH_2{=}CHCH_2Br \xrightarrow{K_2CO_3} ?$

(x) 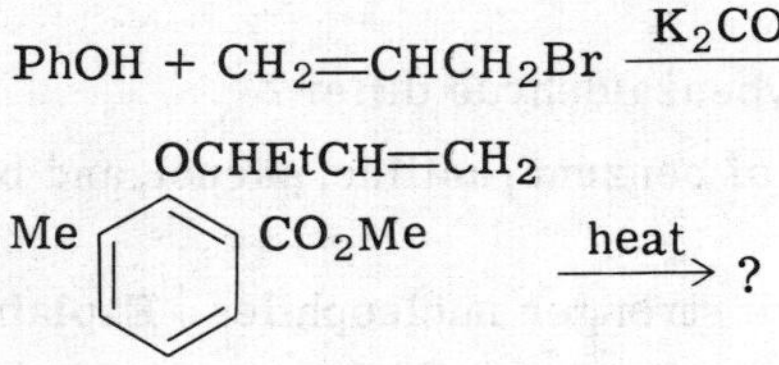

(xi) 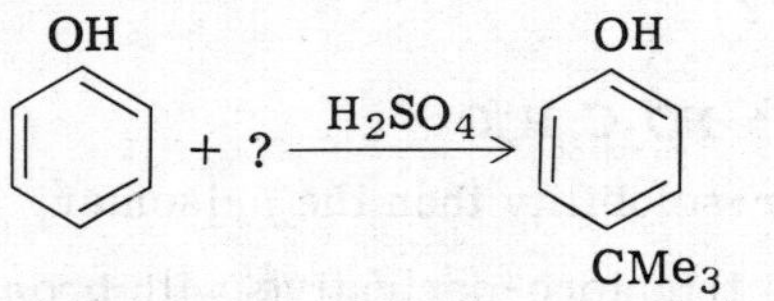

(xii) 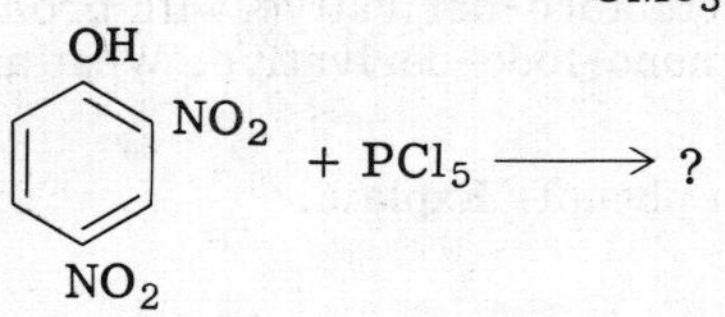

(xiii) 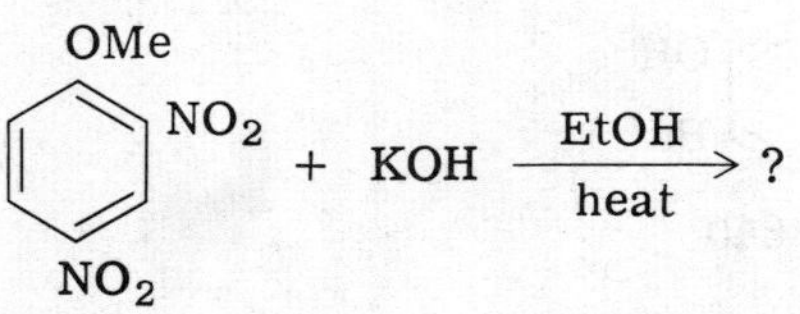

(xiv) 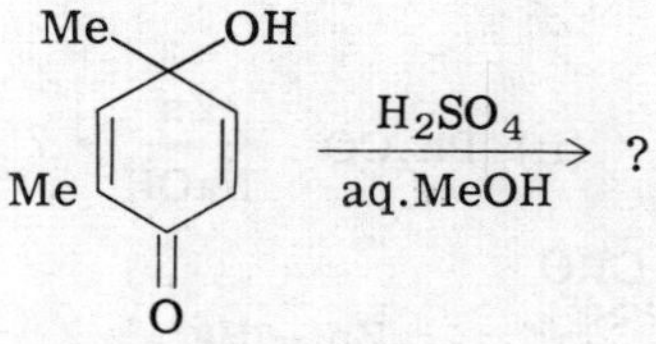

(xv) 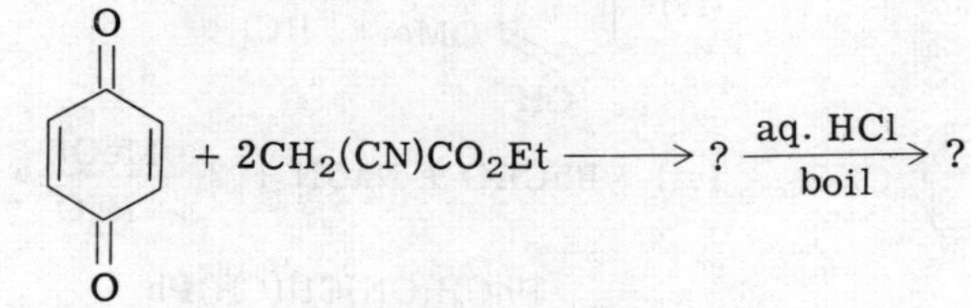

10. Phenyl triphenylmethyl ether, on treatment with acid, gives *p*-triphenylmethylphenol. Discuss.

11. How many N.M.R. signals would you expect from Ph—O—CH_2CH_2—O—Ph?

12. Convert:

(i) phenol ⟶ *p*-fluorophenol; (ii) PhH ⟶ *m*-$HOC_6H_4NO_2$;

(iii) PhCl ⟶ 2, 4-$(NO_2)_2$-phenol; (iv) PhOH ⟶ 2, 6-Br_2-4-NO_2-phenol;

(v) *o*-hydroxyacetophenone ⟶ 2, 5-dihydroxyacetophenone;

(vi) *o*-MeC_6H_4OH ⟶ *o*-$HOC_6H_4CO_2H$.

13. The Wurster salt derived from *p*-$Me_2NC_6H_4NMe_2$ is very stable, whereas that derived from the corresponding 2, 3, 5, 6-tetramethyl compound (durene derivative) cannot be detected. Explain.

14. Account for the following ions in the mass spectra of:

(i) *o*-ethylphenol: 122 (40%), 107 (100%), 79 (9%), 77 (18%), 51 (5%); (ii) n-butyl phenyl ether: 150 (19%), 107 (2%), 94 (100%), 93 (1%), 66 (4%), 65 (4%).

27 Aromatic alcohols, aldehydes and ketones

(Solutions on page 291)

1. Distinguish between PhCOEt and *p*-MeC_6H_4COMe by:

(i) a chemical method; (ii) a physical method.

2. Complete the following equations:

(i) 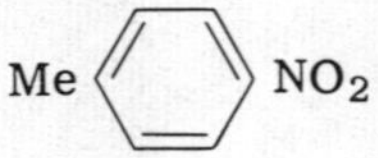 $\xrightarrow[H_2SO_4]{Ac_2O;CrO_3}$? $\xrightarrow{H^+}$?

(ii) $Ph_2CO \xrightarrow[NaOH]{Zn}$?

(iii) $p\text{-}MeC_6H_4CHO \xrightarrow[KOH]{HCHO}$?

(iv) 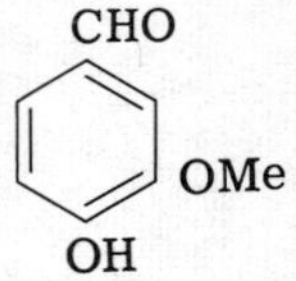 $\xrightarrow[HCl]{Zn-Hg}$?

(v) PhMe + ? $\xrightarrow{?}$ 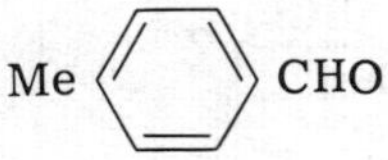

(vi) PhCHO + NaCN + ? $\xrightarrow[H_2O]{MeOH}$ PhCH(CN)CH(CN)Ph

(vii) 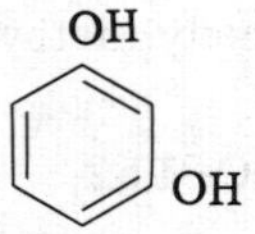 + $MeCO_2H \xrightarrow[heat]{ZnCl_2}$?

(viii) 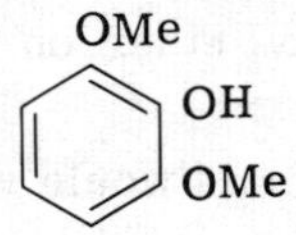 + $(CH_2)_6N_4 \xrightarrow{H_3BO_3}$?

(ix) 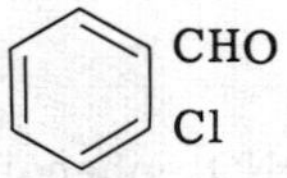 + $Cl_2 \xrightarrow{heat}$?

(x) PhCH=CHCOPh + KCN $\xrightarrow{AcOH}$?

(xi) $PhCH_2OH$ + ? $\longrightarrow PhCH_2OCOCl$

(xii) PhCHBrCHBrCHO $\xrightarrow[heat]{K_2CO_3}$?

(xiii) PhCH(CN)COMe $\xrightarrow[heat]{aq.H_2SO_4}$?

(xiv) PhCOMe $\xrightarrow{?}$ PhCOCHO

(xv) 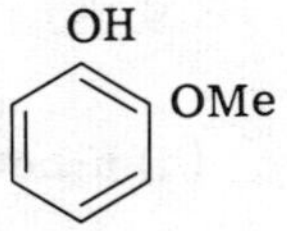 $\xrightarrow[H^+]{CH_2O}$? $\xrightarrow[Cu]{PhNHOH}$? $\xrightarrow{H^+}$?

(xvi) 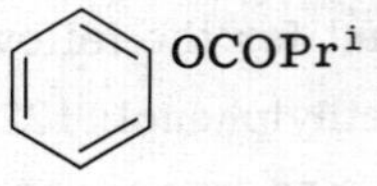 $\xrightarrow{AlCl_3}$?

(xvii) $PhCO_2H \xrightarrow{SOCl_2}$? $\xrightarrow{Me_2NH}$? $\xrightarrow{?}$ PhCHO

3. Which of the carbonyl groups in $p\text{-}MeOC_6H_4COMe$ and $p\text{-}NO_2C_6H_4COMe$ protonates more readily in acid solution and why?

4. Starting with PhH, $EtCO_2H$ and $PhNH_2$, synthesise $PhCH_2CH_2CONHPh$.

5. Convert PhCHO into PhCH=CHCOPh by *two* methods.

6. Convert $Br(CH_2)_3CHO$ into $PhCHOH(CH_2)_3CHO$.

7. What are the products formed when the *syn*-phenyl and *anti*-phenyl monoximes of benzil undergo the acid-catalysed Beckmann rearrangement?

8. Suggest a mechanism for the rearrangement: $Ph_2CHCHO \xrightarrow{H^+} PhCH_2COPh$.

9. Discuss the two reactions:

$$PhCOMe \xrightarrow[NaOH]{NaOCl} PhCO_2^- + CHCl_3$$

$$2,6\text{-}Me_2C_6H_3COMe \xrightarrow[NaOH]{NaOCl} 2,6\text{-}Me_2C_6H_3COCCl_3$$

10. Suggest what A and B could be in the reactions:

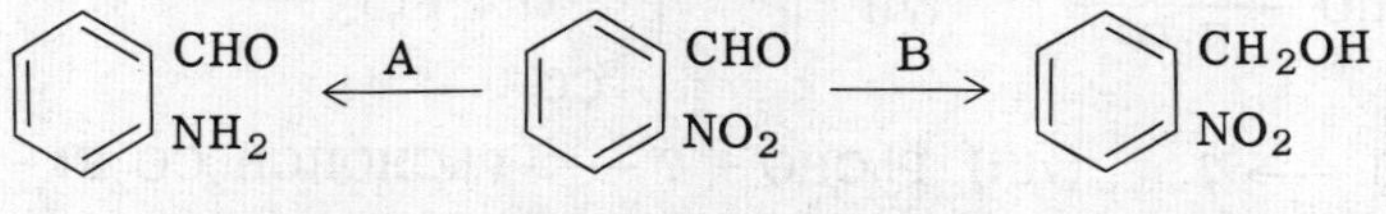

11. The deamination of $Ph_2C(OH)CH_2NH_2$ ($NaNO_2$—HCl) gives a product which, on oxidation, gives benzoic acid only. Formulate the steps involved.

12. Convert:
(i) *p*-xylene ⟶ 2, 5-$Me_2C_6H_3CHO$; (ii) PhH ⟶ *m*-nitroacetophenone;
(iii) $PhNO_2$ ⟶ *m*-bromo-benzaldehyde; (iv) PhH ⟶ *p*-$OHCC_6H_4CHO$;
(v) PhMe ⟶ *p*-$MeC_6H_4CH{=}CHNO_2$.

13. Compound A, $C_8H_6O_2$, on treatment with aq. NaOH followed by acidification, gives B, $C_8H_8O_3$, which, on oxidation, gives benzoic acid only. Suggest structures for A and B, and formulate the reactions.

14. The mechanism of the Cannizzaro reaction involves addition of hydrogen to a benzaldehyde molecule. Devise an experiment to show whether the hydrogen comes from the aqueous solvent or not.

15. Suggest paths whereby *o*-methylbenzyl alcohol can produce the following ions in its mass spectrum: 122 (30%), 120 (1%), 119 (4%), 104 (100%), 91 (36%).

16. Account for the following ions in the mass spectrum of methyl salicylate (methyl *o*-hydroxybenzoate): 152 (41%), 121 (32%), 120 (100%), 93 (13%), 92 (58%), 65 (26%), 64 (17%).

28 Aromatic acids

(Solutions on page 303)

1. Acetyl chloride is rapidly decomposed by water, but benzoyl chloride only very slowly. Suggest an explanation.

2. Complete the equations:

(i) $PhCO(CH_2)_2CO_2H \xrightarrow{?} Ph(CH_2)_3CO_2H$

(ii) $Ph_2C(OH)CO_2H \xrightarrow[\text{heat}]{I_2/P} ?$

(iii) 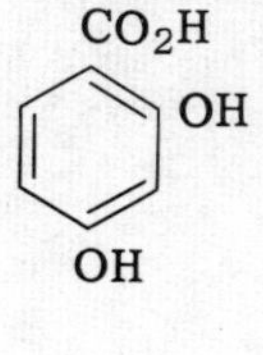

$+ Br_2 \xrightarrow{AcOH} ?$

(iv) 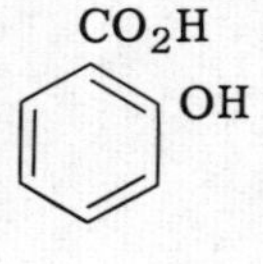

$+ 2ICl \longrightarrow ?$

(v) 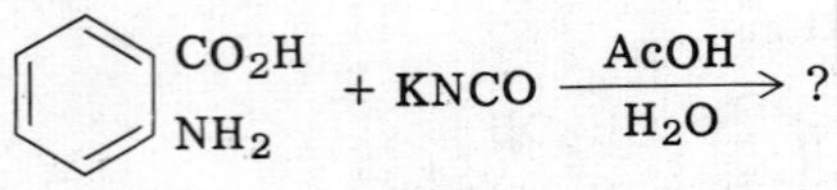

(vi) 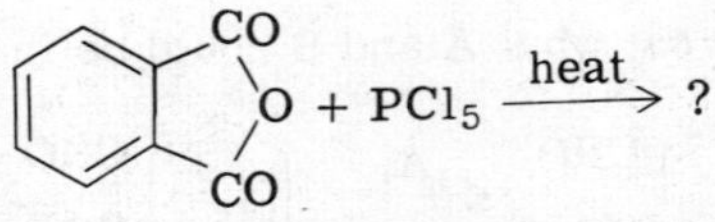

(vii) $PhCOCl + HF\ (gas) \longrightarrow ?$

(viii) $PhCHO + ? \longrightarrow PhCHOHCH_2CO_2Et \xrightarrow{heat} ?$

(ix) 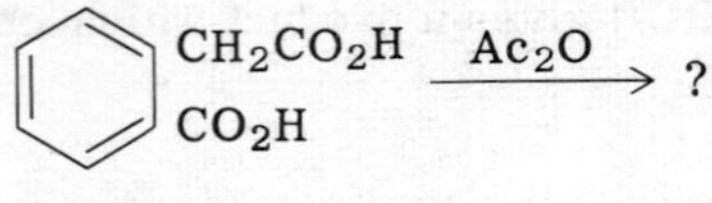

(x) 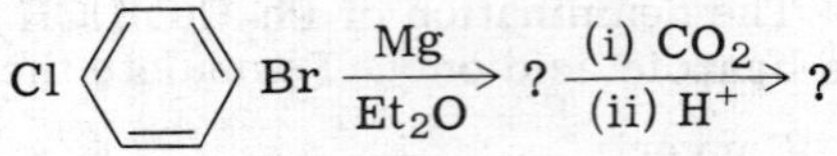

(xi) $CH_2CH_2CH_2CO$ (ring closed through O) $\xrightarrow[(ii)\ SOCl_2]{(i)\ HCl} ? \xrightarrow[AlCl_3]{PhH} ? \xrightarrow[MeOH]{KOH}$ cyclopropane derivative?

(xii) $PhH + ? \xrightarrow{AlCl_3} PhCOCH{=}CHCO_2H$

(xiii) 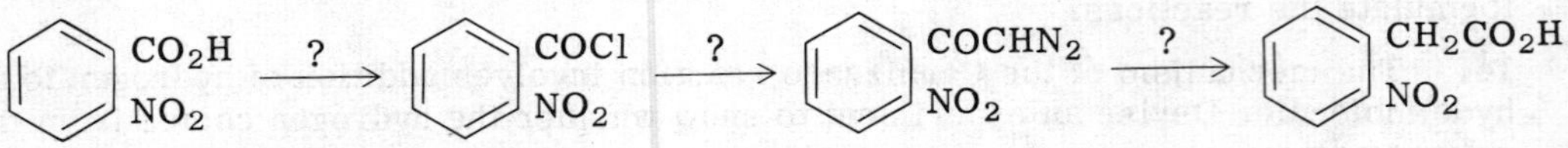

(xiv) $PhCOMe \xrightarrow{HNO_2} ? \xrightarrow{Ac_2O} ? \xrightarrow{H^+} ?$

(xv) 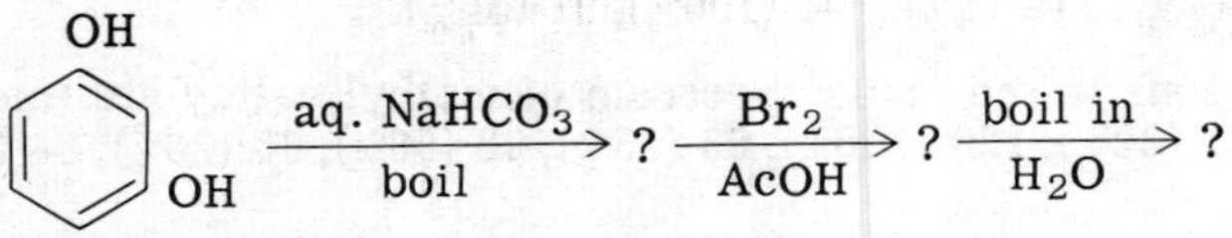

3. Deduce the structures of A, B and C:

$$A, C_9H_{12} \xrightarrow{KMnO_4} B, C_8H_6O_4 \xrightarrow[Fe]{Br_2} C, C_8H_5BrO_4 \text{ (1 product only)}$$

4. Convert:

(i) p-$MeCOC_6H_4CO_2H \longrightarrow p$-$MeCHOHC_6H_4CO_2H$; (ii) $PhMe \longrightarrow o$-BrC_6H_4COCl;
(iii) $PhH \longrightarrow p$-$ClC_6H_4COC_6H_4Br$-p; (iv) $PhH \longrightarrow$ 4-NH_2-2-OH-benzoic acid.

5. Convert toluene into *m*-toluic acid by *two* methods.

6. The sodium salt of 2-OH-*cis*-cinnamic acid (coumarinic acid), on acidification, gives coumarin, whereas the sodium salt of 2-OH-3-nitro-*cis*-cinnamic acid gives the free acid. Suggest an explanation.

7. Starting from benzene and any other compounds you like, synthesise:

(i) $Ph(CH_2)_3CO_2H$; (ii) $PhCH_2COPh$; (iii) 4-OH-3-MeO-cinnamic acid.

29 Polynuclear hydrocarbons and their derivatives

(Solutions on page 312)

1. Arrange the following in order of increasing stability and give your reasons: $PhCH_2^+$, Ph_3C^+, Me^+, Ph_2CH^+.

2. Using PhMgBr, what substrates would lead to Ph_3COH?

3. The rate of reaction of benzhydryl chloride (Ph_2CHCl) in aqueous EtOH with KF to give benzhydryl fluoride is retarded by the addition of NaCl. Explain.

4. Devise an experiment to show whether the rearrangement of naphthalene-1-sulphonic acid to the 2-sulphonic acid is intramolecular or intermolecular.

5. Complete the equations:

(i) 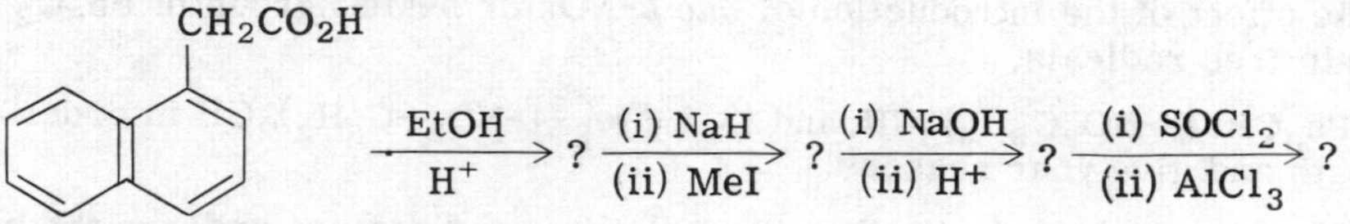

(ii) 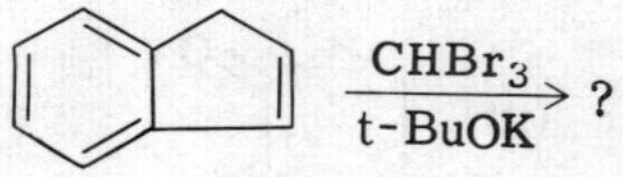

(iii) $PhNH_2 \cdot HCl + Ph_3COH \xrightarrow{AcOH}$? (not an *N*-deriv.)

(iv) 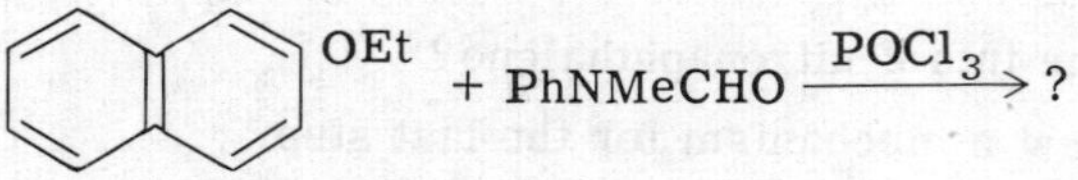

(v) 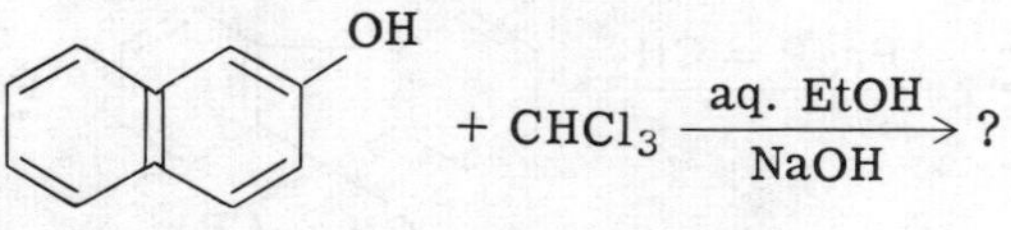

(vi) 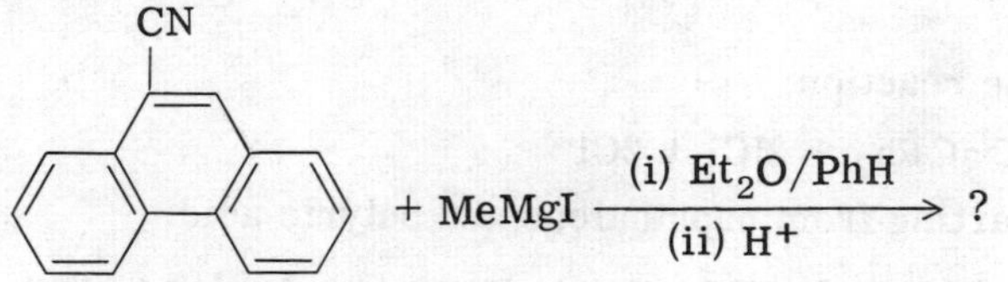

(vii) 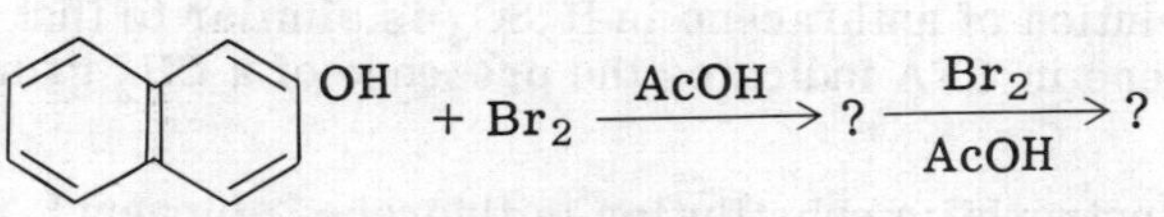

(viii) $PhCHOHCOPh \xrightarrow[\text{aq. NaOH; heat}]{NaBrO_3}$? (rearr.)

(ix) 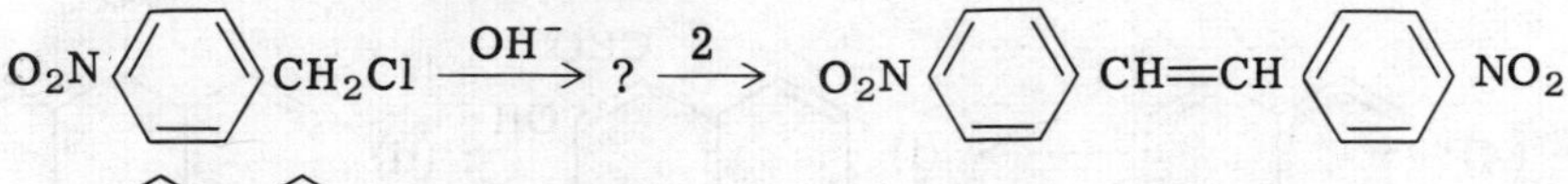

(x) 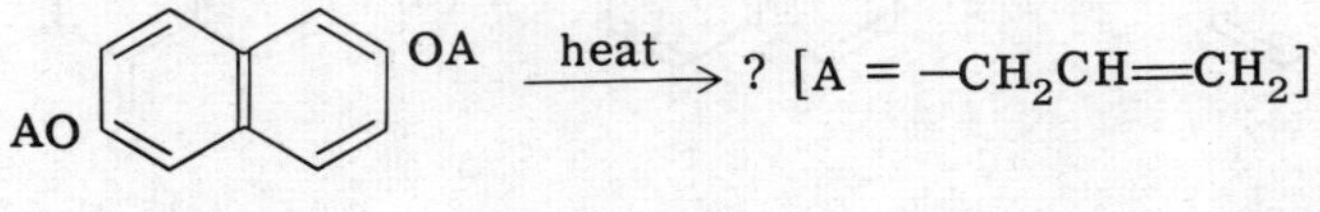

(xi) 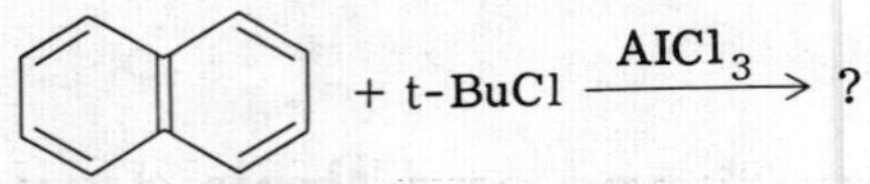

(xii) 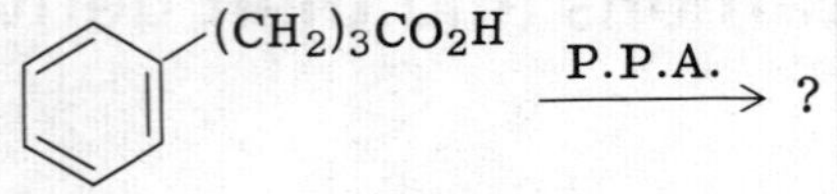

(xiii) ? 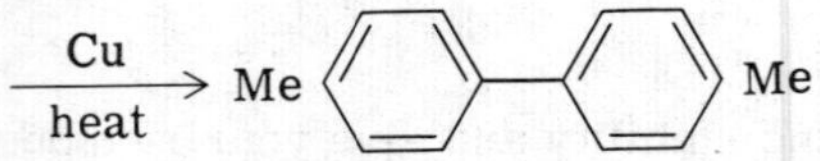

(xiv) 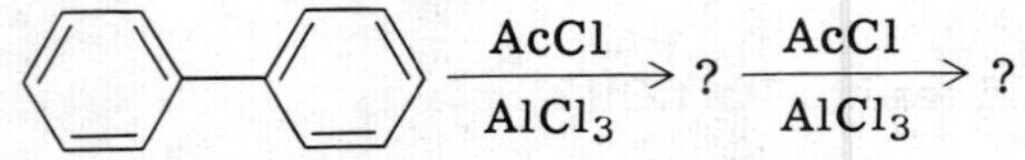

(xv) 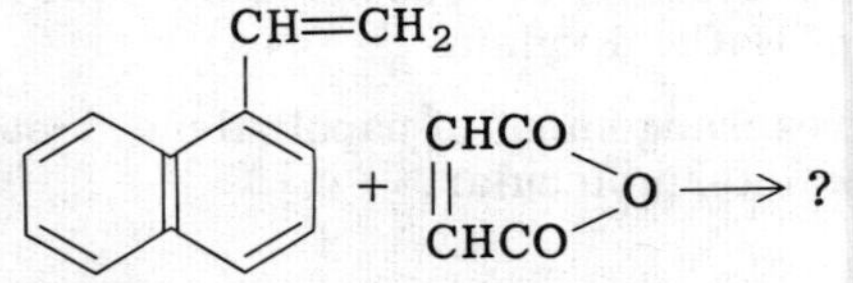

(xvi) 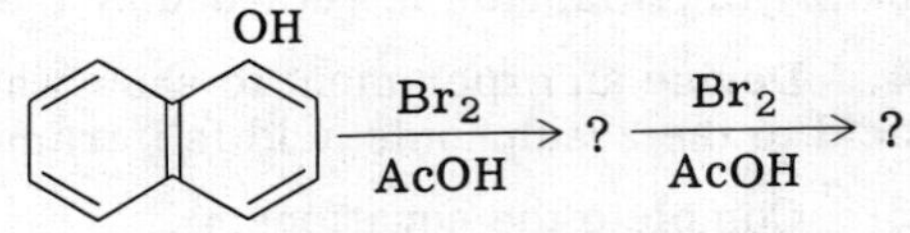

6. Explain the effect of the introduction of *one* p-NO_2 or p-MeO group in Ph_6C_2 on its dissociation into free radicals.

7. Arrange Ph_3CH, $(p\text{-}NO_2C_6H_4)_3CH$, and $(3,5\text{-}Me_2\text{-}4\text{-}NO_2\text{—}C_6H_2)_3CH$ in order of increasing acid strength and give your reasons.

8. What are the expected products when the following α-diketones undergo the benzilic acid rearrangement?

(i) $HO_2CCH_2COCOCH_2CO_2H$; (ii) $MeCOCOCO_2Et$; (iii) 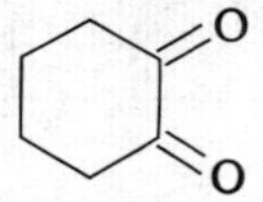

9. How could you convert naphthalene into 2-nitronaphthalene?

10. Complete the equation and suggest a mechanism for the last step.

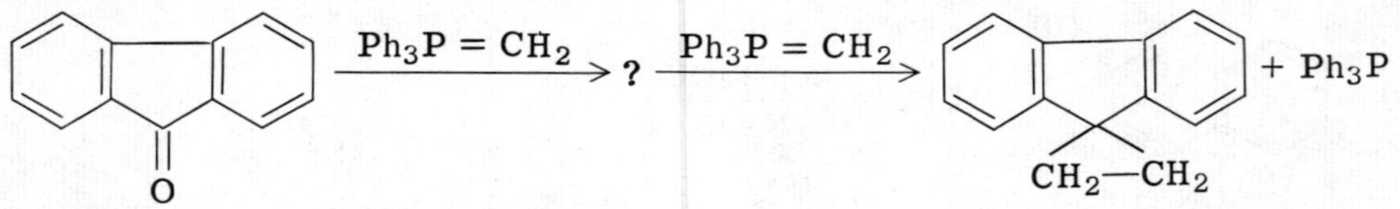

11. Suggest a mechanism for the reaction:

$$2Ph_2CCl_2 + 2NaI \longrightarrow Ph_2C{=}CPh_2 + 2ICl + 2Cl^-$$

12. Synthesise phenanthrene, starting from naphthalene and butyric acid.

13. (i) The u.v. spectrum of a solution of anthracene in H_2SO_4 is similar to that of Ph_2CH^+; (ii) The NMR spectrum of anthracene in TFA indicates the presence of a CH_2 group. Offer an explanation.

14. The order of reactivity to electrophilic substitution is fluorene > biphenyl > benzene. Offer an explanation.

15. Convert: (*A*) 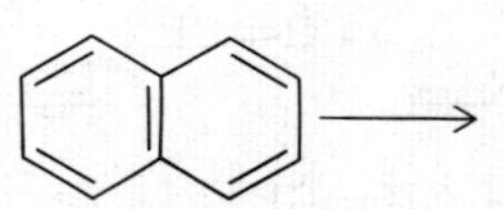 (i) 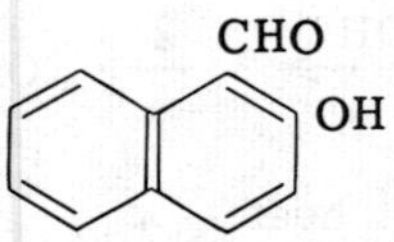

; (ii) 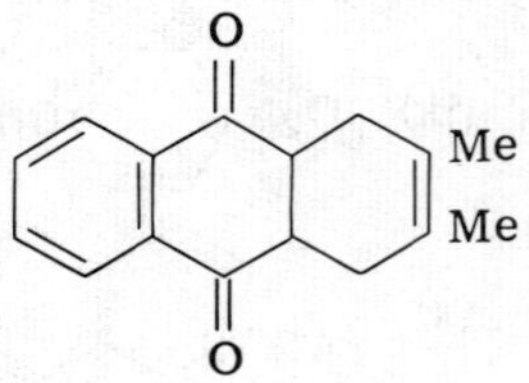

;

(iii) 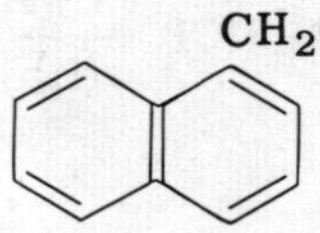 ; (iv) 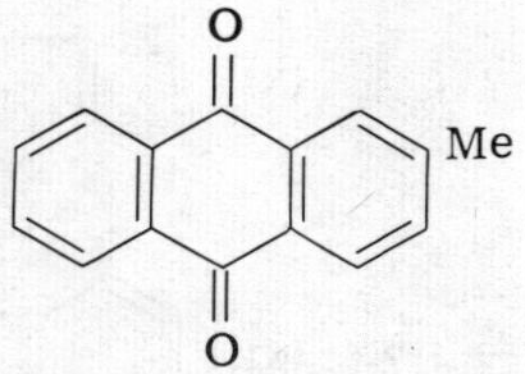 ;

(*B*) (v) PhMe ⟶ $Ph_2CHCHPh_2$; (vi) PhH ⟶ Ph_3CCO_2H;

(vii) PhOMe ⟶ (*a*) 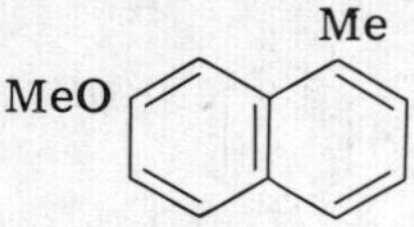 *and* (*b*) 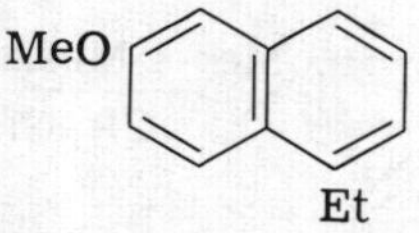

(viii) phenanthrene ⟶ phenanthrene-9-carboxylic acid.

16. Synthesise 2, 9-dimethylanthracene starting with compounds containing one benzene ring.

30 Heterocyclic compounds

(Solutions on page 327)

1. Pyrrole is more reactive than furan. Suggest a reason.

2. Complete the following equations:

(i) 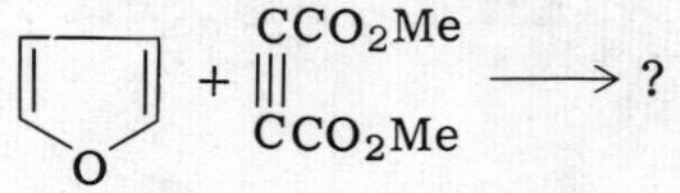 (ii) 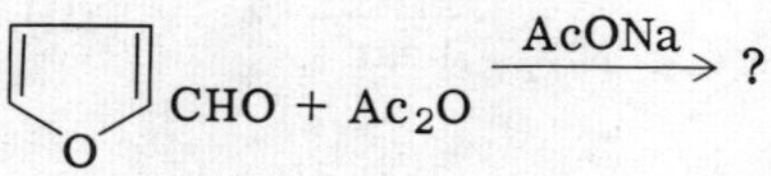

(iii) 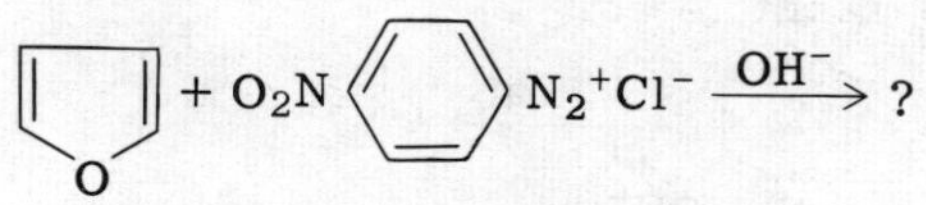 (iv) 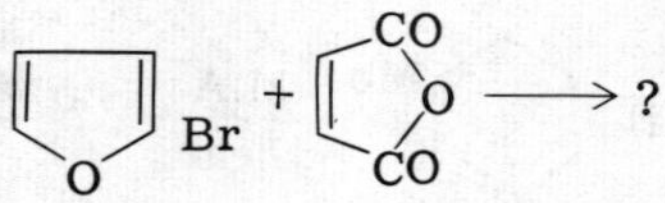

(v) 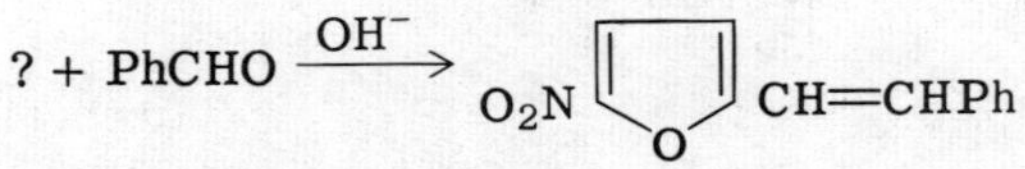 (vi) 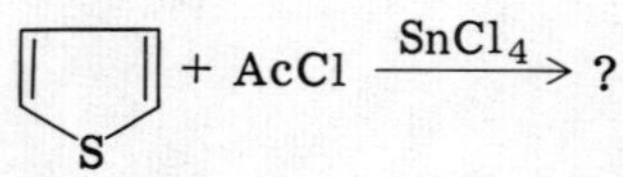

(vii) 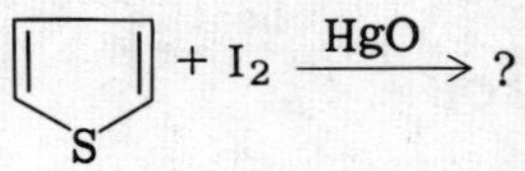 (viii) 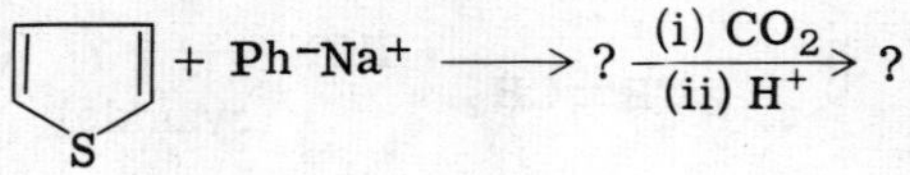

(ix) 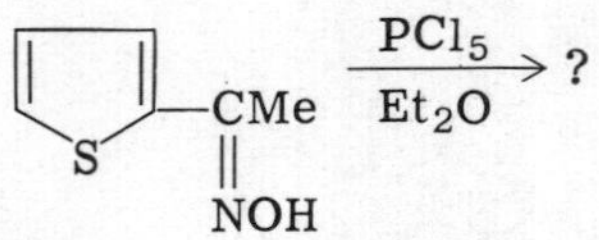

(x) 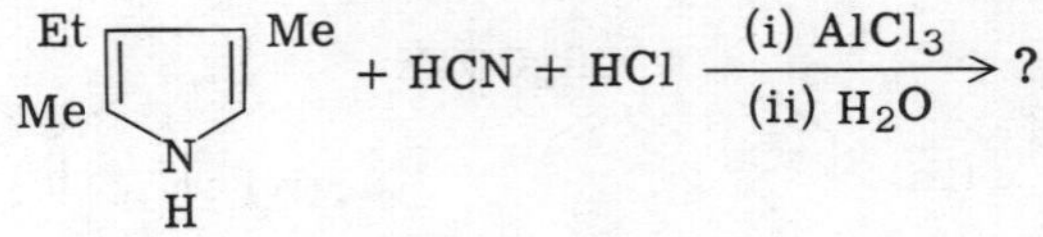

(xi) 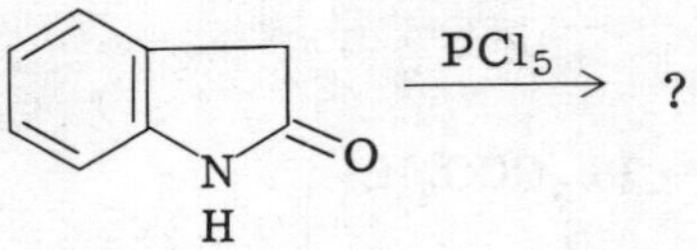

(xii) 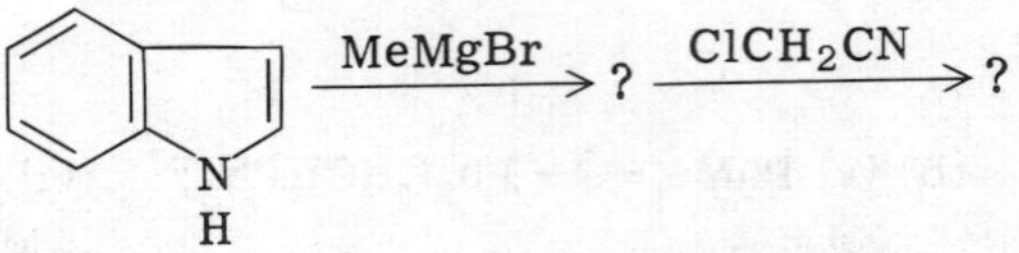

(xiii) 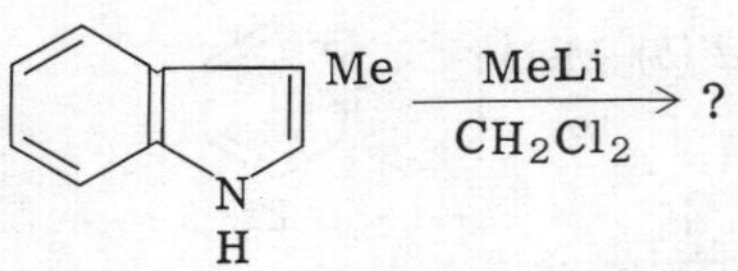

(xiv) 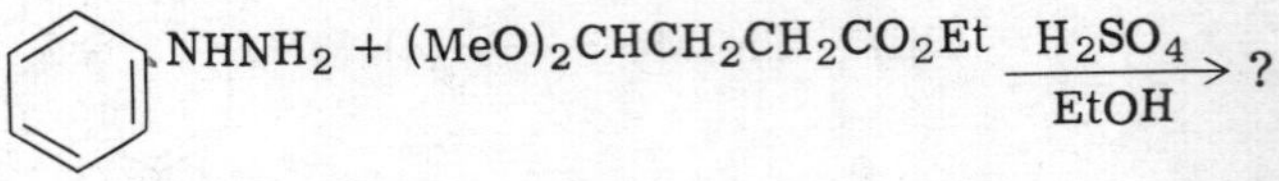

3. Suggest a mechanism for the reaction

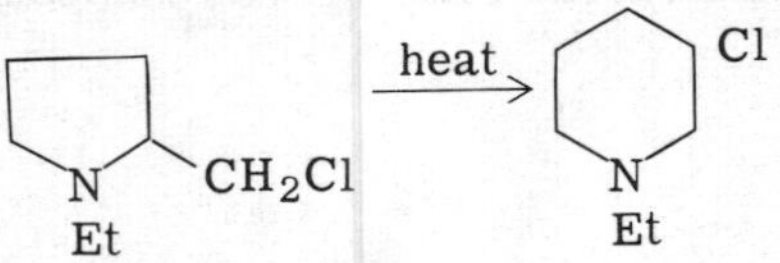

4. Suggest a mechanism for the reaction

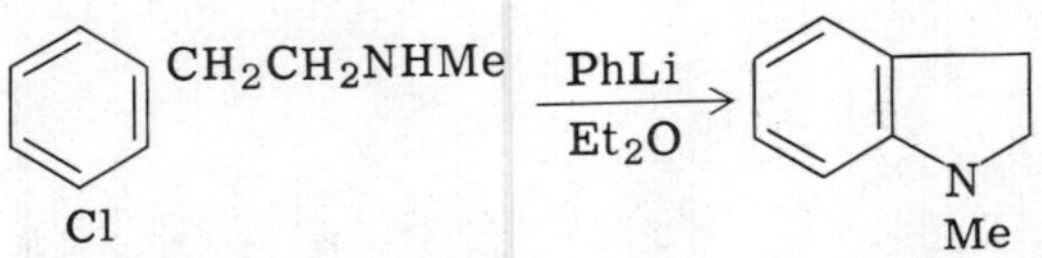

5. The nitration of furan with acetyl nitrate proceeds via a 2, 5-addition intermediate with the reagent. Formulate the complete reaction.

6. Pyrrole is much more acidic than s-alkylamines. Suggest a reason.

7. no action $\xleftarrow{\text{furan}}$ $CH_2O + Me_2NH$ $\xrightarrow{\text{2-Me-furan}}$ A

$\qquad\qquad\qquad\qquad$ $\xrightarrow{\text{pyrrole}}$ B

What are A and B? Explain.

8. Complete the equations:

(i) 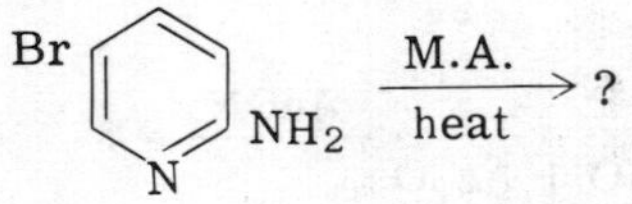

(ii) 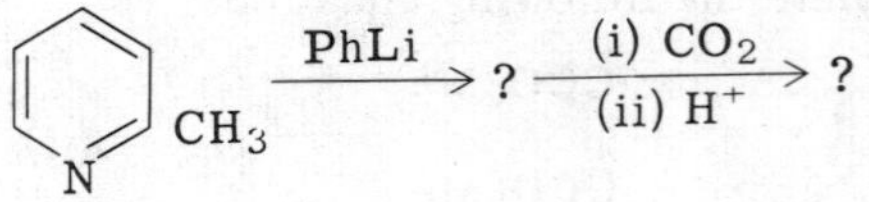

(iii) 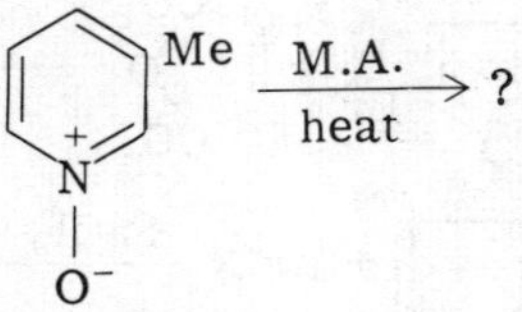

(iv) 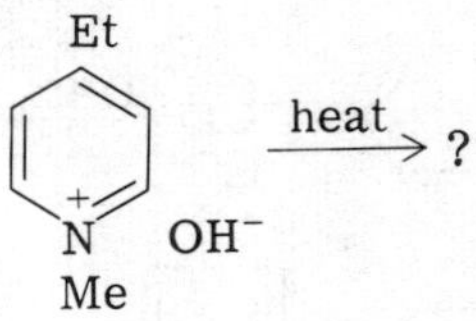

(v) 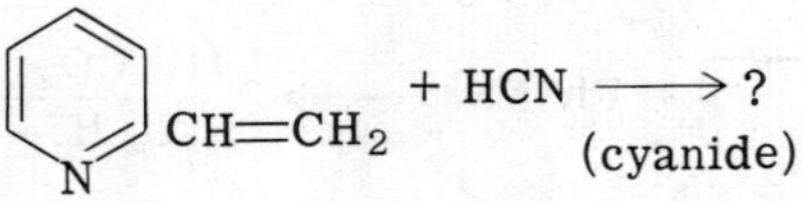

(vi) 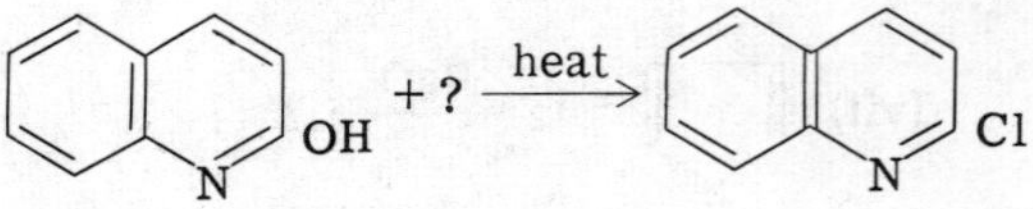

(vii) 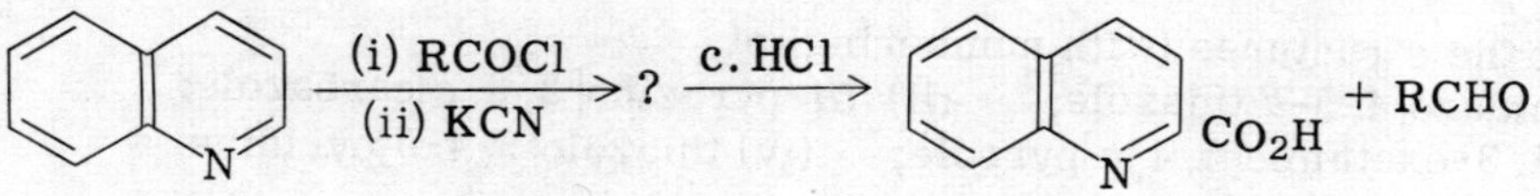

(viii) 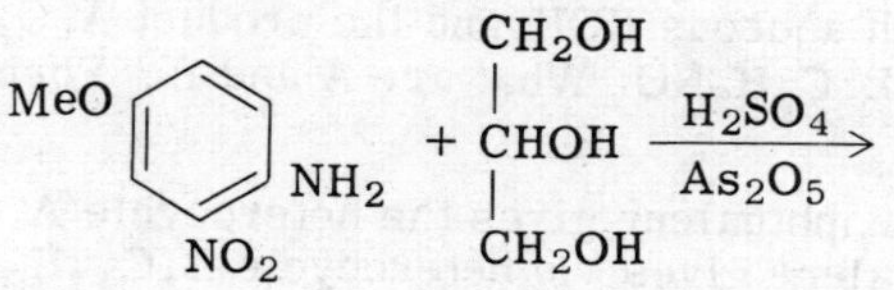

(ix) 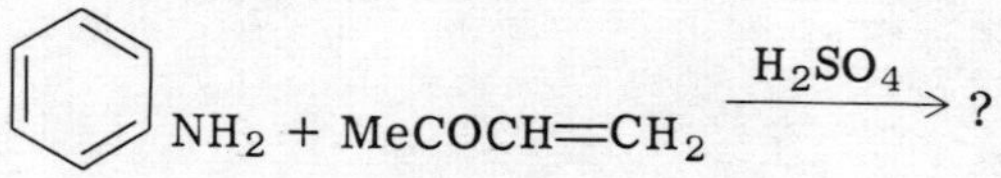

(x) 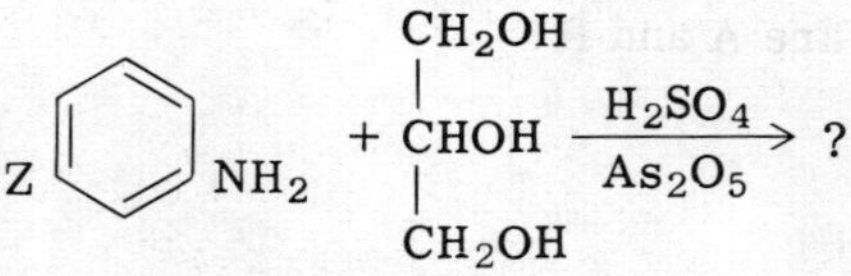

($Z = Cl, NO_2, OMe$)

9. Discuss the behaviour of the three picolines in aqueous NaOH containing D_2O.

10. Suggest a mechanism for the reaction

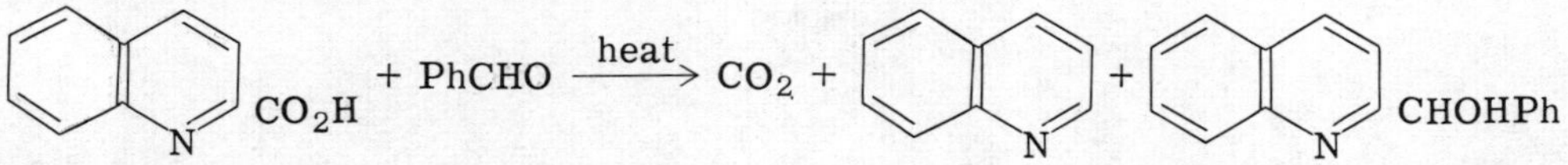

11. What products would you expect to get by the application of Hofmann's exhaustive methylation to:

(i) 3-methylpyrrolidine; (ii) 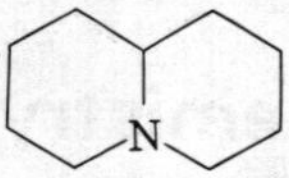 ; (iii) morpholine?

12. Complete:

(i) $2CH_2{=}CHCO_2Et + MeNH_2 \longrightarrow ? \xrightarrow{EtONa} ? \xrightarrow[heat]{H_3O^+} ?$

(ii) 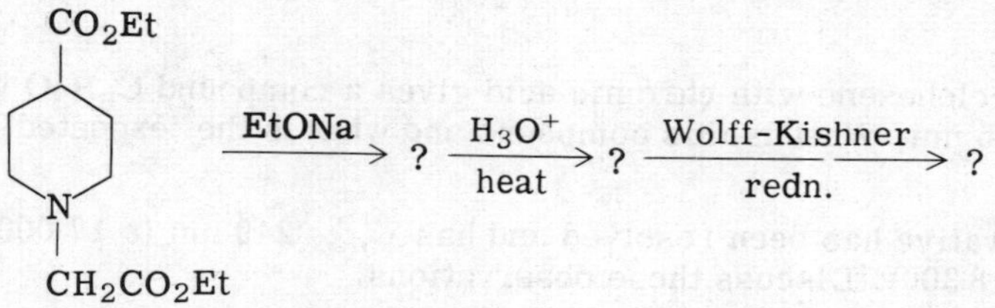

13. Suggest a synthesis for each of the following:
(i) 8-hydroxyquinoline from *o*-aminophenol;
(ii) 2-methyl-3-phenylindole from benzyl methyl ketone;
(iii) 3-aminopyridine from β-picoline; (iv) 2, 6-dimethylpyridine from E.A.A.;
(v) 4-nitroquinoline from quinoline;
(vi) 6-methoxyisoquinoline from *p*-methoxybenzaldehyde;
(vii) 5-nitroquinoline from benzene; (viii) 1-methylisoquinoline from benzene.

14. Arrange the following in order of increasing basicity and give your reasons: pyridine, 4-amino-, 4-methyl-, and 4-cyano-pyridine.

15. Name the following:

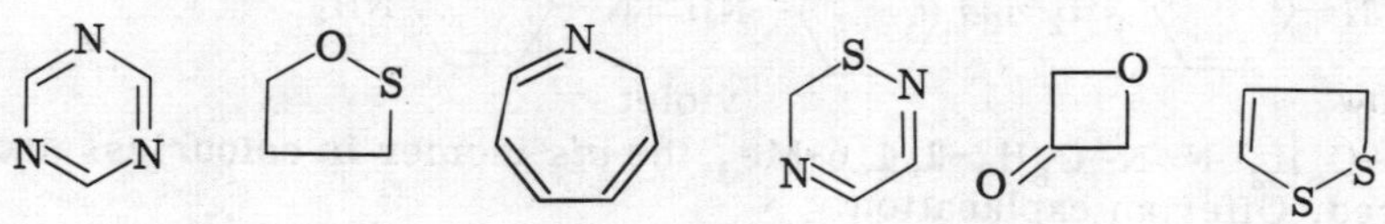

16. Write out the structures (with numbering) of
(i) 4*H*-imidazo[4, 5-*d*]thiazole; (ii) 7*H*-pyrazino[2, 3-*c*]carbazole;
(iii) 4*H*-1, 3-oxathiolo[5, 4-*b*]pyrrole; (iv) thiazolo[5, 4-*b*]pyridine.

17. Pyridine methiodide readily reacts with aqueous KOH, and the product A, C_6H_9NO, on oxidation with potassium ferricyanide, gives B, C_6H_7NO. What are A and B? Formulate the reaction.

18. The Skraup synthesis on 2-NH_2-1-Me-naphthalene gives the heterocycle A, $C_{14}H_{11}N$. The same synthesis on 2-NH_2-1-NO_2-naphthalene gives the heterocycle B, $C_{13}H_9N$. What are A and B?

31 Dyes and photochemistry

(Solutions on page 345)

1. The oxidation of cyclohexene with chromic acid gives a compound C_6H_8O which shows a strong u.v. band at 225 nm. What is this compound and what is the 'expected' product that is also obtained?

2. This biphenyl derivative has been resolved and has λ_{max} 249 nm (ε 17 000). Biphenyl has λ_{max} 251·5 nm (ε 18 300). Discuss these observations.

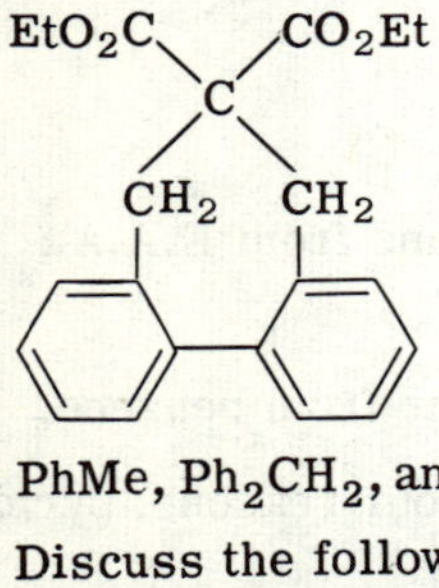

3. PhMe, Ph_2CH_2, and Ph_3CH have similar u.v. spectra (λ_{max} ~262 nm). Explain.

4. Discuss the following spectral data (nm):
(i) PhOH (210), PhOMe (217), PhO^- (235).

(ii) 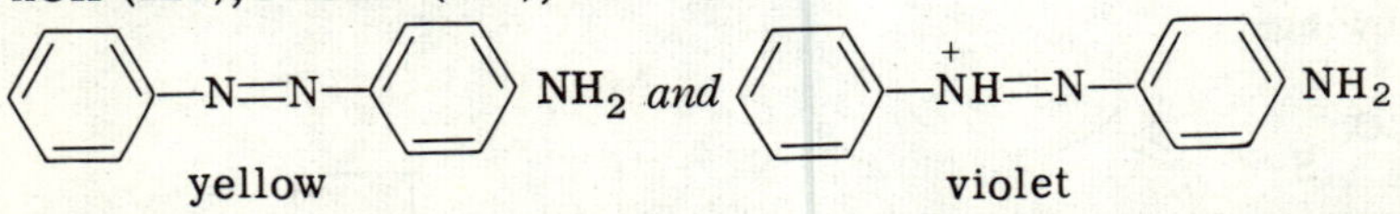

yellow violet

5. In 4-OH-2, 6-Me_2—C_6H_2—N=N—C_6H_2-2, 4, 6-Me_3, the *cis* isomer is colourless and the *trans* isomer is coloured. Offer an explanation.

6. What effect would you expect the polarity of the solvent to have on the position of the absorption maximum of alkenes?

7. Offer an explanation for the following data: λ_{max} (nm): methane, 125; ethane, 135; cyclopropane, 190.

8. Draw diagrams for the relative energies of the various types of π-orbitals of butadiene (which has two bonding and two anti-bonding π-orbitals) to illustrate the ground state and the S_1, T_1, S_2, and T_2 excited states.

9. Complete the equations:

(i) diazotised picramic acid + p-OH-acetanilide ⟶ ?

(ii) diazotised 3-NH_2-4-OH-benzenesulphonic acid + m-phenylenediamine ⟶ ?

(iii) benzenediazonium chloride + resorcinol ⟶ ?

(iv)
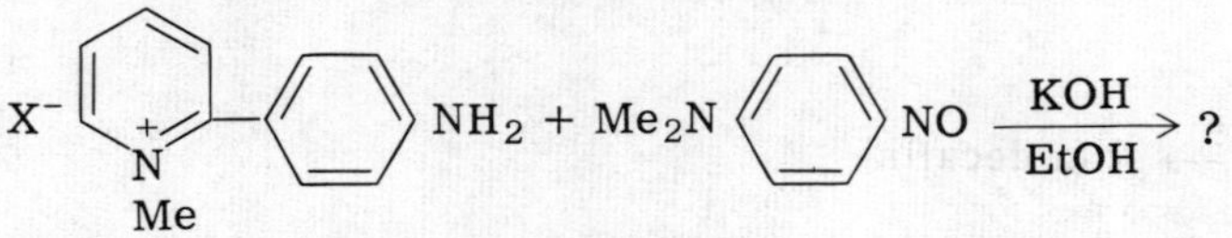

10. Suggest a synthesis of each of the following:

(i)
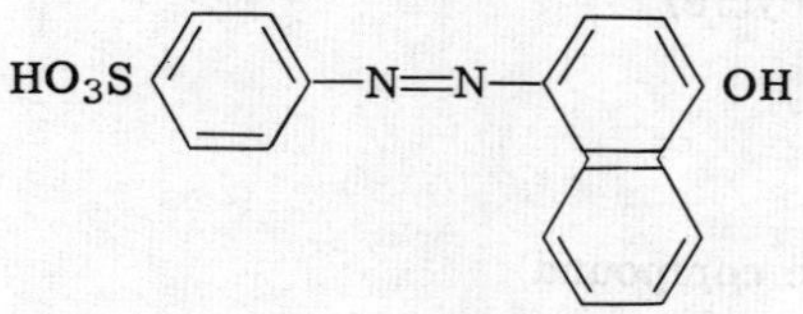

from benzene and naphthalene;

(ii)
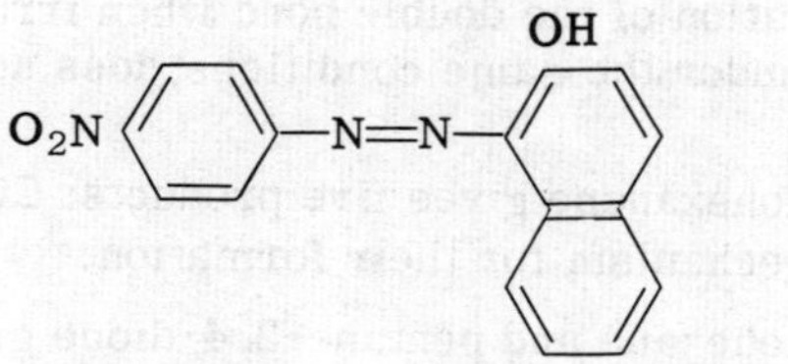

from benzene and naphthalene;

(iii)
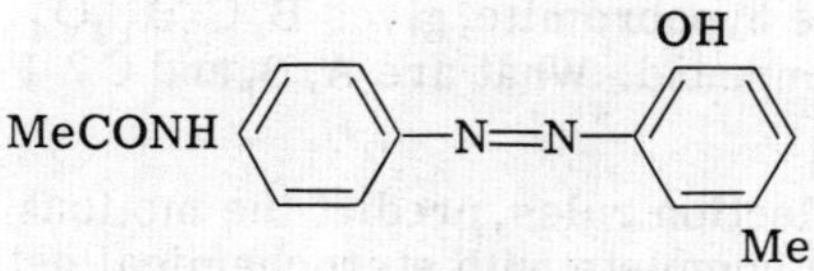

from aniline and p-cresol;

(iv)
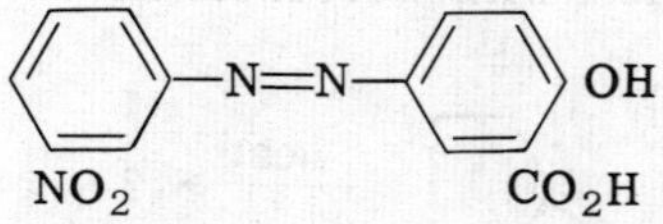

from benzene;

(v)
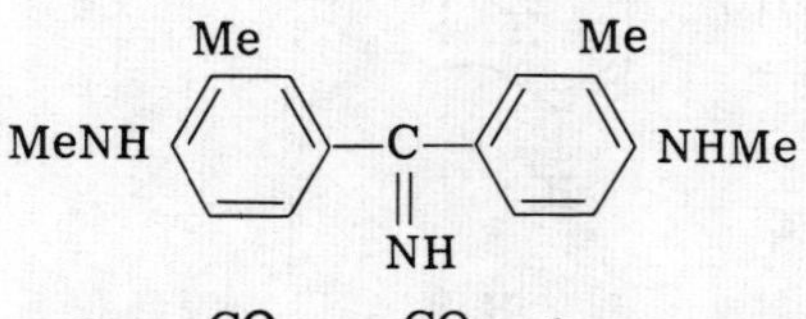

from N-Me-o-toluidine;

(vi)
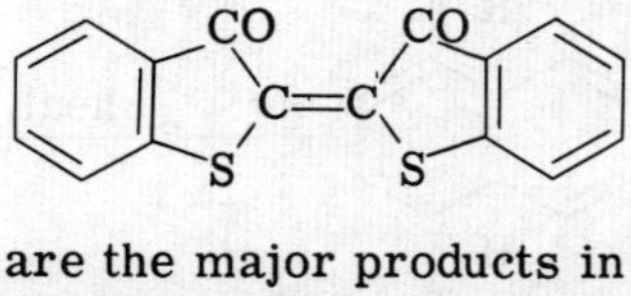

from benzene.

11. What are the major products in the photochemical decomposition of methyl neopentyl ketone? Suggest a mechanism.

12. Complete the following equations:

(i) 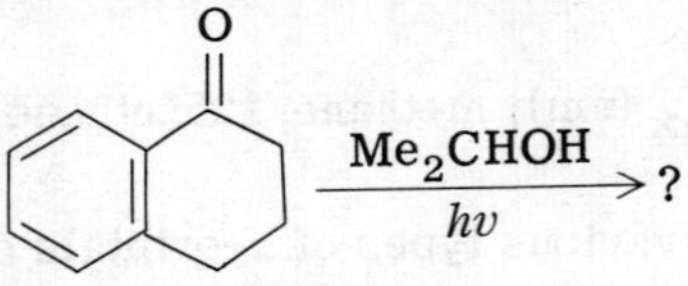

(ii) $Me_2CHONO \xrightarrow{hv}$?

(iii) 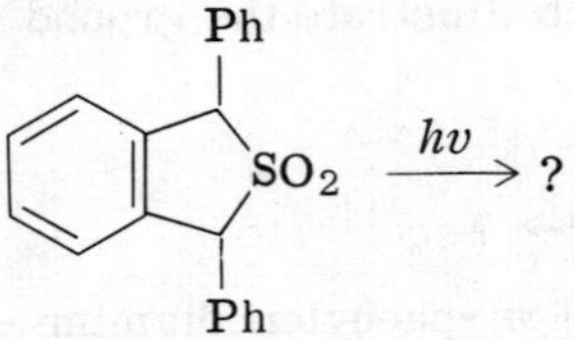

(iv) 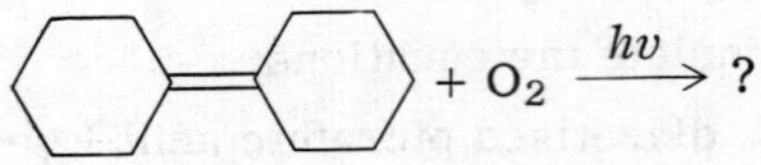

(v) $MeCH_2CH_2COMe \xrightarrow{hv}$? (a cyclobutane is major product)

(vi) 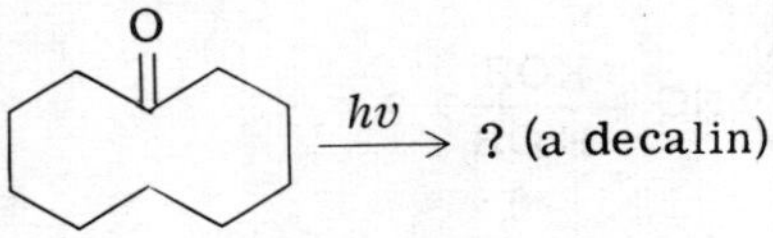

(vii) 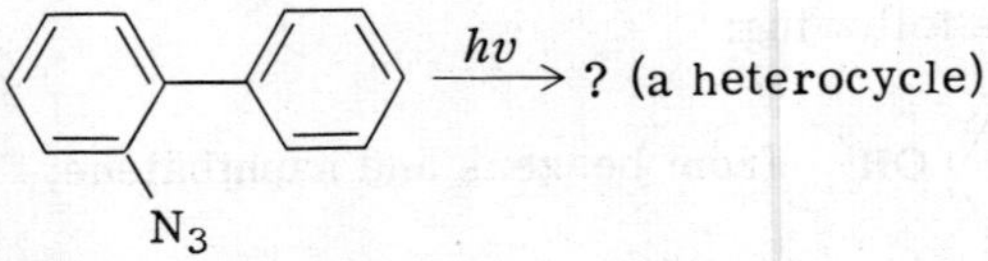

(viii) $CH_3CH_2CH_2CH_2N_3 \xrightarrow{hv}$? (a cyclic compound)

13. 5-Methylhex-3-en-2-one undergoes migration of the double bond when irradiated with light of *low* energy, whereas pent-3-en-2-one, under the same conditions, does not. Suggest a reason.

14. The photochemical decomposition of cyclohexanone gives five products: CO, C_2H_4, C_3H_6, C_5H_{10}, $CH_2{=}CH(CH_2)_3CHO$. Outline a mechanism for their formation.

15. The photochemical reaction between cyclohexane and pentane-2, 4-dione gives a product A, $C_{11}H_{18}O_2$. A, on oxidation with alkaline hypobromite, gives B, $C_9H_{14}O_4$. B, on heating with soda-lime, gives C, C_7H_{14}, a saturated compound. What are A, B, and C? Formulate the reactions.

16. By means of the Woodward-Hoffmann selection rules, predict the motions (conrotatory or disrotatory) for the following reactions, and complete with sterochemical details where necessary.

(i) 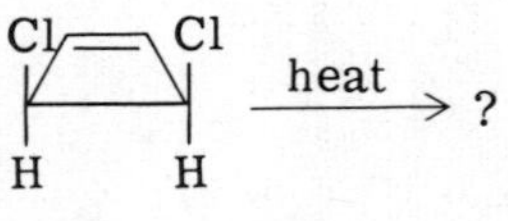

(ii) 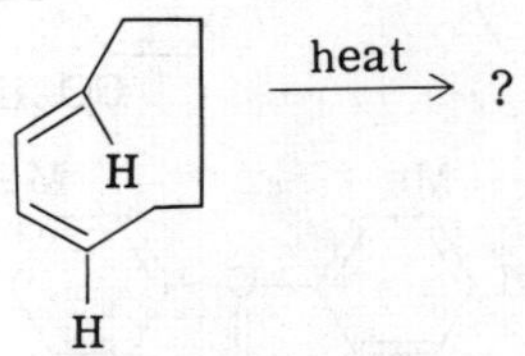

(iii) 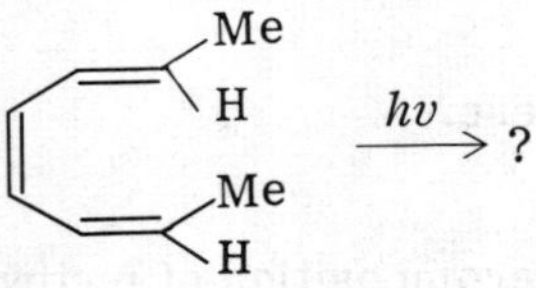

(iv) 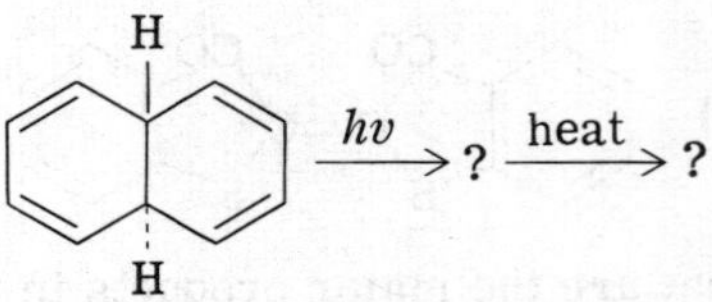

(v) 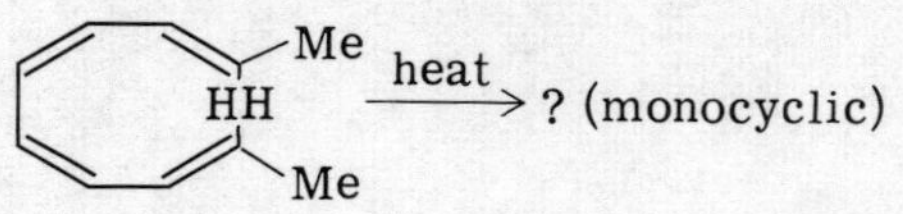

(vi) 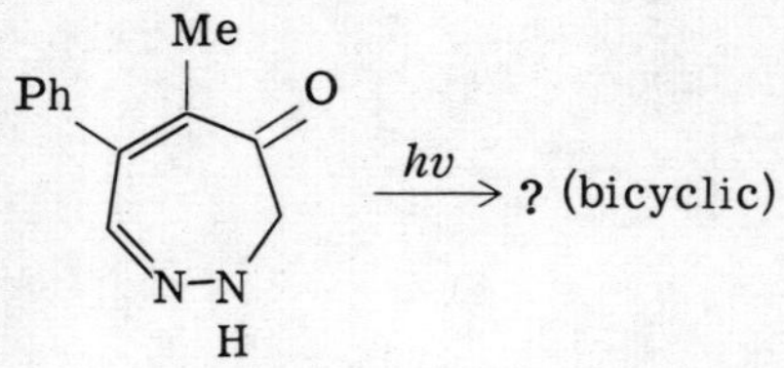

17. Suggest a mechanism for the reaction

$$2PhCH_2CH_2ONO \xrightarrow{h\nu} (PhCH_2NO)_2 + 2CH_2O$$

18. The Norrish type I process is not important for the photolysis of diaryl ketones. Suggest a reason.

SOLUTIONS

1 Determination of structure

(Problems on page 3)

1 In practice, the percentage of oxygen in a compound is determined by difference (from 100%). In the table of results, both the calculated (c.) and the difference result (d.) are given. It should also be noted that values are given to *one* decimal figure unless the *second* decimal figure is actually a five or very close to this value. At. Wts. used are given in brackets.

Molr. form.	M. Wt.	C% (12)	H% (1)	N% (14)	Cl% (35·5)	Br% (80)	S% (32)	O% (16)
$C_{14}H_{14}N_2O_4S$	306	54·9	4·6	9·15	—	—	10·5	c. 20·9 d. 20·85
C_8H_9NO	135	71·1	6·7	10·4	—	—	—	c. 11·85 d. 11·8
$C_8H_8ClNO_2S$	217·5	44·1	3·7	6·4	16·3	—	14·7	c. 14·7 d. 14·8
$C_{15}H_9BrO_2$	301	59·8	3·0	—	—	26·6	—	c. 10·6 d. 10·6

2 Note the remarks made in the answer to Prob. 1.
(i) CH_2O; (ii) $C_5H_5N_2O_3$; (iii) $C_{21}H_{18}N_4O$; (iv) $C_{18}H_{17}N_3O_3S$;
(v) $C_{16}H_{15}BrN_4O_2$; (vi) CH_2Cl.

Worked example: (ii)

	C	H	N	O(diff.)
Atomic ratio:	42·6/12	3·5/1	20/14	33·9/16
=	3·55	3·5	1·43	2·12
Divide by lowest value:	2·48	2.48	1	1·48

Since the values are not integers, multiply by the smallest number to get integers (or very close to them). Use 2.

Hence:	4·96	4·96	2	2·96

∴ Molr. form. is $C_5H_5N_2O_3$

3. Let M be the M. Wt. of cystine. The percentage of S is therefore given by $26{\cdot}7 = (64 \times 100)/M$, since the molecule contains *two* atoms of S (32×2). Hence $M = 240$. [Cystine is $C_6H_{12}N_2O_4S_2$].

4. Unless you are given evidence to the contrary (e.g., low temperature), assume fast rotation about *single* bonds. Then try to decide the number of different environmental protons, ignoring at this stage the number of protons in a given environment and whether there is coupling with protons in other environments.

(i) One signal (only one type of proton); (ii) one signal (again, all protons equivalent); (iii) three signals (Me, CH_2, Me (of Et)); (iv) four signals; (v) one signal; (vi) three signals; (vii) two signals; (viii) three signals; (ix) six signals.

5. (i) CH_3OCH_2Cl. Ignore the Cl (for number of signals). Hence two type of protons, and so two signals. Relative intensities 3: 2 (CH_3: CH_2). There is no *vicinal* coupling (second-order coupling negligible). Hence both signals are singlets. Both types of protons are subject to the —I effect of the oxygen atom, but the CH_2(Cl) protons are also subject to the —I effect of the Cl atom. Hence these CH_2 protons are more deshielded than the CH_3 protons and consequently the signal for CH_2 moves *downfield* (lower τ value) with respect to the CH_3 signal (higher τ value).

(ii) CH_3CHO. Ignore the O (for number of signals). Two type of protons. Therefore 2 signals, relative intensities 3 : 1 (CH_3 : CH). Since there is vicinal coupling, there will be splitting of these signals. The Me protons are coupled to *one* proton ($n = 1$) and so the signal will be a doublet ($n + 1$) of relative intensities 1 : 1 (coefficients of expansion $(1 + x)^1$). The proton of CHO is coupled to three protons (CH_3) and so $n = 3$ and hence the multiplicity is 4, i.e., the signal is a quartet with relative intensities 1 : 3 : 3 : 1 (coeffs. of expansion $(1 + x)^3$). Since the proton of CHO is strongly deshielded by the —I effect of the oxygen atom, its signal will be downfield with respect to the methyl signal.

(iii) $BrCH_2CHBr_2$. There will be two signals, relative intensities 2 : 1. The CH_2 signal is a doublet, 1 : 1, and the CH signal a triplet, 1 : 2 : 1. Because of the 'doubled' —I effect of the two Br atoms attached to the same carbon atom, the triplet will be downfield with respect to the doublet.

(iv) $CHCl_2CHBr_2$. There will be two signals, relative intensities 1 : 1. Both are doublets, 1 : 1, and because the —I effect of Cl > Br, the doublet for $CHCl_2$ will be downfield with respect to that for $CHBr_2$.

(v) $CD_3CH_2CH_3$. Ignore D (no signal and no coupling).

CD_3 —	CH_2 —	CH_3	
	quartet	triplet	2 : 3
	1 : 3 : 3 : 1	1 : 2 : 1	
	downfield		

6. (i) A trace of acid will not affect the NMR spectrum of an ether, but it may show a signal for the acidic proton (low field). The ether will show four signals, and for the simple rules of multiplicity without overlapping:

CH_3 —	O —	CH_2 —	CH_2 —	CH_2Br
singlet		triplet	nonet	triplet
(n=0)		(2 + 1)	(2 + 1)(2 + 1)	(2 + 1)

If overlapping can occur, this can occur only in the nonet. The chemical shift of the protons in —OCH_2 is quite different from that in the 'central' CH_2, and since $\Delta\nu/J \geqslant 6$, no overlapping occurs (see Table 3.2). On the other hand, the chemical shifts of the protons in —OCH_2 and CH_2Br are close enough to give rise to the possibility of overlapping when both are coupled with the central CH_2, i.e., this central CH_2, instead of being a nonet, may now produce a quintet (if the J values of the two pairs are about the same). Thus, there will still be *four* signals, but:

CH_3 —	O —	CH_2 —	CH_2 —	CH_2Br
singlet		triplet	quintet	triplet
		$(2 + 1)$	$(2 + 2 + 1)$	$(2 + 1)$
		1 : 2 : 1	1 : 4 : 6 : 4 : 1	1 : 2 : 1
		$(1 + x)^2$	$(1 + x)^4$	$(1 + x)^2$

The ratios indicated by the integration trace would be expected to be close to (as is also the case without overlapping) 3 : 2 : 2 : 2.

(ii) In the presence of a trace of acid, because of rapid chemical exchange, the proton in OH does not couple with vicinal protons. Since the two methyl groups are in identical environments, they produce *one* signal.
Thus:

$(CH_3)_2$ —	CH —	OH
doublet	16	singlet
$(1 + 1)$	$(3 + 1)(3 + 1)$	$(n=0)$

This would be the situation for no overlapping. However, the two CH_3 groups have the same chemical shifts (and the same *J* values with the CH) and so overlapping occurs to give the spectrum:

$(CH_3)_2$ —	CH —	OH
doublet	septet.	singlet
$(1 + 1)$	$(3 + 3 + 1)$	$(n=0)$
1 : 1	1 : 6 :15 : 20 : 15 : 6 : 1	
	$(1 + x)^6$	

The integration trace ratios will be close to 6 : 1 : 1.

7. One way to get the answers is to look up a table of fragment ions (see Table 1. 6) and choose the appropriate fragments. In the absence of a table, the usual way is to apply known fragmentation patterns. The molecular formula is C_2H_6O and *M* is 46. Then, starting at the 'top end': m/e 42= 46-4. Hence loss of 4H; i.e., C_2H_6O-4H= $C_2H_2O^+$. m/e 28= 46-18. Since 18 is the M. Wt. of H_2O, m/e 28 is $C_2H_4^{+\cdot}$. Hence m/e 26 is $C_2H_2^{+\cdot}$; m/e 25 is C_2H^+.

8. The lowest possible value for M^+ of $CBrCl_3$ is $12 + 79 + 3 \times 35 = 196$, and the highest possible value is $13 + 81 + 3 \times 37 = 205$ (see Table 1. 5). Hence the compound does not show a molecular ion. Since $196 - 167 = 29$ and $205 - 167 = 38$, and both are smaller than 79 (and $12 + 35 = 47$), the only possibility is loss of a chlorine atom to give $CBrCl_2^{+\cdot}$.

This has a minimum M. Wt. $=12 + 79 + 2 \times 35 = 161$. Hence this ion contains higher isotopes and requires the addition of 6 units. The necessary combination can be evaluated as follows.

The maximum addition that can be supplied by C is 1; Br, 2; Cl, 4 (=7). Since 6 is required, the formula that fits is: $^{12}C^{81}Br^{37}Cl_2^{+\cdot}$; M^+167.

9. The most general type of formula is $C_aH_bN_cO_d$. Then D.B.E. $= a + 1 - (b - c)/2$ Oxygen is ignored, and so are other bivalent elements, e.g., S; and univalent elements, e.g., halogen, may be replaced by one hydrogen atom (per halogen atom).

(i) $C_7H_6O_2$ D.B.E. $= 7 + 1 - 6/2 = 5$

(ii) C_3H_7N D.B.E. $= 3 + 1 - (7 - 1)/2 = 1$

(iii) $C_{10}H_7Cl \equiv C_{10}H_8$ D.B.E. $= 10 + 1 - 8/2 = 7$

(iv) $C_{12}H_{16}N_2O$ D.B.E. $= 12 + 1 - (16 - 2)/2 = 6$

(v) $C_{13}H_9BrS \equiv C_{13}H_{10}S$ D.B.E. $= 13 + 1 - 10/2 = 9$

2 Properties of molecules

(Problems on page 4)

1. (i) :Ċ·

$2s^2$, $2p$, $2p$

(ii) H:C:H (with H above and below C)

4σ (sp^3); tetrahedral

(iii) H H / C :: C / H H

5σ (sp^2), 1π ($2p$, $2p$);

trigonal

(iv) H:C⋮⋮C:H

3σ (sp), 2π ($2p$, $2p$);

digonal

(v) :Ö·

$2s^2$, $2p^2$, $2p$, $2p$

(vi) H:Ö:H

2σ (sp^3); tetrahedral;

2 lone pairs (in sp^3 orb.)

(vii) :N·

$2s^2$, $2p$, $2p$, $2p$

(viii) H:N: (with H above and below N)

3σ (sp^3); tetrahedral;

1 lone pair (in sp^3 orb.)

(ix) H:N⁺:H (with H above and below N)

4σ (sp^3); tetrahedral

2. The use of V.B. and M.O. structures comes with experience. In the V.B. method, lone pairs are shown when these are involved. Charges are shown and the V.B. structures chosen are those in which the more electronegative atom (in a bonded pair) acquires the negative charge. Carefully note examples (ii) and (iv); although O and N are more electronegative than C, when they acquire a *positive* charge, C is uncharged (sp^3 hybridised). In general, for the V.B. method, progressively extend the conjugation. For M.O. diagrams, draw the $2p$ orbitals involved in π-bonding.

(i) 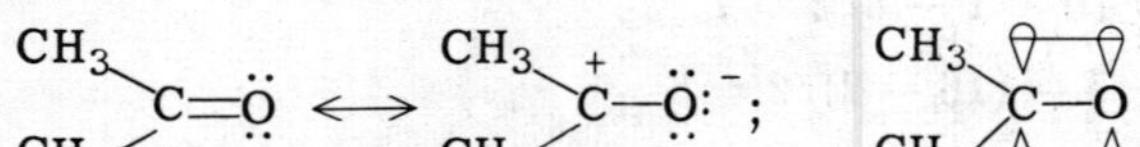

(ii) 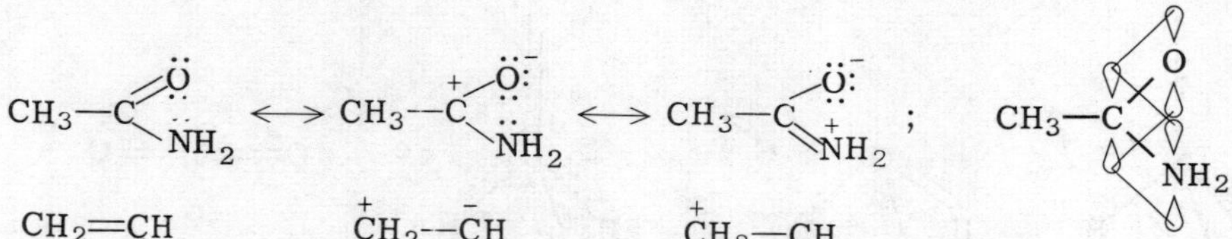

(iii) 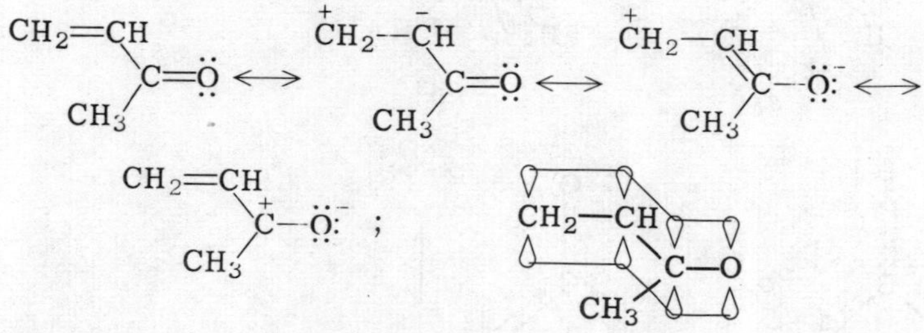

A V.B. structure carrying four charges could have been drawn. What is it and why has it been rejected?

(iv) 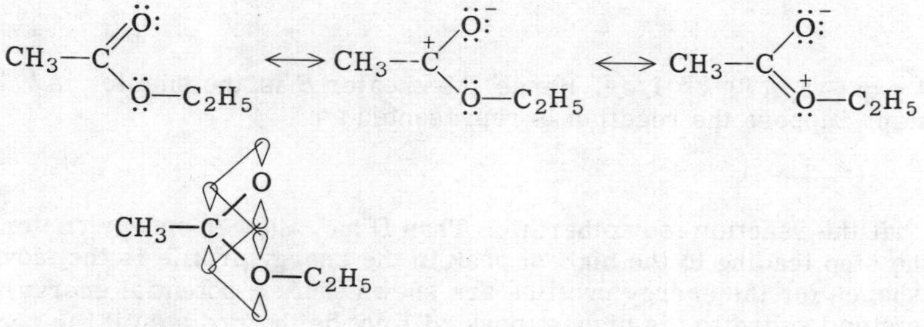

(v) 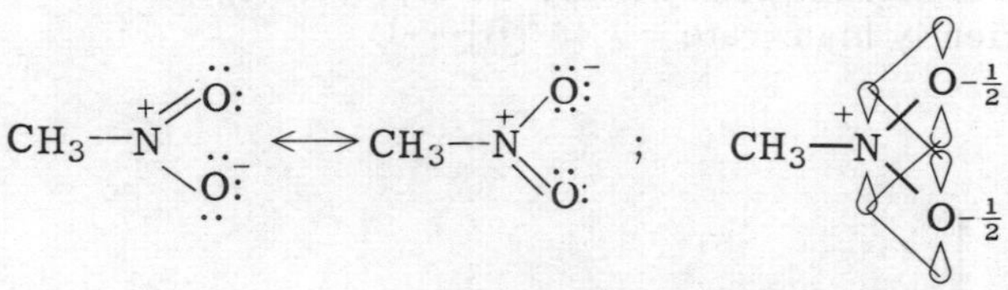

3. The dipole moment of a molecule is the vector sum of the bond dipoles and includes the dipoles due to the presence of lone pairs on the *central* atom (lone pairs elsewhere may be ignored). Symmetrical molecules have zero dipole moment. Bond dipoles are in the direction of the more electronegative atom (of the bonded pair). In the diagrams, the general direction of μ is indicated by the bold arrow (or by zero).

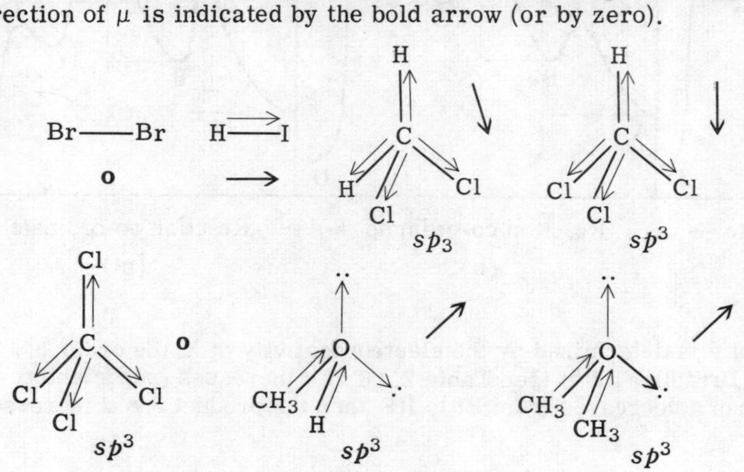

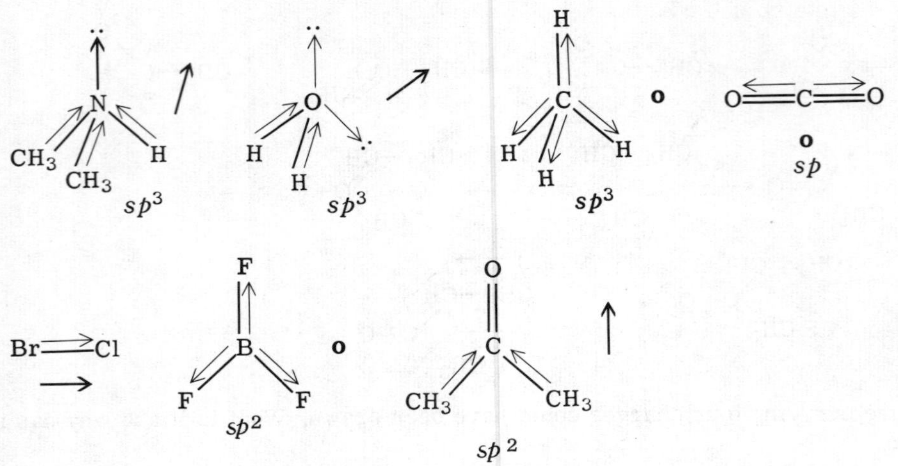

4. Since $k = A \exp(-E/RT)$, $k \propto 1/e^E$. Hence, the greater E is, the smaller is k, i.e., the slower is that step. Suppose the reaction is represented as:

$$A \xrightarrow{1} B \xrightarrow{2} C \xrightarrow{3} D$$

Let us assume that the reaction is exothermic. Then D has a lower energy content than A. Also note that the step leading to the highest peak in the energy profile is the slowest (r/d) step. Possible shapes for the energy profiles are shown (*P.E.* = potential energy). It should be noted that the step leading to the highest peak will not be the r/d step if the concentration used in some other step is sufficiently high (rate = k [A][B]---).

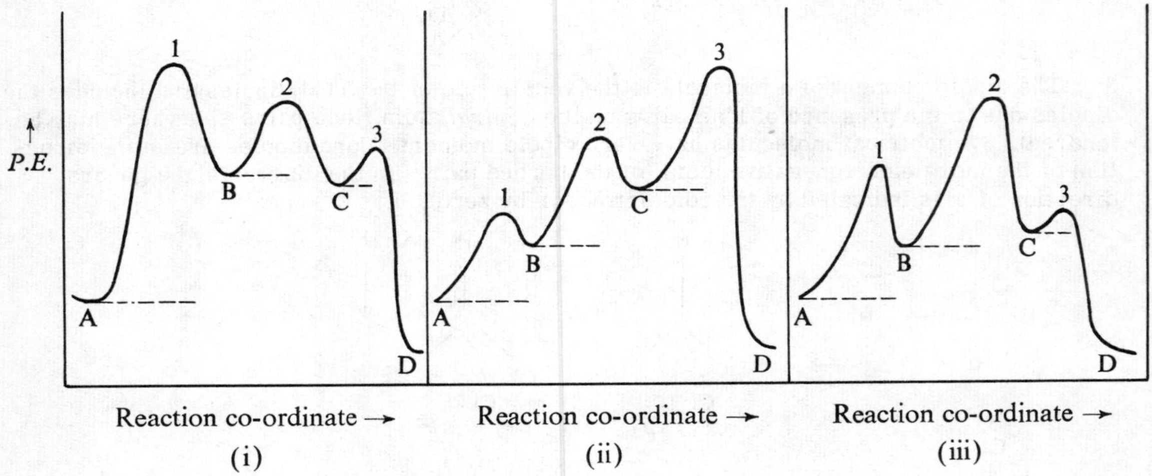

5. $\mu = e \times d$. The value of e is determined by the electronegativity of X, the order of which is F(4·0) > Cl(3·0) > Br(2·8) > I(2·5) (see Table 2.1). If e increases (sufficiently) more rapidly from I to F than d decreases from HI to HF, then the product $e \times d$ increases from HI to HF.

6. The order of acidity of the four halogen acids is : HI > HBr > HCl > HF. Since the stronger an acid is, the weaker is its conjugate base (and vice versa), it therefore follows that the order of basicity of the halide ions (the cB of its acid) is: $F^- > Cl^- > Br^- > I^-$.

7. $\Delta G^\ominus = \Delta H^\ominus - T\Delta S^\ominus$

$\therefore T\Delta S^\ominus = \Delta H^\ominus - \Delta G^\ominus = -802{\cdot}5 + 800{\cdot}8 = -1{\cdot}7$ kJ

$\therefore \Delta S^\ominus = -1700/298$ K $= -5{\cdot}7$ JK^{-1}

This is a very small value, and might have been anticipated on the basis that there has been no change in the number of molecules in going from reactant [3 mol. (g)] to product [3 mol. (g)].

8. Since this problem requires an answer to an equilibrium constant, it will be necessary to determine the value of $\Delta G^\ominus$. Since K is *positive* for a negative $\Delta G^\ominus$, write the chemical equation to give a negative $\Delta G^\ominus$.

$$-RT \ln K = \Delta G^\ominus = \Delta H^\ominus - T\Delta S^\ominus$$

$$A \rightleftharpoons B$$

$$\therefore \Delta G^\ominus = (-25{\cdot}44 + 24{\cdot}27) - 298(252{\cdot}3 - 246{\cdot}4) \times 10^{-3}$$

$$= -1{\cdot}17 - (298 \times 5{\cdot}9 \times 10^{-3})$$

$$= -1{\cdot}17 - 1{\cdot}76 = -2{\cdot}93\text{kJ}$$

Now, $\Delta G^\ominus = -RT \ln K = 5{\cdot}7 \log K$ kJ (at 298*K*)

$$\therefore 5{\cdot}7 \log K = 2{\cdot}93$$

$$\therefore \log K = 2{\cdot}93/5{\cdot}7 = 0{\cdot}51$$

$$\therefore K = 3{\cdot}24$$

i.e., in the equilibrium $A \rightleftharpoons B$

$$\frac{[B]}{[A]} = 3{\cdot}24 \text{ or for } 3{\cdot}24 \text{ parts B there is 1 part A}$$

$$\therefore \%\ A = \frac{1}{4{\cdot}24} \times 100 = 23{\cdot}6$$

$$\%\ B = \frac{3{\cdot}24}{4{\cdot}24} \times 100 = 76{\cdot}4$$

Answer: 23·6% A and 76·4% B.

N.B. Before the author is told that he can subtract 23·6 from 100 to get B%, he has done it this way to keep a check on his arithmetic!

9. General rules: (i) stronger the acid, smaller the pK_a; (ii) stronger the base, smaller the pK_b; (iii) stronger the base, larger the pK_a of its *conjugate acid*.

Order of increasing strength:

acids, pK_a: 15·8, 9·5, 6·3, 4·7, –2·1; *bases,* pK_a: 4·2, 8·1, 10, 16, 25; *bases,* pK_b: 12·8, 11·4, 9·2, 7, 4·7.

3 Alkanes

(Problems on page 5)

1. See also Ch. 2, Prob. 8.

Write equation: n-butane ⇌ isobutane

$\therefore \Delta G^{\ominus} = -17{\cdot}99 + 15{\cdot}69 = -2{\cdot}3$ kJ

$\therefore 5{\cdot}7 \log K = 2{\cdot}3$

$\therefore \log K = 2{\cdot}3/5{\cdot}7 = 0{\cdot}4$

$\therefore K = 2{\cdot}5$

Hence $\dfrac{[\text{isobutane}]}{[\text{n-butane}]} = 2{\cdot}5$

$\therefore$ moles of n-butane $= \dfrac{1}{3{\cdot}5} = 0{\cdot}29$

moles of isobutane $= \dfrac{2{\cdot}5}{3{\cdot}5} = 0{\cdot}71$

Answer: 0·29 mole n-butane and 0·71 mole isobutane.

It might be noted that since $\Delta G_f^{\ominus}$ for isobutane is more negative than that for n-butane, the former is thermodynamically more stable than the latter and consequently it can be expected that the former will predominate in the equilibrium mixture.

2. As the number of carbon atoms in the alkane increases, the number of possible isomers increases far more rapidly, e.g., C_6H_{14}(5), C_7H_{16}(9), C_8H_{18}(18), C_9H_{20}(35); these numbers *exclude* stereoisomers, e.g., *including* stereoisomers: C_7H_{16}(11), C_8H_{18}(24), C_9H_{20}(55). Obviously, the difficulty in writing down structures of isomers of a given alkane is no easy task, even for the members C_7 and C_8.

One way in which the student might tackle the problem is as follows. We shall use C_6H_{14} as our example.

(*a*) Write down the straight chain; only carbon atoms need be indicated.

C—C—C—C—C—C **One form only**

(*b*) Remove a terminal C, treat this as a Me group and progressively move it along the remaining straight chain. To avoid repetition, number the chain in both directions as shown:

$$\overset{1}{\underset{5}{\mathrm{C}}}-\overset{2}{\underset{4}{\mathrm{C}}}-\overset{3}{\underset{3}{\mathrm{C}}}-\overset{4}{\underset{2}{\mathrm{C}}}-\overset{5}{\underset{1}{\mathrm{C}}}$$

Position: 2(4); 3(3).

You cannot start at 1(5), since you will now have the original straight chain. If you go beyond 3(3), you now get 4(2), which has already been recorded as 2(4). Hence there are ***two* methylpentanes.**

(*c*) Remove *two* C from one end of the straight chain and treat them:

(i) as *two* Me groups. Start the substitution at the *same* C, and shift *both* to the next C, etc.

1 2 3 4
C—C—C—C
4 3 2 1

Position: 2, 2(3, 3). **One form only.** *N.B.* 1, 1(4, 4) is 2-Me-pentane.

Now place one Me at 2, the other at 3, and keeping 2 fixed, move 3 along the chain. Then start with one Me at 3, the other at 4, etc.

Position: 2, 3(3, 2). **One form only.**

(ii)
1 2 3 4
C—C—C—C
4 3 2 1

Treat the two C as an Et group. If this is placed at 2(3), you now get the longest chain consisting of 5 C atoms, i.e., you are now back again at the methylpentane. It can be seen that for a substituent Me group, you start from C—2; for an Et group, C—3, etc.

If you are dealing with higher homologues (of C_6H_{14}), you can use the same procedure, but you now introduce, where necessary, 3Me, 1Me + 1Et, n-Pr, isoPr, and so on.

Let us now return to the answer to the problem, C_6H_{14}. As shown above, there are five isomers. Hence:

(i) $MeCH_2CH_2CH_2CH_2Me$ — hexane; 2 prim.-C; 4 s-C.

(ii) $\overset{1}{Me}\overset{2}{C}H\overset{3}{C}H_2\overset{4}{C}H_2\overset{5}{Me}$ (Me on C-2) — 2-methylpentane; 3 prim.-C; 2 s-C; 1 t-C.

(iii) $\overset{1}{Me}\overset{2}{C}H_2\overset{3}{C}H\overset{4}{C}H_2\overset{5}{Me}$ (Me on C-3) — 3-methylpentane; as in (ii)

(iv) $\overset{1}{Me}—\overset{2}{C}(Me)_2—\overset{3}{C}H_2\overset{4}{Me}$ — 2, 2-dimethylbutane; 4 prim.-C; 1 s-C; 1 quat.-C.

(v) $\overset{1}{Me}\overset{2}{C}H\overset{3}{C}H\overset{4}{Me}$ (Me on C-2 and C-3) — 2, 3-dimethylbutane; 4 prim.-C; 2 t-C.

The student should now try to work out the formulae of the *structural* isomers for C_7H_{16}(9), and if possible, also include stereoisomers (11 isomers in all).

3. (i) $Me_2CO \xrightarrow[\text{HI/P}]{\text{W–K } or} Me_2CH_2$

The Wolff-Kishner is $Me_2CO \longrightarrow Me_2C{=}NNH_2 \longrightarrow Me_2CH_2$ (see *Text*). Other methods are also available.

(ii) $PrBr \xrightarrow{LiAlH_4} PrH$

Other methods are available.

(iii) $Me_3CH \xrightarrow{\text{alk. } KMnO_4} Me_3COH$ [*N.B.* t-H]

(iv) $MeCH{=}CHMe \xrightarrow[\text{room temp.}]{\text{Raney Ni/H}_2} MeCH_2CH_2Me$

(v) $MeI + EtI \xrightarrow{Na} Me{-}Me + Me{-}Et + Et{-}Et + C_2H_4$

(vi) $EtMgI \xrightarrow{\cdot H_2O} EtH + Mg(OH)I$

(vii) $EtCO_2^- + e \longrightarrow EtCO_2^{\cdot} \longrightarrow CO_2 + Et\cdot \longrightarrow C_2H_4 + C_2H_6 + C_4H_{10} + EtCO_2Et$

(viii) $EtCO_2^-Na^+ \xrightarrow{heat} C_2H_6 + C_2H_4.$

(ix) $EtCH{=}CH_2 \xrightarrow{B_2H_6} (EtCH_2CH_2{-})_3B \xrightarrow[NaOH]{AgNO_3} Et(CH_2)_4Et$

4. The general formula of the alkanes is C_nH_{2n+2}. Hence, for a M. Wt. of 72, $n = 5$, i.e., $MeCH_2CH_2CH_2Me$; Me_2CHCH_2Me; Me_4C.
Since only *one* monochlorination product is formed, *all* the hydrogens are equivalent. Only neopentane satisfies this requirement (Me_4C).

5.

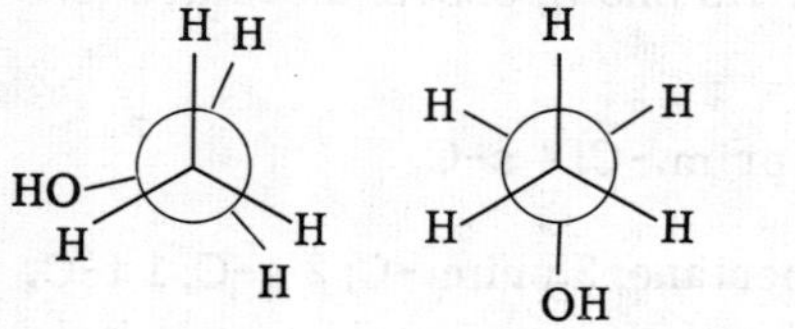

eclipsed staggered

(*a*) **ethanol**
$CH_3{-}CH_2OH$

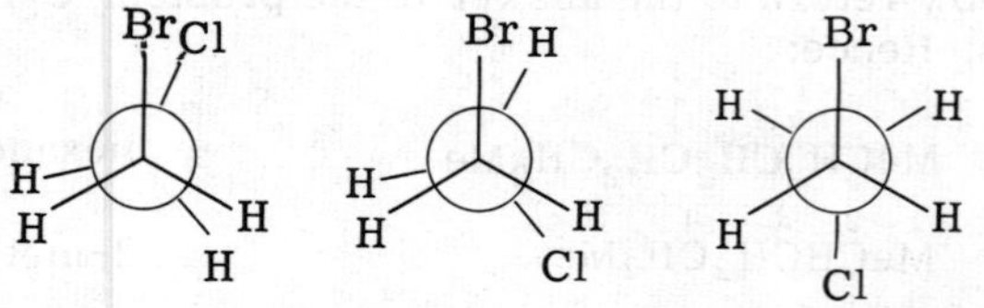

fully ecl. ecl. stag.

(*b*) **1-bromo-2-chloroethane**
$BrCH_2{-}CH_2Cl$

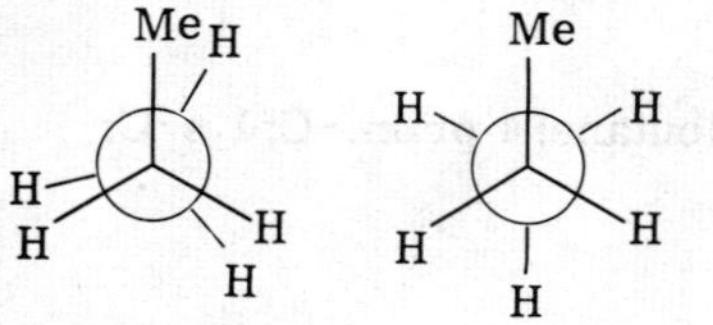

ecl. stag.

(*c*) **propane**
$MeCH_2{-}CH_3$

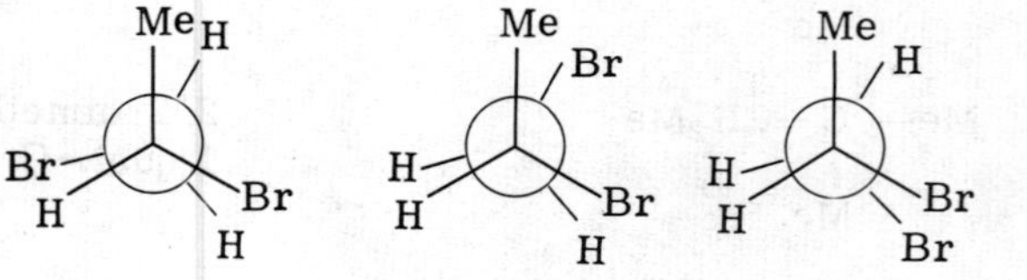

1. 2. 3.
eclipsed

(*d*) **MeCHBr–CH$_2$Br**

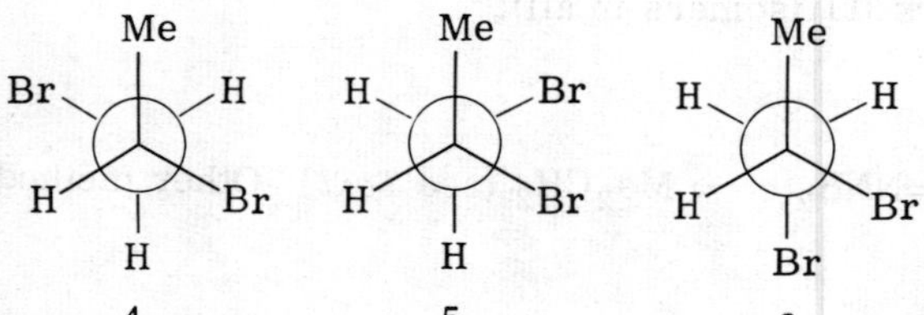

4. 5. 6.
staggered

(*d*) **MeCHBr–CH$_2$Br**

Draw the circle, place the three groups nearer the eye at the radii extremities (their order is arbitrary, but may be arranged in a particular way for some purposes, e.g., elimination reactions; *see Text*). Keep this fixed and rotate the other groups (back C atom) clockwise (or anticlockwise) through 120° at a time to give the three possible positions.

Examples (*a*), (*b*), (*c*) offer no difficulty. Example (*d*) is more complicated. There is a choice of two bonds for rotation: (i) Me—$CHBrCH_2Br$; (ii) MeCHBr—CH_2Br. (i), which would give one eclipsed and one staggered form, may be ignored (write these out). (ii) gives three eclipsed forms (1., 2., 3.) and three staggered forms (4., 5., 6.).

6. The molecular formula C_5H_{12} corresponds to the isomeric pentanes: n-, iso-, neo-. We now proceed by using a table of fragment ions (see Table 1. 6) and the application of the known fragmentation patterns (see text). The M. Wt. of C_5H_{12} (C = 12, H = 1) is 72. One general rule for the fragmentation of alkanes is that the greater the branching, the less likely will there be a peak for the molecular ion, M^+. The order of intensity of M^+ (in the table) is: **B** > **A** > **C**. This *suggests* **B** is n-, **A** is iso-, and **C** is neo-pentane. Supporting evidence for this can be obtained from the following arguments.

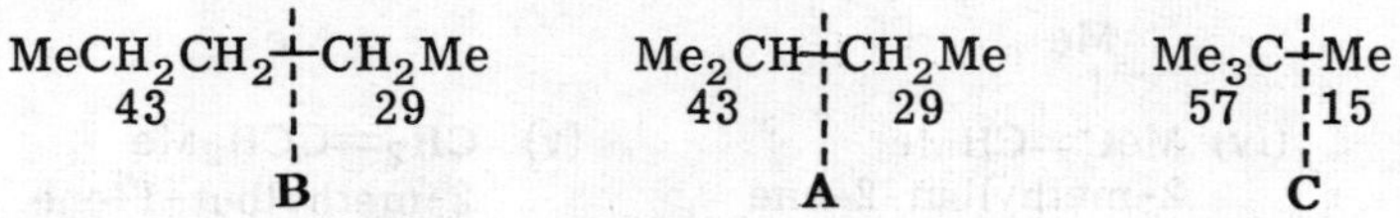

The base peaks of **A** and **B** both have m/e 43. This corresponds to $C_3H_7^+$ (see Table 1. 6). n- and iso-pentane can both give rise to this ion, but not neopentane, unless by rearrangement. In such an event, the intensity of the ion is *usually* weak. Hence **C** is neopentane, and this is supported by the fact that its *B.P.* is 57, which corresponds to $C_4H_9^+$, formed by loss of Me· by fission at the most substituted carbon atom. Also note that the ion m/e 29 corresponds to $C_2H_5^+$, and is absent in **C** (as might be expected).

So far, we are sure that **C** is neopentane, but we have relied only on the intensity of M^+ to distinguish between **A** (isopentane?) and **B** (n-pentane?). To find supporting evidence, let us consider the ion m/e 57. We have chosen this one since, of all the ions listed for **A** and **B**, it has the biggest ratio of relative intensities (55 : 13), and so is worth a closer look. This ion, as we have seen, is formed by loss of Me from M. Inspection of formulae **A** and **B** shows that Me can be lost from a greater branched carbon atom in **A** than in **B**. Hence the intensity of $C_4H_9^+$ for **A** can be expected to be greater than for **B**. This supports designating **A** as isopentane and **B** as n-pentane.

7. MeCHBr—CH_2Br (see also Prob. 5*d*, above).

(*a*) (*b*) (*c*), (*d*)

The most probable conformations are the staggered forms.

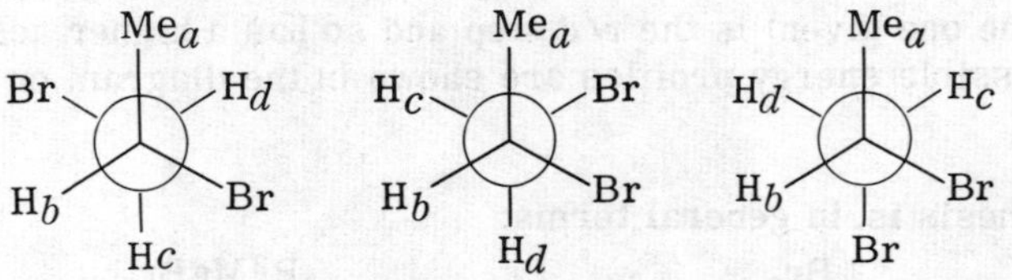

The environments of H_c are different in all three conformations, and is also the case for H_d. Since all three populations are different, H_c and H_d are not equivalent (the student should verify this by the method described in the *Text*, Ch. 3). Hence, if the chemical shifts of H_c and H_d are *sufficiently* different, *four* signals will be observed (H_a, H_b, H_c, H_d).

4 Alkenes and alkynes

(Problems on page 6)

1. See Ch. 3, Prob. 2.

First write out the structures of the isomeric alkanes, C_5H_{12}. Now take each one in turn and introduce one double bond, starting at the least substituted end and shifting the double bond inwards. Thus (see also Prob. 20, below).

(i) $MeCH_2CH_2CH{=}CH_2$ pent-1-ene (ii) $MeCH_2CH{=}CHMe$ pent-2-ene

(iii) $MeCH(Me)CH{=}CH_2$ 3-methylbut-1-ene (iv) $MeC(Me){=}CHMe$ 2-methylbut-2-ene (v) $CH_2{=}C(Me)CH_2Me$ 2-methylbut-1-ene

Note that the lowest number is given to the double bond; this decides which end numbering starts (look carefully at iv).

The products obtained from the ozonide depend on the nature of the reagents used. Here, we shall use zinc and acid to give aldehydes and/or ketones (see *Text*).

(i) $MeCH_2CH_2CHO + CH_2O$; (ii) $MeCH_2CHO + MeCHO$; (iii) $Me_2CHCHO + CH_2O$; (iv) $Me_2CO + MeCHO$; (v) $CH_2O + MeCOCH_2Me$.

2. Reaction (i) produces a primary carbonium ion and reaction (ii) produces a secondary ion.

$$\underset{\substack{+606{\cdot}7\\(C{=}C)}}{MeCH{=}CH_2} + \underset{+x}{H{-}X} \longrightarrow X^- + \underset{\substack{-347{\cdot}3\\(C{-}C)}}{MeCH_2\overset{+}{C}H_2} \; or \; \underset{\substack{-414{\cdot}2\\(C{-}H)}}{Me\overset{+}{C}HCH_3}$$

Hence, $\Delta H = (x - 154{\cdot}8)$ kJ. Since $x > 154{\cdot}8$ for *all* HX bonds (see Table 2.2), ΔH is positive (reaction is endothermic). Thus, the energy level of the carbonium ion is higher than that of the reactants (see also Ch. 8, Prob. 3).

Because bond energies (i.e., average values) have been used, the energy levels of the two carbonium ions are the same. However, since in fact a secondary carbonium ion is more stable than a primary ion, the energy level of the former will be below that of the latter. Also, the first step in the reaction (the one given) is the r/d step and so has a higher activation energy than the second step. Possible energy profiles are shown in the diagram on p. 77 (see also Prob. 14, below).

3. The equation for the Boord synthesis is, in general terms:

$$R^1CH_2CHO \xrightarrow[HCl]{EtOH} R^1CH_2CHCl(OEt) \xrightarrow{Br_2} R^1CHBrCHBr(OEt) \xrightarrow{R^2MgBr}$$

$$R^1CHBrCHR^2OEt \xrightarrow[BuOH]{Zn} R^1CH{=}CHR^2$$

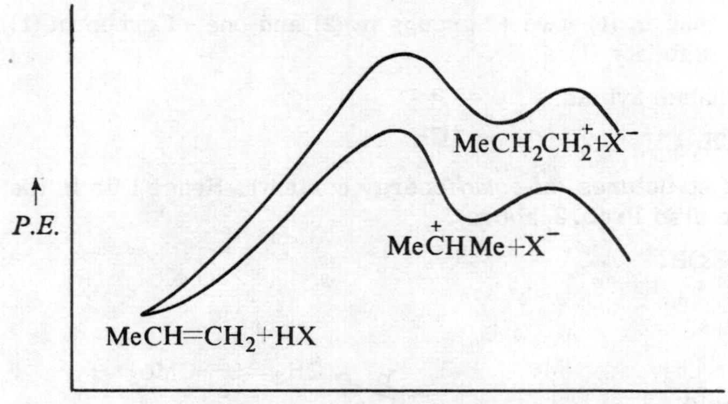

It is important to note that it may be possible to introduce either R^1 or R^2 in the form of the Grignard reagent and so two sets of starting materials may be used.

(i) MeCH=CH⋮Me or Me⋮CH=CHMe. Both require $MeCH_2CHO$ and MeMgBr.

(ii) $MeCH_2CH$=CH⋮CH_2Me or $MeCH_2$⋮CH=$CHCH_2Me$. Both require $MeCH_2CH_2CHO$ and $MeCH_2MgBr$.

Now give the starting materials for:
(*a*) Me_2CHCH=CHMe [2 pairs]; (*b*) $MeCH_2CH_2CH$=CHMe [2 pairs]; (*c*) Me_3CCH=$CHCMe_3$ [1 pair]

4. General principle: each unsaturated carbon atom must have two *different* groups attached to it.

(i) See Prob. 1, above. Hence: 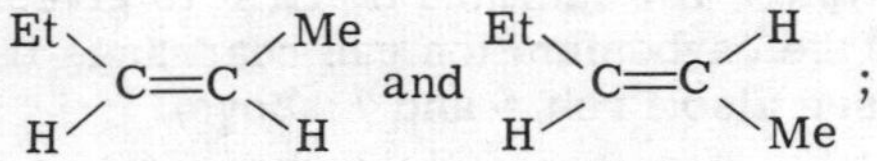 ;

cis *trans*

(ii) See Prob. 1, above. Hence no geometrical isomers.

(iii) Propene is CH_3CH=CH_2, and so the structural isomeric monobromopropenes are: $BrCH_2CH$=CH_2, CH_3CBr=CH_2, CH_3CH=CHBr. Only the last isomer satisfies the requirements for geometrical isomerism.

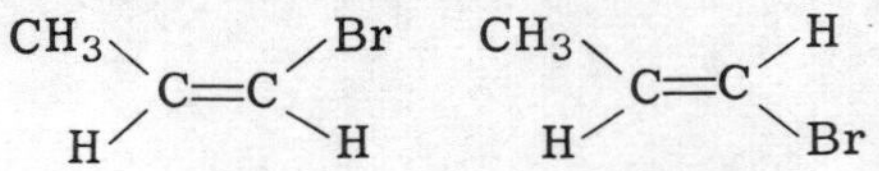

cis *trans*

5. Heat of hydrogenation may be used as a measurement of relative stability provided all the isomeric alkenes are referred to the *same* parent alkane. This is because the heats of formation of isomeric alkanes are different (to varying degrees). Thus, the stabilities of pent-1-ene and *cis*- and *trans*-pent-2-enes may be compared, since *all* give n-pentane on hydrogenation. 3-Methylbut-1-ene gives isopentane, and so cannot be included in the comparison. However, on combustion, all four alkenes give the same products, CO_2 and H_2O (in the same amounts). Hence heats of combustion can be used to compare the stabilities of all four alkenes.

6. General principle: order of stability of carbonium ions is prim. < s < t. This is due to increased spreading of charge, increased number of resonating structures, increased conjugation; which operate depends on the structure of the ion. Order is:

$$\underset{(1)}{MeCH_2CH_2^+} < \underset{(2)}{Me\overset{+}{C}HMe} < \underset{(3)}{CH_2{=}CHCH_2^+}$$

In (2) the charge is spread more than in (1) (two +I groups in (2) and one +I group in (1)). *No bond resonance* also explains stability (1) < (2).

(3) can be represented as a resonance hybrid:

$$CH_2{=}CH{-}CH_2^+ \longleftrightarrow {}^+CH_2{-}CH{=}CH_2$$

Charge is spread; two resonating structures (of *equal* energy content). Hence this is the most stable of all three ions (see also Prob. 2, above).

7. (i) $Me_2C{=}CH_2$; ethanolic KOH.

(ii) 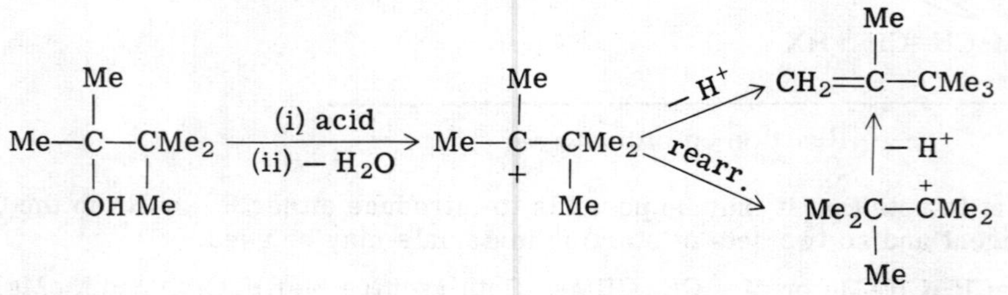

The dehydration occurs in the presence of acid, and only one alkene is formed (since rearrangement produces the original carbonium ion; see also *Text*, Ch. 5; 1, 2-Shifts).

(iii) $Me_2C{=}CH_2$; heat strongly.

(iv) $MeCH_2CH{=}CH_2 + MeCH{=}CHMe$; heat strongly.

In practice, 1-ene > 2-ene, and in latter, *trans* > *cis*.

8. General principles: HX addition; H^+ first to give the more stable carbonium ion, followed by X^-. If the carbonium ion can rearrange (1, 2-shift) to form a more stable ion, this also occurs (see also Prob. 6 and 7, above).

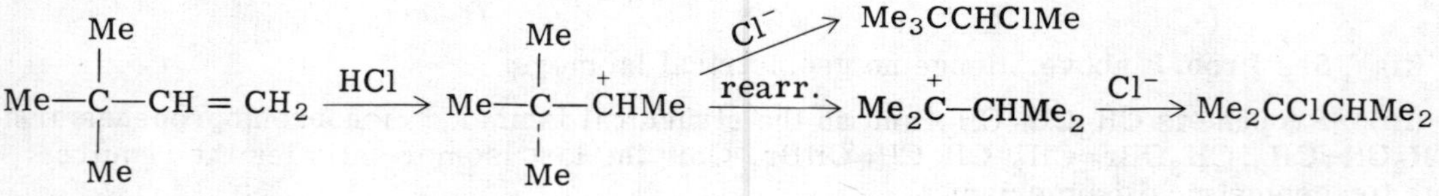

9.

$$\underset{+606{\cdot}7}{>C{=}C<} + \underset{+364{\cdot}0}{H{-}Br} \longrightarrow >CH{-}CBr<$$

$-414{\cdot}2$ (C—H) $-347{\cdot}3$ (C—C) $-284{\cdot}5$ (C—Br)

Thus, $\Delta H^\ominus$ for *both* reactions, using bond energies, is $-75{\cdot}3$ kJ. Equating $\Delta G^\ominus$ with $\Delta H^\ominus$, both reactions would therefore proceed *equally* well. This is *not* the case in practice. The disagreement is due to the use of bond energies (*E*, average values) instead of dissociation energies (*D*, specific values for particular bonds; see *Text*, Ch. 2). Had *D* values been used, $\Delta H^\ominus$ would have been different for the two reactions.

Equation (i) is correct in the absence of oxygen or peroxides (ionic mechanism), and equation (ii) is correct in their presence (peroxide effect; free-radical mechanism).

10. See Prob. 5, above. If we base our argument on the hyperconjugative effect being the predominant factor in determining stability and neglect steric effects (the largest groups involved in steric repulsion are Me groups, which are relatively small), the order of increasing stability is:

$$CH_2{=}CH_2 < MeCH{=}CH_2 < MeCH{=}CHMe < Me_2C{=}CH_2 < Me_2C{=}CHMe < Me_2C{=}CMe_2.$$

11. See also Ch. 2, Prob. 4. Since hydrogenation is exothermic (ΔH is negative), the product system is more stable than its corresponding reactant system. Since the overall activation energy of a multistage reaction is given by the highest peak (with the reactant level as the base line), we can use the simplified energy profiles shown. In this way, we need not be concerned with the mechanisms of the reactions.

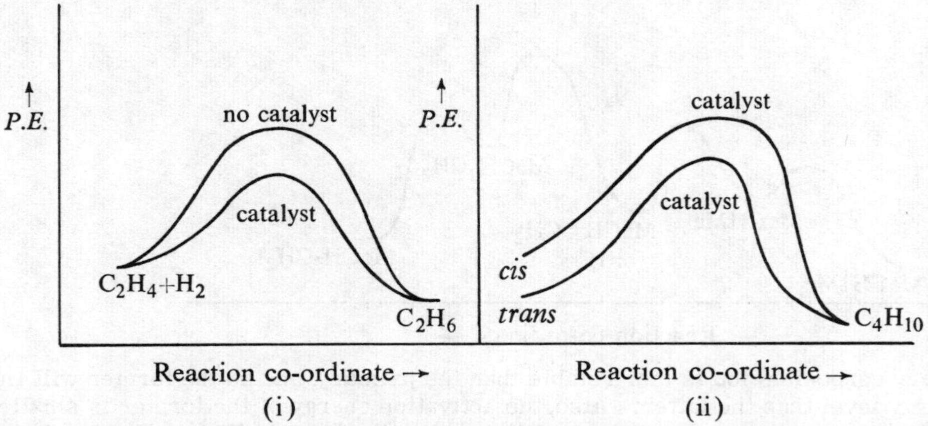

(i) With ethylene and hydrogen, the reactants start at the same energy levels and the product finishes at the same energy level. The paths are different, the one with the catalyst requiring a lower activation energy.

(ii) *trans*-But-2-ene is more stable (lower energy content) then the *cis*-, and both give the same product.

12. $Me_2C{=}CH_2 \xrightarrow{+H^+} Me_3C^+ \quad CH_2{=}CMe_2 \longrightarrow$

$Me_3C{-}CH_2{-}\overset{+}{C}Me_2 \xrightarrow{-H^+} Me_3C{-}CH{=}CMe_2 + Me_3C{-}CH_2{-}CMe{=}CH_2$

Ozonolysis (zinc and acid) would give acetone as one product for the first alkene (2-alkene) and formaldehyde for the second alkene (1-alkene). The Lemieux reagent could also be used (see *Text*).

13. Benzoyl peroxide is a reagent that generates free radicals (phenyl free radicals), and so its presence means we are dealing with a free-radical mechanism.

(i) $(PhCO_2)_2 \longrightarrow 2PhCOO\cdot \longrightarrow 2Ph\cdot + 2CO_2$

$Ph\cdot + CCl_4 \longrightarrow PhCl + \cdot CCl_3$

$EtCH{=}CH_2 + \cdot CCl_3 \longrightarrow Et\dot{C}HCH_2CCl_3 \xrightarrow{CCl_4} EtCHClCH_2CCl_3 + \cdot CCl_3$; etc.

The direction of addition is decided by using the general principle that s-free radicals are more stable than prim.

(ii) In this case, two trihalogeno-methyl radicals are possible.

$PhBr + \cdot CCl_3 \longleftarrow Ph\cdot + CBrCl_3 \longrightarrow PhCl + \cdot CBrCl_2$

Since it is reasonable to assume (but not necessarily true) that the *weaker* bond is more likely to be broken where alternatives exist, the free radical formed will be $\cdot CCl_3$ (see Table 2.2). Hence:

$EtCH{=}CH_2 + \cdot CCl_3 \longrightarrow Et\dot{C}HCH_2CCl_3 \xrightarrow{CBrCl_3} EtCHBrCH_2CCl_3 + \cdot CCl_3$; etc.

This is the product obtained in practice.

14. See Ch. 2, Prob. 4 and Probs. 2 and 6, above. A possible energy profile is shown in the diagram.

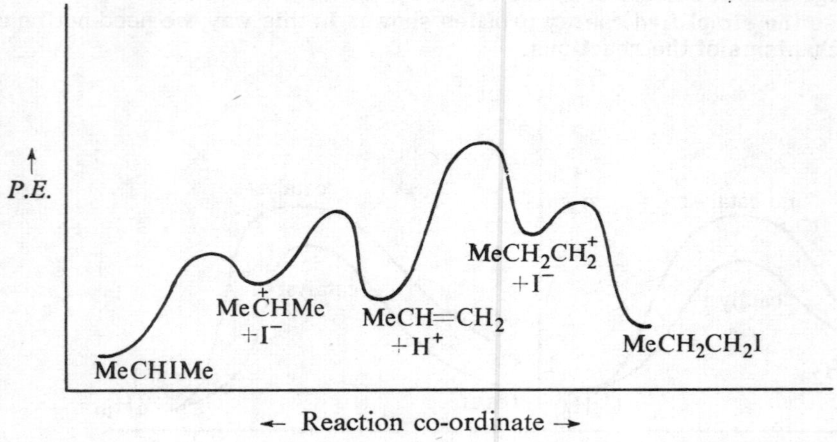

The secondary carbonium ion is more stable than the primary, and so the former will lie at a lower energy level than the latter. Also, the activation energy of the former is smaller than that of the latter (the former is formed faster than the latter). The predominant product is Me_2CHI.

The difficulties encountered with this energy profile are the energy levels of $MeCH{=}CH_2$ (ignoring HI), $MeCH_2CH_2I$, and Me_2CHI. Since the addition reaction is exothermic (see Prob. 9, above), the energy level of propene lies above that of either iodide. The safest way to ascertain the energy contents of the iodides is to look up a table of heats of formation or combustion (according to the 'law of cussedness', you are unlikely to find them!). On the other hand, a rough estimate of their relative stabilities may be calculated from a table of dissociation energies (D) but not from bond energies (E; see Prob. 9, above). If you fail to find the values you want, then you could make use of the rough generalisation: The stabilities of isomers of RZ are in the same order as the isomeric carbonium ions, R^+, derived from them. Thus, Me_2CHI (which gives Me_2CH^+) is more stable than $MeCH_2CH_2I$ (which gives $MeCH_2CH_2^+$).

15. The empirical formula of **A** is C_2H_3. Since its M. Wt. is 54, the molecular formula is C_4H_6 (see also Ch. 1, Prob. 2). The empirical formula of **B** is C_2H_5, and hence its molecular formula is C_4H_{10}. Since 4 H have been added, **A** is an alkadiene, alkyne (and could also be a cycloalkene). The i.r. band 2240 cm^{-1} is in the region of the C≡C bond (str.) for alkynes of the type $R^1C{\equiv}CR^2$. Hence **A** is MeC≡CMe (but-2-yne).

16. The polarisation in HOBr is HO—Br. Hence the Br^+ fragment will add to the unsaturated carbon atom that is the more negatively charged. Me is +I, but CCl_3 is —I (because of combined —I effects of the three chlorine atoms. Hence:

(i) $Me{-}CH{=}CH_2 \;\; Br{-}OH \longrightarrow Me{-}\overset{+}{C}H{-}CH_2Br + OH^-$

$\longrightarrow Me{-}CHOH{-}CH_2Br$

(ii) $CCl_3{-}CH{=}CH_2 \;\; Br{-}OH \longrightarrow CCl_3{-}CHBr{-}CH_2^+ + OH^- \longrightarrow CCl_3{-}CHBr{-}CH_2OH$

Note the situation from the point of view of carbonium-ion stabilities.

Also note that the product in the first reaction is the Markownikoff product, whereas that in the second is the anti-Markownikoff one. The latter is usually associated with the peroxide effect (free-radical mechanism) and so the student might be tempted to formulate the two reactions by different mechanisms (polar and free-radical). Had the peroxide effect been intended as an answer, the question would have had some statement to this effect, explicit or implicit (*cf.* Prob. 9, above). Hence, we accept a polar mechanism, and using the inductive effect to explain the direction of addition in (i), it is reasonable to assume that the inductive effect is acting in the *opposite* direction in (ii). We now can justify this on the basis of the strong electronegativity of a chlorine atom (see above).

17. $\Delta G_f^{\ominus}$ of the *trans* isomer is more negative (less positive) than that of the *cis*. Hence, the former is more stable (thermodynamically) than the latter and therefore can be expected to predominate in the equilibrium mixture.

Write the equation to give a *negative* $\Delta G^{\ominus}$ (see also Ch. 2, Prob. 8). Hence, the equation is written:

cis-but-2-ene $\rightleftharpoons$ *trans*-but-2-ene

$\therefore \Delta G^{\ominus} = 62{\cdot}97 - 65{\cdot}86 = -2{\cdot}89$ kJ

$\therefore -2{\cdot}89 = -RT \ln K = -5{\cdot}7 \log K$ kJ at 298K

$\therefore \log K = 2{\cdot}89/5{\cdot}7 = 0{\cdot}5; K = 3{\cdot}16$

Since $K = [trans]/[cis] = 3{\cdot}16$, the equilibrium mixture contains 3·16 times as much *trans* as *cis*.

At 25°C, and starting with *either* isomer, the interconversion may be said *not* to proceed at all; reaction is immeasurably slow, since the activation energy is too high (remember: $k = A \exp(-E/RT)$). Hence, to obtain an equilibrium mixture, heat (and/or a catalyst) is necessary.

18. Since a hydroxy-compound has been treated with a reagent in the presence of an acid, one possibility is the formation of a carbonium ion intermediate. Now, the product has been formed by rearrangement, and since carbonium ions are known to rearrange to more stable carbonium ions, this line of approach is worth a try. Hence:

$$CH_2{=}CH{-}CHOH{-}CH{=}CHMe \underset{-H^+}{\overset{H^+}{\rightleftharpoons}} CH_2{=}CH{-}C\overset{+}{H}OH_2{-}CH{=}CHMe$$

$$\underset{+H_2O}{\overset{-H_2O}{\rightleftharpoons}} \underset{(I)}{CH_2{=}CH{-}\overset{+}{C}H{-}CH{=}CHMe} \longleftrightarrow \underset{(II)}{CH_2{=}CH{-}CH{=}CH{-}\overset{+}{C}HMe}$$

$$\longleftrightarrow \underset{(III)}{\overset{+}{C}H_2{-}CH{=}CH{-}CH{=}CHMe}$$

Inspection of the carbonium ion first produced, (I), shows that it may be regarded as a resonance hybrid of three resonating structures, (I-III). Since conjugated molecules are more stable than their non-conjugated isomers, (II) and (III) are more stable than (I). Furthermore, because the product is derived from (II) it is reasonable to *assume* that it is more stable than (III). However, it is much better if we can justify this assumption. One argument we may use is that (II) is a secondary carbonium ion, whereas (III) is primary, and so, in accordance with the generalisation of carbonium-ion stabilities (t > s > prim.), (II) is more stable than (III), and therefore makes the larger contribution to the resonance hybrid.

We also know that carbonium ions react with nucleophiles, and so we can complete the equation as follows:

$$CH_2{=}CH{-}CH{=}CH{-}\overset{+}{C}HMe + MeOH \longrightarrow$$

$$\underset{\displaystyle MeO^{+}{-}H}{CH_2{=}CH{-}CH{=}CH{-}CHMe} \xrightarrow{-H^+} CH_2{=}CH{-}CH{=}CH{-}CH(OMe)Me$$

The student should now rewrite the answer to the question, picking out only the necessary statements for this purpose.

19. The empirical formulae of **A** and **B** are C_4H_7 and C_4H_9 respectively (see Ch. 1, Prob. 2). Hence, 'minimum' molecular formulae are **A**, C_8H_{14}, and **B**, C_8H_{18}. Since no M. Wt. has been given, we use these molecular formulae, unless some data given in the question cannot be satisfied by this assumption.

The molecular formula of **A** indicates the possibility of an alkadiene or an alkyne (cyclic isomers, which have been excluded at this stage of the subject in the *Text*, would be rejected on the basis of the products formed by ozonolysis).

Since ozonolysis involves fission at double or triple bonds, the formation of *three* products indicates the presence of *two* sites of attack. Hence, an alkyne is excluded. The positions of the two double bonds may be deduced as follows (on the assumption that they do not migrate in the ozonolysis).

(i) $MeCO_2H \equiv MeCH=$; Therefore occurs at *end* of chain.

(ii) $Me_2CO \equiv Me_2C=$; Therefore occurs at *end* of chain.

(iii) $MeCOCO_2H \equiv =CMeCH=$; Therefore occurs *in* the chain.

These three structural units account for *eight* carbon atoms, the actual number in **A**. If we now put them together we find that *two* possible arrangements give the same products on ozonolysis, viz.

A is $MeCH{=}CMeCH{=}CMe_2$ *or* $MeCH{=}CHCMe{=}CMe_2$

∴ **B** is $MeCH_2CHMeCH_2CHMe_2$ *or* $MeCH_2CH_2CHMeCHMe_2$

20. See also Ch. 3, Prob. 2, and Prob. 1, above.

$MeCH_2CH_2CH_2C\equiv CH$ hex-1-yne	$MeCH_2CH_2C\equiv CMe$ hex-2-yne
$MeCH_2C\equiv CCH_2Me$ hex-3-yne	$Me_2CHCH_2C\equiv CH$ 4-methylpent-1-yne
$Me_2CHC\equiv CMe$ 4-methylpent-2-yne	$MeCH_2CHMeC\equiv CH$ 3-methylpent-1-yne
$Me_3CC\equiv CH$ 3, 3-dimethylbut-1-yne	

21. (i) Ethylene and acetylene add bromine, ethane does not. Acetylene forms silver acetylide with ammoniacal silver chloride, ethylene does not react.

(ii) The i.r. regions of absorption of the C—H bonds are entirely different (—CH, ~2970, 2880 cm^{-1}; =C—H, ~3030 cm^{-1}; ≡C—H, ~3310 cm^{-1}). Also C=C bond (~1650 cm^{-1}) and C≡C (~2130 cm^{-1} in acetylene) will distinguish between C_2H_6, C_2H_4, and C_2H_2.

(iii) All three each show one signal, relative positions being (downfield with respect to TMS): TMS (τ 10), C_2H_6 ($\tau \sim 9{\cdot}15$), C_2H_2 (τ 8·2), C_2H_4 (~4·7).

The student is not expected to memorise τ-values, but should try to remember the *relative* positions.

22. $CH_3COCH_3 \xrightarrow{NaNH_2} CH_3C(O^-Na^+){=}CH_2 \xrightarrow{C_2H_2} (CH_3)_2C(O^-Na^+)C{\equiv}CH \xrightarrow{H^+}$

$(CH_3)_2C(OH)C{\equiv}CH \xrightarrow[Pd]{1H_2} (CH_3)_2C(OH)CH{=}CH_2 \xrightarrow[400°C]{Al_2O_3}$

$CH_2{=}C(CH_3)CH{=}CH_2$

A better understanding of some of these steps will be obtained by reading the appropriate sections of the *Text* (Ch. 6 and 8).

The final product is isoprene and this reaction has been used as a commercial method for its preparation.

23. The shape of a molecule depends on the nature of hybridisation of 'key' atoms.

(i) $MeCH_2—C{\equiv}CH$; *sp*, digonal; may be regarded as *essentially* a linear molecule (the terminal Me group is joined to the CH_2 group by an sp^3 bond, $\angle$ of ~110°).

(ii) $Me—C{\equiv}C—Me$; *sp*, digonal; this *is* a linear molecule.

(iii) This is an alkenyne. Numbering from one end gives the name hept-2-en-5-yne; from the other end hept-5-en-2-yne. The numbers are the same, but since the double bond is given preference (in a case like this), the first name is correct (see *Text*).

Since geometrical isomerism is possible about the double bond (sp^2, trigonal), two stereoisomers are possible, both planar and the *trans* more stable than the *cis*.

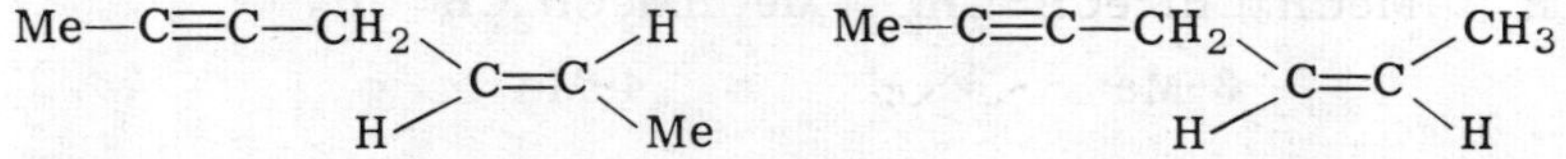

trans *cis*

24. The phrase 'and comment' may be taken to mean 'requires an explanation', and is generally used in connection with reactions that may be associated with descriptions such as unexpected, unusual, rearrangement, peculiar to the given structure, etc. When the question states 'comment where necessary', it is now up to the student to decide what is expected, usual, etc. This differentiation will become easier as the student's knowledge is extended, and for beginners it is probably better to err on the talkative side than to be of the silent type.

(i) $MeCH_2CH_2CMe{=}CHMe \xrightarrow{SeO_2} MeCH_2CHOHCMe{=}CHMe$

This oxidation is typical of CH_2 or CH_3 in the allyl position, and where a choice exists, CH_2 is preferred to CH_3.

(ii) $MeCH_2C{\equiv}CBr + CH{\equiv}CMe \xrightarrow{Cu^+} MeCH_2C{\equiv}C—C{\equiv}CMe$

The cuprous acetylide is formed (to be expected), but the high reactivity of the acetylenic bromine atom, under what appear to be mild conditions, may be regarded as 'unexpected' (*cf.* vinyl bromide).

(iii) $Me_2CHCH{=}CH_2 + HBr \longrightarrow Me_2CHCHBrMe + Me_2CHCH_2CH_2Br$

The first product is the Markownikoff one and the second, the anti-Markownikoff. Unless the reactants are pure and oxygen is excluded, both are formed. When pure and oxygen is excluded, the first product is formed almost exclusively. In the presence of peroxides, the second product is formed exclusively.

(iv) $MeCH_2C{\equiv}CH \xrightarrow[NH_3]{Na} MeCH_2C{\equiv}C^-Na^+ \xrightarrow{EtBr} MeCH_2C{\equiv}CEt$

(v) $BrCH_2CH_2Br \xrightarrow[EtOH]{KOH} [CH_2{=}CHBr] \longrightarrow CH{\equiv}CH$

Two-step elimination. The intermediate can be isolated under suitable conditions.

(vi) $MeCH_2CH_2OH \xrightarrow[\text{(ii)}-H_2O]{\text{(i)}H^+} MeCH_2CH_2^+ \xrightarrow{-H^+} MeCH{=}CH_2$

(vii) $MeCHBrCH_2Br \xrightarrow[MeOH]{Zn} MeCH{=}CH_2$

(viii) 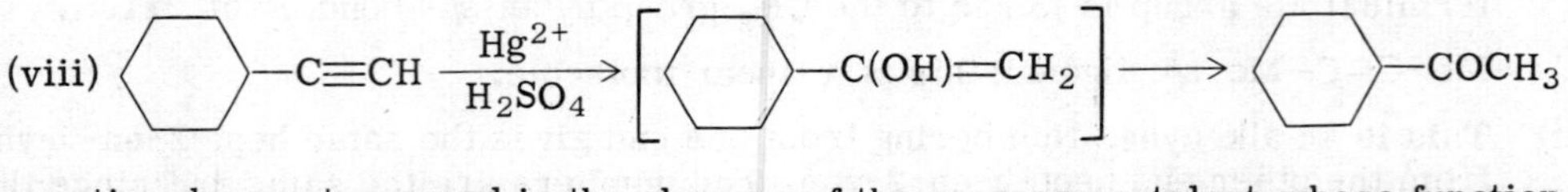

The reaction does not proceed in the absence of the mercury catalyst, whose function is uncertain (see *Text*). Accepting the addition follows the Markownikoff rule, the enol is formed first and then isomerises to the more stable keto form (see *Text*, Ch. 10).

25. The three methylpent-1-enes are:

$MeCH_2CH_2CMe{=}CH_2$	$MeCH_2CHMeCH{=}CH_2$	$MeCHMeCH_2CH{=}CH_2$
2-Me	3-Me	4-Me

M^+ for each is 84. Assuming allylic cleavage:

2-Me ⟶ $C_2H_5^+$ (*m/e* 29) and $C_4H_7^+$ (*m/e* 55)
3-Me ⟶ $C_2H_5^+$ (*m/e* 29) and $C_4H_7^+$ (*m/e* 55)
4-Me ⟶ C_3H_7 (*m/e* 43) and $C_3H_5^+$ (*m/e* 41)

Ion, *m/e* 43, is the base peak (*B.P.*) of **C**; this suggests **C** is the 4-Me isomer. Both 2- and 3-Me isomers given ion, *m/e* 55, by allylic cleavage, and this cleavage is 'assisted' in the 3-Me isomer because the allylic carbon is more branched than in 2-Me. Hence, it might be expected that this ion would be more intense in 3-Me than in 2-Me. In fact, this ion is the *B.P.* of **A**; hence **A** is probably 3-Me. Also, the other ion, *m/e* 29, is more intense in **A**, and further supports **A** is 3-Me. This leaves **B** is 2-Me. Its *B.P.* is ion, *m/e* 56. This is therefore $C_4H_8^+$. It cannot be obtained by direct fragmentation, and so must be obtained through some rearrangement (this is also the case for **A** and **C**). The formation of $C_4H_8^+$ can be explained on the basis of a McLafferty rearrangement. For 2-Me:

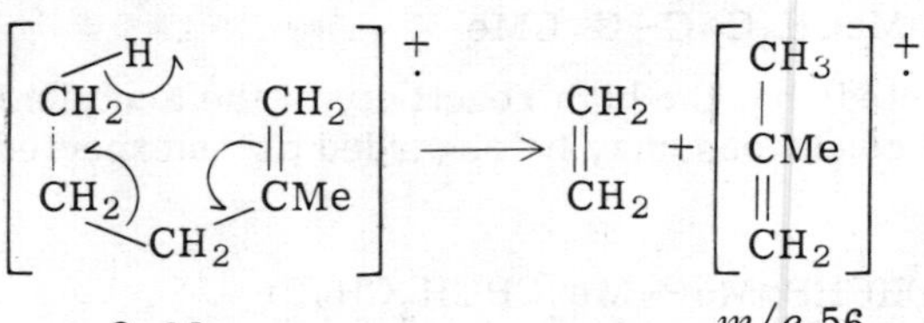

2—Me   *m/e* 56

Thus, the evidence given suggests: **A** is 3-Me, **B** is 2-Me, and **C** is 4-Me-pent-1-ene.

26. (i) One signal; the two protons are equivalent (replacement of each, in turn, by Z, gives the same compound chemically and spatially).

(ii) Three signals and all multiplets, i.e. (1)(m); (2)(m); (3)(m). All three protons are non-equivalent and couple to give multiplets.

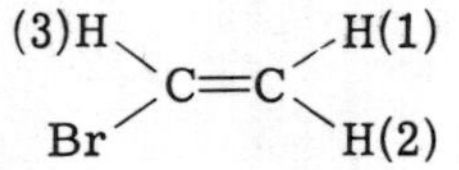

(iii) Two signals: (1)(m); (2)(m). Multiplets arise from long-range coupling (involving the *triple* bond).

$CH{\equiv}CCH_2Br$
(2) (1)

(iv) One signal; all protons equivalent.

(v) Two signals: (1)(m); (2)(m) [*cf.* with (iii)].

$CH_3C{\equiv}CH$
(2) (1)

(vi) Three signals: (1)(m); (2)(m); (3)(m). [*cf.* with (ii)]. Protons (1) and (2) are non-equivalent and are coupled to each other and to the methyl group (3) by long-range coupling.

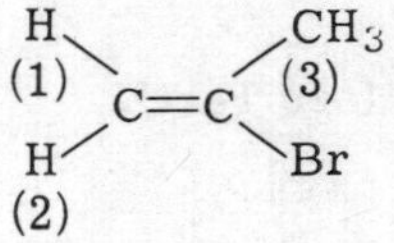

5 Halogen derivatives of the alkanes

(Problems on page 8)

1. See Ch. 2, Prob. 2; Ch. 4, Prob. 1.

(i) $MeCH_2CH_2CHCl_2$
1, 1-dichlorobutane

$MeCH_2CCl_2Me$
2, 2-dichlorobutane

$MeCH_2CHClCH_2Cl$
1, 2-dichlorobutane

$MeCHClCH_2CH_2Cl$
1, 3-dichlorobutane

$ClCH_2CH_2CH_2CH_2Cl$
1, 4-dichlorobutane

MeCHClCHClMe
2, 3-dichlorobutane

(ii)
```
      Me
      |
Me—CH—CHCl2
```
1, 1-dichloro-2-methylpropane

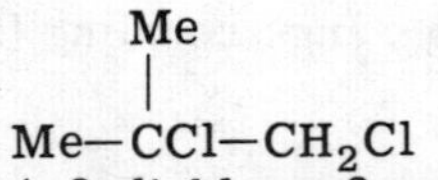

1, 2-dichloro-2-methylpropane

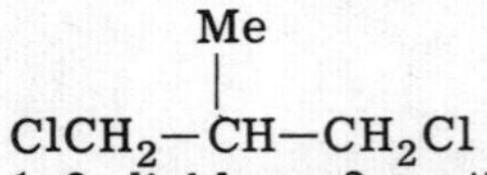

1, 3-dichloro-2-methylpropane

2. (i) Hydrolysis with dilute aqueous alkali:

$$ClCH_2CH_2Cl \xrightarrow{OH^-} HOCH_2CH_2OH \text{ ethylene glycol}$$

$$CH_3CHCl_2 \xrightarrow{OH^-} CH_3CHO \text{ acetaldehyde}$$

The chemical properties of the products are entirely different (see *Text*).

(ii) By means of NMR (i.r. unsatisfactory).

$ClCH_2CH_2Cl$	CH_3CHCl_2
1 signal	2 signals

3. If a structure can exhibit geometrical isomerism, draw these first. Then inspect each geometrical isomer for the possibility of optical isomerism.

(i) No geometrical isomerism possible. Molecule contains one asymmetric carbon atom. Hence, a (+)-form and a (—)-form (and a (±)-form). These optically active forms are represented (see also *Text*, Ch. 17):

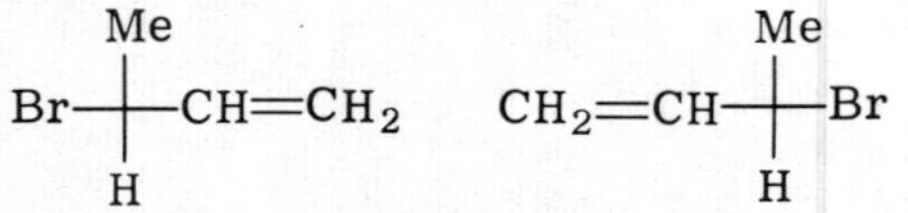

(±)-

These are mirror images (not superimposable). One is (+) and the other is (—), and unless the student knows definitely which is which, it is best to omit the signs.

(ii) Geometrical isomerism possible, and one asymmetric carbon atom present. Hence four forms possible, two geometrical isomers, each existing as a (+)- and (—)-form (and each pair as a (±)-form).

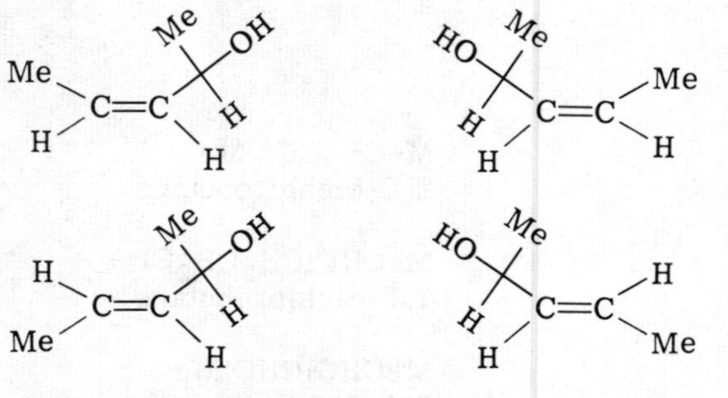

(±)-

(±)-

(iii) One asymmetric carbon atom present.

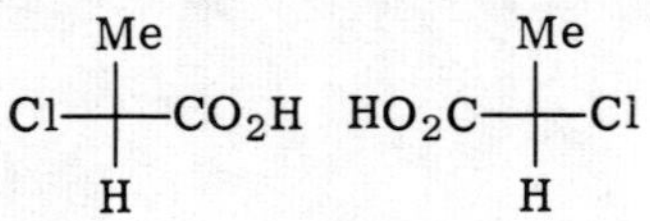

(±)-

(iv) $BrCH_2CH_2CO_2H$. No geometrical or optical isomerism. One form only.

(v) $Me_2C(OH)CO_2H$. One form only (*cf.* iv).

4. The mechanism of (i) is S_N2, whereas that of (ii) is S_N1 (which is characteristic of t-alkyl halides). Furthermore, carbonium ions readily eliminate a proton to form alkenes.

5. (i) HI is a strong acid and can donate a proton to the weak base, EtOH. Thus, the following equilibrium lies to the right.

$$EtOH + HI \rightleftharpoons EtOH_2^+ + I^-$$

Since OH_2^+ is a good leaving group, further reaction occurs:

$$EtOH_2^+ \rightleftharpoons H_2O + Et^+ \xrightarrow{I^-} EtI$$

(ii) HCN is a very weak acid, and because EtOH is too weak a base to accept a proton from HCN, no $EtOH_2^+$ is formed. Also, OH is a very poor leaving group. Hence, there is no reaction.

6. $$\underset{+\,414{\cdot}2}{C_2H_5{-}H} + \underset{+\,188{\cdot}3}{Br{-}Br} \longrightarrow \underset{-284{\cdot}5}{C_2H_5{-}Br} + \underset{-364{\cdot}0}{H{-}Br}$$

Hence, using bond energies (Table 2. 2), $\Delta H^\ominus$ is $-46{\cdot}0$ kJ. Since the number of reactant and product molecules are the same, we shall assume ΔS is ~0 (see *Text*, Ch. 2).

Therefore $\Delta H^\ominus$ may be equated with $\Delta G^\ominus$, and since this is negative (and numerically greater than 40 kJ), the reaction is thermodynamically favourable. However, the activation energy of the reaction is too high to be surmounted at room temperature in the dark, and so does not proceed.

If heated sufficiently or irradiated with u.v. light, the bromine is split into bromine atoms and reaction proceeds by a free-radical mechanism (see *Text*).

7. Carbonium ions can undergo rearrangement, and this is from the less stable to the more stable one. Both compounds give the *same* carbonium ion initially. Hence:

$$Me_3CCHOHMe \xrightarrow[(ii)\ -\ H_2O]{(i)\ H^+} Me_2C(Me){-}\overset{+}{C}HMe \xleftarrow{+\ H^+} Me_3CCH{=}CH_2$$

$$Me_2C(Me){-}\overset{+}{C}HMe \xrightarrow{Cl^-} Me_3CCHClMe$$

$$Me_2C(Me){-}\overset{+}{C}HMe \xrightarrow{rearr.} Me_2\overset{+}{C}CHMe_2 \xrightarrow{Cl^-} Me_2CClCHMe_2$$

Both give the same two isomeric Chlorides in the same proportions, with the tertiary chloride predominating.

8. (i) $$ClCH_2CH_2CH_2Br + KCN \xrightarrow[heat]{EtOH{-}H_2O} ClCH_2CH_2CH_2CN$$

Since equimolecular proportions of reactants are used and bromides are more reactive than chlorides, the main product (if not exclusive) will be the chloro-cyanide.

(ii) $$PhCHO \xrightarrow{SF_4} PhCHF_2$$
Sulphur tetrafluoride replaces oxygen atoms.

(iii) $$BrCH_2CH{=}CHCO_2Me \xrightarrow{AgF} FCH_2CH{=}CHCO_2Me$$

(iv) $Me_2CHBr \xrightarrow{LAH} Me_2CH_2$

(v) $Me_2C{=}CH_2 + HCl \xrightarrow{peroxide} Me_2CClMe$

There is no peroxide effect with HCl.

9. In general terms, the elimination may be written as:

$$\underset{\substack{(I)\\(B + C)}}{EtCH{=}CHMe} \longleftarrow EtCH_2CHBrMe \longrightarrow \underset{\substack{(II)\\(A)}}{EtCH_2CH{=}CH_2}$$

(I) can exist as two geometrical isomers, *cis* and *trans*, and it can be anticipated that the more stable *trans*-form will predominate.

Detailed discussion. If we accept that the elimination of HBr is stereoselectively *trans*, then, according to Saytzeff's rule (based on the hyperconjugative effect), the predominant product will be the most substituted ethylene. Thus, **B** and **C** will be the major products, the *trans*-isomer, say **B**, being formed in larger amount than **C**, the *cis*-isomer. Saytzeff's rule is followed provided steric effects are not large enough to outweigh the hyperconjugative effects to give the 1-alkene (the Hofmann product). Since the order of steric effects is t − R > s − R > prim. R, we shall assume that the elimination follows the Saytzeff rule for the compound under discussion.

On the assumption of an E2 mechanism, we can represent the eliminations as follows:

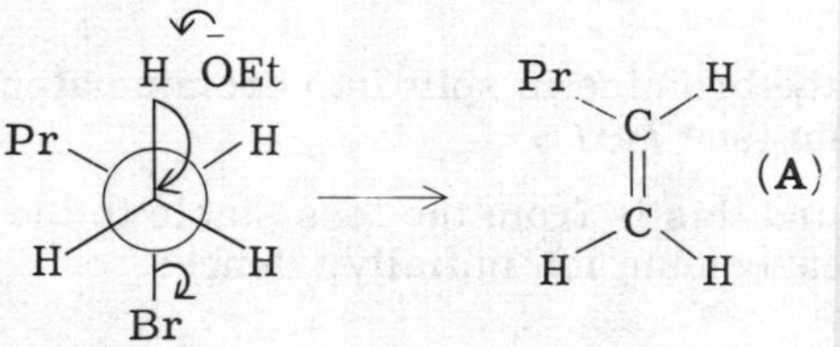

(**A**)

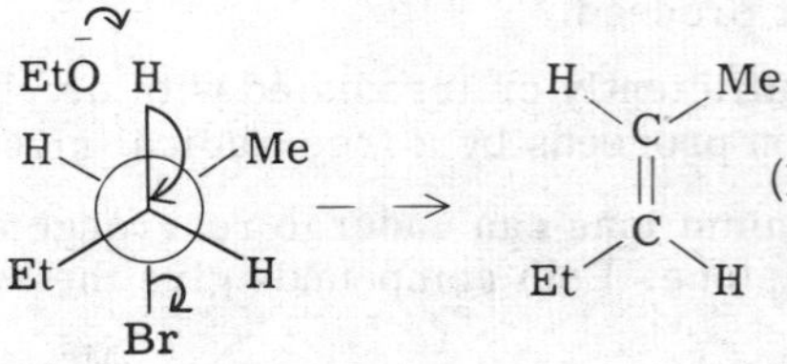

(**B**) *trans*

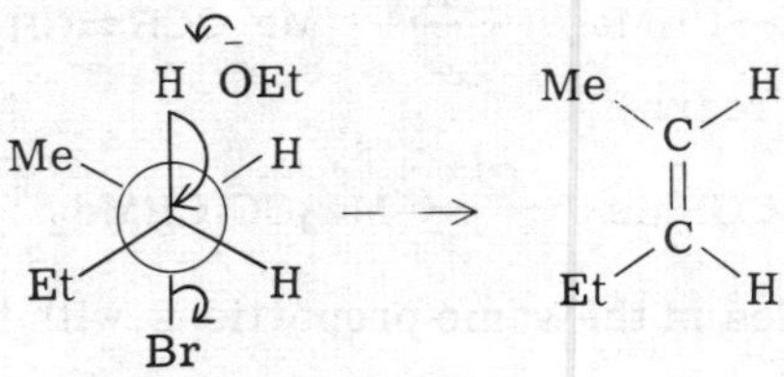

(**C**) *cis*

The T.S. leading to the *cis*-isomer is more crowded than that leading to the *trans*-isomer. Hence, **B** > **C** > **A**.

This order has been based on the assumption of the presence of only *small* steric effects. If this assumption is not justified, then the order given, which is based on the formation of thermodynamically controlled products, may be incorrect. *Rates* of reaction are predominantly controlled by activation energies (concentrations are far less important). The T.S. leading to **C** is crowded and so the activation energy is raised, and may be higher than that for the reaction leading to **A**. In this case, **A** will be formed faster than **C** (kinetic control), and the order will now be **B** > **A** > **C**. More information is required to distinguish between these two orders. In actual fact, the second order has been shown to operate: **B** (~50%) > **A** (~30%) > **C** (~20%).

10. 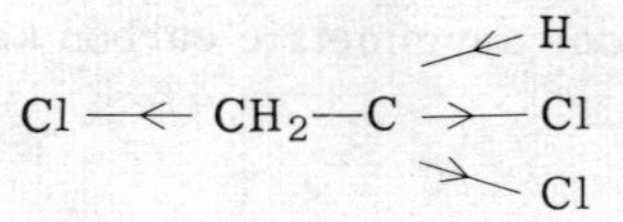

The —I effect of two chlorine atoms is greater than that of one. Consequently, the proton of $CHCl_2$ is removed more readily by OH^- than a proton of CH_2Cl. Hence (for an E2 mechanism):

$$HO^- \; H{-}CCl_2{-}CH_2{-}Cl \longrightarrow H_2O + Cl^- + CCl_2{=}CH_2$$

11. Under these conditions, the mechanism is S_N2. Since the rate changes in the presence of NaI and no temperature change has been mentioned, it may be assumed that the increased rate is brought about by a change in path which requires a lower activation energy than the original path. Since alkyl iodides are more reactive than the corresponding chlorides, then a possible route for the catalysed reaction is:

(i) $I^- \; Bu{-}Cl \longrightarrow I{-}Bu + Cl^-$
$+326{\cdot}4 \qquad -213{\cdot}4 \qquad \Delta H = +113{\cdot}0$ kJ

(ii) $HO^- \; Bu{-}I \longrightarrow HO{-}Bu + I^-$
$+213{\cdot}4 \qquad -334{\cdot}7 \qquad \Delta H = -121{\cdot}3$ kJ

If this *is* the reaction sequence, and because we are told that the reaction is accelerated, it must be assumed that the activation energy in *each* step, (i) and (ii), must be lower than that of the direct replacement, (iii).

(iii) $HO^- \; Bu{-}Cl \longrightarrow HO{-}Bu + Cl^-$
$+326{\cdot}4 \qquad -334{\cdot}7 \qquad \Delta H = -8{\cdot}3$ kJ

In order to draw the energy profiles of these reactions, we need to know the energy levels of the various systems involved (reactants, intermediates and products). Contributions by the ions are difficult to estimate and so the relative energy levels will be interpreted in terms of ΔH of each elementary step (this problem is discussed in some detail in Ch. 8, Prob. 3). On this basis, the two curves of the energy profiles have arbitrarily been brought together at reactants and products. Because reaction (iii) is slightly exothermic, the energy level of BuOH will be only slightly below that of BuCl. Also, because reaction (i) is endothermic, the energy level of BuI will be above those of BuCl and BuOH. Hence, possible profiles are:

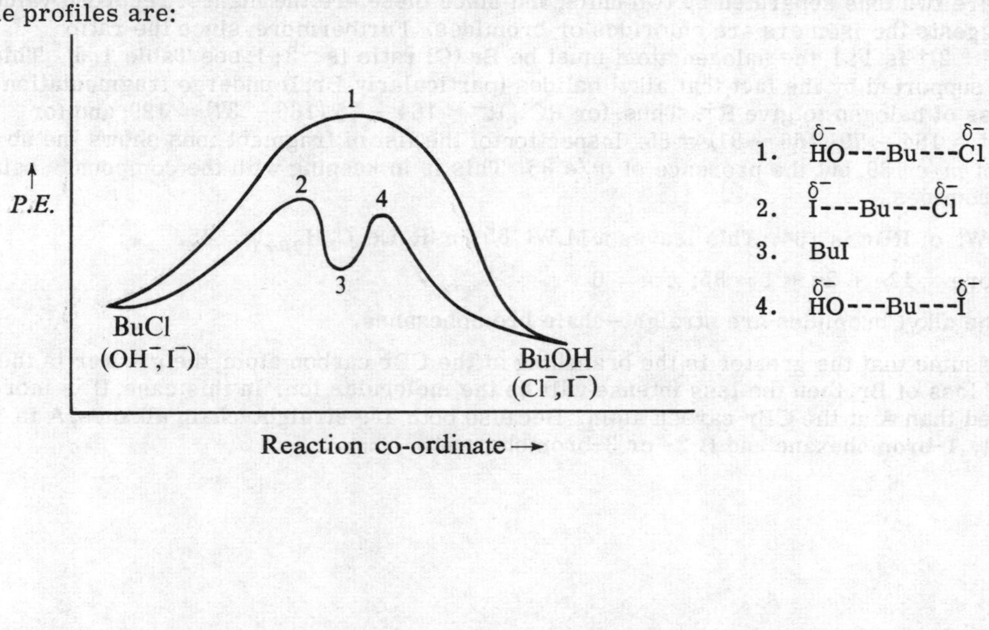

12. To exhibit optical activity, an alkane must contain at least one asymmetric carbon atom. Hence, the simplest optically active alkane is 3-methylhexane.

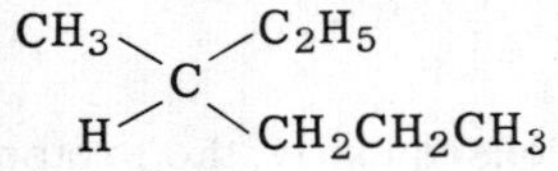

13. Two signals mean two types of protons. Since $C_3H_5Cl_3$ is equivalent to C_3H_8, the compound is a trichloropropane. The presence of a doublet suggests coupling with a CH group. In this case, the remaining four protons must be equivalent. In propane, the maximum number of protons that can be attached to a carbon atom is 3. Hence, the 4 protons must be symmetrically placed with respect to the CH proton. This is satisfied by 1, 2, 3-trichloropropane.

$ClCH_2—CHCl—CH_2Cl$
(1) (2) (1)

The terminal protons (1) (at each end) are equivalent, and would give a doublet when coupled to (2). (2) is coupled to 4 equivalent protons and so will give, because of overlapping, a quintet (2 + 2 + 1).

14. (i) CH_2F_2 One signal; triplet. Each proton will couple equally with the two F atoms (which have nuclear spin; see *Text*). The signal of F does not appear (under the usual working conditions of the instrument).

(ii) Me_3CCH_2Cl

Two signals. Since no coupling (no vicinal protons), each signal is a singlet.

(iii) Me_3CCl One signal; singlet.

(iv) $ClCH_2CHClCCl_3$
(1) (2)

Two signals. (1) doublet; (2) triplet.

(v) $ClCH_2CCl_2CH_3$
(1) (2)

Two signals. Since there is no coupling, each signal is a singlet.

15. Since we are told that the compounds are isomeric halides, we can expect the appearance of two molecular ions for Cl and Br (both have two isotopes separated by two units; see Table 1. 5), and one molecular ion for F and I (both mono-isotopic).

There are two ions separated by two units, and since these are the highest recorded values, this suggests the isomers are chlorides or bromides. Furthermore, since the ratio $M^+:(M+2)^+$ is 1 : 1, the halogen atom must be Br (Cl ratio is ~3 : 1; see Table 1. 5). This is also supported by the fact that alkyl halides (particularly Br, I) undergo fragmentation with loss of halogen to give R^+. Thus, for RCl, $R^+ = 164 - 35\ (166 - 37) = 129$; and for RBr, $R^+ = 164 - 79\ (166 - 81) = 85$. Inspection of the list of fragment ions shows the absence of m/e 129, but the presence of m/e 85. This is in keeping with the compounds being alkyl bromides.

The M.Wt. of RBr is 164. This leaves a M.Wt. 85 for R, i.e. $C_nH_{2n+1} = 85$.

Therefore $12n + 2n + 1 = 85;\ \therefore\ n = 6$

Thus, the alkyl bromides are straight—chain bromohexanes.

If we assume that the greater is the branching at the CBr carbon atom, the greater is the ease of loss of Br, then the less intense will be the molecular ion. In this case, **B** is more branched than **A** at the CBr carbon atom. Because both are straight-chain alkanes, **A** is probably 1-bromohexane and **B** 2- or 3-bromohexane.

A common fragmentation pattern in alkyl bromides is:

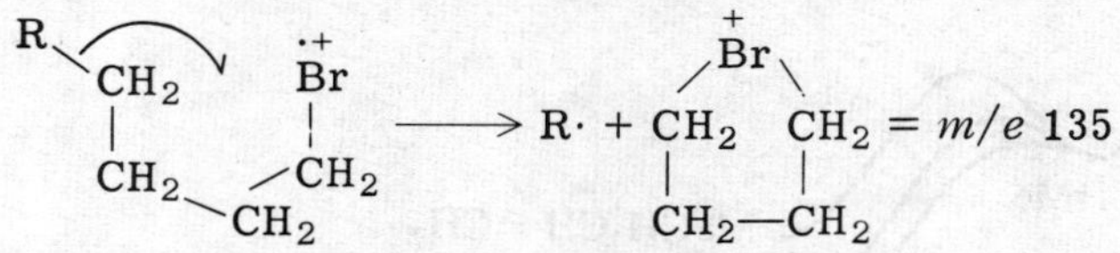

This ion is far more intense in **A** than in **B**, and since it is favoured when the carbon chain (from CBr as 1) contains 6 or more carbon atoms in a straight chain, this supports the argument that **A** is 1-bromohexane.

The base peak for *both* **A** and **B** is m/e 43. This is $C_3H_7^+$, and does not distinguish between the two.

Answer: **A** is 1-bromohexane (**B** is actually 3-bromohexane).

6 Monohydric alcohols

(Problems on page 10)

1. See Ch. 3, Prob. 2; Ch. 5, Prob. 1.

$MeCH(Me)CH_2CH_2CHOHMe$ — 5-methylhexan-2-ol

$MeCH(Me)CH_2CHOHCH_2Me$ — 5-methylhexan-3-ol

$MeCH(Me)CHOHCH_2CH_2Me$ — 2-methylhexan-3-ol

$MeCH_2CH(Me)CH_2CHOHMe$ — 4-methylhexan-2-ol

$MeCH_2CH(Me)CHOHCH_2Me$ — 4-methylhexan-3-ol

$MeCHOHCH(Me)CH_2CH_2Me$ — 3-methylhexan-2-ol

2. See Ch. 4, Probs. 5 and 18; Ch. 5, Prob. 9.

$$MeCH_2CHOHMe \xrightarrow[\text{(ii) } -H_2O]{\text{(i) } H^+} MeCH_2\overset{+}{C}HMe \begin{cases} \xrightarrow{-H^+} MeCH{=}CHMe \\ \xrightarrow{-H^+} MeCH_2CH{=}CH_2 \end{cases}$$

The 2-alkene will predominate (*trans* > *cis*), since this is more stable than the 1-alkene. Thus, this elimination gives the Saytzeff product (predominantly).

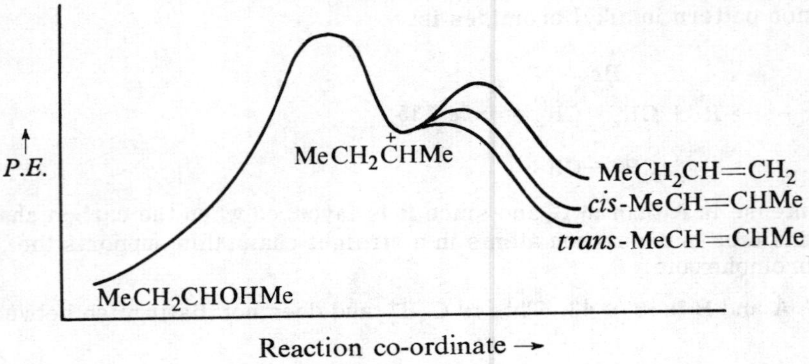

The relative energy levels of reactant and product may be established as follows.

$$\underset{\substack{+414{\cdot}2 \quad +334{\cdot}7\\ +347{\cdot}3}}{\overset{\text{MeCH}-\text{CHMe}}{\text{H} \quad\quad \text{OH}}} \longrightarrow \underset{-606{\cdot}7}{\text{MeCH}{=}\text{CHMe}} + \underset{-464{\cdot}4}{\text{H}-\text{OH}}$$

Hence, $\Delta H^{\ominus} = +25{\cdot}1$ kJ, and so the product contains more energy than the reactant. Because bond energies have been used (Table 2.2), all alkenes have the *same* energy level. It is therefore necessary to use other information to place the three alkenes at their relative energy levels.

3. This is a question of stepping up a series by (i) one carbon atom; (ii) two carbon atoms. The second reaction could be effected by repetition of the first or in one step. The methods used depend largely on the student's knowledge of the techniques of stepping up a series. The syntheses required here are relatively simple, but let us consider a general principle in synthetic work, viz., *work backwards*. PrOH contains one carbon atom more than EtOH. Hence, somewhere along the route this extra carbon atom has to be introduced, but it must be done in a 'form' that will permit ultimate conversion into CH_2OH. This, therefore, requires a knowledge of general methods of converting appropriate groups into CH_2OH. One such group is the carboxyl group; this is converted into the primary alcoholic group by reduction with lithium aluminium hydride (LAH). If, then, we can introduce a carboxyl group, we have a possible synthesis. Now, there are two very useful ways of preparing acids: (*a*) hydrolysis of cyanides ($CN \longrightarrow CO_2H$) and (*b*) Grignard reagent and carbon dioxide ($RMgX + CO_2 \longrightarrow RCO_2H$). Cyanides may be prepared by reaction between alkyl halides and KCN; Grignard reagents from alkyl halides and magnesium. Since alcohols may be used for preparing alkyl halides, we have now reached our starting material. All that has been described may be summarised as follows (write the equations *starting from the left*):

$$RCH_2OH \longleftarrow RCO_2H \longleftarrow RCN; RMgX + CO_2 \longleftarrow RX \longleftarrow ROH$$

Grignard reagents react with ethylene oxide to produce alcohols containing *two* more carbon atoms (than R in RMgX). This reaction is dealt with later (see *Text*, Ch. 15), but because of its usefulness, it is mentioned here. The syntheses required are:

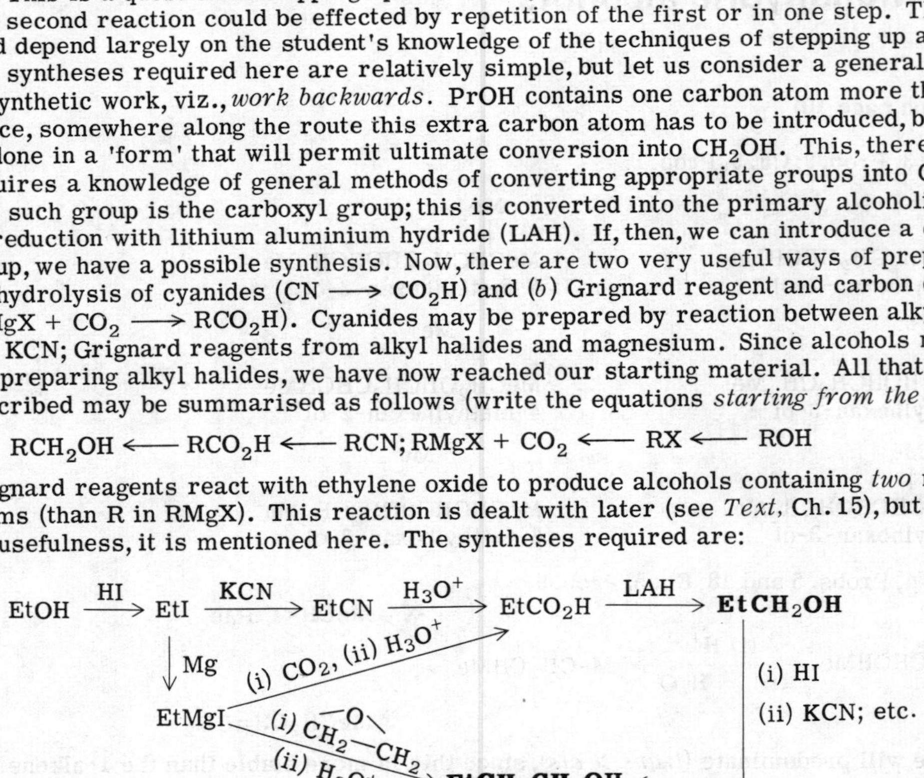

4. $R \leftarrow O \leftarrow H$. The greater the electron-attracting effect of R, the more readily will the proton be released, i.e., the stronger is ROH as an acid. Since the electronegativities of the elements in R are $H < Cl < F$, the predicted order of increasing acid strength is $CH_3CH_2OH < CCl_3CH_2OH < CF_3CH_2OH$.

5. See also Prob. 2, above.

A primary carbonium ion is formed initially, and this undergoes hydride-ion shift to form the more stable secondary carbonium ion. Both ions eliminate a proton to give the 1-alkene (from the primary ion), but the secondary ion gives the 2-alkene predominantly (Saytzeff product) and also a small amount of the 1-alkene. Because of the greater stability of the secondary ion, this will be present in far greater concentration than the primary ion. Hence the predominant product is but-2-ene (*trans* > *cis*).

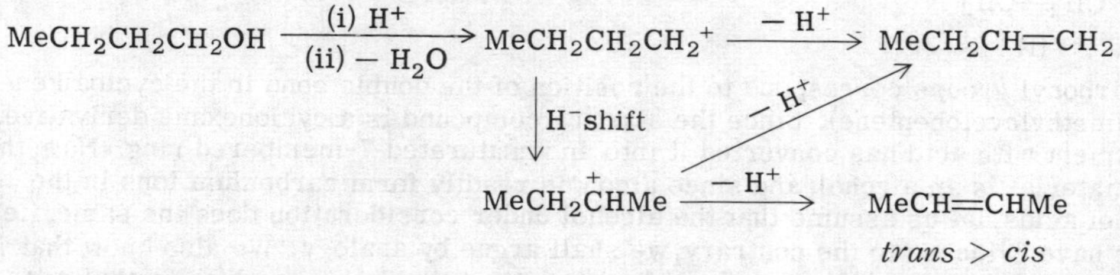

Since these are the observed results, it therefore means that the rearrangement occurs faster than proton elimination (yields: 55% 2-ene and 45% 1-ene).

6. However the reactions are carried out, i.e., directly (if possible) or indirectly (if possible), the *overall* result is the addition of a molecule of water (treated as H^+ and ^-OH fragments) and is anti-Markownikoff in (i) and Markownikoff in (ii). This suggests a peroxide effect for (i), and hence a possibility is (see Ch. 4, Prob. 13).

$$MeCH{=}CH_2 \xrightarrow[(PhCO)_2O_2]{HBr} MeCH_2CH_2Br \xrightarrow{OH^-} MeCH_2CH_2OH$$

There is also the method involving the use of diborane (*Text*, Ch. 4).

$$MeCH{=}CH_2 \xrightarrow{B_2H_6} (MeCH_2CH_2{-})_3B \xrightarrow[NaOH]{H_2O_2} 3MeCH_2CH_2OH + H_3BO_3$$

(ii) $$MeCH{=}CH_2 \xrightarrow{HI} Me_2CHI \xrightarrow{OH^-} Me_2CHOH$$

Isopropanol may also be prepared by absorbing propene in concentrated sulphuric acid, followed by the addition of water. *Direct* hydration requires special conditions (see *Text*).

7. $ROH + H^+ \rightleftharpoons ROH_2^+ \rightleftharpoons R^+ + H_2O$

Since water is *formed* and reacts *reversibly*, then using sulphuric acid in water containing $H_2{}^{18}O$ and *unlabelled* ROH, recovered alcohol will contain labelled oxygen if the equilibrium exists.

The geometry of a carbonium ion is *planar* (sp^2) and is produced by fission of the C—O bond (sp^3). If the alcohol is optically active and is of the type $R^1R^2R^3C{-}OH$, then formation of the carbonium ion results in racemisation. Hence, if we start with one optically active form of the alcohol, racemised alcohol will be recovered if the equilibrium exists.

These answers illustrate the use of isotopes and optical activity in organic chemistry. The student should always bear in mind the possibility of solving problems this way.

8. Both are elimination reactions and both tend to give mixtures of alkenes, particularly by the E1 mechanism.

9. Since ozonolysis converts **A**, C_9H_{16}, into nonane-2, 8-dione *without* loss of carbon, **A** is therefore a ring compound. Since it undergoes ozonolysis, **A** is a monocyclic compound with one double bond [saturated monocyclic: C_9H_{18} (*cf.* alkenes); saturated bicyclic: C_9H_{16} (*cf.* alkadienes, alkynes)]. If we write nonane-2, 8-dione as shown, (I), then the positions of

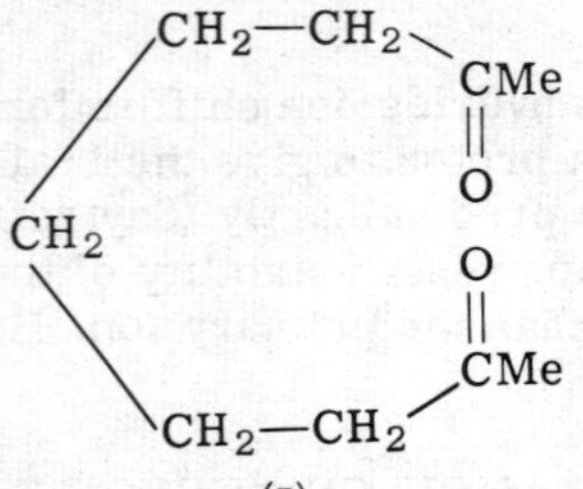

(I)

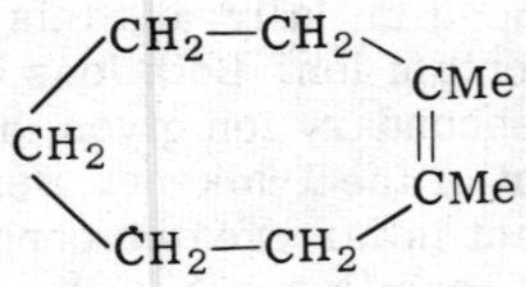

(II) [A]

the two carbonyl groups correspond to the position of the double bond in the cycloalkene, (II; 1, 2-dimethylcycloheptene). Since the starting compound is a cyclohexane derivative, then treatment with acid has converted it into an unsaturated 7-membered ring. Now, the starting material is an alcohol, and since alcohols readily form carbonium ions in the presence of acids, let us assume that the alcohol under consideration does the same, i.e., unless we have evidence to the contrary, we shall argue by analogy. We also know that alkyl carbonium ions can undergo 1, 2-shifts involving hydride ions and/or methyl carbanions. Since *ring expansion* has occurred, the ring in the starting compound must have opened and then closure followed to give the larger ring. Just as the methyl carbanion, Me^-, migrates, let us assume that the section of the ring CH_2—* can do the same if * is the carbon atom *adjacent* to C^+ in the carbonium ion. We can then formulate the steps as follows:

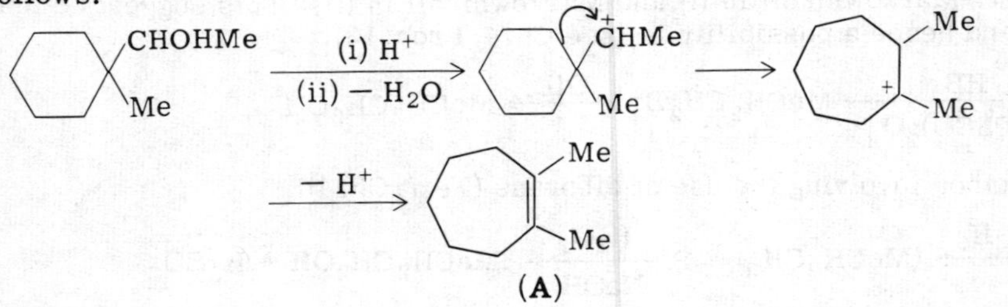

(A)

Thus, we have arrived at an explanation that appears to be eminently reasonable. If, for some reason or other, we are wrong, we are at least entitled to some credit for our ingenuity!

10. 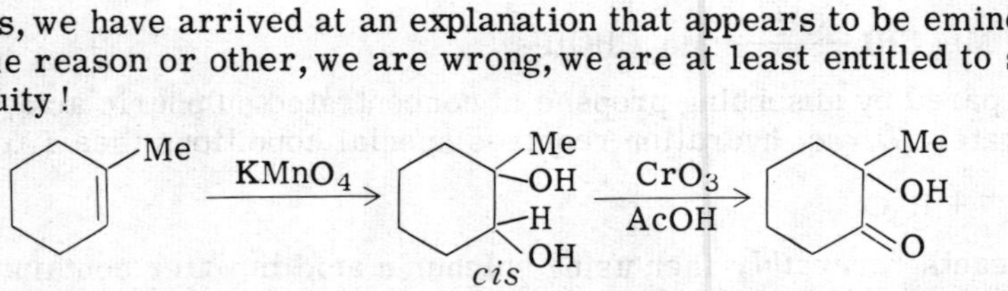

Permanganate hydroxylates a double bond to give the *cis*-1, 2-diol (see *Text*, Ch. 4). t-Alcohols are much more difficult to oxidise than s-alcohols, and since chromium trioxide in acetic acid is a relatively mild oxidising agent, only the secondary alcoholic group in the diol is attacked (see also Ch. 4, Prob. 24).

11. (i) A general scheme for introducing a triple bond into a compound is first the introduction of a double bond followed by its conversion into a triple bond. Thus:

$$MeCH_2CH_2CH_2OH \xrightarrow{HBr} MeCH_2CH_2CH_2Br \xrightarrow[EtOH]{KOH} MeCH_2CH{=}CH_2 \xrightarrow{HI}$$

$$MeCH_2CHIMe \xrightarrow[EtOH]{KOH} MeCH{=}CHMe \xrightarrow[(ii)\,KOH/EtOH]{(i)\,Br_2} MeC{\equiv}CMe$$

Note that a primary halide must first be converted into a secondary halide in order to get the double bond in the 2-position, which is the position of the triple bond.

(ii) The preparation of a particular stereoisomer (geometrical or optical) often involves one (or more) stereospecific or stereoselective reaction (see *Text*, Ch. 4). Two problems are involved in this question: (i) $C_3 \longrightarrow C_6$; (ii) position of the double bond in C_6 and *cis*-isomer. Are the two independent or interrelated? For example

$$\text{PrOH} \xrightarrow{\text{HI}} \text{PrI} \xrightarrow[\text{(Wurtz)}]{\text{Na}} \text{Pr—Pr}$$

Now we would have to introduce a double bond into the 2-position. Conversion of the hexane into its monobromide would result in a number of isomeric monobromides. This would mean a separation and probably a poor yield of the isomer we require. This line of solving our problem does not appear to be profitable. Let us therefore try a different approach. *cis*-Alkenes are usually the predominant products when alkynes are catalytically reduced, and exclusive when diborane or di-imide are used (see *Text*, Ch. 4). Hence, if we could prepare the appropriate hexyne, we can complete our synthesis. The hexyne we require is the 2-isomer, and inspection of the formula shows that it may be formed by joining a propyne to a propyl group.

$$\text{MeC}\equiv\text{C}\,\vdots\,\text{CH}_2\text{CH}_2\text{Me}$$

This can be done as follows (from PrOH):

$$\text{MeCH}_2\text{CH}_2\text{OH} \xrightarrow[\text{(ii) KOH/EtOH}]{\text{(i) HBr}} \text{MeCH=CH}_2 \xrightarrow[\text{(ii) KOH/EtOH}]{\text{(i) Br}_2} \text{MeC}\equiv\text{CH} \xrightarrow[\text{(ii) MeCH}_2\text{CH}_2\text{Br}]{\text{(i) Na/liq. NH}_3}$$

$$\text{MeC}\equiv\text{CCH}_2\text{CH}_2\text{Me} \xrightarrow{\text{redn.}}$$ 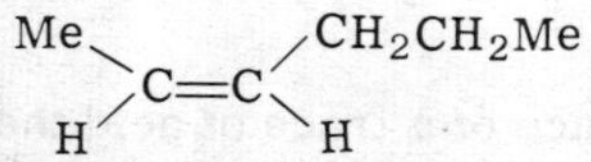

12. See also Ch. 5, Prob. 4.

(i) $\text{EtOH} + \text{HBr} \rightleftharpoons \text{Br}^- \curvearrowright \text{Et—}\overset{\frown}{\text{O}}\text{H}_2^+ \longrightarrow \text{Br—Et} + \text{H}_2\text{O}$

(ii) $\text{Br}^- \curvearrowright \text{Et—OH} \longrightarrow \text{Br—Et} + \text{OH}^-$

We can assume an S_N2 mechanism for (i), and if we apply this to (ii), the leaving group would be the hydroxide ion. Since this is a very poor leaving group (its activation energy is very high), this reaction does not occur. In contrast, all the steps in (i) require relatively low activation energies (OH_2^+ is a good leaving group).

13. If the student has read the question carefully, then the given molecular formulae and their changes, coupled with the nature of the reagent involved, will have suggested the *class* of each compound, e.g., in going from **(A)** to (B), the thionyl chloride has replaced OH by Cl; and so on. Hence, we may write:

alcohol **(A)** ⟶ alkyl chloride **(B)**
↓
acid **(D)** ⟵ alkene **(C)** ⟶ alkyl chloride **(E)**

If we now accept, purely as a working guide, that the smaller the molecule the easier it is to identify it, then the step (or steps) leading to a break-down of the molecule is worth looking at first. Ozonolysis of **(C)** gives acid **(D)** with loss of *one* carbon atom. Therefore **(C)** is a 1-alkene and **(D)**, from inspection of its molecular formula, is propionic acid,

$MeCH_2CO_2H$. Hence, **(C)** is but-1-ene and so **(A)** must be a straight-chain butanol, i.e., 1- or 2-butanol. Since the addition of HCl to but-1-ene gives the Markownikoff product 2-chlorobutane, **(E)** is this compound and therefore **(B)** is the isomer, 1-chlorobutane. Only butan-1-ol would give **(B)** under the given conditions and so **(A)** is this alcohol.

14. (i) $n\text{—}C_3H_7CO_2H \xrightarrow[\textit{or } B_2H_6]{LAH} n\text{—}C_4H_9OH$

Other reagents are also available (see Table 6.1).

(ii) $Me_2CO + EtMgI \longrightarrow Me_2C(OMgI)(Et) \xrightarrow{H^+} Me_2C(OH)Et$

(iii) $EtCO_2Et + 2MeMgI \longrightarrow IMgO\text{—}C(Me)_2(Et) \xrightarrow{H_2O} EtCMe_2OH$

(iv)

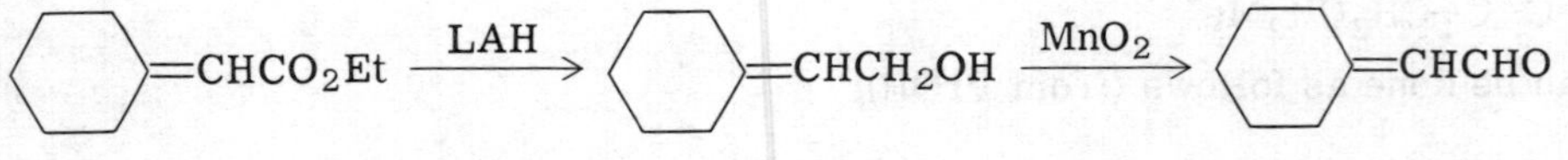

See also *Text*, Ch. 12.

(v) $CH_3CH_2CH_2OH \xleftarrow{LAH} BrCH_2CH_2COCl \xrightarrow[AlCl_3]{LAH} BrCH_2CH_2CH_2OH$

15. In the presence of a trace of acid there is no coupling between the H of OH and H on a vicinal carbon atom (rapid chemical exchange).

(i)	$CH_3\text{—}CH_2\text{—}CH_2\text{—}OH$ (m m m s)	Four signals (s = singlet)
(ii)	$(Me)_2CH\text{—}CH_2\text{—}OH$ (m m m s)	Four signals
(iii)	$Me_3C\text{—}OH$ (s s)	Two signals
(iv)	$Me_3C\text{—}CH_2\text{—}OH$ (s s s)	Three signals
(v)	$Me\text{—}CH_2\text{—}CMe_2\text{—}OH$ (m m s s)	Four signals

16. Since we are told that the compounds are the four isomeric butanols, we have:

$MeCH_2CH_2CH_2OH$	$MeCH_2CHOHMe$	Me_2CHCH_2OH	Me_3COH
(I)	(II)	(III)	(IV)

All show a molecular ion at 74, and on the basis that the intensity of M decreases in the order prim. > s > t, **A** could be (I) or (III); **B**, (III) or (I); **C**, (II); **D**, (IV).

A characteristic fragmentation pattern of alcohols is derived from β-cleavage with preferential elimination of the heaviest alkyl group. Some possibilities are, in general terms:

$$\left[R^1{-}\underset{R^3}{\overset{R^2}{\underset{|}{\overset{|}{C}}}}{-}OH\right]^{+\cdot} \longrightarrow R^1\cdot + \left[\begin{matrix}R^2\\R^3\end{matrix}\!\!>C{-}OH\right]^{+} \quad \text{and } R^{1^+} + \left[\begin{matrix}R^2\\R^3\end{matrix}\!\!>\dot{C}{-}OH\right]; \text{ etc.}$$

Hence:

(I): $R^1=C_3H_7$ (43); $R^2=R^3=H$; CH_2-OH (31).

(II): $R^1=C_2H_5$ (29); $R^2=Me$; $R^3=H$; $MeCH-OH$ (45).

(III): $R^1=C_3H_7$ (43); $R^2=R^3=H$; CH_2-OH (31).

(IV): $R^1=Me$ (15); $R^2=R^3=Me$; Me_2C-OH (59).

The base peak of **A** is *m/e* 31 and there is also a strong peak at *m/e* 43. **B** shows these two peaks reversed. Hence, it is highly probable that **A** is (I) or (III) and **B** is (III) or (I). However, (I) gives the n—Pr carbonium ion and (III) the isoPr carbonium ion, and since the latter is more stable than the former, this would explain the result that *m/e* (isoPr$^+$) is the *B.P.* for (III), i.e. **B** is (III) and so **A** is (I). It should also be noted that isoPr$^+$ is more stable than CH_2-OH^+.

The *B.P.* of **C** is *m/e* 45. Therefore **C** is (II). Also the *B.P.* of **D** is *m/e* 59, and therefore **D** is (IV). It should also be noted that all give the peak *m/e* 31 (CH_2-OH), and this is formed by rearrangement for **C** and **D** (see Text).

Answer: **A** is butan-1-ol (I); **B** is 2-methylpropan-1-ol (III); **C** is butan-2-ol (II); **D** is 2-methylpropan-2-ol (IV).

17. The answers require the use of mainly Table 6.1. Constant mention of many of these reagents in the *Text* will help the student to memorise some of them.

(i) $MeCO(CH_2)_2CO_2Et \xrightarrow{a} MeCHOH(CH_2)_2CO_2Et$
a: $NaBH_4$; $LiAlH(OBu^t)_3$; LAH-pyridine.

(ii) $HO_2C(CH_2)_4COCl \xrightarrow{b} HO_2C(CH_2)_4CH_2OH$
b: $LiBH_4$; $NaBH_4$.

(iii) $O_2N(CH_2)_2CN \xrightarrow{c} O_2N(CH_2)_2CH_2NH_2$
c: $NaBH_4-AlCl_3$; B_2H_6.

(iv) $O_2N(CH_2)_2CH{=}CH_2 \xrightarrow{d} H_2N(CH_2)_2CH{=}CH_2$
d: LAH.

(v) $O_2N(CH_2)CH{=}CH_2 \xrightarrow{e} O_2N(CH_2)_3CH_3$
e: $NaBH_4-AlCl_3$; B_2H_6.

(vi) $Me_2CHCOCl \xrightarrow{f} Me_2CHCHO$
f: $LiAlH(OBu^t)_3$ at −78°C.

(vii) $O_2N(CH_2)_3CHO \xrightarrow{g} O_2N(CH_2)_3CH_2OH$
g: $LiBH_4$; $NaBH_4$; $NaBH_4-AlCl_3$; B_2H_6; LAH-pyridine.

7 Ethers

(Problems on page 12)

1. $\text{EtMeCH—OH} \xrightarrow{\text{Na}} \text{EtMeCH—O}^-\text{Na}^+ \xrightarrow{\text{MeI}} \text{EtMeCH—OMe}$

The asymmetric carbon atom is C of CHOH. Since no bond joined to this C atom is broken during the two steps, no change in configuration is possible. Thus, the configuration of each product is the same as that of the starting material, i.e., the reaction proceeds with retention of configuration.

2. Reaction (i) involves the use of base and a *tertiary* alkyl halide. Under these conditions, the latter readily undergoes elimination to form an alkene. Reaction (ii) also involves the same two types of reagents, but in this case the halide is *primary* and so elimination, if it occurs, does so to a far less extent than in (i). Hence:

(i) $Me_3CBr + MeO^-K^+ \longrightarrow Me_2C{=}CH_2$

(ii) $Me_3CO^-K^+ + MeBr \longrightarrow Me_3COMe$

3. (i) $\text{MeOEt} + \text{HI} \rightleftharpoons \text{I}^- \; \text{Me—}\overset{\text{H}}{\overset{|}{\text{O}}}{}^+\text{—Et} \xrightarrow{\text{cold}}$

$$\text{MeI} + \text{EtOH} \xrightarrow[\text{hot}]{\text{HI}} \text{EtI}$$

The reaction is S_N2, the smaller group forms the iodide (steric effect), and stops at the first stage in the cold. When heated, the reaction proceeds to the second stage.

(ii) $Et_2O \xrightarrow{\text{Na}}$ *no reaction*

(iii) $Me_2C{=}CH_2 \xrightarrow[\text{pressure}]{H_2SO_4\text{—}H_2O} Me_3C^+ \xrightarrow[\text{pressure}]{Me_3COH}$

$$\text{Me}_3\text{C—}\overset{\text{H}}{\overset{|}{\text{O}}}{}^+\text{—CMe}_3 \xrightarrow{-\text{H}^+} \text{Me}_3\text{C—O—CMe}_3$$

Pressure is applied to get a large concentration of the carbonium ion in solution, and this type of reaction works well for alkenes which produce t-carbonium ions.

4. Since there are two signals, there are only two types of protons. The molecular formula indicates a saturated ether, which must therefore by symmetrical, i.e., of the type R—O—R. Hence $R{=}C_3H_7$. This could be $MeCH_2CH_2$ or Me_2CH. The former would give three NMR signals and the latter, two. Hence, the compound is di-isopropyl ether, $(Me_2CH)_2O$.

5. $CH_2{=}CH_2 + HX \longrightarrow CH_3CH_2{}^+ \xrightarrow{X^-} CH_3CH_2X$

$\downarrow$ EtOH

$$\text{CH}_3\text{CH}_2\text{—}\overset{\text{H}}{\overset{|}{\text{O}}}{}^+\text{—Et} \xrightarrow{-\text{H}^+} \text{Et}_2\text{O}$$

The first step is the formation of the ethyl carbonium ion, and this then reacts with the competing nucleophiles I^- and EtOH or Cl^- and EtOH. The relative amounts of products will depend on the relative nucleophilicities. The general order is $I^- > EtO^- > Cl^-$ (see *Text,* Ch. 5). If we *assume* this order holds good for EtOH, i.e., $I^- > EtOH > Cl^-$, then for iodine the predominant product will be EtI and for chlorine, Et_2O.

EtO^- is a stronger nucleophile than EtOH, and so I^- will be the strongest but EtOH and Cl^- might have changed places, i.e., $I^- > Cl^- > EtOH$. However, it follows from the data given in the question that the *assumed* order of nucleophilicity is correct.

6. See also Ch. 6, Prob. 16. α- and β-Cleavage are the main types of fission for ethers, and rearrangements can also occur.

***α*-Cleavage**

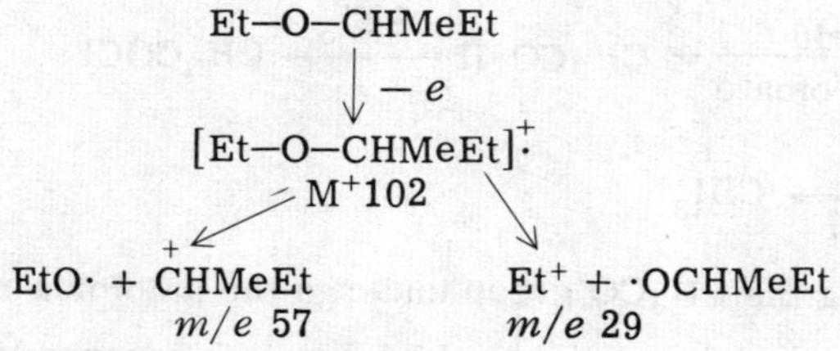

***β*-Cleavage and rearrangements**

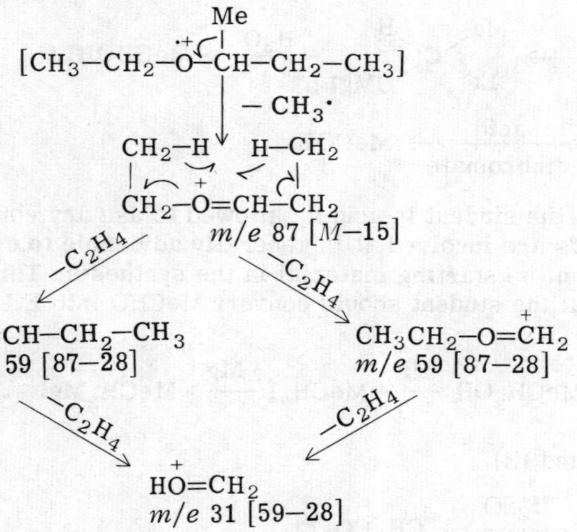

The only ion we have not accounted for is *m/e* 45. This could be EtO^+ or $MeCH{=}\overset{+}{O}H$ (see Table 1.6). The former is not a usual type of product for ethers, but the formation of the latter can be explained as follows, using double cleavage and rearrangement (*cf.* above for *m/e* 59).

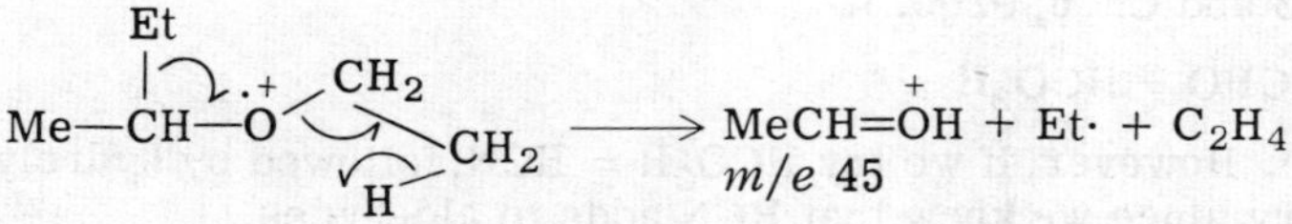

In practice, this ion, *m/e* 45, is the base peak of the ether

8 Aldehydes and ketones

(Problems on page 12)

1. (i) $MeCHO \longrightarrow EtOH$

Almost any reducing agent may be used. Convenient ones are LAH or $NaBH_4$ (see Table 6.1).

(ii) $$CH_3CHO \xrightarrow[\text{dichromate}]{\text{acid}} CH_3CO_2H \xrightarrow{PCl_5} CH_3COCl$$

(iii) $$CH_3CHO \xrightarrow[NaOH]{I_2} CHI_3$$

Most compounds containing the CH_3CO group undergo the haloform reaction (see *Text*).

(iv) The required product is a ketone, which can be prepared by oxidation of a *secondary* alchohol. All aldehydes, except formaldehyde, react with Grignard reagents to form secondary alcohols (see *Text*, Ch. 6). Hence:

$MeCHO + EtMgI \longrightarrow$ 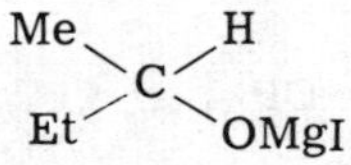 $\xrightarrow{H_3O^+} MeCHOHEt$

$$\xrightarrow[\text{dichromate}]{\text{acid}} MeCOEt$$

In this type of question, the student is usually allowed to use any compounds he wants, but when organic compounds are involved, it is generally advisable to use those which contain *fewer* carbon atoms than the starting material in the synthesis. This means, unless he is told to the contrary, that the student should convert MeCHO into EtI. This is readily done as follows.

$$MeCHO \xrightarrow{NaBH_4} MeCH_2OH \xrightarrow{HI} MeCH_2I \xrightarrow{Mg} MeCH_2MgI$$

(v) See also (i) and (ii)

$$CH_3CO_2H + EtOH \xrightarrow{H_2SO_4} CH_3CO_2Et$$

Alternatively, the ester may be obtained directly from MeCHO by the Tischenko reaction (see *Text*).

$$2MeCHO \xrightarrow{Al(OEt)_3} MeCO_2Et$$

(vi) See also Ch. 4, Prob. 3 and Ch. 6, Prob. 3.

$$CH_3CHOHCO_2H = CH_3CHO + HCO_2H$$

This addition will not take place. However, if we try $HCO_2H = HCN$, followed by hydrolysis, then this approach is satisfactory since we know that HCN adds to aldehydes.

$$CH_3CHO + HCN \longrightarrow CH_3CHOHCN \xrightarrow{H_3O^+} CH_3CHOHCO_2H$$

(vii) This can be done in several ways, e.g.,

(*a*) $MeCHO \dashrightarrow EtI \xrightarrow[\text{press.}]{NH_3} EtNH_2$

(*b*) $EtI + AgONO \longrightarrow EtNO_2 \xrightarrow{LAH} EtNH_2$

(*c*) $CH_3COCl + NH_3 \longrightarrow CH_3CONH_2 \xrightarrow{LAH} CH_3CH_2NH_2$

The relative merits of these preparations require further knowledge (see *Text,* Ch. 13).

(viii) The product contains four carbon atoms; the starting material contains two. The product also contains two bromine atoms in the 1, 2-position, and this is obtained by addition of bromine to a double bond. Hence, if we could synthesise $MeCH{=}CHCO_2H$, we can easily get the dibromide. The fragments on either side of the double bond in the acid each contain two carbon atoms, the number of carbon atoms in the starting material, MeCHO. Hence, a possibility is:

$$MeCH\boxed{O + H_2}CHCHO \longrightarrow MeCH{=}CHCHO \xrightarrow{[O]} MeCH{=}CHCO_2H$$

The first step may be effected by the aldol condensation. Thus:

$$2CH_3CHO \xrightarrow{NaOH} CH_3CHOHCH_2CHO \xrightarrow{heat} CH_3CH{=}CHCHO \xrightarrow[NH_4OH]{AgNO_3}$$
$$CH_3CH{=}CHCO_2H \xrightarrow{Br_2} CH_3CHBrCHBrCO_2H$$

Since the usual oxidising agents, e.g., permanganate, dichromate, etc., attack the double bond, the mild oxidising agent, ammoniacal silver nitrate is used.

2. Two OH groups in the 1, 2-position may be formed by hydroxylation of a double bond with alkaline permanganate. Hence, we require the compound $CH_2{=}CHCHO$. The difficulty here is that ethanolic KOH removes HCl, but at the same time causes condensations involving the aldehyde group. One way out of this difficulty is to try to protect the aldehyde group under these conditions (alkaline solutions). Since acetals are stable in alkali, the synthesis may be carried out as follows.

$$ClCH_2CH_2CHO \xrightarrow[HCl]{EtOH} ClCH_2CH_2CH(OEt)_2 \xrightarrow[EtOH]{KOH} CH_2{=}CHCH(OEt)_2$$
$$\xrightarrow[NaOH]{KMnO_4} HOCH_2CHOHCH(OEt)_2 \xrightarrow{H_3O^+} HOCH_2CHOHCHO$$

3. The accepted mechanism for this reaction is:

(i) $\underset{+414{\cdot}2}{CH_3COCH_2{-}H} + OH^- \overset{slow}{\rightleftharpoons} \underset{-464{\cdot}4}{H{-}OH} + \overline{C}H_2COCH_3$ $\Delta H = -50{\cdot}2$ kJ

(ii) $\overline{C}H_2COCH_3 + \underset{+188{\cdot}3}{Br{-}Br} \xrightarrow{fast} \underset{-284{\cdot}5}{Br{-}CH_2COCH_3} + Br^-$ $\Delta H = -96{\cdot}2$ kJ

To determine the energy levels of each system (reactants and products), we have to estimate the enthalpies of OH^-, Br^-, and $\overline{C}H_2COCH_3$. This involves the use of bond energies, electron affinities and ionisation potentials (see *Text,* Ch. 2). Because of the complexity of these evaluations, we shall ignore the energies of ions and assign relative energy levels based on changes due to fission and formation of bonds. Since both steps (i) and (ii) are exothermic (see above), the energy profile for the complete reaction may be represented by Fig. 1, on p. 102.

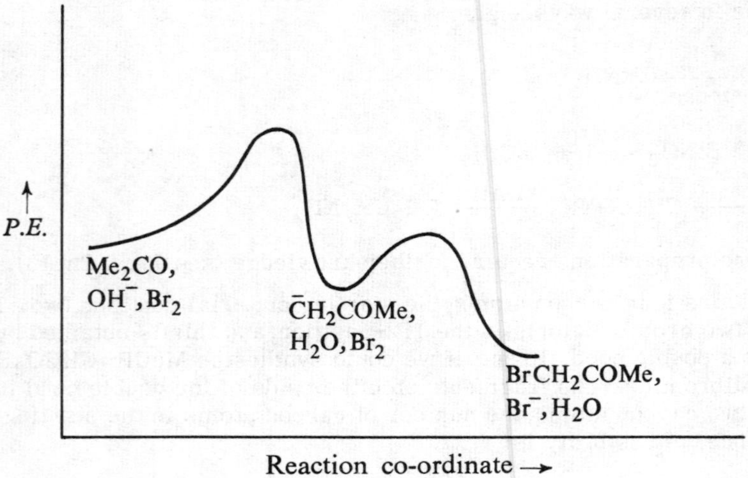

Fig. 1.

Now let us consider the reaction:

$$A + B \longrightarrow A\text{---}B \longrightarrow C + D \longrightarrow C\text{---}D \longrightarrow E + F$$

The products of the first step are the reactants of the second, and so it is possible to plot a *continuous* energy profile with two peaks representing transition states, the rate-determining step being the one leading to the higher peak. Each step in a multistep reaction is represented by its own energy profile, and in the case discussed (A + B, etc.) the two are continuous. We have also drawn this type of curve for the bromination of acetone (Fig. 1). We can, however, look at this reaction from a different point of view. Step (i) involves the formation of H_2O, but this is *not* a reactant for step (ii). Hence, we can treat the overall reaction as consisting of two *separate* systems and so draw two separate energy profiles.

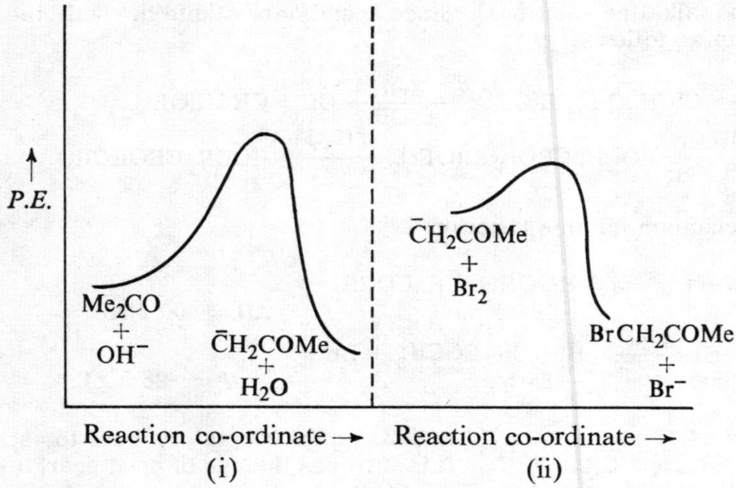

Fig. 2.

Fig. 2 (i) represents the first step. This step is exothermic and this is entirely due to the formation of an O—H bond. The low energy level of the $H_2O + \bar{C}H_2COMe$ system is the *sum* of the very low energy level of H_2O and the high energy level of $\bar{C}H_2COCH_3$. Hence, to draw the energy profile of the second step, the high energy level of $\bar{C}H_2COMe$ will be lowered by the presence of Br_2 in the system, but the sum will be higher than that of the system $H_2O + \bar{C}H_2COMe$ (the O—H bond is much stronger than the Br—Br bond). Thus, the energy profile of the second step may be represented by Fig. 2 (ii). The merit of using Figs. 2(i) and 2(ii) instead of Fig. 1 is that the former show the high reactivity of one of the reactants ($\bar{C}H_2COMe$) in the second step. This might, perhaps, be represented by Fig. 3, the dotted line producing a 'continuous' curve and indicating that the product system of the first step is *not* the reactant system of the second step. In this way, we get a 'continuous' curve which

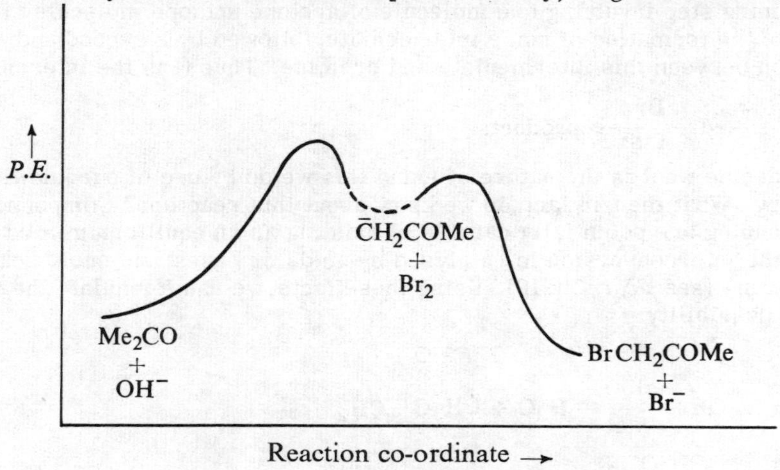

Fig. 3.

does show the relative energy levels of 'reactants' and 'products'.

The student should now be in a position to see that energy profiles are not always what they are meant to represent. Even so, they are still useful, particularly for picking out the r/d step when drawn as in Fig. 3 (with the dotted part or with the curve continuous, if preferred).

We shall now deal with an entirely different problem, and it is discussed here because the reaction we shall investigate is the bromination of acetone in alkaline solution. By now, the student will have dealt with mechanisms of many reactions and the sort of evidence that has led to their formulation. The most general method for determining reaction mechanisms is kinetic studies and it is this method that we shall now consider.

(i) *The determination of the mechanism of the base-catalysed bromination of acetone.*

First, we start with the facts. *Reactants.* Acetone, sodium hydroxide (in aqueous solution) and bromine. *Products.* Tribromoacetone ($MeCOCBr_3$), bromoform ($CHBr_3$) and acetate ion ($MeCO_2^-$).

Kinetic studies show that the rate of reaction is proportional to the concentration of acetone (represented as A), proportional to the concentration of hydroxide ion, and is independent of the concentration of bromine, i.e.,

$$\text{rate} \propto [A][OH^-]$$

Thus, the rate law is first-order in acetone, first-order in hydroxide ion, and zero-order in bromine; and consequently the rate law is second-order (see *Text*, Ch. 2).

$$\therefore \text{rate} = k[A][OH^-]$$

where k is the rate constant. On the other hand, since the rate of a reaction is given by the rates at which the reactants are consumed, we may write (in differential form):

$$\text{rate} = -\frac{d[\text{A}]}{dt} = \frac{d[\text{OH}^-]}{dt} = k[\text{A}]\,[\text{OH}^-]$$

By convention, k (and rate) is always taken to be a positive quantity. Since [A] and $[\text{OH}^-]$ *decrease* with time, their differentials have *negative* values. Hence the use of the negative sign for the differential (to make k positive).

A mechanism consistent with this rate law is at *least* a two-stage process, the first being the rate-determining step involving one molecule of acetone and one molecule of hydroxide ion and leading to the formation of some intermediate, followed by a second and *fast* step involving reaction between this intermediate and bromine. Thus (I is the intermediate):

$$\text{A} + \text{OH}^- \xrightarrow{\text{slow}} \text{I} \xrightarrow[\text{fast}]{\text{Br}_2} \text{products}$$

We now have to decide what is the nature of I, and this we do by use of our knowledge of organic chemistry. What then, in fact, do we know about this reaction? Compounds containing the —CHCO— grouping are potentially capable of existing as an equilibrium mixture of keto and enol forms, the interconversion is catalysed by acids or bases, and enols react very rapidly with bromine (see *Text*, Ch. 10). Using these facts, we can formulate the following mechanism as a possibility:

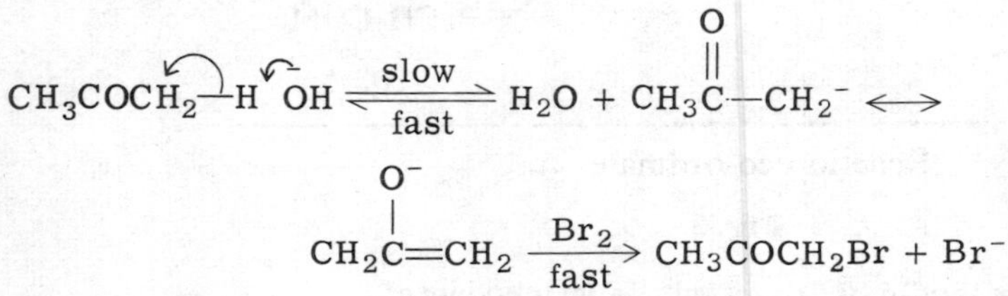

However, since the products isolated are MeCOCBr_3, CHBr_3 and MeCO_2^-, we now assume the following sequence:

$$\text{CH}_3\text{COCH}_2\text{Br} + \text{OH}^- \underset{\text{fast}}{\overset{\text{fast}}{\rightleftharpoons}} \text{H}_2\text{O} + \text{CH}_3\text{CO}\overset{-}{\text{C}}\text{HBr} \xrightarrow[\text{fast}]{\text{Br}_2}$$

$$\text{Br}^- + \text{CH}_3\text{COCHBr}_2 \xrightarrow[\text{(ii)Br}_2\text{; fast}]{\text{(i)OH}^-\text{; fast}} \text{Br}^- + \text{CH}_3\text{COCBr}_3$$

$$\xrightarrow[\text{fast}]{\text{OH}^-} \text{CH}_3\text{CO}_2^- + \text{CHBr}_3$$

We can justify these subsequent steps based on the —I effect of bromine.

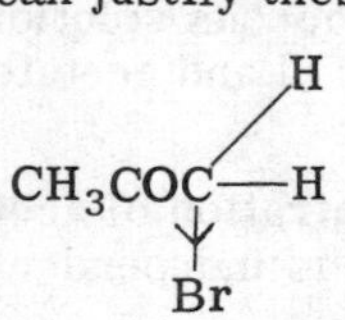

This facilitates easier proton removal (by OH^-) than from acetone, and so the second bromination is faster than the first; and the third is faster than the second. At the same time, this argument offers an explanation why the brominated carbon atom is further brominated rather than the remaining methyl group.

(ii) *Correlation of the mechanism with the rate law.*

What we have done so far is to propose a mechanism based on a rate law (*which is obtained experimentally*). Now we shall discuss the reverse procedure, i.e., determine whether the proposed mechanism satisfies the rate law. The answer may appear obvious, but the method of establishing the correlation (in this case at least) is somewhat complicated. If we represent the enolate anion by En, then we can write:

$$A + OH^- \underset{k_{-1}}{\overset{k_1}{\rightleftharpoons}} En + H_2O \tag{1}$$

$$En + Br_2 \xrightarrow{k_2} MeCOCH_2Br + Br^- \tag{2}$$

Since the rate of reaction is given by the rate at which A or OH^- is consumed, then:

$$\text{rate} = -\frac{d[A]}{dt} = -\frac{d[OH^-]}{dt} = k_1[A][OH^-] - k_{-1}[En] \tag{3}$$

Since water is one of the products (equation 1), its concentration should be included in equation (3). However, because the reaction is carried out in dilute aqueous alkaline solution, $[H_2O]$ is virtually unchanged and the reverse reaction is first-order (in En). To evaluate the rate, it is necessary to know the concentration of En. This cannot be measured, and unless we can overcome this difficulty, we shall not be able to proceed further. Now, the overall rate of the formation of En is given by its actual rate of formation from A and OH^- minus the sum of its rate of return to A and OH^- and the rate of its loss by reaction with bromine, i.e.,

$$\frac{d[En]}{dt} = k_1[A][OH^-] - k_{-1}[En] - k_2[En][Br_2] \tag{4}$$

The value of [En] is obtained from equation (4) by use of the *method of the stationary state* (or the *steady state treatment*). The assumption is made that the concentration of En is always very small and hence it follows that the rate of change of its concentration ($d[En]/dt$) will be smaller still, so much so that we assume it to be zero. Thus putting $d[En]/dt = 0$ in equation (4), we have

$$0 = k_1[A][OH^-] - k_{-1}[En] - k_2[En][Br_2]$$

$$\therefore [En] = \frac{k_1[A][OH^-]}{k_{-1} + k_2[Br_2]} \tag{5}$$

Substituting this value of [En] in equation (3):

$$\text{rate} = -\frac{d[A]}{dt} = -\frac{d[OH^-]}{dt} = k_1[A][OH^-] - k_{-1}\left\{\frac{k_1[A][OH^-]}{k_{-1} + k_2[Br_2]}\right\}$$

$$= k_1[A][OH^-]\left\{1 - \frac{k_{-1}}{k_{-1} + k_2[Br_2]}\right\}$$

Hence:

$$\text{rate} = -\frac{d[A]}{dt} = -\frac{d[OH^-]}{dt} = k_1[A][OH^-] \times \frac{k_2[Br_2]}{k_{-1} + k_2[Br_2]} \tag{6}$$

If we assume that the rate at which En reacts with bromine (equation 2) is very much faster than the rate at which En reforms acetone (equation 1), then

$$k_2[\text{En}][\text{Br}_2] \gg k_{-1}[\text{En}]$$

$$\therefore\ k_2[\text{Br}_2] \gg k_{-1}$$

Hence we may neglect k_{-1} in equation (6) which now becomes

$$\text{rate} = -\frac{d[\text{A}]}{dt} = -\frac{d[\text{OH}^-]}{dt} = k_1[\text{A}][\text{OH}^-]$$

This is the rate law that was obtained *experimentally* (see part (i) of this discussion).

When a mechanism has been proposed, its acceptance is strengthened if further evidence can be obtained to support it. A common approach to this problem is to make predictions based on the mechanism and then try to verify them experimentally. For example, since chlorine, bromine and iodine are chemically similar, it is reasonable to assume that the base-catalysed halogenation of acetone could follow the same path for each halogen. It has, in fact, been shown that all three halogens, under the *same* conditions, halogenate at the *same* rate and that the rate law is zero-order in halogen. This is in keeping with the proposed mechanism.

4. Although acetone is 100% unhydrated in dilute aqueous solution, acetone can react reversibly with water, the equilibrium lying virtually completely on the unhydrated side. Hence, in water enriched with ^{18}O, there are the following equilibria:

$$Me_2CO + H_2\overset{*}{O} \rightleftharpoons Me_2C\begin{matrix}\overset{*}{O}H \\ OH\end{matrix} \rightleftharpoons Me_2C\overset{*}{O} + H_2O$$

Recovered acetone will therefore contain ^{18}O.

5. (i) $EtCH_2CH_2OH \longrightarrow EtCH_2CH_2{-}CHEtCH_2OH$

We can solve this problem by using arguments similar to those used in Prob. 1 (viii). Inspection of the formula of 2-ethylhexan-1-ol shows that it consists of two butane chains joined by C1 of one to C2 of the other. This is the characteristic mode of linkage when aldehydes undergo the aldol condensation, the aldehyde group (C1) of one molecule condensing at the α-carbon atom (C2) of the other. Thus:

$$EtCH_2CH_2OH \xrightarrow[\text{dichromate}]{\text{acid}} EtCH_2CHO \xrightarrow[\text{NaOH}]{\text{2 mol.}} EtCH_2CH(OH)CHEtCHO$$

This aldol now has to be converted into the required compound. The aldehyde group is readily reduced to alcohol. The problem then is to 'replace' the OH in the aldol by H. This is not a direct reaction, but since the aldol is an alcohol as well as an aldehyde, a possibility is that the molecule can be dehydrated by acid to an unsaturated aldehyde and this reduced to the saturated alcohol.

$$EtCH_2CH(OH)CHEtCHO \xrightarrow[-H_2O]{H_2SO_4} EtCH_2CH{=}CEtCHO \xrightarrow[\text{cat.}]{H_2} EtCH_2CH_2CHEtCH_2OH$$

The alkene formed is the Saytzeff product (see *Text*, Ch. 5), and both the double bond and the aldehyde group are reduced together by catalytic hydrogenation. Other reagents will also do this, e.g., diborane (see Table 6.1).

(ii) 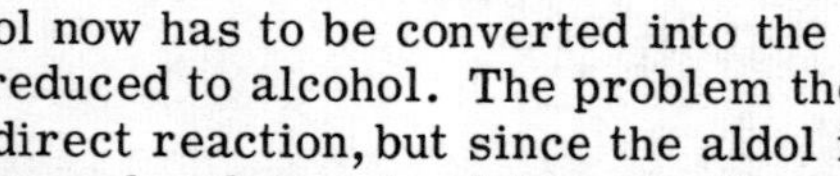

In this synthesis there is a change in ring-size. There are two reasonable possibilities; (i) to deliberately open the 6-membered ring and somewhere along the route to reclose to a 5-membered ring. (ii) The starting material contains 6 carbon atoms and so does the product. This suggests that the synthesis might be performed by a rearrangement of a suitable precursor. We have seen that 1, 2-glycols undergo rearrangement in acid solution (pinacol-pinacone), and argument by analogy suggests that cyclic 1, 2-glycols might do the same (see also Ch. 6, Prob. 9). What we therefore require is cyclohexane-1, 2-diol. Since this type of diol is formed by hydroxylation of an alkene, the synthesis is (see also *Text*, Ch. 19):

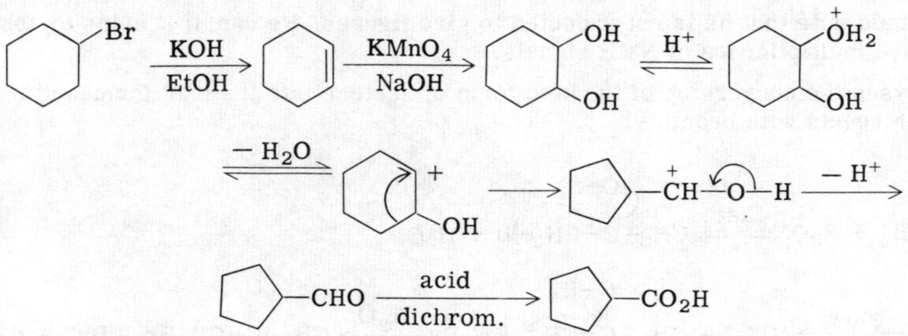

6. $H{-}CH_2CHO + OH^- \rightleftharpoons \overline{C}H_2CHO + H{-}OH$
+414·2 −464·4

$H{-}CH_2COCH_3 + OH^- \rightleftharpoons \overline{C}H_2COCH_3 + H{-}OH$
+414·2 −464·4

Hence, using the appropriate bond energies, ΔH for *both* reactions is −50·2 kJ. If we now equate ΔH with ΔG, then both reactions are *equally* thermodynamically favourable (see also Ch. 5, Prob. 6). This is *not* the case in practice. Acetaldehyde is a stronger acid (pK_a 14·5) than acetone (pK_a 20). Hence, the former equilibrium lies more to the right than does the latter and so the former reaction is the more thermodynamically favourable. Bond dissociation energies would give a more accurate estimate of ΔH (see Ch. 4, Prob. 9).

7. The i.r. band at 1715 cm^{-1} suggests a ketone (there is always the possibility that it might be an aldehyde). The molecular formula, $C_5H_{10}O$, corresponds to a saturated ketone (or aldehyde). Alternatively, we could use the D.B.E. approach (See Ch. 1, Prob. 9).

C_aH_bO: $a = 5$; $b = 10$. $\therefore$ D.B.E. $= a + 1 - b/2 = 5 + 1 - 5 = 1$

Since the carbonyl group contains 1 double bond, the molecule is saturated.

The two NMR signals are a triplet and a quartet. This is characteristic of the ethyl group $CH_3{-}CH_2$ (triplet—quartet). The student should always bear this in mind. Since these are the only two signals, this suggests *two* ethyl groups (C_4H_{10}) symmetrically placed around the CO group (no coupling between the two Et groups). Hence, the compound is diethyl ketone, Et_2CO.

8. D.B.E. of $C_3H_6O = 3 + 1 - 6/2 = 1$ (see Prob. 7 above). Hence the molecule contains 1 double bond.

$MeCH_2CHO$	MeCOMe	$CH_2{=}CHCH_2OH$	$CH_2{=}CHOMe$
(I)	(II)	(III)	(IV)

The shift of an i.r. band to a higher frequency on progressive dilution is characteristic of *decrease* in association due to hydrogen bonding (see *Text*, Ch. 6). Only (III), allyl alcohol, can be expected to exhibit hydrogen bonding.

	(I)	(II)	(III)	(IV)
i.r.	CO group (probably CHO)	CO group (probably >CO)	double bond; OH group	double bond; ether
NMR	3 signals (low field for C*H*O)	1 signal	4 signals (olefinic H)	3 signals (olefinic H)

The two types of spectroscopy, coupled together, distinguish the four structures.

The student should note that he is not expected to give figures. He can, if it helps to answer the question, give multiplicities of NMR signals.

9. Acid catalyses the conversion of the keto form of acetone into the enol form, and it is the latter which reacts with bromine.

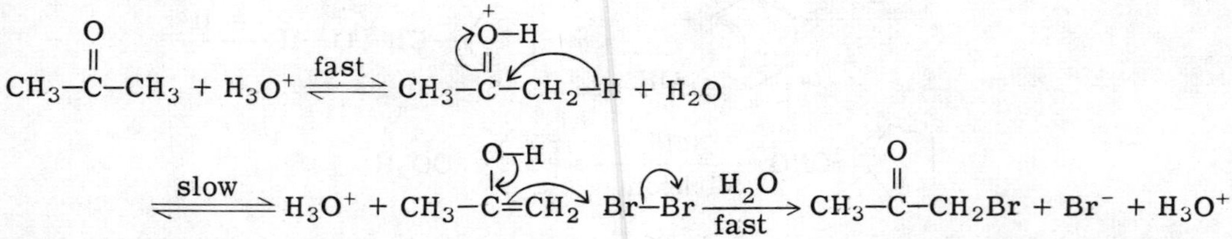

According to this mechanism, the r/d step *precedes* the addition of bromine. Hence, the rate law does not contain the concentration of bromine, i.e., the rate law is zero order with respect to bromine (see *Text*, Ch. 2).

It should be noted that rate laws are determined *experimentally*. The above reaction has been *shown* to be zero order in bromine. Hence, *whatever* is the mechanism, the step involving bromine comes *after* the r/d step in a multistage reaction.

10. The equations have been written down in steps in order to give a clearer picture of the reactions.

(i) Because acetylene is very strongly acidic relative to acetone, sodium acetylide will be formed preferentially. Any sodio-acetone produced will regenerate acetone by reaction with acetylene.

$$CH{\equiv}CH + NaNH_2 \longrightarrow NH_3 + CH{\equiv}C^-Na^+$$

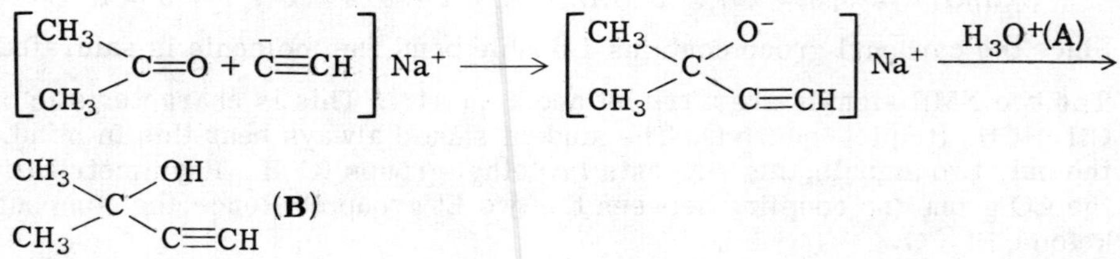

(ii) $ClCH_2CO_2Et \xrightarrow{NaNH_2} Cl\bar{C}HCO_2Et \xrightarrow{PhCOCH_3}$

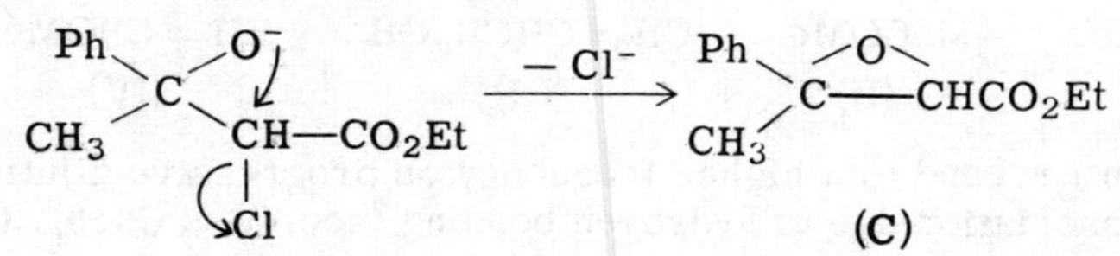

(iii) 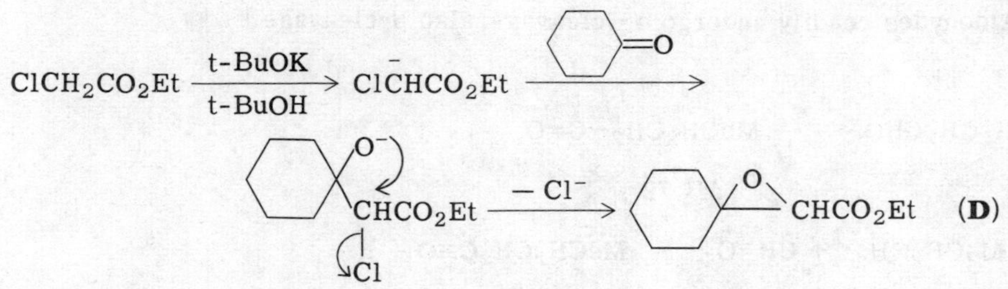

(iv) $EtONO + HCl \longrightarrow EtCl + HONO \overset{H^+}{\rightleftharpoons} H_2\overset{+}{O}\text{–}NO \rightleftharpoons H_2O + NO^+$

$$CH_3COCH_2CH_3 \underset{-\,H^+}{\overset{H^+}{\rightleftharpoons}} CH_3C(OH)\text{=}CHCH_3 \xrightarrow{NO^+} CH_3C(\text{=}O)\text{–}CH(N\text{=}\overset{+}{O}H)CH_3 \xrightarrow{-\,H^+} CH_3COC(\text{=}NOH)CH_3 \quad \textbf{(E)}$$

(v) $$Et_2CO + NH_3 \longrightarrow \left[Et_2C\begin{matrix}OH\\NH_2\end{matrix}\right] \xrightarrow{-H_2O} Et_2C\text{=}NH \xrightarrow{CN^-} Et_2C\begin{matrix}\bar{N}H\\CN\end{matrix} \xrightarrow{H_2O} Et_2C\begin{matrix}NH_2\\CN\end{matrix} \quad \textbf{(F)}$$

11. In both cases, the substrate is a 1, 2-glycol, and these are known to undergo the pinacol-pinacone rearrangement in acid solution. The general rules are: (i) The OH group lost (as H_2O) is the one that leads to the more stable carbonium ion; (ii) When there is competition for migration, it is the more electron-releasing group that migrates (predominantly or exclusively).

(i) $$Me_2C(OH)\text{–}CH_2OH + H^+ \rightleftharpoons Me_2C(\overset{+}{O}H_2)\text{–}CH_2OH \rightleftharpoons$$

$$H_2O + Me_2\overset{+}{C}\text{–}CH(H)\text{–}OH \xrightarrow[\text{shift}]{H^-} Me_2CH\text{–}\overset{+}{C}H\text{–}O\text{–}H \xrightarrow{-\,H^+} Me_2CHCHO$$

(ii) $$PhMeC(OH)\text{—}CMePh(OH) + H^+ \rightleftharpoons PhMeC(\overset{+}{O}H_2)\text{—}CMePh(OH) \rightleftharpoons$$

$$H_2O + PhMe\overset{+}{C}\text{—}CMe(Ph)(OH) \longrightarrow Ph_2MeC\text{—}\overset{+}{C}Me(O\text{—}H) \xrightarrow{-\,H^+} Ph_2MeCCOMe$$

12. (i) Aldehydes readily undergo α- cleavage; also β-cleavage.

(*a*) $MeCH_2CH_2CHO \xrightarrow{-e} MeCH_2CH_2{-}\overset{\displaystyle H\atop |}{C}{=}\overset{\cdot +}{O}$ (M^+ 72)

$\xrightarrow{\alpha}$ $MeCH_2CH_2\cdot + CH{\equiv}\overset{+}{O}$ (*m/e* 29)

$\xrightarrow{\alpha}$ $MeCH_2CH_2C{\equiv}\overset{+}{O} + H\cdot$ (*m/e* 71)

$$MeCH_2{-}CH_2CH\overset{\cdot +}{O} \xrightarrow{\beta} MeCH_2\cdot + [CH_2{=}CHO]^+ \quad m/e\ 43$$

We have now accounted for all fragments except *m/e* 44 (the *B.P.*). Since a γ—H is present, the McLafferty rearrangement can occur.

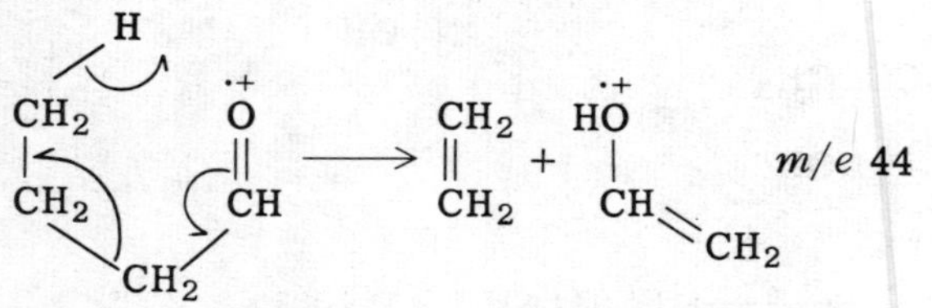

This rearrangement is suggested by the fact that $m/e\ 44 = M{-}28$ (28 = C_2H_4 or CO; see Table 1.6).

$Me_2CHCHO \xrightarrow{-e} Me_2CHCH\overset{\cdot +}{O}$ (M^+ 72)

$\xrightarrow{\alpha}$ $H\cdot + Me_2CHC{\equiv}\overset{+}{O}$ (*m/e* 71)

$\xrightarrow{\alpha}$ $Me_2CH^+ + \cdot CHO$ (*m/e* 43)

$\xrightarrow{\alpha}$ $Me_2CH\cdot + CH{\equiv}\overset{+}{O}$ (*m/e* 29)

This alternative α-cleavage to give *m/e* 43 (*B.P.*) is favoured because of the stability of the isopropyl carbonium ion (note the absence of *m/e* 44; there is no γ—H).

(ii) The fragmentation patterns of ketones are similar to those of aldehydes.

(*c*) $MeCH_2CH_2COMe \xrightarrow{-e} MeCH_2CH_2\overset{\overset{\cdot +}{O}\atop \|}{C}Me \xrightarrow{\alpha}$ (M^+ 86)

$$MeCH_2CH_2\cdot + MeC{\equiv}\overset{+}{O} \text{ (m/e 43) } \textit{and } MeCH_2CH_2C{\equiv}\overset{+}{O} + Me\cdot \text{ (m/e 71)}$$

Loss of the heavier alkyl group occurs preferentially; $MeC{\equiv}\overset{+}{O}$ is the *B.P.*

$$[MeCH_2{-}CH_2COMe]^{\overset{+}{\cdot}} \xrightarrow{\beta} MeCH_2^+ + \cdot CH_2COMe \text{ (m/e 29) } \textit{and } MeCH_2\cdot + {}^+CH_2COMe \text{ (m/e 57)}$$

Because of the presence of a γ–H, the McLafferty rearrangement is possible [m/e 58 = M–28].

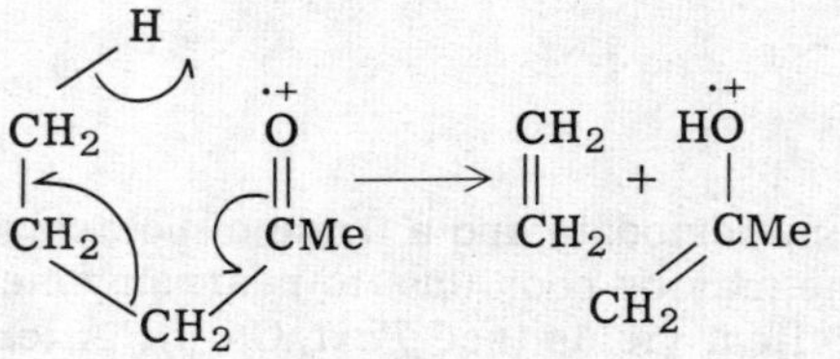

m/e 58

(*d*) $[MeCH_2COCH_2Me]^{+\cdot} \xrightarrow{\alpha} MeCH_2^+ + MeCH_2CO\cdot$ *and*

M^+ 86 $\quad$ m/e 29

$\downarrow \beta$ $\quad$ $MeCH_2\cdot + MeCH_2C{\equiv}\overset{+}{O}$

$\quad$ m/e 57 (*B.P.*)

$MeCH_2COCH_2^+ + Me\cdot$

m/e 71

Note the absence of m/e 58 (no γ–H) and of m/e 43 (no MeCO group).

9 Saturated monocarboxylic acids

(Problems on page 14)

1. If the student wishes to be on the safe side, he can always give *every* step separately in a synthesis. However, when various steps are standard reactions, it is permitted to combine appropriate steps labelling the reagents in order of use as (i), (ii), etc. Even so, the student should not be so enthusiastic about this approach that he uses, say (i)-(vi) with *one* arrow. Until he has more experience, the student should not exceed *three*, and will probably be safer with *two*.

The conversion requested is a stepping down of the acid series. This can be done by a number of *standard* methods. The student would be well-advised to master the following three:

(i) $C_4H_9CH_2CO_2H \xrightarrow[\text{(ii) } Br_2/CCl_4]{\text{(i) HgO or Ag salt}} C_4H_9CH_2Br$

$\xrightarrow[\text{(ii) acid dichr.}]{\text{(i) } OH^-} C_4H_9CO_2H$ [*N.B.* Hunsdiecker reaction]

(ii) $$C_4H_9CH_2CO_2H \xrightarrow{\text{LAH}} C_4H_9CH_2CH_2OH \xrightarrow[-H_2O]{\text{acid}}$$

$$C_4H_9CH{=}CH_2 \xrightarrow[\text{reagent}]{\text{Lemieux}} C_4H_9CO_2H$$

The Lemieux reagent is an aqueous solution of sodium periodate and a trace of potassium permanganate (see *Text*, Ch. 4). The yield of 1-alkene may be poor (due to rearrangement); pyrolysis of the acetate (or xanthate) would give excellent yields (see *Text*, Ch. 4). Alternatively, we could use HCl and then treat the chloride with ethanolic KOH.

(iii) $$C_4H_9CH_2CO_2H \xrightarrow[\text{HCl}]{\text{MeOH}} C_4H_9CH_2CO_2Me$$

$$\xrightarrow[\text{(ii) } H_3O^+]{\text{(i) 2PhMgBr}} C_4H_9CH_2C(OH)Ph_2 \xrightarrow[-H_2O]{\text{acid}} C_4H_9CH{=}CPh_2$$

$$\xrightarrow[\text{reagent}]{\text{Lemieux}} C_4H_9CO_2H + Ph_2CO$$

For both (ii) and (iii), other ways of oxidizing the alkene are ozonolysis (silver oxide or hydrogen peroxide), chromium trioxide, etc.

2. If we accept that the $B_{AC}2$ mechanism operates in all cases, we have, as the r/d step:

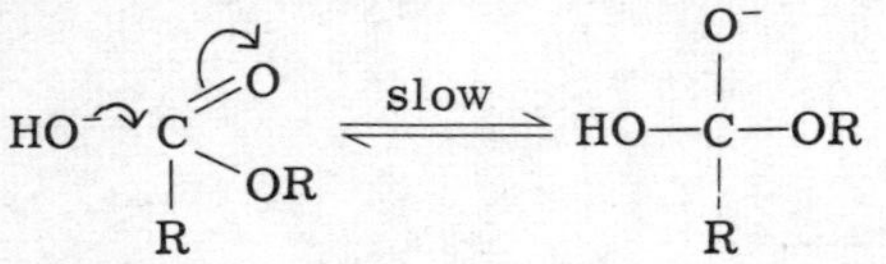

The larger the size of the R group, the greater would be the steric effect. Hence, it is to be expected that $MeCO_2Me$ would be hydrolysed most readily and $MeCO_2Bu^t$ least readily.

3. (i) $$\overset{*}{\mathrm{M}}\mathrm{eI} + \mathrm{NaCN} \xrightarrow{\text{EtOH}} \overset{*}{\mathrm{M}}\mathrm{eCN} \xrightarrow{H_3O^+} \overset{*}{\mathrm{M}}\mathrm{eCO_2H}$$

(ii) $$\mathrm{MeI} + \mathrm{Na}\overset{*}{\mathrm{C}}\mathrm{N} \xrightarrow{\text{EtOH}} \mathrm{Me}\overset{*}{\mathrm{C}}\mathrm{N} \xrightarrow{H_3O^+} \mathrm{Me}\overset{*}{\mathrm{C}}\mathrm{O_2H}$$

(iii) $$\overset{*}{\mathrm{M}}\mathrm{e\,I} + \mathrm{Na}\overset{*}{\mathrm{C}}\mathrm{N} \dashrightarrow \overset{*}{\mathrm{M}}\mathrm{e}\overset{*}{\mathrm{C}}\mathrm{O_2H}$$

(iv) $$2\overset{*}{\mathrm{M}}\mathrm{eCO_2H} \xrightarrow{Ca(OH)_2} (\overset{*}{\mathrm{M}}\mathrm{eCO_2})_2\mathrm{Ca} \xrightarrow{\text{heat}} \overset{*}{\mathrm{M}}\mathrm{eCO}\overset{*}{\mathrm{M}}\mathrm{e}$$

(v) $$\overset{*}{\mathrm{M}}\mathrm{e}\overset{*}{\mathrm{C}}\mathrm{O_2H} \xrightarrow{\text{LAH}} \overset{*}{\mathrm{M}}\mathrm{e}\overset{*}{\mathrm{C}}\mathrm{H_2OH} \xrightarrow[\text{dichr.}]{\text{acid}} \overset{*}{\mathrm{M}}\mathrm{e}\overset{*}{\mathrm{C}}\mathrm{HO}$$

This oxidation must be carried out under controlled conditions, otherwise much of the aldehyde will be further oxidised (to acid).

4. (i) $$CHCl_2COCl \xrightarrow{\text{LAH}} CHCl_2CH_2OH$$

LAH reduces $COCl \longrightarrow CH_2OH$ and $RX \longrightarrow RH$; reduction of RCl is fairly difficult and far more so is $RCHCl_2$.

(ii) $$HOCH_2CH_2CN \xrightarrow[H_2O]{\text{HBr}} BrCH_2CH_2CO_2H$$

The cyano group is hydrolysed to carboxyl and the alcoholic group is converted to bromide. Intermediates could be $BrCH_2CH_2CN$ and/or $HOCH_2CH_2CO_2H$.

(iii) $B_{AC}2$ mech. $\quad Me_3CCH_2OAc + OH^- \longrightarrow Me_3CCH_2OH + AcO^-$

Since this is acyl-oxygen fission, the CH_2—O bond remains intact and no rearrangement is possible.

(iv) $B_{AL}1$ mech. Since this is alkyl-oxygen fission with formation of a carbonium ion, a number of products are formed as follows:

$$Me_3CCH_2\text{—}OAc \longrightarrow AcO^- + Me_3CCH_2^+ \xrightarrow{OH^-} Me_3CCH_2OH$$

$$\downarrow \text{rearr.}$$

$$Me_2C(OH)Et \xleftarrow{OH^-} Me_2\overset{+}{C}CH_2Me \xrightarrow{-H^+} Me_2C{=}CHMe$$

(v) $ClCH_2CO_2H + NaI \xrightarrow{\text{acetone}} ICH_2CO_2H + NaCl$

This is an example of the Finkelstein reaction and takes place because NaCl is far less soluble in acetone than NaI (see *Text,* Ch. 3).

(vi) $Me_2CBrCONH_2 \xrightarrow[NaOH]{Br_2} Me_2CO + 3NaBr + Na_2CO_3$

Had H been present instead of Br, the reaction would have been a typical example of the Hofmann primary amine preparation to give Me_2CHNH_2. A possible explanation for the formation of acetone is that the substrate leads to the formation of (I) or (II).

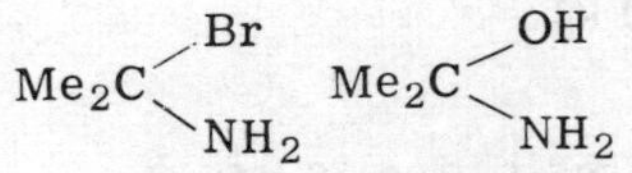

(I) (II)

If (I), this is readily converted to (II), which then readily eliminates NH_3 to form acetone.

5. $\Delta G^\ominus = -RT \ln K = -5{\cdot}7 \log K$ (kJ at 298 K)

$$K = \frac{[AcOEt][H_2O]}{[AcOH][EtOH]} = \frac{\frac{2}{3}\times\frac{2}{3}}{\frac{1}{3}\times\frac{1}{3}} = 4$$

Since we are dealing with the log of K, it is better to write the equilibrium equation to give $K > 1$, i.e., a positive value for log K. It so happens that the equation is already written to do this. Had this not been the case, the equation would have been 'turned round'.

Hence $\Delta G^\ominus = -5{\cdot}7 \times 0{\cdot}6021 = -3{\cdot}43$ kJ

The value of $\Delta G^\ominus$ gives *no* information on the time for equilibrium to be reached (see *Text,* Ch. 2).

6. The accepted mechanism of $A_{AC}2$ esterifications is:

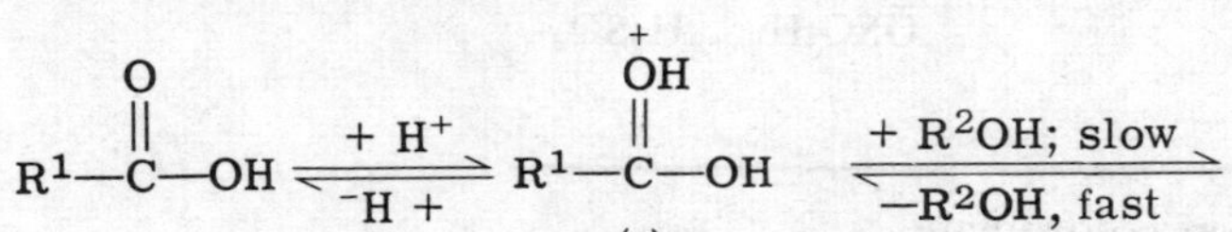

(I)

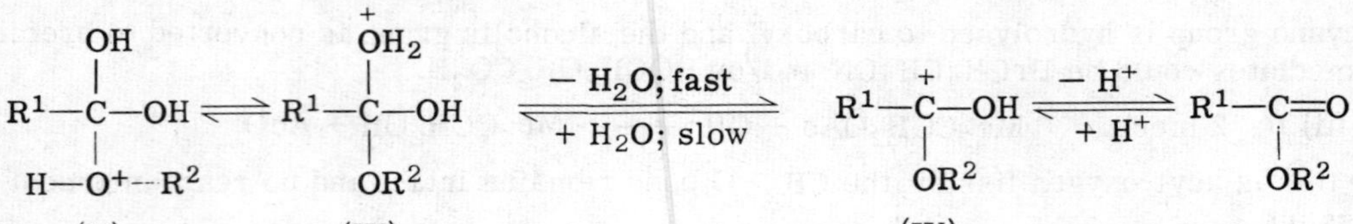

(II) (III) (IV)

All proton transfers are fast (not indicated in the equations), and the r/d step is (I) → (II). This step will therefore lead to the highest peak in the energy profile.

Now let us work out the energy changes of the various steps in order to determine the relative energy levels. We shall use sulphuric acid as the catalyst (see also Table 2.2).

(i) $R^1{-}C({=}O){-}OH + H{-}OSO_3H \rightleftharpoons (I) + \bar{O}SO_3H$ $\Delta H = 0$ kJ
$\qquad + 464{\cdot}4 \qquad - 464{\cdot}4$

(ii) (I) + $R_2OH \rightleftharpoons$ (II)
$+ 694{\cdot}5 \qquad - 2\times 334{\cdot}7$ $\Delta H = +25{\cdot}1$ kJ

(iii) (II) $\rightleftharpoons$ (III)
$+ 464{\cdot}4 \quad - 464{\cdot}4$ $\Delta H = 0$ kJ

(iv) (III) $\rightleftharpoons H_2O$ + (IV)
$+ 2\times 334{\cdot}7 \qquad - 694{\cdot}5$ $\Delta H = -25{\cdot}1$ kJ

(v) (IV) + $\bar{O}SO_3H \rightleftharpoons R^1C({=}O)OR^2 + H{-}OSO_3H$
$+ 464{\cdot}4 \qquad - 464{\cdot}4$ $\Delta H = 0$ kJ

The enthalpy change for the overall reaction is: $0 + 25{\cdot}1 + 0 - 25{\cdot}1 - 0 = 0$ kJ

This enthalpy change can also be obtained from the overall equation:

$R^1{-}C({=}O){-}OH + R^2O{-}H \longrightarrow R^1{-}C({=}O){-}OR^2 + H{-}OH$
$+ 334{\cdot}7 \qquad + 464{\cdot}4 \qquad - 334{\cdot}7 \qquad - 464{\cdot}4$ $\Delta H = 0$ kJ

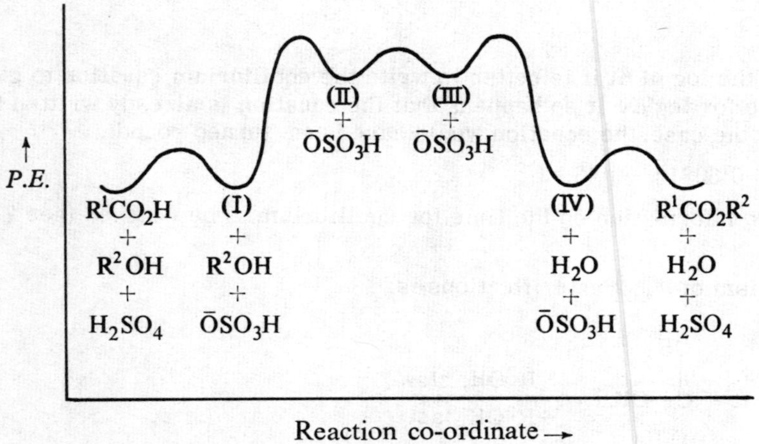

The values are the same (this follows from Hess's law; sulphuric acid is a catalyst). The second calculation is also useful as a check on the stepwise calculation.

We are now in a position to draw an energy profile, and it should be noted that the curve obtained is symmetrical (see also Ch. 8, Prob. 3).

7. Under the conditions described, sterically hindered acids usually give esters by the $A_{AC}1$ mechanism, i.e., an acylium ion is an intermediate. Thus:

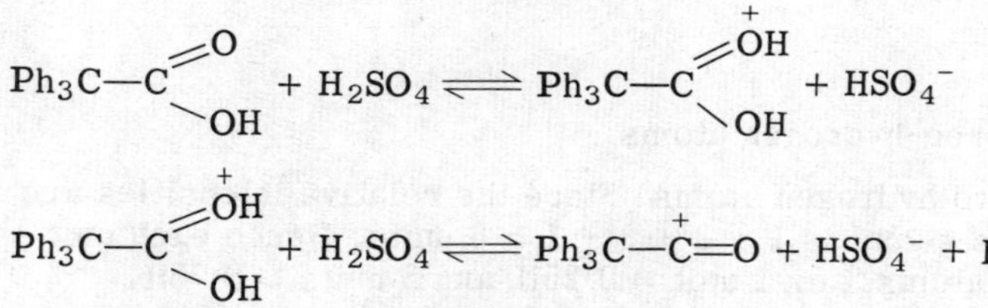

This acylium then reacts with the alcohol to form an ester (see *Text*). In the case under discussion, however, the product is an ether. Inspection of its formula shows that the carbonyl group of the acylium ion has been lost. Hence the further steps can be formulated as follows:

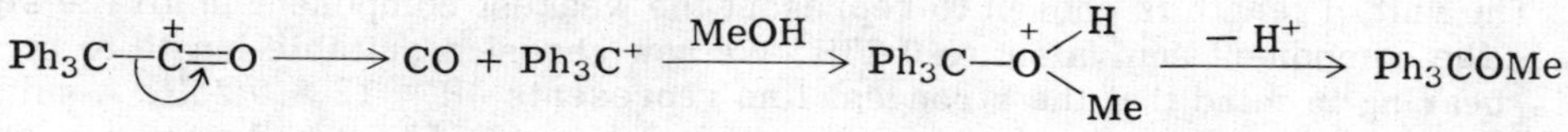

What we now have to try to do is offer an explanation for this result. The fact that the acylium ion is extremely crowded suggests that the carbonyl group is 'squeezed out' to relieve the steric strain. The triphenylmethyl carbonium ion is a very stable species (stabilised by resonance). This ion is sp^2 hybridised and so steric strain is operating to a far less extent than in the acylium ion which is sp^3 hybridised, and because of the resonance stabilisation of the carbonium ion, this is an additional 'driving force' for the elimination of the carbonyl group from the acylium ion.

8. (i) n-BuOH has lost a carbon atom to form a primary amine. This suggests the use of the Hofmann primary amine preparation (see also Prob. 1.).

$$MeCH_2CH_2CH_2OH \xrightarrow[\text{dichr.}]{\text{acid}} MeCH_2CH_2CO_2H$$

$$\xrightarrow[\text{(ii) }NH_3]{\text{(i) }SOCl_2} MeCH_2CH_2CONH_2 \xrightarrow[NaOH]{Br_2} MeCH_2CH_2NH_2$$

(ii) This is $CO_2H \longrightarrow CHO$. In a case like this, it is always worth checking whether a complex metal hydride will do the trick (the conversion is a *reduction*). Inspection of Table 6.1 leads to the following method:

$$Et_2CHCO_2H \xrightarrow{SOCl_2} Et_2CHCOCl \xrightarrow[-78^\circ C]{LiAl(OBu^t)_3} Et_2CHCHO$$

An alternative procedure is catalytic reduction of the acid chloride (Pd—$BaSO_4$-S-quinoline; Rosenmund Reduction).

9. The answer is yes, and the reasons are shown in the following chart (s = singlet; d = doublet; t = triplet; q = quartet). One compound gives two signals and the other two give three signals of different multiplicities which can be predicted for each compound.

$CH_3{-}CH_2{-}COCl$ (t, q) $CH_3{-}CHCl{-}CO_2H$ (d, q, s)

$ClCH_2{-}CH_2{-}CO_2H$ (t, t, s)

10. (i) $CH_3{-}CO{-}O{-}CH_2{-}CH_3$

	CH_3	CH_2	CH_3
	s	q	t
	1	1:3:3:1	1:2:1
	3H	2H	3H

(*a*) The singlet corresponds to three-hydrogen atoms.

(*b*) The quartet corresponds to two hydrogen atoms. Since the relative intensities are 1:3:3:1, the total intensity of this signal is 1 + 3 + 3 + 1 = 8 units. Hence each unit corresponds to 1/8 of 2 hydrogen atoms, i.e., 1 unit ≡ 0·25H, and 3 units ≡ 0·75H.

(*c*) The triplet corresponds to three hydrogen atoms. Using the argument given in (*b*), we see that the total intensity is 4 and so 1 unit ≡ 0·75H and 2 units ≡ 1·5H.

Each signal may be drawn as a vertical line if a singlet or as a multiplet of vertical lines if a multiplet, the lengths of the lines being proportional to the number of protons they represent. The unit of length is chosen to represent the weakest component in all the signals. This is the component equivalent to 0·25H. We now choose a suitable length to represent this, bearing in mind that the strongest line represents 3H ≡ 12 × 0·25H. A suitable value is 0·25H represented by a line 5 mm in length (minimum is 5 mm and maximum is 60 mm).

To make a reasonable job of the diagram, we should also have some idea of the appropriate τ-values. These can usually be estimated from tables (see, e.g., Table 3.2). Thus:

$CH_3{-}CO{-}O{-}CH_2{-}CH_3$

τ ~8·0 (CH₃), ~5·9 (CH₂), ~9·0 (CH₃)

If tables are not available, then an attempt might be made to work out the *relative* positions of the signals with respect to TMS. A methyl group, unless *directly* attached to an electron-attracting atom or group, usually gives a signal closest to the TMS signal. In the same circumstances a methylene group gives a signal a little downfield and a methine group further downfield. When *directly* attached to an electron-attracting group, the signals of CH_3, CH_2, and CH are all shifted downfield, the amount of the shift depending on the electron-attracting power of the attached atom or group. A rough order is: $>C{=}O < -NH_2 < -X(I < Br < Cl) < -OH < -OCOR$ (this is equivalent to $OH + >C{=}O$) $< -F < -NO_2$.

The diagrammatic NMR spectrum of $MeCO_2Et$ may be drawn as shown (based on *relative* positions). If τ-values are estimated, then the distances between signals should be drawn to scale, the centre of a multiplet being taken as the τ-value of that signal.

(ii) Using the same method as before, we get the results summarised as shown.

$CH_3{-}CH_2{-}CO_2{-}CH_3$

	CH_3 (1)	CH_2 (2)	CH_3 (3)
	t	q	s
	1:2:1	1:3:3:1	1
	3H	2H	3H
	4 units = 3H	8 units = 2H	1 unit = 3H
	1 unit = 0·75H	1 unit = 0·25H	1 unit = 3H
estimated τ	~9·0	~7·9	~6·35

The diagrammatic NMR spectrum of $EtCO_2Me$ may be drawn as shown (0·25H is represented by 5 mm).

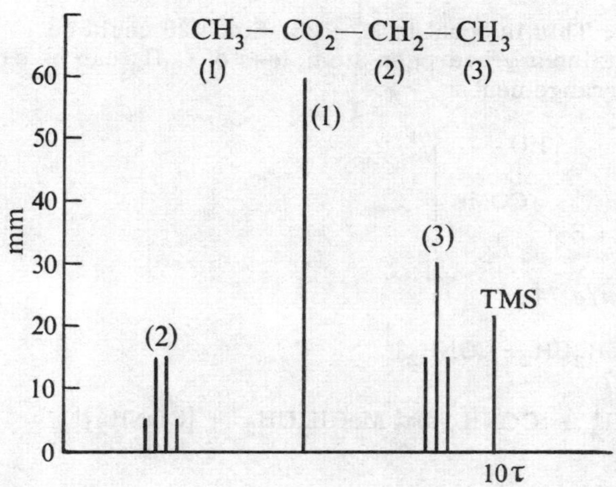

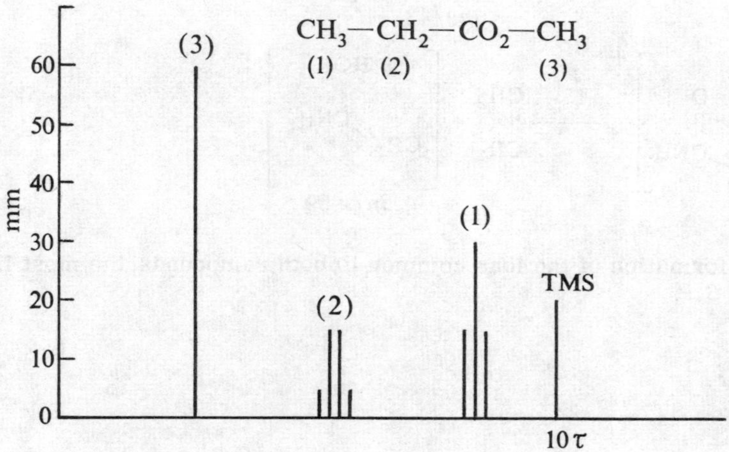

11. Two characteristic fragmentation patterns of esters and amides are: (*a*) α-cleavage; (*b*) the McLafferty rearrangement if they contain a γ-hydrogen atom.

(i) $$\mathrm{MeCH_2CH_2{-}CO_2Me} \xrightarrow{-e} \underset{M^+\,102}{[\mathrm{MeCH_2CH_2{-}CO_2Me}]^{+\cdot}}$$

$$\longrightarrow \underset{m/e\ 43}{\mathrm{MeCH_2CH_2^+}} + \cdot\mathrm{CO_2Me}\ \textit{and}\ \mathrm{MeCH_2CH_2}\cdot + \underset{m/e\ 59}{[\mathrm{CO_2Me}]^+}$$

From Table 1.6, it can be seen that m/e 31 could be OMe. Hence (see also Ch. 8, Prob. 12):

$$[\mathrm{MeCH_2CH_2CO{-}OMe}]^{+\cdot} \longrightarrow \mathrm{MeCH_2CH_2CO}\cdot + \underset{m/e\ 31}{\mathrm{OMe^+}}$$

$$\downarrow$$

$$\mathrm{MeO}\cdot + \underset{m/e\ 71}{\mathrm{MeCH_2CH_2{-}C{\equiv}\overset{+}{O}}} \longrightarrow \mathrm{CO} + \underset{m/e\ 43}{\mathrm{MeCH_2CH_2^+}}$$

The only ion not accounted for is m/e 74. This is equal to $M - 28$. Since 28 could be ethylene (or CO), and since the ester contains a γ-hydrogen atom, loss of C_2H_4 can be explained on the basis of a McLafferty rearrangement.

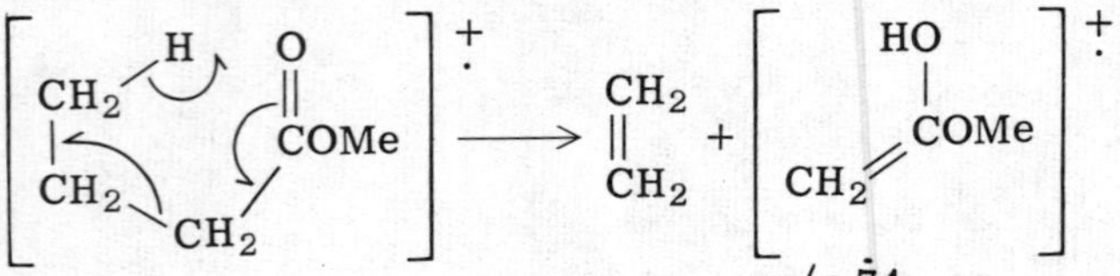

m/e 74

(ii) $MeCH_2CH_2{-}CONH_2 \xrightarrow{-e} [MeCH_2CH_2{-}CONH_2]^{+\cdot}$
$M^+ 87$

$\longrightarrow MeCH_2CH_2^+ + {}^{\cdot}CONH_2$ *and* $MeCH_2CH_2^{\cdot} + [CONH_2]^+$
m/e 43 $\qquad$ m/e 44

$[MeCH_2CH_2CO - NH_2]^{+\cdot} \longrightarrow MeCH_2CH_2CO\cdot + NH_2^+$
m/e 16

$\downarrow$

$NH_2\cdot + MeCH_2CH_2 - C{\equiv}\overset{+}{O} \longrightarrow CO + MeCH_2CH_2^+$
m/e 71 $\qquad$ $me/43$

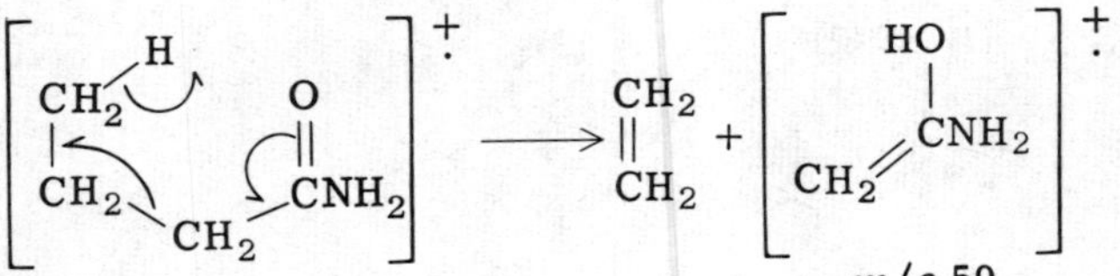

m/e 59

The student should note the formation of the ions common to both compounds, the most important difference being the *B.P.*.

10 Polycarbonyl compounds

(Problems on page 15)

1. If **A** and **B** are tautomers, then their enthalpies (heat contents) will be different. Whatever the type of tautomerism involved, chemical reaction occurs in the interconversion.

Since this requires energy of activation, each step (if a multistep reaction) will pass through a T.S. If, for simplicity, we ignore the actual mechanism for a particular system, we can write the equation in the following general terms:

$\mathbf{A} \rightleftharpoons \text{T.S.} \rightleftharpoons \mathbf{B}$

In these circumstances, a possible energy profile is *T*.

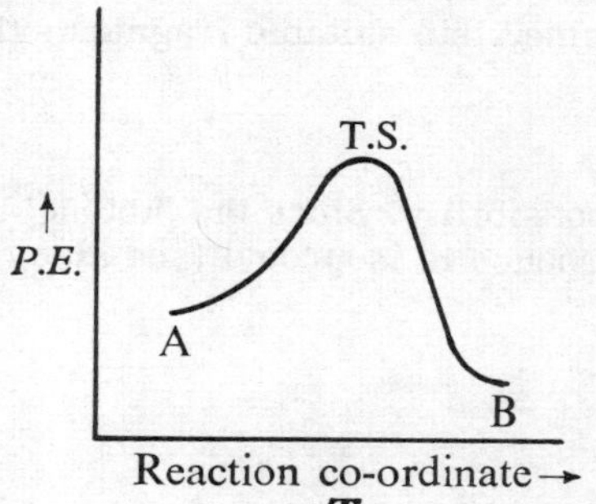

T

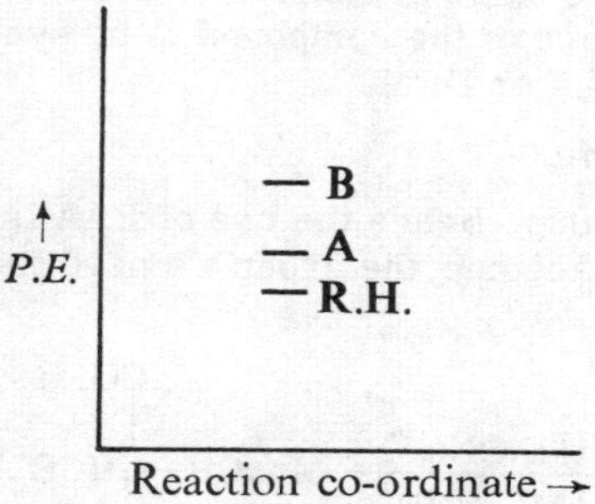

R

If there is resonance between **A** and **B** this means that **A** and **B** are resonating structures, the actual molecule being a resonance hybrid of the two contributions. If we assume that the enthalpy of **A** is lower than that of **B**, then we may accept that **A** makes the greater contribution (see *Text*, Ch. 2). In these circumstances, the structure of **R.H.** (resonance hybrid) will be closer to **A** than to B. Now, in energy profiles, the *r.c.* is taken to mean the progress of the reaction, i.e., 'something' is taking place and is progressing from reactants to products. This situation is *not true* for resonance. A resonance hybrid molecule is a *single* species and is *not* formed by some dynamic equilibrium between **A** and **B**. **A** and **B** are fictitious structures (classical structures); **R.H.** is the *real molecule*. Hence, we cannot draw an energy profile to represent a resonance hybrid, since there is no *r.c.* One way to represent diagrammatically what has been described is to indicate the energy levels of **A**, B, and **R.H.** as shown in diagram ***R***.

The student, having read this discussion, may think a 'catch' question has been asked. This has not been the intention. The object of this exercise is to help him realise that tautomerism and resonance are entirely different phenomena.

2. 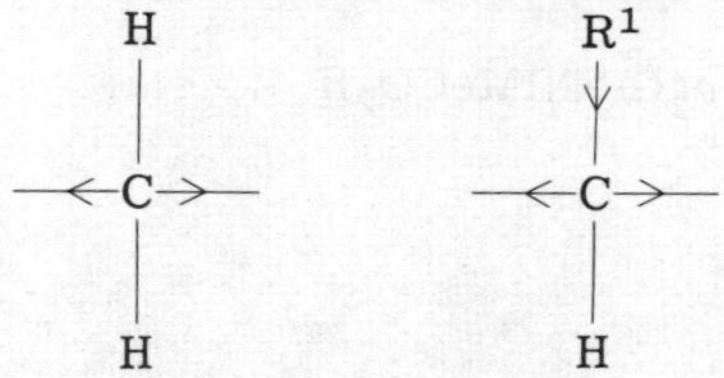

R^1 (taken to mean an alkyl group) has a + I effect. This tends to nullify the electron-withdrawing effects of the CO and CO_2Et groups, the result being that the H in the R^1CH group is less acidic than that in the CH_2 group, and so the former is the more difficult to remove. At the same time, the presence of R^1 creates steric hindrance at this C atom and consequently decreases the tendency to attack by R^2X.

3. **(A)** $MeCOCH_2CH_2COMe$ **(B)** $MeCOCH_2COCH_2Me$. **i.r.** There is very little enolisation in **(A)**, and so the i.r. spectrum will show a strong CO band [and CH_2 and Me bands]. On the other hand, **(B)** is considerably enolised, and so the i.r. spectrum will show bands for CO, C=C, OH [and CH_2 and Me bands]. **(B)** can exist in two enol forms.

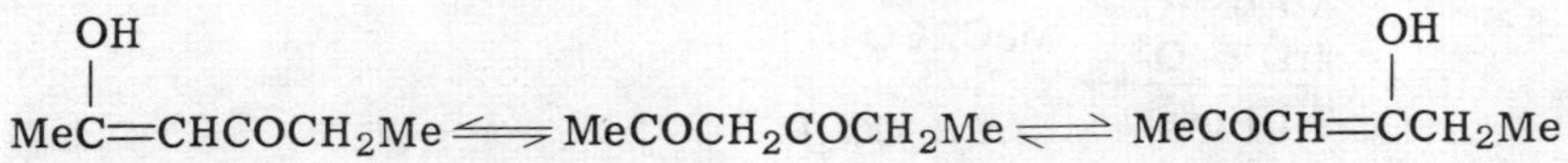

NMR Because of symmetry, the two CH_2 groups in **(A)** are equivalent. **(A)** will therefore show two NMR signals only. On the other hand, because of enolisation, **(B)** will show a complex NMR spectrum which includes OH and olefinic protons.

4. In general, DEM should be used for synthesising acids. EAA can generally be used for ketones containing a CH_3CO group.

The student *must* know the basic principles of EAA and DEM syntheses. It is then a question of breaking up the structure of the compound to be synthesised into suitable fragments that can be introduced into EAA or DEM.

(i) $Me_2CH \vdots CH_2COMe$

This contains a MeCO group. Hence the use of EAA is a possibility. Since the 'ketone' hydrolysis of EAA gives acetone, the group we need to introduce is isopropyl (see also Ch. 9, Prob. 1).

$$MeCOCH_2CO_2Et \xrightarrow[\text{(ii) } Me_2CHBr]{\text{(i) EtONa}} Me_2CH{-}\overset{\displaystyle COMe}{\overset{|}{C}}HCO_2Et$$

$$\xrightarrow{\text{dil. KOH}} Me_2CHCH_2COMe$$

(ii) $\overset{\displaystyle Me}{\overset{|}{Me}CH} \vdots \overset{\displaystyle Me}{\overset{|}{C}}HCO_2H$

This is an acid. Hence the use of DEM is a possibility. Since DEM itself leads to acetic acid, the groups we need to introduce are Me and Me_2CH. To keep steric effects to a minimum, the larger group is introduced first (see *Text*).

$$CH_2(CO_2Et)_2 \xrightarrow[\text{(ii) } Me_2CHBr]{\text{(i) EtONa}} Me_2CHCH(CO_2Et)_2$$

$$\xrightarrow[\text{(ii) MeI}]{\text{(i) EtONa}} Me_2CHCMe(CO_2Et)_2 \xrightarrow[\text{(ii) } H_3O^+]{\text{(i) KOH}}$$

$$Me_2CHCMe(CO_2H)_2 \xrightarrow[-CO_2]{\text{heat}} Me_2CHCHMeCO_2H$$

(iii) $\underset{\displaystyle CH_2CO_2H}{Me \vdots \underset{|}{C}HCO_2H}$

$$CH_2(CO_2Et)_2 \xrightarrow[\text{(ii) MeI}]{\text{(i) EtONa}} MeCH(CO_2Et)_2 \xrightarrow{Br_2}$$

$$MeCBr(CO_2Et)_2 \xrightarrow{\text{NaDEM}} \underset{\displaystyle CH(CO_2Et)_2}{Me\underset{|}{C}(CO_2Et)_2}$$

$$\xrightarrow[\text{(iii) heat}]{\text{(i) KOH, (ii) } H_3O^+} \underset{\displaystyle CH_2CO_2H}{Me\underset{|}{C}HCO_2H}$$

(iv) 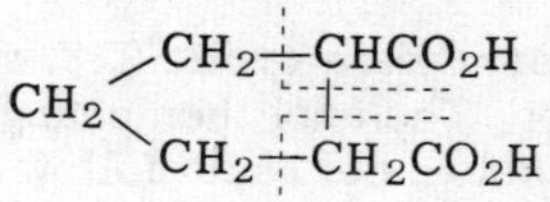

This contains two DEM residues. The question is: Do we join them together first and then add $-CH_2CH_2CH_2-$, or do we prepare the chain compound with a DEM residue at each end and then ring-close? In practice we use the second method (see *Text*, Ch. 19, for a discussion on cyclisation reactions).

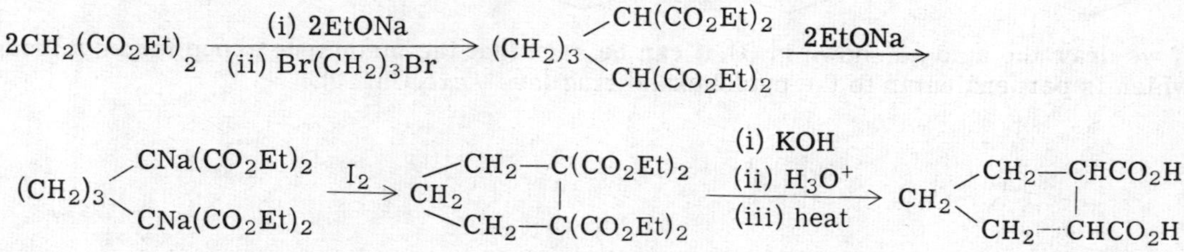

Note the use of iodine (or bromine) to join together two DEM (or EAA) residues. Also note that the behaviour of each DEM residue in the *ring compound* is the same as that in acyclic compounds.

(v) $PhCO{-}CH_2CO_2Et$

If we try the DEM synthesis, we shall get to $PhCOCH_2CO_2H$, which then has to be esterified. This intermediate is a β-keto-acid and so (probably) will be decarboxylated during the heating of its dicarboxylic acid precursor. In practice, one could try to carry out the synthesis, in spite of the anticipated difficulty. You never know your luck! Some molecules have a habit of not playing the game (in practice, decarboxylation does occur).

Since the student is asked for a synthesis, it being understood that the one given has a reasonable chance of success, then it would be unwise to give the one described unless he *knows* it will work and can give references. Since we are asked to start from EAA or DEM, we shall assume that the questioner knows what he is talking about (whatever the student himself may think!). We have seen that the DEM approach is not promising. This leaves us with EAA. An interesting reaction of this compound is that the acetyl group may be replaced by another acyl group under suitable conditions:

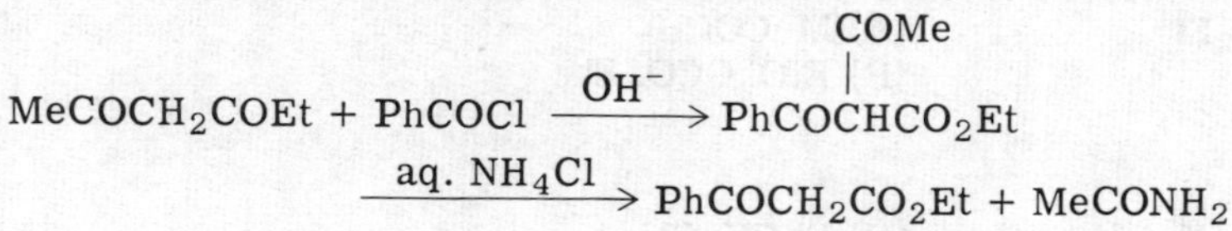

(vi) 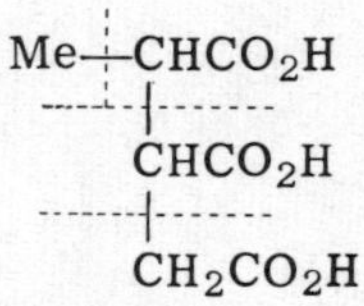

See (iii) above, and carry on as follows:

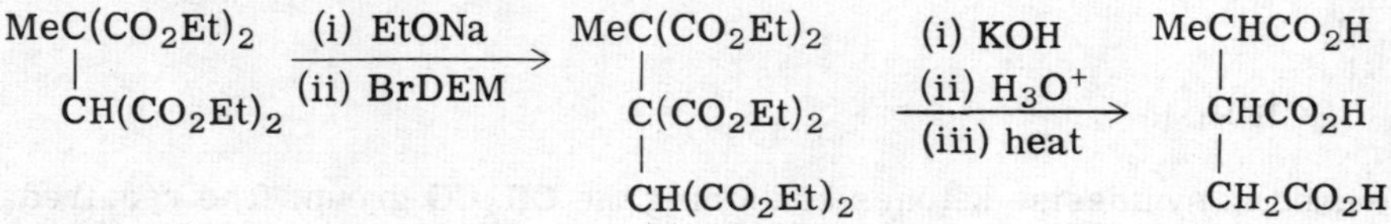

5. β-Keto-acids undergo decarboxylation through the keto form via a cyclic T.S. which results in the formation of the *enol* form of the ketone product. The enol then changes to the more stable keto form. If decarboxylation of the given β-keto-acid could follow this route, then:

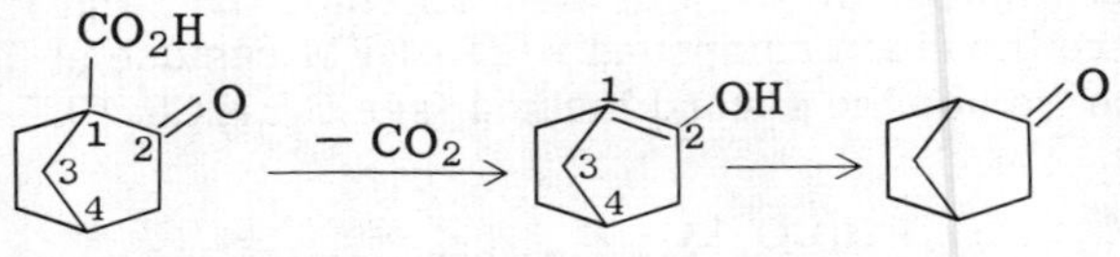

If we draw the acid as shown in (I), it can be seen that the methylene group forms a bridge, which is perpendicular to the cyclohexane ring (see *Text*, Ch. 19).

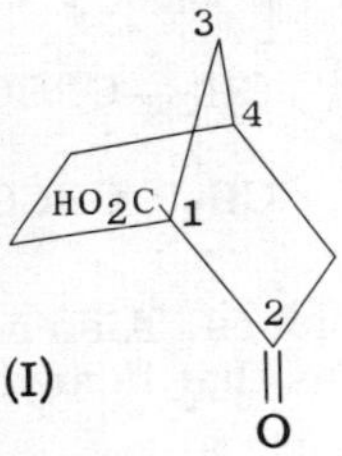

(I)

The enol produced would have a double bond in the 1, 2-position. Thus, C—1 is sp^2 hybridised and therefore the bond leading to the methylene bridge (1, 3) would have to be coplanar with the groups joined to C—2 in the enol, i.e., the bridge would have to flatten. This would require the normal valency angles (~109·5°) to change to ~120°. Even if a compromise situation occurred, i.e., all bond angles involved (∠s 1, 3, and 4) underwent change, the resulting strain would be very large, too large to be expected for it to occur. Hence, it is reasonable to argue that since decarboxylation does occur, it does so through a route requiring a much higher activation energy than that required for the 'usual' route. Hence the difficulty of decarboxylation.

6. The use of the word *condense*, the use of EtONa as the condensation reagent, and the fact that the products are keto-esters or ketones, all suggest that a Claisen condensation is involved. If we try this as our line of approach, then we require two molecules, one containing a CO_2Et group and the other 'active' hydrogen atoms. Since the CO group in the product is derived from the CO_2Et group with elimination of EtOH, it is now easy to decide the structure of each compound required.

(i) $MeCOCH_2$ ┆ $COCO_2Et$
      H ┆ OEt
(A) $MeCOCH_3$
(B) EtO_2CCO_2Et

(ii) PhCH(CN) ┆ $COCO_2Et$
      H ┆ OEt
(C) $PhCH_2CN$
(D) EtO_2CCO_2Et

(iii) PhCH(CN) ┆ COMe
      H ┆ OEt
(E) $PhCH_2CN$
(F) EtO_2CMe

7. $MeCH_2$ ┆ CH(Me)CHOHMe

EAA may be used to synthesise ketones containing the CH_3CO group. The required alcohol contains this group reduced to alcohol. Hence, we prepare the appropriate ketone and reduce it (see also Prob. 4 ii).

$$\mathrm{MeCOCH_2CO_2Et \xrightarrow[\text{(ii) EtI}]{\text{(i) EtONa}} MeCOCHEtCO_2Et}$$

$$\mathrm{\xrightarrow[\text{(ii) MeI}]{\text{(i) EtONa}} MeCOCEtMeCO_2Et \xrightarrow{\text{dil. KOH}} MeCOCHMeEt}$$

$$\mathrm{\xrightarrow{NaBH_4} MeCHOHCHMeEt}$$

8. $$\mathrm{CH_2(CO_2Me)_2 \xrightarrow{Na} NaCH(CO_2Me)_2 \xrightarrow[\text{(ii) AcOH}]{\text{(i) ClCO}_2\text{Me}} CH(CO_2Me)_3}$$

Chloroformates, $ClCO_2R$, are very useful for introducing a CO_2R group.

(ii) $$\mathrm{MeCH(CO_2Et)COCO_2Et \xrightarrow{heat} CO + MeCH(CO_2Et)_2}$$

This elimination is characteristic of α-keto-esters.

(iii) 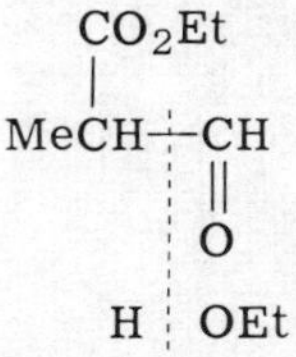

The Claisen condensation is a useful means of introducing the formyl group into a molecule containing active hydrogen atoms. Note that only α-hydrogens are 'active'. Hence:

$$\mathrm{MeCH_2CO_2Et + HCO_2Et \xrightarrow{EtONa} MeC(CO_2Et){=}CH{-}\bar{O}\overset{+}{N}a}$$

$$\mathrm{\xrightarrow{AcOH} MeCH(CHO)CO_2Et}$$

(iv) $$\mathrm{MeCH_2CO_2Et + EtO_2CCO_2Et \xrightarrow[\text{(ii) AcOH}]{\text{(i) EtONa}}}$$

$$\mathrm{MeCH(CO_2Et)COCO_2Et \xrightarrow{heat} CO + MeCH(CO_2Et)_2}$$

9. 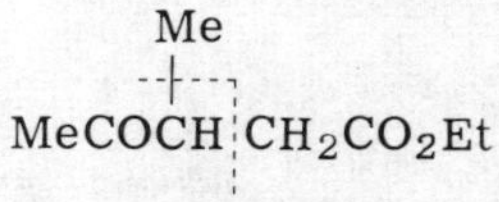

MeCO group; hence try EAA synthesis.

$$\mathrm{MeCOCH_2CO_2Et \xrightarrow[\text{(ii) ClCH}_2\text{CO}_2\text{Et}]{\text{(i) EtONa}} MeCOCH(CO_2Et)CH_2CO_2Et}$$

$$\mathrm{\xrightarrow[\text{(ii) MeI}]{\text{(i) EtONa}} MeCOCMe(CO_2Et)(CH_2CO_2Et) \xrightarrow[\text{(ii) H}_3\text{O}^+]{\text{(i) dil. KOH}}}$$

$$\mathrm{MeCOCHMeCH_2CO_2H \xrightarrow[HCl]{EtOH} MeCOCHMeCH_2CO_2Et}$$

Chloroacetic *acid* cannot be used, since this will decompose Na—EAA to EAA. Esterification 'masks' the carboxyl group. 'Ketonic hydrolysis' also hydrolyses the CO_2Et group. Hence the last esterification step is necessary.

10. The D.B.E. of $C_4H_6O_2$ is $4 + 1 - 6/2 = 2$.

The bond at 1721 cm^{-1} suggests the presence of a keto group. This accounts for 1 double bond, and so there is present another double bond or a ring system. The structural formula may be written as $(C_3H_6O)CO$. Since the keto band is said to *very strong,* this could mean that *two* keto groups are present, and this would account for the other double bond. Hence, a tentative structural formula is $(C_2H_6)(CO)_2$. The (C_2H_6) part can only be two methyl groups. Hence the compound could be butane-2, 3-dione, $CH_3COCOCH_3$. This is confirmed by the fact that the NMR spectrum consists of a single-line signal.

11 Polyhydric alcohols

(Problems on page 16)

1. See Ch. 3, Prob. 2 and Ch. 5, Prob. 1.

Since two OH groups attached to the *same* carbon atom form an unstable arrangement except in certain circumstances, e.g., chloral hydrate, 1, 1-diols may be disregarded.

$MeCH_2CH_2Me$ (i) $MeCH_2CHOHCH_2OH$; (ii) $MeCHOHCH_2CH_2OH$; (iii) $HOCH_2CH_2CH_2CH_2OH$; (iv) MeCHOHCHOHMe.

$MeCH(Me)Me$ (v) $MeC(Me)(OH)CH_2OH$; (vi) $HOCH_2CH(Me)CH_2OH$

Only 1, 2-diols are oxidised by periodic acid. Hence, the butanediols oxidised are: (i), (iv), and (v).

2. When hydrogen bonding is possible between molecules, the b.p. of that compound will (usually) be higher than expected. Glycol can form hydrogen bonds on 'both sides', the monomethyl derivative from one side only, and the dimethyl derivatives cannot form hydrogen bonds at all. Hence the order of increasing b.p. will be expected to be: $(CH_2OMe)_2 <$ $HOCH_2CH_2OMe < (CH_2OH)_2$. This is the case in practice ($84° < 125° < 197°C$).

3. (i) Two H atoms have been removed and both oxygen atoms are joined to the same carbonyl group. This, coupled with the fact that the reaction is carried out in the presence of pyridine (a *base*) suggests the compound required is of the type COX_2.

Hence, carbonyl chloride could be used (this is the acid chloride of carbonic acid and behaves like acid chlorides of the monocarboxylic acids; see *Text,* Ch. 16).

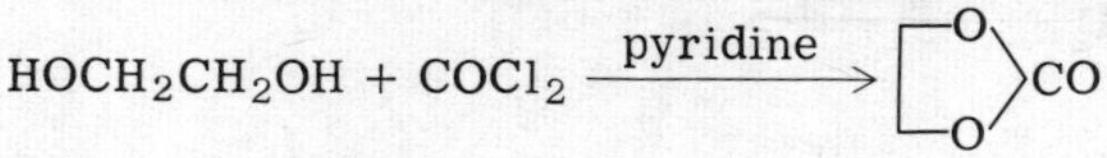

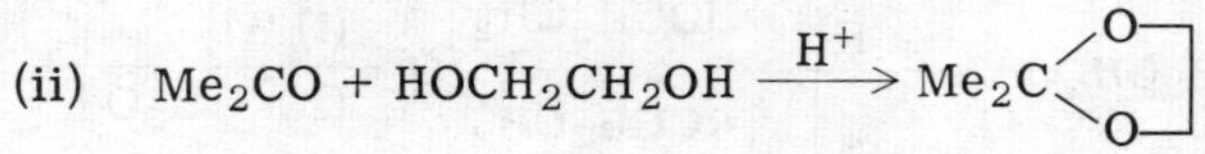

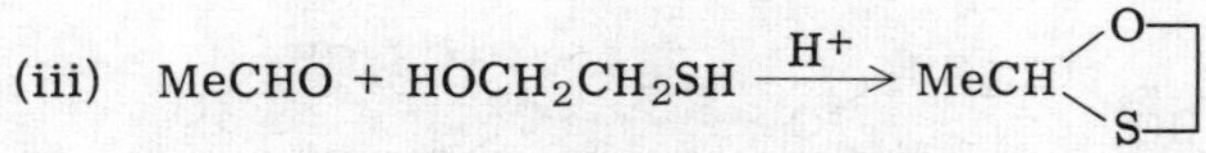

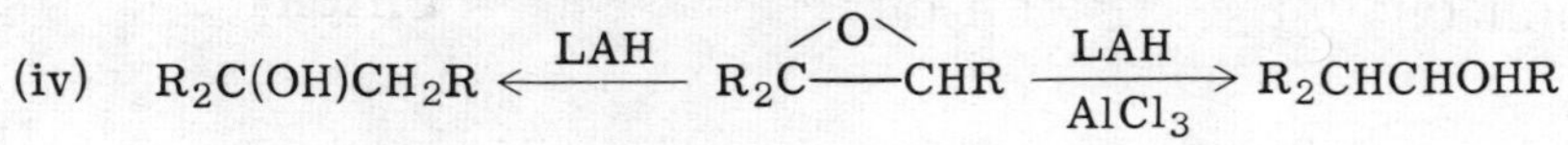

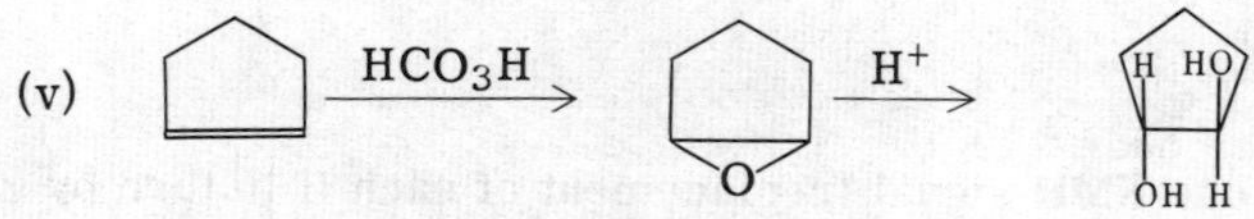

The hydrolysis of epoxides results in the formation of the *trans*-diol.

(vi) $CH_2{=}CH(CH_2)_2CH{=}CH_2 \xrightarrow[\text{(ii) } H_2O_2;\ OH^-]{\text{(i) } B_2H_6} HOCH_2(CH_2)_4CH_2OH$

(See *Text,* Ch. 4. Hydroboronation).

(vii) $(CH_2)_8(CO_2Et)_2 \xrightarrow{LAH} (CH_2)_8(CH_2OH)_2$

Other reagents may also be used (see Table 6.1).

4. The oxidation of 1, 2-diols by periodic acid has been shown to be subject to steric hindrance; the greater the hindrance, the slower is the oxidation. Hence, the expected rates are:

$CH_2OH{-}CH_2OH > MeCHOH{-}MeCHOH > Me_2COH{-}CH_2OH > Me_2COH{-}MeCHOH > Me_2COH{-}Me_2COH$

5. Since 1, 2-diols are conveniently prepared from alkenes, the first step is to convert cyclohexanol (an alcohol) into cyclohexene. Alkenes give *cis*-1, 2-diols with alkaline permanganate. On the other hand, *trans*-1, 2-diols are formed by hydrolysis of epoxides. Hence:

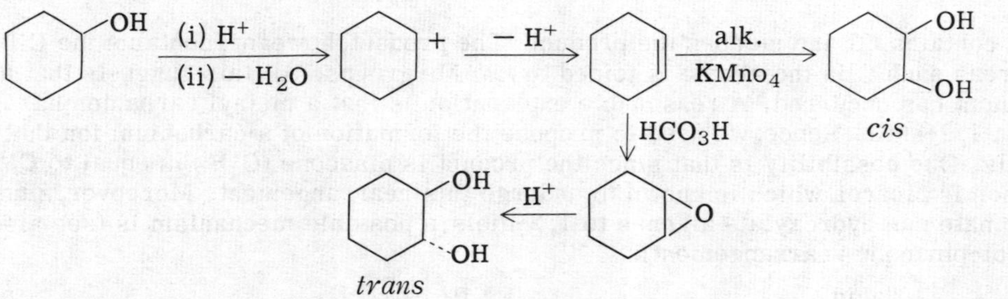

6. Dioxan is a di-ether. Ethers may be prepared from alcohols by the action of acids, and the mechanism of this reaction is accepted to be via a carbonium-ion intermediate.

Hence, arguing by analogy, we have:

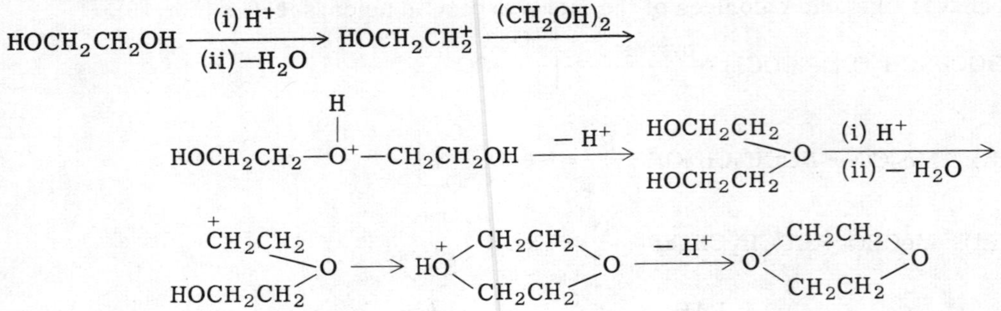

7. (i)

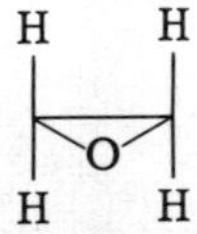

All hydrogens are equivalent; hence one NMR signal (replacement of each H in turn by Z gives the same compound).

(ii)

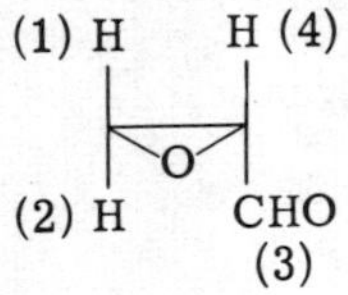

Replacement of each H in turn (1, 2, and 3) by Z gives a different compound for each. Hence each H gives its own signal. The H in CHO is different from the other three, and so also gives its own signal. Hence, the compound gives four NMR signals.

(iii)

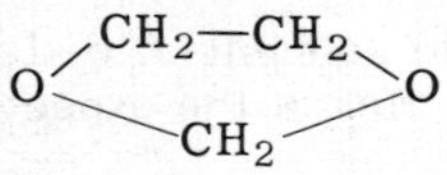

The upper two methylene groups are equivalent, but the lower one, being attached to two oxygen atoms, has a different environment. Hence, this compound gives two NMR signals.

8. The presence of the 1 700 cm^{-1} band suggests the presence of a carbonyl group, probably keto. The positive haloform reaction indicates the presence of a CH_3CO group. Hence:

$$Me_2C{=}CMe_2 \xrightarrow[\text{dichr.}]{\text{acid}} C_6H_{12}O = CH_3CO{-}C_4H_9$$

The alkene contains 6C and so does the product. The product, however, contains the $CH_3{-}C$ group, whereas each C in the alkene is joined to *two* Me groups. All this suggests that a rearrangement has occurred. A reasonable explanation is that a methyl carbanion has undergone a 1, 2-shift. Hence, we need to propose the formation of a carbonium ion that could do this. One possibility is that since the product is pinacone (C_4H_9 is equal to CMe_3), its precursor is pinacol, which is known to undergo this rearrangement. Moreover, since acid dichromate can hydroxylate alkenes to 1, 2-diols, a possible mechanism is (see also *Text*, pinacol-pinacone rearrangement):

$$Me_2C{=}CMe_2 \xrightarrow[\text{dichr.}]{\text{acid}} Me_2C(OH)C(OH)Me_2 \xrightarrow{H^+} MeCOCMe_3$$

Thus, the structure suggested for $C_6H_{12}O$ is $CH_3COC(CH_3)_3$.

The question did not ask for a mechanism for its formation, but one has been suggested because it helps to explain the structure assigned. This now raises an interesting point. Let us, for the moment, ignore the mechanism. The structure assigned is in agreement with the i.r. data and the haloform reaction. The presence of the C_4H_9 part of the ketone is reasonably expressed as t-butyl, since we have then accounted for the four Me groups in tetramethylethylene.

Now let us return to the mechanism proposed. Even if this is wrong, the structure is still correct and almost certainly was formed via a 1, 2-shift. Can we then propose other reasonable mechanisms? Since many alkenes are hydrated to alcohols in acid solution, a possible route to the product is:

$$Me_2C{=}CMe_2 \xrightarrow{H^+} Me_2CH{-}\overset{+}{C}Me_2 \longrightarrow Me\overset{+}{C}H{-}CMe_3$$

$$\xrightarrow[\text{(ii)} -H^+]{\text{(i)} H_2O} MeCHOHCMe_3 \xrightarrow{[O]} MeCOCMe_3$$

Supporting evidence for this mechanism could be found by determining experimentally whether the alkene can be hydrated under acid conditions.

We have now proposed two mechanisms for which some supporting evidence may be obtained by specially chosen experiments. The final question is: Which of these two is correct (in the sense of being acceptable), or are both wrong? As the student will appreciate, the proposed mechanisms have been derived from a knowledge of known reactions and arguing that these can reasonably be expected to operate in the case of the compound under discussion. However, the most satisfactory way is to actually investigate the mechanism of this particular reaction. As a matter of fact, this has been done by Hickinbottom (1954), and he showed that neither mechanism is correct. According to him, the reaction proceeds via the epoxide.

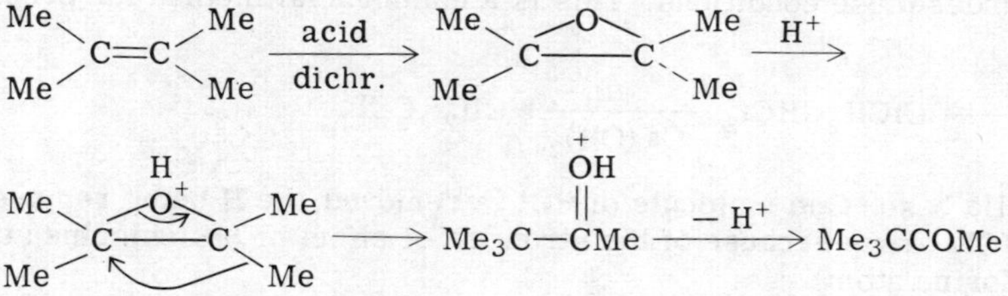

The student will now realise that however reasonable is a proposed mechanism based on analogy, that reaction should be investigated in its own right to obtain evidence that the mechanism is in keeping with the facts. The student should also note that he will always be given credit for showing that he is capable of applying principles (even though his proposals may prove to be wrong).

12 Unsaturated alcohols, ethers, carbonyl compounds and acids

(Problems on page 18)

1.

(i) $$\mathrm{Me(C{\equiv}C)_3COCH{=}CH_2 \xrightarrow{NaBH_4} Me(C{\equiv}C)_3CHOHCH{=}CH_2}$$

$$\mathrm{\xrightarrow[(ii)\,-H_2O]{(i)\,H^+} Me(C{\equiv}C)_3\overset{+}{C}HCH{=}CH_2 \longleftrightarrow Me(C{\equiv}C)_3CH{=}CHCH_2^+}$$

$$\mathrm{\xrightarrow[(ii)\,-H^+]{(i)\,H_2O} Me(C{\equiv}C)_3CH{=}CHCH_2OH}$$

Sodium borohydride reduces only the CO group and the resulting alcohol undergoes the allylic rearrangement to give the more stable alcohol (in which there is increased conjugation).

(ii) $$\mathrm{\overbrace{CH_2{-}CH_2}^{O} + HCN \longrightarrow HOCH_2CH_2CN \xrightarrow[300^\circ C]{Al_2O_3} CH_2{=}CHCN}$$

Alcohols are dehydrated under these conditions. This is a commercial method for preparing vinyl cyanide.

(iii) $$\mathrm{CH_2{=}CHCl + Cl_2 \longrightarrow ClCH_2CHCl_2 \xrightarrow[Ca(OH)_2]{aq.} CH_2{=}CCl_2}$$

Calcium hydroxide is a mild base. One molecule of HCl is removed, the H being removed more readily from the $CHCl_2$ group because of the stronger –I effect of *two* chlorine atoms (as compared with *one* chlorine atom).

(iv) $$\mathrm{CH_2{=}CHCN + EtOH \xrightarrow[H_2O]{H_2SO_4} CH_2{=}CHCO_2Et + NH_4HSO_4}$$

The cyanide is hydrolysed to amide, then to acid, and finally esterified.

(v) $$\mathrm{CH_2{=}CHCOCH_3 \xrightarrow{NaOCl} CH_2{=}CHCO_2H + CHCl_3}$$

This is an example of the haloform reaction.

(vi) $$\mathrm{MeCH{=}CHCOMe \xrightarrow[H_2\,(1\,atm.)]{Pd{-}C} MeCH_2CH_2COMe}$$

Under these conditions, the olefinic bond is reduced preferentially.

(vii) $$\mathrm{MeCH{=}CHCOEt \xrightarrow[press.]{Ni;\,H_2} MeCH_2CH_2CHOHEt}$$

Nickel is an extremely efficient catalyst for hydrogenation and so is less selective than Pd—C.

(viii) $$\mathrm{MeCH{=}CHCO_2Et + HCN \longrightarrow MeCH(CN)CH_2CO_2Et}$$

This is an example of 1,4-addition. HCN does not normally add to an olefinic bond, but in this case the double bond is 'activated' by the adjacent CO_2Et group.

(ix) $CH_2{=}CHCO_2Et + EtO^- \longrightarrow EtOCH_2\bar{C}HCO_2Et$

$\xrightarrow{EtOH} EtOCH_2CH_2CO_2Et + EtO^-$

Compare with (viii).

(x) $CH_2{=}CHCOMe \xrightarrow[OH^-]{H_2O_2} CH_2\!-\!\!\overset{O}{\frown}\!\!-CHCOMe$

It is usually more difficult to convert α,β-unsaturated carbonyl compounds than alkenes into their corresponding epoxides. Hence the use of alkaline hydrogen peroxide.

The mechanism is believed to be:

$HOOH + OH^- \longrightarrow H_2O + HOO^-$

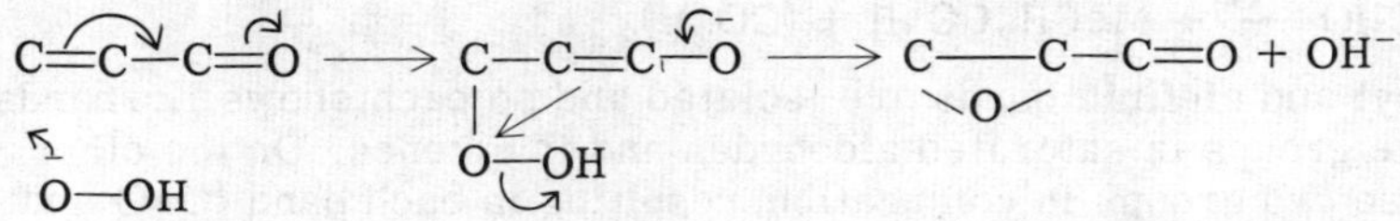

(xi) $HC{\equiv}CNa + Cl(CH_2)_3I \xrightarrow[NH_3]{liq.} HC{\equiv}C(CH_2)_3Cl \xrightarrow{NaCN}$

$HC{\equiv}C(CH_2)_3CN \xrightarrow{H^+} HC{\equiv}C(CH_2)_3CO_2H$

Alkyl iodiodes are more reactive than alkyl chlorides, and so, in competitive experiments, the iodide reacts preferentially.

(xii) $MeCH{=}CHCOMe + NH_3 \longrightarrow MeCH(NH_2)CH_2COMe$

See (viii) and (ix), above.

(xiii) $MeCH{=}CHCH_2Cl + NaI \xrightarrow{Me_2CO} MeCH{=}CHCH_2I + NaCl$

This is the Finkelstein reaction (see *Text*, Ch. 3).

(xiv) $MeCH{=}CHCO_2Et + CH_3NO_2 \xrightarrow{EtONa} MeCH(CH_2NO_2)CH_2CO_2Et$

This is an example of the Michael condensation (see *Text*).

(xv) $MeCH{=}CHCHO + HCN \xrightarrow{OH^-} MeCH{=}CHCH(OH)CN$

1, 2-Addition is favoured because of the high reactivity of the aldehyde group (see *Text*).

(xvi) $MeCH{=}CHCHO \xrightarrow{LAH} MeCH{=}CHCH_2OH$

(xvii) $PhCHO + CH_2(CO_2H)_2 \xrightarrow{pyridine} PhCH{=}CHCO_2H$

This is an example of the Knoevenagel reaction (see Text).

(xviii) $EtCHO + 2CH_2(CO_2Et)_2 \xrightarrow{piperidine} EtCH[CH(CO_2Et)_2]_2$

This type of condensation is favoured when DEM is used in excess and piperidine is the base.

(xix) $EtCHO + CH_2(CO_2H)_2 \xrightarrow{pyridine} EtCH{=}CHCO_2H$

See (xvii), above.

(xx) $EtCH{=}CHCH_2CONH_2 \xrightarrow{NaOCl/NaOH} EtCH{=}CHCH_2NH_2 \xrightarrow{HCl}$

$EtCH{=}CHCH_2NH_3^+Cl^-$

This is an example of the Hofmann amine preparation. It does not work for α,β- unsaturated amides (see *Text*).

(xxi) 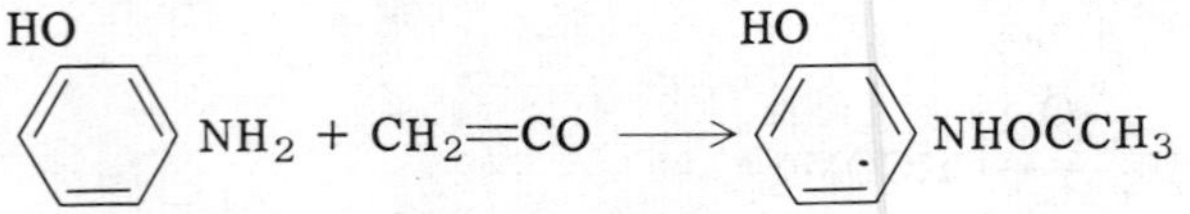

The amino group is more reactive than the hydroxy group.

2. (i) On oxidation by the Lemieux reagent, different products are obtained.

(A) $MeCH{=}CHCH_2CHO \xrightarrow{[O]} MeCO_2H + CH_2(CO_2H)_2$

(B) $MeCH_2CH{=}CHCHO \xrightarrow{[O]} MeCH_2CO_2H + (CO_2H)_2$

(ii) In **(A),** the carbonyl and olefinic bonds are isolated and so each shows i.r. bands characteristic of these groups in saturated aldehydes and in alkenes. On the other hand, **(B)** contains these two groups in conjugation, resulting in each band (C=O and C=C) being shifted to a lower frequency (longer wavelength).

(iii) Both NMR spectra are complicated because of long-range coupling, but only **(B)** will show a triplet ($\sim\tau$ 9) for the methyl group (see also Ch. 4, Prob. 26).

3. (i) $CH_2{=}CH{-}CH{=}O \longleftrightarrow \overset{+}{C}H_2{-}CH{=}CH{-}\overline{O}$

$CH_3{-}CH_2{-}CH{=}O \longleftrightarrow CH_3{-}CH_2{-}\overset{+}{C}H{-}\overline{O}$

The dipole moment, μ, is equal to $e \times d$ (electronic charge × distance). Because of conjugation in acraldehyde, the charges and the distance between them are both greater than the corresponding values in propionaldehyde. Hence the former has the greater μ.

(ii) $\overset{\delta+}{CH_3} \rightarrow CH_2 \rightarrow \overset{\delta-}{Br}$ $\qquad \overset{\delta+}{CH_2}{=}CH \rightarrow \overset{\delta-}{Br} \longleftrightarrow \overset{-}{C}H_2 - CH = \overset{+}{Br}$

μ: $\mapsto$ $\qquad\qquad \mapsto \qquad \leftarrow\!\!\!+$

The electron displacement in ethyl bromide is towards the Br atom. In vinyl bromide, there are two electron displacements which are acting in opposite directions in the two canonical forms. The dipole moment of the resonance hybrid may be regarded as the sum of these two contributions. As the matter stands, it is difficult to say which is the larger contribution. However, in actual fact, the direction of the dipole moment in vinyl bromide has the negative end at the bromine atom. Hence, it is reasonable to argue that, since we are asked to explain why the dipole moment in EtBr is the greater, the charge on the bromine atom in the resonance hybrid must be smaller than that in ethyl bromide. It could also be argued, however, that because the bonds in the resonance hybrid have partial double-bond character, the distance between the unlike charges is shorter than that in EtBr. This factor would also tend to increase the dipole moment of EtBr with respect to the resonance hybrid.

4. (i) This is base-catalysed bimolecular acyl-oxygen fission. Since the C—O bond is not broken, no allylic rearrangement is possible. Hence:

$CH_2{=}CHCHMeO{-}OCCH_3 \xrightarrow{OH^-} CH_2{=}CHCHMe{-}OH + CH_3CO_2^-$

(ii) This is base-catalysed unimolecular alkyl-oxygen fission. Since the C—O bond is broken, a carbonium-ion intermediate is formed. Hence, the allylic rearrangement is possible and so we can expect a mixture of alcohols.

$$CH_2{=}CHCHMe{-}OOCCH_3 \longrightarrow CH_3CO_2^- + CH_2{=}CH\overset{+}{C}HMe$$

$$\longleftrightarrow \overset{+}{C}H_2 - CH{=}CHMe \equiv \overset{\delta+}{C}H_2 \overset{...}{-} CH \overset{...}{-} \overset{\delta+}{C}HMe$$

OH^- ↙ ↘ OH^-

$$HOCH_2CH{=}CHMe \qquad CH_2{=}CHCH(OH)Me$$

(iii) This is acid-catalysed unimolecular acyl-oxygen fission, and so no allylic rearrangement is possible (see i, above).

$$CH_2{=}CHCHMeO - OCCH_3 \xrightarrow{H^+} CH_2{=}CHCHMeOH + CH_3CO_2^-$$

See *Text*, Ch. 9, for details.

5. $CH_2{=}CHCH_2OH + HCl \longrightarrow CH_2{=}CHCH_2Cl + H_2O$

The S_N1' mechanism means nucleophilic substitution by a unimolecular mechanism resulting in rearrangement. Hence, a carbonium ion is formed as an intermediate and because it is a resonance hybrid, contains *two* centres for subsequent attack.

$$CH_2{=}CH{-}CH_2OH \xrightarrow[\text{(ii)} -H_2O]{\text{(i)} H^+} CH_2{=}CH{-}\overset{+}{C}H_2 \longleftrightarrow \overset{+}{C}H_2{-}CH{=}CH_2$$

Since this resonance hybrid would result in two allyl chloride molecules which are identical, we need some scheme to differentiate between them. This can be done by labelling the C of the CH_2OH group with ^{14}C, since the resulting chlorides will be differently labelled. Thus:

$$CH_2{=}CH - \overset{*}{C}H_2^+ \longleftrightarrow \overset{+}{C}H_2 - CH{=}\overset{*}{C}H_2$$

$$\downarrow Cl^- \qquad\qquad \downarrow Cl^-$$

$$CH_2{=}CH - \overset{*}{C}H_2Cl \qquad ClCH_2 - CH = \overset{*}{C}H_2$$

If, then, we find the product contains the CH_2Cl group, the S_N1' mechanism must be operating.

6. 1, 2-*Addition* (see Table 2. 2)

$$C{=}C{-}C{=}O + H - Z \longrightarrow C{=}C{-}C\begin{matrix} OH \\ Z \end{matrix}$$

$$+ 694{\cdot}5\,(C{=}O) + x \qquad \underset{(C-O)}{-334{\cdot}7} \underset{(O-H)}{-464{\cdot}4} \underset{(C-Z)}{-y}$$

Since Z is unknown, we can represent bond energies H—Z and C—Z as x and y kJ, respectively. Hence, $\Delta H^\ominus$ for 1, 2-addition is $(x{-}y{-}104{\cdot}6)$ kJ.

1, 4-*Addition*

$$\underset{+606{\cdot}7}{C{=}C{-}C{=}O} + \underset{+x}{H - Z} \longrightarrow [Z{-}C{-}C{=}C{-}OH] \longrightarrow \underset{-y-347{\cdot}3-414{\cdot}2}{Z{-}C{-}CH{-}C{=}O}$$

Hence, $\Delta H^\ominus = (x{-}y{-}154{\cdot}9)$ kJ. If we equate $\Delta H^\ominus$ with $\Delta G^\ominus$ (assuming entropy changes about the same; see Ch. 5, Prob. 6), then 1, 4-addition is favoured (more thermodynamically favourable) since its $\Delta G^\ominus$ is more negative than that for 1, 2-addition.

7. 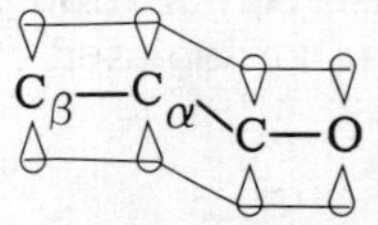

All atoms are sp^2 hybridised and overlap of all $2p_z$ orbitals occurs to give an MO embracing all four atoms. Because oxygen is strongly electronegative, the MO is 'fat' at the oxygen atom and 'thin' at the terminal carbon atom (C-β). Hence, the terminal carbon atom (C-β)

carries a small positive charge (the volume of the MO at an atom is a measure of the electron density). Therefore C-β is susceptible to nucleophilic attack, resulting in 1, 4-addition.

8. $MeCH{=}CHCO_2H \longrightarrow \underset{\displaystyle CH_2CO_2Et}{MeCH}{-}CH_2CO_2Et$

Inspection of the structure of the product shows that the original 'framework' is still present, but there is an additional acetic ester group. This, coupled with the fact that the starting material is an α, β- unsaturated acid, suggests that one route could be via a Michael condensation. Hence:

$$MeCH{=}CHCO_2H \xrightarrow[HCl]{EtOH} MeCH{=}CHCO_2Et \xrightarrow[EtONa]{CH_2(CO_2Et)_2}$$

$$\underset{\displaystyle CH(CO_2Et)_2}{MeCH}{-}CH_2CO_2Et \xrightarrow[(ii)\ H_3O^+]{(i)\ KOH} \underset{\displaystyle CH(CO_2H)_2}{MeCH}{-}CH_2CO_2H \xrightarrow[-\ CO_2]{heat}$$

$$MeCH(CH_2CO_2H)_2 \xrightarrow[HCl]{EtOH} MeCH(CH_2CO_2Et)_2$$

It is possible that under the conditions used (heating to eliminate CO_2), the cyclic anhydride rather than the dicarboxylic acid is the product (see *Text*, Ch. 16 and Ch. 19). This, however, would still give the di-ester on esterification.

9. $EtO_2CCH{=}CHCH_2 \vdots COCO_2Et$

The presence of the $COCO_2Et$ group suggests a condensation involving $(CO_2Et)_2$. This Claisen condensation requires an active methylene group. The remaining fragment of the product is $EtO_2CCH{=}CHCH_3$. If the CH_3 group were active, then the synthesis can be carried out. According to the principle of vinology (see *Text*), the presence of the CH=CH group between CH_3 and CO_2Et still allows the CH_3 group to behave like the CH_3 group in ethyl acetate. Hence:

$$CH_3CH{=}CHCO_2H \xrightarrow[HCl]{EtOH} CH_3CH{=}CHCO_2Et$$

$$EtO_2CCH{=}CHCH_3 + EtO_2CCO_2Et \xrightarrow[(ii)\ AcOH]{(i)\ EtONa}$$

$$EtO_2CCH{=}CHCH_2COCO_2Et$$

10. The product obtained almost exclusively is in accordance with the equation:

$$MeCHOHCH_2CHO \xrightarrow{-H_2O} MeCH{=}CHCHO$$

The mechanism involves the removal of an adjacent proton (to the OH group). In the CH_2CHO group, because of the presence of the CHO group (strongly $-I$), the CH_2 is an active methylene group. Because there is no activation for the terminal Me group, proton loss occurs predominantly from the CH_2 group to form crotonaldehyde. Another supporting factor for this direction of elimination is that crotonaldehyde has extended conjugation ($-CH{=}CH{-}CH{=}O$), whereas the alternative product, but-3-enaldehyde, has an isolated double-bond system and so is less stable than its conjugated isomer.

11. $CH_2{=}CH{-}CMe_2{-}CO_2H \xrightarrow{-CO_2} MeCH{=}CMe_2$

The double bond has migrated and the H of CO_2H has become attached to the *terminal* CH_2 group. A possible explanation is that the reaction proceeds via a cyclic T.S. in which this transfer occurs. Hence, a possible mechanism is:

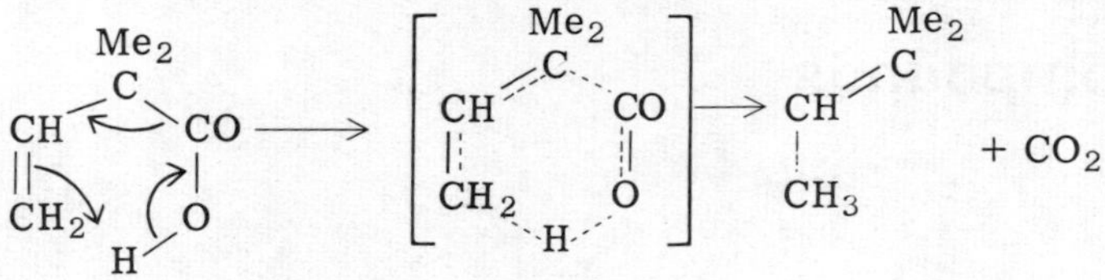

12. (i)

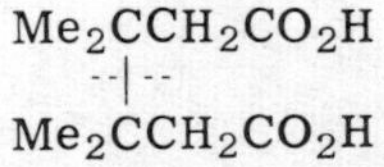

The two 'halves' are identical. This suggests the possibility of a 'dimerisation' reaction. This type of reaction is characteristic of compounds containing the α,β-unsaturated carbonyl group, CH=CHCHO. Furthermore, this 'dimerisation' is a bimolecular *reduction* product and is common with α,β- unsaturated ketones. Since the product is an acid, we require a ketone that is readily oxidised to an acid. A ketone of this type is one containing a CH_3CO group, since this forms acid by means of the haloform reaction. Thus (note the use of magnesium amalgam):

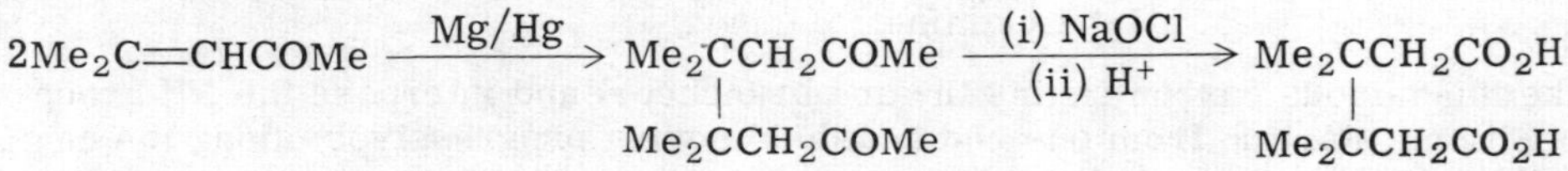

(ii)

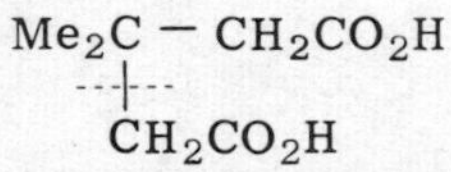

See Prob. 8 for the discussion leading to the following synthesis.

$$Me_2CH{=}CHCO_2H \xrightarrow[HCl]{EtOH} Me_2C{=}CHCO_2Et$$

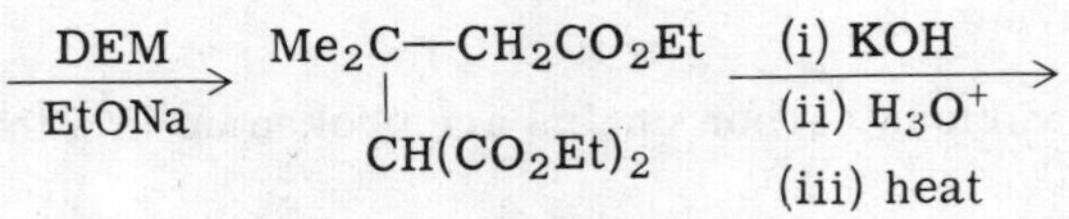

$$Me_2C(CH_2CO_2H)_2$$

Whereas acids of the type $RCH{=}CHCO_2H$ are readily prepared by the Knoevenagel reaction (see *Text*), it is more difficult to prepare acids of the type $R_2C{=}CHCO_2H$. The general method is to condense a ketone with cyanoacetic ester in the presence of an aminoacid (or acetamide; see *Text*).

(iii) $CH_3CH{=}CHCH_2CH_2OH$ This is a β,γ-unsaturated alcohol, and a possible precursor is α β,γ-unsaturated acid. Hence, a synthesis is (see also *Text*)

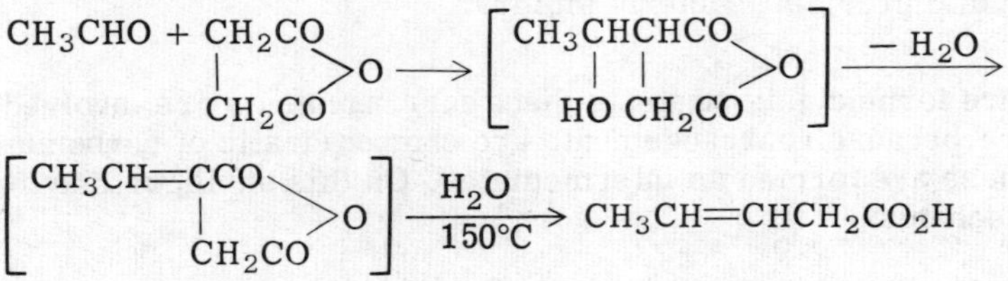

$$\xrightarrow{LAH} CH_3CH{=}CHCH_2CH_2OH$$

13 Nitrogen compounds

(Problems on page 19)

1. See also Ch. 3, Prob. 2.

$CH_3CH_2CH_2CH_2NH_2$
butylamine

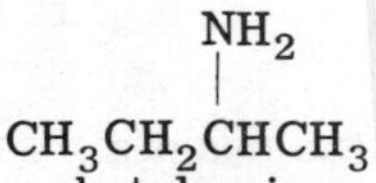
s-butylamine

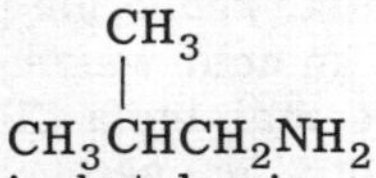
isobutylamine

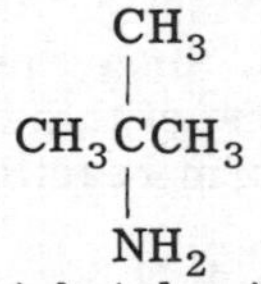
t-butylamine

We now take the various carbon chains (n- and branched-) and interpose the NH group at different positions, starting from one end and then moving progressively along the chain. This gives the isomeric secondary amines.

$CH_3CH_2CH_2NHCH_3$
methylpropylamine

$CH_3CH_2NHCH_2CH_3$
diethylamine

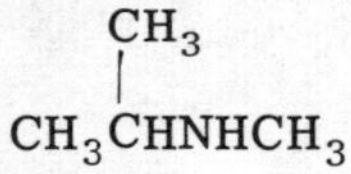
isopropylmethylamine

$(CH_3)_2NC_2H_5$
ethylolimethylamine

To obtain the isomeric tertiary amines, the various carbon chains are broken up into three fragments:

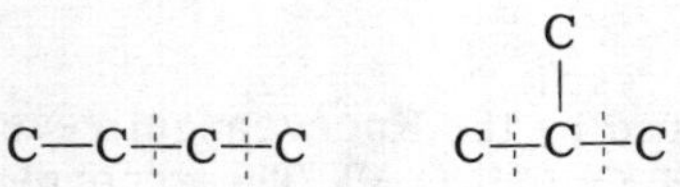

In each case, only one set is possible and both are identical.

2. Hydrogen bonding between molecules raises the b.p. of that compound. Primary (NH_2) and secondary amines (NH) can form hydrogen bonds; tertiary amines (> N—) cannot. Hence, t-amines have the lowest b.p. of a group of isomeric amines.

3. Since pairs of isomers are formed, this suggests that rearrangements are involved in their formation. Furthermore, because rearrangements are characteristic of carbonium ions, we may conclude that these are formed as intermediates. On this basis, we may formulate the following possible mechanism.

$$MeCH_2CH_2CH_2NH_2 \xrightarrow[HCl]{NaNO_2} MeCH_2CH_2CH_2\text{—}N_2^+ \xrightarrow{-N_2}$$

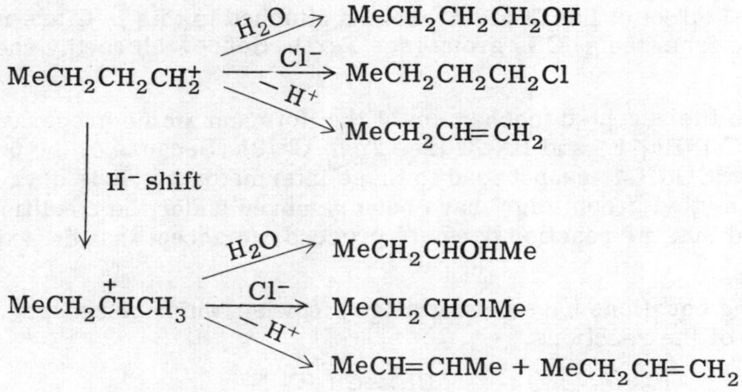

4. (i) $Me_2CHCH_2CH_2NH_2 \xrightarrow[\text{(ii) } OH^-]{\text{(i) MeI}}$

$$Me_2CH\overset{\beta}{C}H_2\overset{\alpha}{C}H_2\overset{+}{N}Me_3OH^- \xrightarrow[-H_2O]{\text{heat}} Me_3N + Me_2CHCH{=}CH_2$$

The general rules of the Hofmann exhaustive methylation method are (i) loss of a β-proton, and (ii) when alternative β-protons are present, the one lost preferentially is that which leads to the least substituted ethylene (the Hofmann product).

(ii) $Me_2CHCH_2NHMe \xrightarrow[\text{(ii) } OH^-]{\text{(i) MeI}} Me_2\overset{\beta}{C}H\overset{\alpha}{C}H_2\overset{+}{N}Me_3OH^-$

$$\xrightarrow[-H_2O]{\text{heat}} Me_3N + Me_2C = CH_2$$

(iii) $Me_2CHCH_2NHCH_2CH_2Me \xrightarrow[\text{(ii) } OH^-]{\text{(i) MeI}}$

$$Me_2\overset{\beta}{C}H\overset{\alpha}{C}H_2\overset{+}{N}Me_2\overset{\alpha}{C}H_2\overset{\beta}{C}H_2MeOH^- \xrightarrow[-H_2O]{\text{heat}}$$

$$Me_2CHCH_2NMe_2 + \underset{\text{major product}}{MeCH{=}CH_2} + MeCH_2CH_2NMe_2$$

$$+ \underset{\text{minor product}}{Me_2C{=}CH_2}$$

Propene is a less substituted ethylene than isobutene (see *Text* for an explanation).

(iv) $CH_3CH_2NHCH_2CH_2Cl \xrightarrow[\text{(ii) } OH^-]{\text{(i) MeI}}$

$$\overset{\beta}{C}H_3\overset{\alpha}{C}H_2\overset{+}{N}Me_2\overset{\alpha}{C}H_2\overset{\beta}{C}H_2ClOH^- \xrightarrow[-H_2O]{\text{heat}}$$

$$CH_3CH_2NMe_2 + \underset{\text{major product}}{CH_2{=}CHCl} + Me_2NCH_2CH_2Cl +$$

$$\underset{\text{minor product}}{CH_2{=}CH_2}$$

Because of the —I effect of the Cl atom, protons attached to this β—C are more easily removed than those from the β—CH_3 group (see *Text*). Hence, chloroethylene is the major product.

5. According to the accepted mechanism of the Hofmann amine preparation, two intermediates are $[RCONBr]^- K^+$ and RNCO (see *Text*, Ch. 9). Because of the presence of the *N*-methyl group, RCONHMe cannot lead to these intermediates. It is always possible, however, that the *N*-methyl group might have been removed under the conditions of the reaction. Since we are told that the reaction does *not* proceed, we accept that the explanation offered is satisfactory.

6. The following equations have been written stepwise, where necessary, in order to explain the course of the reactions.

(i) $PhCH_2OH \xrightarrow{H_2SO_4} PhCH_2^+ \xrightarrow{CH_2=CH-C\equiv N}$

$PhCH_2-\overset{+}{N}\equiv CCH=CH_2 \xrightarrow{H_2O} PhCH_2NHCOCH=CH_2$

This is an example of the Ritter reaction (see *Text*).

(ii) $PhCH_2COCHN_2 + HCl(g) \xrightarrow{Et_2O} PhCH_2COCH_2Cl$

(iii) $NH_2CONH_2 \xrightarrow{heat} HNCO \xrightarrow{BuOH}$

$NH_2CO_2Bu \xrightarrow{HNCO} NH_2CONHCO_2Bu$

Urea forms cyanic acid which reacts with the alcohol to form a urethan. If excess of acid is present, the final product is an allophanate (see *Text*).

(iv) $2MeNH_2 + ClCO_2Et \longrightarrow MeNHCO_2Et + MeNH_3^+Cl^-$

(v) $HOCH_2CH_2CN \xrightarrow[heat]{aq.\ HBr} BrCH_2CH_2CO_2H$

(vi) 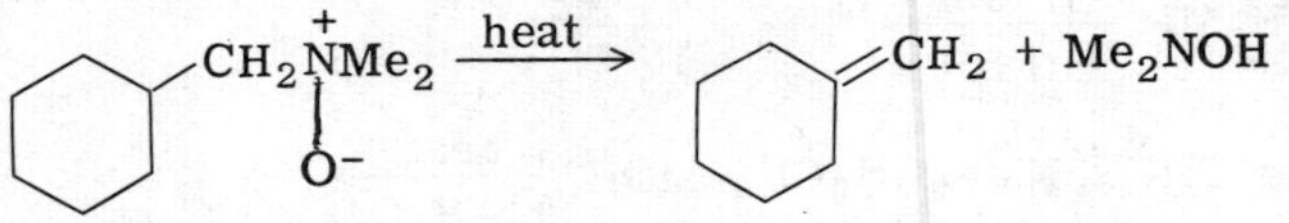

This is an example of the Cope reaction (see *Text*, Ch. 4).

(vii) $PhCH(CN)CH(CN)CO_2Et \xrightarrow[heat]{aq.\ HCl}$

$PhCH(CO_2H)CH(CO_2H)_2 \xrightarrow{-CO_2} PhCH(CO_2H)CH_2CO_2H$

The intermediate acid is a malonic acid derivative and readily eliminates carbon dioxide under the conditions of the experiment (*cf.* DEM syntheses).

(viii) This is an example of the enamine synthesis.

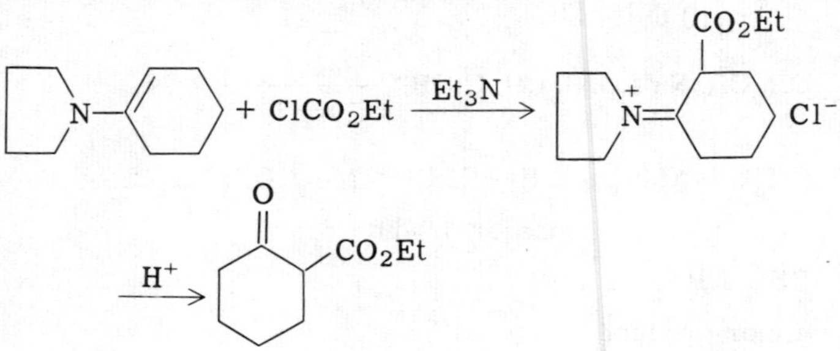

(ix)

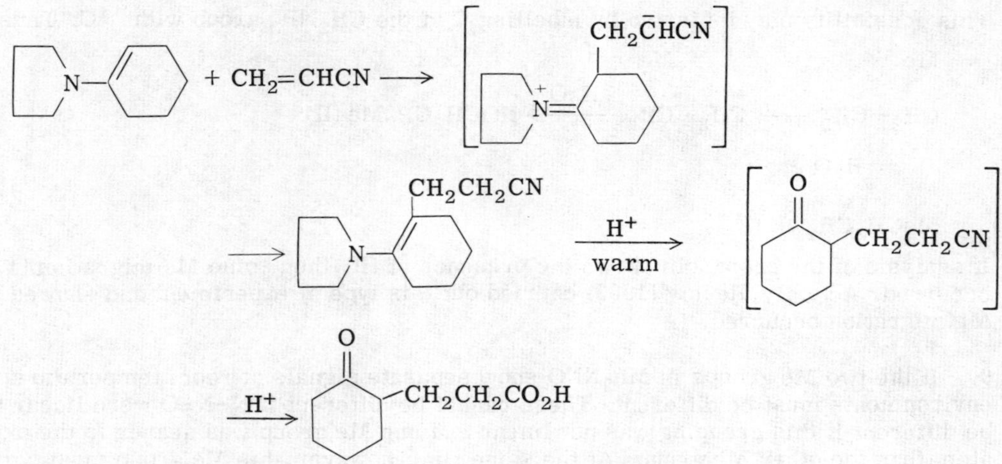

(x) This is an example of the Mannich reaction (see *Text*).

$$EtNH_2 + CH_2O \overset{H^+}{\rightleftharpoons} H_2O + Et\overset{+}{N}H{=}CH_2 \longleftrightarrow EtNH{-}\overset{+}{C}H_2$$

$$Me_2CH{-}\overset{+}{N}(=O)O^- \rightleftharpoons Me_2C{=}\overset{+}{N}(OH)O^-$$

$$Me_2C{=}\overset{+}{N}(O{-}H)O^- + \overset{+}{C}H_2{-}NHEt \xrightarrow{-H^+} Me_2C(NO_2)CH_2NHEt \xrightarrow{HCl} Me_2C(NO_2)CH_2\overset{+}{N}H_2Et\,Cl^-$$

7. (i) At first sight, the expected reaction would be:

$$R_2C(OH)CH_2CR_2NH_2 \xrightarrow{HNO_2} R_2C(OH)CH_2CR_2{-}N_2^+ \xrightarrow{-N_2}$$

$$R_2C(OH)CH_2CR_2^+ \xrightarrow{-H^+} R_2C(OH)CH{=}CR_2$$

Since we are told that the products are R_2CO and $R_2C{=}CH_2$, it therefore follows that fission of the molecule takes place more easily than removal of a proton. An electronic redistribution that would facilitate fission is:

$$R_2C(O{-}H){-}CH_2{-}CR_2^+ \xrightarrow{-H^+} R_2C{=}O + R_2C{=}CH_2$$

(ii) $PhCH{=}NR \xrightarrow{MeI} PhCH{=}\overset{+}{N}MeR\,I^- \xrightarrow{H^+} PhCHO + MeNHR$

Nitrogen, because of its lone pair, is readily attacked.

8. If we represent the deamination as shown, then it appears possible that a methyl group could undergo a 1,2-shift.

$$MeCH_2CH_2NH_2 \xrightarrow[\text{(ii) }-N_2]{\text{(i) }HNO_2} CH_2(Me){-}CH_2^+ \longrightarrow \overset{+}{C}H_2{-}CH_2(Me)$$

This possibility may be tested by labelling C of the CH_2NH_2 group with ^{14}C. Thus:

$$Me{-}CH_2{-}\overset{*}{C}H_2^+ \longrightarrow \overset{+}{C}H_2{-}\overset{*}{C}H_2{-}Me \xrightarrow{H_2O} HOCH_2\overset{*}{C}H_2Me \text{ (II)}$$

$$Me{-}CH_2{-}\overset{*}{C}H_2^+ \xrightarrow{H_2O} MeCH_2\overset{*}{C}H_2OH \text{ (I)}$$

If analysis of the propanols shows the presence of (II), then some Me migration (1,2—) has occurred. Actually, Reutov (1963) carried out this type of experiment and showed that no Me migration occurred.

9. If the two Me groups in Me_2NNO show separate signals at room temperature, then their environments must be different. These cannot be different if N—N=O were linear but could be different if this grouping was not linear *and* one Me group was nearer to the oxygen atom than the other Me group. At the same time, however, this Me group must *stay* nearer.

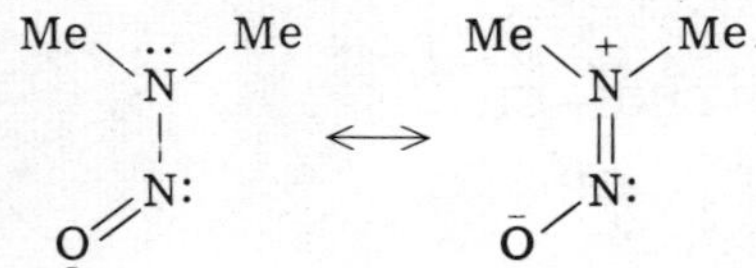

Since N has a tetrahedral distribution of its valancies (see *Text*; also Ch. 17), then if Me_2NNO is represented as a resonance hybrid, the N—N bond has some double-bond character and so there will be hindered rotation about this bond. Consequently, one Me group will be *cis* to the oxygen and the other *trans*. Hence, the environments of the two Me groups are different and so two signals are observed. This is true only at room temperature (and at lower temperatures). As the temperature is raised, the rate of internal rotation increases, and at a sufficiently high temperature (~ 200°C), the rotation is sufficiently fast for an 'average' signal to be observed. Only if the rotation is slow enough will the environments be sufficiently different for each Me to give its own signal.

10. Tertiary alkyl halides readily eliminate hydrogen halide by the EI mechanism to form alkenes. Hence:

$$Me_3CBr \xrightarrow{-HBr} Me_2C{=}CH_2$$

Neopentyl bromide is a *primary* alkyl halide, but because of the large steric effect of the Me_3C group, S_N2 reaction is difficult. Furthermore, because NH_3 is such a poor nucleophile (see *Text*, Ch. 5) S_N2 reaction is made even more difficult. Hence rearranged alkene is the major product (via EI mechanism).

$$Me_3CCH_2Br \xrightarrow{-Br^-} Me_3CCH_2^+ \longrightarrow Me_2\overset{+}{C}CH_2Me \xrightarrow{-H^+} Me_2C{=}CHMe$$

A general method of converting acids into primary amines is the Hofmann amine preparation. Since the rearrangement step is intramolecular, there is no chance of alkene formation occurring.

$$\text{(i)}\quad Me_3CCO_2H \xrightarrow[\text{(ii) } NH_3]{\text{(i) } SOCl_2} Me_3CCONH_2 \xrightarrow[KOH]{Br_2} Me_3CNH_2$$

$$\text{(ii)}\quad Me_3CCH_2CO_2H \xrightarrow[\text{(iii) } Br_2/KOH]{\text{(i) } SOCl_2,\ \text{(ii) } NH_3} Me_3CCH_2NH_2$$

The Schmidt reaction could also be used on these acids (see *Text*, Ch. 9); this is also intramolecular. As a matter of choice, however, since the use of hydrazoic acid can be hazardous, the Hofmann method is preferable.

11. (i) $Me_2C(CN)NHNHC(CN)Me_2 \xrightarrow{HOCl} Me_2C(CN)N{=}NC(CN)Me_2$

This oxidation is characteristic of 1,2-disubstituted hydrazines.

(ii) $MeNHNHMe \xrightarrow{HgO} MeN{=}NMe$

Mercuric oxide is the most commonly used oxidising agent for hydrazines (to azo-compounds).

(iii) $Me(CH_2)_2Me \xrightarrow[\sim 500°C]{HNO_3} Me(CH_2)_2CH_2NO_2 +$

$MeCH(NO_2)CH_2Me + MeCH_2CH_2NO_2 + MeCH_2NO_2 + MeNO_2$

All possible isomeric nitro-compounds are obtained from the alkane, together with all possible fission products.

(iv) $3RCH{=}CH_2 \xrightarrow{B_2H_6} (RCH_2CH_2^-)_3B \xrightarrow{ClNH_2} 3RCH_2CH_2NH_2$

(v) $Me_3N + \text{D}{-}EtMeCHCl \longrightarrow \text{L}{-}EtMeCH\overset{+}{N}Me_3\,Cl^-$

This takes place by the S_N2 mechanism and so inversion occurs (see *Text*, Ch. 5. Also Ch. 17).

(vi) $PhNCO \xrightarrow{H_2O} PhNHCO_2H \xrightarrow{-CO_2} PhNH_2$

$\xrightarrow{PhNCO} PhNHCONHPh$

(vii) $CH_2(CO_2Et)_2 \xrightarrow[(ii)\ BrCN]{(i)\ EtONa} NCCH(CO_2Et)_2$

(viii) $KCN + Br_2 \longrightarrow BrCN + KBr$

$EtNH_2 + BrCN \xrightarrow{KOH} EtNHCN + KBr$

12. The formation of these products is typical of the reaction between diazomethane and ketones. To formulate the structures, it is best to work them out on the basis of the accepted mechanisms of this reaction. We are compelled to use this approach since we are given no information (chemical or otherwise) for elucidating the structures.

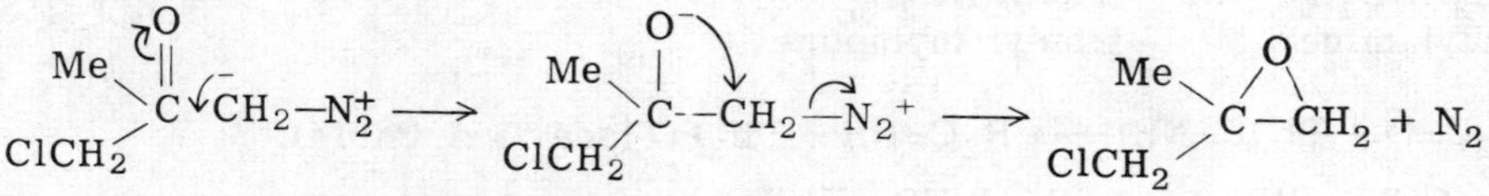

This epoxide is C_4H_7ClO, and so is **(B)**.

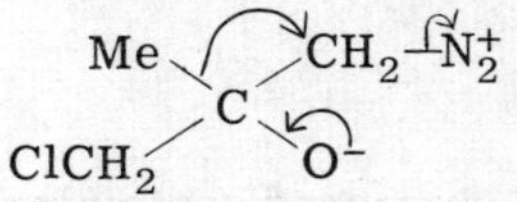

$\longrightarrow N_2 + ClCH_2COCH_2Me$ **(A)**

We have given **(A)** the ketone structure derived by migration of the Me group. However, had the $ClCH_2$ migrated, the ketone obtained would have been $MeCOCH_2CH_2Cl$. Obviously, then, we must give a reason for choosing Me as the migrating group. It has been pointed out (see *Text*, Ch. 8) that migratory aptitude increases with increasing electron-releasing property of that group. Since Me has a +I effect and $ClCH_2$ a −I effect, the former therefore migrates preferentially.

13. $RMgBr + \overset{*}{C}O_2 \longrightarrow R\overset{*}{C}(=O)OMgBr \xrightarrow{H^+} R\overset{*}{C}O_2H \xrightarrow[\text{(ii) } CH_2N_2]{\text{(i) } SOCl_2}$

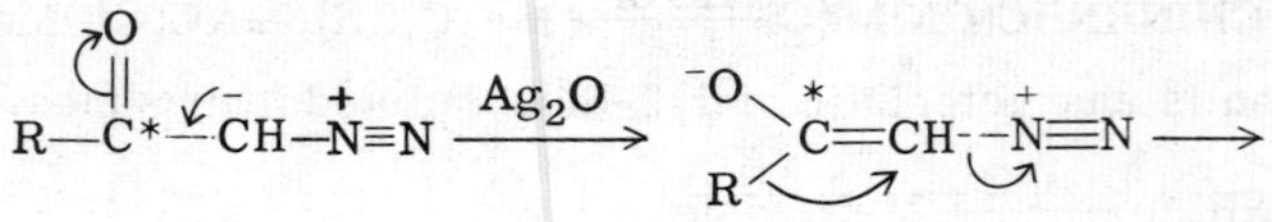

$$N_2 + O{=}\overset{*}{C}{=}CHR \xrightarrow{H_2O} RCH_2\overset{*}{C}O_2H$$

The labelled carbon atom ends up in the carboxyl group because the diazoketone, under the influence of silver oxide, undergoes the Wolff rearrangement to form the keten (see *Text*).

14. (**A**) $C_6H_{15}N \xrightarrow[\text{(ii) KOH}]{\text{(i) MeI}} C_8H_{20}N^+OH^-$ (**B**)

Now, on methylation, $-NH_2 \longrightarrow -NMe_3^+$; $>NH \longrightarrow >NMe_2^+$; $\equiv N \longrightarrow \equiv NMe^+$. Since methylation of (**A**) results in the addition of *two* carbon atoms to give (**B**), (**A**) must be a *secondary* amine:

$C_6H_{14}NH \longrightarrow C_6H_{14}NMe_2^+OH^- \equiv C_8H_{20}N^+OH^-$ (**B**)

We also have:

$C_8H_{20}N^+OH^- \xrightarrow{\text{heat}} H_2O + (CH_3)_2C{=}CH_2 + C_4H_{11}N$ (**C**)

Since (**C**) is a tertiary amine and contains two Me groups (which were introduced by methylation), its structure is $(CH_3)_2NC_2H_5$, i.e., one alkyl group in (**A**) is ethyl. Hence the other group in (**A**) is $C_4H_9(C_4H_9 + C_2H_5 = C_6H_{14})$. There are three isomeric butyl groups: $MeCH_2CH_2CH_2-$; Me_2CHCH_2-; Me_3C-. n-Butyl cannot produce isobutene, but the other two could:

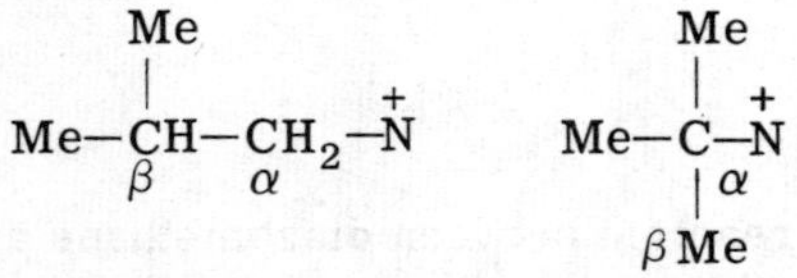

Since no data are given in the question that will enable us to distinguish between the two, either group could be present. Thus, (**A**) is

$Me_2CHCH_2NHC_2H_5$ or $Me_3CNHC_2H_5$
ethylisobutylamine t-butylethylamine

15. (i) $R_2C{=}O + Ph_3P{=}NH \longrightarrow R_2C{=}NH + Ph_3PO$ (see *Text*, Ch. 14).

(ii) $Ph_3CCl + RNH_2 \longrightarrow Ph_3CNHR + HCl$

(iii) $EtNO_2 + 2CH_2O \xrightarrow{KHCO_3} MeC(NO_2)(CH_2OH)_2$

This is an example of the condensation of formaldehyde with a compound containing an active methylene group.

(iv) $RCH_2NH_2 + 2CH_2O + 2HCO_2H \longrightarrow RCH_2NMe_2 + 2CO_2 + 2H_2O$

This reaction is an example of the Eshweiler-Clark methylation (see *Text*, Ch. 8).

(v) $CH_2{=}CO + HCN \longrightarrow CH_2{=}C(OH)CN \xrightarrow{CH_2{=}CO} CH_2{=}C(OAc)CN$

The compound expected could have been CH_3COCN (see *Text*, Ch. 12). The enol form is more stable because it is stabilised by conjugation and by hydrogen bonding, and

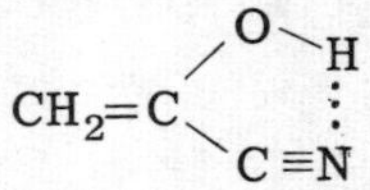

so further reaction is possible.

(vi) $Me_2C{=}CHCOMe + RCH_2NH_2 \longrightarrow Me_2\underset{\displaystyle NHCH_2R}{\underset{|}{C}}{-}CH_2COMe$

$$\xrightarrow{HNO_2} Me_2\underset{\displaystyle ON{-}N{-}CH_2R}{\underset{|}{C}}{-}CH_2COMe \xrightarrow{EtONa} Me_2C{=}CHCOMe + RCHN_2 + H_2O$$

Note the 1,4-addition. This is a general method for preparing diazoalkanes (*cf.* with methods described for CH_2N_2 in the *Text*).

(vii) $EtCHO + CH_2N_2 \longrightarrow EtCOMe + \overset{\;O}{EtCH{-}CH_2}$

(viii) $MeCOEt + CH_2N_2 \longrightarrow EtCOEt + \overset{\;O}{MeEtC{-}CH_2} + MeCOCH_2Et$
(See Prob. 12, above).

(ix) $Et_4N^+Br^- \xleftarrow{MeBr} Et_3\overset{+}{N}{-}CH_2^- \xrightarrow{Me_2CO} Et_3\overset{+}{N}CH_2CMe_2O^-$

(x) 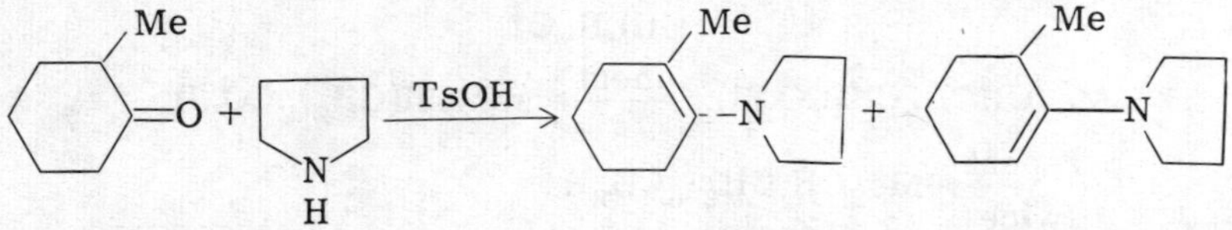

(xi) 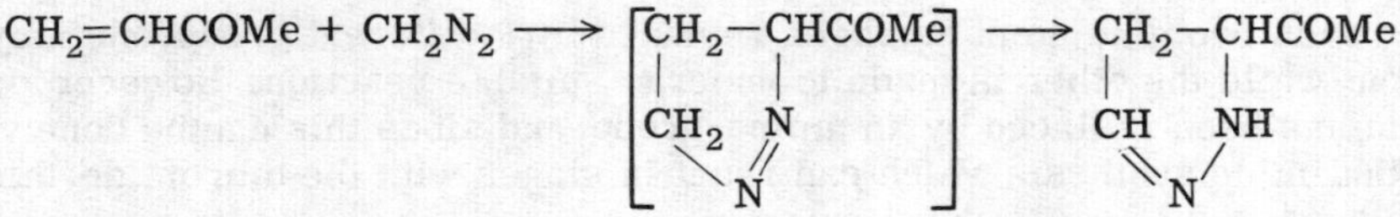

The product is a pyrazoline; the double bond is attacked rather than the carbonyl group.

16. $\underset{(1)}{CH_3}{-}\underset{(2)}{CH_2}{-}\underset{(3)}{CH_2}{-}NO_2$

The NMR spectrum shows three signals.

(1) **CH_3.** This is a triplet and occurs upfield; $\tau \sim 9$ (see Table 3.2).

(3) **CH_2.** This is a triplet and occurs considerably downfield due to deshielding caused by the strong —I effect of the nitro-group; $\tau \sim 5{\cdot}6$

(2) **CH_2.** This is coupled to both CH_3 (1) and CH_2 (2). Hence, if there is no overlapping, the multiplicity of this signal is given by $(3 + 1)(2 + 1) = 12$ lines. On the other hand, if the coupling constants $J_{1,2}$ and $J_{2,3}$ are almost the same, overlapping will occur and then the total number of lines in the signal will be $(3 + 2 + 1) = 6$. Whichever of these two signals is the actual one, it will be found *between* the other two signals. Finally, both the CH_3 (1) and CH_2 (2) signals will be further downfield with respect to the $MeCH_2$ group in an alkane because the —I effect of the nitro-group is transmitted along the chain.

17. (i) The length of the chain has been increased by six carbon atoms. Since this is the number of carbon atoms in cyclohexanone, one possible synthesis could start with a cyclohexeno-enamine. Thus:

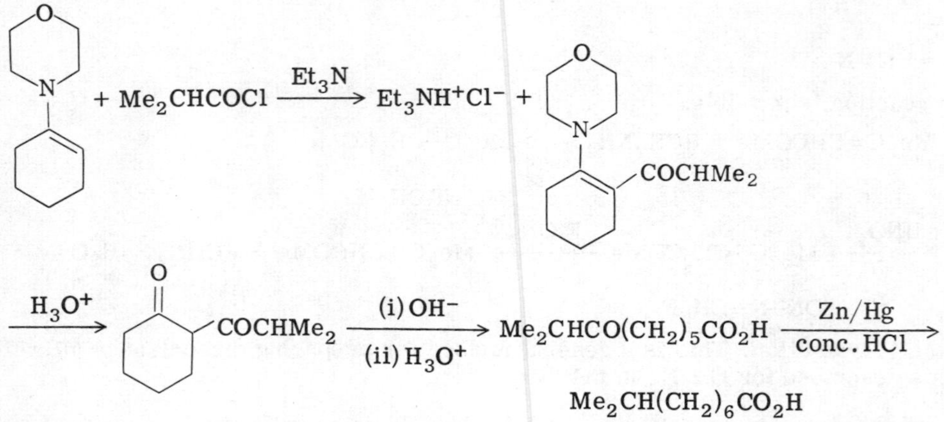

The acid chloride is prepared from the acid by the action of thionyl chloride; β-diketones are split by alkali; the Clemmensen reduction converts $CO \longrightarrow CH_2$.

If the student did not remember this method of stepping up the acid series, he can always fall back on the standard method of stepping up the acid series by two carbon atoms at a time using DEM:

$$Me_2CHCO_2H \xrightarrow[\text{(ii) HBr}]{\text{(i) LAH}} Me_2CHCH_2Br \xrightarrow[\text{(ii) KOH (iii) } H_3O^+]{\text{(i) NaDEM}}$$

$$Me_2CHCH_2CH(CO_2H)_2 \xrightarrow{\text{heat}} Me_2CH(CH_2)_2CO_2H$$

$$\xrightarrow[\text{twice}]{\text{repeat}} Me_2CH(CH_2)_6CO_2H$$

(ii) Since each bromine atom has been replaced by a different group, one may need 'protection' while the other is made to undergo suitable reaction. However, one bromine atom has been replaced by an amino-group, and since this can be done via Gabriel's phthalimide synthesis which can react in stages with the dibromide, this route should be successful.

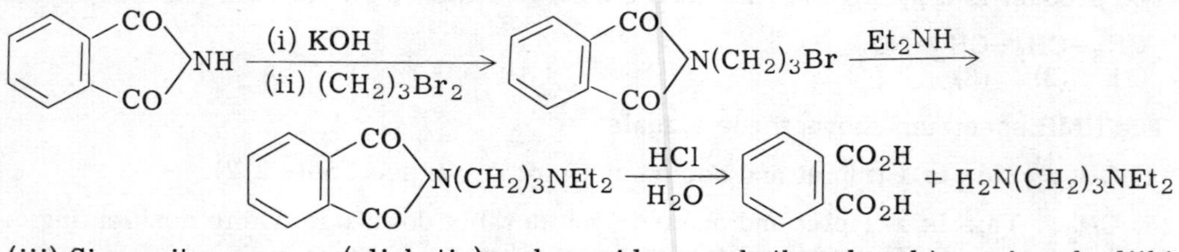

(iii) Since nitro-groups (aliphatic) and cyanides are both reduced to amines by lithium aluminium hydride, this reduction can readily be done (see Table 6.1).

$$O_2NCH_2CH_2CN \xrightarrow{\text{LAH}} H_2NCH_2CH_2CH_2NH_2$$

(iv) For this reduction, the reagent is sodium borohydride—aluminium chloride (see Table 6.1).

$$O_2NCH_2CH_2CN \xrightarrow[AlCl_3]{NaBH_4} O_2NCH_2CH_2CH_2NH_2$$

18. (i) Four signals. All groups of protons different.

(ii) Three signals. Whether *cis* or *trans*, the two olefinic protons are not equivalent.

(iii) Five signals. All groups of protons different.

(iv) Two signals. The Me groups not equivalent, i.e., Me in $MeCO_2$ and MeN.

19. The most characteristic fragmentation pattern of amines is via β-cleavage. Also common is the McLafferty rearrangement of *N*-containing fragments.

(i) From Table 1.6, we have m/e 30 = CH_2NH_2 and m/e 58 = C_3H_8N. Hence:

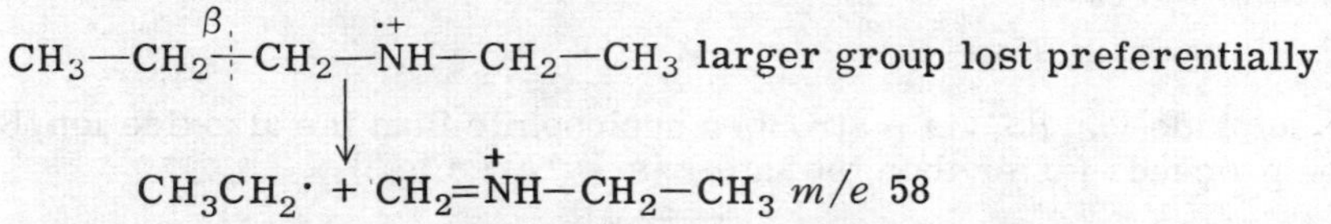

McLafferty rearrangement. 58 − 28 = 30. Suggests loss of ethylene. Hence:

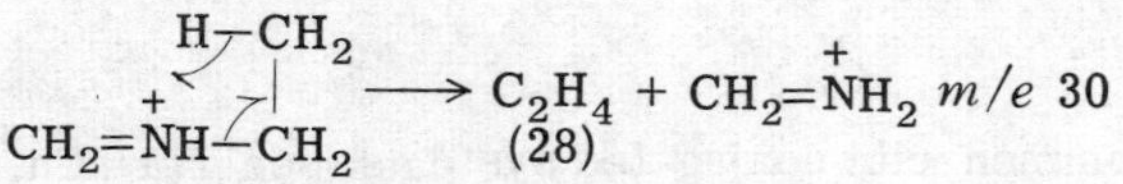

(ii) m/e 30 = CH_2NH_2; m/e 58 = C_3H_8N; m/e 86 = 58 + 28 = $C_5H_{12}N$. This suggests *two* McLafferty rearrangements.

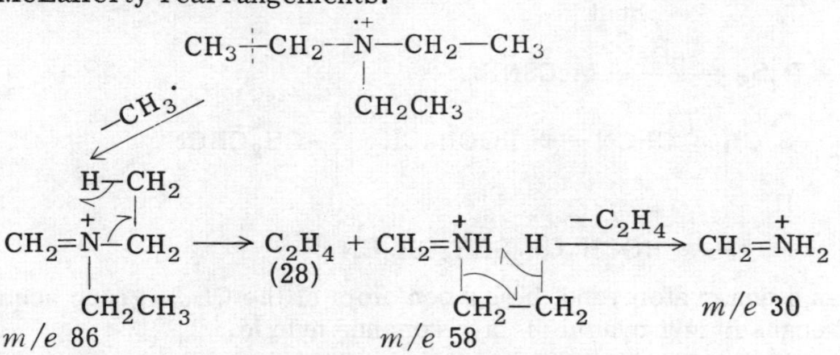

(iii) The M.Wt. of butyl cyanide (C_5H_9N) is 83. Now, 41 = 83 − 42. Since 42 could be C_3H_6, i.e., propene, this suggests a McLafferty rearrangement.

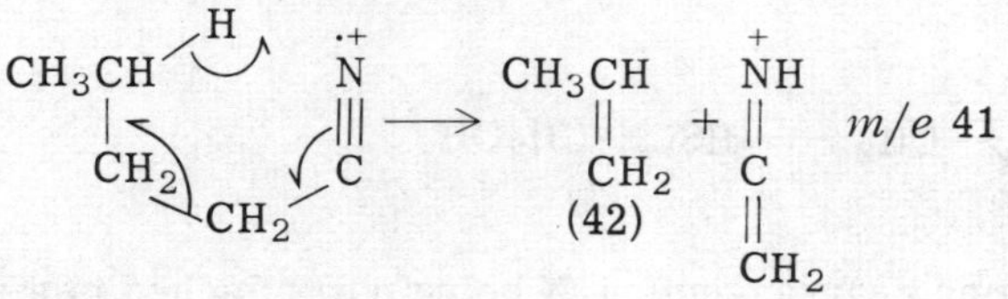

14 Aliphatic compounds of sulphur, phosphorus, silicon and boron

(Problems on page 22)

1. The mechanism is S_N2:

$$RZ^- \; R{-}X \longrightarrow R{-}Z{-}R + X^-$$

Since the alkyl sulphide ion, RS^-, is a stronger nucleophile than the alkoxide ion, RO^-, the former reaction proceeds faster than the latter (see *Text*, Ch. 5).

2. (i)

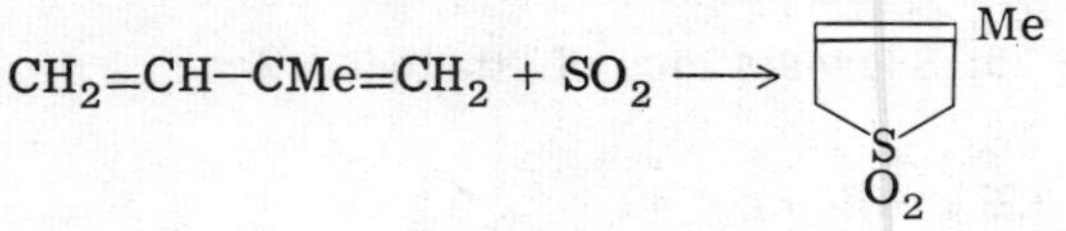

This is a 1,4-addition and is common with conjugated systems (see *Text*, Ch. 4).

(ii) $$MeNH_2 + CS_2 + NaOH \xrightarrow[\text{heat}]{H_2O} MeNHCS_2^-Na^+$$

(iii) $$MeCONH_2 + P_2S_5 \xrightarrow{C_6H_6} MeCSNH_2$$

(iv)

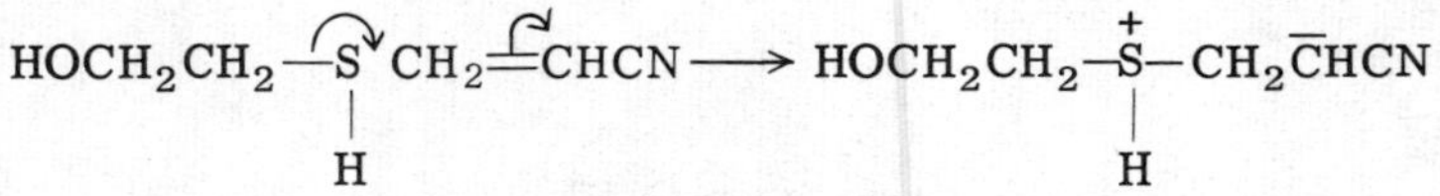

$$\longrightarrow HOCH_2CH_2SCH_2CH_2CN$$

Sulphur can act as a donor atom, and the carbon atom of the $CH_2{=}$ group acquires a positive charge because vinyl cyanide is a resonance hybrid.

$$CH_2{=}CH{-}C{\equiv}N \longleftrightarrow \overset{+}{C}H_2{-}\overset{-}{C}H{-}C{\equiv}N \longleftrightarrow \overset{+}{C}H_2{-}CH{=}C{=}\overline{N}$$

(v)

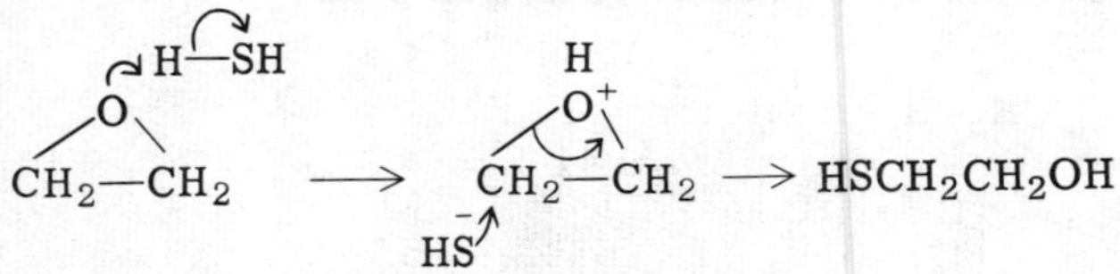

(vi) The molecule has undergone rearrangement, N being now attached to a carbon atom instead of S. A possible explanation is that the rearrangement proceeds by the S_N1' mechanism (see *Text*, Ch. 12).

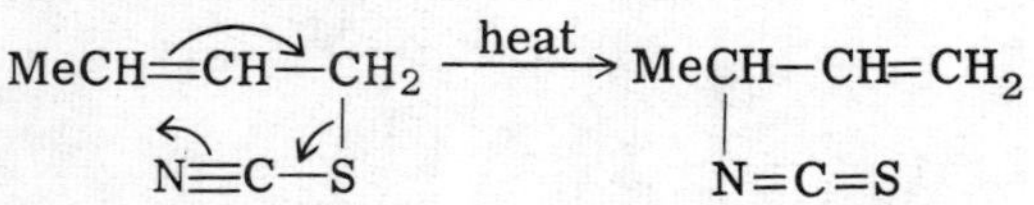

Alternatively, the mechanism could be:

$$MeCH{=}CHCH_2SCN \xrightarrow{\text{heat}} [SCN]^- + MeCH{=}CHCH_2^+$$

$$MeCH{=}CHCH_2^+ \longleftrightarrow Me\overset{+}{C}H{-}CH{=}CH_2 \xrightarrow{[SCN]^-} MeCH(NCS)CH{=}CH_2$$

The carbonium ion is a resonance hybrid and the secondary ion can be expected to make a larger contribution than the primary ion. Recombination results in the formation of the more thermodynamically stable isothiocyanate. If the mechanism were S_N1', then this stability is also a contributing factor to the 'driving force' of the rearrangement.

(vii) $CH_2{=}CH_2 + (SCN)_2 \longrightarrow NCSCH_2CH_2SCN$

The S—S bond is relatively weak (225·9 kJ) and so undergoes fission:

$$\text{—S—S—} \longrightarrow 2\ \text{—S}\cdot \ (\textit{or}\ \text{—S—S—} \longrightarrow \text{—S}^+ + \text{—S}^-)$$

In either case, since S is a much stronger nucleophile than N, the reaction is kinetically controlled, resulting in the formation of the thiocyanate (see Prob. vi, above).

(viii) $Me_2C(SEt)_2 \xrightarrow{KMnO_4} Me_2C(SO_2Et)_2$

(ix) $PhSO_2Me \xrightarrow{EtONa} PhSO_2CH_2^- \xrightarrow{PhCO_2Me}$

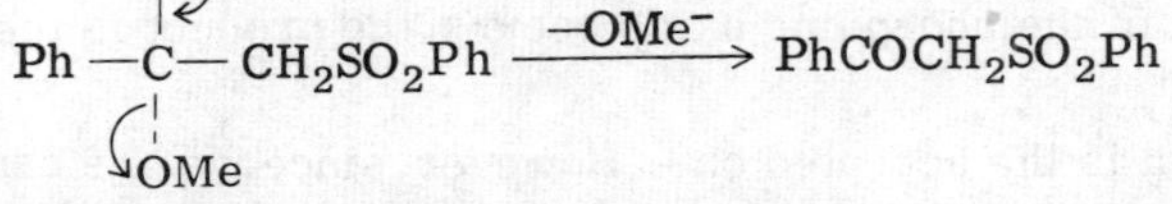

Because of the powerful —I effect of the sulphone group, the Me group is 'activated' and proton removal by the strong base, EtO^-, is made possible.

The student should note that the use of EtONa as a reagent usually means a condensation is being effected via the removal of an 'activated' α-hydrogen atom.

(x) $MeCOCl + EtSH \longrightarrow Me\ COSEt + HCl$

Compare this reaction with alcohols as substrate.

(xi) $(RS)_2Hg \xrightarrow{heat} R_2S + HgS$

This reaction proceeds via free radicals (RS·).

(xii) $Et_2S + H_2SO_4 \longrightarrow Et_2SH^+HSO_4^- \xrightarrow{H_2O} Et_2S + H_2SO_4$

Dialkyl sulphides are weak bases and so their salts are readily hydrolysed (*cf.* with ethers).

(xii) 

Cyanogen halides readily react with compounds containing a donor atom to form cyano-compounds as their halide salts (*cf.* R_3N + BrCN; see *Text*, Ch. 13).

3. See *Text* (not worth learning by heart)

(i) Phenylphosphonic acid; (ii) Diethyl hydrogen phosphite; (iii) Ethylphosphonous dichloride; (iv) Methylphosphonic dichloride; (v) 2-Chloroethylphosphine; (vi) Iminophosphorane; (vii) Phosphorodichloridic acid; (viii) Phosphoramidic dichloride; (ix) Phosphonochloridic acid; (x) Phosphinic chloride; (xi) Methylphosphinic amide; (xii) Diethylphosphinous chloride.

4. Phosphorus, in its tervalent state, is a donor atom. Hence, a possible mechanism is:

This mechanism has been described in its simplest form; the geometries of some of the intermediates have been neglected. In order to appreciate this aspect, we shall replace each CH_2 by C*ab* (dotted lines represent groups behind the plane of the paper).

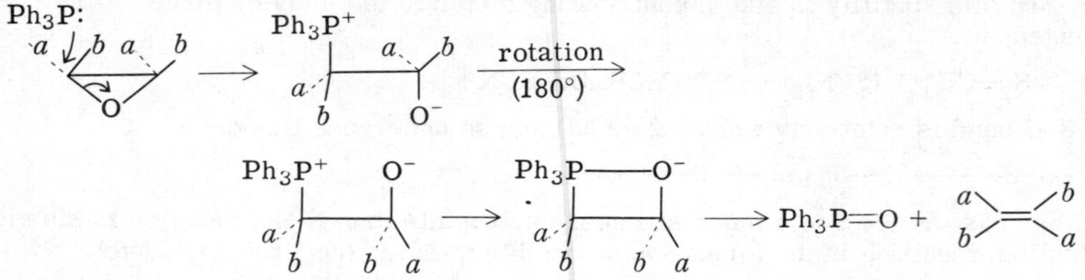

The phosphine attacks on the *opposite* side and along the C—O bond (see ethylene oxide; *Text*, Ch. 11), resulting in inversion at this carbon atom. In order to form the cyclic intermediate, rotation occurs about the single bond to bring the P and O atoms close together to form a bond between them. Inspection of the formulae shows that the epoxide has been derived from a *cis*-alkene, but after rotation to form the cyclic intermediate, the 'alkene' fragment is *trans*, and when elimination of the phosphine oxide occurs, the product is the *trans*-alkene.

It appears that the mechanism described is the accepted one. However, since we are asked to suggest a mechanism, then, if we were not aware of the one described, we are entitled to suggest *any* mechanism that we think is reasonable. The following is an alternative, and is based on the assumption that the first step involves attack on oxygen (see also Prob. 13 (ii), below).

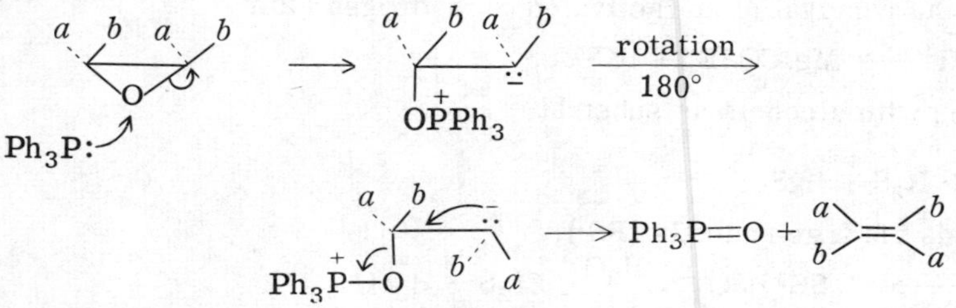

For the electron-pair set free (in the first step) to assist the removal of the phosphine oxide molecule, its 'best' geometry is to be in such a position that it attacks at the C of C—O from the back of the C—O bond. This it can do if rotation occurs as shown, and at the same time explains the formation of the *trans*-alkene (see also *Text*, Ch. 17).

5. (i) $3PhBr + PCl_3 + 6Na \longrightarrow Ph_3P + 3NaCl + 3NaBr$

(ii) $EtPH_2 + Na \xrightarrow[NH_3]{liq.} NaPHEt \xrightarrow{MeI} MePHEt$

(iii) (*a*) $Et_3P \xrightarrow[or\ H_2O_2]{HNO_3} Et_3PO$

(*b*) $3EtMgBr + POCl_3 \longrightarrow Et_3PO + 3MgBrCl$

(iv) $PCl_3 + Et_4Pb \longrightarrow Et_3PCl + Et_2PbCl_2$

(v) The Arbusov reaction is the most satisfactory method.

$P(OEt)_3 + MeI \longrightarrow MePO(OEt)_2 + EtI$

(vi) $Me_2PCl + Cl_2 \longrightarrow Me_2PCl_3 \xrightarrow{SO_2} Me_2POCl + SOCl_2$

(vii) $EtPOCl_2 + 4NH_3 \longrightarrow EtPO(NH_2)_2 + 2NH_4Cl$

Since the question does not state the starting materials, the student can start where he likes. This, no doubt, will mean he will use one-step reactions where possible. For those who are interested in phosphorus chemistry, it would be worthwhile finding out how the starting materials used in the above syntheses may be prepared (this does *not* include Na, PCl_3 and $POCl_3$!).

6. (i) $PhCH{=}CHCH_2PPh_3^+Cl^- \xrightarrow[-HCl]{EtOLi}$

$$PhCH{=}CHCH{=}PPh_3 \xrightarrow{PhCHO} Ph_3P{=}O + PhCH{=}CHCH{=}CHPh$$

(ii) 

(iii) $Ph_3P{=}CCl_2 + p\text{-}Me_2NC_6H_4CHO \longrightarrow Ph_3PO + p\text{-}Me_2NC_6H_4CH{=}CCl_2$

(iv) $CHCl_3 \xrightarrow[-HCl]{t\text{-BuOK}} CCl_2 \xrightarrow{Ph_3P} Ph_3P{=}CCl_2$

(v) $Ph_3P + ClCH_2COMe \longrightarrow Ph_3\overset{+}{P}CH_2COMeCl^-$

$$\xrightarrow{Na_2CO_3} Ph_3P{=}CHCOMe \xrightarrow{PhCHO} Ph_3PO + PhCH{=}CHCOMe$$

(vi) $PhCO(CH_2)_4Br \xrightarrow{Ph_3P} PhCO(CH_2)_4\overset{+}{P}Ph_3Br^-$

$$\xrightarrow[EtOH]{EtONa} PhC(=O){-}(CH_2)_3{-}CH{=}PPh_3 \longrightarrow \text{PhC=CH–CH}_2\text{–CH}_2\text{–CH}_2 \text{ (ring)} + Ph_3PO$$

Five-membered rings are formed readily, and this is the final product since its precursor contains *both* the carbonyl group and the alkylidenephosphorane fragment suitably placed with resepct to each other to form the ring compound.

(vii) $PhN{=}O + Ph_3P{=}CHPh \longrightarrow PhN{=}CHPh + Ph_3PO$

(viii) $Et_2PCl + Et_2PNa \longrightarrow Et_2P{-}PEt_2 + NaCl$

This product is a diphosphine derivative.

(ix) $Ph_3P + Et_3\overset{+}{N}{-}\overline{O} \longrightarrow Et_3N + Ph_3PO$

Triphenylphosphine removes oxygen from many types of oxide.

(x) $R^1_3P + R^2_2SO \longrightarrow R^2_2S + R^1_3PO$

(xi) $Ph_3P{=}CH_2 + MeCH{=}CHCOMe \longrightarrow MeCH{=}CHCMe{=}CH_2 + Ph_3PO$

(xii) $(EtO)_3P + Cl_2 \longrightarrow ClPO(OEt)_2 + EtCl$

This is a special case of the Arbusov reaction (see *Text*).

(xiii) $(EtO)_3P + Me_3\overset{+}{N}{-}\overline{O} \longrightarrow (EtO)_3PO + Me_3N$

Phosphites are powerful reducing agents.

(xiv) $Me_2SO + Ph_3P \longrightarrow Me_2S + Ph_3PO$

(xv) $Ph_4P^+I^- + PhLi \longrightarrow Ph_5P + LiI$

7. See *Text* (not worth learning by heart).

(i) Tetrasilane; (ii) Methylsilanediamine; (iii) Ethylsilanetriol;

(iv) Dimenthyldichlorosilane; (v) Acetoxydiethylsilane;

(vi) 8-Ethoxypentasiloxanamine (numbering starts at the amino-group; the chain consists of oxygen and silicon atoms);

(vii) 3-Ethoxy-2-methyltrisilazane.

8. See the remarks made in Prob. 5, above.

(i) $SiCl_4 + 4n\text{-}PrMgBr \longrightarrow n\text{-}Pr_4Si + 4MgBrCl$

(ii) $SiCl_4 + MeMgI \longrightarrow MeSiCl_3 \xrightarrow{LAH} MeSiH_3$

(iii) $2Et_3SiCl + 2Na \longrightarrow Et_3Si\text{—}SiEt_3 + 2NaCl$

(iv) $Me_4Si \xrightarrow[Cl_2]{1\ mol.} Me_3SiCH_2Cl$

(v) $Me_3SiCl + 2Et_2NH \longrightarrow Me_3SiNEt_2 + Et_2NH_2^+Cl^-$

(vi) $Me_3SiNHMe + Me_3SiCl \longrightarrow Me_3SiNMeSiMe_3 + HCl$

(vii) $Me_3SiCl + EtMgI \longrightarrow Me_3SiEt + MgClI.$

9. (i) $MeCH{=}CHMe + Me_2SiHCl \xrightarrow{cat.} MeCH_2CHMeSiMe_2Cl$

This is an example of hydrosilation.

(ii) $2CH_2{=}CHCH_2Cl + Si \xrightarrow{Cu} (CH_2{=}CHCH_2)_2SiCl_2$

This is a free-radical reaction and probably proceeds via $CH_2{=}CHCH_2Cu$ (see *Text*).

(iii) $Me_3SiCl + BrCl \longrightarrow BrCH_2SiMe_2Cl + HCl$

BrCl is a brominating reagent.

(iv) $Me_3SiCl \xrightarrow{H_2O} Me_3SiOH \xrightarrow[-H_2O]{2\ mol.} Me_3SiOSiMe_3$

(v) $Et_3SiCl + Me_3SiOH \longrightarrow Et_3SiOSiMe_3 + HCl$

(vi) $Me_3SiH + NaOH \xrightarrow{H_2O} Me_3SiOH + H_2$

(vii) $Me_3SiCH_2CHClCH_3 + OH^- \longrightarrow Me_3SiOH + MeCH{=}CH_2 + Cl^-$

10.

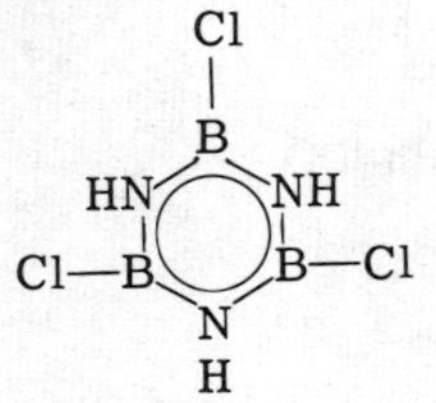

(i)

$HB(OMe)_2$ (ii)

Me_2BOMe (iii)

$MeB(OEt)_2$ (iv)

$B(OMe)_3$ (v)

$(EtO)BH(OMe)$ (vi)

CH_3COOBH_2 (vii)

11. The accepted structure of diborane has a bridged structure.

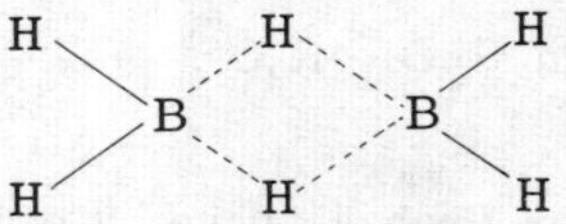

Since the bridged hydrogen atoms cannot be replaced, penta- and hexa-alkyldiboranes cannot be prepared (see *Text*). The five possible methyldiboranes are:

$MeBH_2BH_3$, Me_2BHBH_3, $MeBH_2BH_2Me$, Me_2BHBH_2Me, $Me_2BHBHMe_2$.

12. (i) $B(OMe)_3 + 3EtMgBr \longrightarrow Et_3B + 3MgBrOMe$

(ii) $3PhMgBr + BF_3 \longrightarrow Ph_3B + 3MgBrF$

(iii) $Et_3B + Br_2 \longrightarrow Et_2BBr + EtBr$

Compare this with the reaction of Et_4Si.

(iv) $2(MeO)_3B + NaH \longrightarrow (MeO)_2BH + NaB(OMe)_4$

(v) $Me_3B + HCl \longrightarrow Me_2BCl + CH_4$

(vi) $EtBCl_2 + 2MeOH \longrightarrow EtB(OMe)_2 + 2HCl$

(vii) $B(OMe)_3 + 2EtMgBr \longrightarrow Et_2BOMe \xrightarrow{H_2O} Et_2BOH$

(viii) $MeB(OH)_2 \xrightarrow{\text{heat}} (MeBO)_3$

13. (i) Since the substrate is an alcohol, we can expect the formation of a carbonium ion in the presence of acid.

$$EtSCHMeCH_2OH \xrightarrow[\text{(ii) } -H_2O]{\text{(i) } H^+} EtSCHMeCH_2^+$$

This is a primary carbonium ion and so we might have expected that a 1,2-methyl shift would occur to give the more stable secondary carbonium ion. Thus:

$$EtSCH(Me)CH_2^+ \longrightarrow EtS\overset{+}{C}HCH_2Me \xrightarrow{Cl^-} EtSCHClCH_2Me$$

This, however, is *not* the observed product. We must therefore reconsider the situation. A rearrangement has definitely occurred, and if we look at the product (more carefully than before), we see that the EtS group ends up on the CH_2 group. Since a methyl shift gives the wrong answer, and since we know that S can act as a donor atom, a possible mechanism can be formulated on the basis of neighbouring group participation as follows (see *Text*, Ch. 5; also Ch. 17):

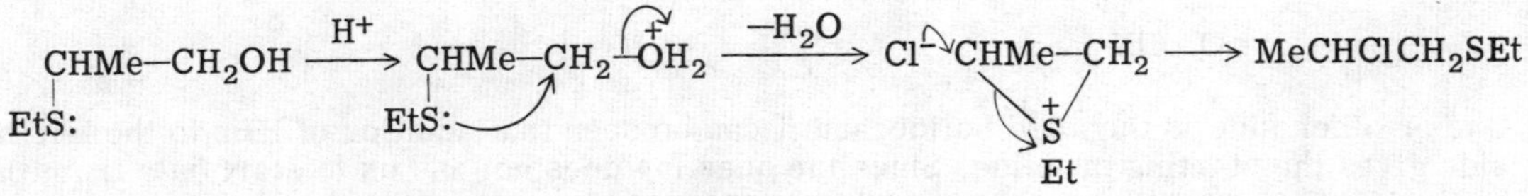

This is the observed product, but in view of the fact that the more sterically hindered carbon atom is attacked by the chloride ion, this requires an explanation. A reasonable argument is that since Me has a +I effect, electron-transfer of the C—S bond is easier from CHMe (as shown) than from the CH_2—S bond. In this case, the inductive effect outweighs the opposing steric effect.

(ii) Since we have been told that the mechanism is not free-radical, we must look for an ionic mechanism. If the phosphorus atom (a donor) attacked an alkyl group, we get a quaternary phosphorus ion. Thus:

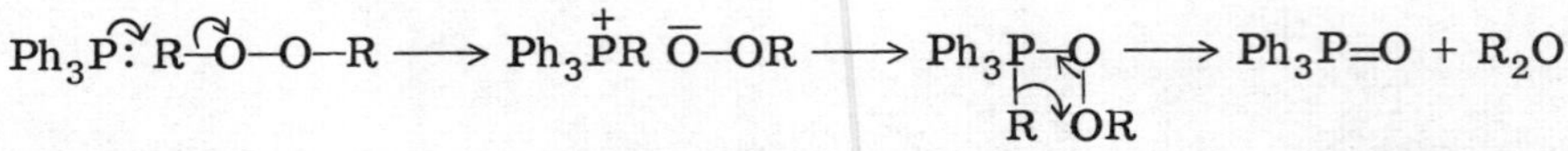

This mechanism bears a formal resemblance to that proposed in Prob. 4, above. It could be argued that there is also a formal resemblance to the reaction:

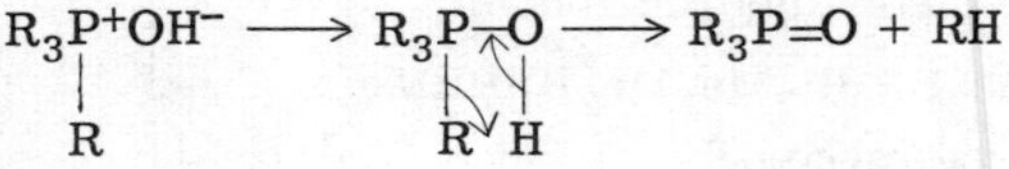

This latter reaction is known to occur, but its mechanism is uncertain.

We can also propose an alternative mechanism in which the phosphine attacks an oxygen atom directly. Thus:

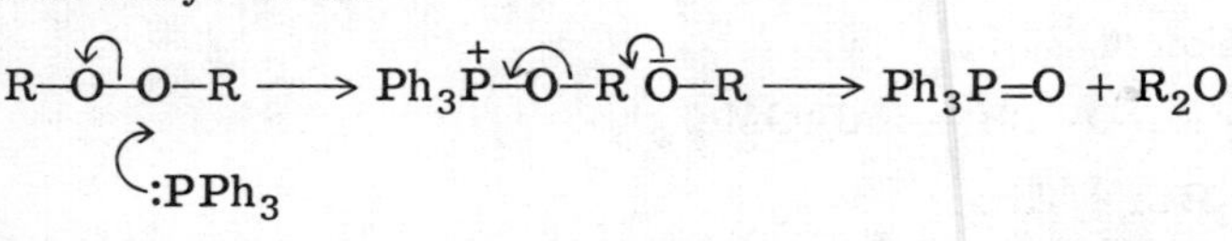

It appears from the literature that only the second mechanism has been proposed.

14. (i) The presence of a double bond should always recall the possible use of an alkylidenephosphorane if one fragment of the alkene product has been added to the starting material. The question then is whether the added fragment is the phosphorane and the starting material the carbonyl compound, or vice versa. Since the phosphorane is prepared from an alkyl halide then, for the case under consideration, we have the following synthesis:

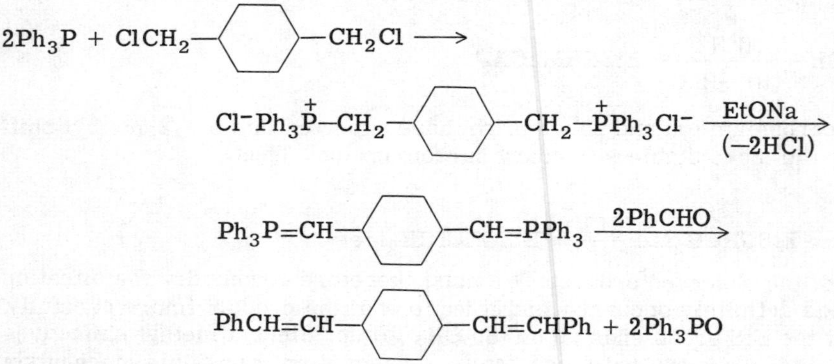

(ii)

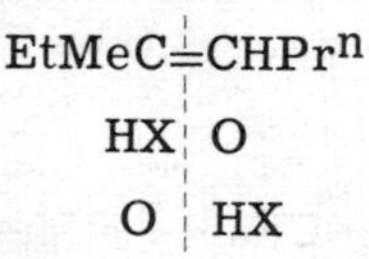

One or other side is the alkyl halide, and it can be seen that addition of HBr to the left-hand side gives the starting bromide. Since the question does not ask us to start *only* from the bromide, we are entitled to use any other compounds we want. Hence:

$$\mathrm{EtMeCHBr + Ph_3P \longrightarrow EtMeCH\overset{+}{P}Ph_3Br^- \xrightarrow{EtONa}}$$

$$\mathrm{EtMeC{=}PPh_3 \xrightarrow{Pr^nCHO} EtMeC{=}CHPr^n + Ph_3PO}$$

Had we been expected to start from the bromide only (and this would have been clearly stated), it means we have also to convert the bromide into n-butyraldehyde. This may be done in several ways, e.g. (see *Text*, Ch. 4):

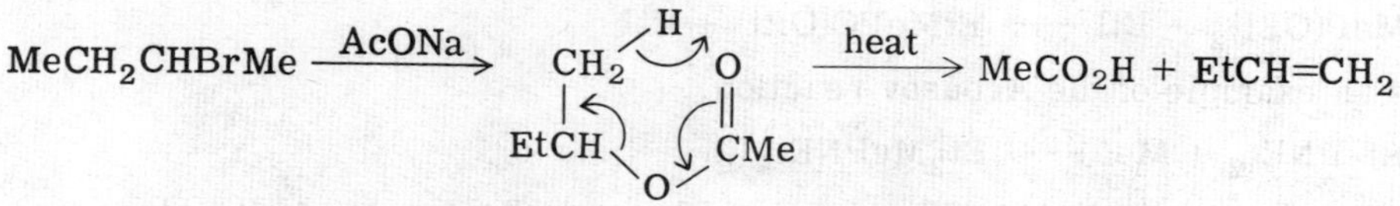

$$\xrightarrow[\text{peroxide}]{HBr} EtCH_2CH_2Br \xrightarrow{OH^-}$$

$$EtCH_2CH_2OH \xrightarrow[\text{dichr.}]{\text{acid}} EtCH_2CHO$$

The major product of the pyrolysis of acetates of secondary alcohols is the 1-alkene. In the presence of peroxides, addition of HBr gives the anti-Markownikoff product.

15. (i) $Ph_2CO + Ph_3P{=}CHOMe \longrightarrow Ph_3PO + Ph_2C{=}CHOMe$

$\xrightarrow{H^+} [Ph_2C{=}CHOH] \longrightarrow Ph_2CHCHO$

The unstable enol tautomerises to the more stable keto form.

(ii) 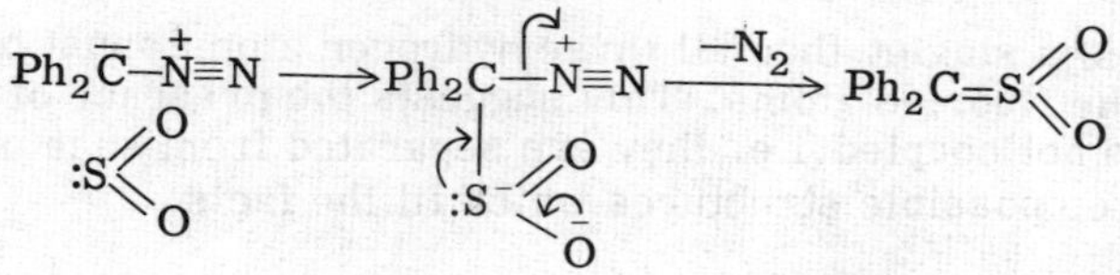

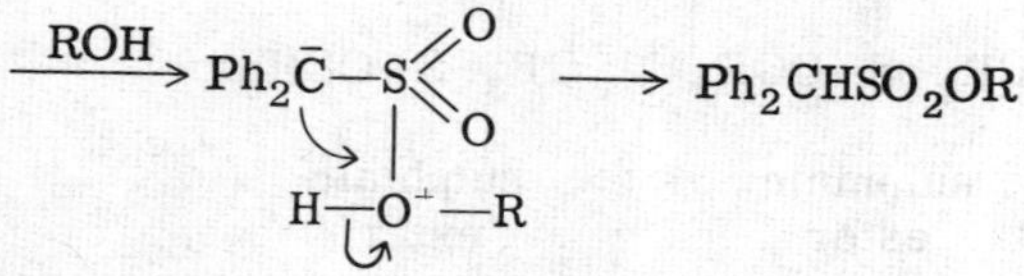

(iii) 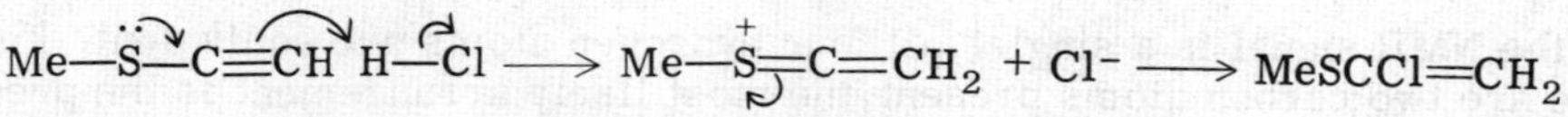

(iv) $(R^1O)_3P + R^2COCl \longrightarrow R^2COPO(OR^1)_2 + R^1Cl$

The product is a ketophosphonate, and the reaction is a special case of the Arbusov reaction (see *Text*).

(v) $Et_3B + Et_3N \longrightarrow$ no reaction

No complex is formed since steric effects (in the complex) make it too unstable.

(vi) $Et_3B + O_2(H_2O) \longrightarrow Et_2BOEt$

(vii) $B(OMe)_3 + 2EtMgBr \longrightarrow Et_2BOMe \xrightarrow{H_2O} Et_2BOH$

(viii) $EtB(OH)_2 \xrightarrow{\text{heat}} (EtBO)_3$

(ix) 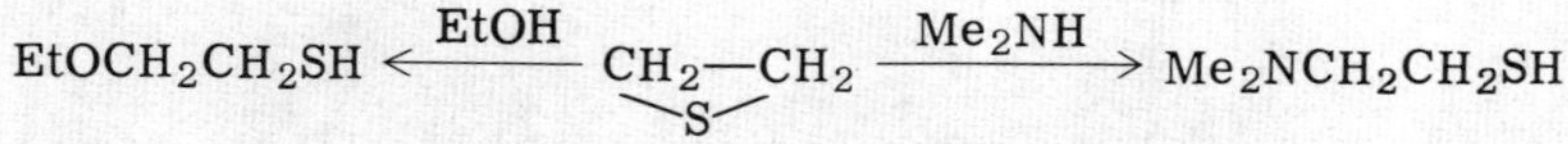

Compare these reactions with those of ethylene oxide (see *Text*, Ch. 11).

(x) 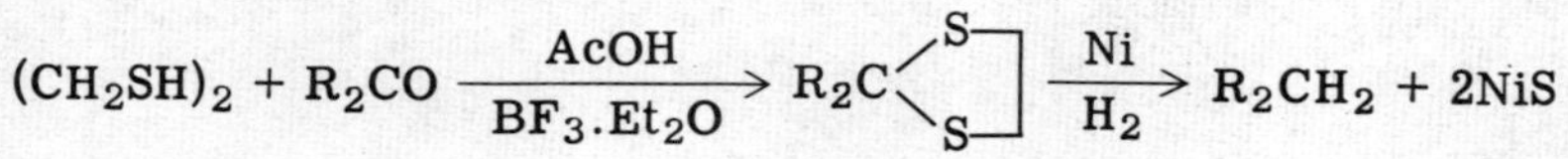

This is an example of desulphurisation.

(xi) $MeP(OEt)_2 + EtI \longrightarrow EtMePO(OEt) + EtI$

This is an example of the Arbusov reaction.

(xii) $Et_2PNEt_2 + MeI \longrightarrow Et_2Me\overset{+}{P}NEt_2I^-$

Phosphorus is attacked rather than nitrogen because the former is the stronger nucleophile.

(xiii) $R^1SSR^2 \xrightarrow[\text{Ni}]{\text{Raney}} R^1H + R^2H + 2NiS$

This is an example of desulphurisation; the hydrogen required is adsorbed on the surface during the preparation of Raney nickel (see *Text*, Ch. 3). Compare with (x), above.

(xiv) $3BCl_3 + 3EtNH_2 \xrightarrow{\text{heat}}$ 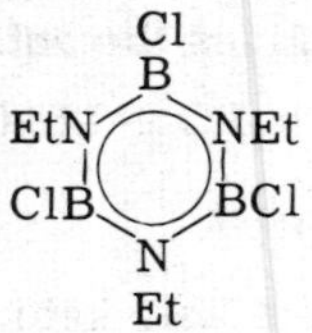

16. (i) Since the NMR signal is a singlet, then all three hydrogen atoms must be equivalent and cannot be coupled with the fluorine atoms. This suggests the presence of a CH_3 group and a CF_3 group which are not coupled, i.e., they are separated from each other by at least one atom. There are three possible structures which fit the facts:

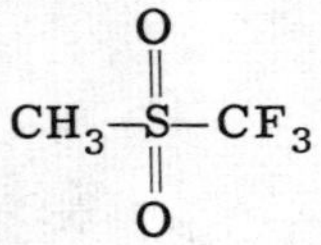

sulphone

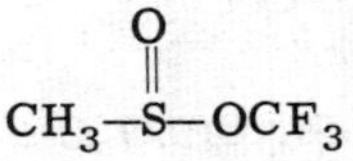

sulphinic ester

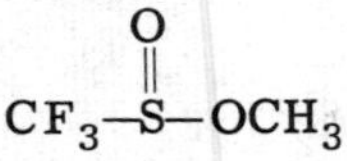

sulphinic ester

(ii) Since the NMR signal is a singlet, all four hydrogen atoms are equivalent. Furthermore, since there are two carbon atoms present, the most likely arrangement is the presence of a $-CH_2-CH_2-$ group which is symmetrically placed in the molecule. Only one structure, ethylene sulphate, satisfies the facts.

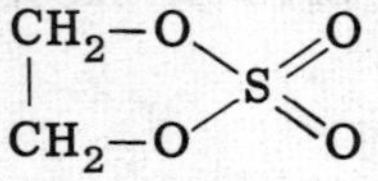

15 Organometallic compounds

(Problems on page 25)

1. Since the order of reactivity is usually: active hydrogen $\gg$ CHO $>>$CO $>$ CO_2R the equations will be:

$$HOCH_2CH_2COCH_2CH_2CO_2Et \xrightarrow{1RMgX}$$

$$RH + XMgOCH_2CH_2CO\ CH_2CH_2CO_2Et \xrightarrow[\text{(predominantly)}]{1RMgX}$$

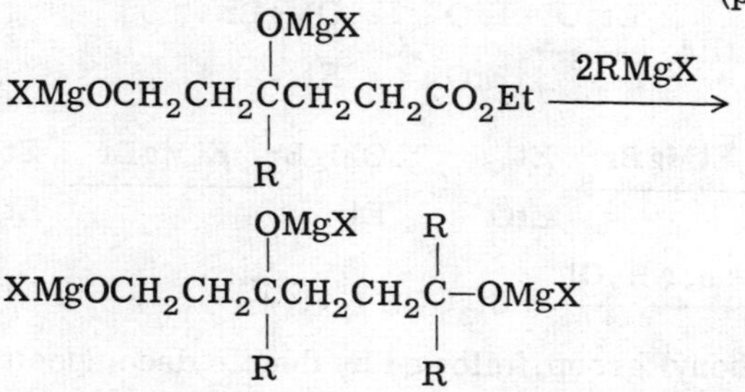

2. The most important functional group that reacts with Grignard reagents is the carbonyl group, and the addition may be represented (see *Text*):

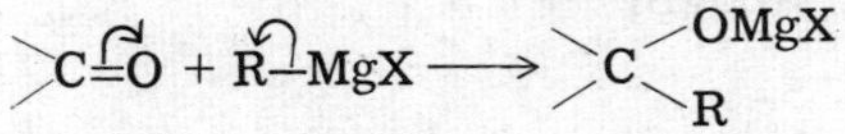

The general rule is that whatever the nature of the carbonyl group (i.e., $>$ CO, CO_2R, COCl), the final product is an oxygenated compound. This type of product, however, may also be obtained from cyanides:

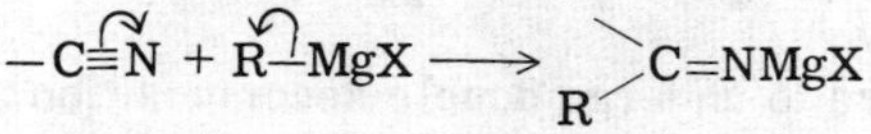

Another important type of reaction with Grignard reagents is double decomposition with compounds containing an active hydrogen atom, a reactive halogen atom, and a reactive alkoxyl group (in ortho-esters, i.e., compounds of the type $-C(OR)_3$). In general, if a carbonyl group is also present in the last two types of compound, this reacts first with the Grignard reagent.

(i) $n\text{-}C_4H_9MgBr \longrightarrow n\text{-}C_4H_9-CH_2CH_2CH_2OH$

The final product is a primary alcohol and contains three more carbon atoms than the starting material. Hence, we have to step up the alcohol series (the starting material is derived from a primary alcohol: $PrCH_2OH \longrightarrow PrCH_2Br \longrightarrow PrCH_2MgBr$).

(*a*) We can always try stepping up by *one* carbon atom at a time. Since the product is oxygenated, we can try a carbonyl compound, and because only one carbon atom is involved, the compound must be formaldehyde. Hence:

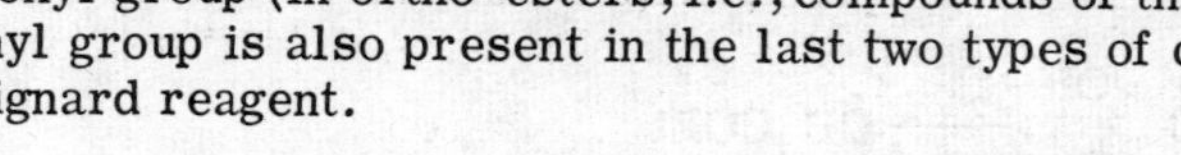

$$\xrightarrow{H^+} C_4H_9CH_2OH \xrightarrow[\text{(iii) } CH_2O \text{ (iv) } H^+]{\text{(i) HBr (ii) Mg}} C_5H_{11}CH_2OH$$

$$\xrightarrow{\text{repeat}} C_6H_{13}CH_2OH$$

(*b*) It often happens that stepping up a series can be done in fewer steps than by one carbon atom at a time, the method used depending on the nature of the series. Grignard reagents react with epoxides to form primary alcohols. For the problem in hand, we require trimethylene oxide.

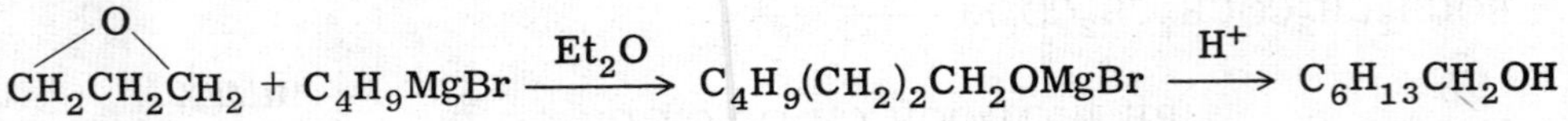

(ii)

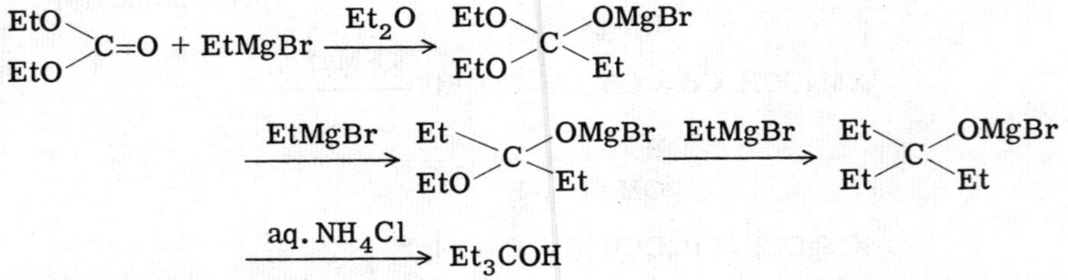

Note the addition to the carbonyl group, followed by double decomposition involving the ethoxy-groups.

(iii)

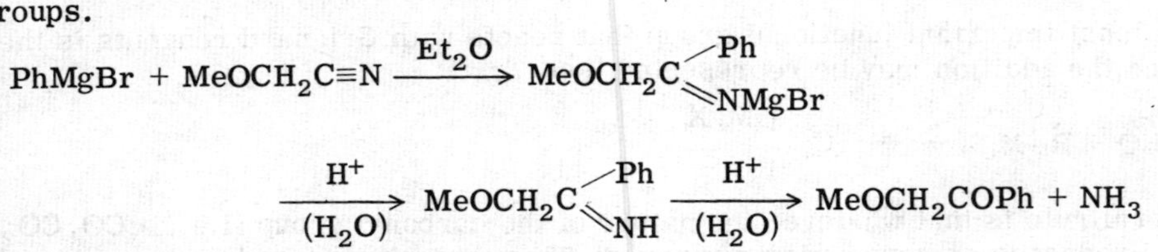

(iv) $$C_6H_{11}MgBr + BrCH_2CBr{=}CH_2 \xrightarrow{Et_2O} MgBr_2 + C_6H_{11}CH_2CBr{=}CH_2$$

Only the allylic bromine atom is sufficiently reactive to undergo double decomposition.

(v)

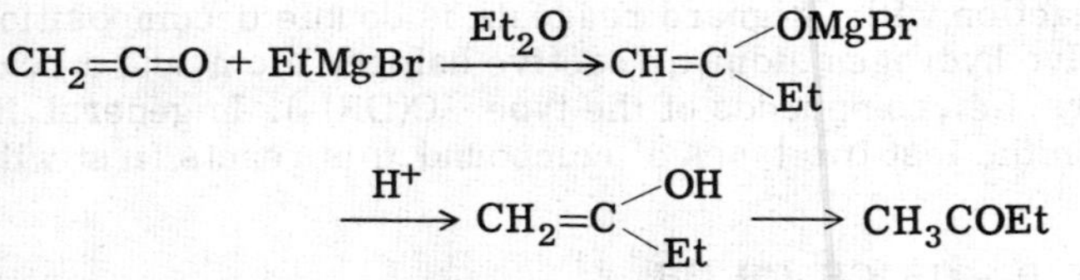

Addition occurs at the carbonyl group, and the enol tautomerises to the more stable keto-form.

(vi)

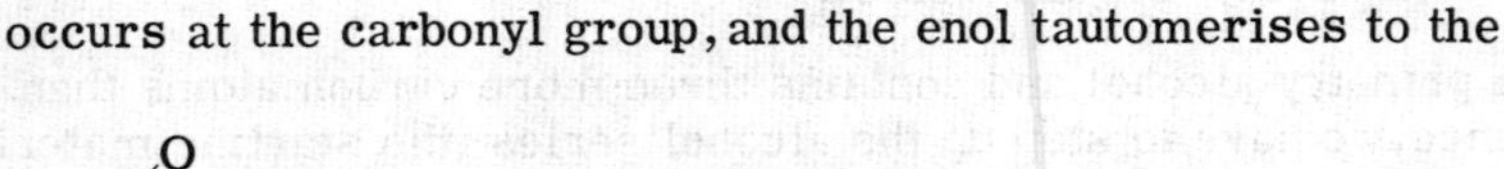

This direction of addition can be attributed to steric effects (see also (x), below).

(vii) The use of cuprous bromide suggests that the reaction proceeds by a free-radical mechanism, which is *known* to result in the formation of predominantly the 1,4-addition product. Hence:

$$\mathrm{EtCH{=}CHCOEt} \xrightarrow[\text{(ii) } \mathrm{H^+}]{\text{(i) EtMgBr + CuBr}} \mathrm{Et_2CHCH_2COEt}$$

The actual details of the mechanism are uncertain.

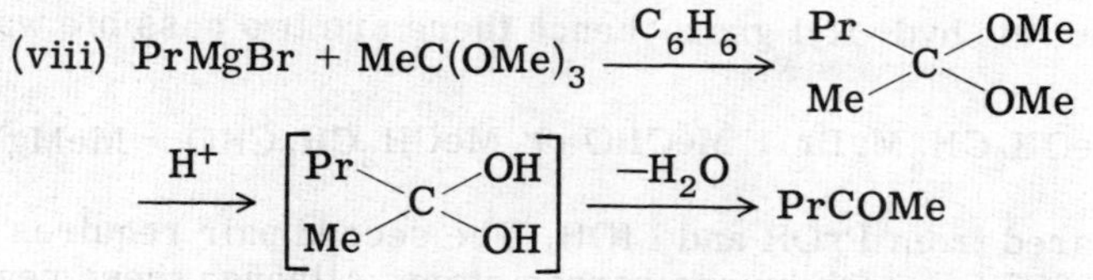

This is an example of double decomposition, and the use of methyl ortho-acetate offers a good method of preparing ketones containing the acetyl group.

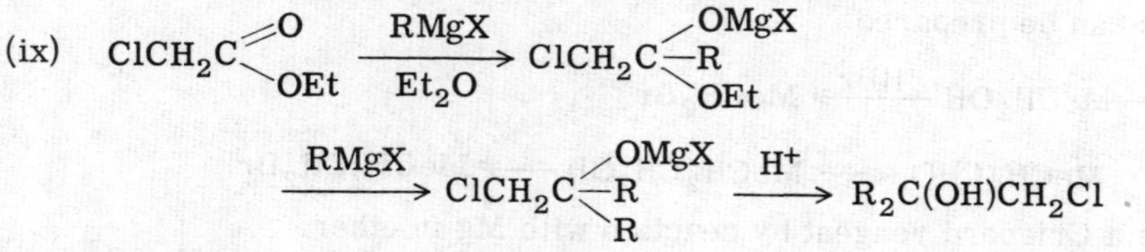

The attack occurs at the carbonyl group (this is more reactive than the chlorine atom), and the sequence is believed to be as shown (involving double decomposition).

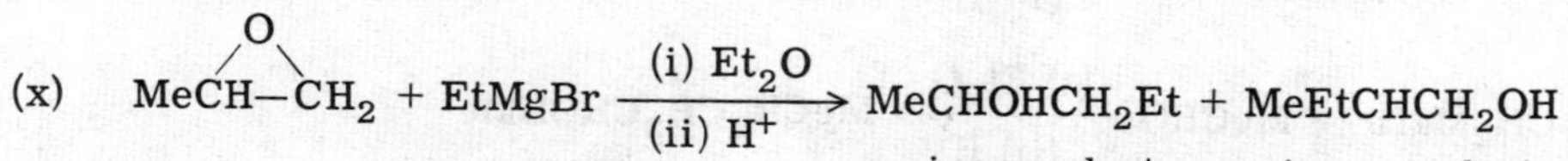

major product minor product

Steric effects operate, but since they are comparatively small, *both* products are formed (see also (vi), above).

$$\text{(xi)}\quad \mathrm{(CH_2)_5Br_2} \xrightarrow[\mathrm{Et_2O}]{\mathrm{2Mg}} \mathrm{(CH_2)_5(MgBr)_2} \xrightarrow[\text{(ii) } \mathrm{H^+}]{\text{(i) } \mathrm{2MeONH_2}} \mathrm{(CH_2)_5(NH_2)_2}$$

The di-Grignard reagent is formed only if $n \geqslant 4$ (see *Text*).

$$\text{(xii)}\quad \mathrm{MeCOCH_2CO_2Et + Me_2Cd} \xrightarrow[\text{(ii) } \mathrm{H^+}]{\text{(i) } \mathrm{Et_2O}} \mathrm{Me_2C(OH)CH_2CO_2Et}$$

Cadmium alkyls are less reactive than Grignard reagents and so are better to use for attacking a ketone group in a compound which also contains a carbethoxy group (the latter is also attacked, but much less than the ketone group, by Grignard reagents).

$$\text{(xiii)}\quad \mathrm{Et_2Cd + CCl_3CHO} \xrightarrow[\text{(ii) } \mathrm{H^+}]{\text{(i) } \mathrm{Et_2O}} \mathrm{CCl_3CHOHEt}$$

Compare with (ix), above.

3. Since alkenes may be prepared by the acid-catalysed dehydration of alcohols, a possible synthesis involves the formation of a suitable alcohol. Because pent-2-ene is unsymmetrical about the double bond, the addition of a molecule of water (working backwards) can be done in two ways.

$$\underset{\text{(I)}}{\mathrm{MeCH_2CHOHCH_2Me}} \longleftarrow \mathrm{MeCH_2CH{=}CHMe} \longrightarrow \underset{\text{(II)}}{\mathrm{MeCH_2CH_2CHOHMe}}$$

Now let us consider the preparations of (I) and (II).

(I) Since this is symmetrical about the hydroxyl group, there is only one way of preparing it (from EtOH and PrOH):

$$MeCH_2 \vdots CHOH \vdots CH_2Me \equiv MeCH_2MgBr + MeCH_2CHO$$

(II) This is not symmetrical about the hydroxyl group; hence there are two possible ways of synthesising it:

$$MeCH_2CH_2 \vdots CHOH \vdots Me \equiv MeCH_2CH_2MgBr + MeCHO \text{ or } MeCH_2CH_2CHO + MeMgBr$$

The first pair can be readily prepared from PrOH and EtOH. The second pair requires stepping up PrOH and stepping down EtOH, each by one carbon atom. Although these reactions may be carried out by standard procedures, there is no point in doing them, since the first pair, which is readily available, leads to the same alcohol.

The necessary pairs can be prepared:

$$MeCHO \xleftarrow[\text{dichr.}]{\text{acid}} MeCH_2OH \xrightarrow{HBr} MeCH_2Br$$

$$MeCH_2CHO \longleftarrow MeCH_2CH_2OH \longrightarrow MeCH_2CH_2Br$$

Each bromide forms a Grignard reagent by reaction with Mg in ether.

(I) $$MeCH_2CHO + MeCH_2MgBr \xrightarrow[\text{(ii) } H^+]{\text{(i) } Et_2O} MeCH_2CHOHCH_2Me$$

(II) $$MeCH_2CH_2MgBr + MeCHO \xrightarrow[\text{(ii) } H^+]{\text{(i) } Et_2O} MeCH_2CH_2CHOHMe$$

Acid-catalysed dehydration

(I) $$MeCH_2CHOHCH_2Me \xrightarrow[\text{(ii) } -H_2O]{\text{(i) } H^+} MeCH_2CH{=}CHMe$$

Only one alkene is produced.

(II) $$MeCH_2CH_2CHOHMe \xrightarrow[\text{(ii) } -H_2O]{\text{(i) } H^+} \underset{\text{major product}}{MeCH_2CH{=}CHMe} + \underset{\text{minor product}}{MeCH_2CH_2CH{=}CH_2}$$

Two products are possible, the major product being formed according to Saytzeff's rule (see *Text*, Ch. 5)

Thus, the best way of preparing pent-2-ene is via alcohol (I).

4. This involves stepping down the acid series by one carbon atom, and since no restrictions are mentioned we can use any methods we like.

(i) $$C_4H_9CH_2CO_2H \xrightarrow[\substack{\text{(ii) 2 PhMgBr} \\ \text{(iii) aq. } NH_4Cl}]{\text{(i) MeOH/HCl}} C_4H_9CH_2C(OH)Ph_2$$

$$\xrightarrow[(-H_2O)]{H_3O^+} C_4H_9CH{=}CPh_2 \xrightarrow{CrO_3} C_4H_9CO_2H + Ph_2CO$$

(ii) $$C_4H_9CH_2CO_2H \xrightarrow[\text{(ii) } Br_2/CCl_4]{\text{(i) Ag salt or HgO}} C_4H_9CH_2Br$$

$$\xrightarrow[\text{(ii) acid dichr.}]{\text{(i) OH}^-} C_4H_9CO_2H$$

See also Ch. 9, Prob. 1.

5. $$R^1CH{=}CHCOR^2 + R^3MgX \longrightarrow \underset{1,2\text{-}}{R^1CH{=}CHC(OH)R^3R^3} + \underset{1,4\text{-}}{R^1R^3CHCH_2COR^2}$$

The general rule is that steric effects cause a decrease in 1,2-addition. Hence, as R^2 changes from Et → isoPr → t-Bu, 1,4-addition increases in this order.

6. (i) $$CHD_2I \xrightarrow[Et_2O]{Mg} CHD_2MgI \xrightarrow[\text{(ii) } H_3O^+]{\text{(i) } CH_2O} CHD_2CH_2OH$$

(ii) Dehydration of this alcohol can occur in two ways:

$$CD_2{=}CH_2 \xleftarrow[\text{(ii) } -H_2O]{\text{(i) } H^+} CHD_2CH_2OH \xrightarrow[\text{(ii) } -H_2O]{\text{(i) } H^+} CHD{=}CH_2$$

Since deuterium shows a kinetic isotope effect, i.e., undergoes reaction more slowly than protium, the predominant product is $CD_2{=}CH_2$ (see *Text*, Ch. 2).

7. (i) The product contains two more carbon atoms than the starting material. This suggests the use of EtMgI. There are two reasonable routes:

(*a*) $$MeCO_2H \xrightarrow{PCl_5} MeCOCl \xrightarrow{EtMgI} MeCOEt$$

(*b*) $$MeCO_2H \xrightarrow{LAH} MeCH_2OH \xrightarrow[\text{dichr.}]{\text{acid}} MeCHO \xrightarrow[\text{(ii) acid dichr.}]{\text{(i) EtMgI}} MeCOEt$$

The difficulty with (*a*) is that t-alcohol is also formed, and so (*b*) may be better.

(ii) The product is a ketone which contains three more carbon atoms. Since we can use any compounds we like, some possible routes are:

(*a*) $$EtCHO + EtMgBr \longrightarrow EtCHOHEt \xrightarrow[\text{dichr.}]{\text{acid}} EtCOEt$$

(*b*) $$EtC{\equiv}N + EtMgBr \longrightarrow EtC(Et){=}NMgBr \xrightarrow{H_3O^+} EtCOEt$$

(iii) $$Me_2CHOH \xrightarrow{HCl} Me_2CHCl \xrightarrow[Et_2O]{Mg} Me_2CHMgCl \xrightarrow[\text{(ii) } H_3O^+]{\text{(i) } MeONH_2} Me_2CHNH_2$$

See *Text* for details.

(iv) $$MeCOEt \longrightarrow Me{-}CH(CHBrMe){-}Et$$

Direct reaction between a ketone and a Grignard reaction normally gives a t-alcohol. Inspection of the product shows that the ketone has been converted into a secondary alkyl group. This suggests reduction of the ketone to secondary alcohol. Also, if we assume that the Br has been derived from OH (action of HBr), the product can be broken down into the fragments shown.

MeCHOHEt ⋮ CH(OH)Me ≡ MeCHBrEt (Grignard) ⋮ OCHMe (aldehyde)

A possible synthesis is:

$$\text{MeCOEt} \xrightarrow{\text{NaBH}_4} \text{MeCHOHEt} \xrightarrow[\text{(ii) Mg}]{\text{(i) HBr}} \text{MeEtCHMgBr}$$

$$\xrightarrow[\text{(ii) H}_3\text{O}^+]{\text{(i) MeCHO}} \underset{\text{CHOHMe}}{\underset{|}{\text{MeCHEt}}} \xrightarrow{\text{HBr}} \underset{\text{CHBrMe}}{\underset{|}{\text{MeCHEt}}}$$

8. (i) $4PhLi + (CH_2{=}CH)_4Sn \longrightarrow 4CH_2{=}CHLi + Ph_4Sn$

Because of the ease with which the vinyl group is removed, this reaction may be used to prepare vinyl derivatives of metals more electropositive than tin.

(ii) $2EtNa + Hg \longrightarrow$ no reaction

Metal exchange occurs only between a metal higher in the electrochemical series and an organo compound of a metal lower in the series.

(iii) $BuLi + MeONH_2 \longrightarrow BuNHLi \xrightarrow{H_2O} BuNH_2$

Compare the analogous reaction with Grignard reagents.

(iv) $Me_2CHLi + C_2H_4 \longrightarrow Me_2CHCH_2CH_2Li$

$$\xrightarrow[\text{(ii) H}_2\text{O}]{\text{(i) CO}_2} Me_2CHCH_2CH_2CO_2H$$

Some $Me_2CH(CH_2)_4CO_2H$ is also formed. This type of reaction is the main difference between organolithium compounds and Grignard reagents.

(v) $Me_3CCl + Me_2Zn \longrightarrow Me_4C + MeZnCl$

The yield of neopentane is better than when a Grignard reagent is used. The method is satisfactory only with t-alkyl halides, but organozinc compounds are difficult to handle.

(vi) $2EtI + Na_2/Hg \longrightarrow Et_2Hg + 2NaI$

Mercury, in the form of its amalgam, is extremely reactive.

(vii) $Et_2Hg + 2K \longrightarrow 2Et^-K^+ + Hg$

See the remarks in (ii), above.

(viii) $4RMgCl + SnCl_4 \longrightarrow R_4Sn + 4MgCl_2$

Mg is higher in the electrochemical series than Sn and so this exchange occurs (see *Text*).

(ix) $Me_2Hg + Zn \longrightarrow Me_2Zn + Hg$

See (ii), above.

(x) $2CH_2(CN)COCl + Et_2Cd \longrightarrow 2CH_2(CN)COEt + CdCl_2$

Cadmium dialkyls react very slowly with the carbonyl group and not at all with the cyano-group. This offers a means of preparing ketones in high yield from acid chlorides (*cf.* with Grignard reagents).

(xi) $3Ph_2Hg + 2Al \xrightarrow{\text{heat}} 2Ph_3Al + 3Hg$

See (ii), above.

(xii) $CH_2{=}CHCOMe \xrightarrow{Me_2SnH_2} CH_2{=}CHCHOHMe$

(xiii) $R_2C{=}O + BrCH_2CH{=}CHCO_2Et \xrightarrow{Zn}$

$[R_2C{=}O + BrZnCH_2CH{=}CHCO_2Et] \longrightarrow$

$R_2C(OZnBr)CH_2CH{=}CHCO_2Et \xrightarrow{H^+}$

$R_2C(OH)CH_2CH{=}CHCO_2Et$

Normally, α-bromo-esters are used in the Reformatsky reaction, and the organozinc compound does not react with the carbethoxy group (*cf.* with Grignard reagents). In the bromoester given in the equation, the bromine atom is sufficiently active for reaction to occur; it is an example of vinylogy (see *Text*, Ch. 12).

(xiv) $EtMgX \xrightarrow{PbCl_2} Et_4Pb$

Note the formation of tetra-alkyl-lead from Pb(II) chloride

(xv) $(PhCH_2)_4Sn \xrightarrow{heat} 4PhCH_2\cdot + Sn$

The generation of free radicals is characteristic of many metal alkyls.

(xvi) $2MeI + Zn/Cu \longrightarrow Me_2Zn + Cu$

Compare with (vi), above.

(xvii) $EtLi + Br_2 \longrightarrow EtBr + LiBr$

9. $$Me_2CHNO_2 \xrightarrow{OH^-} Me_2\bar{C}{-}\overset{+}{N}\begin{matrix}{=}O\\ {-}\bar{O}\end{matrix} \longleftrightarrow Me_2C{=}\overset{+}{N}\begin{matrix}{-}\bar{O}\\ {-}\bar{O}\end{matrix}$$

$$\xrightarrow{Br_2} Br^- + \underset{(A)}{Me_2CBrNO_2} \xrightarrow{Me_2Zn} \underset{(B)}{Me_3CNO_2}$$

Removal of a proton by hydroxide ion from 2-nitropropane is made possible by the strong —I effect of the nitro-group (active hydrogen). Zinc alkyls replace halogen but have no effect on the nitro-group. On the other hand, Grignard reagents tend to reduce a nitro-group (see *Text*).

10. The simplest way to solve this problem is to try to formulate the steps on the basis that each step goes 'according to plan'. Should this fail, it is most likely that any unusual step (or steps) will be identified and this will give some idea of what to try next.

$$EtMeCHCOPh \xrightarrow[\text{(ii) aq. } NH_4Cl]{\text{(i) } PhMgBr} \underset{(A)}{EtMeCHC(OH)Ph_2}$$

$$\xrightarrow[(-H_2O)]{H^+} \underset{(B)}{EtMeC{=}CPh_2} \xrightarrow{O_3} \underset{(C)}{EtCOMe} + \underset{(D)}{Ph_2CO}$$

$$EtCHO \xrightarrow{CH_2N_2 (E)} (C)$$

11. $$R_2C{=}O + Et_3Al \longrightarrow R_2C\begin{matrix}{-}H\\ {-}OAlEt_2\end{matrix} + C_2H_4$$

Ethylene has been derived from one ethyl group by loss of a hydrogen atom. This can be formulated as:

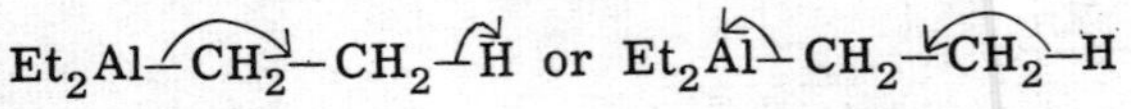

(I) (II)

Since the group Et_2Al is in the product, it is reasonable to argue that the hydrogen atom transferred to the CO group to give $C\begin{smallmatrix}H\\O-\end{smallmatrix}$ is the one lost in (I) or (II). Furthermore, since the CO group polarises $C{=}O$, the carbon atom acquires a positive charge and so the added hydrogen atom can be expected to be in the form of a hydride ion. This indicates the electron shifts are those in (I), and is supported by the fact that this leaves the Al with a positive charge and so the aluminium atom can readily combine with the negatively charged oxygen atom. We now have worked out the way the reaction is going to proceed, but this leaves us with one other question: Is the reaction one-step or two-step? Now, a one-step mechanism is readily interpreted in terms of a cyclic T.S., and it can be seen that this will be six-membered, which is a very stable type, i.e.,

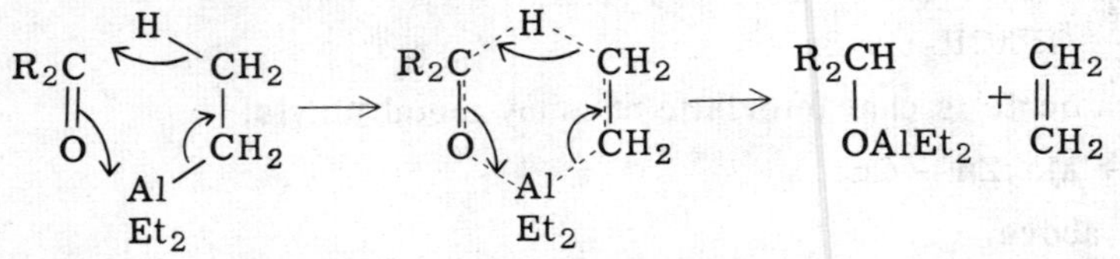

Many students, when they read the question, may immediately have thought of the M.P.V. reduction, and remembering that the accepted mechanism involves a cyclic T.S., would have argued by analogy that the reaction under consideration proceeds in a like manner. This, of course, will lead to the same answer in a shorter time. The object of the approach described above has been to give the student some idea how to tackle a problem that may not appear to lend itself to argument by analogy.

12. (i) The preparation of a β-hydroxy-ester from a carbonyl compound suggests the use of the Reformatsky reaction. The difficulty with this method, apart from often obtaining poor yields, has been the hydrolysis of the ester to acid. The yield is usually low when carried out with alkali followed by acid. Several by-products are obtained, e.g., the α,β-unsaturated acid (by dehydration).

$$Me_2CO + BrCH_2CO_2Et \longrightarrow Me_2C(OZnBr)CH_2CO_2Et$$

$$\xrightarrow{H_3O^+} Me_2C(OH)CH_2CO_2Et \xrightarrow[\text{(ii) } H_3O^+]{\text{(i) } OH^-} Me_2C(OH)CH_2CO_2H$$

$$+ Me_2C{=}CHCO_2H$$

Improved yields of hydroxy-acid have been obtained by various modifications (see *Text*).

(ii) The first thing to note is that the product contains a branched chain, which is isobutyl. The functional group is aldehyde, and since this may be prepared by oxidation of a primary alcohol, a possible synthesis is to prepare isobutanol.

$$Me_2CH \mid CH_2OH \equiv Me_2CHMgBr + CH_2O$$

The carbon skeleton of isopropyl is the same as that of acetone. Now, one way of preparing acetone is to heat calcium acetate, and since ethanol is readily oxidised to acetic acid, we now have a synthetic route.

$$MeCH_2OH \xrightarrow[\text{dichr.}]{\text{acid}} MeCO_2H \xrightarrow[\text{heat}]{\text{Ca salt}} Me_2CO \xrightarrow{NaBH_4} Me_2CHOH \xrightarrow[\text{(ii) Mg}]{\text{(i) HBr}}$$

$$Me_2CHMgBr \xrightarrow[\text{(ii) } H_3O^+]{\text{(i) } CH_2O} Me_2CHCH_2OH \xrightarrow[\text{dichr.}]{\text{acid}} Me_2CHCHO$$

Acid dichromate oxidises primary alcohols to aldehydes or acids according to the conditions used. LAH may be used instead of sodium borohydride, but the former is more difficult to handle.

An alternative route is to oxidise the butanol to the corresponding acid and then (see Table 6.1):

$$Me_2CHCO_2H \xrightarrow{SOCl_2} Me_2CHCOCl \xrightarrow[-78°C]{LiAlH(OBu^t)_3} Me_2CHCHO$$

The acid may also be prepared as follows:

$$Me_2CHMgBr \xrightarrow[(ii)\ H_3O^+]{(i)\ CO_2} Me_2CHCO_2H$$

(iii) $Me_3COH \longrightarrow Me_3C \vdots CH_2CH{=}CH_2 \equiv Me_3CMgCl + ClCH_2CH{=}CH_2$

Since allyl halides are reactive, they undergo double decomposition with Grignard reagents. Hence:

$$Me_3COH \xrightarrow[(ii)\ Mg]{(i)\ HCl} Me_3CMgCl \xrightarrow{ClCH_2CH=CH_2} Me_3CCH_2CH{=}CH_2$$

(iv) This problem is similar to (ii), above, but in this case we have $MeCHO \longrightarrow EtMeCHMgBr$. Hence, we have to convert MeCHO into MeCHOHEt.

$$MeCHO \xrightarrow[(ii)\ H_3O^+]{(i)\ EtMgBr} MeCHOHEt \xrightarrow[(ii)\ Mg]{(i)\ HBr} Me{-}CHMgBr{-}Et$$

Then we proceed as in (ii), above.

(v) $HO(CH_2)_4OH \longrightarrow Me_2C(OH)(CH_2)_4C(OH)Me_2$

It can be seen that the product contains two tertiary alcoholic groups and an extra isopropyl group at each end of the chain. These facts suggest a Grignard reagent and a ketone, acetone, since these give tertiary alcohols. Since the two bromine atoms are separated by *four* carbon atoms, this dibromide can form a di-Grignard reagent (see Prob. 2(xi), above). Hence:

$$HO(CH_2)_4OH \xrightarrow[(ii)\ 2Mg]{(i)\ HBr} BrMg(CH_2)_4MgBr$$

$$\xrightarrow[(ii)\ aq.\ NH_4Cl]{(i)\ 2Me_2CO} Me_2C(OH)(CH_2)_4C(OH)Me_2$$

Because the product is a tertiary alcohol, it is better to play safe and use aqueous ammonium chloride to decompose the complex (acid may cause dehydration).

16 Saturated dicarboxylic acids

(Problems on page 28)

1. See Ch. 3, Prob. 2; Ch. 5, Prob. 1.

$C_6H_{10}O_4 = C_4H_8(CO_2H)_2$. Thus, we are dealing with the isomeric dicarboxy-derivatives of (i) butane, and (ii) isobutane.

$MeCH_2CH_2CH(CO_2H)_2$ butane-1, 1-dicarboxylic acid; 2-propylpropanedioic acid.

$MeCH_2C(CO_2H)_2Me$ butane-2, 2-dicarboxylic acid; 2-ethyl-2-methylpropanedioic acid.

$MeCH_2CH(CO_2H)CH_2CO_2H$ butane-1, 2-dicarboxylic acid; 2-ethylbutanedioic acid.

$MeCH(CO_2H)CH_2CH_2CO_2H$ butane-1, 3-dicarboxylic acid; 2-methylpentanedioic acid.

$HO_2CCH_2CH_2CH_2CH_2CO_2H$ butane-1, 4-dicarboxylic acid; hexanedioic acid.

$MeCH(CO_2H)CH(CO_2H)Me$ butane-2, 3-dicarboxylic acid; 2, 3-dimethylbutanedioic acid.

$Me{-}CH(Me)CH(CO_2H)_2$ 2-isopropylpropanedioic acid; 2-methylpropane-1, 1-dicarboxylic acid.

$MeC(Me)(CO_2H)CH_2CO_2H$ 2, 2-dimethylbutanedioic acid.

$HO_2CCH_2CH(Me)CH_2CO_2H$ 3-methylpentanedioic acid.

2. (i) $(CH_2CO)_2O + MeOH \xrightarrow{\text{heat}} HO_2CCH_2CH_2CO_2Me$

The 'half-ester' is the main product under these conditions. With excess of MeOH, the di-ester is obtained.

(ii) 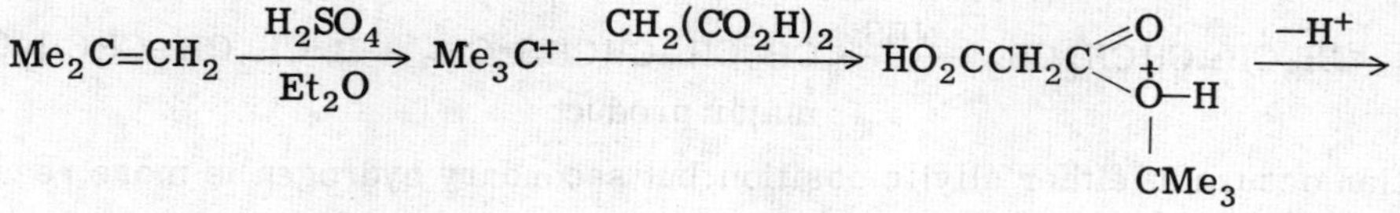

$$HO_2CCH_2CO_2CMe_3 \xrightarrow[\text{(ii) }-H^+]{\text{(i) }Me_3C^+} CH_2(CO_2CMe_3)_2$$

The alkylating agent is the t-butyl carbonium ion.

(iii) $$HO_2C(CH_2)_4CO_2Me \xrightarrow{AgNO_3} AgO_2C(CH_2)_4CO_2Me \xrightarrow[CCl_4]{Br_2} Br(CH_2)_4CO_2Me$$

This is an example of the Hunsdiecker reaction (see *Text*, Ch. 5).

(iv) 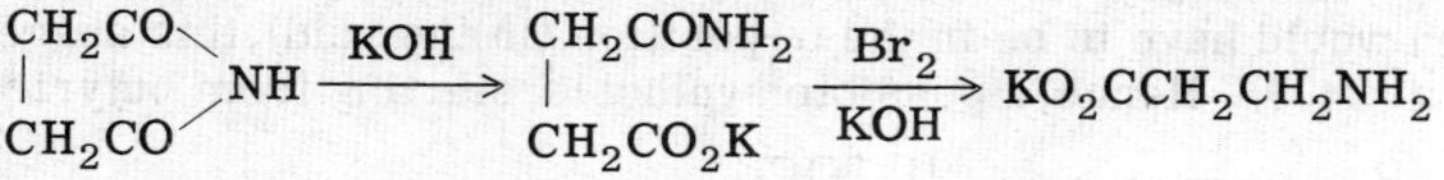

The imide is hydrolysed to the 'half-amide', which then undergoes the Hofmann reaction to form the amine (see *Text*, Ch. 9).

(v) 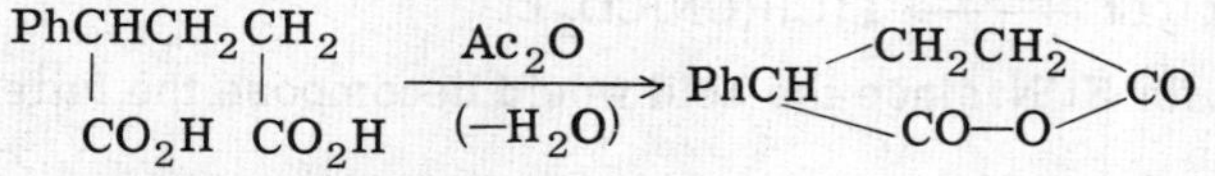

Cyclic anhydrides are readily formed from 1,4- and 1,5-dicarboxylic acids.

(vi) $$CH_2{=}CH(CH_2)_nCO_2R^1 \xrightarrow{O_3} [HO_2C(CH_2)_nCO_2R^1] \xrightarrow{R^2OH} R^2O_2C(CH_2)_nCO_2R^1$$

See ozonolysis (*Text*, Ch. 4).

(vii) 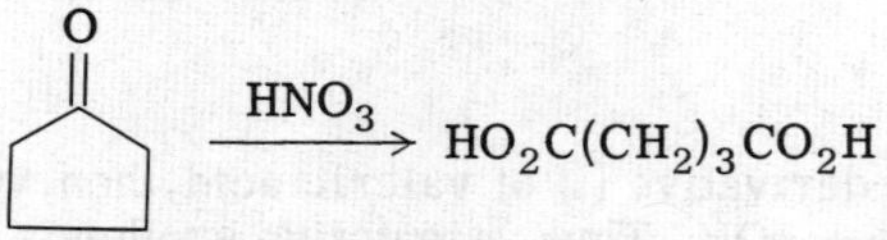

Ketones of the type RCH_2COCH_2R undergo oxidation to give a mixture of acids, RCH_2CO_2H and RCO_2H. Cyclic ketones give a dicarboxylic acid because the 'two R groups are joined together'.

(viii) $$Bu^tOH \xrightarrow[\text{(ii) }-H_2O]{\text{(i) }H^+} Bu^{t+} \xrightarrow{NH_2CONH_2} Bu^t{-}\overset{+}{N}H_2CONH_2 \xrightarrow{-H^+} Bu^tNHCONH_2$$

This type of reaction is characteristic for *tertiary* alcohols.

(ix) $$KNCS \xrightarrow[H_2O]{H_2SO_4} H{-}N{=}C{=}S$$

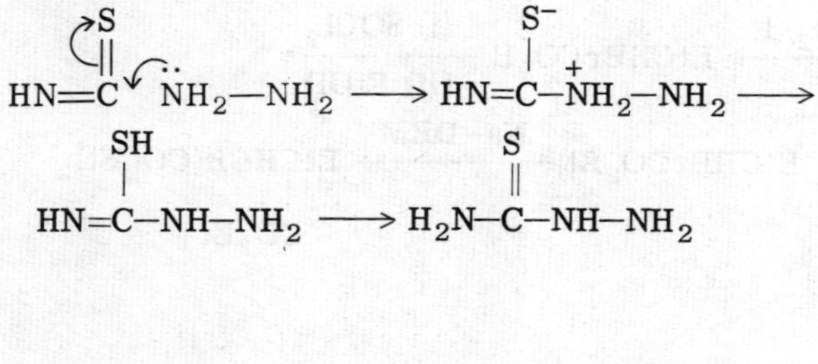

(x) $$CH_3CH{=}CHCH_2CH_3 \xrightarrow{NBS} \underset{\text{major product}}{CH_3CH{=}CHCHBrCH_3} + BrCH_2CH{=}CHCH_2CH_3$$

Substitution occurs at either allylic position, but secondary hydrogen is more readily substituted than primary.

(xi) $$HOCH_2(CH_2)_nCH_2OH \xleftarrow{LAH} HO_2C(CH_2)_nCO_2Et$$

$$\xrightarrow[\textit{or}\ Na/EtOH]{LiBH_4} HO_2C(CH_2)_nCH_2OH$$

Sodium borohydride in excess may reduce CO_2R to CH_2OH (see Table 6.1).

3. (i) Let us try working backwards. The ester can be prepared from its acid by esterification. A cyano-group can usually be introduced by reaction between alkyl halide and KCN. Since the halogen would have to be in the α-position (in the acid), this suggests the HVZ reaction (see *Text*, Ch. 9). Hence, a possible synthesis, starting from butyric acid, is:

$$EtCH_2CO_2H \xrightarrow{Br_2/P} EtCHBrCO_2H \xrightarrow[\text{(ii) EtOH}]{\text{(i) }SOCl_2}$$

$$EtCHBrCO_2Et \xrightarrow{KCN} EtCH(CN)CO_2Et$$

The ester is used for reaction with KCN, since the acid would decompose the latter (into HCN).

(ii) Here we have two carbethoxy-groups attached to nitrogen. A standard method for attaching CO_2R to N is the use of ethyl chloroformate in the presence of sodium.

$$EtNH_2 + 2ClCO_2Et + 2Na \xrightarrow{Et_2O} EtN(CO_2Et)_2 + 2NaCl + H_2$$

(iii) $$\underset{\underset{CO_2H}{|}}{EtCH}{-}CH_2CO_2H$$

(disconnection (ii) at the $CH{-}CH_2CO_2H$ bond; disconnection (i) at the $CH{-}CO_2H$ bond)

If we regard this compound as the β-carboxy-derivative (i) of valeric acid, then, working backwards, we have: $CO_2H \longleftarrow CN \longleftarrow Br \longleftarrow OH$. Thus, a tentative synthesis requires the use of β-hydroxyvaleric acid. This suggests the use of the Reformatsky reaction (see *Text*, Ch. 15). Hence:

$$EtCHO + BrCH_2CO_2Et + Zn \longrightarrow EtCH(OZnBr)CH_2CO_2Et \xrightarrow{H_3O^+} EtCHOHCH_2CO_2Et$$

$$\xrightarrow[\text{(ii) KCN}]{\text{(i) HBr}} \underset{\underset{CN}{|}}{EtCH}CH_2CO_2Et \xrightarrow{H_3O^+} \underset{\underset{CO_2H}{|}}{EtCH}CH_2CO_2H$$

An alternative approach is to regard the compound as containing an acetic acid residue joined to the α-carbon atom of butyric acid (ii). This suggests the use of the HVZ reaction and a DEM synthesis:

$$EtCH_2CO_2H \xrightarrow{Br_2/P} EtCHBrCO_2H \xrightarrow[\text{(ii) EtOH}]{\text{(i) }SOCl_2}$$

$$EtCHBrCO_2Et \xrightarrow{Na{-}DEM} \underset{\underset{CO_2Et}{|}}{EtCH}CH(CO_2Et)_2$$

$$\xrightarrow[\text{(iii) heat}]{\text{(i) KOH, (ii) } H_3O^+} EtCH(CO_2H)CH_2CO_2H$$

4. The presence of a nitro-group in the product and the presence of nitric acid in the starting material suggests that the mechanism involves the formation of the nitronium ion (see *Text*, Ch. 22).

$$HNO_3 + 2H_2SO_4 \rightleftharpoons NO_2^+ + H_3O^+ + 2HSO_4^-$$

5.

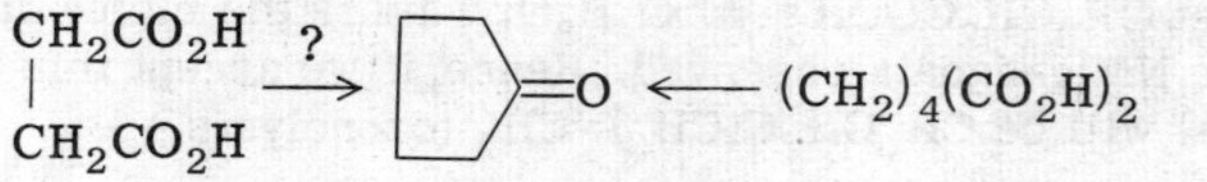

The solution to this problem is really a question of whether we know how to prepare cyclic ketones. Now, it has been mentioned in the *Text* that cyclopentanone may be prepared from adipic acid (hexanedioic acid). Hence, our problem is to convert succinic acid into adipic acid. This can be done by stepping up each end by one carbon atom by the following standard method.

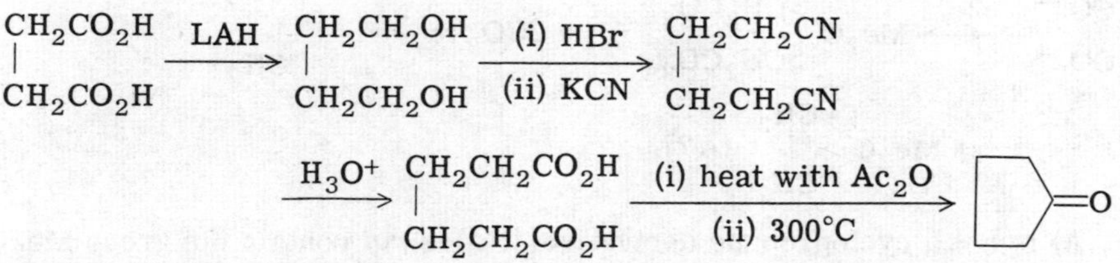

Alternatively, the calcium salt of adipic acid can be heated (see *Text*, Ch. 19).

6.

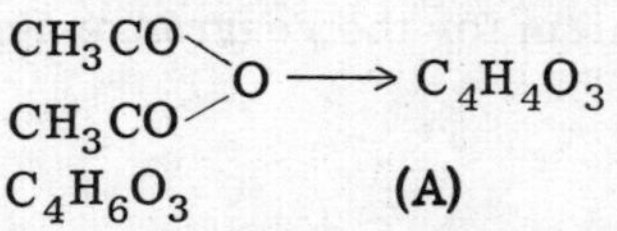

Since the i.r. bands occur in the regions 1 870–1 830 and 1 800–1 760, this suggests that (A) is a five-membered ring anhydride. On this basis, (A) could be succinic anhydride.

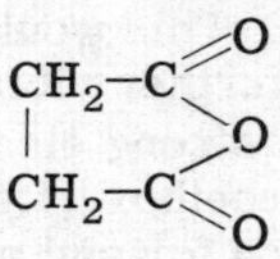

(A)

This is in keeping with:

D.B.E. $= 4 + 1 - 4/2 = 3$, i.e., two double bonds (C=O) and one ring (see Ch. 1, Prob. 9).

The formation of (A) from acetic anhydride under the influence of potassium permanganate involves the loss of two hydrogens and the formation of a C—C bond. One possibility is that the two hydrogens are removed as hydrogen atoms, thereby producing a diradical which then ring-closes:

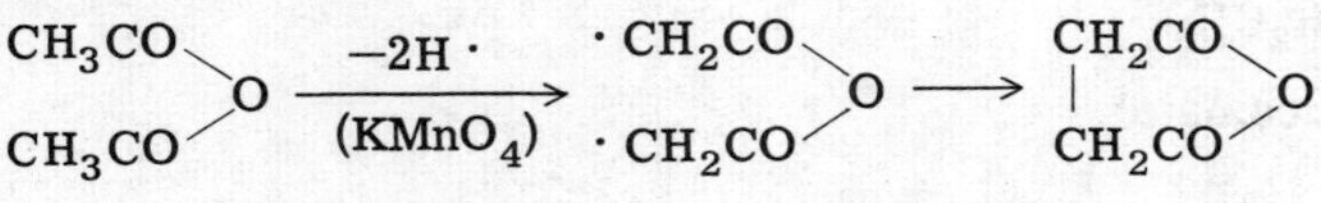

7. 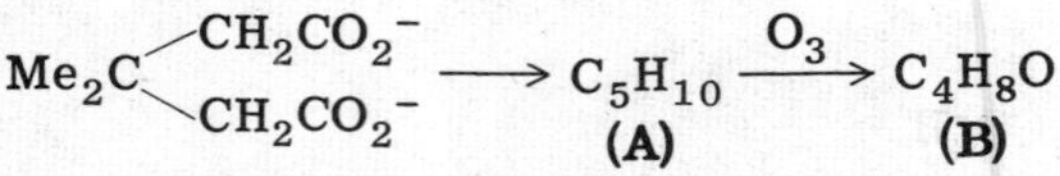

Since (A) has the molecular formula C_5H_{10}, it is an alkene (or a saturated monocyclic system). Since it loses one carbon atom on ozonolysis, (A) must therefore be a 1-alkene and so (B) is an aldehyde or a ketone. The i.r. band at 1 720 cm^{-1} for (B) strongly suggests (B) is a ketone; aldehyde is not completely excluded because this band occurs at the lower end of the aldehyde region (see *Text*, Ch. 8). However, the τ value of an aldehydic hydrogen is ~0·0 – 1·0. Since a signal in the NMR spectrum of (B) is not given in this region, (B) is therefore a ketone. If we now refer to Table 3.2, we can make the following tentative assignments:

τ 9·1, CH_3—C; τ 7·8, CH_3—CO; τ 7·6—CH_2—CO.

Putting these groupings together, we get $CH_3CH_2COCH_3$ = C_4H_8O. This is the molecular formula of (B) and would give the three NMR signals observed. Hence, if we accept this as the structure of (B), the structure of (A) will be $CH_3CH_2C(CH_3)$=CH_2 (ozonolysis ⟶ $CH_3CH_2COCH_3$).

When the salt of a monocarboxylic acid is electrolysed, a free radical is formed first, and then eliminates CO_2 to form an alkyl free radical (Kolbe's electrolytic method; see *Text*, Ch. 3). Extending this to dicarboxylic acids, we would expect the salt of β,β-dimethglutaric acid to behave as follows:

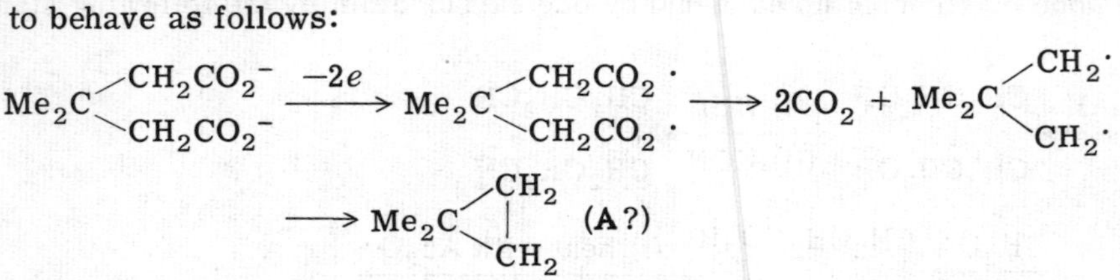

As we have seen, (A) is not a cyclopropane derivative; it does *not* contain the group Me_2C, and actually contains a straight chain of four carbon atoms. Hence we must conclude that some rearrangement of the hydrocarbon diradical has occurred to give (A). As we now know the structure of (A), we can propose the following mechanism for the rearrangement:

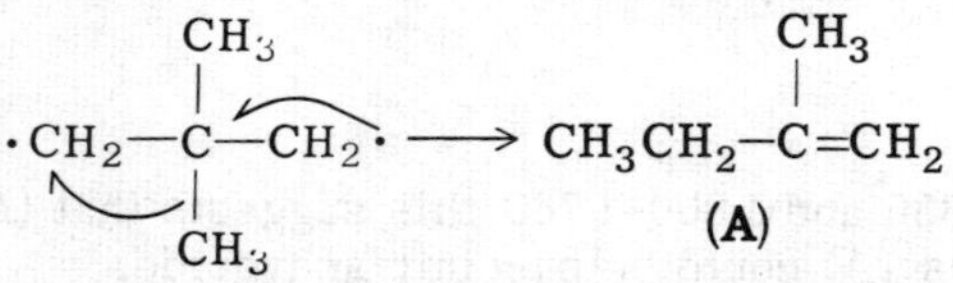

8. The presence of a triplet and a quartet suggests an ethyl group, CH_3CH_2. This would account for two signals, and because the CH_2 signal is a quartet, there is no further vicinal coupling. Also because of the low τ-values, the CH_2 group is deshielded by a strong electron-attracting group. Inspection of Table 3.2 shows that such a group is carbethoxy, i.e., CH_3CH_2OOC—. This fragment, written as $C_3H_5O_2$, leaves $C_5H_9O_2$ unaccounted for and must be explained on the basis that it gives *one* signal only. The fact that 14 hydrogens produce three signals leads to the conclusion that the molecule has repeated units which, because of their equivalence, produce one set of signals. Hence, we may write the formula as $(CH_3CH_2OOC)_2C_2H_4$. This means that the fragment C_2H_4 gives a singlet and so must be symmetrical. The combination that satisfies this requirement is ethylene, i.e., if the compound has the structure

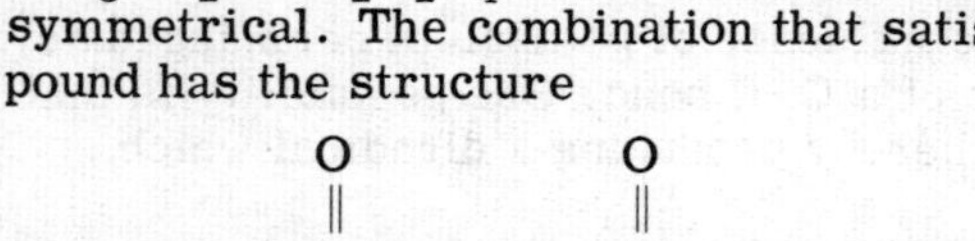

This is diethyl succinate (ethyl succinate).

The student might think that this structure could have been deduced from the fact that the NMR spectrum contains three signals (quartet, triplet and singlet), *without* their respective τ-values being given. This is true, but in this case there would have been an alternative structure which also gives three signals, *viz.*, ethylene glycol dipropionate.

$$CH_3CH_2\overset{\overset{\large O}{\|}}{C}{-}O{-}CH_2CH_2{-}O{-}\overset{\overset{\large O}{\|}}{C}CH_2CH_3$$

Reference to Table 3.2 shows that the τ-values of the signals will be different from those given in the question.

17 Hydroxyacids, stereochemistry, unsaturated dicarboxylic acids

(Problems on page 29)

1. See also Ch. 5, Prob. 3.

(i)

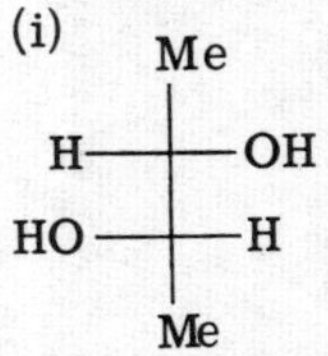

(I)

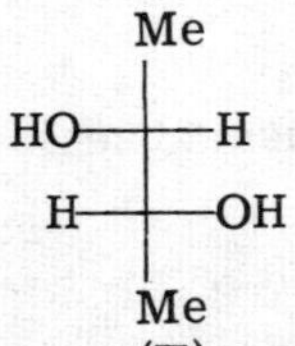

(II)

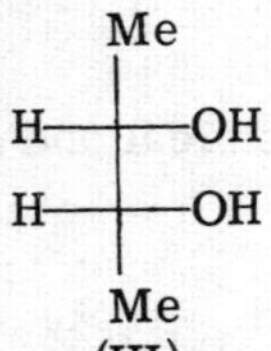

(III)

(I) and (II) are optically active and are related to each other as a pair of enantiomers. (III) is a diastereoisomer of (I) or (II) and is a *meso* form (optically inactive by internal compensation; it has a plane of symmetry).

(ii)

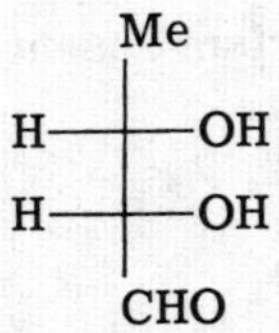

(I)

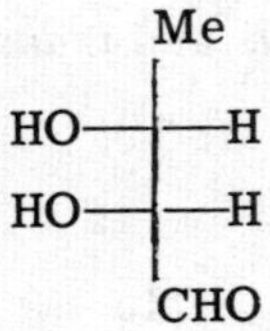

(II)

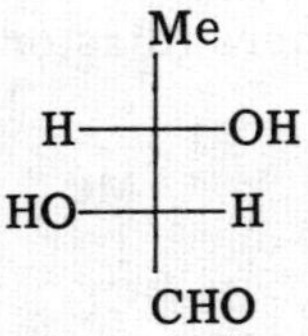

(III)

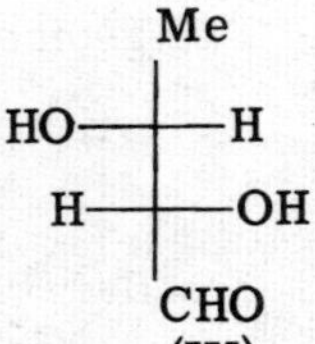

(IV)

(I) and (II) are optically active (pair of enantiomers). (III) and (IV) are optically active (pair of enantiomers). The pairs are related as diastereoisomers.

(iii)

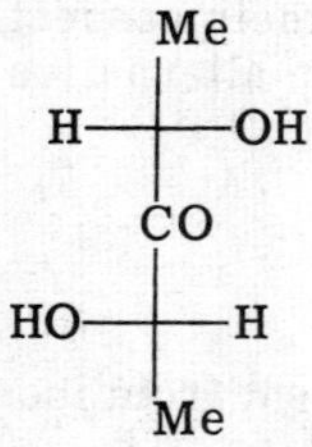
(I)

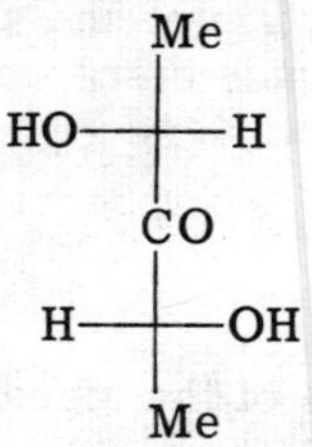
(II)

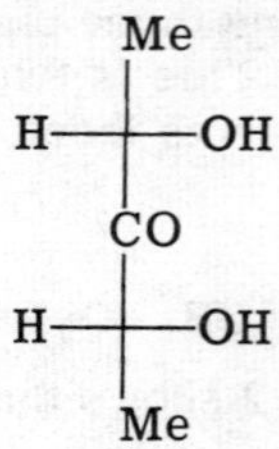
(III)

(I) and (II) are optically active (pair of enantiomers); (III) is the *meso* form (optically inactive). The molecule contains two structurally identical carbon atoms (see (i), above).

(iv)

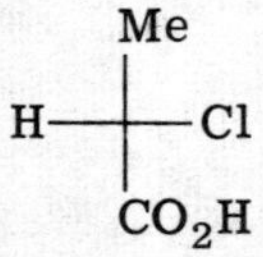

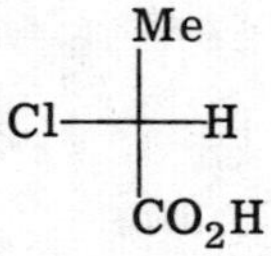

One pair of enantiomers (only one asymmetric carbon atom present).

(v) Only one asymmetric carbon atom present; hence one pair of enantiomers.

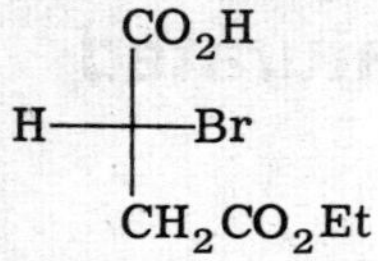

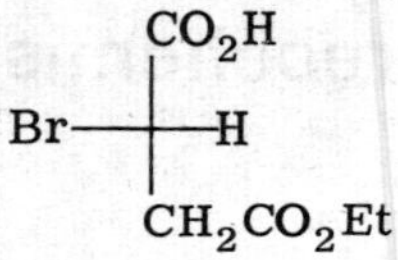

(vi)

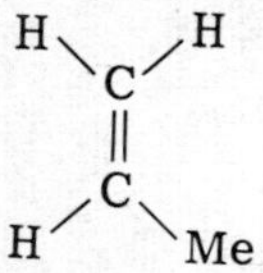

One form only; geometrical and optical isomerism absent.

(vii)

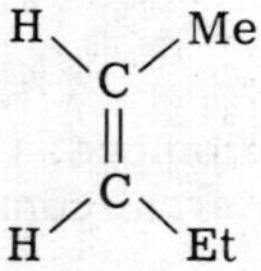
cis

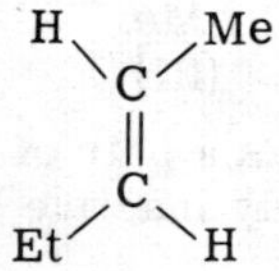
trans

Two geometrical isomers (diastereoisomers), *cis* and *trans*. Optical isomerism absent.

(viii)

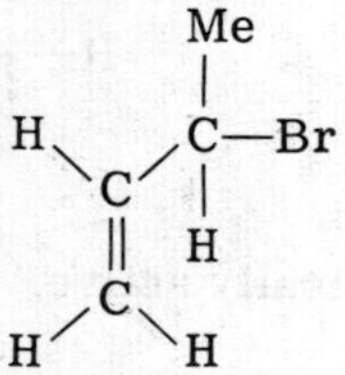

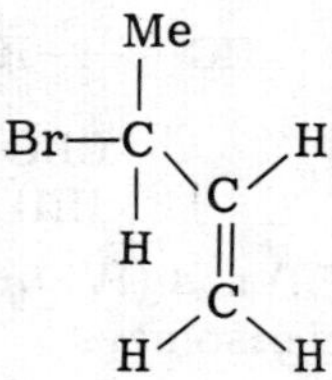

or

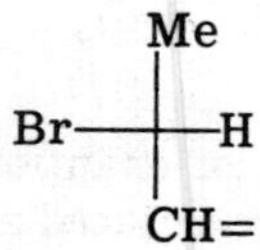

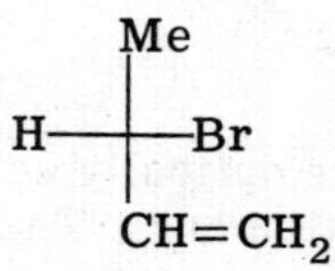

Geometrical isomerism (diastereoisomerism) absent, but the presence of one asymmetric carbon atom gives rise to a pair of enantiomers.

(ix)

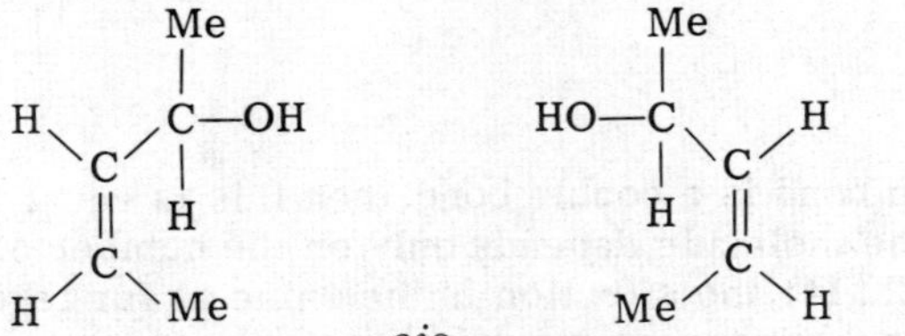

cis

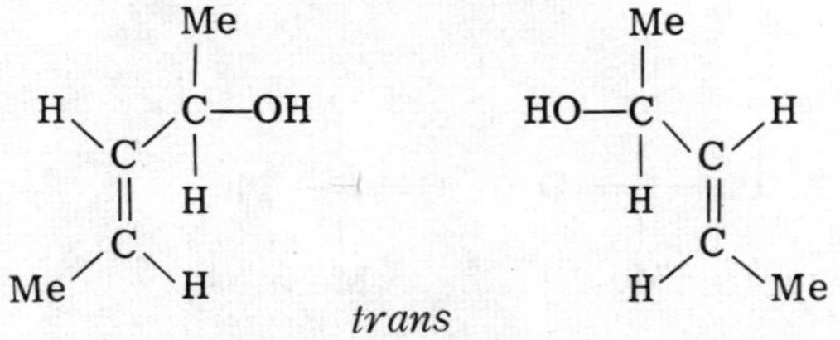

trans

Write down the geometrical isomers (diastereoisomers) first and then look for optical isomerism. Two geometrical isomers (diastereoisomers) are possible, *cis* and *trans*. The presence of one asymmetric carbon atom in *each* geometrical isomer (diastereoisomer) gives rise to a pair of enantiomers. Hence, there are four optically active forms.

(x)

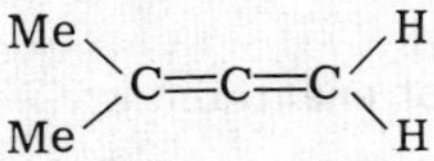

The two Me groups lie in the plane of the paper and the two hydrogens in a plane perpendicular to the paper. Hence, the molecule has a plane of symmetry (in the plane of the paper) and so is not optically active.

(xi)

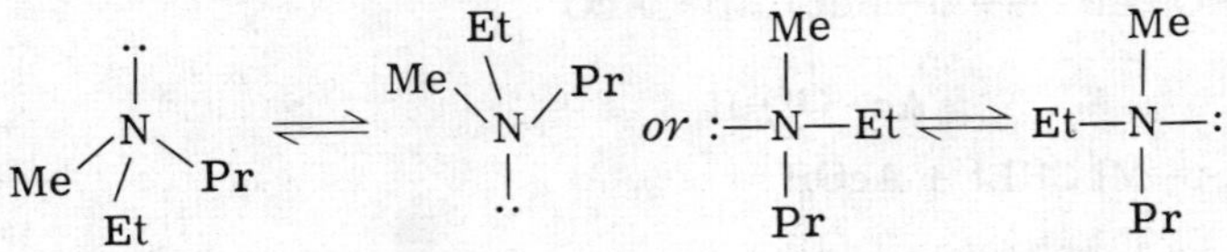

The valencies of tercovalent nitrogen are distributed tetrahedrally, with the lone pair at one corner. If this were a 'rigid' structure, then MeEtPrN would be capable of optical isomerism (four different 'groups' attached to the N atom). In practice, however, the N atom is undergoing rapid oscillation through the plane of the three groups Me, Et and Pr, the result being that the molecule exists as a 'pair of enantiomers' rapidly changing from one to the other. Hence, the compound is not optically active (see also, *Text*).

(xii) In $MeEtPrPr^{i}N^{+}Br^{-}$, the four valencies are distributed tetrahedrally and, ignoring the positive charge on the nitrogen atom, the situation is the same as that for an asymmetric carbon atom, i.e., the nitrogen compound can exist as a pair of enantiomers (*note:* Pr and Pr^{i} are different groups).

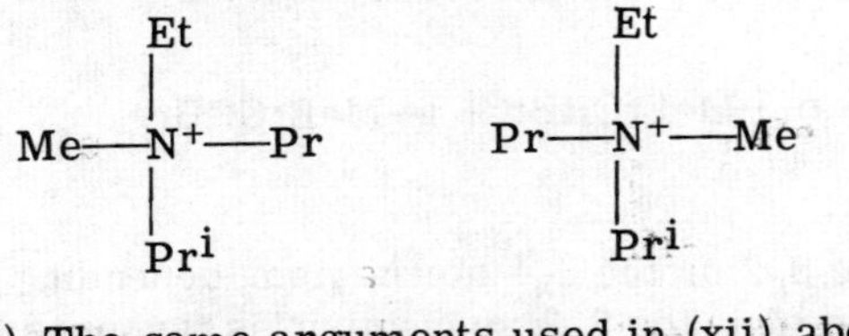

(xiii) The same arguments used in (xii), above, also apply here. Thus, there is a pair of enantiomers.

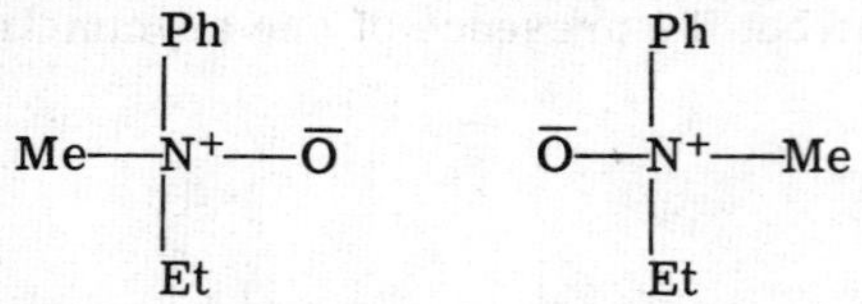

(xiv) If we accept that the sulphur-oxygen bond is a double bond, then this is equal to a σ- and a π-bond, and since the geometry of the molecule depends only on the number of σ-bonds and lone pairs (see *Text*, Ch. 2; see also Ch. 14), the situation is the same as for tercovalent nitrogen, but there is no oscillation with the sulphur compound. Hence, there is a pair of enantiomers

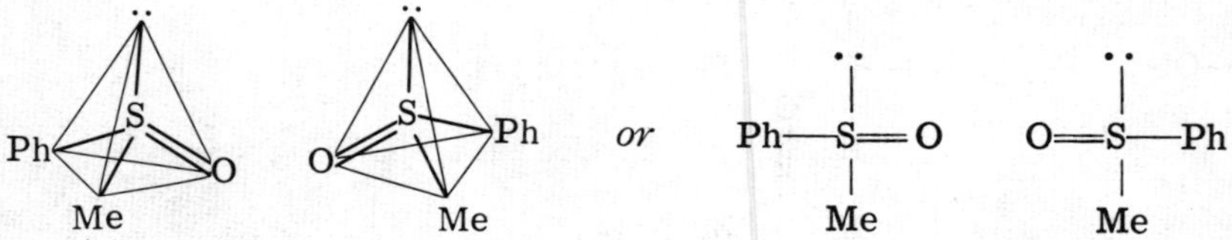

(xv) See also (x), above.

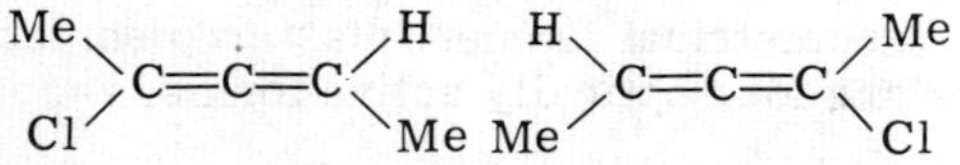

This molecule has no plane of symmetry and so exists as a pair of enantiomers.

2. The general principles are: (*a*) If no bond to the asymmetric carbon atom is broken, optical inversion is impossible; (*b*) If such a bond is broken then, for an S_N2 mechanism there is inversion and for an S_N1 mechanism there are (usually) retention and racemisation.

(i)

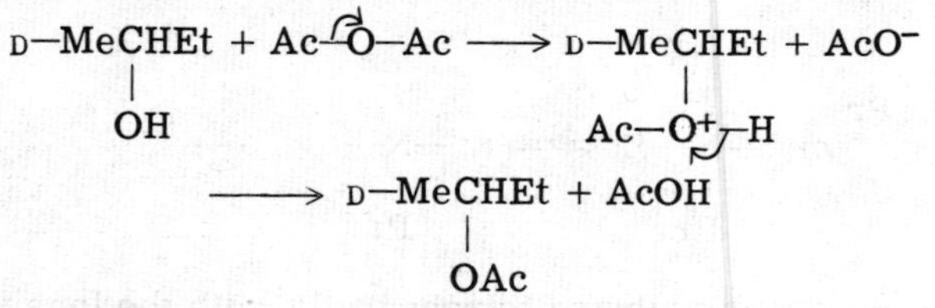

There is no change in configuration, since the C—O bond is not broken.

(ii) If we accept that this hydrolysis proceeds via the common mechanism, $B_{AC}2$, i.e., bimolecular acyl-oxygen fission (see *Text*, Ch. 9), then retention of configuration is obtained.

$$CH_3CO{-}OCHMeEt{-}\text{L} + OH^- \longrightarrow CH_3COOH + \text{L}{-}MeEtCHO^- \longrightarrow \text{L}{-}MeEtCHOH + CH_3CO_2^-$$

(iii) The substrate is a secondary alcohol and in the presence of acid will (most probably) form a carbonium ion (sp^2; flat). Hence, for this S_N1 mechanism, retention and racemisation will be the result.

$$\text{L}{-}MeEtCHOH \xrightarrow[\text{(ii) } -H_2O]{\text{(i) } H^+} \underset{\text{(flat) } sp^2}{MeEtCH^+} \xrightarrow{Br^-} \text{DL}{-}MeEtCHBr + \text{L}{-}MeEtCHBr$$

(iv) This reaction can proceed either by the S_N2 or the S_N1 mechanism, depending on the conditions. If the concentration of the alkali is high, the S_N2 mechanism is favoured (see *Text*, Ch. 5). In these circumstances, inversion will occur.

$$\text{D–MeCHClCO}_2\text{H} + \text{NaOH} \longrightarrow \text{D–MeCHClCO}_2^-\text{Na}^+ + \text{H}_2\text{O}$$

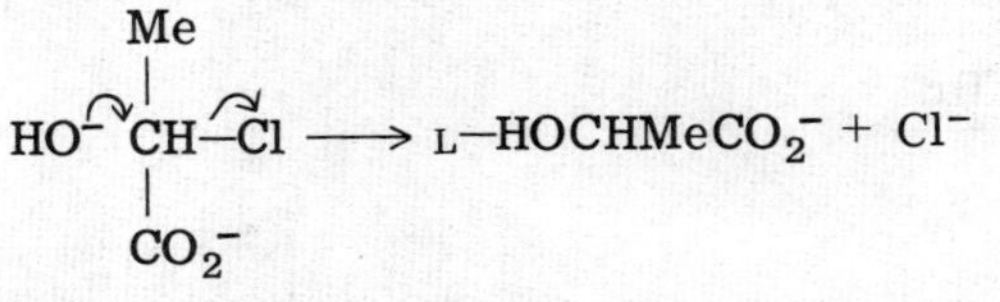

Since dilute alkaline solution favours the S_N1 mechanism, at first sight the expected result might have been retention and racemisation. However, a closer look shows that neighbouring group participation can operate in this molecule between the carboxylate ion and the carbonium ion centre that is formed.

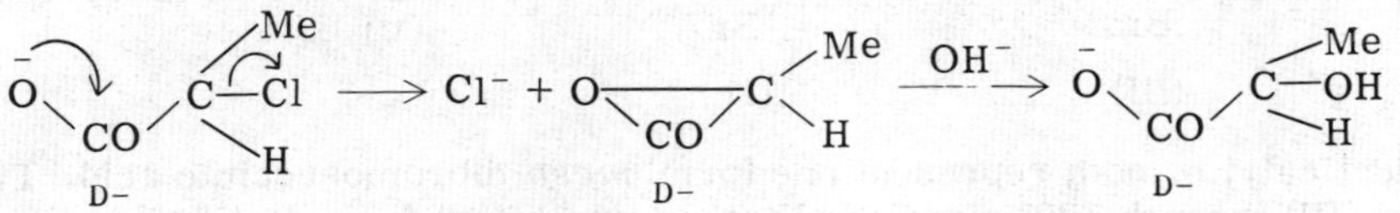

Hence, because of this 'protection', the hydroxide ion is forced to attack on the *same* side as the ejected chlorine atom, thereby resulting in retention of configuration. This is actually the predominant product obtained in practice.

(v) This either contains a t-alkyl group and so the S_N1 mechanism is favoured (see *Text*, Ch. 7). Hence retention and racemisation can be expected.

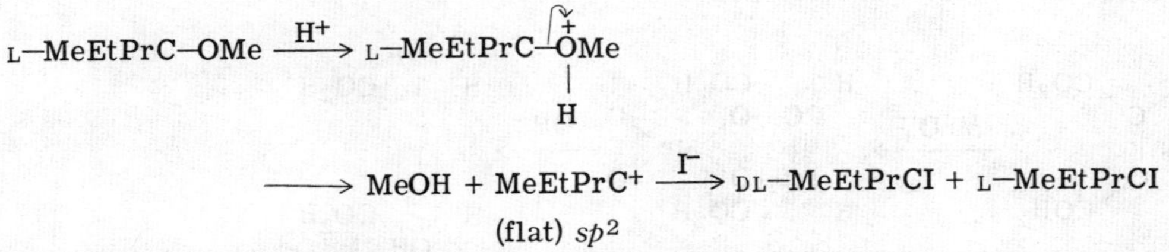

Because of the relatively high stability of t-carbonium ions, the product could well be only the racemised iodide.

3. (i) The addition of bromine is stereoselectively *trans* (via a brominium ion). Hence, the predominant products are: maleic acid gives (±)-dibromosuccinic acid, and fumaric acid gives *meso*-dibromosuccinic acid (see also below).

maleic acid

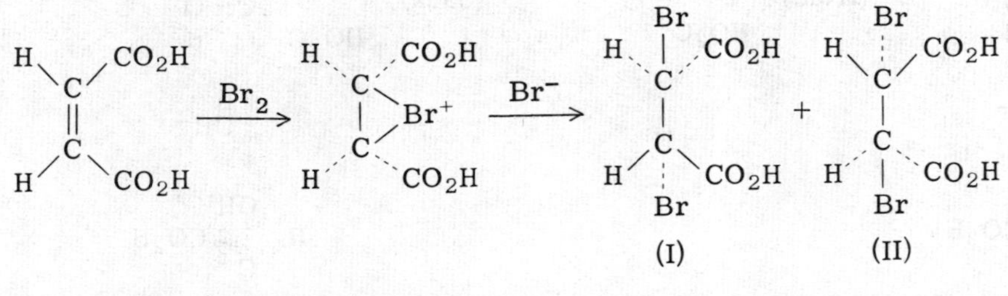

(I) (II)

Because of the symmetry of the brominium ion, the bromide ion can attack, from the back, either carbon atom equally well, thereby resulting in the formation of (I) and (II) in *equal* amounts. Hence, the racemic modification of the dibromosuccinic acid is produced.

fumaric acid

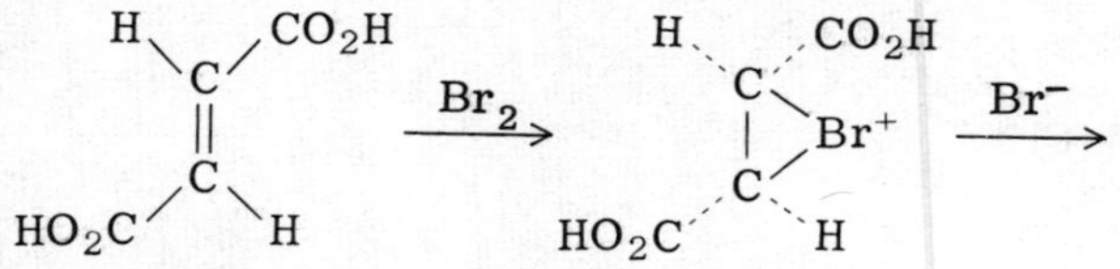

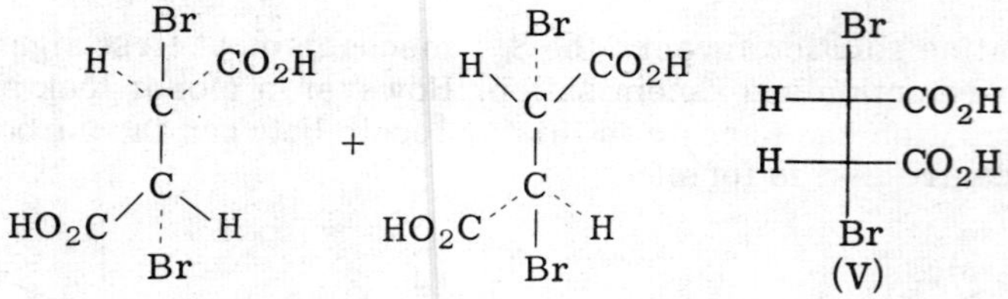

(III) (IV) (V)

(III) and (IV) are identical, i.e., both represent one form, *meso*-dibromosuccinic acid. Twisting the lower half of (IV) through 180° gives the Fischer projection formula (V); this has a plane of symmetry. Twisting the top half of (III) gives the mirror image of (V), and this is superimposable on (V).

(ii) Hydroxylation of a double bond proceeds via a *cyclic* intermediate (ester) which, on hydrolysis, gives the *cis*-diol (see *Text*, Ch. 4). Hence, maleic acid gives *meso*tartaric acid, and fumaric acid gives (±)-tartaric acid.

maleic acid

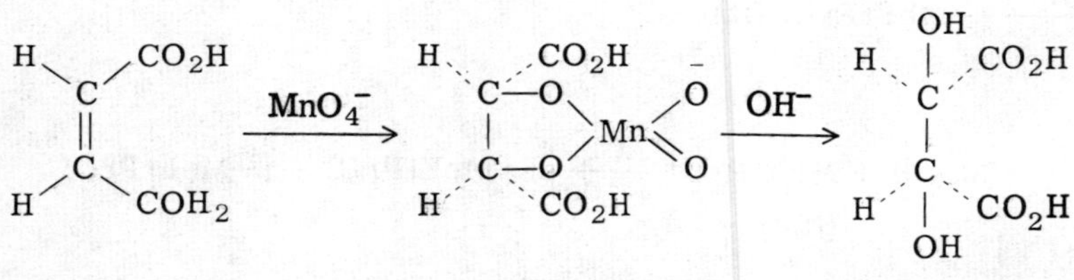

meso

Attack from front or back gives the same product (see also below).

fumaric acid

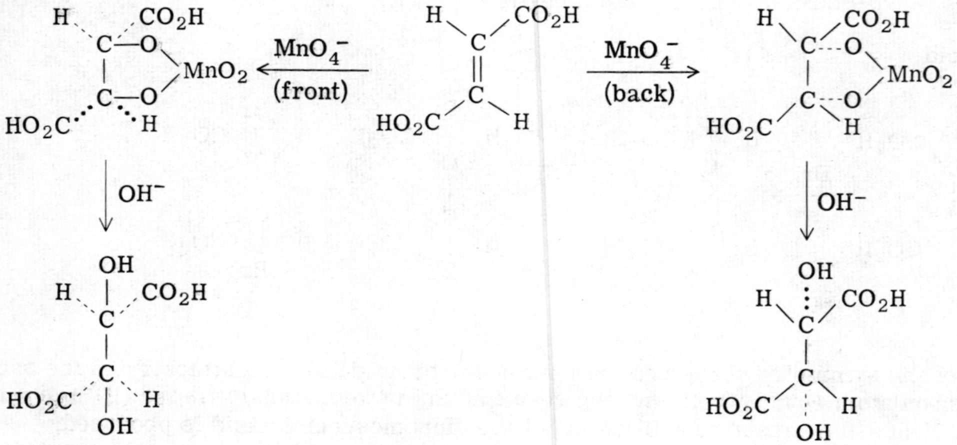

The fumaric acid molecule is flat and can be attacked equally well from the front or from the back. Hence, the products are a pair of enantiomers in equal amount.

(iii) Perbenzoic acid converts the double bond into the epoxide and this, on hydrolysis, gives the *trans*-diol (see *Text*, Ch. 11). Hence, maleic acid gives (±)-tartaric acid, and fumaric acid gives *meso*tartaric acid.

maleic acid

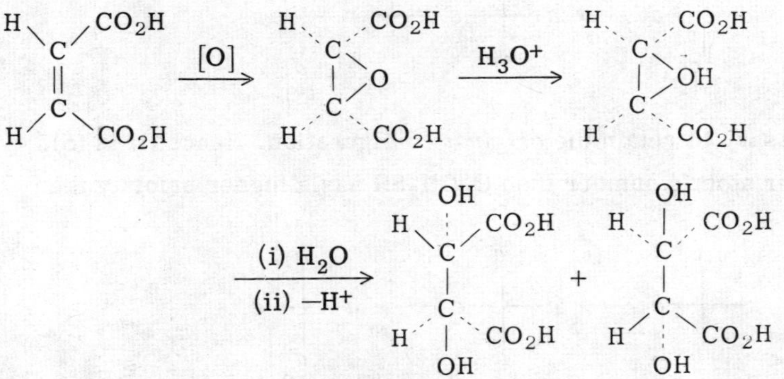

Attack at the back to form the epoxide finally results in the same pair of enantiomers.

fumaric acid

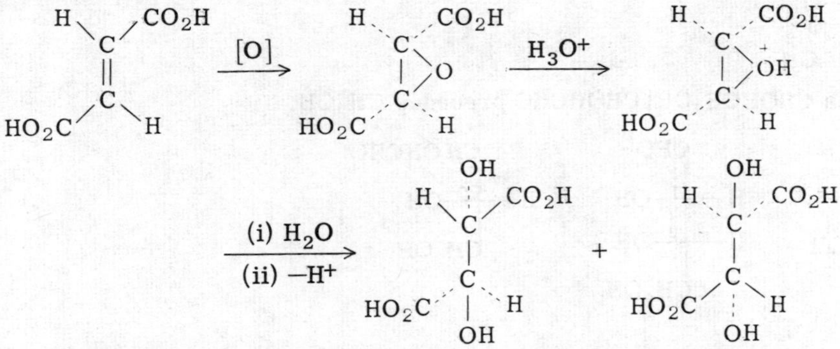

The two products are identical, both being the same *meso*tartaric acid.

(iv) Catalytic hydrogenation usually results in *cis*-addition (see *Text*, Ch. 4), but since the product, succinic acid, contains no asymmetric carbon atoms, both maleic and fumaric acid give the same product.

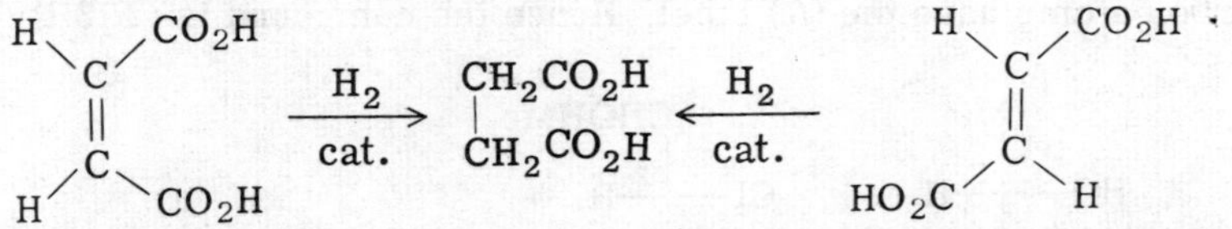

4. If in molecule *Cabcd* the four groups have the priority *a*, *b*, *c* and *d* according to the sequence rule, then labels *R* and *S* are assigned as shown.

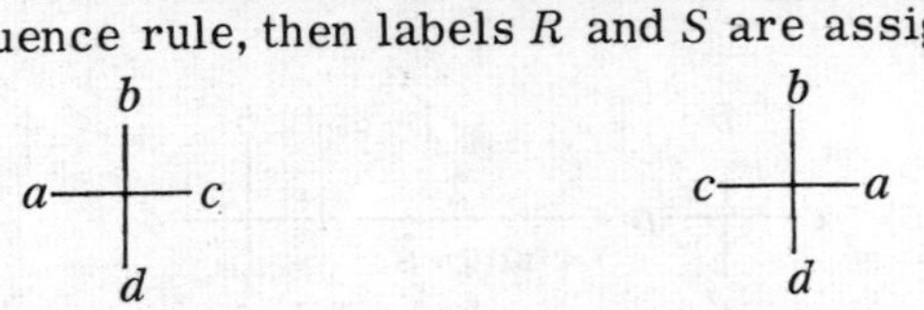

R (clockwise) *S* (anticlockwise)

(i) 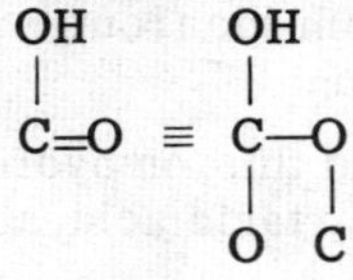 ∴ CO_2H precedes Me.

Hence: 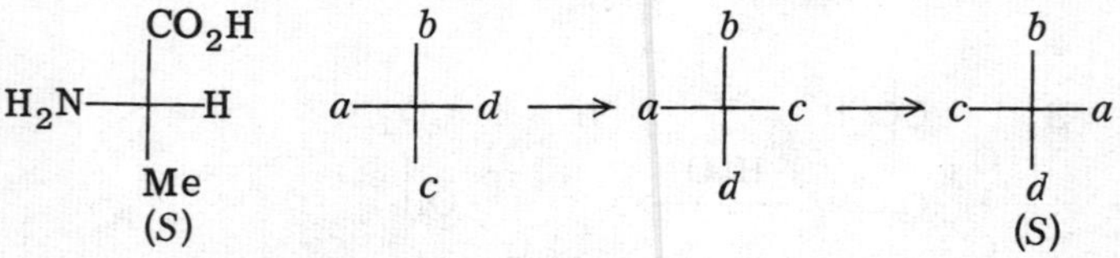

Two interchanges are necessary to retain the *original* configuration. Hence (i) is (*S*).

(ii) Since S has a higher atomic number than O, CH_2SH has a higher priority than CO_2H.

Hence: 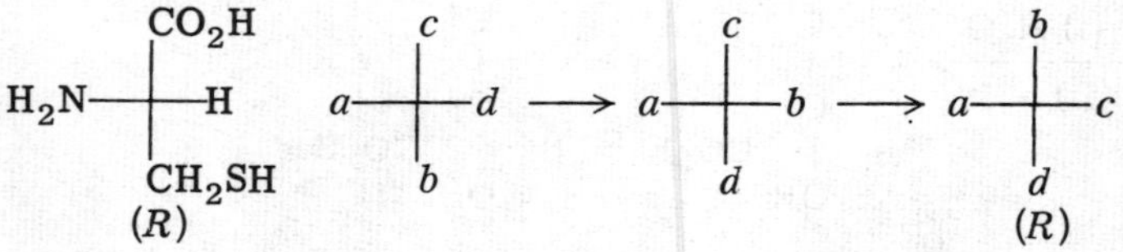

(iii) 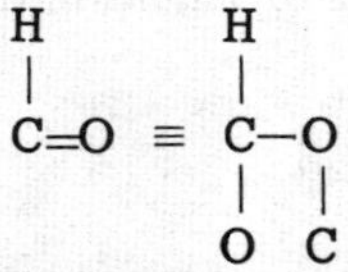

∴ CHO precedes $CHOHCH_2OH$; CHOHCHO precedes CH_2OH.

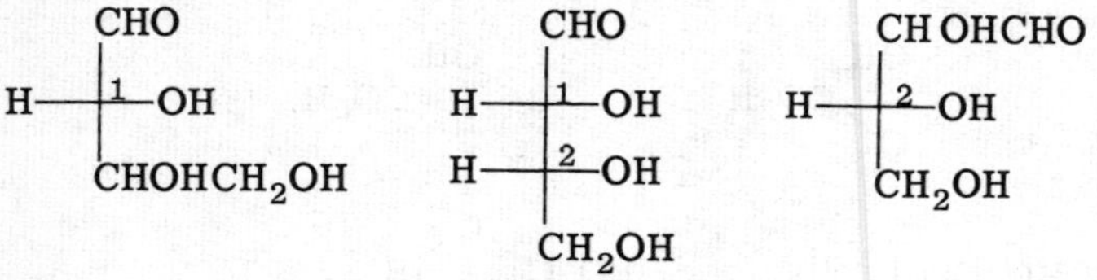

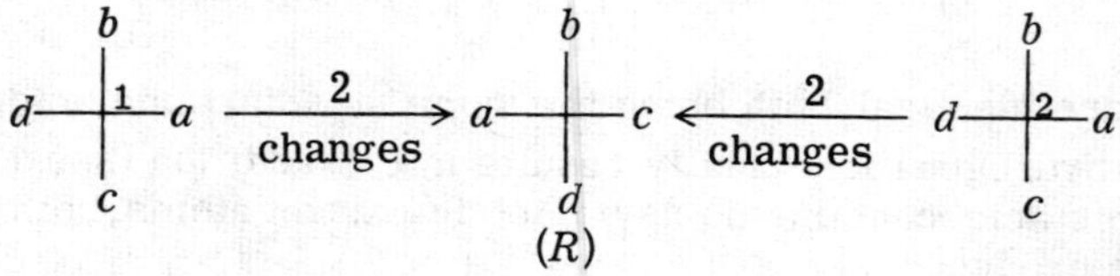

Thus both asymmetric carbon atoms have the (*R*) label. Hence the compound is (2*R*3R)-erythrose.

(iv) 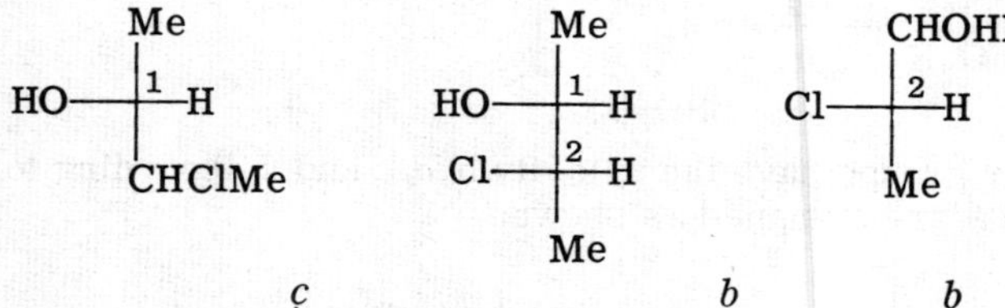

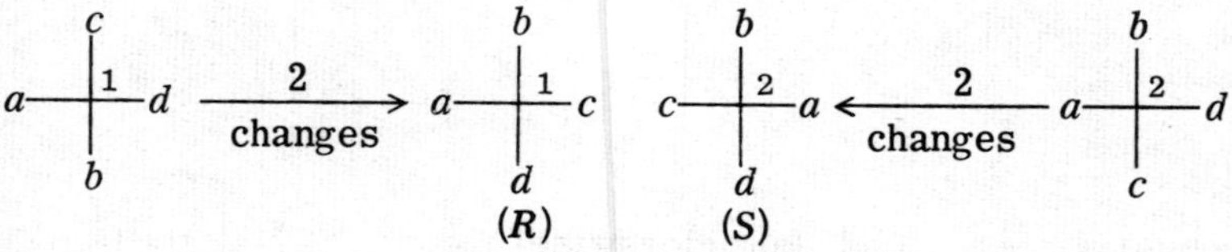

The compound is (2*R*3*S*)-3-chlorobutan-2-ol.

(v) CH_2OH has oxygen attached to C, whereas phenyl in its replica form has only carbon atoms. Hence CH_2OH precedes Ph.

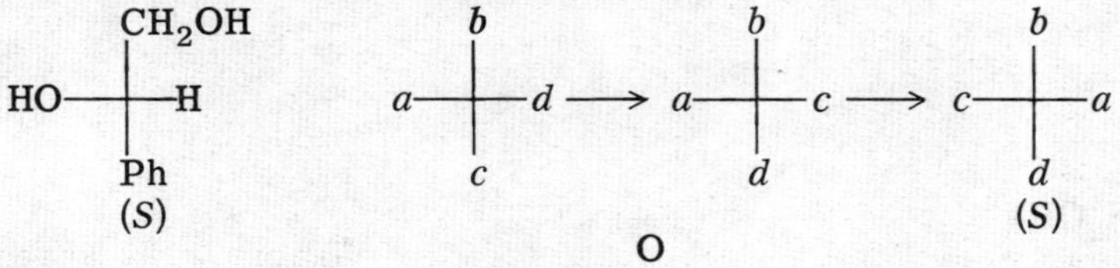

(vi) Order of priority is $CH_2{-}O{-}\overset{\overset{\displaystyle O}{\|}}{C}CH_3 > CH_2OCH_3$

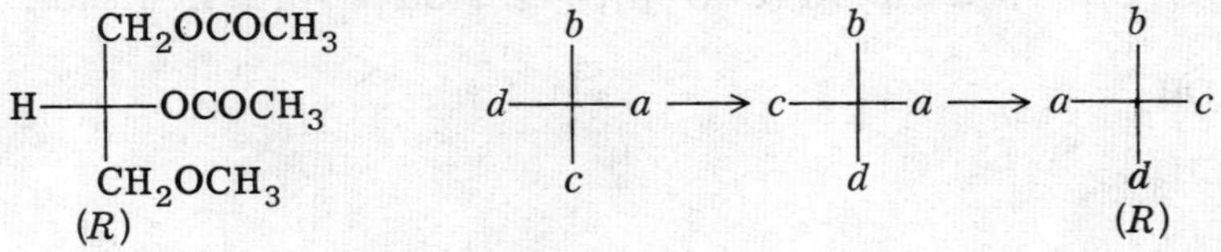

(vii)

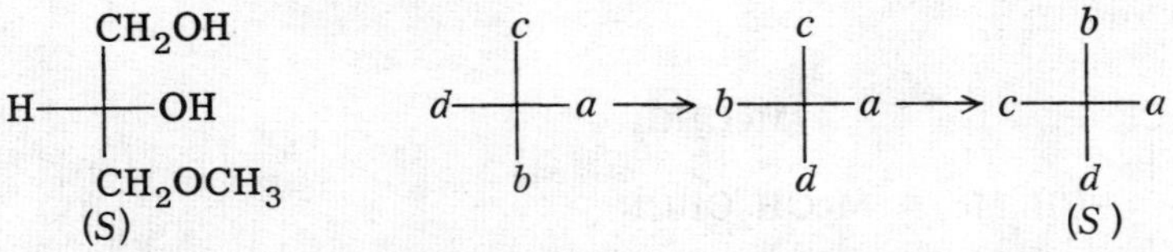

5. If the four groups in C*abcd* have the priority $a > b > c > d$, then by assigning these letters to the groups in the compound under discussion, we can write down the *R*- or *S*-form (see also Prob. 4, above).

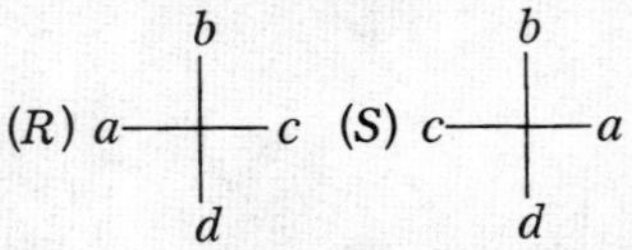

(i)

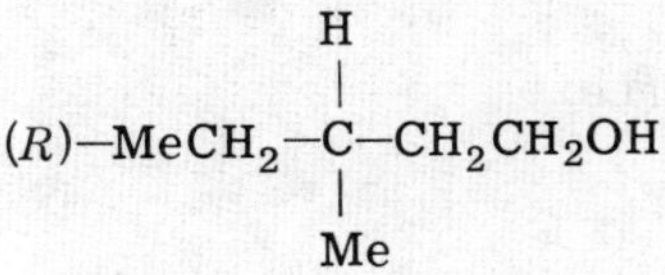

$a = CH_2CH_2OH(C{-}C{-}O)$; $b = CH_2CH_3(C{-}C{-}H)$; $c = CH_3$; $d = H$. Hence:

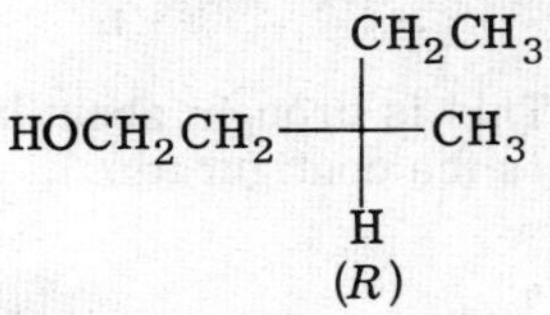

(ii)

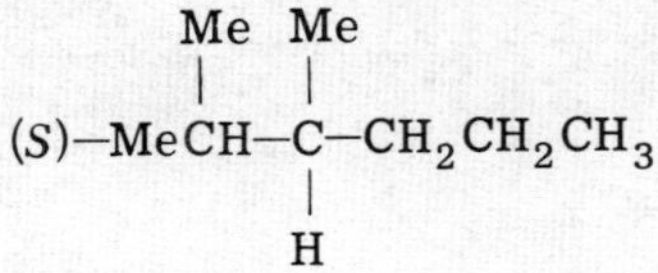

$a = Me_2CH\left(C\begin{smallmatrix}C\\C\end{smallmatrix}\right)$; $b = CH_2CH_2CH_3(C{-}C{-}C)$; $c = CH_3$; $d = H$. Hence:

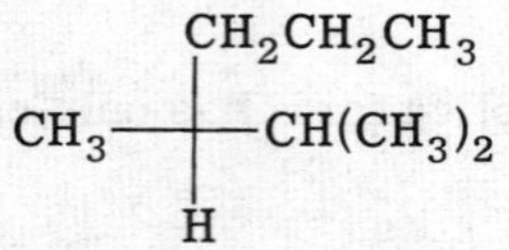

(S)

(iii)

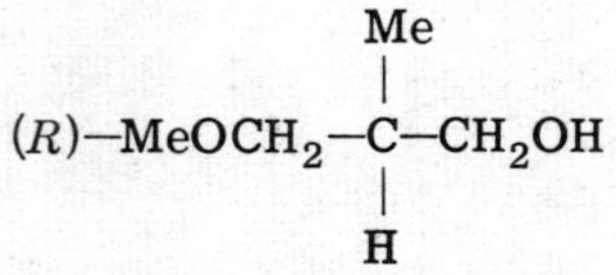

$a = CH_2OCH_3$(C–O–C); $b = CH_2OH$ (C–O–H); $c = CH_3$; d = H. Hence:

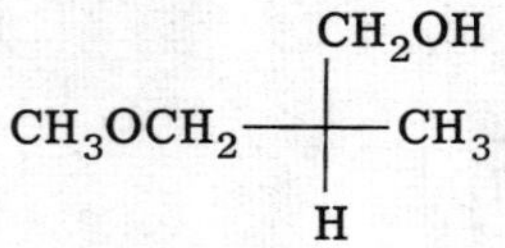

(R)

(iv)

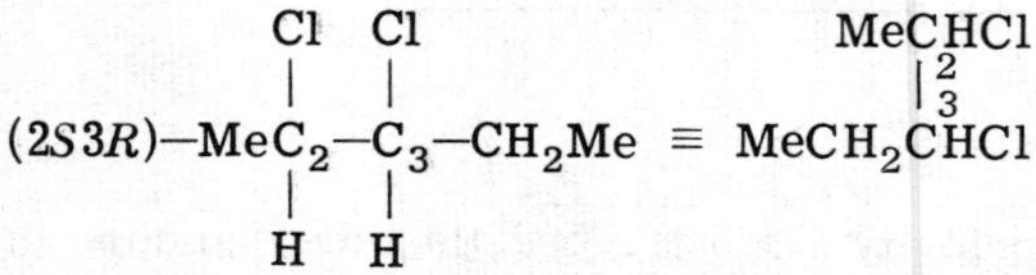

The configurations of the two asymmetric carbon atoms will be, according to the rules:

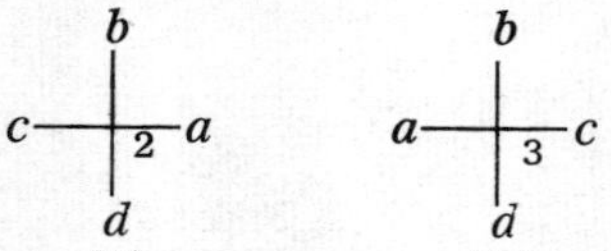

(S) (R)

(S): a = Cl; b = CHClCH$_2$Me; c = Me; d = H. Hence:

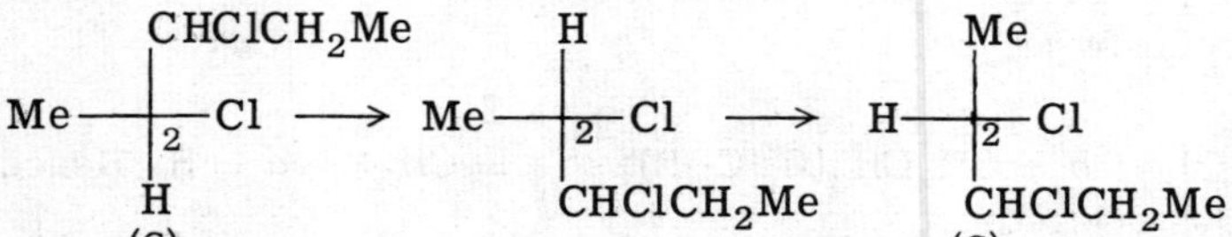

(S) (S)

When C_2 is 'joined' to C_3, the bottom group is CHClCH$_2$Me. This is brought about by *two* interchanges as shown (remember that one interchange inverts the configuration).

(R): a = Cl; b = CHClMe; c = CH$_2$Me; d = H. Hence:

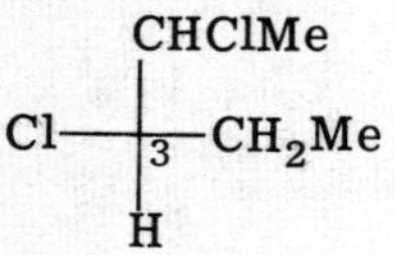

(R)

Since the CHClMe group is at the top (to join with C_3, this arrangement may be used. Thus, (2S3R)-2,3-Dichloropentane is:

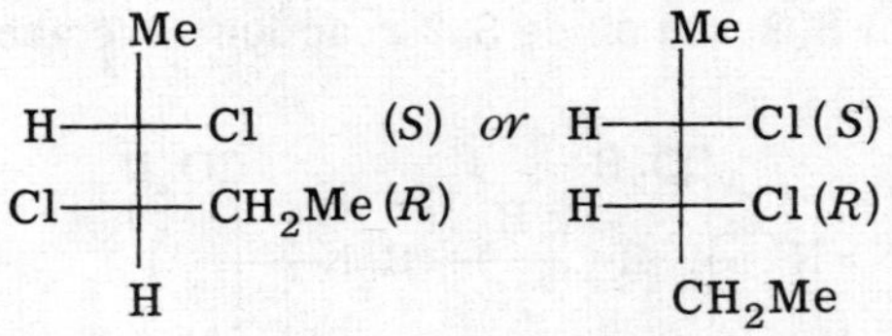

The second formula is obtained from the first by carrying out two interchanges on the lower asymmetric carbon atom. The first formula is acceptable, but the second clearly shows that the compound is the *erythro*-isomer (see *Text*).

6. See also Ch. 3, Prob. 2 and Ch. 5, Prob. 1.

Treat the carboxyl group as a substituent in propane, and since it is the principal function, move the hydroxyl group with respect to it.

(i) $CH_3CH_2CHOHCO_2H$ 2-hydroxybutanoic acid

(ii) $CH_3CHOHCH_2CO_2H$ 3-hydroxybutanoic acid

(iii) $CH_2OHCH_2CH_2CO_2H$ 4-hydroxybutanoic acid

(iv)
OH
|
CH_3CCH_3 2-hydroxy-2-methylpropanoic acid
|
CO_2H

(v) CH_3CHCH_2OH 3-hydroxy-2-methylpropanoic acid
|
CO_2H

Heating: α-Hydroxyacids form *lactides:*

(i) 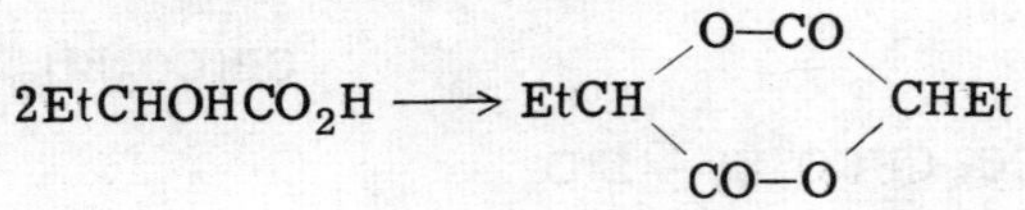

(iv) 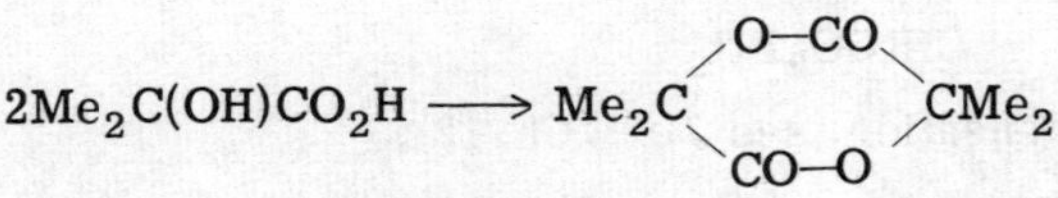

β-Hydroxyacids form *α,β-unsaturated acids:*

(ii) $CH_3CHOHCH_2CO_2H \longrightarrow CH_3CH{=}CHCO_2H$

(v) 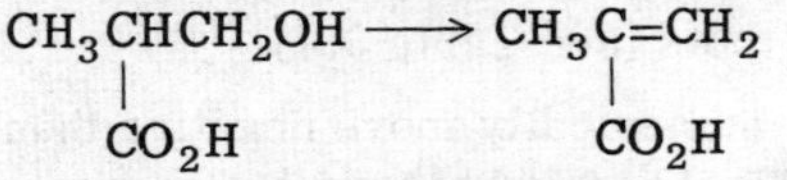

γ- (and δ-) Hydroxyacids form *lactones:*

(iii) 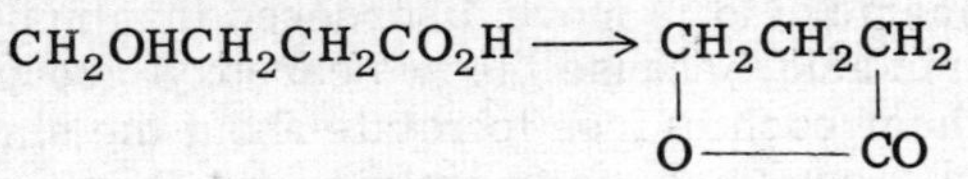

Dilute sulphuric acid on warming

α-Hydroxyacids form aldehydes or ketones, depending on their structure:

(i) $CH_3CH_2CHOHCO_2H \longrightarrow CH_3CH_2CHO + HCO_2H$

(iv) $Me_2C(OH)CO_2H \longrightarrow Me_2CO + HCO_2H$

β- and γ-Hydroxyacids behave in the same way as on heating (see above).

7. Catalytic hydrogenation is the step which is not S_N2, and since S_N2 reactions are accompanied by inversion, the equations are:

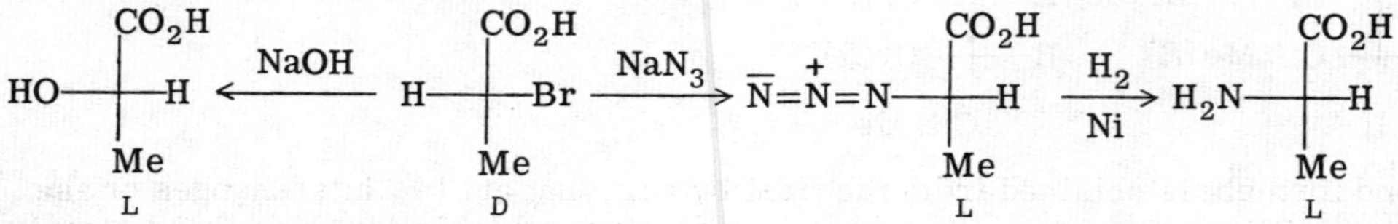

The azide ion is $:\ddot{N}{=}\overset{+}{N}{=}\ddot{N}:$, and azides are reduced to amines. Since the C—N bond in the above azide is not broken during the reduction, there is no change in configuration.

8. (i)

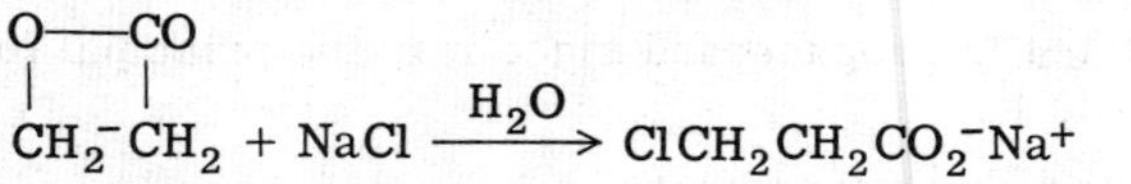

Attack occurs by the chloride ion, and the 'driving force' of the reaction may be attributed to the strain experienced by the four-membered ring (see *Text*, Ch. 19).

(ii) $$(HO_2CCH_2)_2C(OH)CO_2H \xrightarrow[H_2SO_4]{\text{fuming}} (HO_2CCH_2)_2CO$$

$$\xrightarrow[HCl]{EtOH} (EtO_2CCH_2)_2CO$$

Fuming sulphuric acid converts α-hydroxyacids into carbonyl compounds (*cf.* hot dilute sulphuric acid; see Prob. 6, above).

(iii) $$CH_2(CO_2Et)_2 \xrightarrow{EtONa} Na^+\overline{C}H(CO_2Et)_2$$

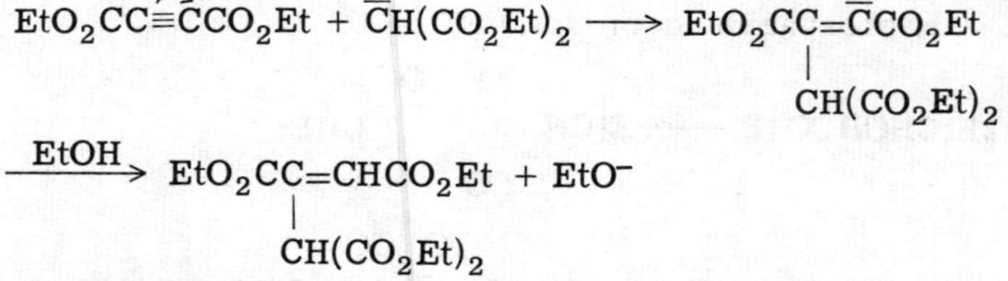

This is an example of the Michael condensation (see *Text*, Ch. 12).

(iv)

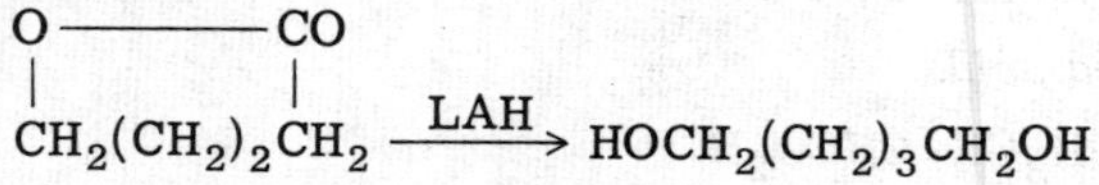

LAH or catalytic reduction converts lactones into diols (see Table 6.1).

9. Since the heat of formation of fumaric acid is numerically more negative than that of maleic acid, the former is thermodynamically more stable than the latter.

If we equate entropy with disorder, then fumaric acid is more disordered than maleic acid. This may be explained as follows. In fumaric acid, because the two carboxyl groups are *trans* and there is no interaction between them, each is free to rotate about the single bond. On the other hand, because the two carboxyl groups are *cis* in maleic acid, they are close enough to form a hydrogen bond (see *Text*). This restricts their freedom to rotate about the single bonds and so the entropy is lowered (*cf.* negative entropy of activation for cyclic transition states).

10. When acyl-oxygen heterolysis occurs, no inversion is possible, but inversion does occur when the mechanism is bimolecular and takes place with alkyl-oxygen heterolysis. Hence (—o→ means inversion):

$$PhCH_2CHMe{-}\ddot{O}{-}H + TsCl \xrightarrow{-HCl} PhCH_2CHMe{-}OTs$$

$$\xrightarrow{AcO^- Na^+} PhCH_2CHMe{-}O{-}Ac \xrightarrow{OH^-}$$

$$PhCH_2CHMe{-}\bar{O} + AcOH \longrightarrow PhCH_2CHMeOH + AcO^-$$

11. The configurations of the 2,3-dibromobutanes are:

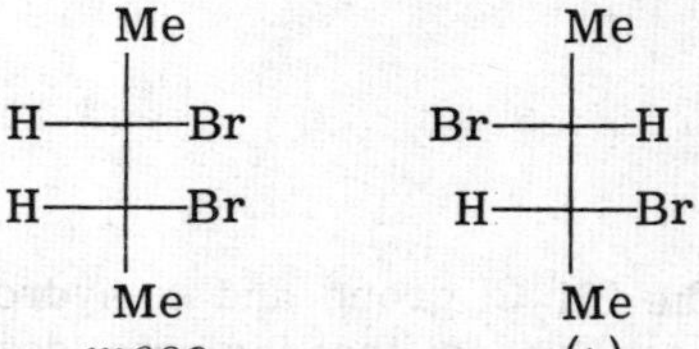

meso (±)

On the basis that *trans*-elimination is preferred, we may formulate the reactions as follows (see also *Text*, Ch. 4).

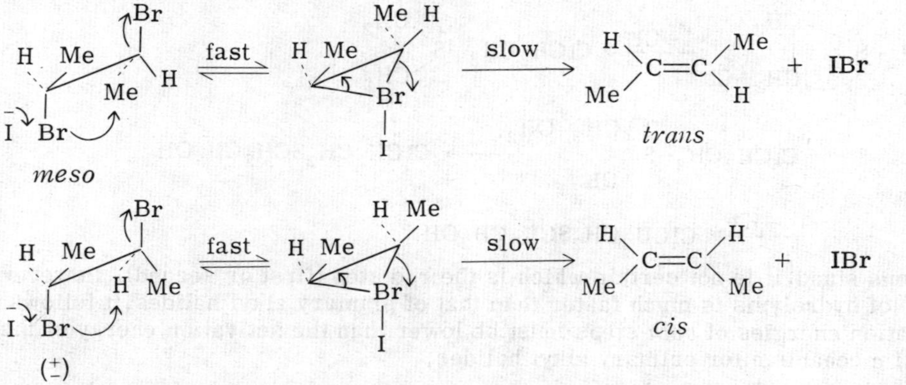

A brominium intermediate is formed in each case, but that from the *meso*-compound experiences less strain due to steric interaction of the *trans* methyl groups than does that from the (±)-compound due to the *cis* methyl groups. Hence, the transition states (for both first and second steps) leading to *trans*-but-2-ene contain less energy than those leading to *cis*-butene. Therefore the activation energy of each T.S. of the former reaction is lower than that of the corresponding T.S.s of the latter reaction, i.e., the *meso*-compound undergoes elimination more rapidly than does the (±)-compound.

12.

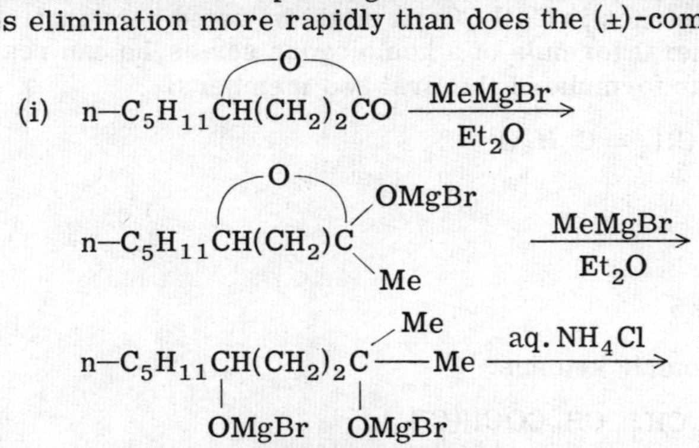

$$n{-}C_5H_{11}CH(OH)CH_2CH_2CMe_2OH$$

Epoxide oxygen is less readily attacked by Grignard reagents than is a carbonyl group. Hence, we can assume the sequence given.

(ii)

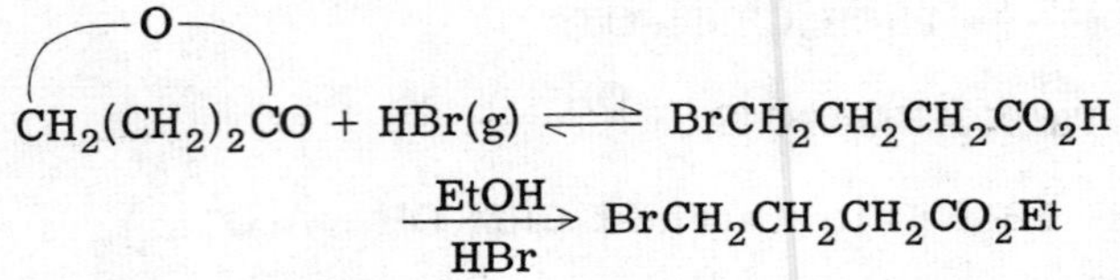

The first step is reversible, but the second one is not. Hence, the overall reaction proceeds to completion.

(iii) $CH_3CHOHCO_2H \xrightarrow{Na} CH_3CH(\bar{O}Na^+)CO_2^-Na^+$

Both OH groups are attacked by sodium.

13. Both mustard gas and primary alkyl halides contain the CH_2Cl group, and so hydrolysis might have been expected to occur at comparable rates. The fact that the rate is *much faster* than expected for mustard gas suggests the possibility of anchimeric assistance, i.e., neighbouring group participation. Since sulphur can act as a donor atom (*note:* this atom is *not* present in alkyl halides), this suggests the following possible mechanism:

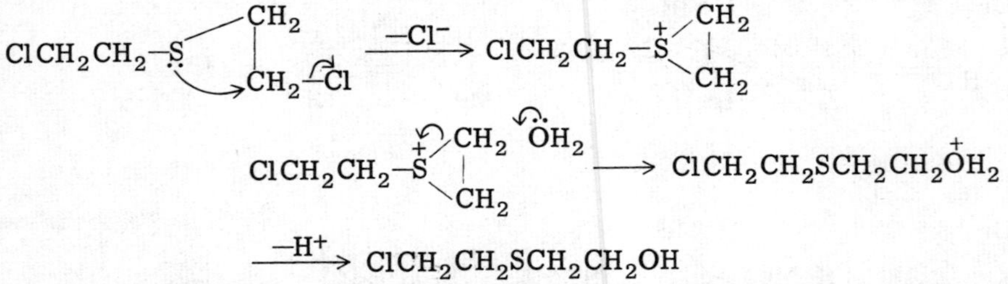

As the equations stand, it is not certain which is the r/d step (first or second). However, since the rate of hydrolysis is much faster than that of primary alkyl halides, it follows that the activation energies of *both* steps must be lower than the activation energy of the one-step (S_N2) mechanism for primary alkyl halides.

The student should also note that inductive effects can change rates of reaction (occurring by a *given* mechanism).

Actually, the electronegativities of C and S are the same (see Table 2.1), and so we can discount the inductive effect of sulphur affecting the rate of hydrolysis.

14. Since we are given the M.W. of the ketone, we can obtain its molecular formula as follows. The general formula of aliphatic ketones is $C_nH_{2n}O$. It might be noted here that should the student have forgotten the general formula of a homologous series, he can readily work it out by writing down the molecular formulæ of the first two members:

$$CH_3COCH_3 = C_3H_6O; \quad CH_3COCH_2CH_3 = C_4H_8O$$

Hence, for n C there are $2n$ H.

Now, returning to our problem.

$$C_nH_{2n}O\text{: } 12n + 2n + 16 = 86; \therefore n = 5$$

$C_5H_{10}O$ gives rise to three possible isomeric ketones:

$$CH_3COCH_2CH_2CH_3 \quad CH_3CH_2COCH_2CH_3 \quad CH_3COCH(CH_3)_2$$

Formation of the amine may be represented as:

$$R^1COR^2 \xrightarrow{NH_2OH} R^1C({=}NOH)R^2 \xrightarrow{\text{redn.}} R^1CH(NH_2)R^2$$

Since the amine can be resolved, the C of the $CHNH_2$ group must be an asymmetric carbon atom (chiral centre). Hence R^1 and R^2 must be different. This eliminates diethyl ketone (pentan-3-one) and so the ketone is either methyl propyl ketone (pentan-2-one) or isopropyl methyl ketone (3-methylbutan-2-one).

These two alternatives may be distinguished by NMR spectroscopy.

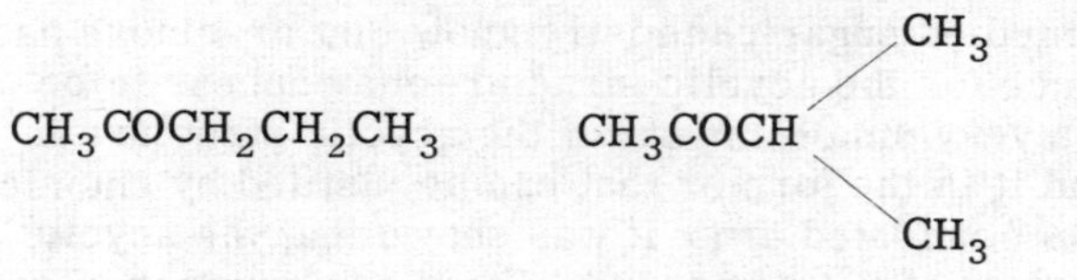

four signals three signals

15.

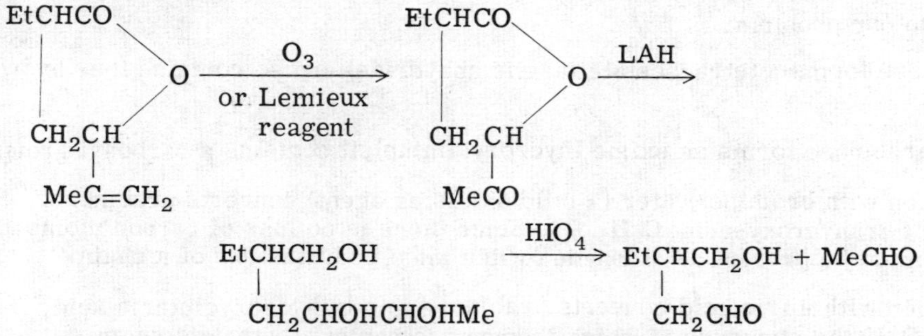

For a discussion of ozonolysis and the Lemieux reagent, see *Text*, Ch. 4. Lithium aluminium hydride reduces ketones to alcohols and lactones to diols (see also Table 6.1). Periodic acid oxidises only the 1,2-glycol structure in the compound (see *Text*, Ch. 11).

18 Carbohydrates

(Problems on page 31)

1. Since we are *given* the structure of arabinose, what we have to do is to show the presence of an aldehyde group, four hydroxyl groups, and that the carbon skeleton consists of five carbon atoms in a straight chain. Furthermore, because we have been given the structure, the approach to *justify* it is not necessarily the same as that if we had to *elucidate* it. For example, if we were starting from scratch, we would first show the presence of a carbonyl group by routine chemical tests and *then* show whether it is an aldehyde or a ketone. In our example, we know it is an aldehyde and so we are justified to show this group is present without, if it is possible, first showing that a carbonyl group (aldehyde *or* ketone) is present. It so happens that aldoses do not give some of the usual reactions of an aldehyde, e.g., Schiff's reaction. Hence, a student who showed the 'presence' of an aldehyde group by means of Schiff's reaction would be wrong even though his method was reasonable. Obviously, then, to get the right answer, the student would have to have some specific knowledge about sugars. Those students who have studied sugar chemistry know that arabinose has a *cyclic* structure, and so to ask for evidence for the acyclic structure may appear to be unjustified. However, in solution, there is a very small amount of the acyclic structure in equilibrium with the cyclic structure, and it is the former that can be justifed by chemical evidence. The structure of the latter was elucidated *after* it was shown that the acyclic structure did not satisfy all of the properties of the compound. There is a number of cases of this kind where a structure has been modified in the light of later information (this excludes cases where the completely *wrong* structure had been assigned in the first place).

We now return to our problem.

(i) Arabinose forms a tetra-acetate (acetic anhydride) and so contains four hydroxyl groups.

(ii) Since arabinose forms an oxime (hydroxylamine), it contains a carbonyl group.

(iii) Oxidation with bromine-water (a mild oxidising agent) converts arabinose, $C_5H_{10}O_5$, into a tetrahydroxy-acid, $C_5H_{10}O_6$. Since there is no loss of carbon atoms, the carbonyl group must be present as aldehyde (which *must* be at the *end* of a chain.

(iv) Oxidation with nitric acid converts arabinose into trihydroxyglutaric acid, $HO_2C(CHOH)_3CO_2H$, the structure of which is known to contain a straight chain.

2. The general rule is that groups of the following type undergo fission by the action of periodic acid:

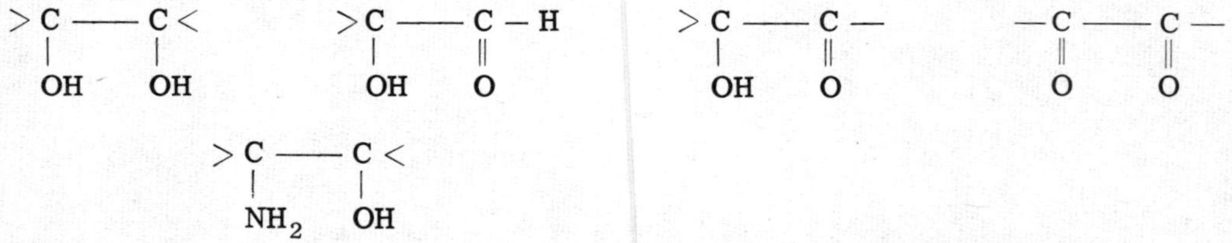

(i) $CH_3CHOH{-}CH_2OH \xrightarrow{1\ HIO_4} CH_3CHO + CH_2O$

(ii) $HOCH_2{-}CO{-}CH_2OH \xrightarrow{2HIO_4} CH_2O + CO_2 + CH_2O$

(iii) $HOCH_2{-}CHOH{-}CHOH{-}CH_2OH \xrightarrow{3HIO_4} CH_2O + HCO_2H + HCO_2H + CH_2O$

(iv) $HOCH_2CH_2CHOHCO_2H \xrightarrow{HIO_4}$ no reaction

(v) $HO_2CCHOH{-}CHOH{-}CHO \xrightarrow{2HIO_4} HO_2CCHO + HCO_2H + HCO_2H$

(vi) $OCHCH_2CHOMeCHOH{-}CH_2OH \xrightarrow{1HIO_4} OCHCH_2CHOMeCHO + CH_2O$

(vii) $OCHCHOMeCHOHCH_2OEt \xrightarrow{HIO_4}$ no reaction

The equations would not normally be written as shown; it has been done this way to show the fate of each carbon atom that has been oxidised.

3. (i)

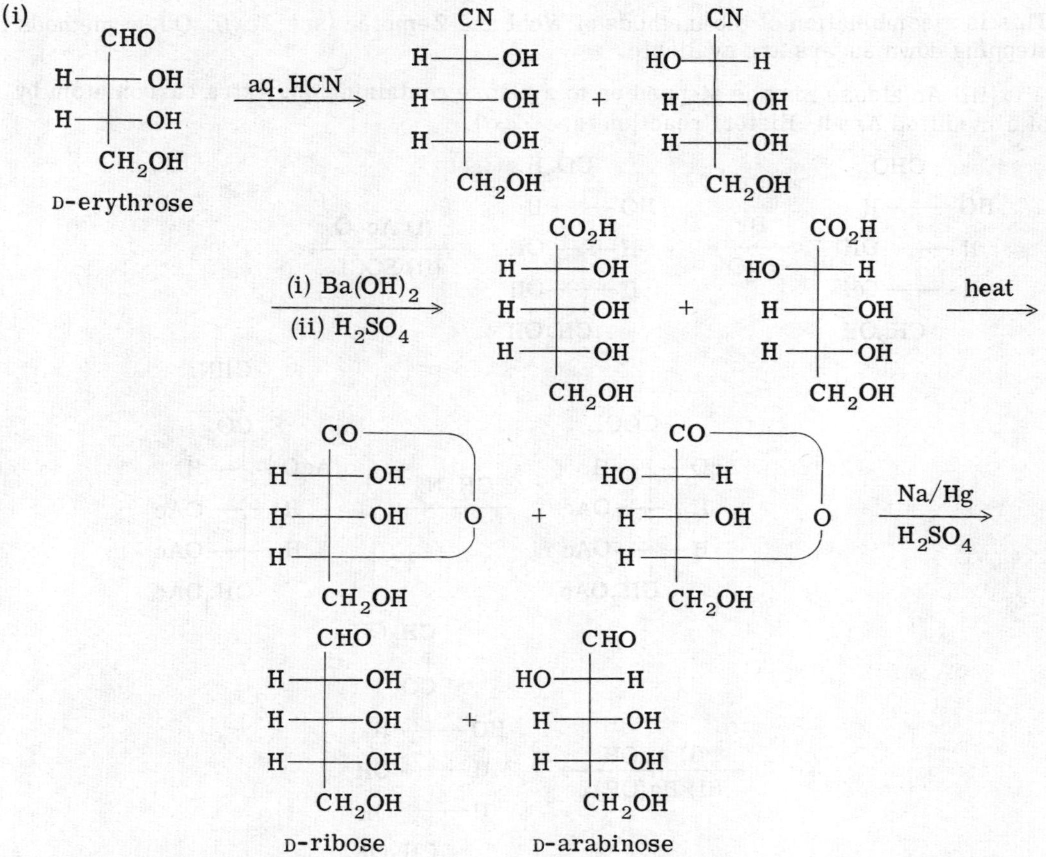

This is an example of the Kiliani reaction. The addition of HCN creates a new asymmetric carbon atom (chiral centre), finally resulting in a mixture of epimeric aldopentoses in unequal amounts. Also note the reduction of the γ-hydroxylactones to aldehydes (see also *Text*, Ch. 17).

Other methods for stepping up sugars are also available.

(ii)

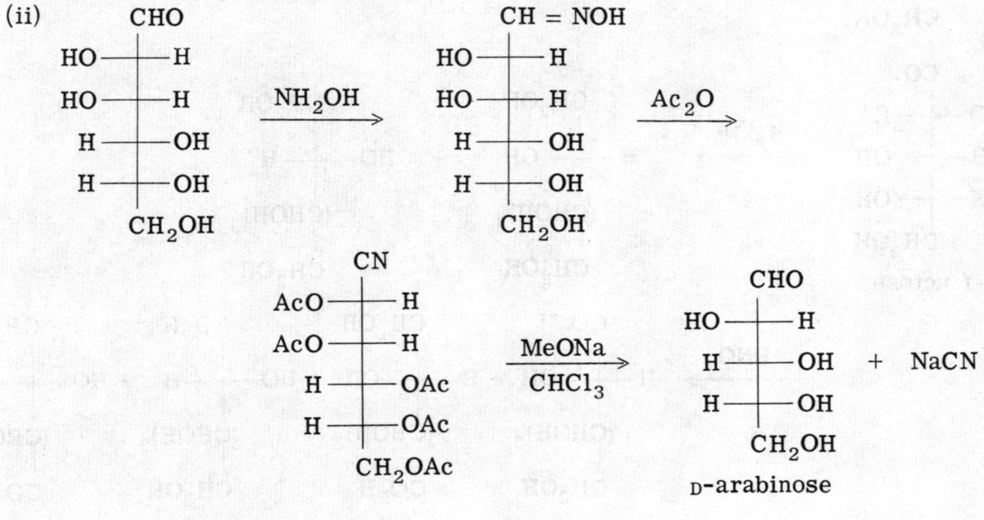

This is a combination of the methods of Wohl and Zemplén (see *Text*). Other methods for stepping down sugars are available.

(iii) An aldose may be stepped up to a ketose containing one extra carbon atom by means of a modified Arndt–Eistert reaction (see *Text*).

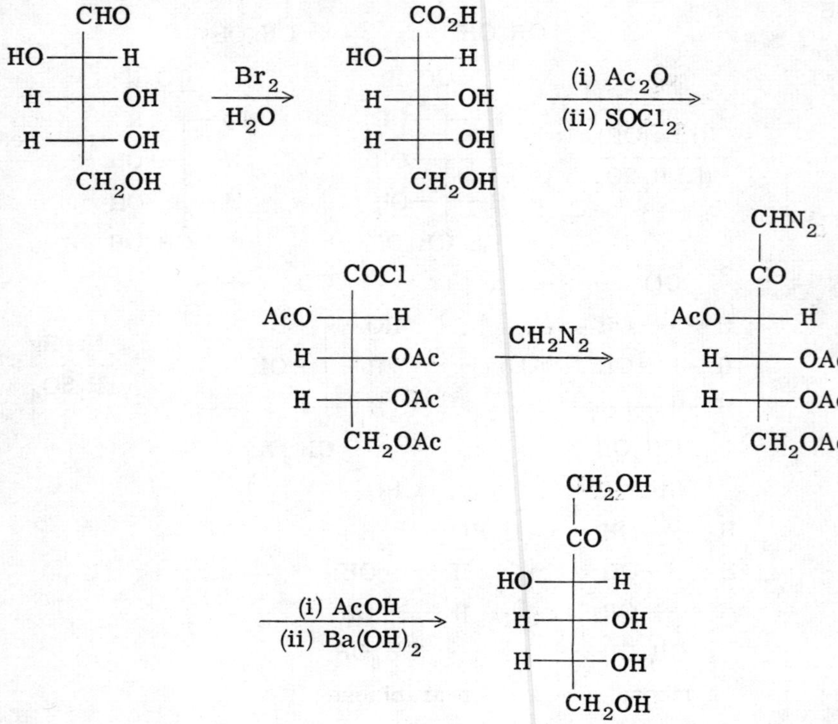

D-fructose

(iv) Conversion of a ketose into the next lower aldose requires the transformation of the ketose into an aldose with the same number of carbon atoms. This may be done as follows.

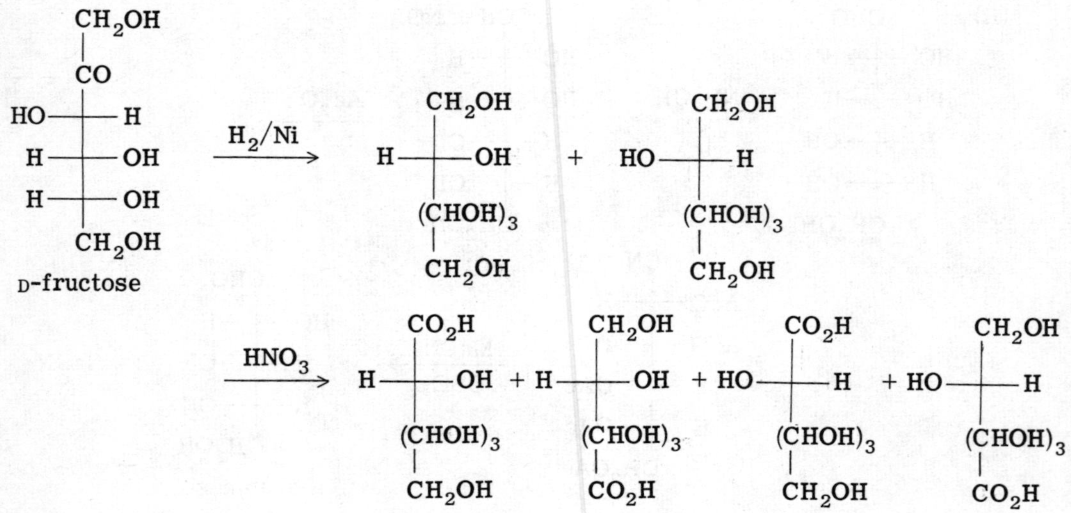

Reduction produces two hexitols, and each of these gives two carboxylic acids arising from oxidation at either end. Thus, the product is a mixture of four carboxylic acids (not present in equal amounts). Each acid may be converted into its γ-lactone which, on reduction with sodium amalgam in faintly acid solution, will give an aldose (see (i), above). The final product will therefore be a mixture of four aldohexoses. These, on stepping down to the next lower aldose (see (ii), above), will result in a mixture of four aldopentoses.

Since pentoses and hexoses actually have cyclic structures of the pyranose type, their formulæ should be written as such. Which type the student uses, planar or conformational, will probably depend on what his lecturer recommends.

Many students find it difficult to remember the configurations of the various pentoses and hexoses. The following scheme is offered as a possible aid.

First write down the configurations of the four aldopentoses as shown. The student *must* know the names of these pentoses.

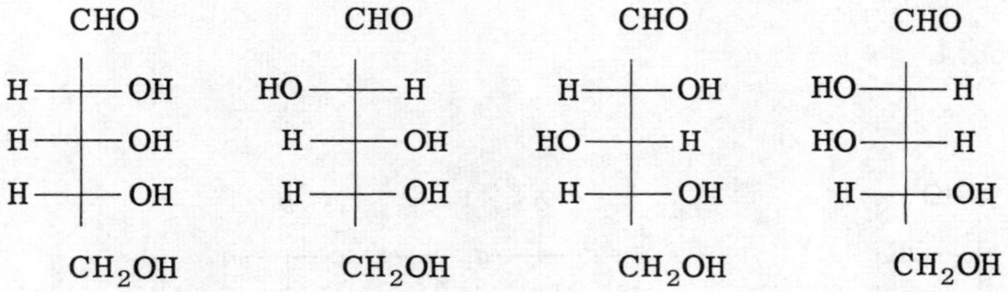

D-Ribose **D-Arabinose** **D-Xylose** **D-Lyxose**

The first pentose is written with all hydrogens to the left. The next formula is written with the first hydrogen to the right. The next formula is written with the second hydrogen to the right. So far, we have inverted the configuration, one at a time, working downwards. If we do this with the lowest asymmetric carbon atom, we get another pentose, but it will be the L-form. Hence, to have this in the D-form, we take its mirror image. The mnemonic is then **RAXL**.

Now we come to the aldohexoses. This also requires the names of the eight pairs of enantiomers and the pentoses from which they may be derived, the latter being part of the second nmemonic.

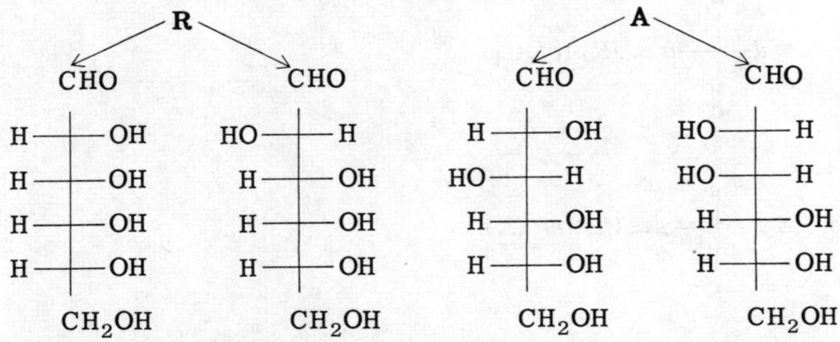

D-Allose **D-Altrose** **D-Glucose** **D-Mannose**

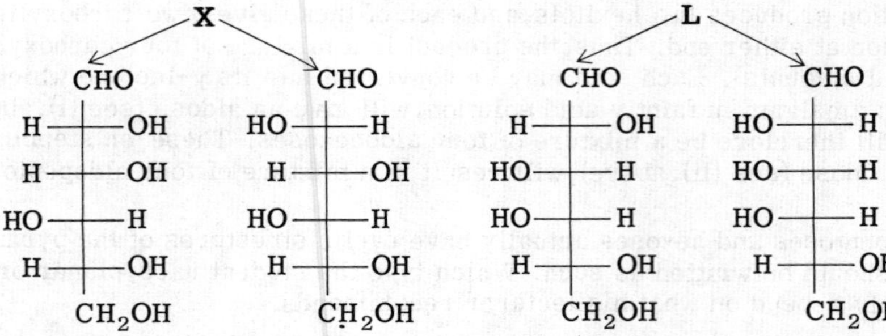

D-Gulose **D-Idose** **D-Galactose** **D-Talose**

Write down the configurational formula of each pentose twice and then step them up by one carbon atom, the formula on the *left* having the added hydrogen to the *left* and the formula on the *right* having the added hydrogen to the *right*. The mnemonics are: aRa (*allose, ribose, altrose*. Allose precedes altrose alphabetically); *glAm* (*glucose, arabinose, mannose*); guXi (*gulose, xylose, idose*); gaLYt; (*galactose, lyxose, talose*). The mnemonics should be pronounced as words (guxi is a beauty!)

4. See also Ch. 17, Prob. 4.

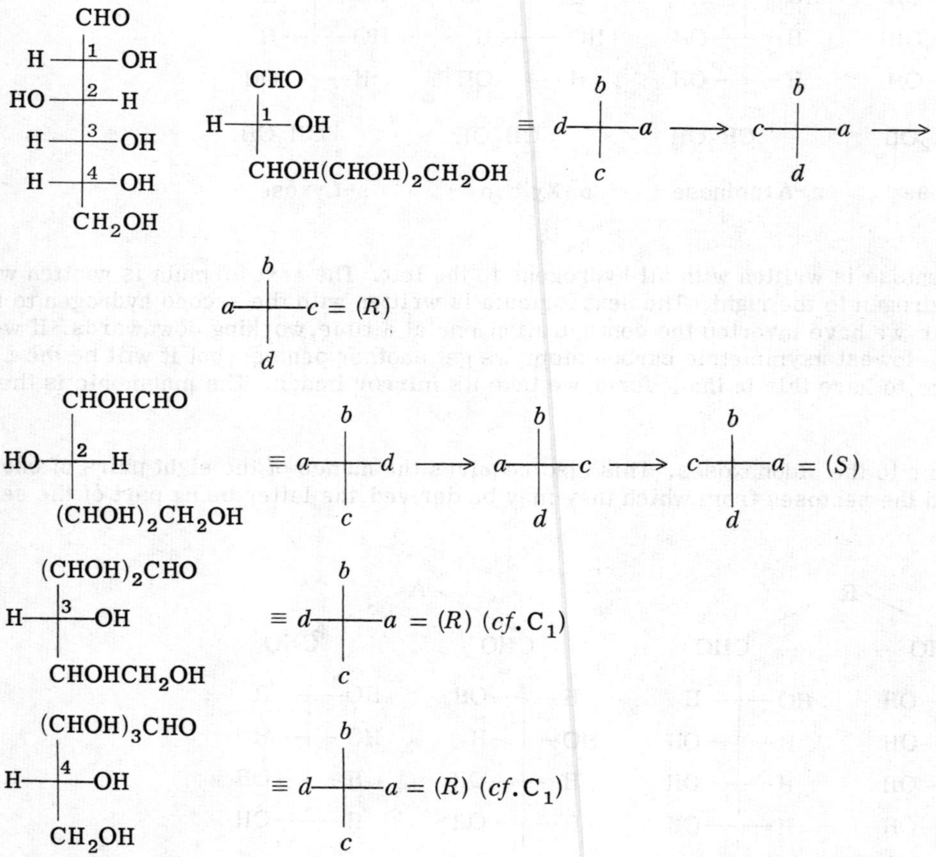

Hence, D(+)-glucose is:

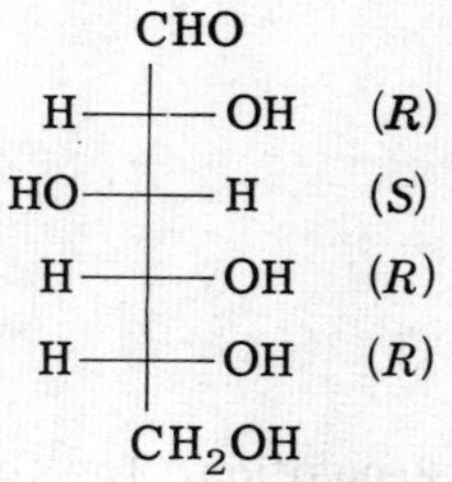

$(2R3S4R5R)$-2, 3, 4, 5, 6-pentahydroxyhexanal

5. (i)

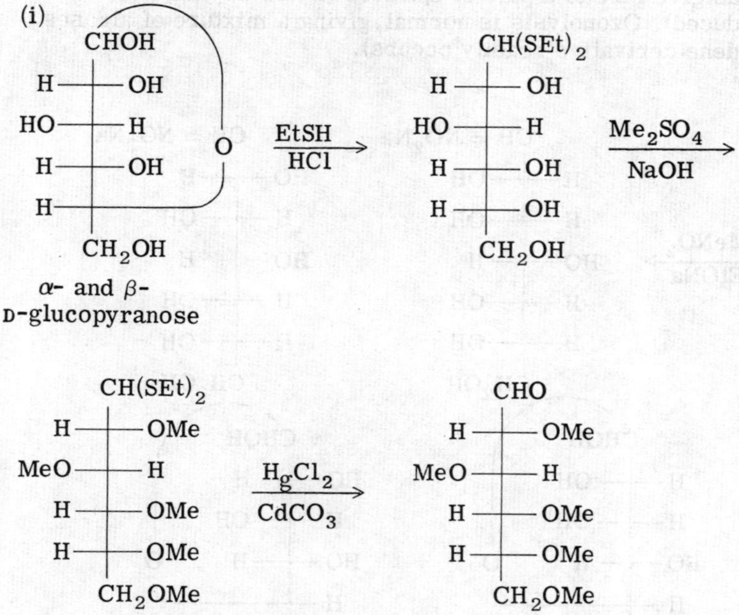

D-Glucose has a ring structure (pyranose), which is opened to give the acyclic mercaptal. Methylation, followed by treatment with mercuric chloride and cadmium carbonate, regenerates the aldehyde group (see *Text*, Ch. 14). Since all hydroxyl groups are methylated, ring-closure is prevented, resulting in the acyclic product, 2, 3, 4, 5, 6-penta-*O*-methyl-D-glucose. Methylation of D-glucose itself gives the cyclic product, methyl 2, 3, 4, 6-tetra-*O*-methyl-D-glucoside (see *Text*).

(ii)

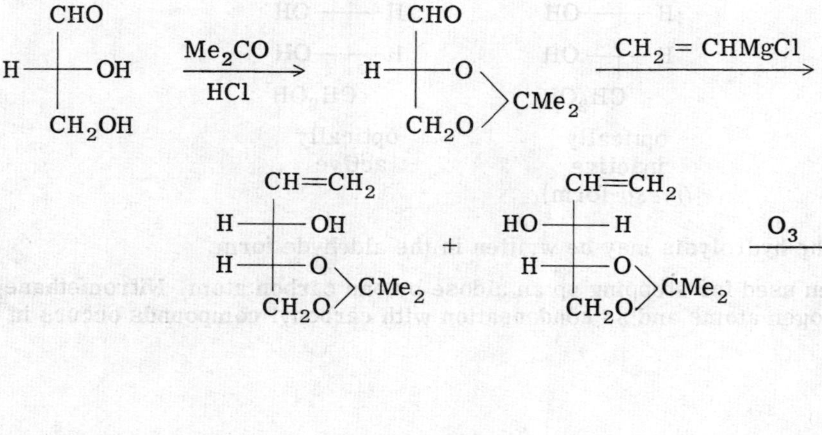

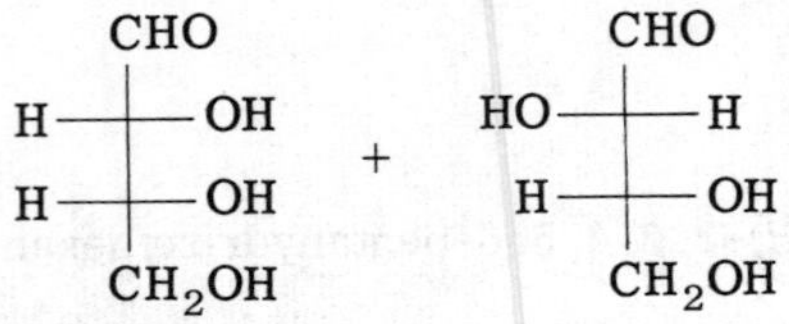

D-erythrose D-threose

Isopropylidene formation normally involves hydroxyl groups in the 1, 2-position. The Grignard reacts with the aldehyde group (the hydroxyl groups are protected) to give a secondary alcohol. However, addition can give rise to a pair of epimers (a new asymmetric carbon atom (chiral centre) is introduced). Ozonolysis is normal, giving a mixture of aldoses (hydrolysis of the isopropylidene derivative readily occurs).

(ii)

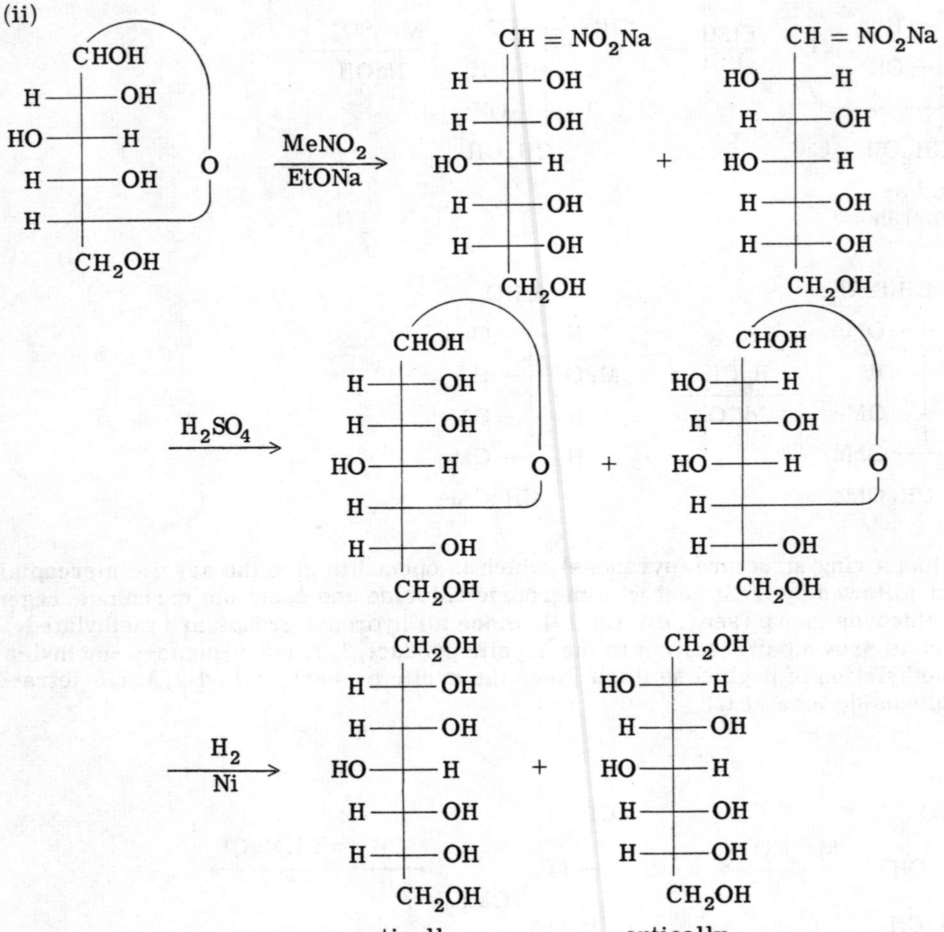

optically inactive (*meso*-form) optically active

The products from the hydrolysis may be written in the aldehyde form.

This method has been used for stepping up an aldose by one carbon atom. Nitromethane contains active hydrogen atoms and so condensation with carbonyl compounds occurs in the presence of sodium ethoxide (see *Text*, Ch. 8). A possible mechanism (involving the acyclic form) is:

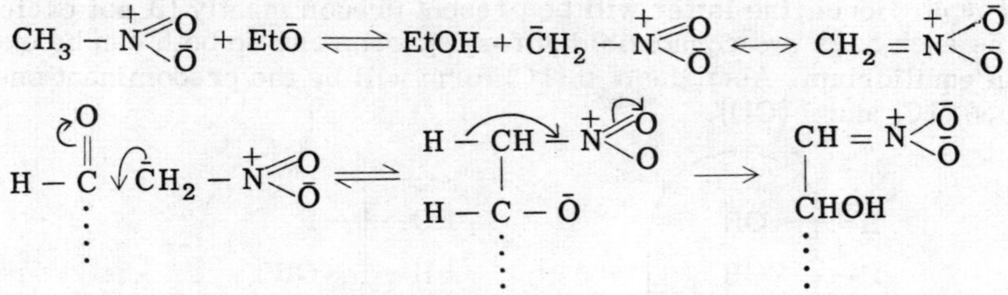

6. This question requires some knowledge of conformational analysis of ring systems (see *Text*, Ch. 19). There are two major conformations for sugar rings, C1 and 1C, one changing into the other by 'flipping'.

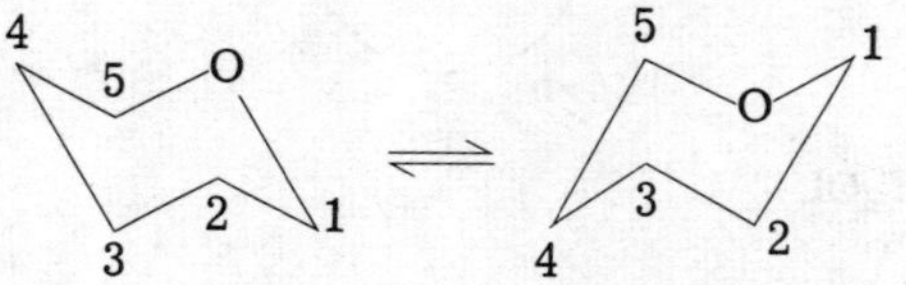

C1 1C

The general principle for unsubstituted sugars is that the form with the largest number of hydroxyl groups equatorial (including the CH_2OH when present) is the more stable one. This is usually the factor that decides whether the predominant ring is C1 or 1C.

(i)

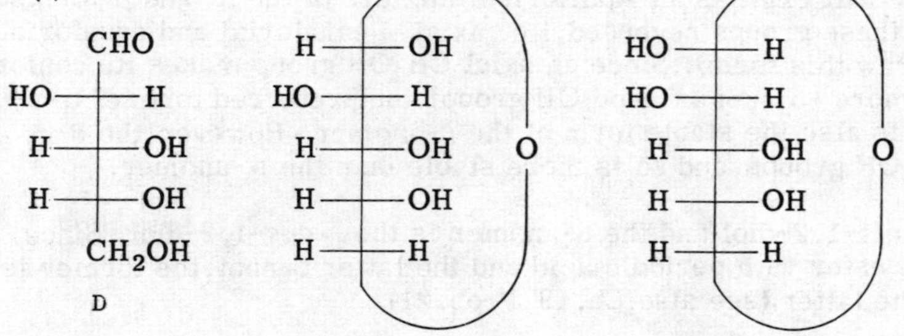

D α-D β-D

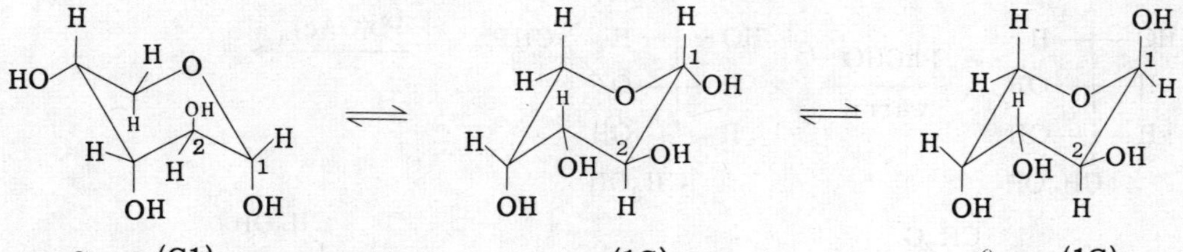

α – D (C1) α – D (1C) β – D (1C)

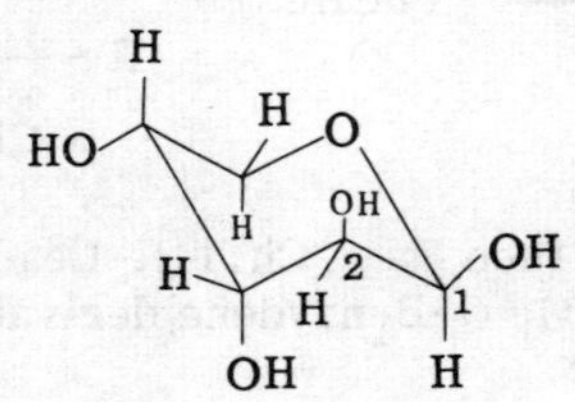

β – D (C1)

In solution, D-arabinopyranose will exist as an equilibrium mixture of the α- and β-forms. The α-D (C1) form has three axial hydroxyl groups, whereas the α-D (1C) form has three equatorial hydroxyl groups. Hence, the latter will be present predominantly (if not exclusively). Both β-D forms each have two equatorial hydroxyl groups and so both can be expected to be present in equilibrium. Also, the α-D (1C) form will be the predominant one in the mixture (of α (1C), β (1C) and β (C1)).

(ii)

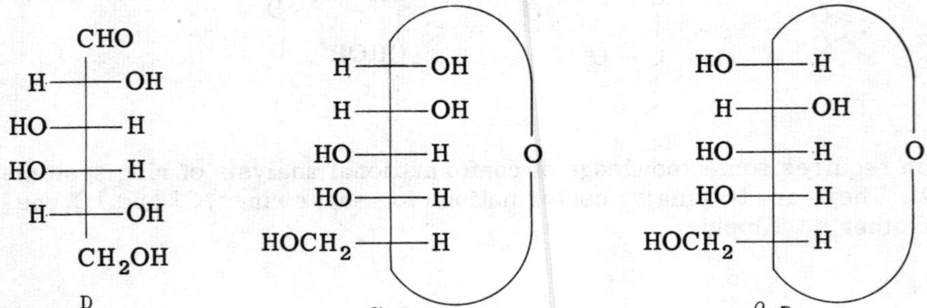

D α-D β-D

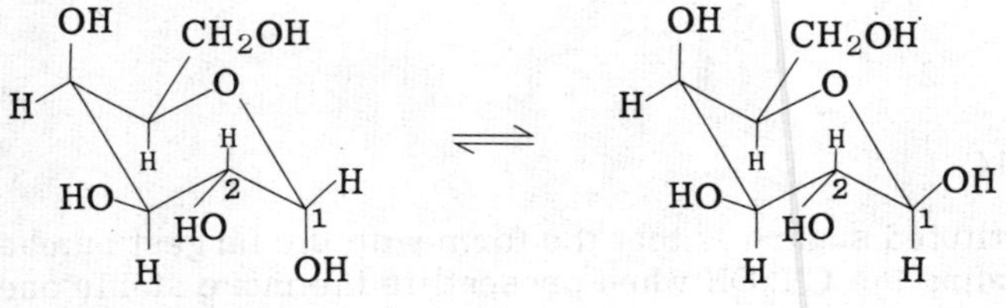

α — D (C1) β — D (C1)

In solution, D-galactopyranose will exist as an equilibrium mixture of the α- and β-forms. The α-D (1C) form will have these groups reversed, i.e., axial → equatorial and equatorial → axial (the student should draw this form). Since an axial CH_2OH group makes its conformation very unstable (much more so than an axial OH group), the preferred form of α-D-galactopyranose is C1. This is also the stable form of the β-anomer. However, the β-anomer has three equatorial OH groups, and so is more stable than the α-anomer.

7. α-Glucopyranose is the *cis*-1, 2-diol and the β-anomer is the *trans*-1, 2-diol. Since the former can form a *cyclic* ester with periodic acid and the latter cannot, the former is oxidised more rapidly than the latter (see also Ch. 19, Prob. 21).

8.

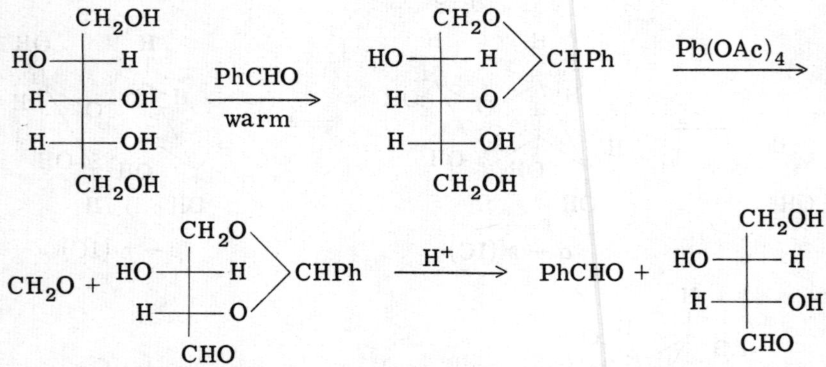

Benzaldehyde combines with *cis*-1, 3-diols (see *Text*, Ch. 11). Lead tetra-acetate oxidises 1, 2-diols (*cf.* periodic acid; see *Text*, Ch. 11). *O*-Benzydene derivatives are readily hydrolysed by acids.

The aldose formed above is D-threose. An alternative benzylidene derivative is also formed, and this finally gives D-erythrose.

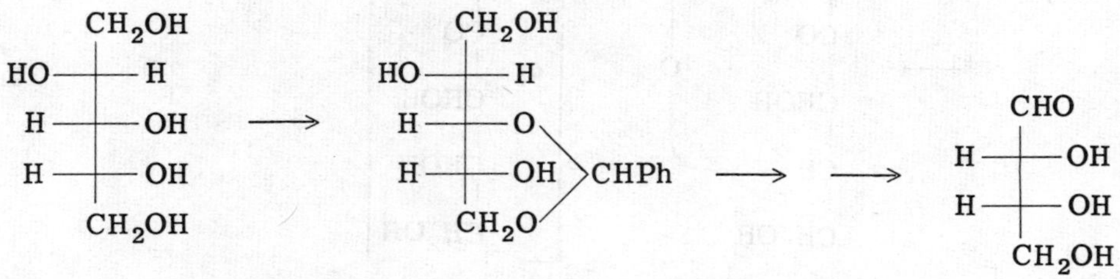

9. In order to oxidise C-3, all other CHOH groups must be protected. Now, acetone forms isopylidene derivatives with *cis*-1, 2-diols. Furthermore, because formation of the di-isopylidene derivative is so highly favoured, this is the usual product from the reaction between a sugar and acetone. Also, to form this di-isopropylidene derivative, the ring will, if necessary, change from pyranose to furanose. If we consider the D-aldohexoses, then all eight diastereoisomers can form the 1, 2-derivative, since the α- or β-anomer will be *cis* with respect to the hydroxyl group on C-2. If the hydroxyl groups on C-3 and C-4 are *cis*, then these form the 3, 4-isopropylidene derivative. In this case, the C-3 group is protected, and so the oxidation to the 3-ketone will not proceed. If, however, the hydroxyl groups on C-3 and C-4 are *trans*, the 5, 6-isopropylidene derivative is formed, the ring changing from pyranose to furanose. This has the C-3 hydroxyl group free and this can be oxidised to the 3-ketone. Thus:

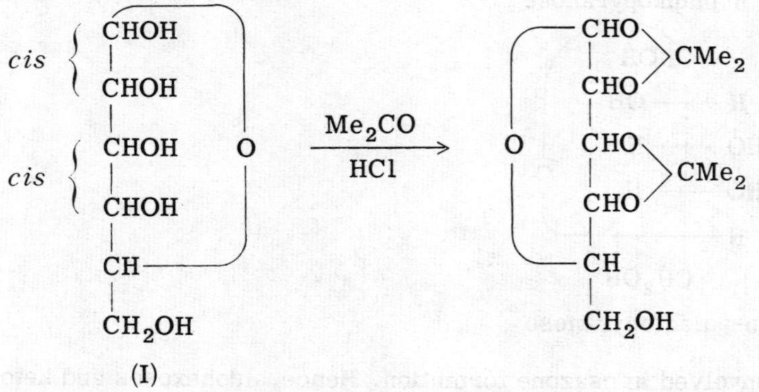

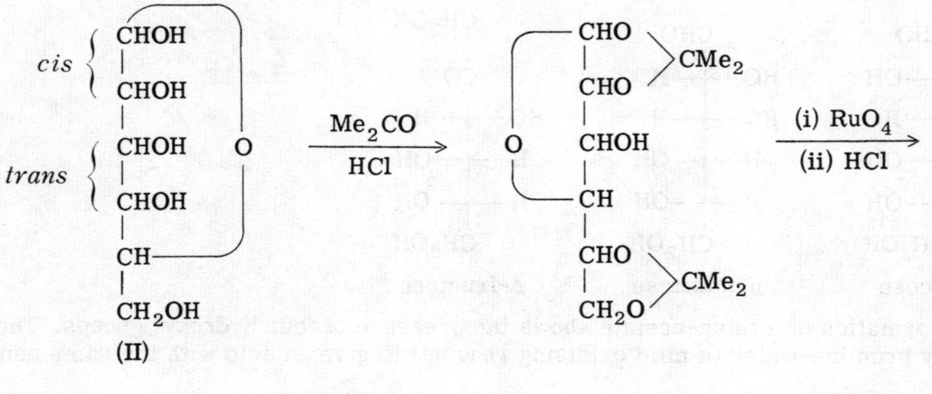

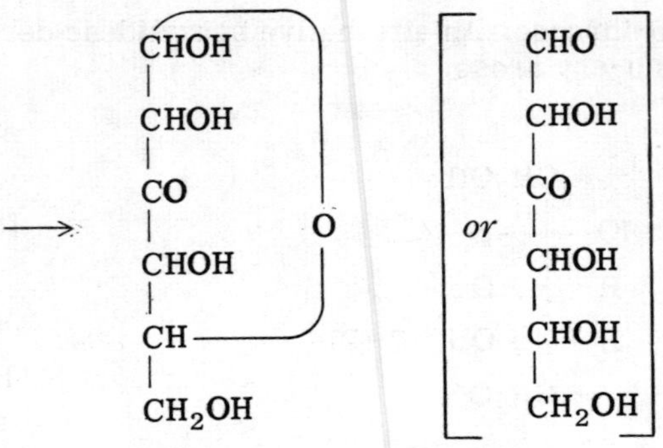

Aldohexoses of type (I) are: allose, altrose, galactose, and talose; and those of type (II) are: glucose, mannose, gulose, and idose. Only those of type (II) will form the 3-ketone via their di-isopropylidene derivatives.

10.

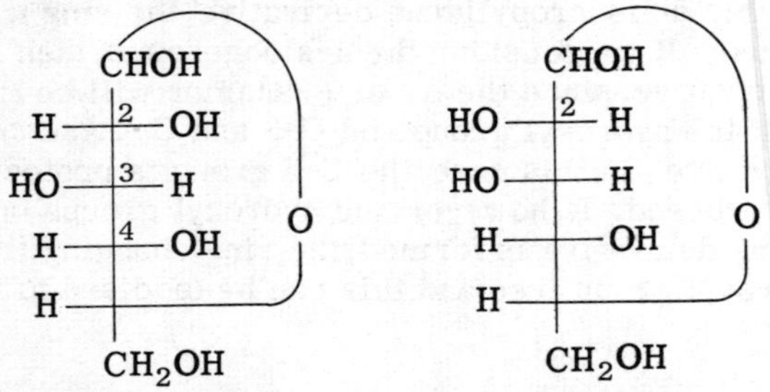

D-glucopyranose D-mannopyranose

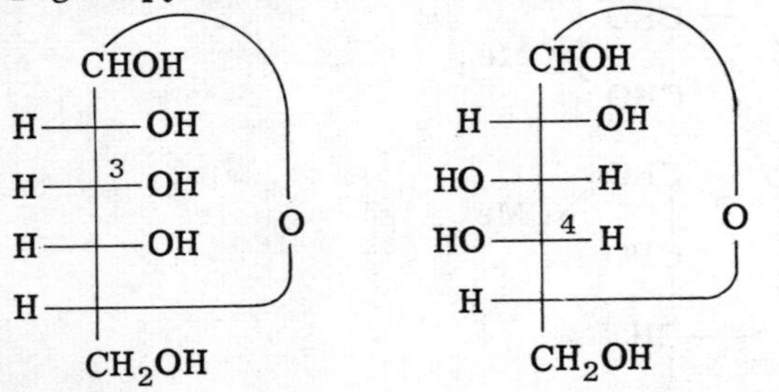

D-allopyranose D-galactopyranose

11. Only C-1 and C-2 are involved in osazone formation. Hence, aldohexoses and keto-hexoses which have the same configurations at C-3, C-4, and C-5, give the same osazone.

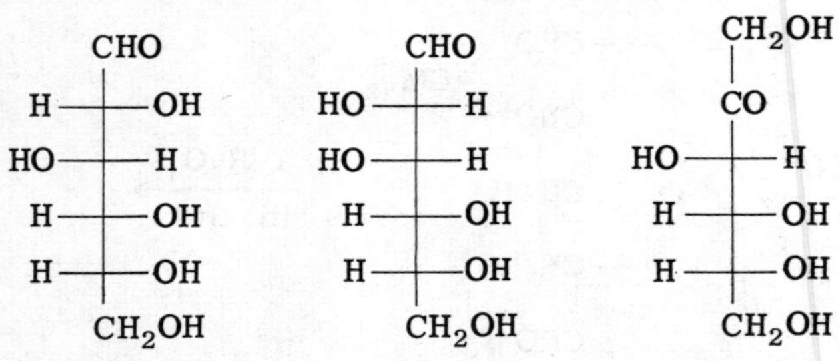

D-glucose D-mannose D-fructose

12. The formation of a tetra-acetate shows the presence of four hydroxyl groups. The oxidation by bromine-water (a mild oxidising reagent) to give an acid with the same number of carbon atoms and the fact that **A**, $C_5H_{10}O_4$, adds oxygen to give $C_5H_{10}O_5$ (the acid) indicate that **A** is an aldehyde. Reduction with HI to isopentane shows that **A** has this carbon skeleton. Hence, to obtain possible structures of **A**, we write down the isopentane skeleton, place the aldehyde group at a terminal carbon and then insert four hydroxyl groups, one to each remaining carbon atom. This is repeated if the carbon skeleton is unsymmetrical with respect to its ends (the longest carbon chain). In this way, it will be found that two structural isomers are possible.

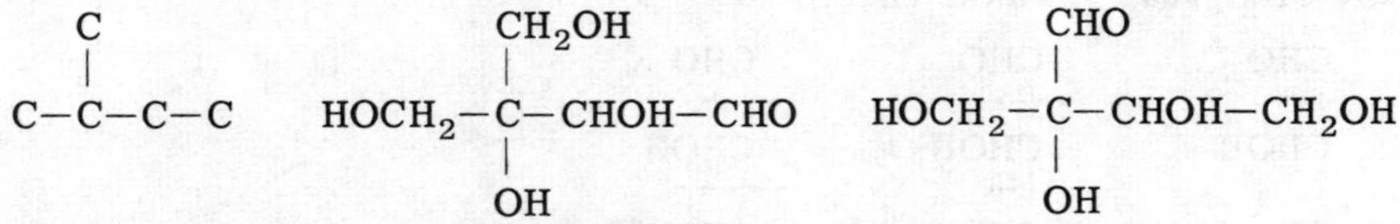

(I) [apiose] (II)

Because of the complexity of these structures and their close similarity, it would be extremely difficult if not impossible (except for an expert) to differentiate between the two structures by spectroscopic methods (i.r. and NMR). This therefore leaves us in the dreadful situation of having to resort to chemical methods! With so many hydroxyl groups in vicinal positions, it is always worth trying the periodate oxidation. In this particular case, the method fails; both structures would consume four molecules of reagent and both would give $2CH_2O$, $2HCO_2H$ and $1CO_2$. In actual fact, since we are told that **A** is reduced by HI, this leads to a possible method. Alcoholic groups are far more readily reduced by HI—P than carboxyl groups. Inspection of (I) and (II) shows that reduction of the corresponding hydroxy-acids would give different products. Hence:

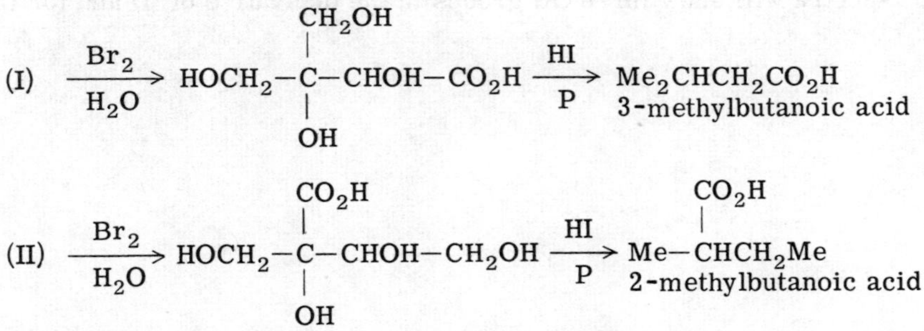

It might be noted that the two hydroxy-acids readily form their γ-lactones, but these with HI—P give the acids shown above.

13. **A** contains three hydroxyl groups and an aldehyde group (see Prob. 12, above). Formation of a lactone shows that one hydroxyl group is in the γ- or δ-position with respect to the carboxyl group. Since **A** contains four oxygen atoms, and these have been accounted for as three hydroxyl groups and an aldehyde group, the formation of an osazone shows the presence of the group —CHOHCHO (see *Text*). Since only one molecule of periodic acid is consumed, **A** contains only one set of adjacent oxidisable groups. This must therefore be the CHOHCHO group, and so the other CHOH groups must be 'separated' from each other and from the CHOH of CHOHCHO. Also, the absence of *four* hydroxyl groups in a five-carbon chain suggests **A** is a deoxy-compound. This 'deoxy-carbon atom' is therefore the one that 'separates' all the three CHOH groups. A structure which fits all the facts is:

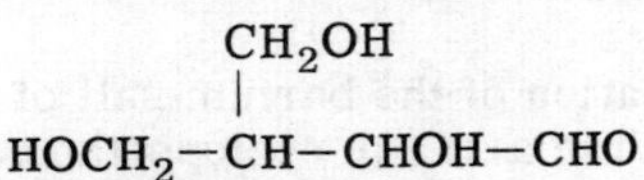

A (cordycepose)

This gives a γ-lactone (on oxidation to the acid).

14. **A** is an aldehyde, contains three hydroxyl groups, and the carbon skeleton consists of five carbon atoms in a straight chain (see Prob. 12, above). Also, the formula $C_5H_{10}O_4$ therefore suggests that **A** is a deoxy-sugar (see Prob. 13, above). If we now try to work out the possibilities based directly on the periodic oxidation of **A**, we shall find it difficult to get an answer. In a case like this, it is simpler to use what we have deduced so far and then see what can be deduced from the periodic acid oxidation. This approach is reasonable provided that the number of the possibilities is relatively small. Hence, we have (by systematically shifting the CH_2 group down the chain).

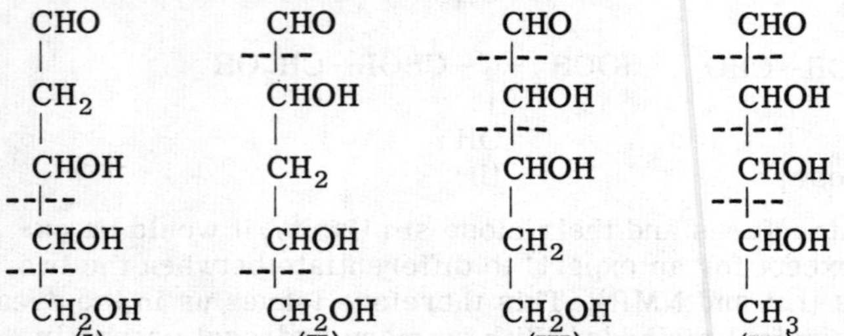

(I) (II) (III) (IV)

Now, the periodic acid oxidation produces one molecule of CH_2O and one molecule of HCO_2H. Inspection of the structures shows that (I) and (II), but not (III) and (IV), can give the required products. Hence, **A** is (I) or (II).

(I) and (II) may readily be distinguished by reaction with phenylhydrazine. (I) forms a phenylhydrazone, but (II) forms an osazone (the two derivates will have different elemental analyses; their NMR spectra will show three OH groups in the derivative of (I) and for (II), two OH groups).

19 Alicyclic compounds

(Problems on page 33)

1. Since the ring has changed size, one approach will be to deliberately open the ring of the starting material and reclose to get the product. The word *deliberately* is used to distinguish this method from reactions in which ring expansion or contraction occurs as part of those reactions.

Since 5- to 7-ring ketones are readily prepared by the distillation of the barium salt of a dicarboxylic acid (with loss of carbon dioxide), we require adipic acid for cyclopentanone. Since this acid contains six carbon atoms and so does cyclohexanone, this suggests a direct oxidation of the latter. Hence:

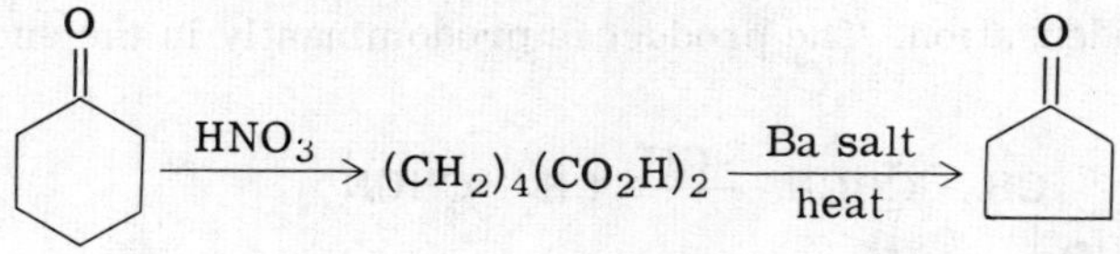

2.

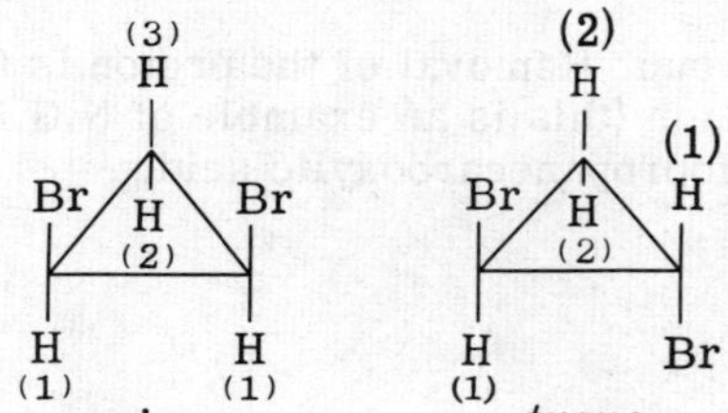

cis *trans*

In the *cis* compound, hydrogens labelled (1) are in identical environments and so show only one signal. H(2) and H(3) are not in identical environments, and so each will show its own signal (provided their τ-values are sufficiently separated; in this case, they are). Hence the *cis*-isomer will give three NMR signals (H(1) at lowest field).

In the *trans*-isomer, the hydrogens labelled (1) are equivalent, and those labelled (2) are also equivalent. Hence, this isomer gives two NMR signals (H(1) at lower field).

3.

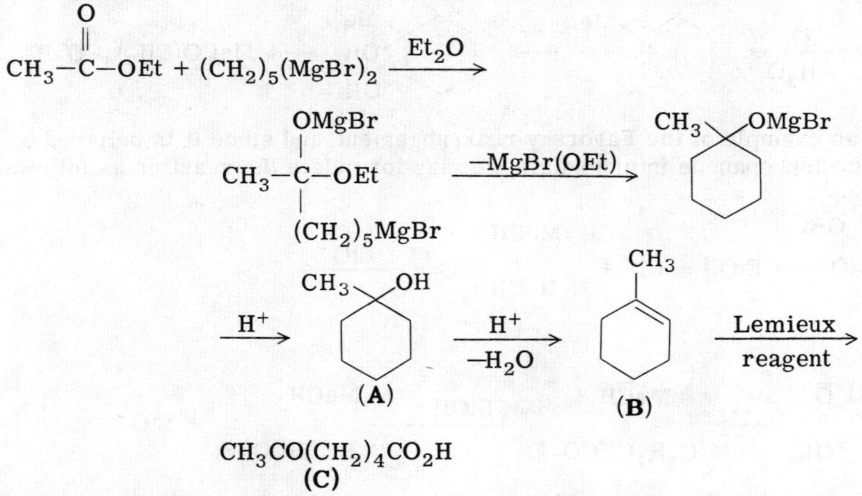

Esters react with two molecules of alkyl Grignard reagents to form alcohols. The di-Grignard reagent above reacts in an analogous manner to form a six-membered ring. The product, a tertiary alcohol, is readily dehydrated to the cycloalkene, and opening of the ring is effected by the Lemieux reagent (see *Text,* Ch. 4).

4. (i)

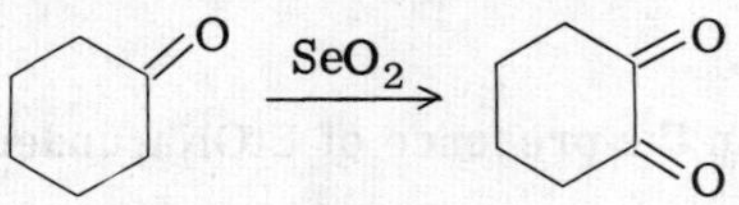

SeO_2 readily oxidises an active methylene group to CO.

(ii)

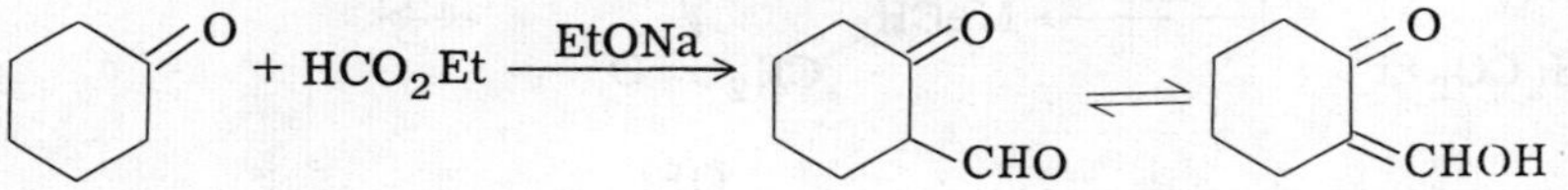

This is an example of the Claisen condensation. The product is predominantly in the enol form.

(iii) 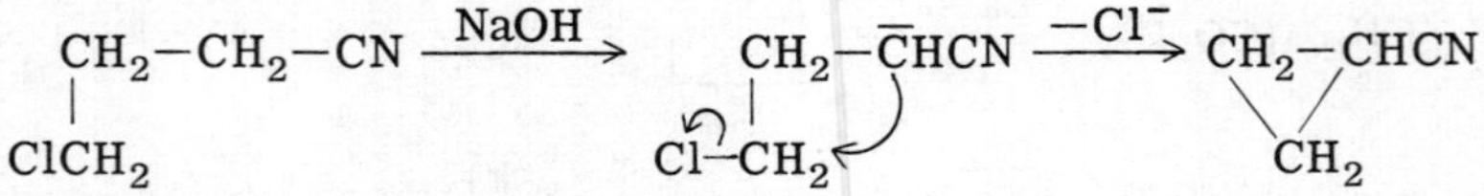

The cyano-group activates the adjacent methylene group. Removal of the proton is followed by ring-closure with explusion of the chloride ion (this is an example of N.G.P.; see *Text*, Ch. 17). The final product could be the cyclopropanecarboxylic acid.

(iv) 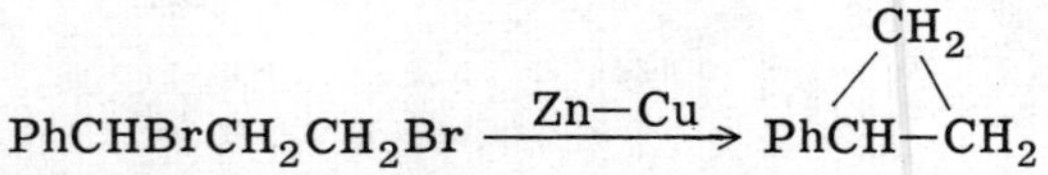

This is an example of ring-closure of α,ω-dihalogeno-alkanes (Freund reaction).

(v) 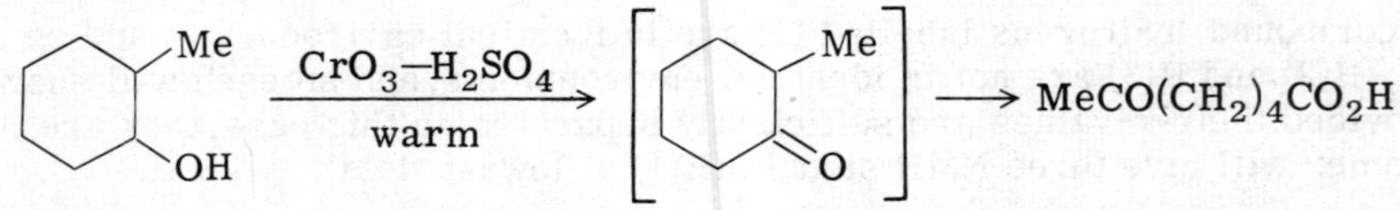

In view of the fact that the product, a keto-compound, is not further oxidised by the reagent, reaction through the keto-intermediate would appear to be unlikely. A better sequence is:

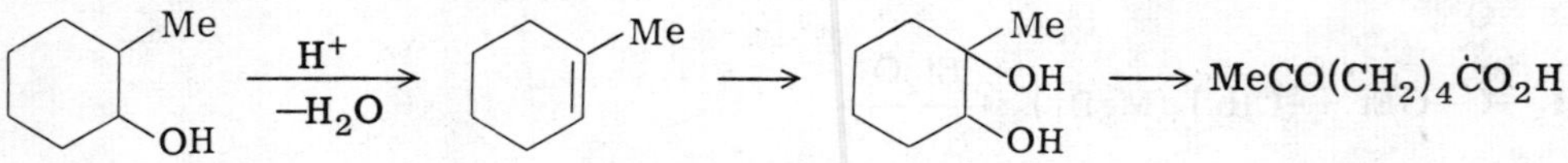

(vi) This is an example of the Favorsky rearrangement, and since it is believed to occur via a cyclopropanone intermediate, we may formulate the reaction as follows:

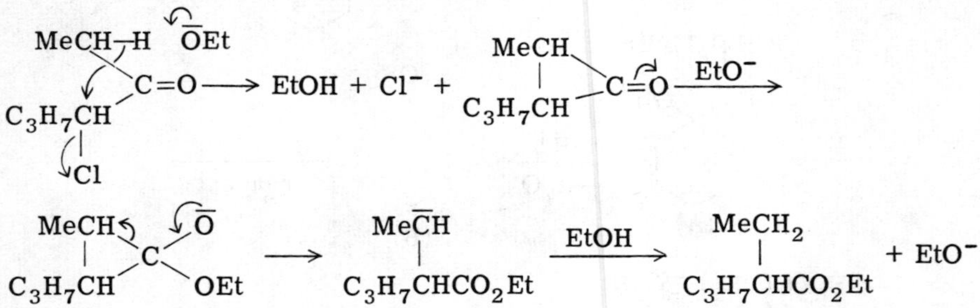

It can also be expected that the ring could open on the other side, i.e.,

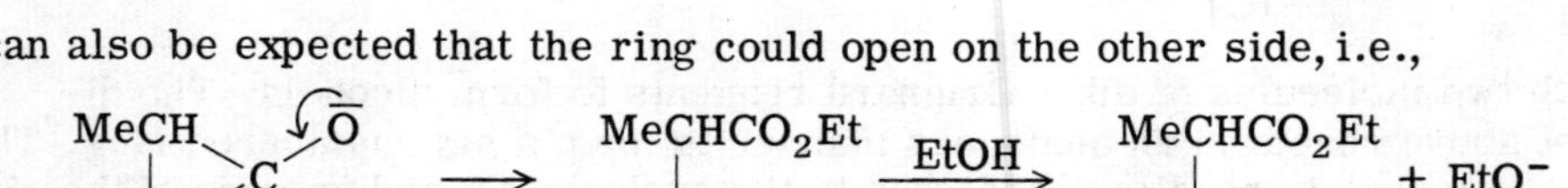

Hence, two products are possible.

(vii) The substrate is a 1,6-diester and so, in the presence of EtONa, undergoes the Dieckmann reaction.

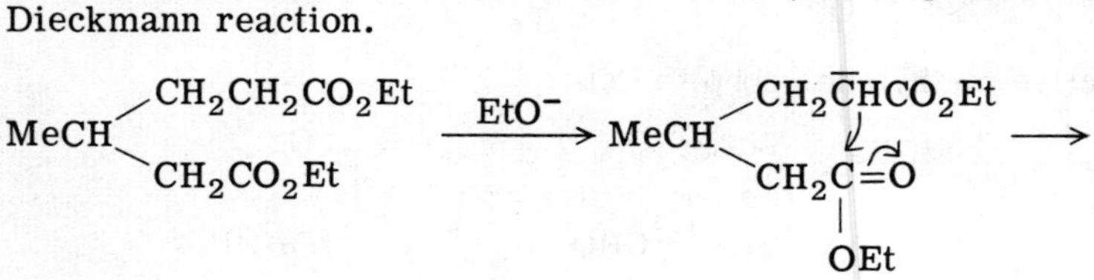

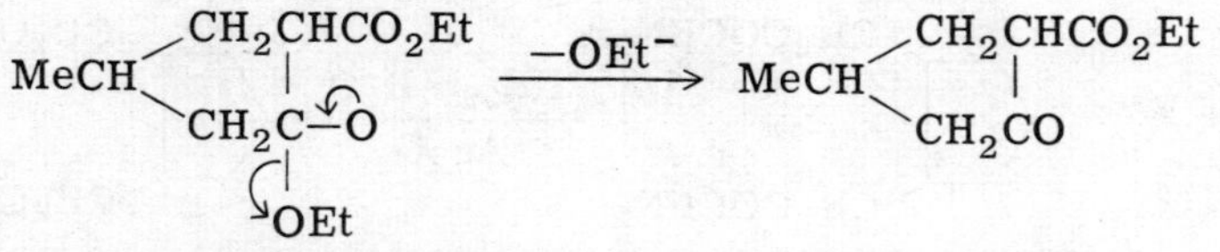

It is also possible for the other α-methylene group to be attacked to give a different (isomeric) product.

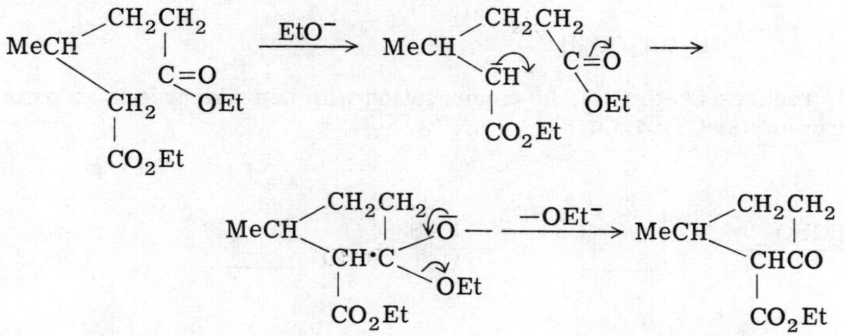

The first isomer is ethyl 4-methylcyclopentan-2-one-1-carboxylate, and the second isomer is ethyl 5-methylcyclopentan-2-one-1-carboxylate.

(viii) This is an example of the Dieckmann reaction.

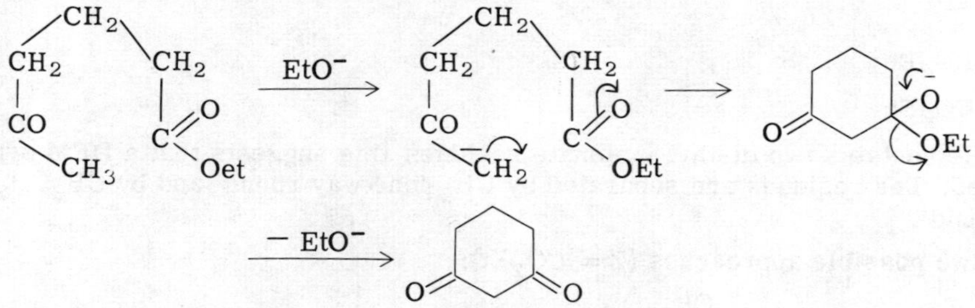

If the methylene group in the α-position to the CO group became the carbanionic centre, the product would be a four-membered ring ketone. If the methylene group in the α-position to the CO_2Et group became the carbanionic centre, the product would be a four-membered ring alcohol. Since four-membered rings experience large strain and six-membered rings do not, the latter are produced in preference to the former.

(ix) This is an example of the Schmidt reaction with ketones and acids (see *Text*, Ch. 8 and Ch. 9 for mechanisms).

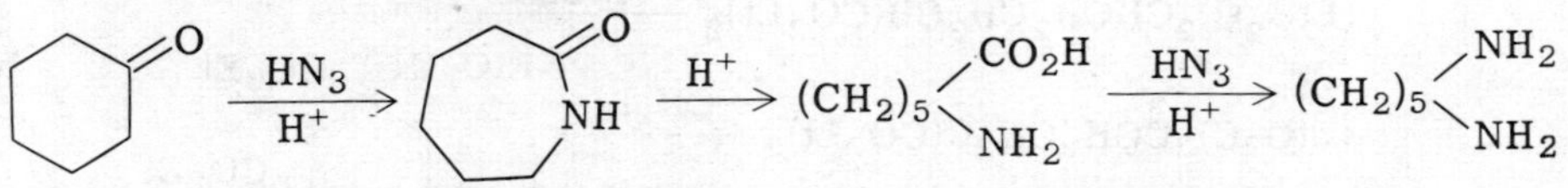

(x) This sequence of reactions involves the Arndt-Eistert synthesis (see *Text*, Ch. 13).

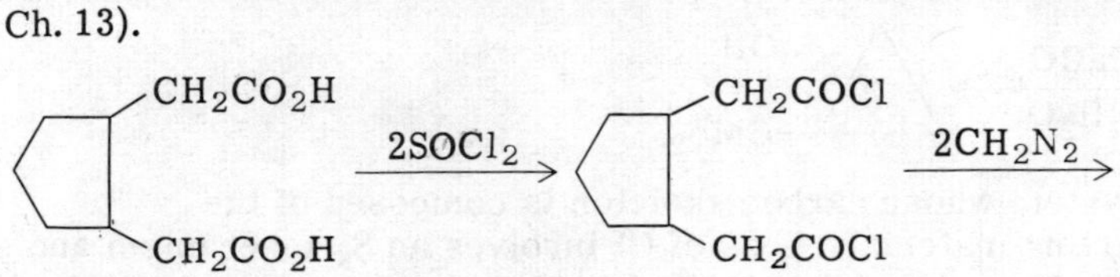

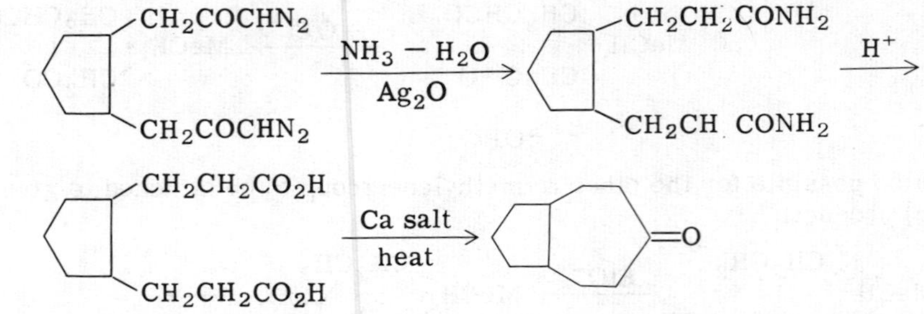

(xi) LAH—$AlCl_3$ reduces CO to CH_2; the condensation with benzaldehyde is an example of the Claisen reaction (see *Text*, Ch. 8).

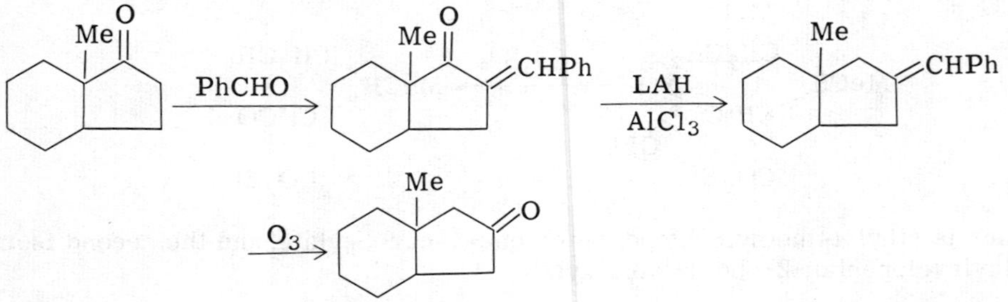

5.

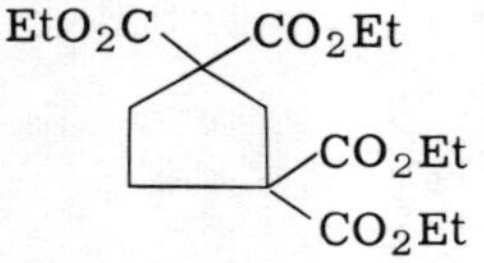

Since this molecule contains two diethyl malonate residues, this suggests that a DEM synthesis may be used. The residues are separated by CH_2 'one way round' and by CH_2CH_2 the 'other way round'.

Hence, there are two possible approaches ($Z = CO_2Et$):

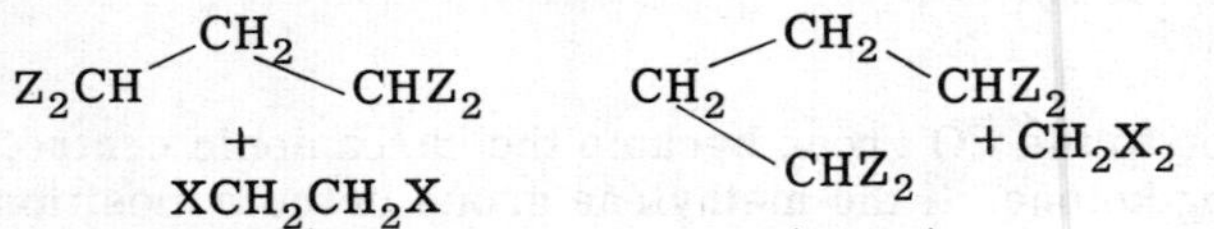

(3 + 2) (4 + 1)

We shall use the (4 + 1) to illustrate the synthesis.

$$2CH_2(CO_2Et)_2 \xrightarrow{2EtONa} 2\overset{+}{Na}\overset{-}{C}H(CO_2Et)_2 \xrightarrow{(CH_2)_2Br_2}$$

$$(EtO_2C)_2CHCH_2CH_2CH(CO_2Et)_2 \xrightarrow{2EtONa}$$

$$(EtO_2C)_2\bar{C}CH_2CH_2\bar{C}(CO_2Et)_2 \xrightarrow{CH_2I_2}$$

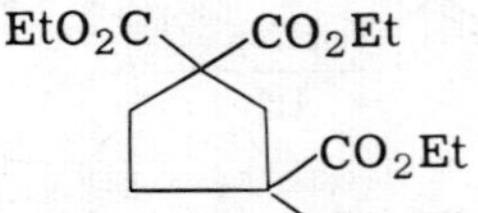

6.

$$Me_2C = CHCH_2CH_2Cl \xrightarrow[H_2O]{CaCO_3}$$

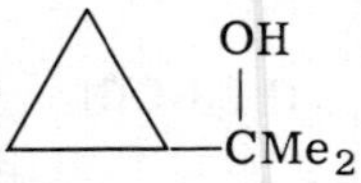

The product contains a ring system whose carbon skeleton is composed of the $=CHCH_2CH_2Cl$ part of the starting material. Loss of Cl involves an S_N mechanism and so the carbon atom bonding with the CH_2Cl carbon atom (to form the ring) must behave as the nucleophile. Since the 'nucleophilic carbon atom' is linked to its neighbour by a double bond, then a possible mechanism is one in which N.G.P. involves this double bond. Hence:

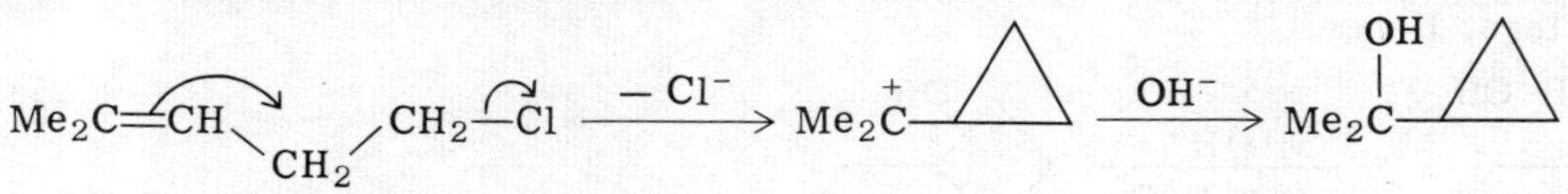

7. (i) The substrate is a 1, 2-diol, the reaction is carried out in acid solution, and the product is a carbonyl compound. These facts lead to the suggestion that a pinacol-pinacone rearrangement is involved (see *Text*, Ch. 8). Since the ring expands, this means that it is the carbon atom of the CPh_2OH group that becomes the carbonium ion centre (I). In the

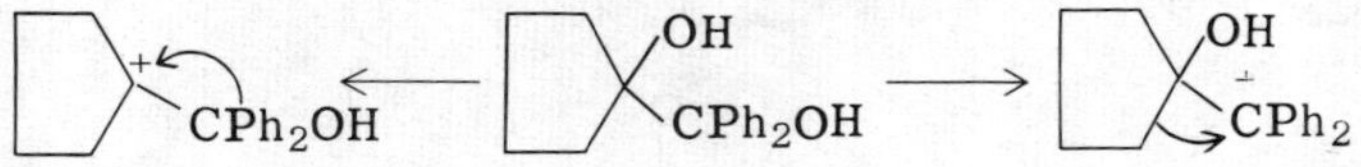

(II) (I)

alternative carbonium ion (II), a phenyl group would migrate and there would be no ring expansion.

What we have done so far is to recognise that the reaction is an example of the pinacol-pinacone rearrangement, and deduce which group migrates from the nature of the product. What we should now do is justify this preferential elimination of the OH group. This we can do quite readily on the basis of the relative stabilities of the two alternative carbonium ions (see also Ch. 8, Prob. 11). (I) is a tertiary ion which is stabilised by conjugation with the phenyl groups, e.g.,

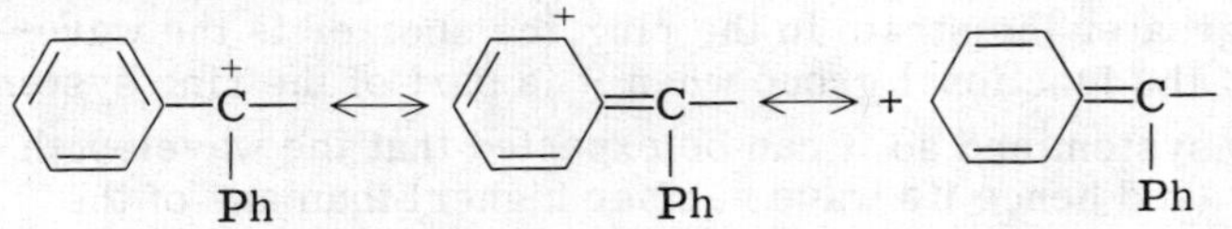

Since no such stabilisation is possible for the tertiary carbonium ion (II), (I) is formed. We may now write the mechanism as:

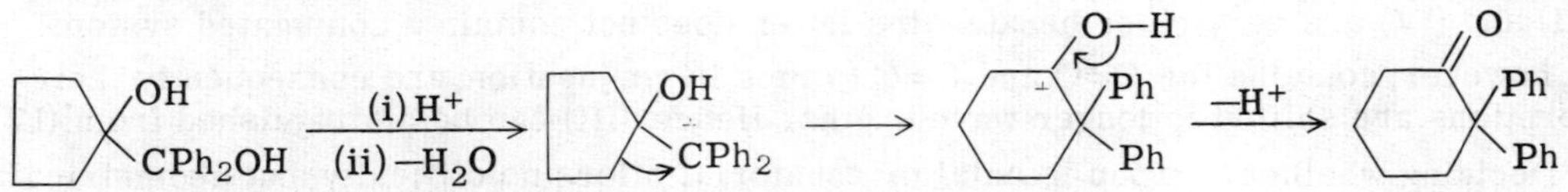

(ii) Since two products are formed, this means that the stabilities of the two possible carbonium ions are about the same. Hence:

(*a*)

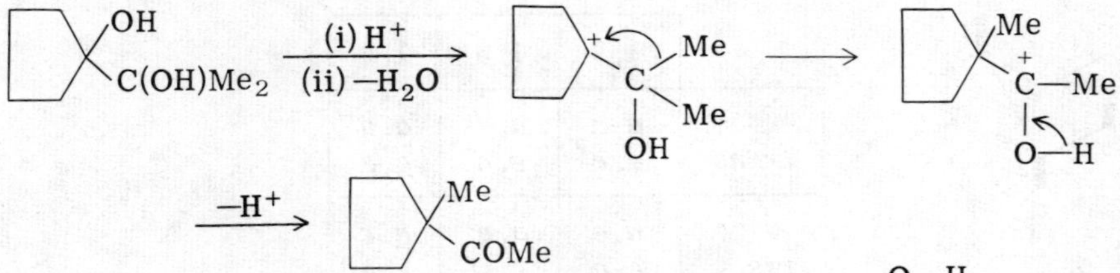

(*b*)

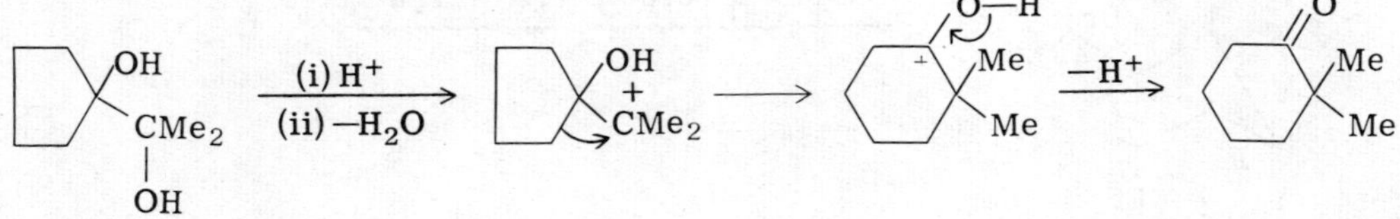

This similarity in stability may be attributed to the similarity between the inductive effects of alkyl and cycloalkyl groups.

(iii) The substrate is symmetrical and the product contains a carbon atom that is common to the two rings. Hence:

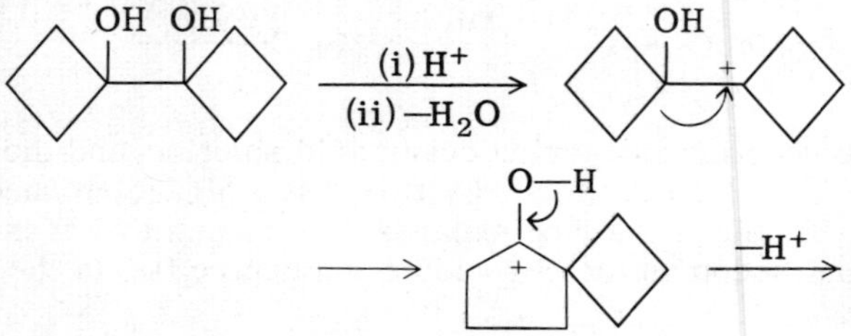

8.

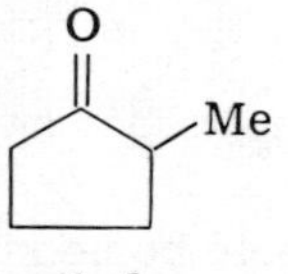

2-methyl-cyclopentanone (I)

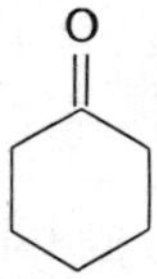

cyclohexanone (II)

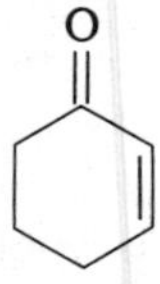

cyclohex-2-enone (III)

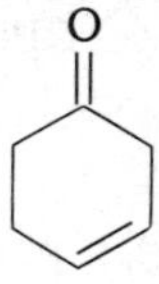

cyclohex-3-enone (IV)

Since a student is not expected to memorise various wave numbers, he has to deal with this question from a descriptive point of view. This essentially means that since he has been *given* the structures, he is expected to discuss *relative* positions of absorption bands and how these positions may vary with changes in structure. One important point about cyclic structures is that, in general, the greater the strain in the ring, the shorter is the wavelength (higher the wave number) of the functional group when it is part of the ring system. (I) is the only five-membered ring system and so it can be expected that the wavelength of the carbonyl group will be shorter (and hence its wave number higher) than any of the others. The carbonyl band in (II) will occur at about the same wavelength as that in (IV), but since the latter also has a double bond in the ring, this will be seen in the spectrum. Consequently, (II) and (IV) may be distinguished. The wavelengths of the carbonyl groups in (II) and (IV) are very close because the latter does not contain a conjugated system. (III), however, contains the C=O and C=C groups in conjugation, and consequently their absorptions are shifted to longer wavelengths. Hence, (III) can be distinguished from (IV).

9. Deciding whether a group is axial or equatorial offers no difficulty, but geometrical relationships may be confusing. The following table shows these relationships and these are readily obtained by choosing the diaxial positions *first* and assigning the other accordingly.

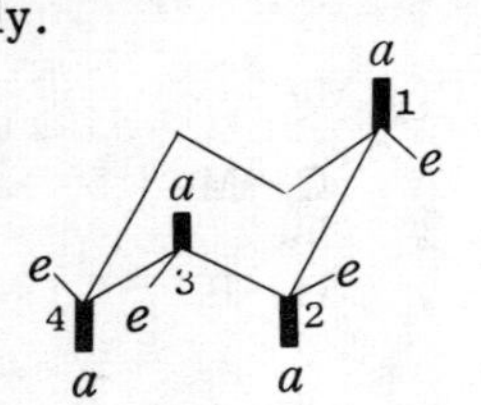

	1, 2	1, 3	1, 4
trans	***a, a*** *e, e*	*a, e* *e, a*	***a, a*** *e, e*
cis	*a, e* *e, a*	***a, a*** *e, e*	*a, e* *e, a*

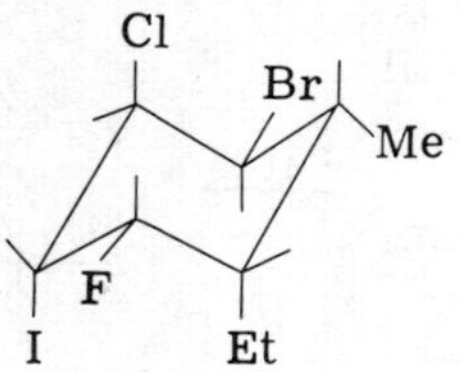

(i) 1*e*Br, 2*a*Cl; *cis*. (ii) 1*a*Et, 3*a*I; *cis*. (iii) 1*a*Cl, 4*a*Et; *trans*. (iv) 1*e*Br, 3*a*Et; *trans*. (v) 1*e*F, 3*e*Me; *cis*. (vi) 1*a*Cl, 3*e*Me; *trans*. (vii) 1*e*Br, 4*e*F; *trans*. (viii) 1*a*Et, 2*e*Me; *cis*. (ix) 1*a*Cl, 2*a*I; *trans*. (x) 1*a*I, 4*e*Me; *cis*.

10. If we accept that the E2 mechanism occurs most readily when the two groups which are eliminated are coplanar and *trans*, then two halogen atoms, Cl and I, satisfy these requirements (see *Text*, Ch. 17). Both are axial and each has an adjacent axial hydrogen atom. Since iodine is more readily eliminated than chlorine (see *Text*, Ch. 5), the reaction will follow this path predominantly (if not exclusively). Also note that fluorine is more electronegative than bromine and so the H of CHF can be expected to be more labile than the H of CHBr.

11. Unless we have good reason to suppose otherwise, we usually accept that the chair conformation is the most stable one (i.e., compared with boat and twist-boat). We also accept, unless we have reasons to suppose otherwise, that the conformation with the largest number of substituents equatorial is the most stable one. To decide whether the compound is optically active, it is much simpler to examine the *planar* ring formula for elements of symmetry (the results agree with experiment).

(i) *cis*

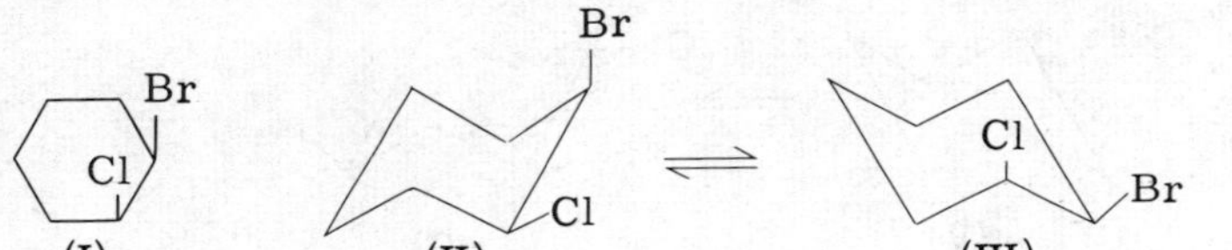

(I) (II) (III)

Since bromine is a larger atom than chlorine, the predominant form will be (III). (I) has no elements of symmetry, and so the compound is optically active.

trans

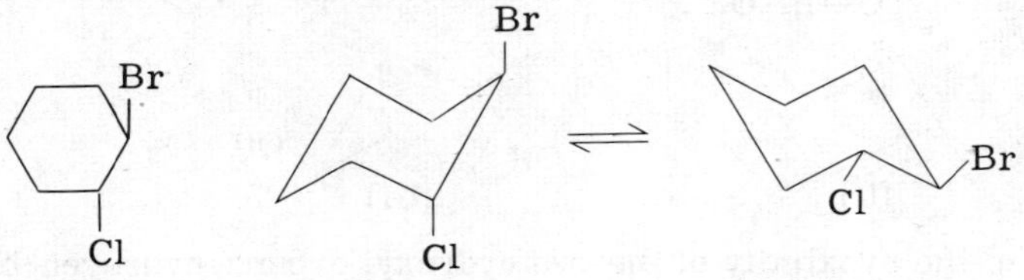

(IV) (V) (VI)

The predominant form is (VI), and the compound is optically active. A point to notice here is that the C—X bond is polar. In (V), the dipole-dipole interaction is a minimum, but in (VI) there is some interaction. If this interaction outweighs 1, 3-interactions, then (V) would be more stable than (VI). In practice, (VI) is the predominant form. This illustrates the use of generalisations followed by a further consideration of other factors. Unfortunately, it is often difficult to say which is the predominant factor; experimental results are usually required to decide the answer. What the student is expected to do is to show he is aware of the problem and apply his knowledge accordingly.

In view of what has been said, (VI) will be the most stable conformation of the four (II, III, V, VI).

(ii) *cis*

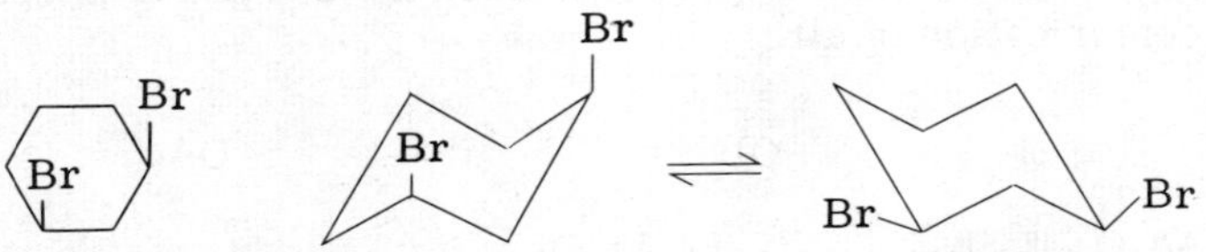

(I) (II) (III)

The predominant form is (III); in fact, virtually exclusively. The compound is not optically active ((I) has a plane of symmetry).

trans

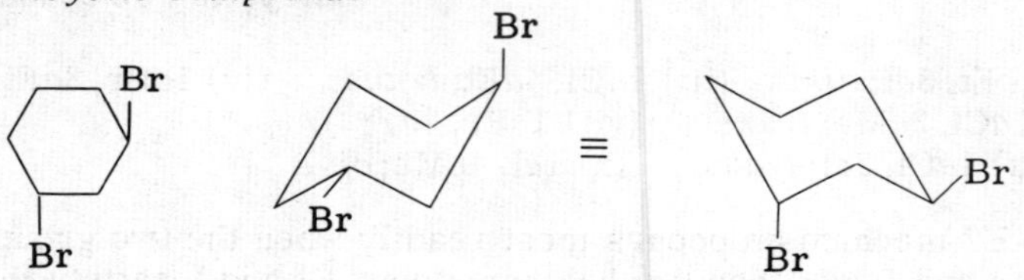

(IV) (V)

The two *trans*-conformations are identical; the compound is optically active. (III) is the most stable conformation of all.

(iii)

cis

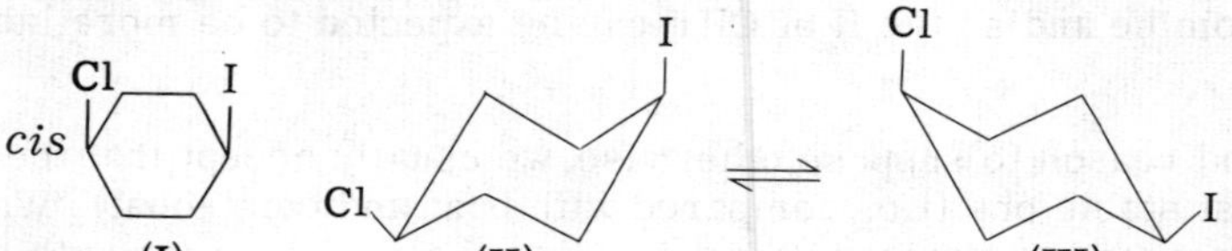

(I) (II) (III)

(III) is the predominant form (I is larger than Cl) and the compound is optically inactive

trans

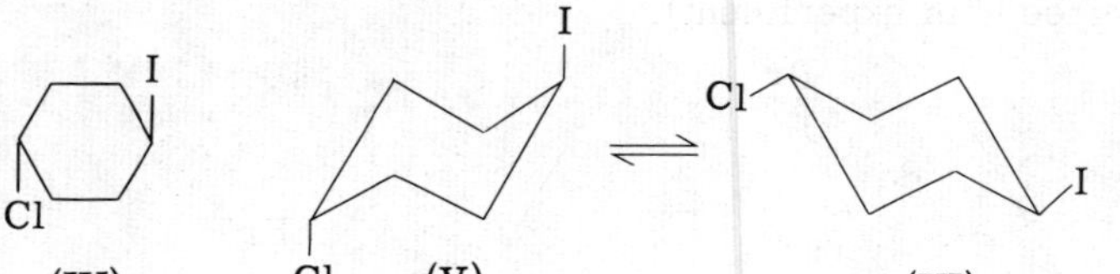

(IV) (V) (VI)

(VI) is present virtually exclusively, and the compound is optically inactive. (VI) is the most stable conformation of all.

(iv)

cis

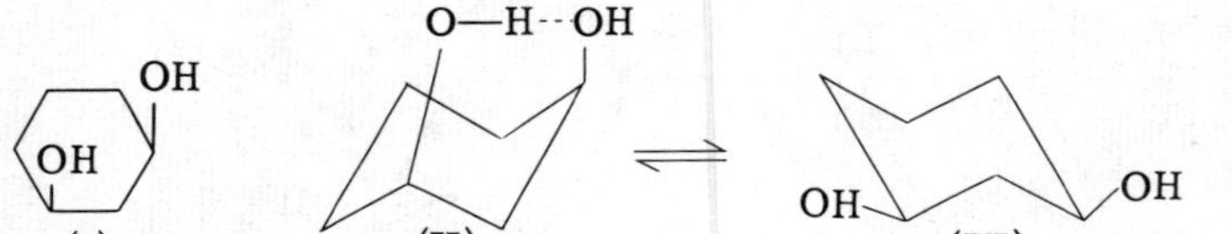

(I) (II) (III)

(II) is diaxial, but because of the proximity of the two hydroxyl groups, hydrogen-bonding occurs between them. If this 'bonding' energy is high enough, it will outweigh the 1, 3-interactions and so (II) will be more stable than the diequatorial form (III). Experimental results have shown that (II) is more stable than (III). The compound is not optically active.

trans

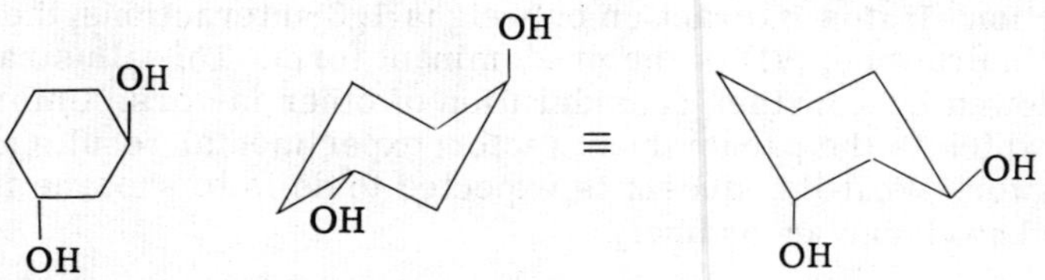

(IV) (V)

The two *trans*-conformations are identical (*cf.* (ii), above); the compound is optically active. (II) is the most stable conformation of all.

12.

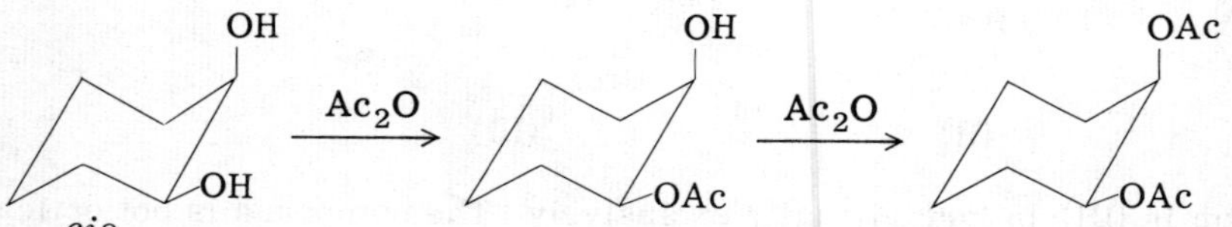

cis

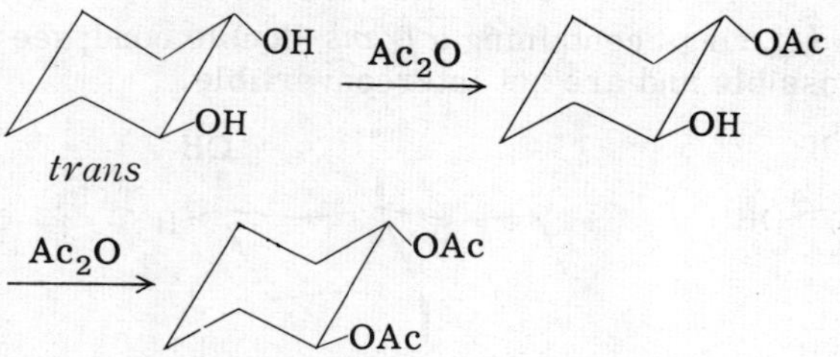

The C—O bond is not broken in the acetylation and so attack at the oxygen atom increases the 'volume' with consequent increase in 1, 3-interactions if the hydroxyl group is axial. Hence, for the *cis*-compound, the equatorial OH is acetylated first, and this is followed by acetylation of the axial OH, this step being slower than the first. This is the result of an increase in activation energy due to the increased 1, 3-interactions in the second step.

For the *trans*-compound, since both OH groups are equatorial, either is attacked equally well, and is followed by the second step at almost the same rate (probably a little slower due to the larger bulk of the acetylated group).

13. (i) In *cis*-decalin, the two cyclohexane rings are fused together by an axial and an equatorial bond, both rings being in the chair form. Because of the type of fusion, the molecule can flip to the other *cis*-form ($a \rightarrow e$ and $e \rightarrow a$). The two forms are identical and in equilibrium. When, however, there is a hydroxyl group in the 2-position, it is axial in one conformation and equatorial in the 'flipped' conformation. Hence four conformations for *cis*-decal-2-ol are possible.

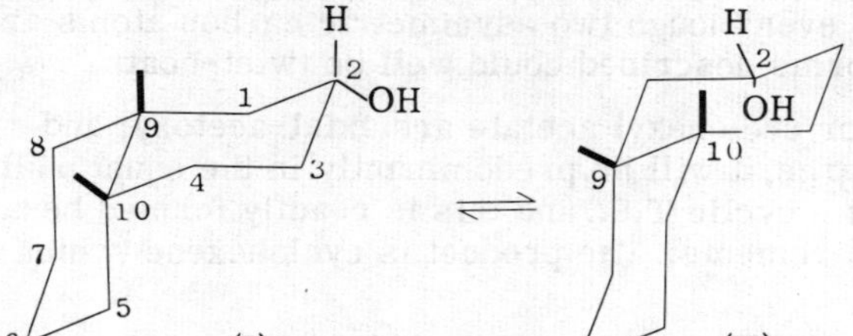

(I) (II)

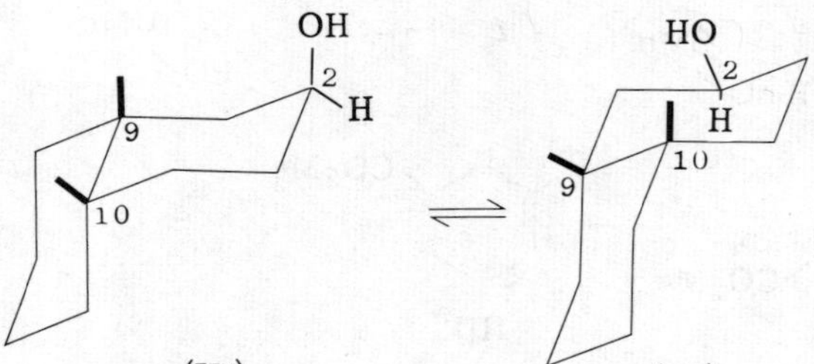

(III) (IV)

(I), in which the OH is equatorial, flips to give (II). (III), which is the same decalin conformation as (I), has the OH axial, and when this flips, (IV) is obtained. Inspection of these formulae shows that all four conformations are different:

(I) 2*e*OH, 9*a*H; (II) 2*a*OH, 9*e*H.

(III) 2*a*OH, 9*a*H; (IV) 2*e*OH, 9*e*H.

(ii) In *trans*-decalin, the two cyclohexane rings are fused together by equatorial bonds. This cannot flip because the other form would require two axial bonds of fusion, which would require the bonds pointing in *opposite* directions. Large rings could take up this strain, but not six-membered rings (*cf.* rings containing a *trans*-double bond; see *Text*). Hence, two *trans*-decal-2-ols are possible and are not interconvertible.

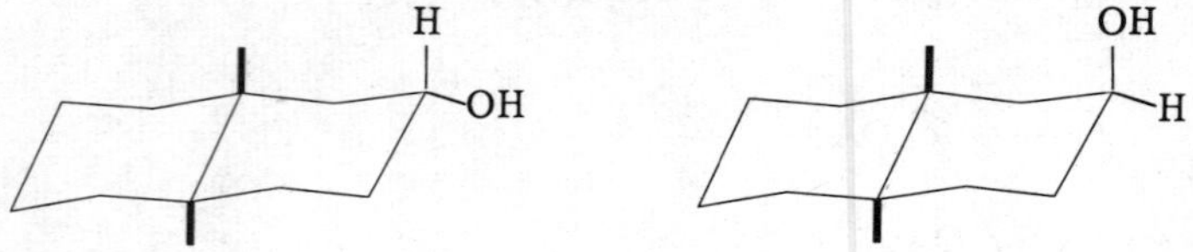

14.

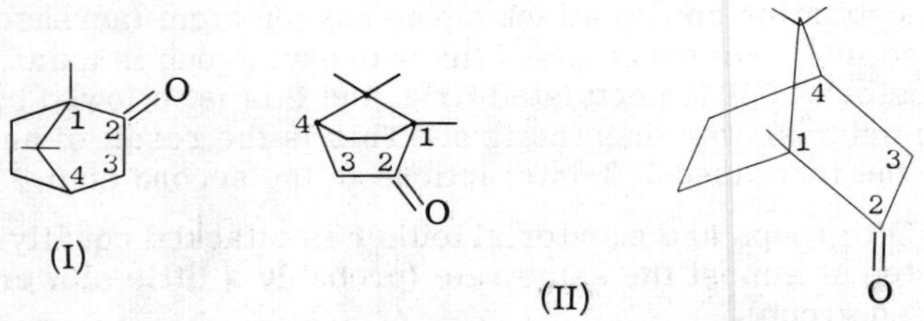

(I) (II)

In the chair form, the positions of 1 and 4 are too far apart to be a possible conformation of camphor, (I) [*see* Prob. 13, above]. The bridge, composed of two C—C bonds is not long enough. On the other hand, the distance between the 1, 4-positions in the boat conformation, (II), can be covered by the bridge. Hence, although the boat-form is energetically unfavourable, it is maintained in this conformation by the bridge. Also, because of the length of the bridge, its two bonds must be *cis*. Since this is a rigid structure, *both* asymmetric carbon atoms at 1 and 4 cannot act independently, i.e., only this *one* structure is possible. Hence, there can be only one pair of enantiomers even though two asymmetric carbon atoms are present. It might be noted that the boat-forms described could well be twist-boats.

15. Two possible chair conformations for cyclohexyl acetate are axial-acetoxyl and equatorial-acetoxyl. Since this group is large, it will be predominantly in the equatorial form. Pyrolysis involves the formation of a cyclic T.S., and this is readily formed because the geometry of the molecule fits the requirements. The product is cyclohexene (which may be drawn in the half-chair form).

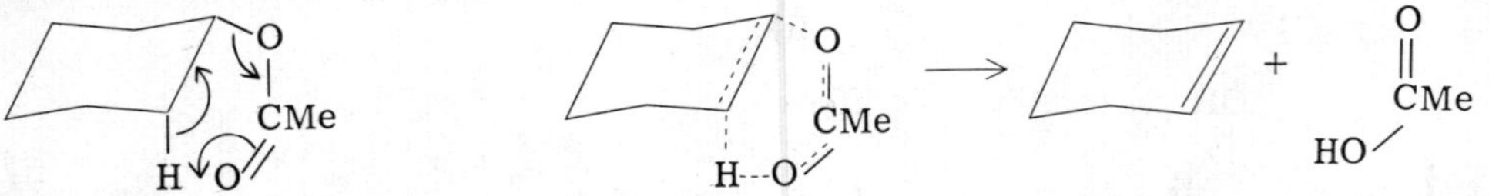

16. (i)

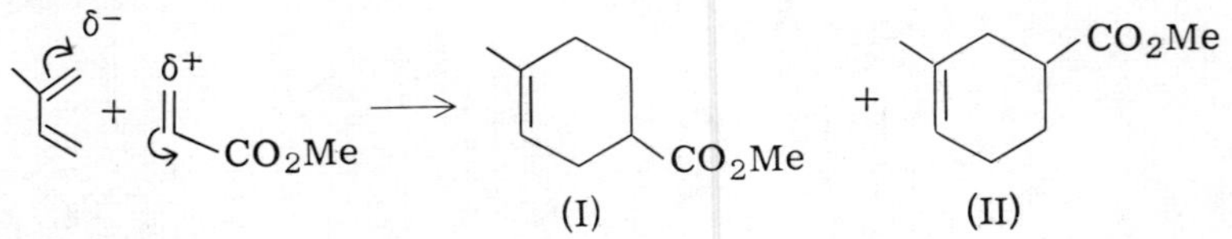

(I) (II)

(I) [methyl 4-methylcyclohex-3-enecarboxylate] is the major product and (II) [methyl 3-methylcyclohex-3-enecarboxylate] is the minor product. Because of the inductive effects of the methyl and carbomethoxy groups, the reactants carry charges (as shown) and hence the orientation of the two reactants as they approach each other favours the formation of (I).

(ii)

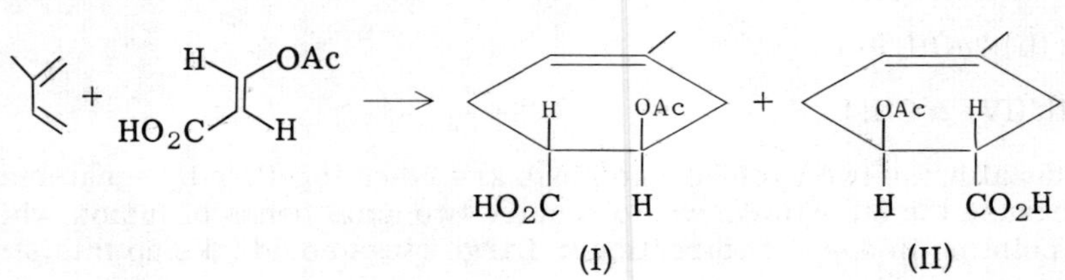

(I) (II)

The addition is sterospecifically *cis*. Because both groups have —I effects and their relative values are not easy to estimate, it is difficult to say which product predominates. If, however, we refer to Table 3.2, we see that OCOR has a greater deshielding effect than CO_2H, i.e., the former has the greater —I effect. If we assume this holds good for the molecule under consideration, then (II) would be the predicted major product (*cf.* with (i), above).

(iii)

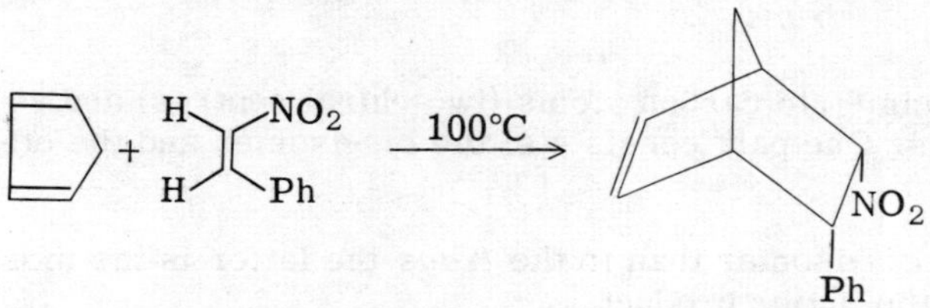

(boat-form)

The addition is stereospecifically *cis* and steric effects will favour the formation of the isomer with the bridge and substituent groups on opposite sides of the ring (the *endo*-compound).

17. If we assume that *trans*-elimination is easier than *cis*, then we must examine the geometry of all the possible (geometrical) isomers of hexachlorocyclohexane to see if there is one in which all the chlorine atoms are *trans* to adjacent chlorine atoms, i.e., there is no alternate pair of hydrogen and chlorine in the *trans* position. The one shown fits these requirements, and its conformation has all chlorine atoms equatorial (*trans: e, e* or *a, a*). The axial configuration is very highly unstable with respect to its equatorial conformation.

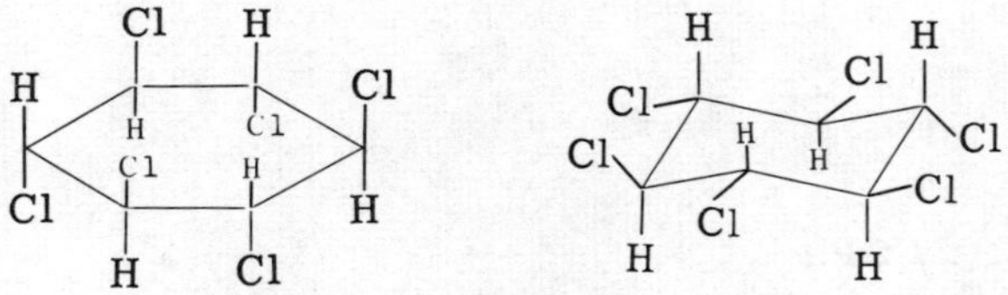

Since all the protons are equivalent, the NMR spectrum will consist of one signal (singlet).

18. Since the substrate is an α,β-unsaturated ketone, 1,2- and/or 1,4-additions with Grignard reagents are possible. Also, it is known that the presence of cuprous chloride favours 1,4-addition (see *Text*, Ch. 15). The i.r. band at about 1 680 cm^{-1} suggests the presence of a ketone (possibly of the alkaryl type). This also supports the probability of 1,4-addition, since 1,2-addition would give an alcohol. Hence, we may formulate the reaction as:

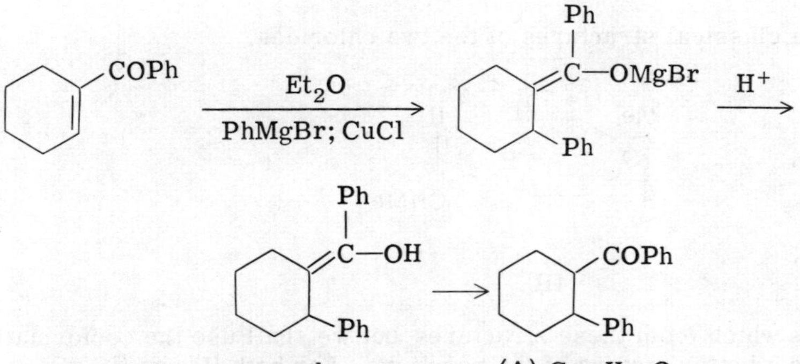

enol (A) $C_{19}H_{20}O$

Two geometrical isomers are possible for the product:

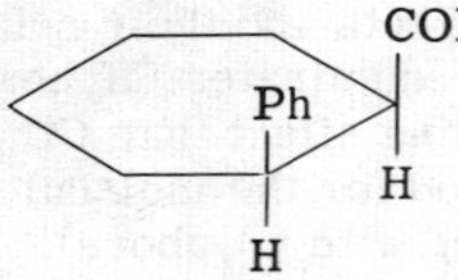

cis

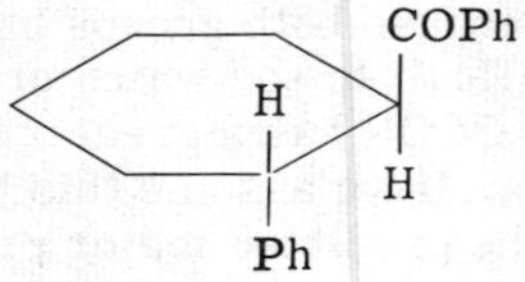

trans

This compound contains two asymmetric carbon atoms (two chiral centres) and so can exist as two pairs of enantiomers. One pair consists of the *cis*-isomer and the other pair of the *trans*-isomer.

Because strain is greater in the *cis*-isomer than in the *trans*, the latter is the more stable and hence can be expected to be the major product.

Finally, let us examine these isomers from the conformational point of view.

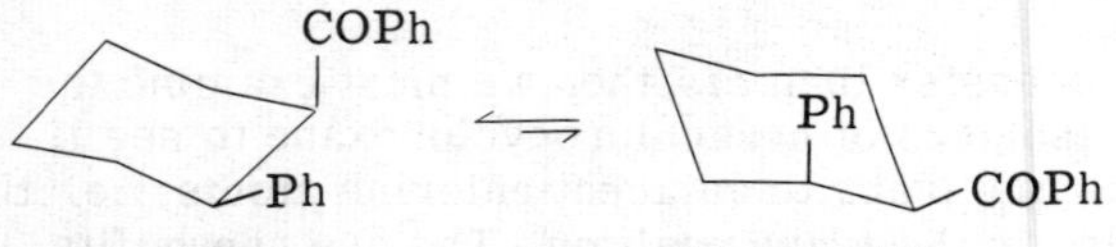

*e*Ph, *a*COPh   *cis*   *a*Ph, *e*COPh

The Ph group is larger than the COPh group (in the latter, the Ph is separated from the ring by the carbonyl carbon atom). Therefore the preferred form is the one with the phenyl group equatorial.

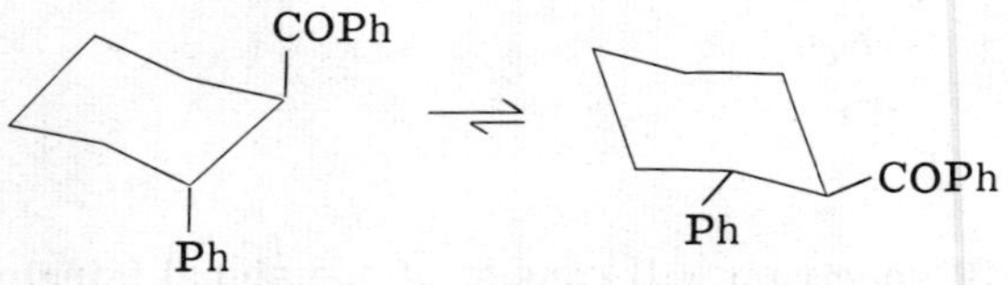

*a*Ph, *a*COPh   *trans*   *e*Ph, *e*COPh

The *e*, *e*-conformation is far more stable than the *a*, *a*, and so the molecule is virtually exclusively in this form. Also, since the *e*, *e*-form is the most stable one of all four forms, this is the one which will predominate in the product mixture, **(A)**. It has already been predicted that the *trans*-isomer would be the predominant product (using the classical structures; see above).

19. First, let us draw the classical structures of the two chlorides.

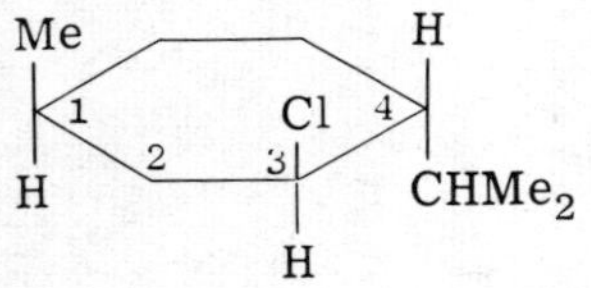

(I)

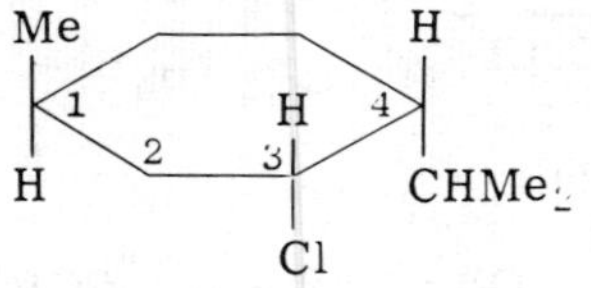

(II)

We could work out which is which from these structures, but we shall use the conformational structures since these give a better picture of the reactions. For both (I) and (II), the *trans* alkyl groups (1, 4) will be *a*, *a* or *e*, *e* (see also Prob. 9, above). Hence (*trans*-1, 2 is *a*, *a* or *e*, *e*).

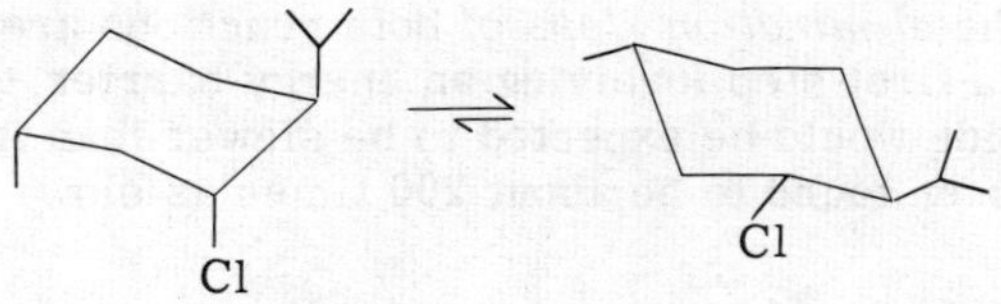

(I*a*) (I*b*)

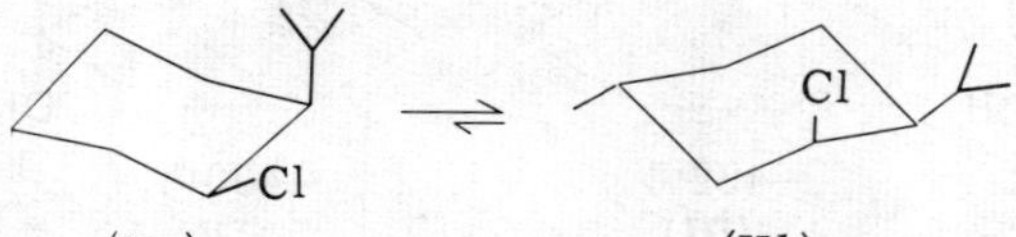

(II*a*) (II*b*)

On the basis that the conformation with the larger number of groups equatorial is the more stable one, then (I*b*) is preferred to (I*a*) and (II*b*) to (II*a*).

The use of EtONa suggests an E2 mechanism (very strong base; see *Text*, Ch. 5), and we shall assume that *trans*-elimination is easier than *cis*. For *trans*-elimination, the chlorine and hydrogen are both axial (to satisfy the spatial requirements). This means that elimination from (I) occurs better in conformation (I*a*), even though this is not the preferred conformation. Since the activation energy of the elimination reaction is far higher than the energy barrier between the two forms, interconversion to the less stable form occurs, followed by elimination (compare with a two-step reaction in which the second step is r/d).

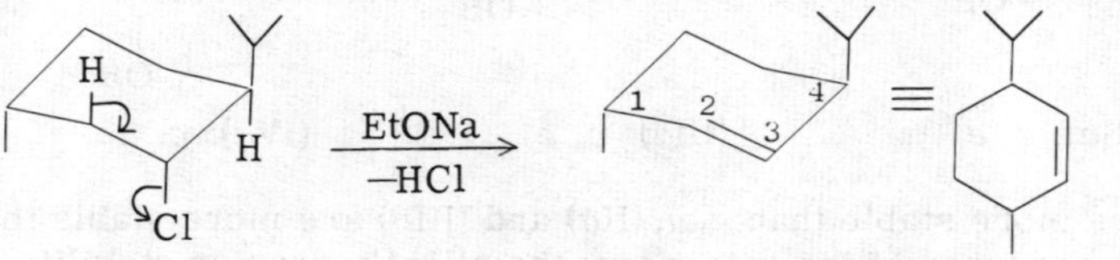

(I*a*) menth–2–ene

Since there is only *one* H *trans* to the Cl, only one product is formed, menth-2-ene. Hence (I) is menthyl chloride, and the chlorine atom is equatorial in the stable form (I*b*) and the molecule undergoes elimination in its less stable form (I*a*; chlorine axial).

Similarly, elimination from (II) also occurs from the conformation with axial chlorine, and in this case this is the preferred form, (II*b*).

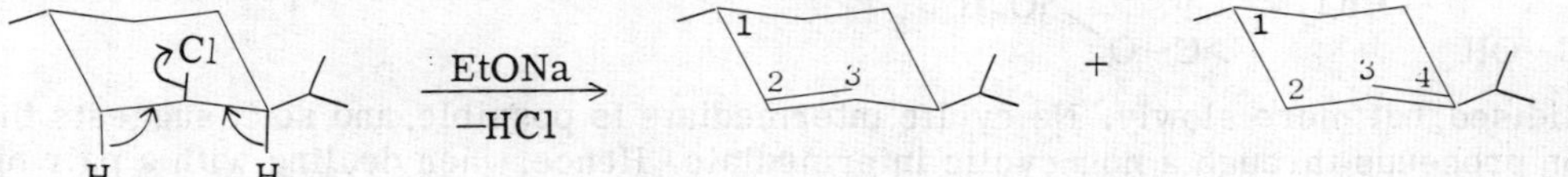

(II*b*) *or*

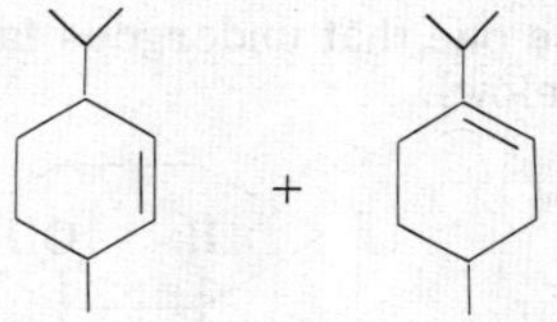

Inspection of (II*b*) shows that two hydrogen atoms are axial with respect to the chlorine atoms, and hence two products are formed, menth-2-ene and menth-3-ene. Therefore (II) is neomenthyl chloride, and the chlorine atom is axial in the preferred conformation (II*b*).

It might be noted that the 3-ene is more highly substituted than the 2-ene and so is the expected predominant product (Saytzeff's rule; see *Text*, Ch. 5).

If we assume that the activation energies for the *elimination steps* of both reactions are close, then because menthyl chloride requires a first step involving an energy barrier, the *overall* rate of elimination from menthyl chloride would be expected to be slower than that from neomenthyl chloride. In practice, it has been found to be about 200 times as slow.

20.

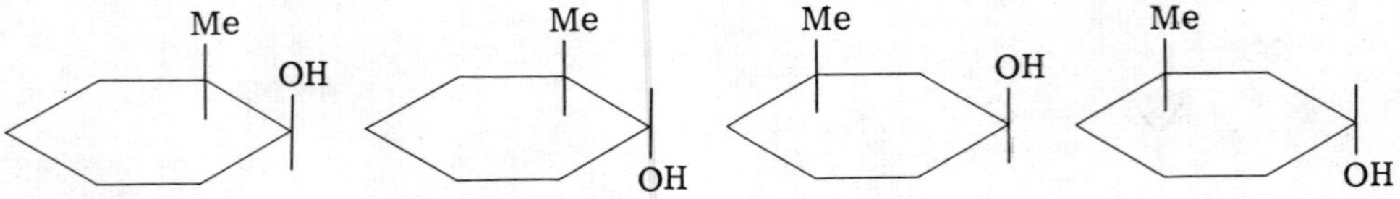

kJ —396·6 (I) —422·2 (II) —422·6 (III) —400·8 (IV)

Since the greater $-\Delta H_f^{\ominus}$ is numerically, the more stable is the compound (the enthalpies of the constituent atoms are arbitrarily put equal to zero), the order of stability is: (III) > (II) > (IV) > (I).

Now let us consider these compounds from the conformational point of view. Since Me has a larger volume than OH, the former will tend to be equatorial rather than the latter when these alternatives exist. Hence, the preferred conformations are:

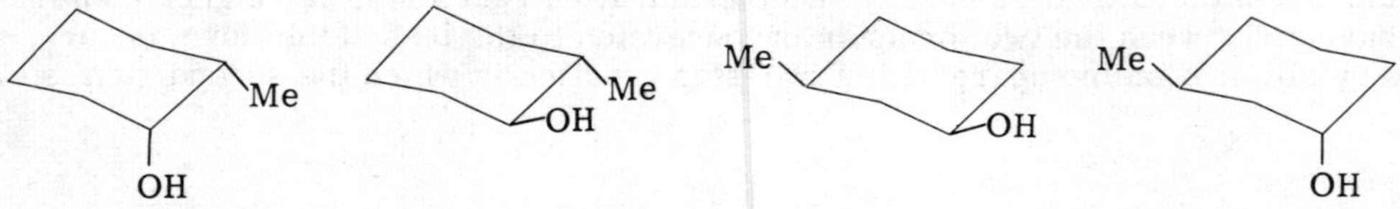

(I*a*): *e*, *a* (II*a*): *e*, *e* (III*a*): *e*, *e* (IV*a*): *e*, *a*

Because *e*, *e*-conformations are more stable than *e*, *a*, (II*a*) and (III*a*) are more stable than the other two. If we accept the accuracy of the data given, the slightly greater stability of (III*a*) with respect to (II*a*) could be attributed to the fact that the two groups are further apart in the former than in the latter. This argument could also be used to explain why (IV*a*) is more stable than (I*a*). Thus, we get the same order of stability as before.

21. Experimental evidence is in favour of the formation of a cyclic intermediate when *cis*-1, 2-glycols are oxidised with periodic acid (see *Text*, Ch. 11). *trans*-1, 2-Diols are

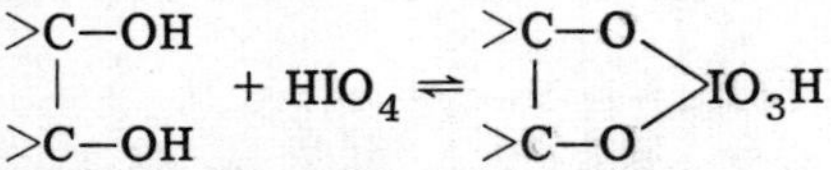

also oxidised, but more slowly. No cyclic intermediate is possible, and so it suggests that reaction proceeds through a non-cyclic intermediate. Hence, when dealing with a pair of *cis*- and *trans*- 1, 2-diols, the one that undergoes fast oxidation is identified as the *cis*-isomer (see also Prob. 26, below).

cis *trans*

cyclic intermediate possible cyclic intermediate not possible

22. We can write the equation as shown (the conformations are boat-forms; see also Prob. 14, above):

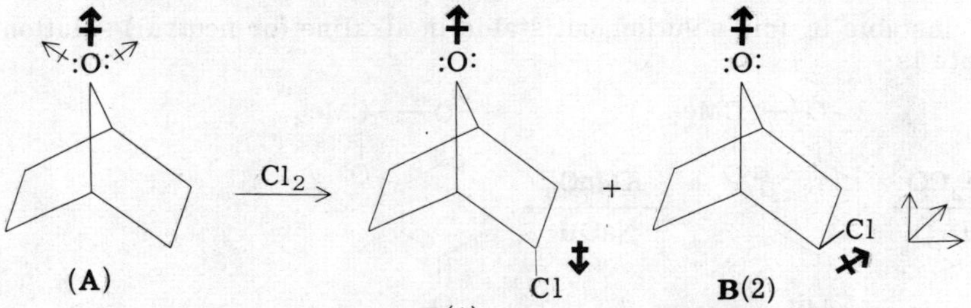

(A) **B**(1) *axial* **B**(2) *equatorial*

The two lone pairs of electrons on the oxygen atom (tetrahedral distribution) have a resultant dipole moment which points vertically. The C—Cl bond is polar, and in the axial conformation **(B1)** the resultant dipole moment is obtained by *substracting* the two vectors (since they are parallel and pointing in opposite directions). In the equatorial conformation **(B2)**, the resultant dipole moment is obtained by the vector *summation* of the two separate dipole moments. Hence, the resultant of **(B2)** must be greater than that of **(B1)**. Therefore the isomer with $\mu = 1{\cdot}07$D is **(B1)** and that with $\mu = 3{\cdot}06$D is **(B2)**.

23.

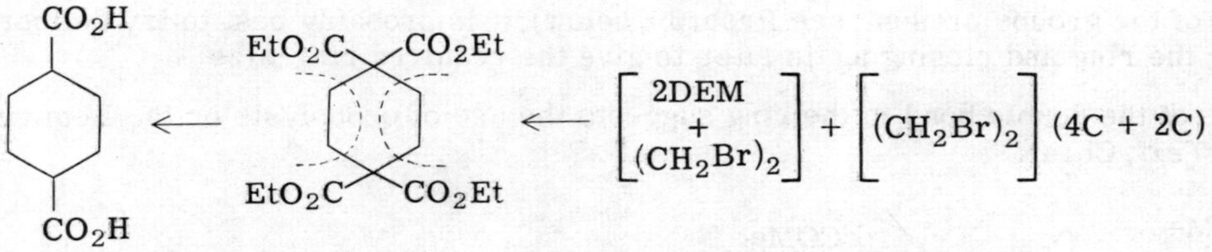

Since we are told to start from DEM, we can work backwards as shown. Hence:

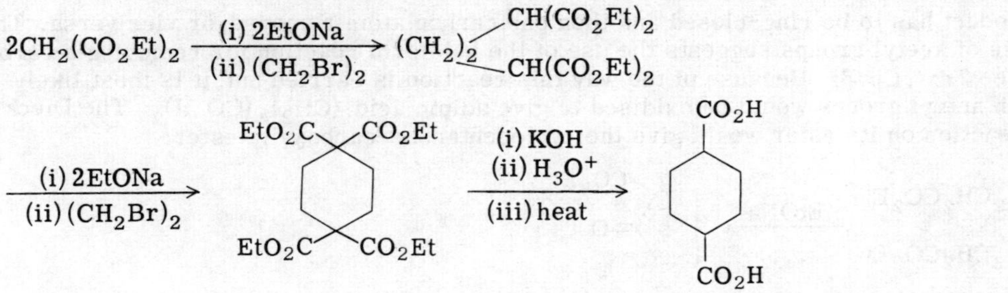

24.

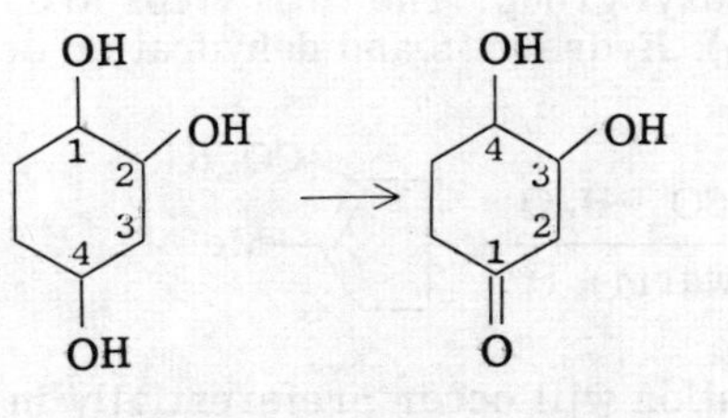

All are secondary alcoholic groups and this suggests protecting two (1 and 2) and oxidising the third (4). Furthermore, since the two to be protected are in the 1, 2-position, a possibility is the formation of the isopropylidene derivative (see *Text*, Ch. 18).

This is usually formed from *cis*-1, 2-diols, and since we have not been given the geometry of the triol, we are entitled to assume this configuration for our purpose. Now, isopropylidene derivatives are unstable in acid solution, but stable in alkaline (or neutral) solution. Hence, a possible route is:

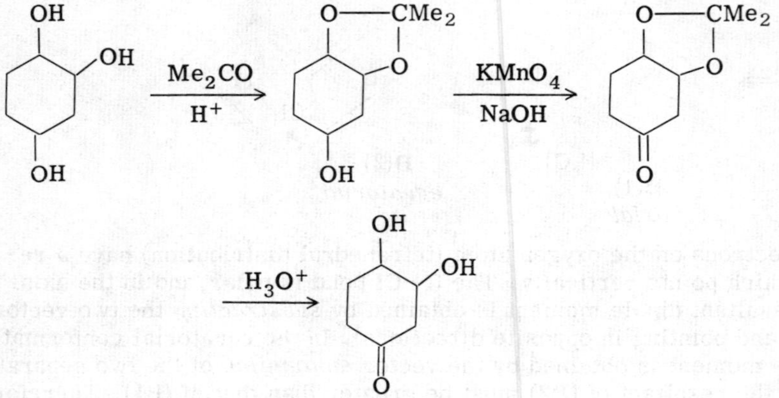

25. This reaction involves a change in ring-size and loss of one carbon atom. Unless the student can see an obvious route in which a 'spontaneous' change in ring-size occurs (due to the nature of the groups present; see Prob. 30, below), it is probably best to try deliberately opening the ring and closing again later to give the required ring-size.

The presence of the double bond in the ring suggests the use of ozonolysis or the Lemieux reagent (see *Text*, Ch. 4).

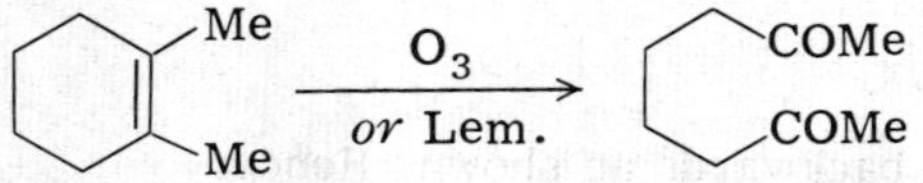

This product has to be ring-closed and then one carbon atom removed, or vice versa. The presence of acetyl groups suggests the use of the haloform reaction for removing a carbon atom (see *Text*, Ch. 8). Because of the way this reaction is carried out, it is most likely that both acetyl groups would be oxidised to give adipic acid, $(CH_2)_4(CO_2H)_2$. The Dieckmann reaction on its ester would give the cyclopentanone-carboxylic ester:

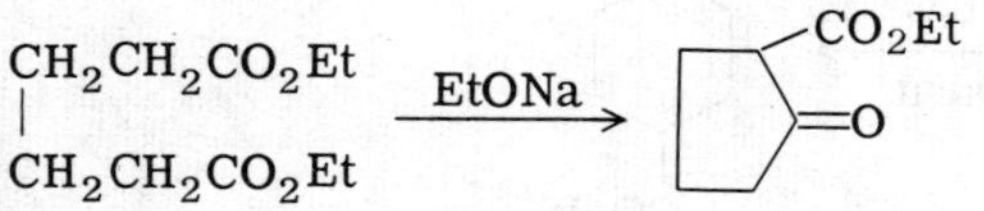

The product contains the ring we want and also the carboxyl group. The final steps are (note the use of one molecule of MeMgI; see *Text*, Ch. 15): Hydrolysis and dehydration occur,

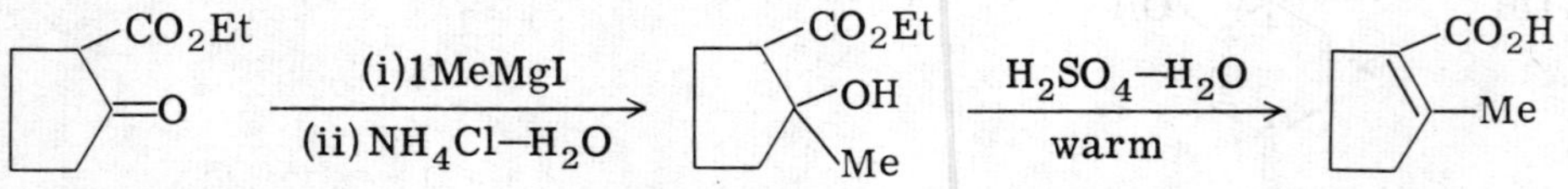

and because the α-methylene group is activated, elimination will occur preferentially in the direction shown (to give the desired product).

An alternative route is as follows. The synthesis given uses the Dieckmann reaction, and this involves an active methylene group and a suitably placed carbonyl group (of CO_2Et). Inspection of the diketone shows the situation is analogous to that of a di-ester. Hence, we could *expect* (but not necessarily get) a Dieckmann reaction.

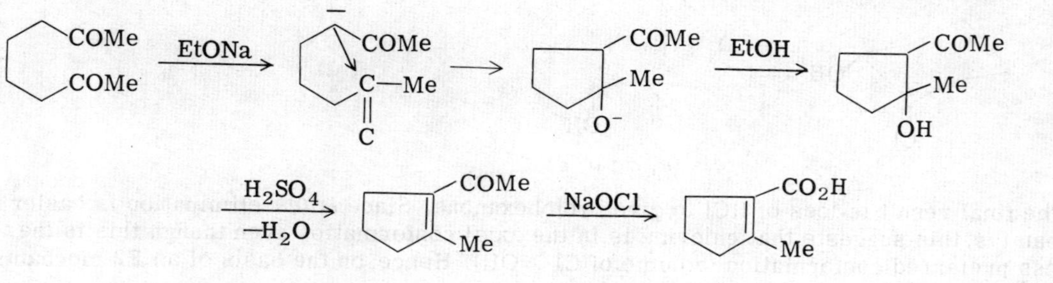

This synthesis is based on analogy with a known reaction, and whether it works or not can only be decided by experiment. In practice, it does work.

26. In alicyclic chemistry, a clearer picture of the course of a reaction is obtained by use of conformational formulae than by planar (classical) formulae. However, if the planar formulae illustrate quite clearly the points under discussion, it is reasonable to use these (*cf.* Prob. 21, above).

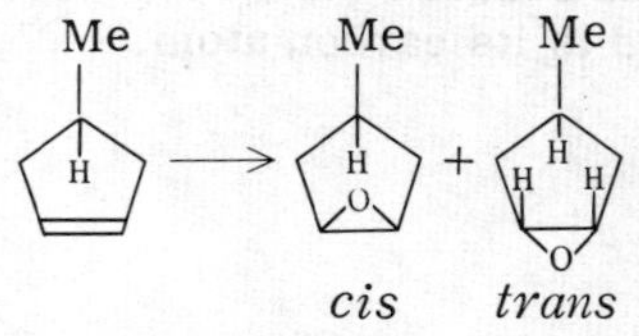

The two isomeric epoxides are the *cis*- and *trans*-products. Because strain (due to steric effects) is smaller in the *trans*- isomer than in the *cis*, the former can be expected to predominate.

27. Since 1, 2-diols are readily prepared by hydroxylation of a double bond, this suggests first preparing cyclohexene.

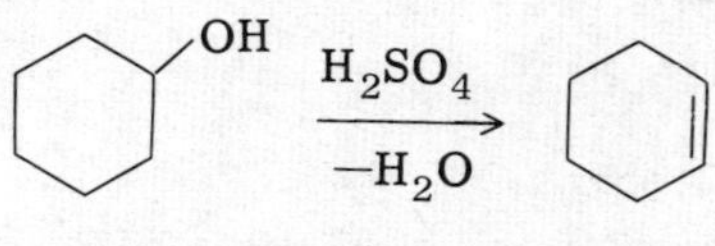

The next step involves stereospecific reactions and so it is now a question of knowing which reagents to use. Addition with alkaline permanganate (or with osmium tetroxide) is stereospecifically *cis* (proceeds via a cyclic intermediate; see *Text*, Ch. 4). On the other hand, *trans*-addition may be achieved by first preparing the epoxide (*cis*-addition) and then hydrolysing this with acid (*trans*-addition; see *Text*, Ch. 11). Thus:

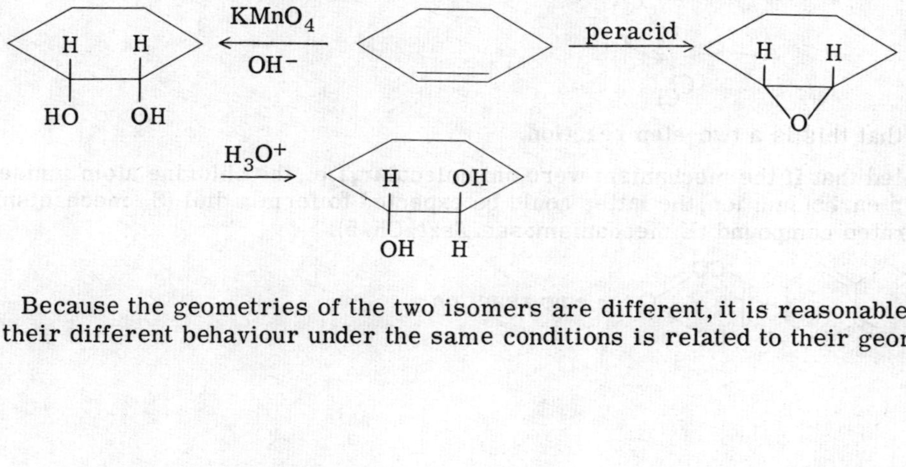

28. Because the geometries of the two isomers are different, it is reasonable to expect that their different behaviour under the same conditions is related to their geometries.

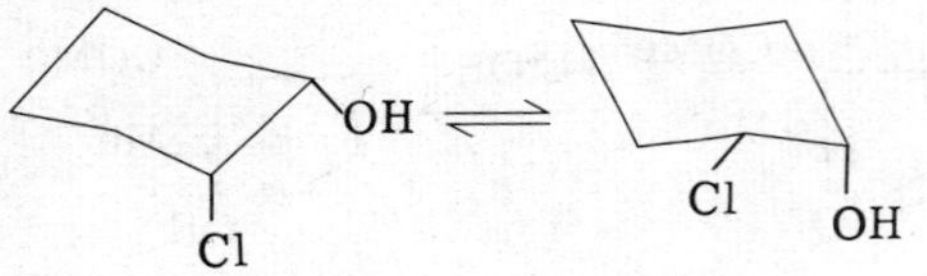

cis (*e, a* ; *a, e*)

The final result is loss of HCl to give cyclohexanone. Since *trans*-elimination is easier than *cis*, this suggests that chlorine is in the axial conformation even though this is the less preferred conformation (volume of Cl > OH). Hence, on the basis of an E2 mechanism:

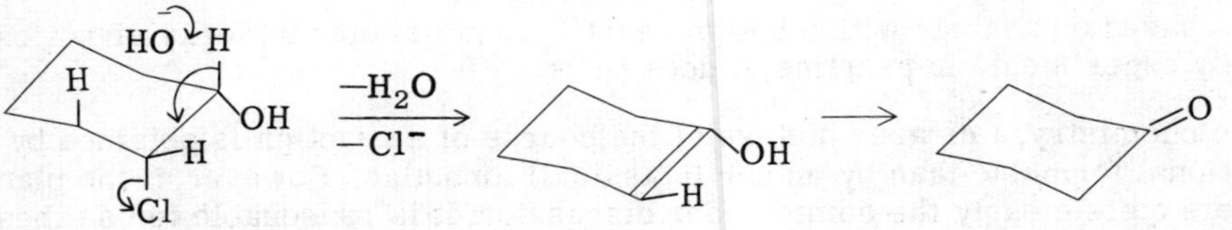

Removal of the other adjacent proton *trans* to Cl is not favoured because the strong –I effect of the OH group facilitates release of the proton attached to its carbon atom.

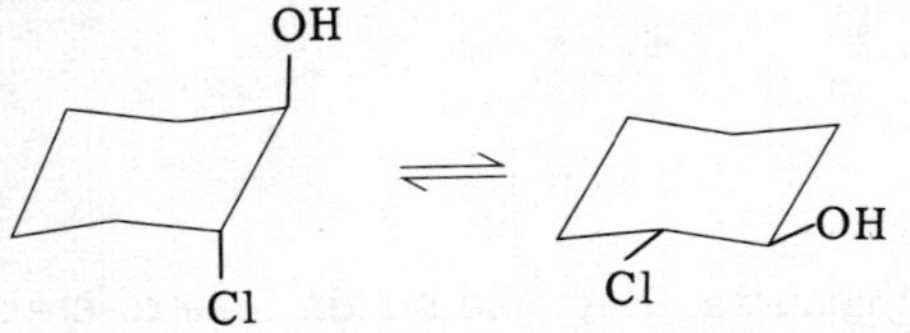

trans (*a, a*; *e, e*)

As before, the final results is the loss of HCl, but in this case the product is the expoxide. Hence, the oxygen atom of OH has linked up with the carbon atom of CCl. This suggests the possibility of N.G.P. and that the geometrical requirements of O and Cl are *trans*-diaxial (the less stable form; see also Prob. 19, above). Thus:

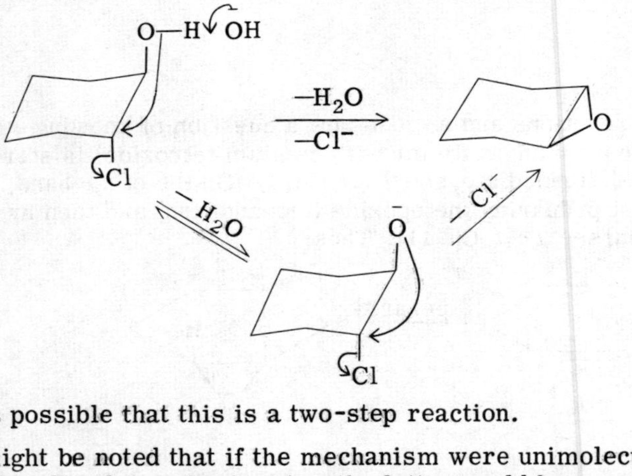

It is possible that this is a two-step reaction.

It might be noted that if the mechanism were unimolecular, i.e., the chlorine atom ionised and produced a carbonium ion, the latter could be expected to form a diol (S_N mechanism) and an unsaturated compound (E mechanism; see *Text*, Ch. 5).

29. (i)

Conjugated systems undergo the Diels-Alder reaction in the cisoid form. If the system is held rigidly in the transoid form, as is the case here, the ends are too far apart to form bonds with the dienophile.

(ii)

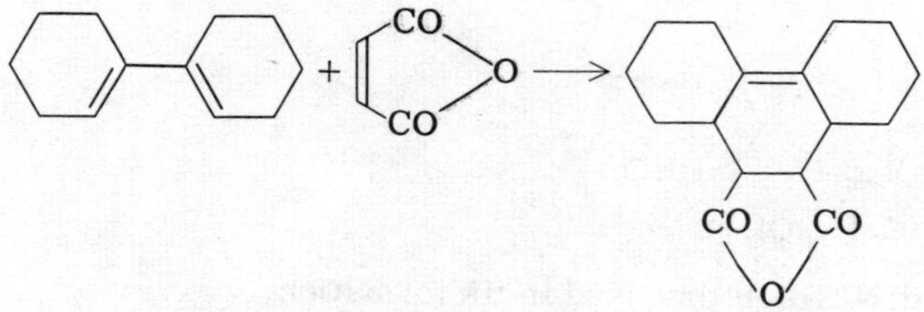

(iii)

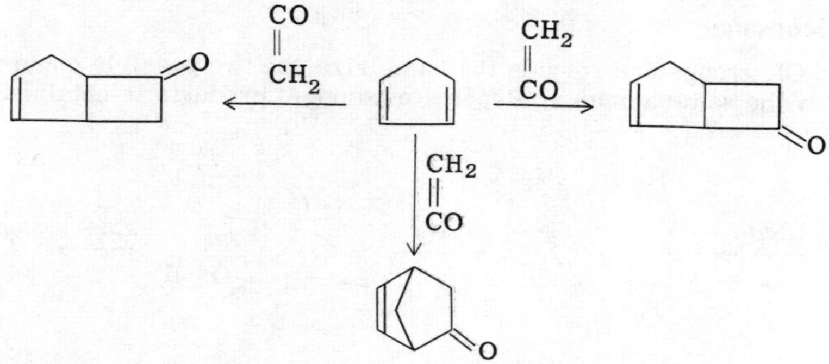

Without any prior knowledge, we would have to argue that the three products shown are possibilities. In practice, it has been found that keten apparently undergoes only 1, 2-additions. Hence, the 1, 4-adduct is ruled out.

(iv)

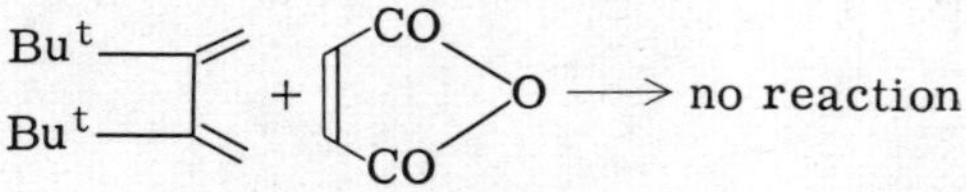

The Diels-Alder reaction is susceptible to steric effects. In the diene given, the steric effect of the large t-Bu groups prevents the reactants from approaching sufficiently closely to form the lateral bonds.

30. The general principles for rearrangements in cyclohexane derivatives are: (i) When the leaving group is axial, only an axial group on an adjacent carbon atom can migrate; the usual migrating groups are hydride ion and the methyl carbanion: (ii) When the leaving group is equatorial, only an adjacent carbon atom in the *ring system* can migrate. These principles are based on the argument that *trans*-migration, i.e., attack from *behind,* is the required geometry.

trans-**2-aminocyclohexanol**

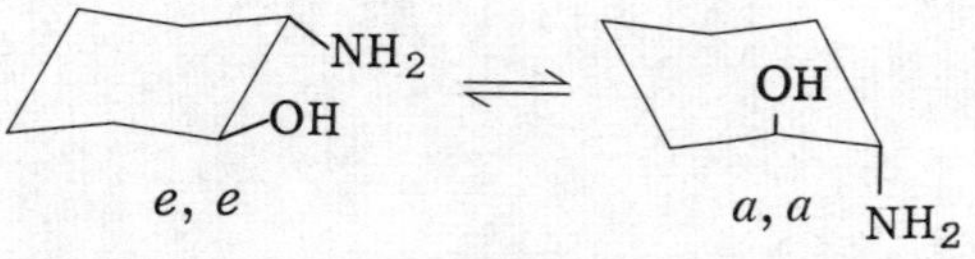

The conformation is virtually exclusively *e*, *e*, and so the compound reacts in this form. If the diaxial form could undergo reaction by some different path which required a lower activation energy (than that for the diequatorial path), it can be expected that this would be followed (provided that activation energy of the r/d step is greater than the energy barrier of interconversion; see Prob. 19, above). Hence:

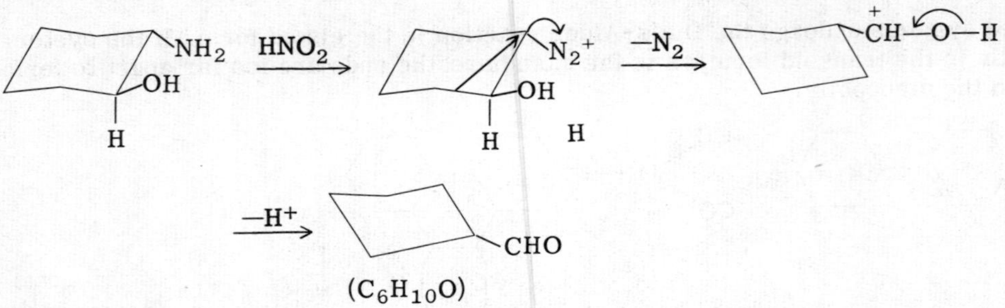

Formylcyclopentane is produced; N.G.P. is involved in its formation.

cis-**2-aminocyclohexanol**

Because the NH_2 and OH groups are roughly the same size, the two possible conformations are present in roughly the same amounts. Hence a mixture of products is obtained, one from each conformation.

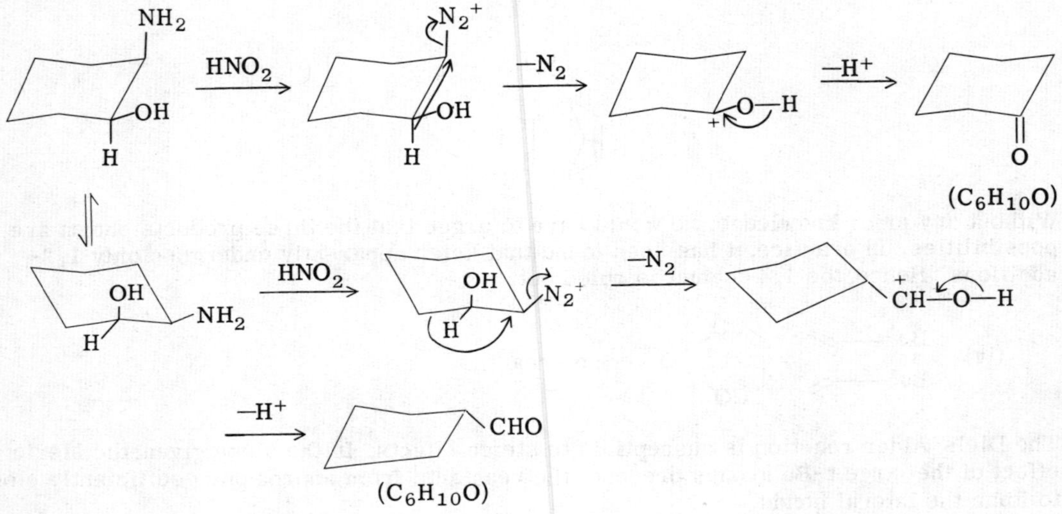

The formylcyclopentane is the same product as that from the *trans*-aminoalcohol.

20 Monocyclic aromatic hydrocarbons

(Problems on page 36)

(*a*) $C_9H_{12} \equiv C_6H_5^- + C_3H_7$

Hence:

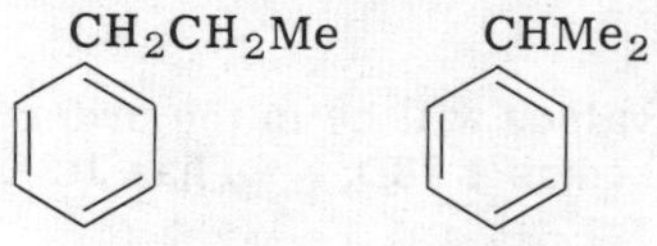

(*b*) $C_9H_{12} \equiv C_6H_4 < + C_3H_8$

$C_3H_8 \equiv Et + Me$

Hence:

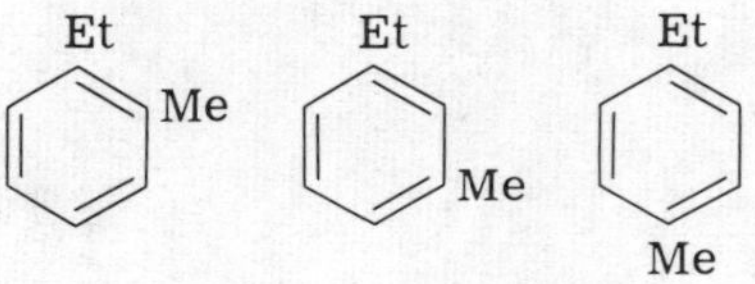

(*c*) $C_9H_{12} \equiv C_6H_3 \leftarrow + C_3H_9$

$C_3H_9 \equiv Me + Me + Me$

Hence:

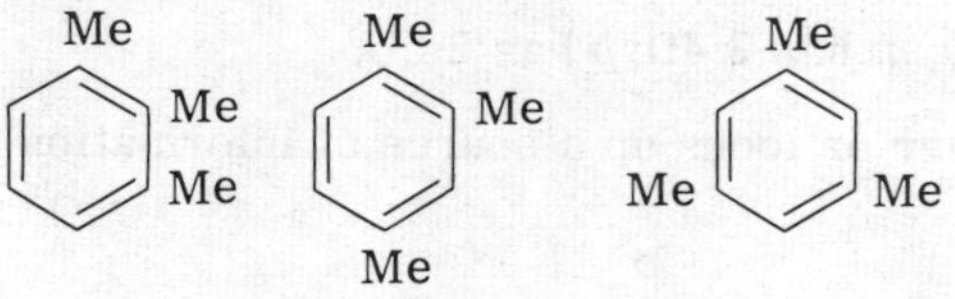

1, 2, 3 1, 2, 4 1, 3, 5

In this case, it is best to number the isomers in order to avoid repeating some as if they were different.

2. The student can, if he wishes, use the formulæ for the vector additions of the dipole moments in the *o*-, *m*-, and *p*-isomers. This has not been done here; all the student has to remember is that a vector can be resolved into two components at right angles.

First, we decide the directions of the dipoles (based on I and R effects of the group; see also Prob. 6, below).

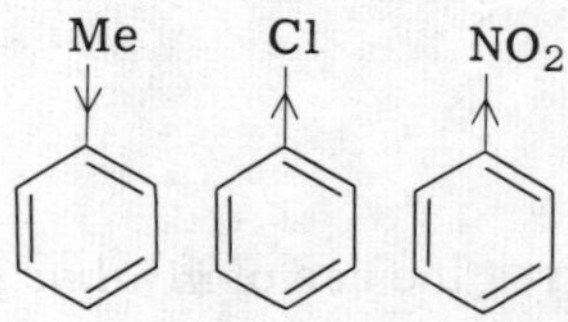

We now draw the isomers with one group at the top. This dipole will be vertical and so we resolve the other vertically and horizontally. Inspection of these readily gives their *relative* values (angles with the horizontal are the same for *o*- and *m*-). Actual values and the angles made with, e.g., the vertical axis, can only be obtained by calculation and a knowledge of the values of the individual dipoles. In the following discussion, *v.c.* means vertical components.

(i)

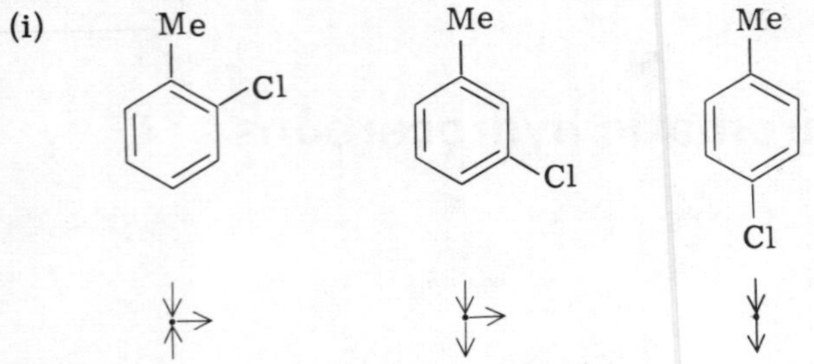

Subtract *v. c.* Add *v.c.* Add vectors

From these diagrams it can be seen that the numerical values will be in the order: $p > m > o$. Hence, we can make the assignments: p has 1·9D; m has 1·78D; o has 1·3D.

(ii)

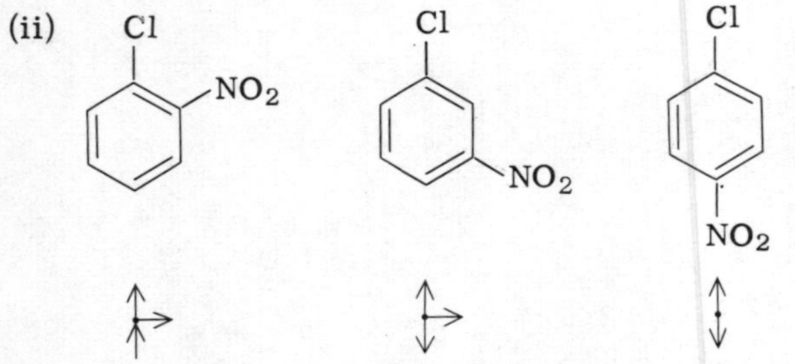

Add *v. c.* Subtract *v.c.* Subtract vectors

The order is: $o > m > p$. Hence: o has 4·3D; m has 3·4D; p has 2·5D.

3. Here, the student either knows the answer or looks up a source of information. Possible oxidising agents are:

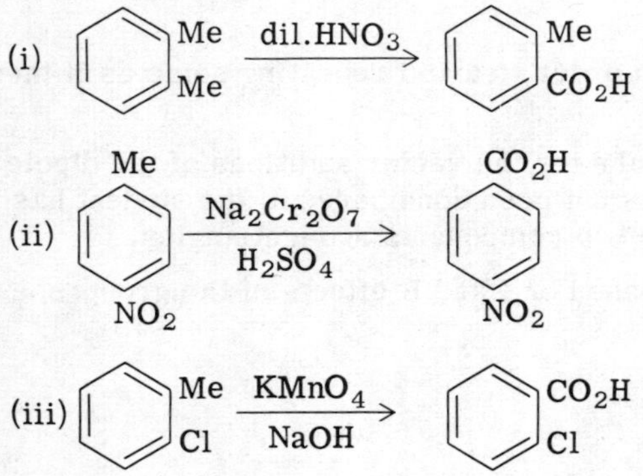

The presence of a –I group *ortho* to the side-chain usually requires the use of *alkaline* permanganate.

4. In the following, each arrow corresponds to an isomer produced, and the number of arrow heads indicates the relative amounts of the isomers. The general rules are that there is reinforcement when the substituent groups direct to the same positions, and precedence of o/p over m when the groups have different directing power (if not directing to the same positions).

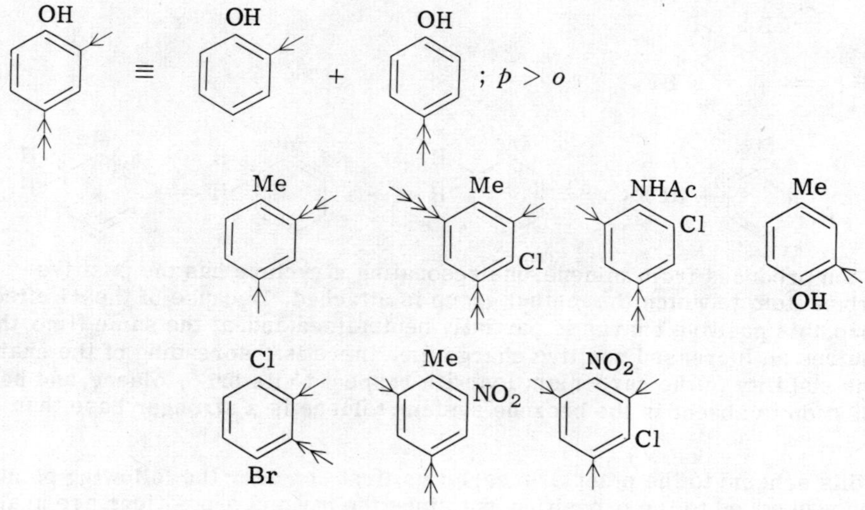

5. 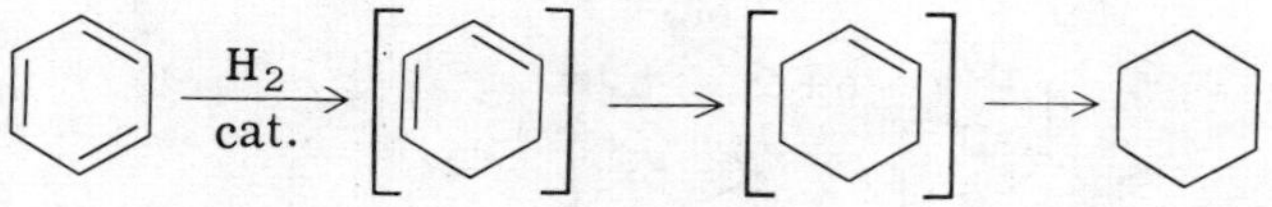

Addition of one molecule of hydrogen produces cyclohexadiene, in which almost all of the resonance energy of benzene has been lost. Hence the activation energy of this step is much greater than that required for each succeeding step, in which the double bonds behave like their acyclic analogues. Thus we have a multi-step reaction in which the first step is rate-determining. This means that the conditions required for the first step are more vigorous than those required for the successive steps (*all* steps involve the addition of one molecule of hydrogen). Because of this, it is not possible to stop the reaction proceeding to complete reduction of benzene to cyclohexane, and consequently it is not possible to isolate the intermediates.

6. NO_2: —I,—R; NH_2: —I, + R; Br: —I, + R; COMe: —I,—R; NMe_3^+: + I (no R effect possible since the nitrogen atom has no lone pair of electrons); SO_3H: —I,—R; OMe: —I, + R.

In all cases, except Br, $\pm R > \mp I$. For the halogens, $\pm R < \mp I$.

It should be remembered that R effects act only from *o*- and *p*-positions with respect to another substituent; they do not act from the *m*-position. On the other hand, the I effect acts from all positions (through the σ-bonds of the benzene ring).

7. The reaction is reversible:

$$ArH + HBF_4 \rightleftharpoons [ArH_2]^+ BF_4^-$$

The more stable the arene carbonium ion, the more will the equilibrium lie on the right, i.e., the more basic is the arene. Since we are given the order of basicity, we must find some factor that increases the stability of the carbonium ion with increase in the number of methyl groups. Because methyl exerts a +I effect, it is worth trying to find out whether a relationship can be formulated between this +I effect and the stability of the ion.

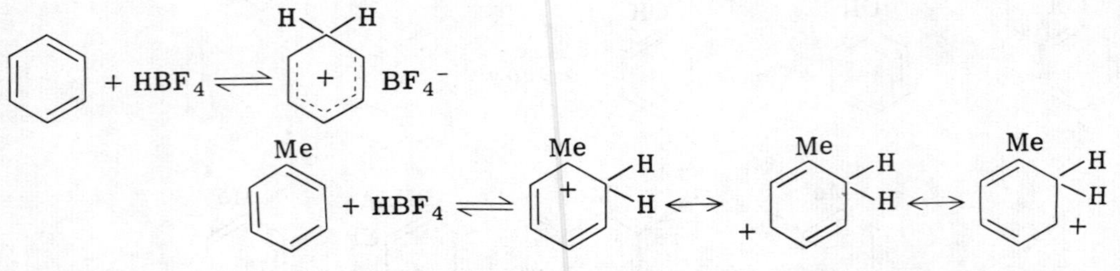

In the carbonium ion produced from toluene, one resonating structure has the positive charge on the carbon atom to which the methyl group is attached. Because of the +I effect of the methyl group, this positive charge is partially neutralised and, at the same time, the methyl group acquires an increased positive charge, i.e., there is a spreading of the charge. This increases the stability of the carbonium ion with respect to its base, toluene, and because this stabilisation is absent in the benzene system, toluene is a stronger base than benzene.

Before we apply this scheme to the other arenes, let us first consider the following point. In toluene, the proton was added to the *o*-position, but since the *m*- and *p*-positions are available, these must also be considered.

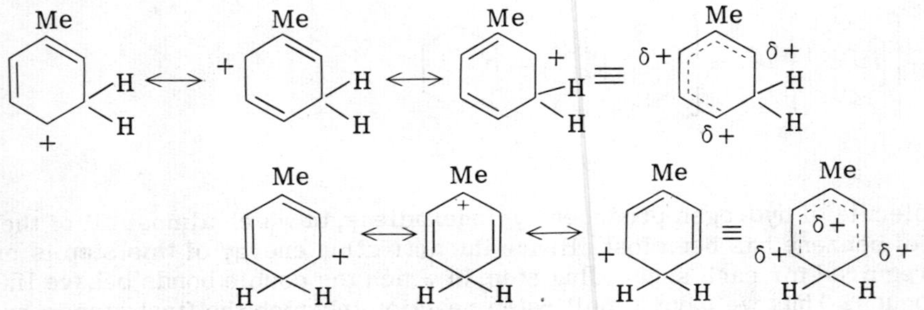

The *m*-carbonium ion has no spreading of charge due to the contribution of a resonating structure with the carbon atom attached to the methyl group carrying a positive charge. Hence, this carbonium ion is less stable than the *o*-ion. The *p*-ion, however, will be as stable as the *o*-ion, and so both of these could be formed equally well. This does not alter the fact that the basicity of toluene would be expected to be greater than that of benzene; all it means is that two carbonium ions of equal stability are present.

The important point to remember when drawing the resonance hybrid molecule is that the charges are in the *o*- and *p*-positions with respect to the attacked carbon atom (now sp^3 hybridised and no longer part of the conjugated system).

If we apply these arguments to the other arenes, we have:

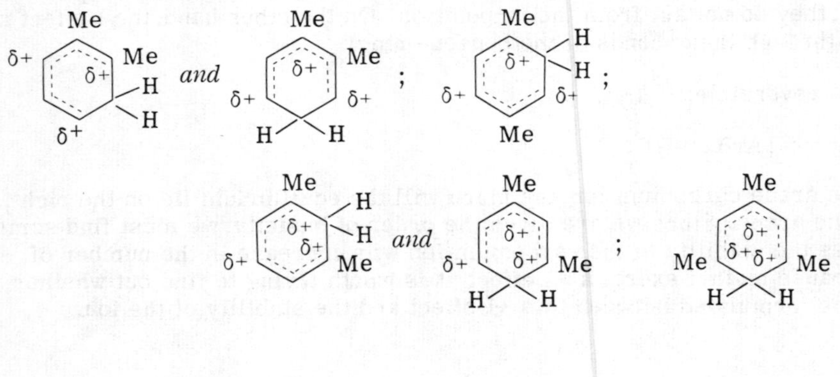

Toluene, *o*- and *p*-xylene each have *one* methyl group stabilising the carbonium ion, and this would suggest that all three have about the same basicity. Since we are told that toluene is less basic than the other two, we should try to explain it. A possibility is as follows. In the *o*- and *p*-xylenes, *one* methyl behaves as the stabilising factor as in toluene, but the other methyl group, because it is adjacent to two positively charged carbon atoms, exerts an increased +I effect (than had these charges been absent) and consequently increases (to a small extent) the spreading of charge. Hence, the carbonium ions of these xylenes are more stable than that of toluene. Thus, the order of basicity is: mesitylene > *m*-xylene > *o*- and *p*-xylene > toluene > benzene.

8. (i)

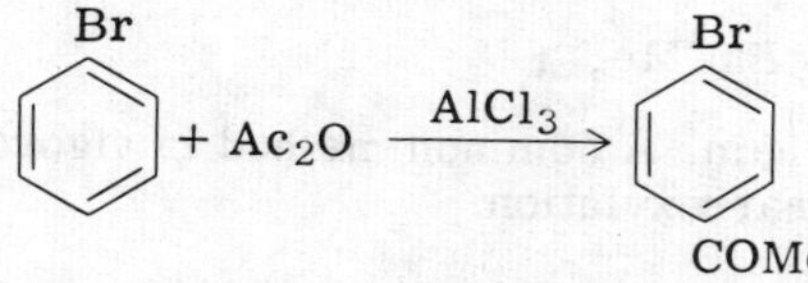

Two molecules of aluminium chloride must be used, since the acylating species is acetyl chloride. The formation of the *p*-product is probably due to steric effects.

(ii)

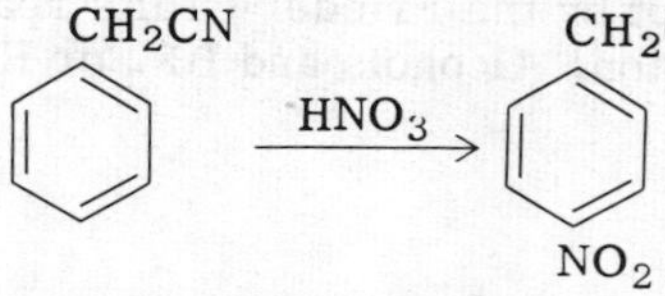

The cyano-group directly attached to the benzene ring has a very strong —R effect and so is *m*-orienting. When 'isolated' from the benzene ring, it exerts only a —I effect on the ring. This effect must be weak, otherwise *m*-substitution would have occurred. At the same time, the CH_2 group exerts a hyperconjugative effect, which favours *o*/*p*-substitution. Presumably the steric effect largely decides *p*-substitution rather than *o*-.

(iii)

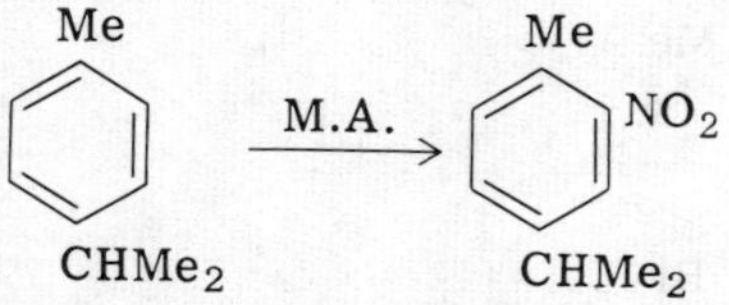

Both alkyl groups have a +I effect, Me_2CH > Me. Both also have a hyperconjugative effect, Me > Me_2CH. Since these effects are in opposition, it is not easy to predict the orientation. However, steric effects due to Me_2CH are far larger than those due to Me, and this decides (presumably) in favour of substitution *ortho* to Me.

(iv)

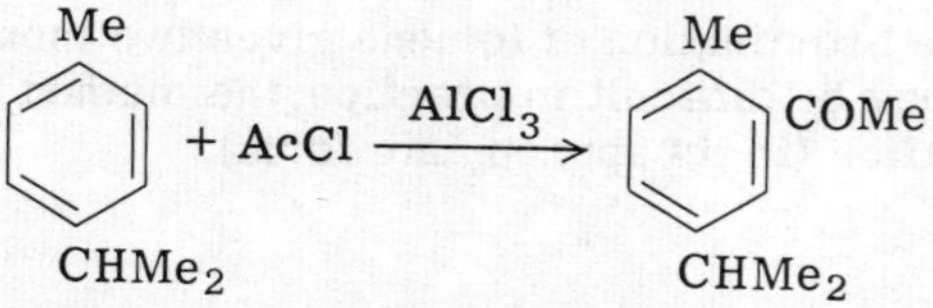

The explanation is similar to that given for (iii).

(v) $Ph(CH_2)_3Me \xrightarrow{KMnO_4} PhCO_2H$

No matter how long the side-chain is, oxidation with strong oxidising reagents results in benzoic acid.

(vi) $PhNO_2 + EtCl \xrightarrow{AlCl_3}$ no reaction

The presence of a strongly *m*-orienting group prevents the Friedel-Crafts reaction. If, however, an activating group (*o*, *p*-orienting) is *also* present, reaction can take place (see *Text*).

(vii) Alkyl halides of suitable structures undergo isomerisation in the presence of aluminium chloride (see *Text*, Ch. 5). Hence:

$$Me_3CCH_2Cl \xrightarrow{AlCl_3} AlCl_4^- + Me_2\overset{\overset{\displaystyle Me}{|}}{C}CH_2^+$$

$$\longrightarrow Me_2\overset{+}{C}CH_2CH_3 \xrightarrow{PhH} PhCMe_2Et$$

9. (i) The final result is removal of the side-chain. A common method is vigorous oxidation to the benzoic acid derivative, followed by decarboxylation.

$$PhEt \xrightarrow{KMnO_4} PhCO_2H \xrightarrow[\text{heat}]{\text{soda-lime}} PhH$$

(ii) Introduction of alkyl groups is often carried out by the Friedel-Crafts reaction. The difficulty that may arise is the problem of orientation. Alcohols and BF_3 or H_2SO_4 give predominantly *o*/*p*-products.

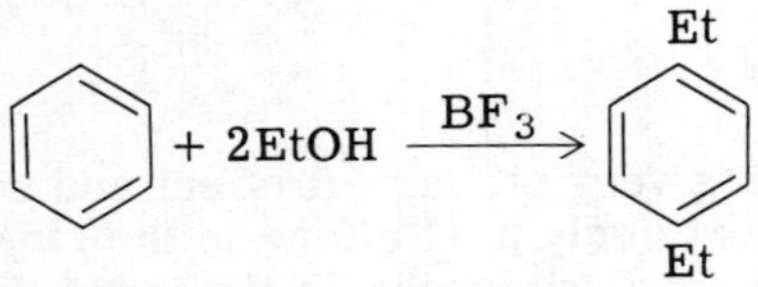

(iii) When the two alkyl groups present are different, their introduction in the correct orientation is usually difficult by the Friedel-Crafts reaction. One way is to use the Wurtz-Fittig reaction.

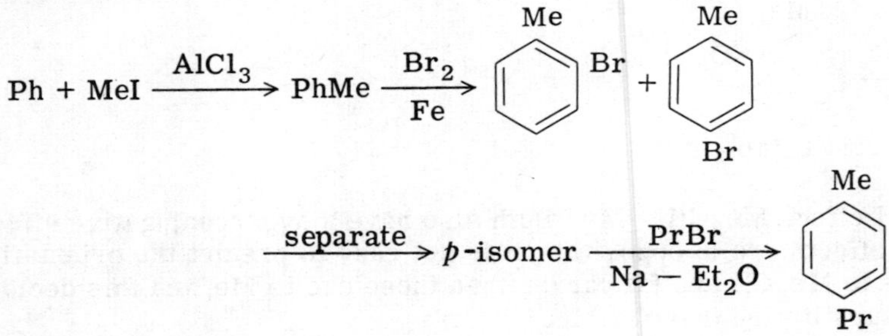

The student should show he is aware that the bromination of toluene gives two isomers and hence necessitates their separation. This may be difficult in practice; the method most likely to succeed is chromatographic separation (in its appropriate form).

(iv) $PhH \longrightarrow PhCH_2CO_2H$

In general, it is inadvisable to try to oxidise a side-chain (alkyl-type) to get a *terminal* carboxylic acid group. At this stage, we shall consider the general approach of working backwards (see also the Willgerodt reaction; *Text*, Ch. 27).

$$PhCH_2CO_2H \longleftarrow PhCH_2CN \longleftarrow PhCH_2X \longleftarrow PhH\ (a)$$

$$PhCH_2CO_2H \longleftarrow PhCH_2CH_2OH \longleftarrow PhH\ (b)$$

The problem is now to be able to synthesise these precursors.

(*a*) Benzyl chloride may be prepared by the chloromethylation of benzene. The student must realise that somewhere along the line he must make use of factual knowledge.

$$PhH + CH_2O + HCl \xrightarrow{ZnCl_2} PhCH_2Cl \xrightarrow{KCN} PhCH_2CN \xrightarrow{H_3O^+} PhCH_2CO_2H$$

(*b*) The compound required is a primary alcohol containing two more carbon atoms than the starting material. This suggests the possibility of the reaction between a Grignard reagent and ethylene oxide (see *Text*, Ch. 15).

$$PhH \xrightarrow[Fe]{Br_2} PhBr \xrightarrow[Et_2O]{Mg} PhMgBr \xrightarrow[(ii)\ H_3O^+]{(i)\ C_2H_4O}$$

$$PhCH_2CH_2OH \xrightarrow[H_2SO_4]{dichr.} PhCH_2CO_2H$$

(v)

$$PhH \xrightarrow[Fe]{Br_2} PhBr \xrightarrow{2Li} PhLi + LiBr$$

(vi)

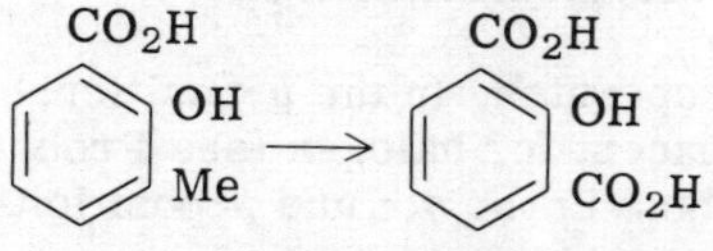

When the benzene ring contains OH or NH_2, the ring is readily broken down by oxidising reagents. The general procedure is to protect the group, oxidise, and then remove the protecting group. A useful 'protecting' method is the preparation of the tosyl (*p*-toluenesulphonyl) derivative. Thus:

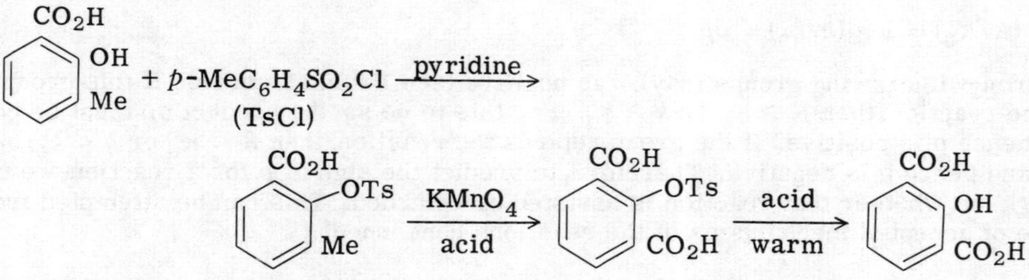

It so happens, however, that hydroxybenzenes can be oxidised without protection if an alkaline solution of lead dioxide is used (for oxidising a *methyl* side-chain).

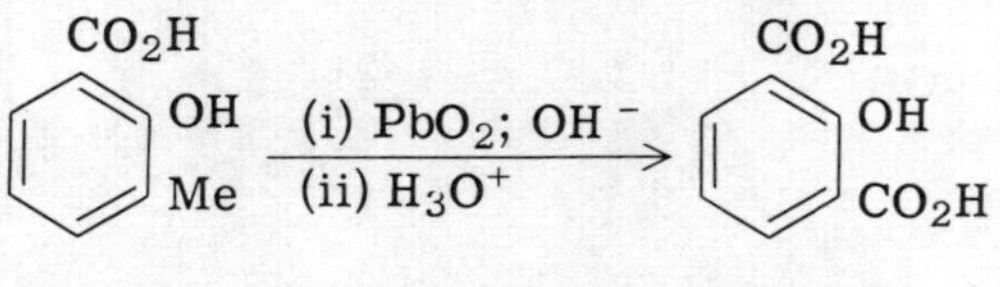

10. The addition of HI is a two-step reaction. The first step is the addition of the proton to give a carbonium ion. There are, however, two possibilities:

$$PhCH{=}CHMe \xrightarrow{HI} Ph\overset{+}{C}H{-}CH_2Me \text{ (I)} \quad or \quad PhCH_2\overset{+}{C}HMe \text{ (II)}$$

(I) can be written as a resonance hybrid:

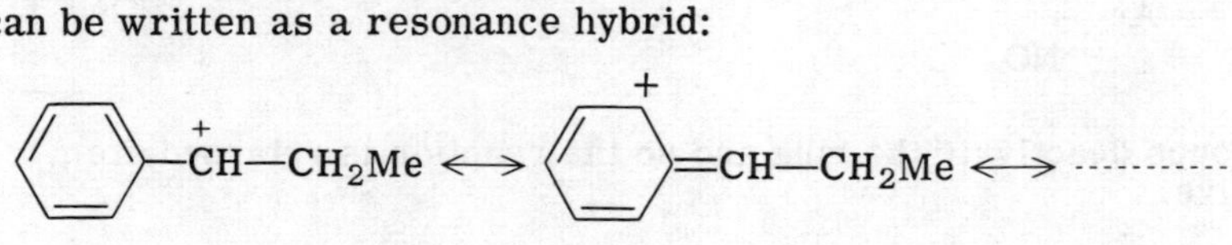

(II) cannot be written as a resonance hybrid (excluding hyperconjugation), and so (I) is far more stable than (II). Hence, the reaction can be expected to proceed predominantly (if not exclusively) through (I).

$$PhCH{=}CHMe \xrightarrow{HI} Ph\overset{+}{C}H{-}CH_2Me + I^- \longrightarrow PhCHICH_2Me$$

11. The σ-constant is a quantitative measure of the polar effect of a *m*- or *p*-substituent and is, by definition:

$$\sigma = \log (K/K_0)$$

where K and K_0 are the ionisation constants of the substituted and unsubstituted acids (see *Text*). The more positive σ is, the greater is the electron-withdrawing power (—I and/or —R) of that substituent, and the more negative σ is, the greater is the electron-releasing power (+I and/or +R) of that substituent. It should also be noted from the definition of σ that if it is positive, $K > K_0$, i.e., the substituted acid is stronger than the unsubstituted one; if negative, $K_0 > K$, and the substituted acid is weaker than the unsubstituted one.

In the *m*-position, only the I-effect of the substituent operates. In the *p*-position, both I and R effects can operate, and when in opposition, R > I except for halogen (see Prob. 6, above). Hence, σ for a given substituent may have opposite signs in the *m*- and *p*-positions.

m-NO_2: [+](—I); *p*-MeO: [—](—I, + R); *p*-Me_2S^+: [+](—I); *m*-Et: [—](+I); *p*-CF_3: [+](—I); *m*-OH: [+](—I); *p*-OH: [—](—I, + R); *m*-NMe_3^+: [+](—I); *p*-CO_2H: [+](—I,—R); *m*-CN: [+](—I); *p*-Me_2N: [—](—I, + R); *p*-SO_2NH_2: [+](—I,—R); *m*-IO_2: [+](—I).

12. $\log (K/K_0) = \log (k/k_0) = \sigma\rho$

For electron-withdrawing groups (say), σ is positive (see Prob. 11, above). If this group assists the reaction, then $K > K_0$ or $k > k_0$. For this to be so, the product σρ must be positive, and hence ρ is positive. If the group retards the reaction, then $K < K_0$ or $k < k_0$; σρ is negative and hence ρ is negative. Therefore, to predict the sign of ρ for a reaction, we must try to work out whether that reaction is assisted or retarded. This can be attempted from a knowledge of accepted mechanisms of the reactions concerned.

(i)

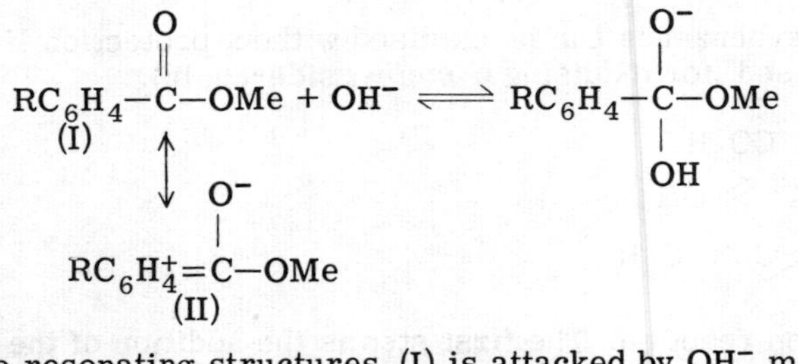

Of the two resonating structures, (I) is attacked by OH^- more readily than (II). If R has a —I effect, then the contribution by (II) is diminished, and so the reaction is facilitated, i.e., $k > k_0$. Hence ρ is positive.

(ii)

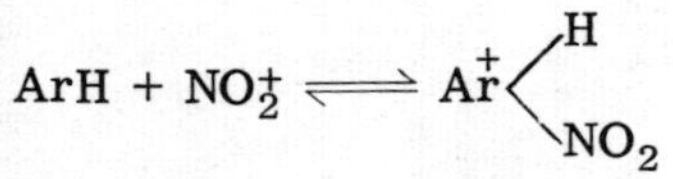

Electron-withdrawing groups deactivate the ring and so the reaction is retarded, i.e., $k < k_0$. Hence ρ is negative.

(iii)

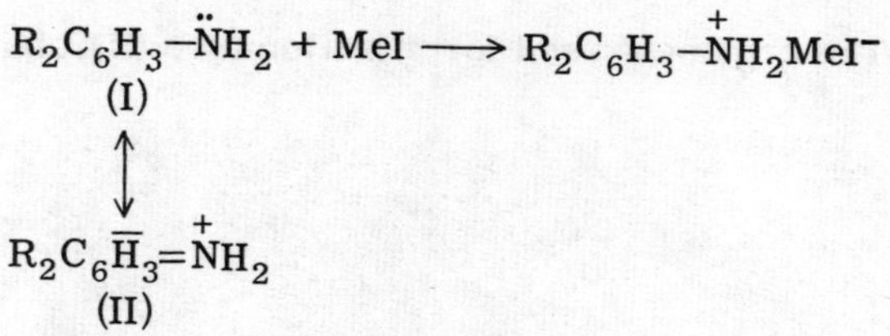

(I) undergoes reaction more readily than (II) because in the former the lone pair of electrons on the nitrogen atom is available for forming the bond with the methyl group. Since electron-withdrawing groups favour (II), the reaction is retarded, i.e., $k < k_0$. Hence ρ is negative.

(iv) Let us assume the r/d step is the formation of a carbanion:

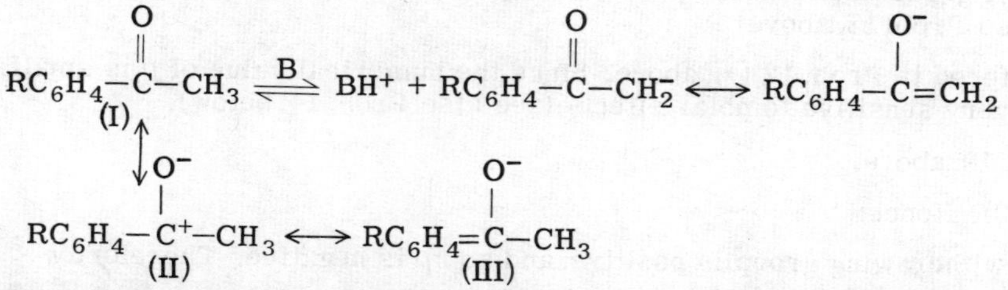

Removal of the proton is easiest from (II) [this is also true for the protonated species in acid solution], and since an electron-withdrawing group favours (I) and (II), the reaction is facilitated, i.e., $k > k_0$. Hence ρ is positive.

(v)

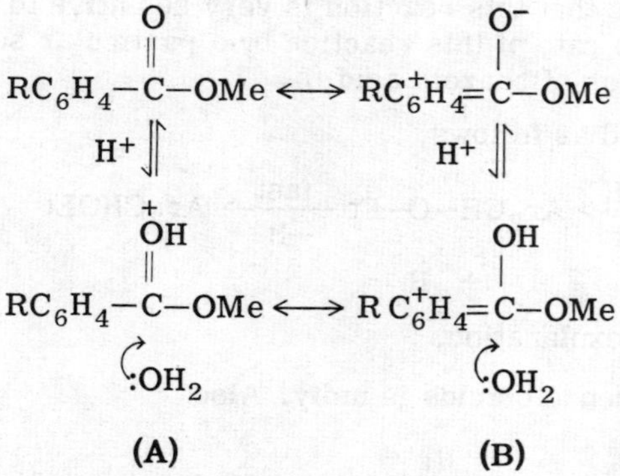

Attack by water on the protonated species is easier with **(A)** than with **(B)**. Since **(A)** is favoured by an electron-withdrawing group, the reaction is facilitated, i.e., $k > k_0$. Hence ρ is positive.

(vi)

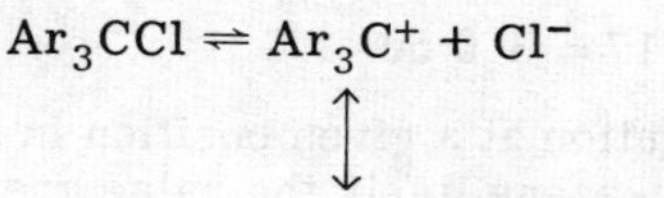

Since the C—Cl bond undergoes heterolytic fission, any effect that withdraws the covalent pair of electrons from the chlorine atom and consequently inhibits stabilisation of the carbonium ion when it is formed, will retard the reaction. Because an electron-withdrawing group behaves this way, $K < K_0$. Hence ρ is negative.

(vii) $ArH + Br_2 \longrightarrow ArBr + HBr$

Bromine is an electrophilic reagent. The argument used for nitration (see (ii), above) also applies here. Hence ρ is negative.

(viii)

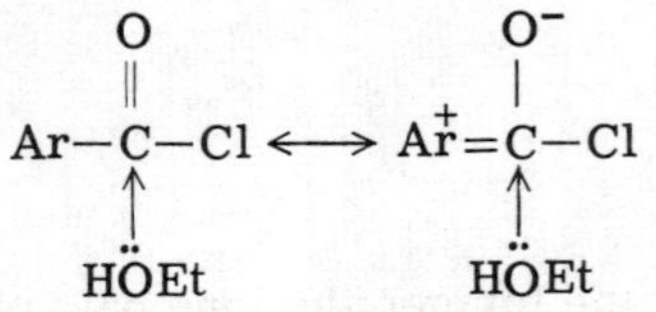

(I) (II)

Attack by ethanol is easier in (I) than in (II). Since electron-withdrawing groups favour (I), reaction is facilitated, i.e., $k > k_0$. Hence ρ is positive.

13. Since σ for a —I group is positive, and ρ is also positive (+0·144), $k > k_0$, i.e., the rate is increased (see also Prob. 12, above).

An explanation is offered in Prob. 12 (v), above. Since the numerical value of ρ is small, this reaction is not very sensitive to polar effects (see also Prob. 14, below).

14. See also Prob. 12, above.

ρ is negative (—5·090). Hence:

(*a*) σ of an electron-withdrawing group is positive, and so $\sigma\rho$ is negative. Therefore $k < k_0$, and so the reaction is retarded.

(*b*) σ of an electron-releasing group is negative, and so $\sigma\rho$ is positive. Therefore $k > k_0$, and so the reaction is accelerated.

(*c*) The large numerical value of ρ indicates that this reaction is very sensitive to polar effects. Thus, for example, the effect on the rate of this reaction by a particular substituent is far greater than its effect on the ionisation of benzoic acid ($\rho = 1$).

The effects of substituents may be explained as follows.

$$Ar_2CHCl \underset{}{\overset{slow}{\rightleftharpoons}} Ar_2CH^+ + Cl^- \xrightarrow[fast]{EtOH} Ar_2CH{-}\overset{+}{\underset{\mid \atop H}{O}}{-}Et \xrightarrow[-H^+]{fast} Ar_2CHOEt$$

See Prob. 12 (vi), above, for the rest of the explanation.

15. ρ for the ionisation equilibria of the benzoic acids is unity. Also

$\log (K/K_0) = \sigma\rho \;\therefore\; \sigma = \log (K/K_0)\rho$

(i) $\sigma_m = \log (K_m/K_0) = \log [(14{\cdot}8 \times 10^{-5})/(6{\cdot}3 \times 10^{-5})]$
$= \log 2{\cdot}35 = +\,0{\cdot}37$

(ii) $\sigma_p = \log (K_p/K_0) = \log [(2{\cdot}0 \times 10^{-4})/(6{\cdot}3 \times 10^{-5})]$
$= \log [(20 \times 10^{-5})/(6{\cdot}3 \times 10^{-5})] = \log 3{\cdot}17 = +\,0{\cdot}50$

16. Since the partial rate factor is the rate of substitution at a given position in a substituted benzene relative to that at any one position in benzene itself, the value greater than unity for p-substitution in fluorobenzene means that p-substitution is faster than in benzene. Hence, the ring in fluorobenzene is activated in the p-position. As we have seen (see *Text*), halogens, although o/p-orienting deactivate the ring; fluorine is the only exception and this applies only to the p-position.

17. Polar effects (I and R) are transmitted strongly to o- and p-positions, and I effects are transmitted weakly to the m-position. Hence, on the assumption that the m-position is very little affected, the m/p ratio reflects the polar influences of the substituent at the p-

position; it is independent of steric effects of the substituent. In the *o*-position, there is always the possibility of a steric effect (large, small, or negligible), and because the *o*- and *p*-positions reflect polar effects, *o/m* and *o/p* ratios therefore reflect *both* polar and steric effects.

18. $PhCH_2CH{=}CH_2 \xrightarrow[\text{heat}]{KOH} PhCH{=}CHMe$

The rearranged product has increased conjugation with the ring; it is the thermodynamically controlled product (T.C.P.) and is the result of an equilibrium controlled reaction (see *Text*, Ch. 2). Hence, the starting material is the kinetically controlled product (K.C.P.).
The conversion of the K.C.P. into the T.C.P. requires removal of a hydrogen atom from the α-carbon, and the use of a base suggests this is effected as proton-removal. We must therefore offer an explanation for this removal being able to occur. One possibility is that the α-proton is sufficiently acidic to be removed because the 'driving force' is the resonance stabilisation of the carbanion thereby produced. On this basis we can formulate the reaction as follows:

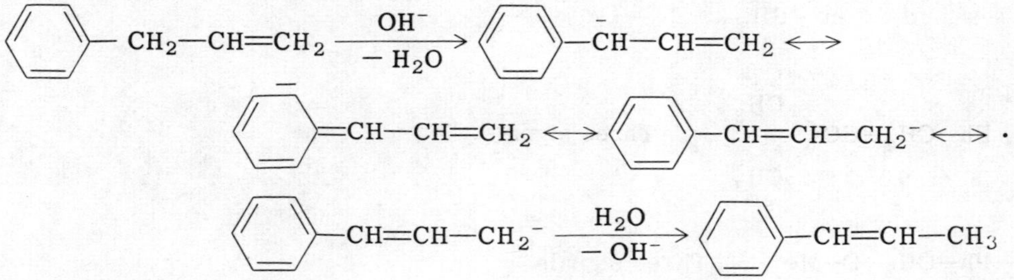

If a proton added to a ring-carbon atom, the R.E. of the benzene ring is lost. If it adds to the terminal methylene group, the product has the maximum amount of stabilisation and so this is the one formed.

19. Hückel's rule requires a monocyclic ring compound to contain $(4n + 2)$ π-electrons for it to exhibit aromaticity. $C_{10}H_{10}$: 10 π-electrons, $n = 2$, $(4n + 2)$ molecule.

$C_{12}H_{12}$: 12 π-electrons, $n = 3$, $4n$ molecule.
$C_{12}H_{12}^{2-}$: 14 π-electrons, $n = 3$, $(4n + 2)$ molecule.
$C_{12}H_{12}^{2+}$: 10 π-electrons, $n = 2$, $(4n + 2)$ molecule.
$C_{20}H_{20}$: 20 π-electrons, $n = 5$, $4n$ molecule.
$C_{20}H_{20}^{-}$: 21 π-electrons, $n = 5$, $(4n + 1)$ molecule.
$C_{20}H_{20}^{2-}$: 22 π-electrons, $n = 5$, $(4n + 2)$ molecule.
$C_{20}H_{20}^{+}$: 19 π-electrons, $n = 4$, $(4n + 3)$ molecule.
$C_{20}H_{20}^{2+}$: 18 π-electrons, $n = 4$, $(4n + 2)$ molecule.

20. On the basis that the r/d step is the formation of the carbonium ion, then, with benzene as the reference compound, the rate is faster when the substituent activates the ring (+I and/or +R; *o/p*) and the rate is slower when the substituent deactivates the ring (–I and/or –R; *m*). Halogen is an exception; it deactivates the ring (–I, +R; –I > +R), but the orientation is *o/p* (see also Prob. 6, above). Hence, the order of increasing rate of nitration is:

$PhNO_2$	< PhCl	< PhH = C_6D_6	< PhMe
(–I, –R)	(–I, +R; –I > +R)		(+I)

Since proton removal from the carbonium ion is fast, no kinetic isotope effect is observed when H is replaced by D. Hence, the rates of nitration of PhH and C_6D_6 are the same. Had a kinetic isotope effect been observed, this would mean that the second step had now become r/d.

21. The following symbols have been used:

s = singlet; d = doublet; q = quartet;

m = multiplet (>4).

Since we are told to assume that the signal of the benzene ring is a singlet, we may assume that there is no coupling between ring and side-chain (i.e., no long-range coupling; see *Text*, Ch. 4).

(i) Ph—CMe_3 (s, s) Two signals

(ii) 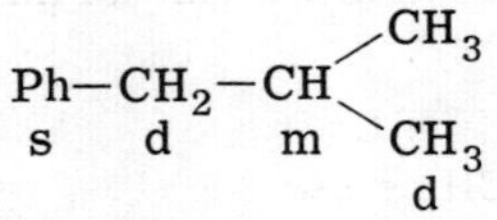 Four signals

(iii) 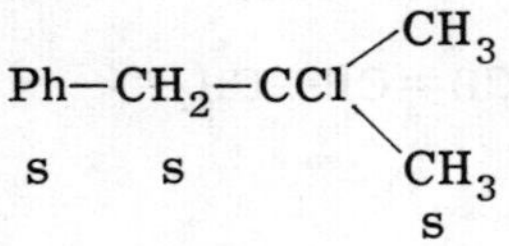 Three signals

(iv) Ph—CH_2—O—Me (s, s, s) Three signals

(v) Ph—CHBr—Me (s, q, d) Three signals (see also (vi), below)

(vi)  Four signals

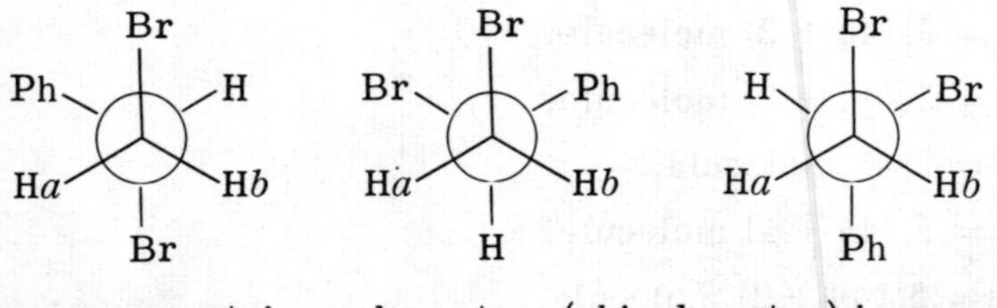

Whenever an asymmetric carbon atom (chiral centre) is present, the various conformations should be investigated. Fast rotation is assumed, and the most probable conformations are taken as the staggered forms (see *Text*, Ch. 3). Inspection of the three conformations shows that the environments of H_a and H_b are different in each form. Also, the three populations are different and hence the chemical shifts of H_a and H_b will be different. Assuming that the difference is sufficient, H_a and H_b each give a signal. Hence, four signals will appear in the NMR spectrum.

The phenyl signal is a singlet. The signal of H (of CHBr) will be a quartet. H is coupled to H_a to give a doublet and each component of this is coupled to H_b, resulting in each being split into a doublet. Hence, a quartet is the result (provided there is no overlapping; see *Text*, Ch. 1). Similarly, H_a and H_b could give rise to quartets.

22. $C_6H_5C_4H_9 = C_{10}H_{14}$ $\therefore$ M.W. = 134

This has been calculated on the basis of C = 12 and H = 1. Hence, the M.W. = 135 is due to 13C or D.

$C_6H_5-CH_2CH_2CH_2CH_3$ (n-) $C_6H_5-CH(CH_3)-CH_2CH_3$ (s-)

M^+ is stronger in **A** than in **B**, and this suggests that the former has the straight chain and the latter the branched chain.

The base peak (*B.P.*) for **A** is *m/e* 91. This is the tropylium cation and its formation is readily explained on the basis of β-cleavage, which is general for alkylbenzenes.

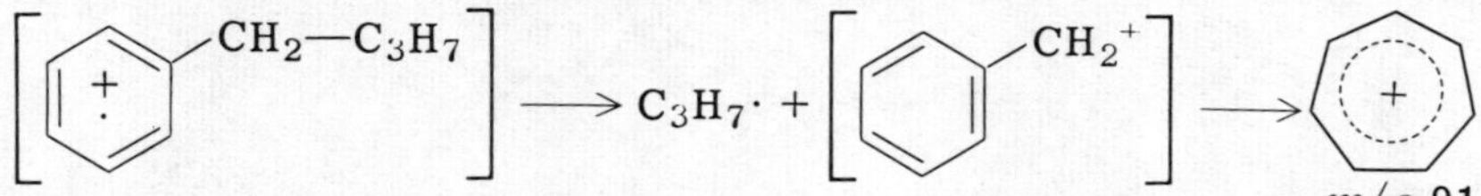

m/e 91

The weak peak at *m/e* 91 in **B** again supports the argument that **B** has the branched chain. The fact that it is present indicates it has been formed by a rearrangement (see *Text*).

The *B.P.* of **B** is *m/e* 105. This can be interpreted as M^+—29, which is equivalent to M^+—C_2H_5 (see Table 1. 6). Here again, this can be expected on the basis of β-cleavage, with the larger group leaving preferentially.

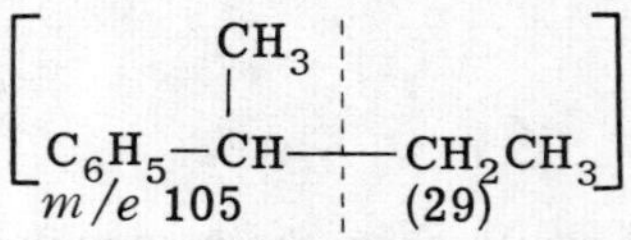

Finally, the strong peak at *m/e* 92 in **A** is readily explained by a McLafferty rearrangement.

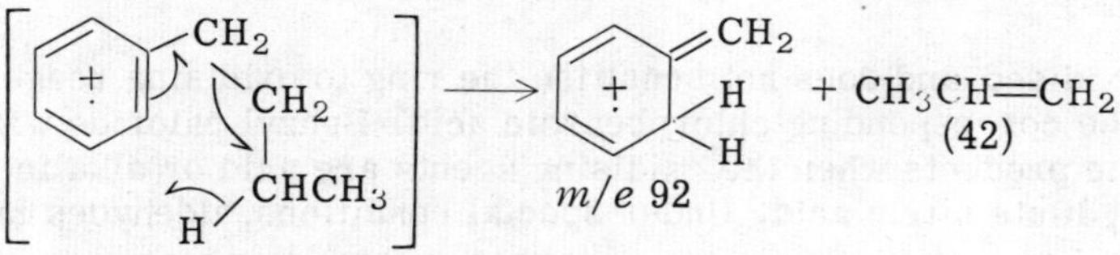

Thus, all the evidence is in favour of **A** being n-butylbenzene and **B** being s-butylbenzene.

21 Aromatic halogen compounds

(Problems on page 39)

1. $C_7H_7Cl = C_6H_5^- + CH_2Cl$

Hence: 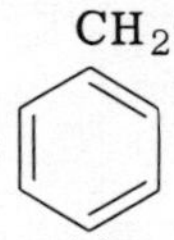

(I)

$C_7H_7Cl = C_6H_4< + CH_3Cl$

$CH_3Cl = CH_3 + Cl$

Hence: 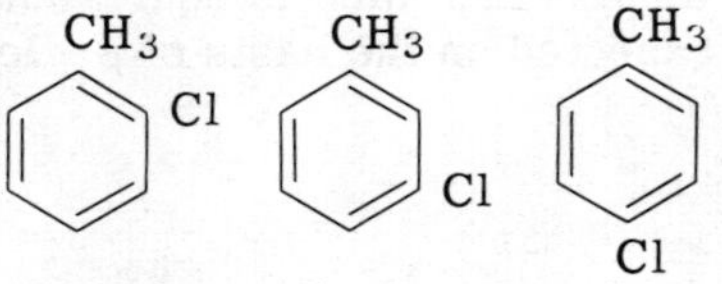

(II) (III) (IV)

I.r. spectroscopy could be used to distinguish between them; (I) would show the monosubstitution pattern and the other three would show the *o*-, *m*-, and *p*-disubstitution pattern, respectively.

Chlorine is a —I group, is not oxidised, and does not sensitise the ring to oxidising reagents. Each chlorotoluene will given the corresponding chlorobenzoic acid. Benzyl chloride will give benzoic acid. These are the products when the oxidising agents are acid or alkaline permanganate, acid dichromate, dilute nitric acid. Under special conditions, aldehydes may be obtained (see *Text*, Ch. 27).

2. (i) **One chlorine atom has been replaced by a carboxyl group. Hence, controlled conditions have to be used, and if we work backwards, two immediate possibilities are: (*a*) CN; (*b*) Grignard and carbon dioxide.**

Aryl halides are far less reactive than alkyl halides, and so to get replacement reactions, certain conditions have to be used.

(*a*) 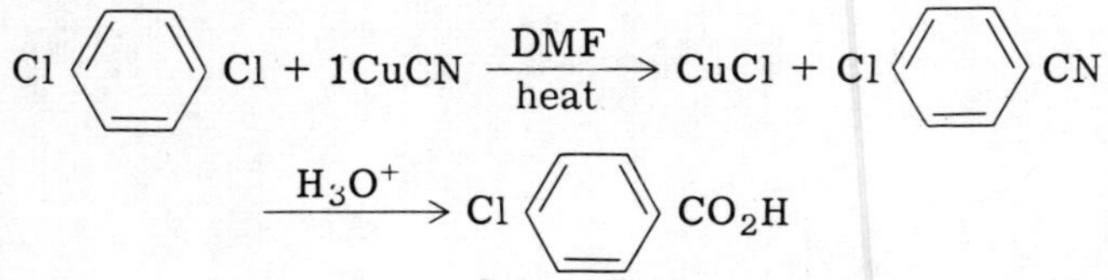

(*b*) 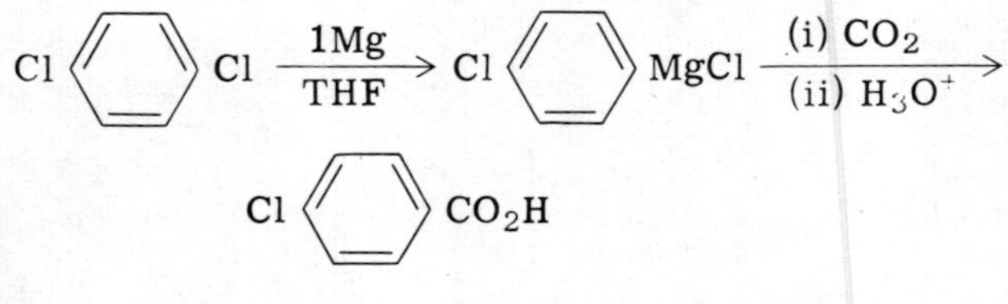

(ii) Since the product is a derivative of *p*-bromoacetophenone, this suggests a Friedel-Crafts reaction. Because the acyl group is *m*-orienting (and readily brominated), the order is bromination followed by acylation.

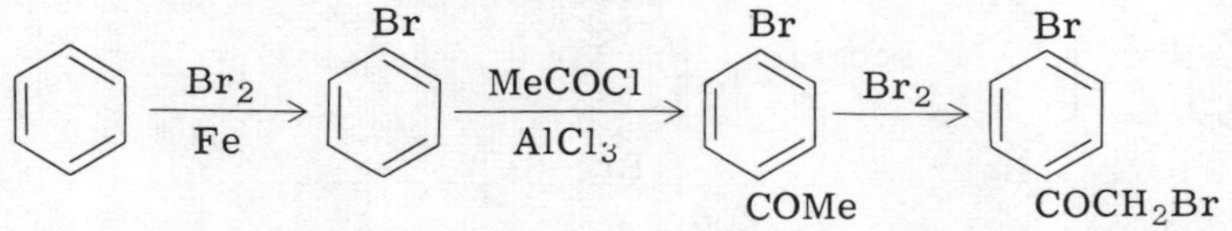

Steric effects are probably the predominant factors that prevent the formation of the *o*-derivative.

(iii) When an aromatic ring is to be alkylated, a common method is the Friedel-Crafts alkylation, and this is usually the first step.

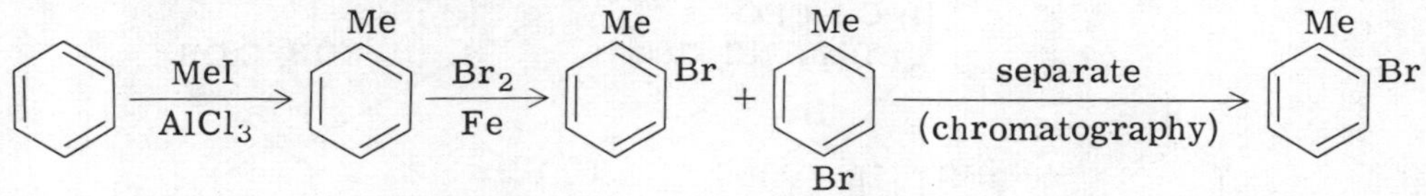

3. The presence of an unsaturated side-chain and bromine in the ring suggests that bromine is introduced first since, if vice versa, bromine addition to the double bond would be expected. The vinyl side-chain suggests the possibility of removal of HX from C_2H_4X. There are two possibilities here, $-CHXCH_3$ or $-CH_2CH_2X$. Both are formed when ethylbenzene is chlorinated at its b.p. in the presence of light and both give the required product when HCl is eliminated. Hence (see also Prob. 2 (iii), above):

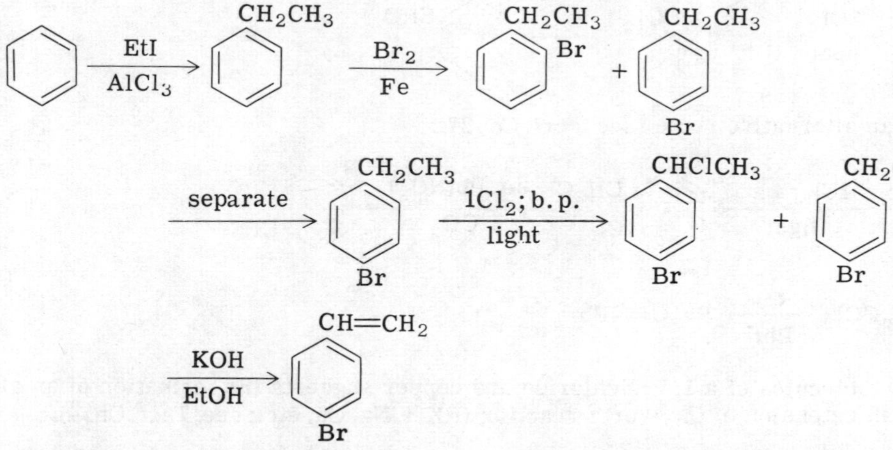

It should be remembered that bromine is not replaced (unless activated).

It is also instructive to tackle this problem in a different way. We could use the method of working backwards as follows:

(*a*) $-CHXCH_3 \longleftarrow -CHOHCH_3$

(*b*) $-CH_2CH_2X \longleftarrow -CH_2CH_2OH$

(*a*) Secondary alcohol: Grignard + aldehyde.

There is also an alternative route. Ketones, on reduction, give secondary alcohols, and if the student refers to Prob. 2 (ii), above, he will see that the ketone precursor is readily prepared.

(*b*) Primary alcohol: two extra carbon atoms; therefore Grignard + ethylene oxide (see *Text*, Ch. 15). Hence:

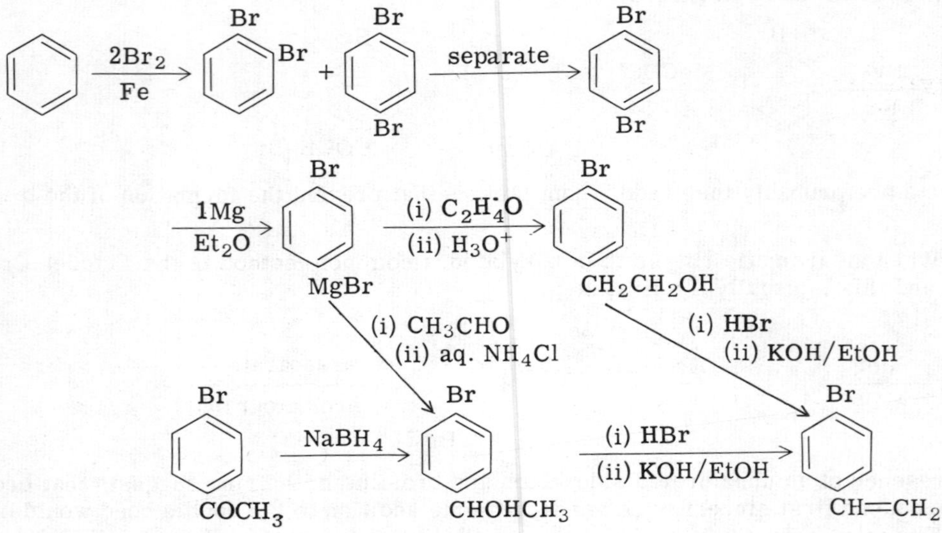

4. (i) Since the final product is an aldehyde, one precursor could be the 1, 1-dichloride; these compounds are readily hydrolysed to carbonyl compounds.

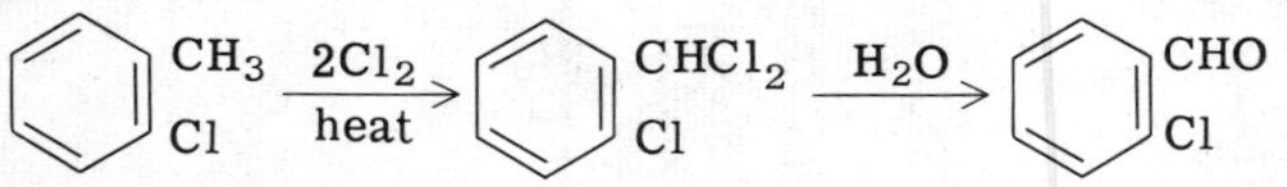

There is also an alternative route (see *Text*, Ch. 27):

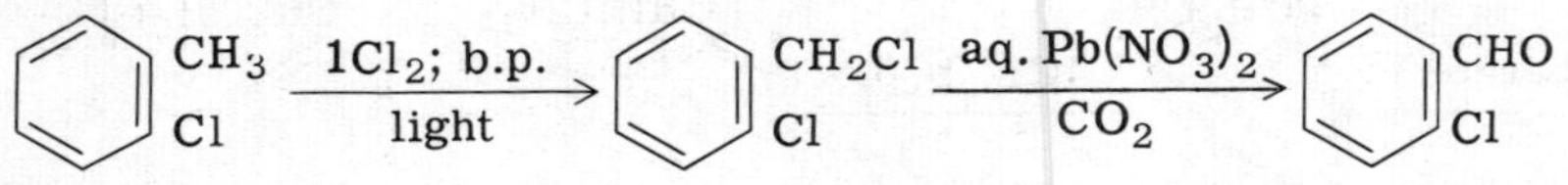

(ii)

$$2Ph_2CCl_2 \xrightarrow[PhH]{Cu} Ph_2C{=}CPh_2$$

The use of *two* molecules of a 1, 1-dichloride and copper suggests the formation of an alkene. This is an extension of the Wurtz reaction (RX + Na, Cu, etc.; see *Text*, Ch. 4).

(iii)

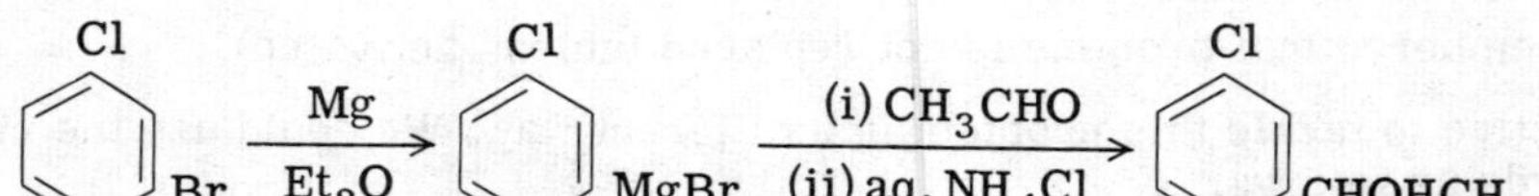

The use of the reagents given clearly indicates that a Grignard reaction is involved. However, aryl *chlorides* do not form Grignard reagents in *ether*; THF is necessary as solvent. On the other hand, aryl *bromides* (and iodides) readily react in ether solution.

The use of aqueous ammonium chloride is a safety precaution to prevent the possibility of dehydration of the alcohol (which might occur if acid is used).

(iv)

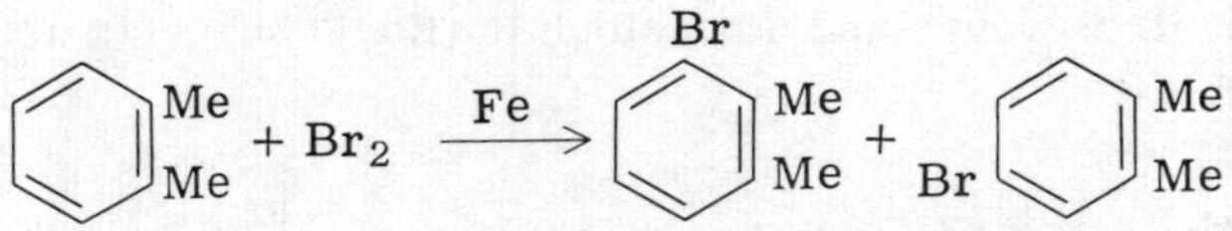

The use of iron indicates *nuclear* substitution. Me is *o*/*p*-orienting, and since there is always the possibility of a steric effect at the *o*-position, all we can suggest is that the 4-Br product will predominate. Experimental work would have to be done to find out what are the actual results. From the literature it appears that the 4-Br compound is formed exclusively.

(v)

Unless the chlorine atom is activated (e.g., by *o*- and/or *p*-NO_2), it is not reactive enough to undergo nucleophilic substitution.

(vi) The use of light suggests a free-radical mechanism. This means that the methane derivative will undergo homolytic fission. Since the C—Br bond is weaker than the C—Cl bond (see Table 2.2), it is reasonable to suppose that the former will be broken. Hence:

$$CBrCl_3 \xrightarrow{h\nu} Br\cdot + \cdot CCl_3$$

$$PhCH_3 \xrightarrow{\cdot CCl_3} CHCl_3 + PhCH_2\cdot \xrightarrow{CBrCl_3} PhCH_2Br + \cdot CCl_3; \text{etc.}$$

Attack by the free radical on toluene occurs at the methyl side-chain and not in the ring because the C—H bond in Me is weaker than that of a ring-hydrogen atom and the benzyl free radical is far more stable than an aryl free radical.

The other point that requires explanation is why toluene is attacked by the CCl_3 free radical and not by the bromine atom. Activation energies involving free radicals are usually very low and so the controlling factor is the heat of reaction (or, more correctly, the free energy of reaction). The more exothermic the reaction (greater is ΔG), the more favoured is that reaction (see *Text*, Ch. 2). If the bromine atom attacks, the result is HBr, the bond of which is much weaker than the C—H bond formed when $\cdot CCl_3$ attacks. Hence, reaction proceeds by the latter route (see also *Text*, Ch. 5).

(vii)

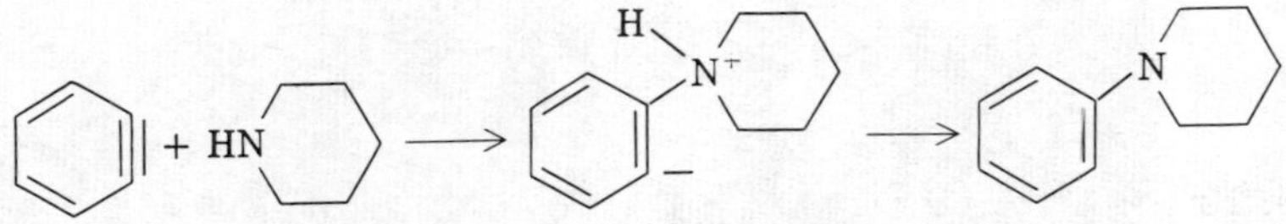

This reaction with piperidine is analogous to that with ammonia (see *Text*).

(viii) The use of an aryl halide and sodamide in liquid ammonia suggests the benzyne mechanism. Since benzyne itself is formed, addition in either direction leads to the same product.

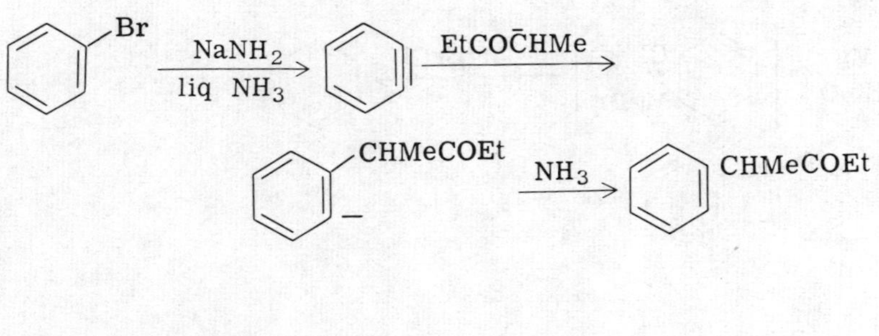

(ix) The presence of *o*/*p*-nitro-groups and a nucelophile (EtO^-) suggests activated nucleophilic aromatic substitution.

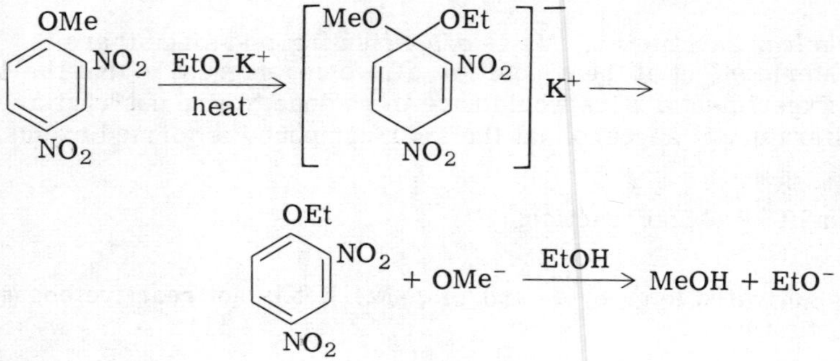

(x) The substrate is a dinitro-compound; this suggests the possibility of activated nucleophilic aromatic substitution. Since only nitro-groups are present in the ring, then, if reaction is going to occur, a hydrogen atom *o*- and *p*- to the nitro-groups will be most activated and so is the most likely one to be attacked. Hydroxide ion is the nucleophile. What, then, is the purpose of potassium ferricyanide? Since this is an oxidising agent, its purpose could be to remove the proton from the complex (see *Text*, Ch. 22).

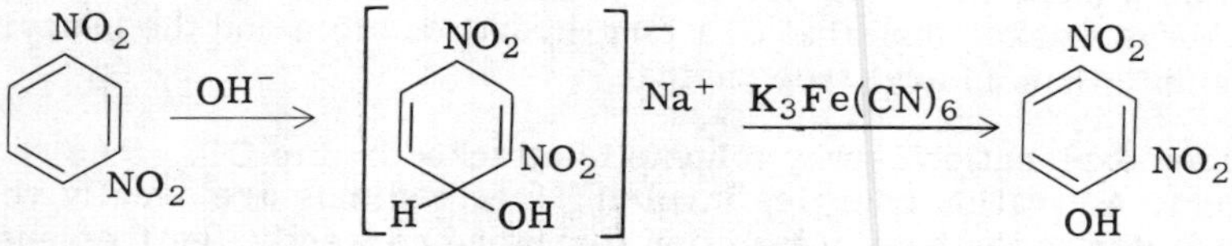

This, 2, 4-dinitrophenol, is the major product; a small amount of 2, 6-dinitrophenol is also produced.

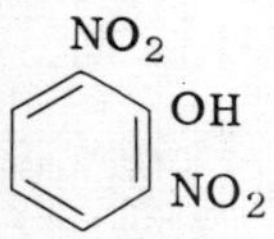

(xi)

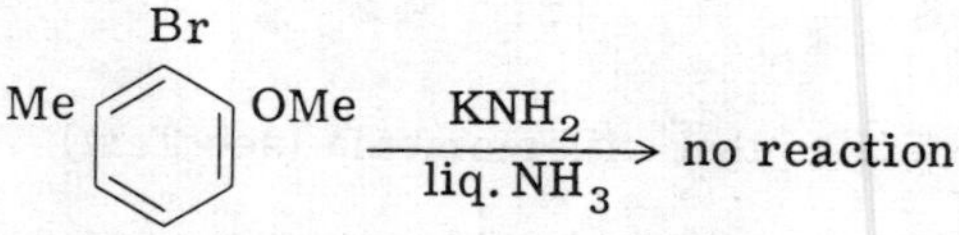

The reagents used suggest the benzyne mechanism, but since an *o*-hydrogen atom is involved, its absence in the substrate leads to the conclusion that this mechanism cannot operate.

(xii)

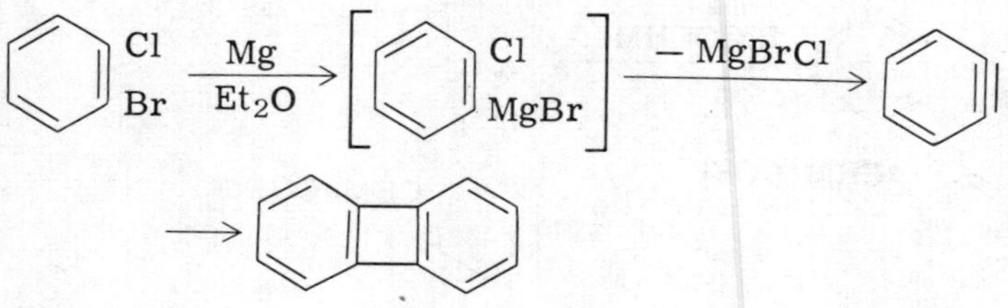

Grignard reagents of this type (with *o*-X atom) are unstable at room temperature; they readily undergo elimination to form benzyne, which then forms biphenylene.

(xiii)

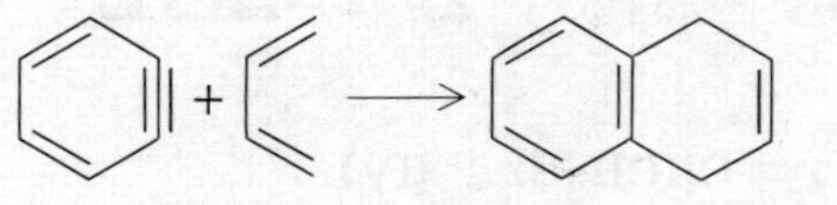

Benzyne behaves as a dienophile in this reaction (see *Text*, Ch. 19).

5. Let us first write down the possible intermediates that we could expect from nuclear substitution and addition to the double bond.

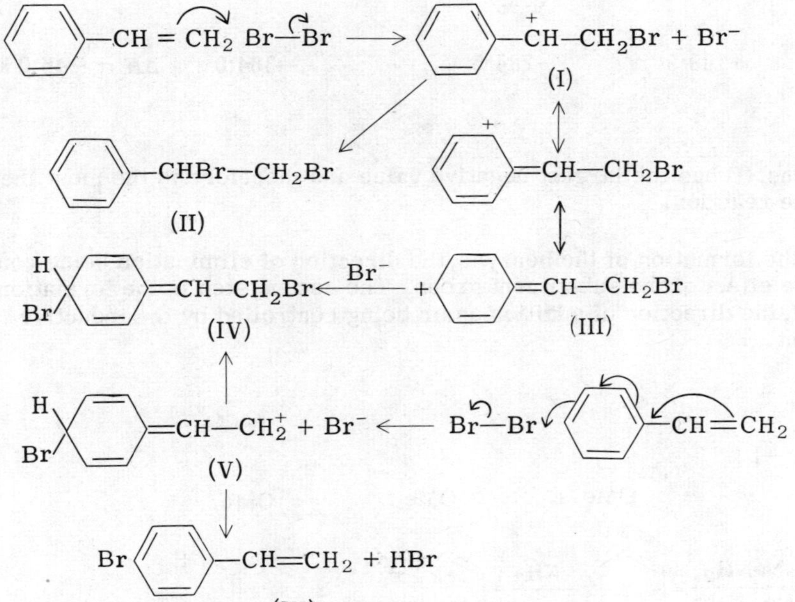

Since we are *told* that (II) is the only product (none of (VI) is formed), we must try to explain these *observed* results. Unless the benzene ring is fairly highly activated by some substituent (usually OH, NH_2, NR_2), a catalyst has to be used in bromination. Hence, the vinyl group, if it activates at all, does so weakly. On the basis that the relative stabilities of the intermediate carbonium ions is the deciding factor for the route to be followed, we therefore have to examine the ions (I), (III) and (V) from this point of view. Of these three, only (I) has retained the R.E. of the benzene ring and hence its contribution will be much greater than that of (III). Therefore the resonance hybrid ion will behave more like (I) than like any other contributing structure, thereby resulting in the formation of (II). Similarly, since carbonium ion (V) is much less stable than (I), this path is not followed. Hence, neither (IV) nor (VI) will be formed.

Now let us look at the problem from the point of view of the free energies of the reactions; we shall equate ΔG with ΔH (see *Text*, Ch. 2). We use Table 2.2.

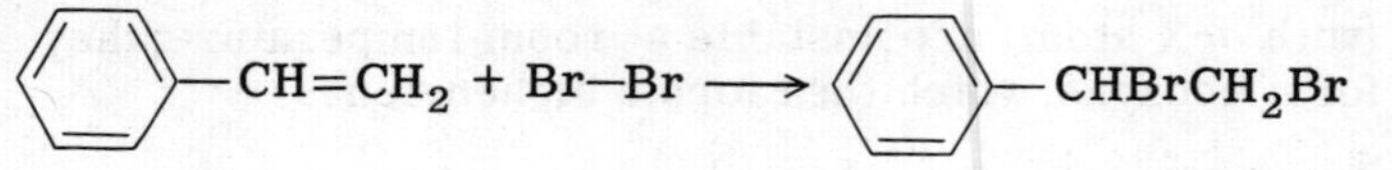

(I)

+606·7   +188·3   −347·3 − 2 × 284·5   **Δ*H* = −121·3 kJ**

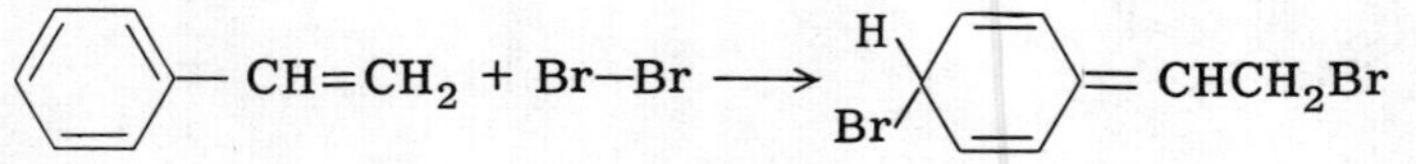

(IV)

+606·7   +188·3   −347·3 (C—C bond formed)
+606·7 (1 double bond lost)   −2 × 284·5
+150·6 (R.E.)   −15 (~R.E.)   **Δ*H* = +14·3 kJ**

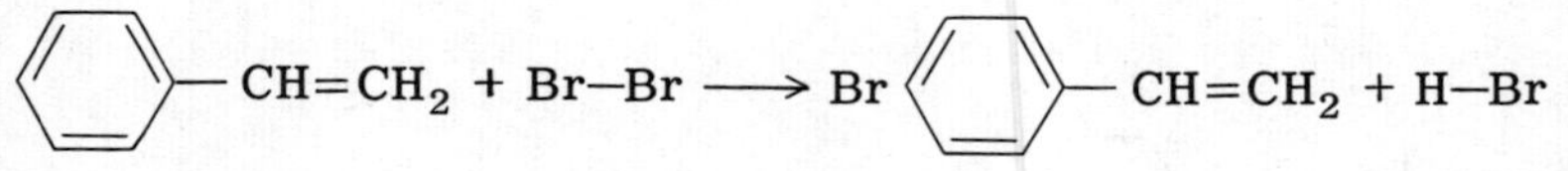

(VI)

+414·2 (C—H)   +188·3   −284·5   −364·0   **Δ*H* = −46·0 kJ**

Of these three reactions, (I) has the largest negative value and therefore is the most thermodynamically favourable reaction.

6. The first step is the formation of the benzyne, the direction of elimination being controlled by the inductive effect of the substituent group. The second step is the formation of the addition product, the direction of addition again being controlled by the inductive effect of the substituent.

(i) R = *o* -OMe; —I

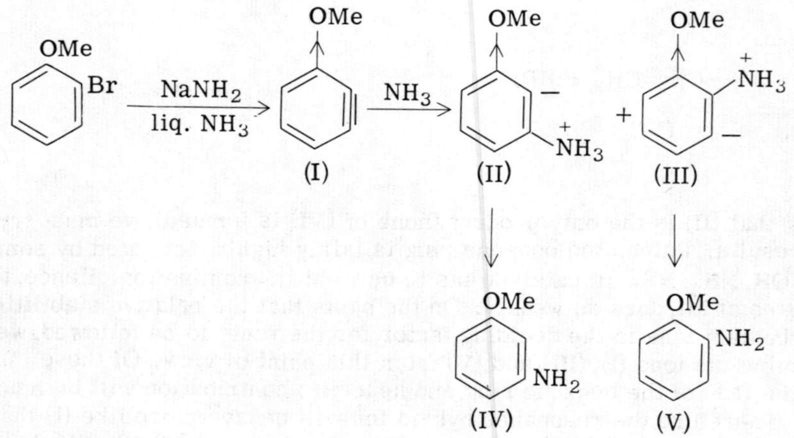

Only one benzyne derivative, (I), is possible. Because of the —I effect of OMe, (II) is preferred to (III), and so the expected predominant product is the *m*-product, (IV). In practice, this product is obtained in 100% yield.

(ii) R = *m*-CF_3; –I.

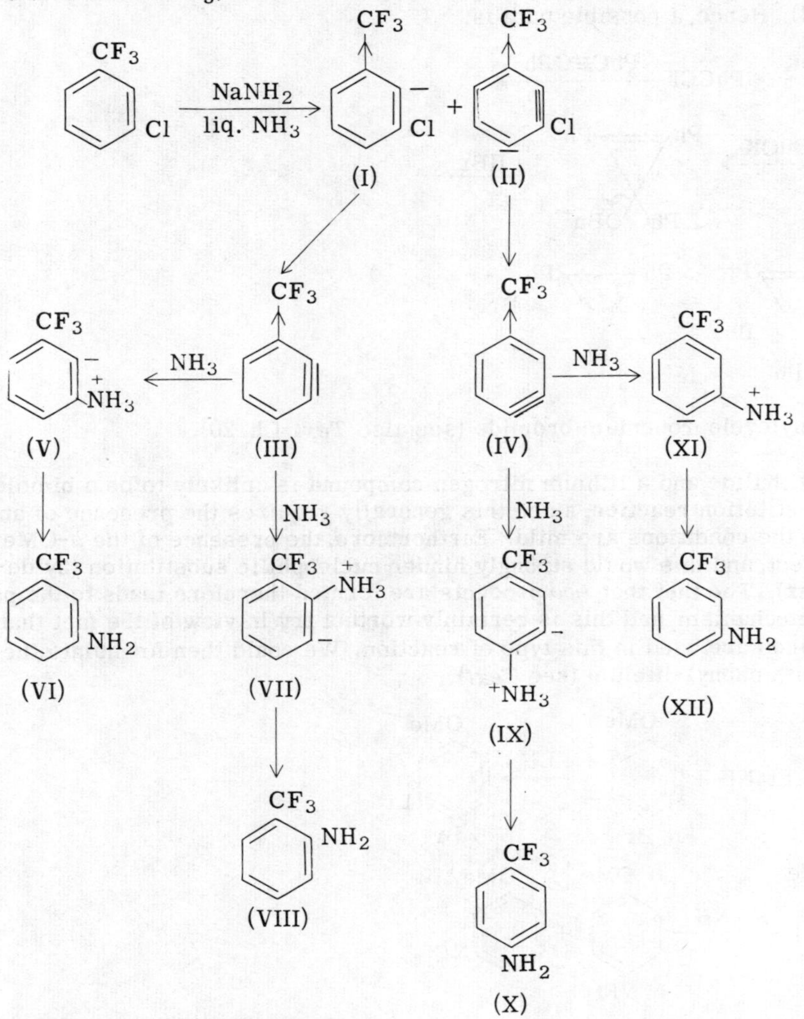

(I) is more stable than (II), because the negative charge is closer to CF_3 in the former than in the latter. Hence benzyne (III) predominates. The preferred direction of addition of NH_3 to (III) is (V) to give (VI). This, therefore, is the expected predominant product. In practice, it is obtained in 100% yield.

(iii) R = *m*-Me; +I.

We shall use the formulæ in (ii), replacing CF_3 by Me and Cl by Br. Because the I effect is in the opposite direction (for Me), (II) is more stable than (I), since the negative charge is further from the Me group in the former than in the latter. Hence benzyne (IV) predominates. The preferred direction of addition of NH_3 to (IV) is (XI) to give (XII). This, therefore, is the expected predominant product. In practice, the percentage yields are *m*-(XII, VI), 56; *o*-(VIII), 22; *p*-(X), 22.

7. The use of a strong base and a dihalogen derivative suggests the possibility of the formation of a methylene intermediate (compare with $CH_2Cl_2 \longrightarrow CHCl$). This is supported by the formation of a *three-membered ring* compound by reaction with an unsaturated compound (see *Text*, Ch. 4). Hence, a possible path is:

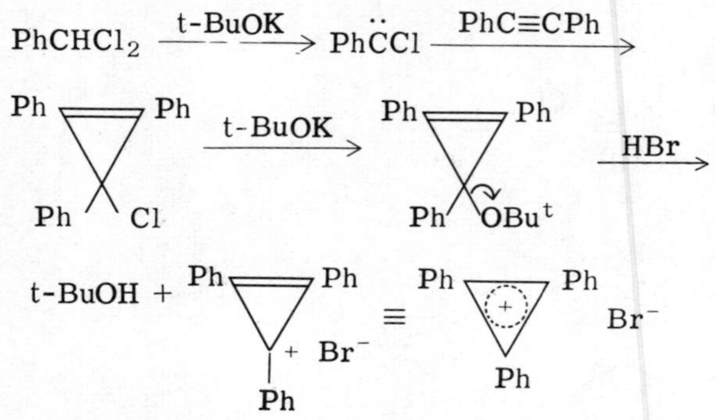

The product is triphenylcyclopropenium bromide (see also *Text*, Ch. 20).

8. The use of an aryl halide and a lithium nitrogen compound is unlikely to be a bimolecular nucleophilic substitution reaction, since this generally requires the presence of an activating group when the conditions are mild. Furthermore, the presence of the *p*-OMe group exerts a +R effect, and this would strongly hinder nucleophilic substitution (by destabilising the complex). The fact that *two* products are formed therefore leads to the possibility of a benzyne mechanism, and this is certainly worth a try in view of the fact that organolithium compounds are used in this type of reaction. We could then formulate the reaction by analogy with phenyl-lithium (see *Text*).

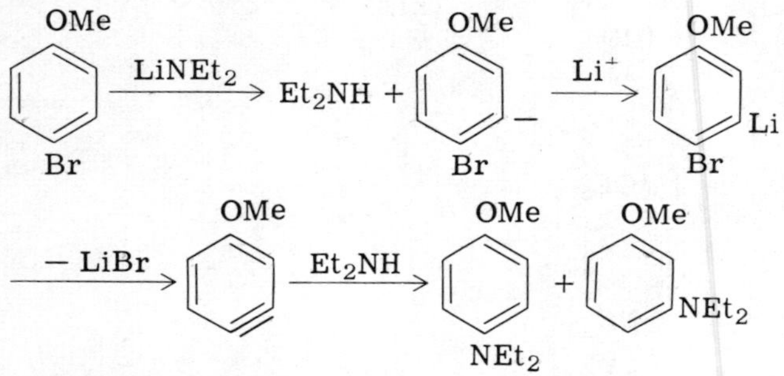

The *p*-product would be the expected predominant product (see Prob. 6, above).

9. Since for a given Z, the σ value is fixed, we may write

$Y = Br: \log (k/k_0) = 2{\cdot}1\sigma$

$Y = Me_2S^+: \log (k/k_0) = 2{\cdot}7\sigma$

Because k_0 is a fixed value, it follows that k undergoes a greater change with Me_2S^+ for a given σ value than it does with Br. Hence, the elimination reaction for Me_2S^+ is more sensitive to the polar effects of Z (measured by σ) than is the corresponding reaction for Br.

10. $PhMe + Br_2 \longrightarrow 0{\cdot}3\%\ m + 67\%\ p$

$PhMe + Cl_2 \longrightarrow 0{\cdot}5\%\ m + 40\%\ p$

Since a halogen carrier is absent, the reactive species is *molecular* halogen. This suggests that the two halogens may behave differently (toluene is usually halogenated in the *presence* of a halogen carrier). Also, this difference will manifest itself from a study of the yields of *m*- and *p*-derivatives. The ratio *m*/*p* indicates the polar influences of the substituent group

(Me in this case) at these positions. This is a property of the *substituent*. On the other hand, the ratio p/m is a measure of the selectivity of the *reagent* in choosing between the m- and p-positions in toluene. The greater the selectivity, the less reactive is the reagent (see *Text*, Ch. 20).

Since there are *two* m-positions and only *one* p-position, a factor of 2 is used in the calculations.

$$\mathbf{Br_2}\ p/m = 2 \times 67/0{\cdot}3 = 447$$

$$\mathbf{Cl_2}\ p/m = 2 \times 40/0{\cdot}5 = 160$$

Hence, since bromine has a greater selectivity than chlorine, the former is less reactive than the latter.

11. The mechanism is S_N1, and so the reaction may be written:

$$\mathrm{Me_3CCHOHPh} \xrightarrow[-\mathrm{H_2O}]{\mathrm{H^+}} \mathrm{Me_3\overset{+}{C}HPh} \xrightarrow{\mathrm{Cl^-}} \mathrm{Me_3CHClPh}$$

The fact that we are asked to discuss this reaction suggests there may be something 'odd' about it. This may not be the case, but there is no harm in taking a second look. A common feature of S_N1 reactions is that they are often accompanied by rearrangements, the driving force being the formation of the most stable carbonium ion (see *Text*, Ch. 5). Let us try this:

$$\mathrm{Me_2\overset{\overset{\large Me}{|}}{C}{-}\overset{+}{C}HPh} \longrightarrow \mathrm{Me_2\overset{+}{C}{-}\overset{\overset{\large Me}{|}}{C}HPh} \xrightarrow{\mathrm{Cl^-}} \mathrm{Me_2CCl\overset{\overset{\large Me}{|}}{C}HPh}$$

Rearrangement would result in the s-carbonium ion being converted into the t-ion. Hence, at first sight, the expected product would have been the t-chloride. Therefore, some factor must operate to make the s-ion more stable than the t-ion, thereby inhibiting the rearrangement. Since I and/or R effects are usually involved in stabilising carbonium ions, we now see whether we can explain the result in these terms. Inspection of the s-ion shows it can be represented as a resonance hybrid (conjugation with Ph).

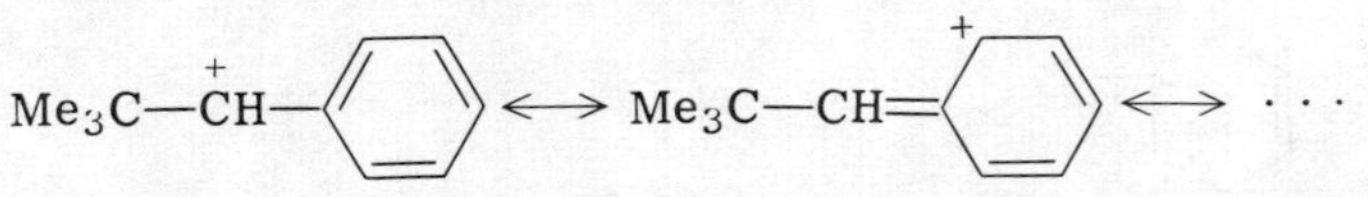

No such stabilisation is possible for the t-ion; partial neutralisation of the positive charge by the +I effect of the Me groups is a much less stabilising factor than resonance. Hence, the s-ion is more stable than the t-ion, and so the s-chloride and not the t-chloride is the product.

12. $\mathrm{PhCH_2MgCl + MeCHO \longrightarrow PhCH_2CHOHMe} + \mathbf{A}$

A is an isomer of the alcohol product; hence its molecular formula is $C_9H_{12}O$.

Now:

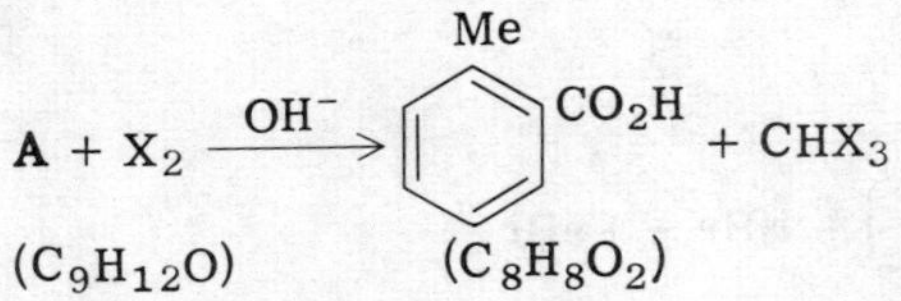

∴ **A** must have the skeleton

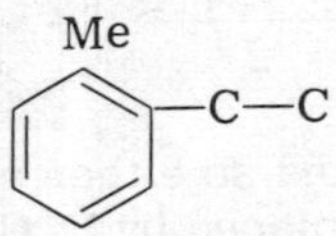

Since **A** undergoes the haloform reaction, the side chain must be $COCH_3$ or $CHOHCH_3$ (see *Text*, Ch. 8).

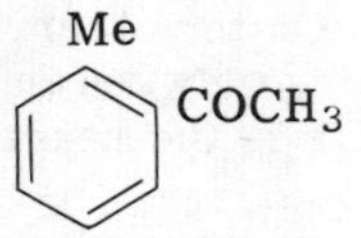

(I) $C_9H_{10}O$

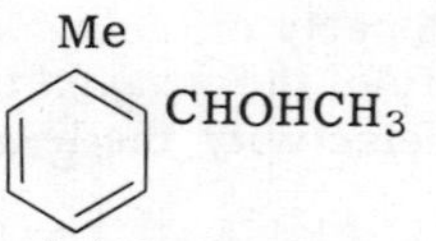

(II) $C_9H_{12}O$

Hence **A** is (II). Its formation must involve a rearrangement, the acetaldehyde residue becoming attached to the ring (*o*-position) and the benzyl residue becoming the tolyl group.

Since the carbonyl group is polarised as: $C{=}O \longrightarrow \overset{+}{C}{-}\overset{-}{O}$ the *o*-position must behave as a nucleophilic centre, i.e.,

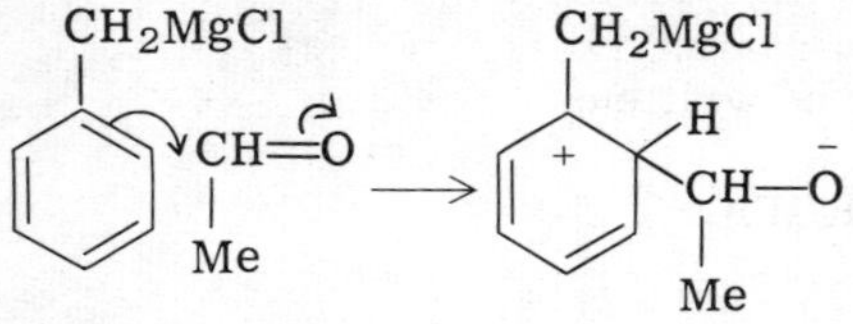

This could be continued as follows:

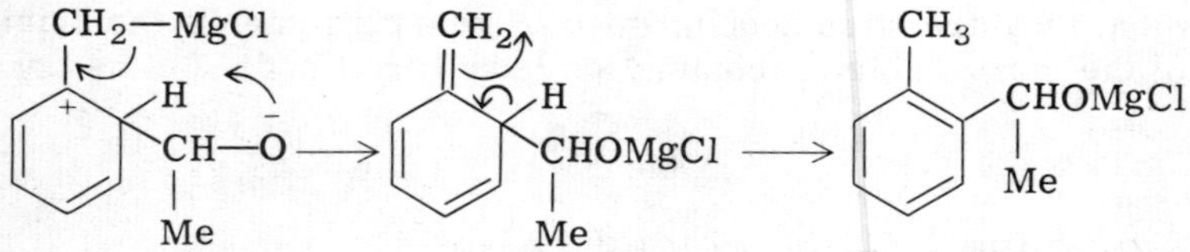

The α,γ-proton shift occurs because the rearranged product is now benzenoid (which is far more stable than its precursor).

The question now is: What makes the *o*-position a nucleophilic centre? A possible explanation is that because the benzyl cation can be stabilised by resonance, the C—Mg bond has a

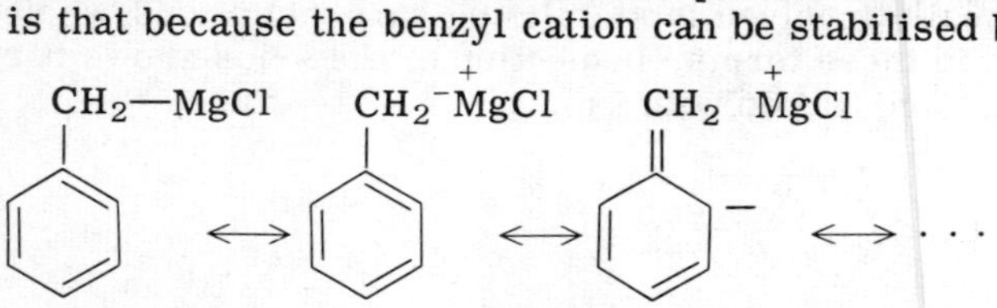

large ionic character. The benzyl cation therefore carries negative charges at the *o*- and *p*-positions. Because of the possibility of a steric effect at the *o*-position, it might have been argued that the rearrangement with *p*-substitution would have been preferred. If, however, we assume that cyclic transition states are involved, then only *o*-substitution satisfies these requirements.

13. All the evidence favours a bimolecular mechanism for bromination; this may be written:

$$Br_2 + FeBr_3 \overset{\text{fast}}{\rightleftharpoons} Br^+ + FeBr_4^-$$

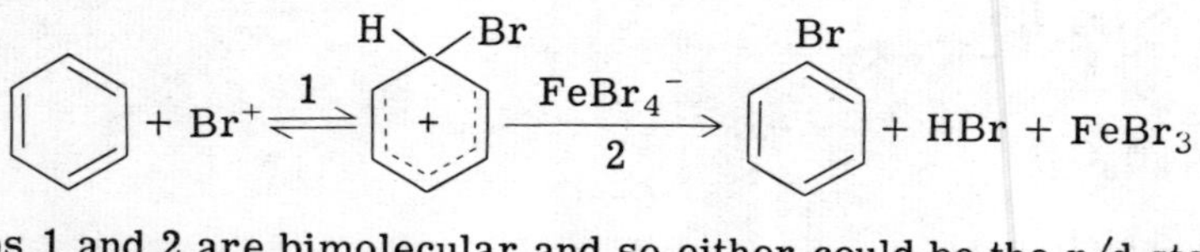

Steps 1 and 2 are bimolecular and so either could be the r/d step. Since no kinetic isotope effect is observed when H is replaced by D, C—H or C—D bond fission cannot be involved in the r/d step. Hence, 1 is the r/d step and so the rate of bromination is the same for C_6H_6 and C_6D_6.

14. The two possible mechanisms are: (i) bimolecular, and (ii) elimination-addition (benzyne mechanism). An experiment that could be used to demonstrate that both are acting simultaneously is to label PhCl at C of C—Cl with ^{14}C.

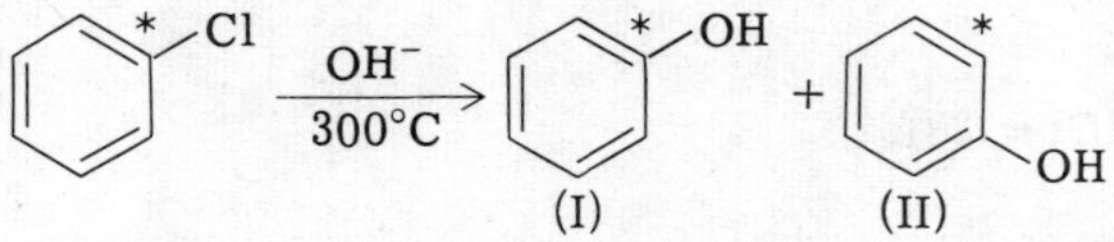

If the mechanism is bimolecular nucleophilic substitution, the product will be 100% of (I). If the benzyne mechanism is operating, then the product will be ~50% of (I) and about ~50% of (II). If (I) and (II) are formed in quite different amounts, then both mechanisms occur together.

In practice, it has been shown that (I) ≫ (II). Hence, the major path is bimolecular substitution.

15. (i)

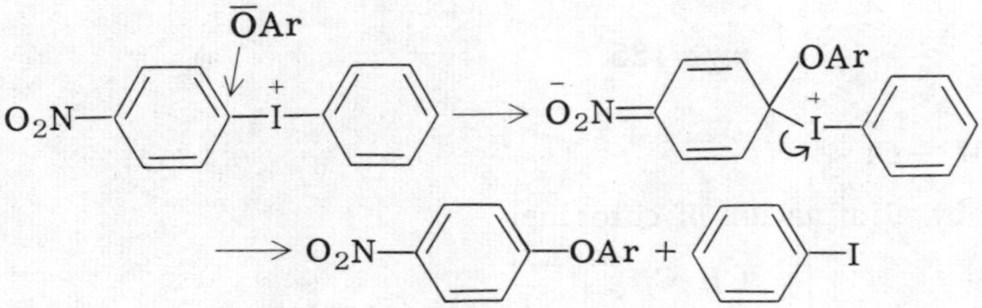

Attack at the benzene ring containing the *p*-NO_2 group gives a complex that is more stable than the one formed by attack at the other ring, i.e., the reaction may be treated as an example of activated nucleophilic aromatic substitution.

(ii)

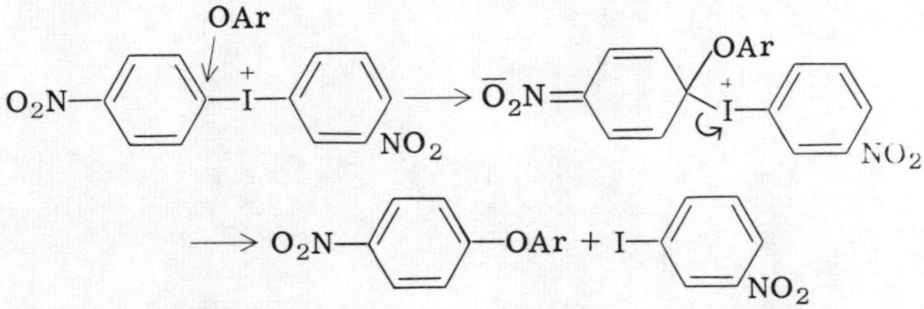

Since a nitro-group activates from the *o*- and *p*-positions, the complex shown is the stabilised one.

(iii)

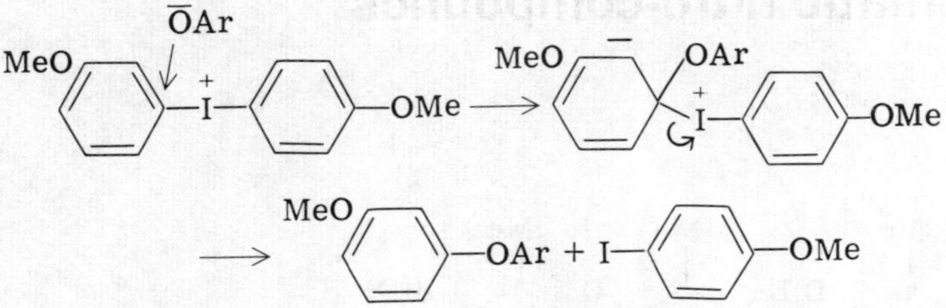

The methoxyl group has a —I effect from the *m*-position and a —I + R effect from the *p*-position. In the latter, +R > —I, and so the complex formed from the ring with the *p*-OMe group is destabilised with respect to that formed from the ring with the *m*-OMe group. Hence, the product is the *m*-methoxy ether.

16. 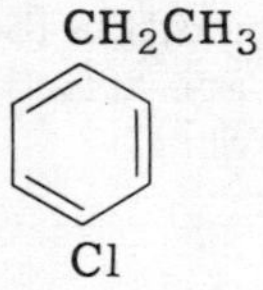 $= C_8H_9Cl$. M.W. = 140 and 142. (C = 12; H = 1)

Hence: $M^+ = 140$ (Cl = 35) and $M^+ = 142$ (Cl = 37).

m/e 125 = $M^+ - 15$. ∴ M^+—Me

This can be accounted for by β-fission in the side-chain.

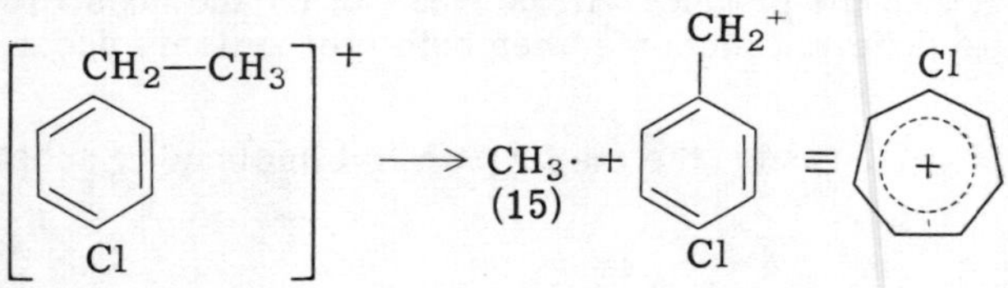

m/e 125

This is actually the *B.P.*

m/e 105 = $M^+ - 35 = M^+$—Cl

This can be accounted for by elimination of chlorine.

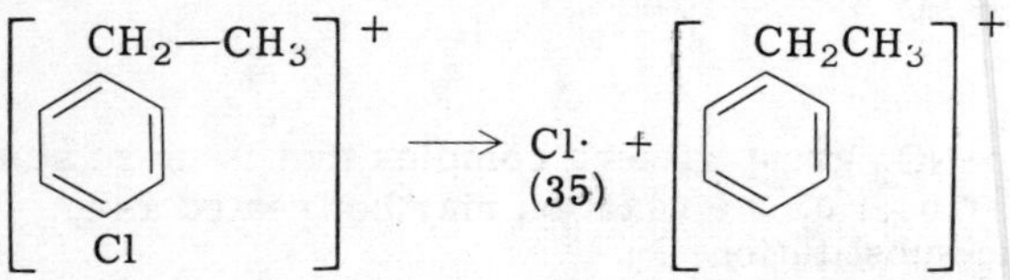

m/e 105

22 Aromatic nitro-compounds

(Problems on page 41)

1.

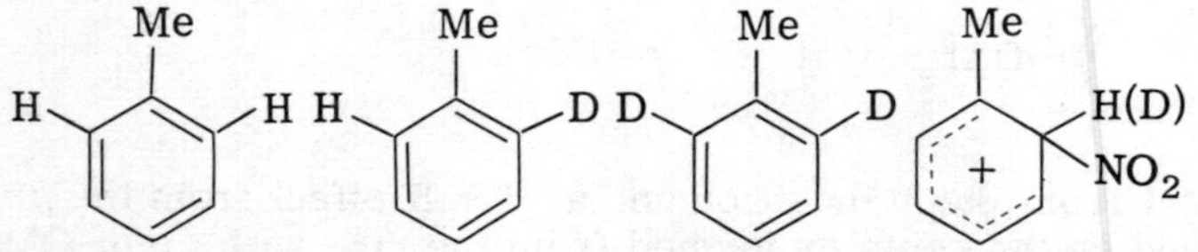

The r/d step is the formation of the σ-complex (with the nitronium ion). Hence, there will not be a kinetic isotope effect, since C—H or C—D bond fission is not involved in the r/d step. The methyl group exerts a steric effect, but this will be the same for all three substrates (the attacking species is the same for all three). Therefore, the rates of nitration of all three will be the same.

2. Since we cannot buy bottles of nitronium ions, we must include the reaction by which they are generated (see also Ch. 8, Prob. 3; Ch. 9, Prob. 6). Table 2.2 has been used to calculate the heat of reaction of each step (these have been formulated according to the accepted mechanisms).

(i)

$$HO{-}NO_2 + 2H{-}OSO_3H \rightleftharpoons NO_2^+ + H{-}\overset{H}{\underset{H}{O^+}} + 2\bar{O}SO_3H$$

+208·8 +2 × 464·4 −2 × 464·4

Hence, ΔH = +208·8 kJ

(ii)

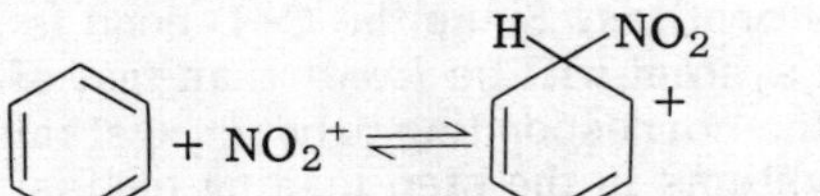

+606·7 (C=C) −184·5 (C—N)
+150·6 (R.E.) −347·3 (C—C)
−15 (R.E.)

Hence, ΔH = +110·5 kJ

(iii)

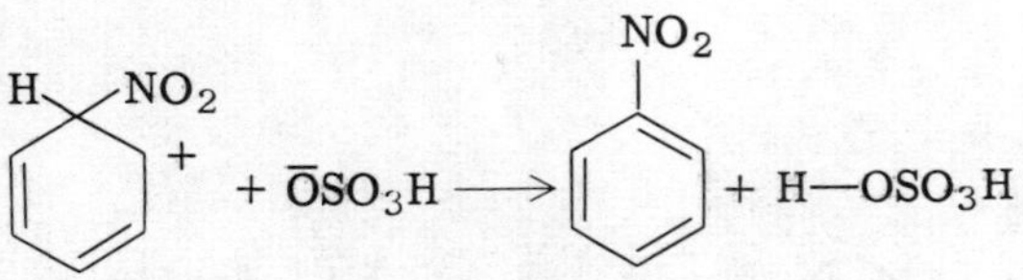

+414·2 (C—H) −606·7 (C=C)
+347·3 (C—C) −464·4 (H—O)
+15 (R.E.) −150·6 (R.E.)

Hence, ΔH = −445·2 kJ

It therefore follows that ΔH for the *overall* reaction is:

ΔH = +208·8 + 110·5 − 445·2 = −125·9 kJ

The overall heat of reaction can also be obtained from the equation:

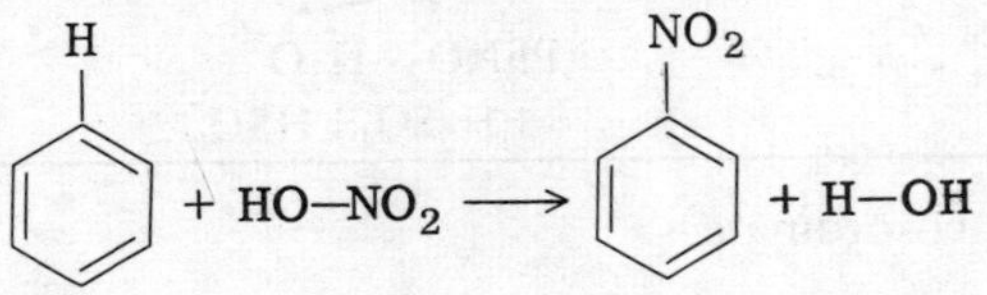

+414·2 +208·8 −284·5 −464·4

∴ ΔH = −125·9 kJ

This is the same value as that obtained from the addition of the heats of reaction of the individual steps. This, of course, follows from Hess's law. This calculation should always be done, since it gives the relative energy levels of the reactants and products (and also acts as a check on the calculations). *N.B.* ($H_3O^+ + HSO_4^-$) = ($H_2O + H_2SO_4$); ΔH = 0.

The energy changes for C_6D_6 as substrate differ from those of C_6H_6 only in the last step (iii).

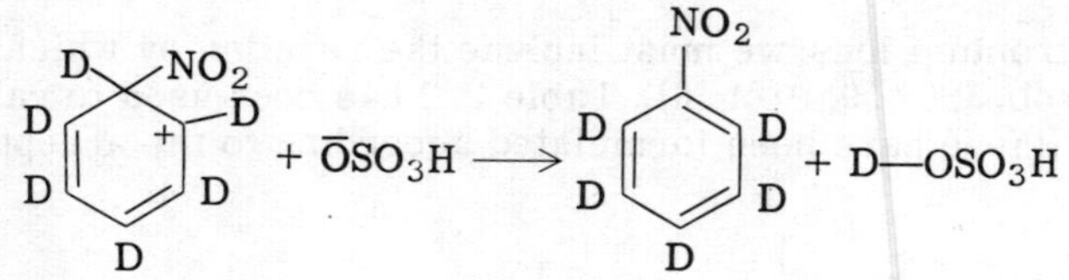

The difficulty here is to find a list of bond energies that gives the values we require (remember the law of cussedness). However, we do know that the C—D bond is stronger than the C—H bond (this must be so from the observations of the kinetic isotope effects involving C—D bonds). Hence, we can say that the energy of activation leading to the T.S. containing a C—D bond is higher than that for the corresponding C—H bond.

We are now in a position to draw possible energy profiles. Since the C—D bond is stronger than the C—H bond, the P.E. of the C_6D_6 reactant system will be *lower* than that of the C_6H_6 reactant system. This will also be the case for the corresponding product systems. Because the most important difference between the two systems is the step leading to fission of the C—H and C—D bonds, both energy profiles may be combined and only this difference is then shown. The student should carefully compare the energy profile drawn here with the one given in the *Text* (Fig. 20.13). This again illustrates the point that energy profiles are not always what they are meant to represent (see Ch. 8, Prob. 3).

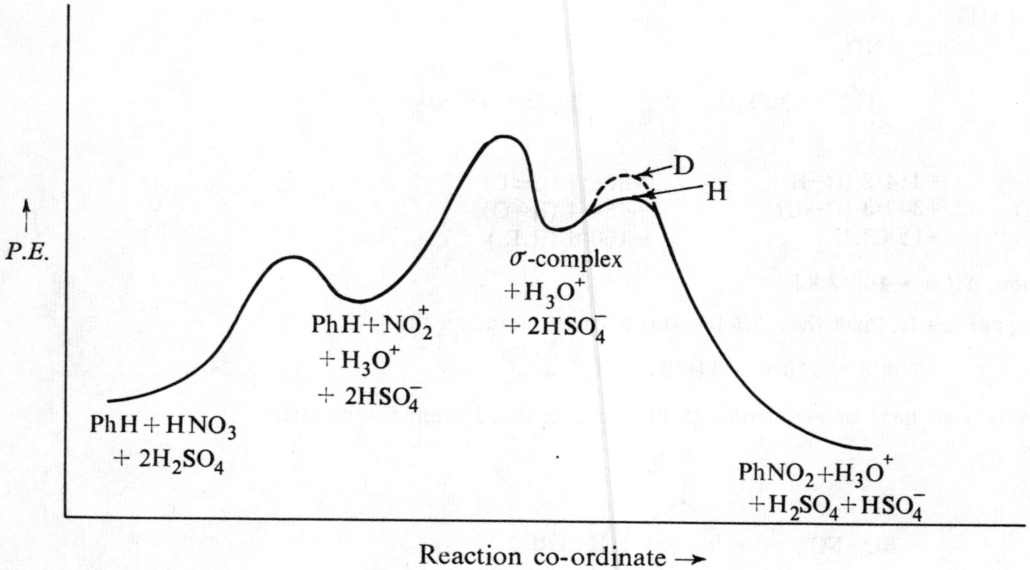

3. Whenever a substituted benzene has to be prepared from benzene, the student should always consider whether the groups can be introduced *directly* and then work out the *order* of substitution in order to get the correct orientation. Should this approach fail to achieve the desired results, the student should then work backwards to discover the substituent that must ultimately be introduced directly, e.g., NH_2 is usually obtained by reduction of a nitro-group, which can be introduced directly.

(i)

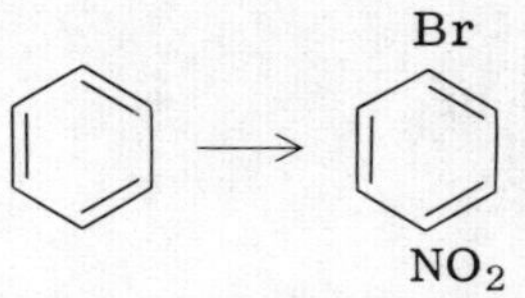

Br and NO_2 can be introduced directly. Block out NO_2. Br is *o/p*-directing and if introduced first, can give the desired product. NO_2 is *m*-directing and if introduced first, will not give the desired product. Hence (M.A. is mixed acid):

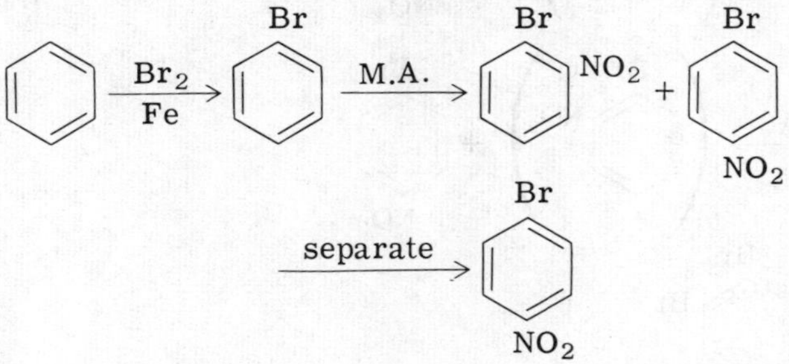

The student should always indicate the isomers formed (or, if there are more than two, then usually the more important ones).

(ii)

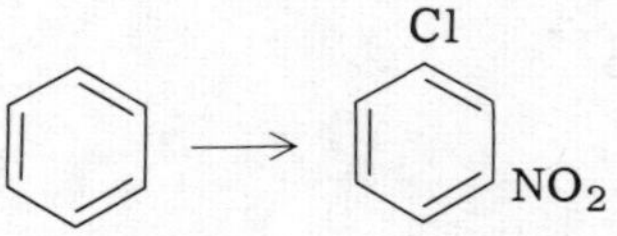

Both Cl and NO_2 can be introduced directly. Cl is *o/p*-directing; NO_2 is *m*-directing. Hence:

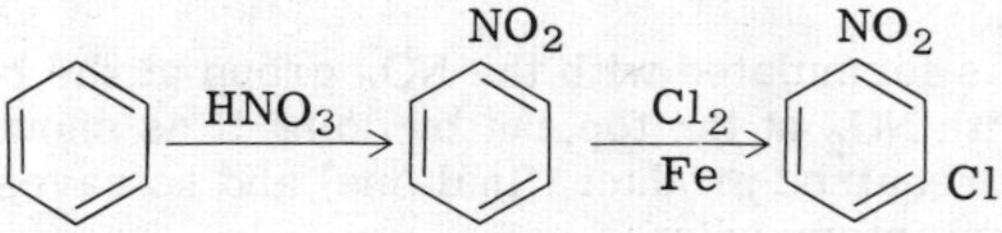

(iii)

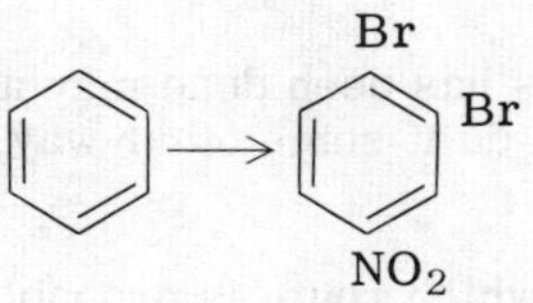

Both Br and NO_2 can be introduced directly; Br is *o/p*-directing and NO_2 is *m*-directing. Starting with the top Br atom, block out the *o*-Br and NO_2 groups. If the top Br is introduced first, *o*-Br can then be introduced, and since NO_2 is *para* to *one* Br, it can also be introduced into the *o*-diBr derivative. On the other hand, the monobromo-compound can be nitrated to give *p*-NO_2, and this, because Br and NO_2 direct to *same* position (*o* to Br and *m* to NO_2), will give the desired product on bromination.

If we block out the two Br atoms, then nitration of benzene gives $PhNO_2$, bromination of this gives *m*-Br, and this compound, on further bromination, gives the desired product.

Now that we have considered all the possibilities, we write out the appropriate equations.

(*a*)

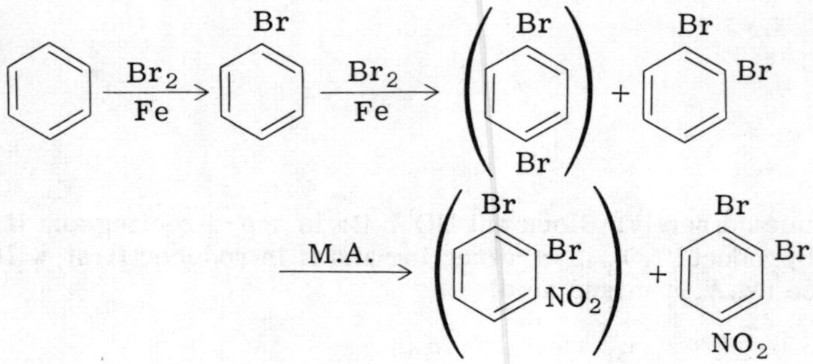

(*b*)

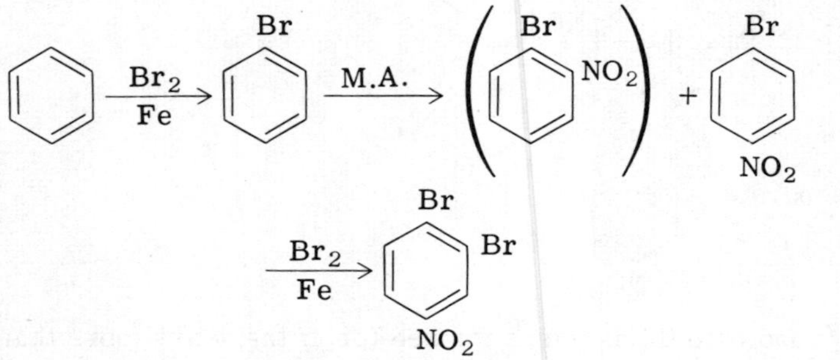

(*c*)

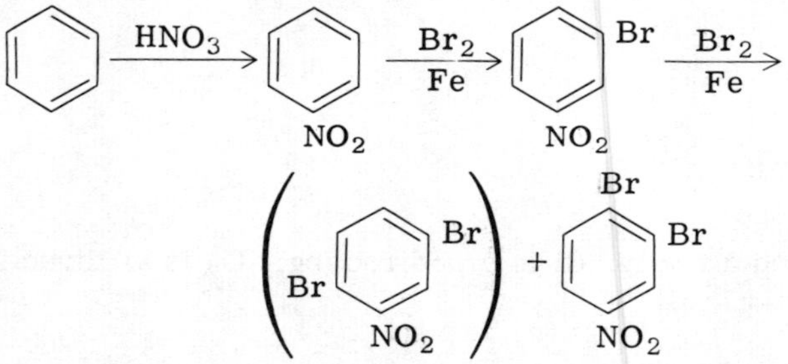

It should be noted that in (*c*), the first product is formulated with the NO_2 group at the bottom. Normally, we would draw nitrobenzene with NO_2 at the top, but by doing it as shown, the nitro-group is in the 'right' position in the required product (final one) and so saves a lot of bother in recognising the various required intermediates.

We now have developed three synthetic routes. The way this has been done may appear somewhat lengthy, but unless the student can be sure he can do it some other way, the systematic approach should be used.

The final point to be considered is whether we can suggest which route is probably the best. Unfortunately, this may be quite difficult, since various factors have to be taken into consideration, and this may require a great deal of experience. One factor is the number of separations of the isomers that must be performed. In general, the fewer the better, particularly when the separations are difficult. Even so, the problem of separation may not be the dominant factor. Thus, if a method involved several separations, but in each case the desired isomer far outweighed the undesired isomers, this method could give a better overall yield than other methods involving fewer separations.

Another factor is the actual number of steps used in the synthesis. This has been partly covered by the previous discussion, but the student must appreciate that the shortest synthesis does not necessarily produce the best yield. Other factors that may have to be considered are cost of materials used, time taken to carry out the synthesis, etc.

The final point we make here regarding the choice of the synthesis is that if we have access to a list of partial rate factors due to various substituent groups, we may be in a position to estimate (approximately) the yields of isomers and so be in a better position to choose the best synthesis (see *Text,* Ch. 20).

If we now return to the syntheses proposed above, we see that all involve three steps (of similar type, i.e., substitution); (*a*) requires two separations, and (*b*) and (*c*) each require one separation. Therefore, unless we can readily see some reason to assume otherwise, we can expect synthesis (*a*) to give the poorest yield.

If only *one* synthesis is asked for, then only *one* should be described. In this case, as long as the method is reasonable, no time should be spent on devising a number of syntheses and then trying to decide which may be the best. The latter approach is worthwhile when the student is solving problems in his own time, since he can learn a great deal this way.

(iv)

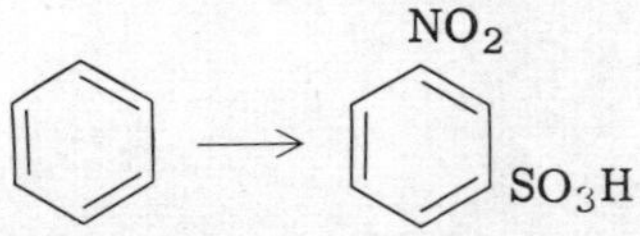

Both groups can be introduced directly and both are *m*-directing. Hence, the order of introducing the groups is immaterial. This is true in this case, but since sulphonation is reversible and the sulphonic acid group can be displaced in some reactions, it may be better to sulphonate as late as possible (see *Text,* Ch. 25).

(v)

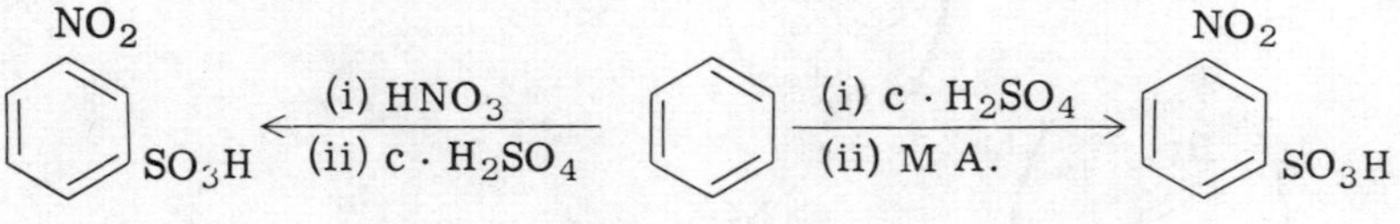

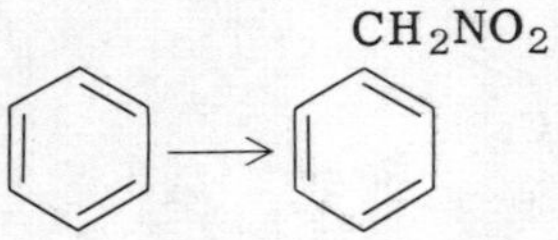

The CH_2NO_2 group cannot be introduced directly. Therefore, working backwards:

$CH_2NO_2 \longleftarrow CH_2Cl \longleftarrow ?$

Since the chloromethyl group can be introduced directly, a possible route is:

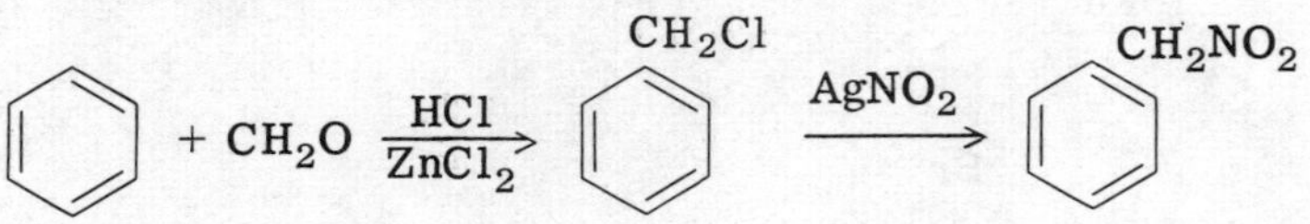

An alternative route is:

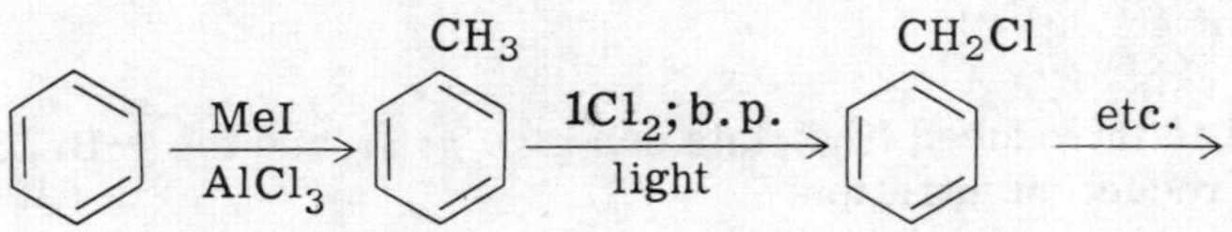

(vi)

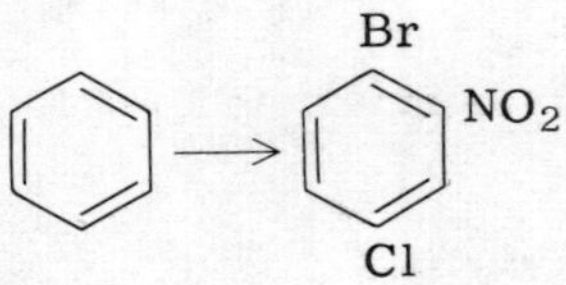

We shall again go through the systematic approach. All groups can be introduced directly; Br and Cl are *o*/*p*-directing and NO_2 is *m*-directing. There are several possibilities (*cf.* with (iii), above).

(*a*) Block out NO_2 and Cl. If Br is introduced first, this can give *o*-NO_2, and the *o*-Br, NO_2-derivative can give the required product on chlorination.

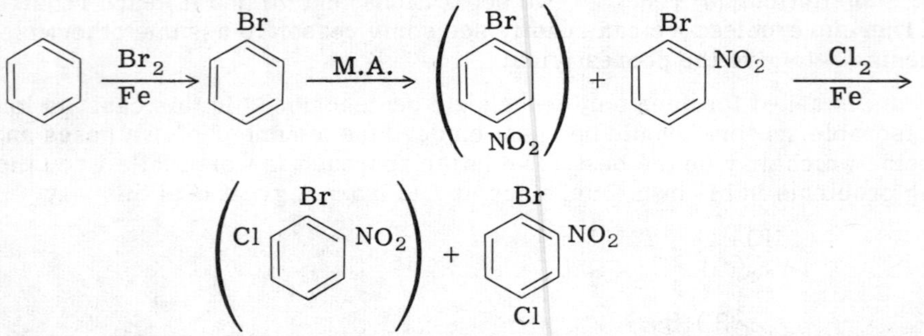

(*b*) An alternative to (*a*) is bromination first (as before), then chlorination (*p*-) and then nitration.

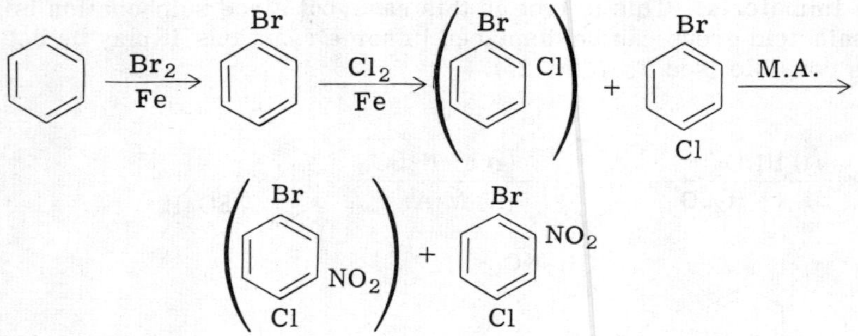

(*c*) Block out Br and Cl. If NO_2 is introduced first, this can give *m*-Cl, and the *m*-Cl, NO_2-derivative can give the required product on bromination.

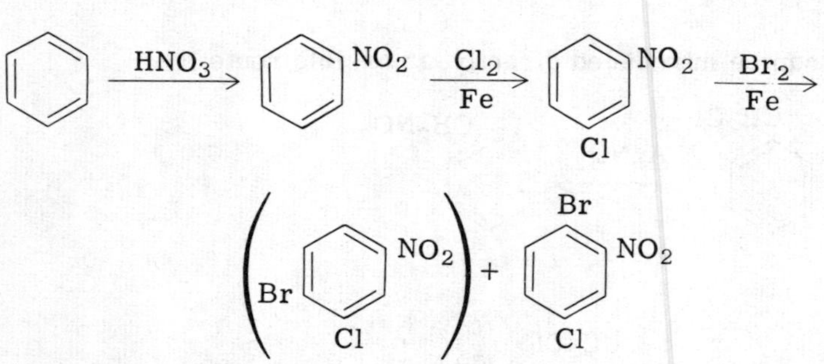

(*d*) Block out Br and NO_2. If Cl is introduced first, this can give *p*-Br, and the *p*-Br, Cl-derivative can give the required product on nitration.

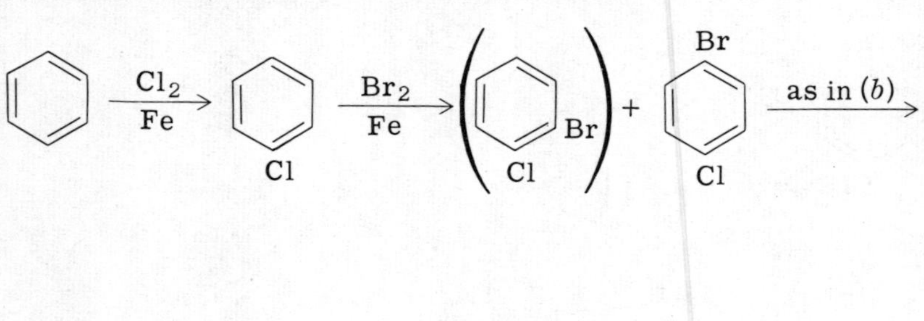

Thus, there are four possible routes. All four involve three steps and all, except (*c*), involve two separations. Since (*c*) involves one separation only, this is likely to be the best route (see also (iii), above).

(vii)

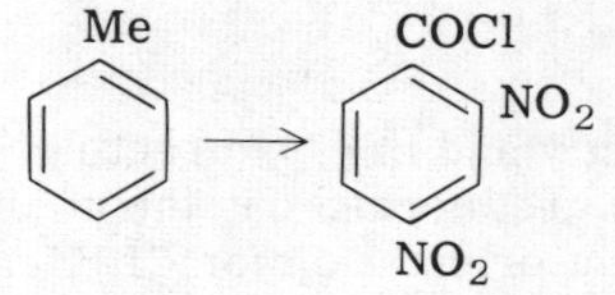

The methyl group is *o/p*-directing, and the dinitro-compound so obtained can be oxidised to the acid without affecting the nitro-groups. Oxidation of toluene to the acid is unsatisfactory, since the carboxyl group is *m*-orienting. Hence:

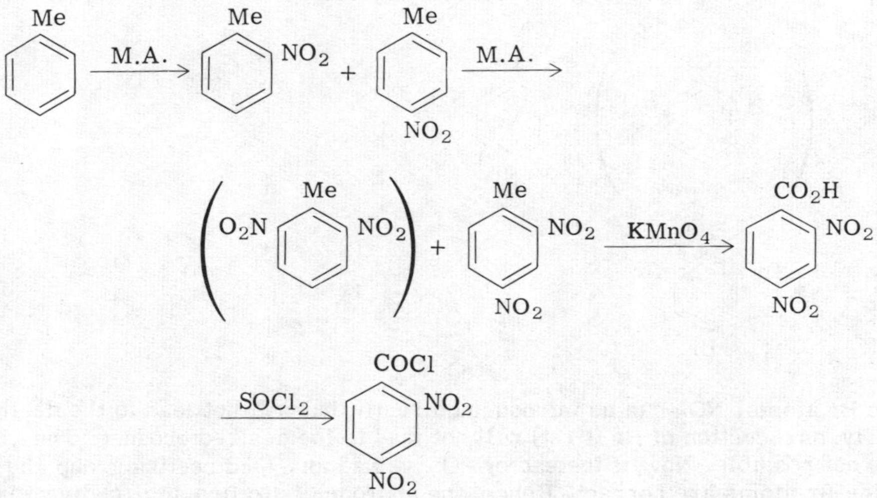

(viii)

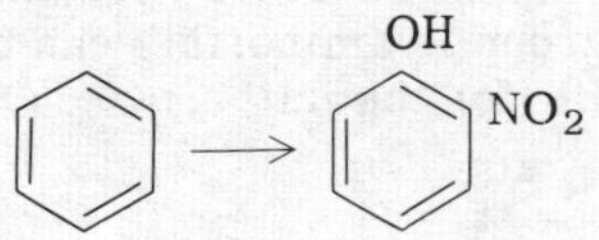

A common method of introducing a hydroxyl group is via the diazonium salt (see *Text*, Ch. 24). Another common method is by activated nucleophilic substitution. This approach is possible for the compound we require since there is a nitro-group *ortho* to the hydroxyl group. Hence (bromination *first*):

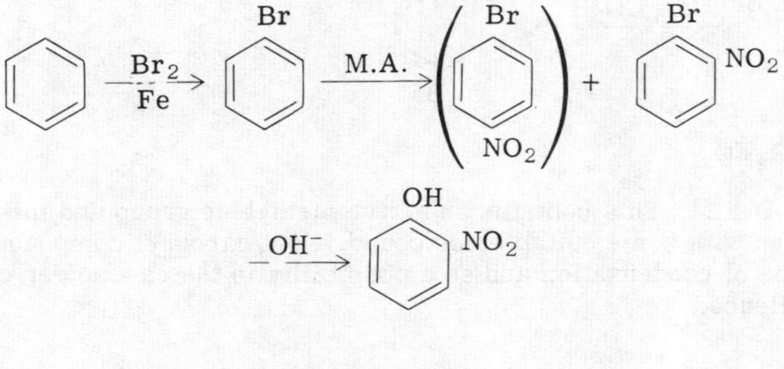

(ix)

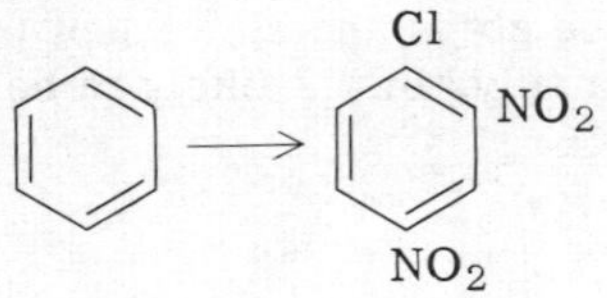

Block out the two NO_2 groups. Cl can be introduced directly and is *o/p*-directing. Hence, we can get the required product. If Cl and one NO_2 group are blocked out, the *m*-dinitro-compound can be obtained, but this will not give the correct orientation for Cl. Hence:

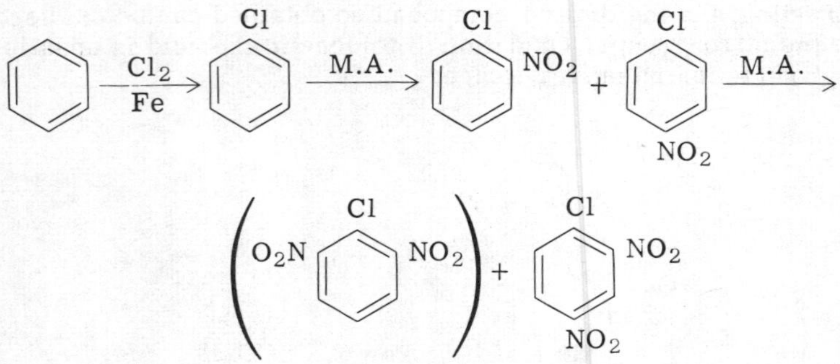

(x)

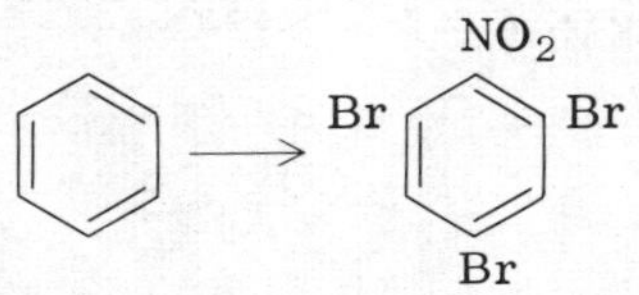

Block out all the Br atoms. NO_2 can be introduced directly, but will not lead to the desired product. Similarly, introduction of Br (first) will not lead to the desired product. Thus, a direct method is not possible. Now, if instead of NO_2 we had an *o/p*-directing group, the orientations of the Br atoms are correct. Hence, one approach is to find an *o/p*-directing group which can be subsequently converted into a nitro-group; at the same time, this group has to be introduced, the more readily the better. Such a group is amino; this can be oxidised directly to nitro or replaced by nitro via the diazonium salt (for the latter, see *Text*, Ch. 24).

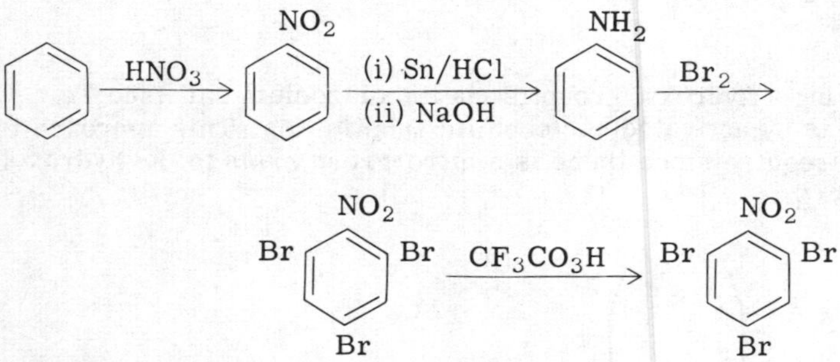

(xi) $PhH \longrightarrow PhN{=}C(CO_2Et)_2$

One fragment of the product is D.E.M. This contains an active methylene group and this suggests a condensation reaction with some suitable compound. Now, carbonyl compounds are commonly used for this type of condensation and so a possibility in the case under consideration is nitrosobenzene. Hence:

$$PhH \xrightarrow{HNO_3} PhNO_2 \xrightarrow[aq.\ NH_4Cl]{Zn} PhNHOH \xrightarrow[H_2SO_4]{dichr.} PhNO \xrightarrow{CH_2(CO_2Et)_2} PhN{=}C(CO_2Et)_2$$

4.

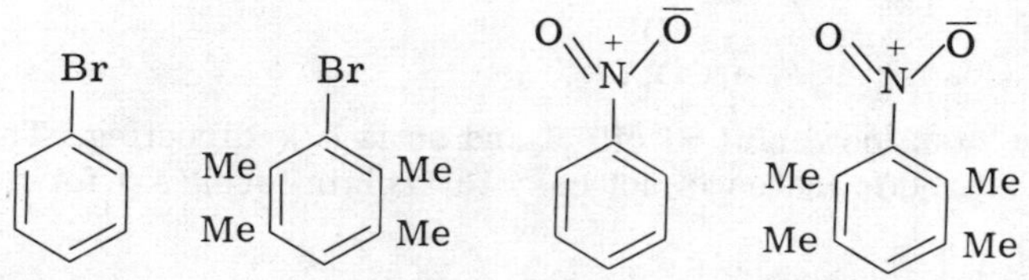

	(I)	(II)	(III)	(IV)
μ(D)	1·53	1·55	3·9	3·39

Introduction of the methyl groups into (I) to give (II) has very little effect on the dipole moment, but very much affects the dipole moment in the conversion of (III) into (IV). Since *o*-Me groups have been introduced, there is now the possibility of steric effects in (II) and (IV), and these effects are far greater in (IV) than in (II). Hence, a reasonable explanation lies in the phenomenon of steric inhibition of resonance (see *Text*).

Bromine has a —I and a +R effect, with —I > +R. The bromine atom in (I) lies in the plane of the ring, and is not displaced from this position (to any appreciable extent, if at all) by the *o*-Me groups in (II). Because of the +I effects of the Me groups, the —I effect of Br is aided and its +R effect is opposed. The net result is a very slight increase in dipole moment in (II).

The nitro-group (—I, —R) is large and is coplanar with the ring in (III), thereby exerting a maximum resonance effect. Thus, electron-withdrawal is a maximum, thereby resulting in the maximum dipole moment. Now, an essential requirement for resonance is the planarity of the system involved. In (IV), because of the steric effects of the *o*-Me groups, the NO_2 group can no longer lie in the plane of the ring. Consequently, resonance is considerably diminished. This results in less electron-withdrawal from the ring and so the dipole moment is decreased in (IV).

5. The formation of the *m*-derivative with M.A. at —20°C implies that the $B(OH)_2$ group has a —I and/or —R effect. The electron configuration of boron is a sextet, but the shell can be expanded to an octet (see *Text*, Ch. 14). This can be achieved by electron-withdrawal from the ring and would explain *m*-substitution.

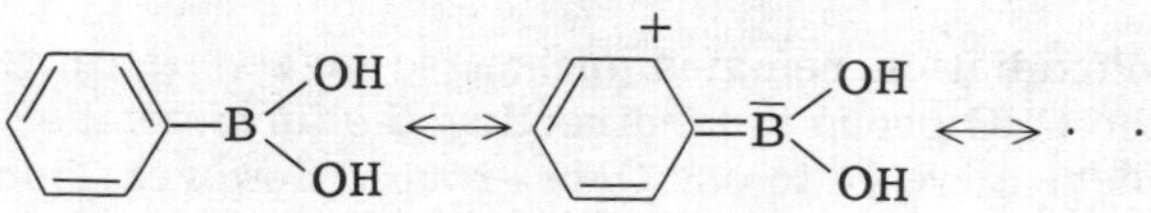

The formation of the *o*-derivative with HNO_3—Ac_2O implies that the electron flow is now reversed, i.e., it is *into* the ring. Since boron with a sextet of electrons cannot do this (it actually *withdraws* electrons from the ring), it would now appear that under the conditions of this experiment the boron group has become +I and/or +R. As we have seen above, if boron expands its shell to an octet, it acquires a negative charge at the same time. Since the difference between the two reagents M.A. and HNO_3—Ac_2O is mainly the presence of Ac_2O in the latter, a possible explanation is that this is involved. Now, a characteristic reaction of acetic anhydride is acetylation, the mechanism of which is believed to be (see *Text*, Ch. 9):

$$MeCO{-}O{-}COMe \longrightarrow MeCOO^- + MeCO^+$$

The acetyl carbonium ion then attacks, e.g., alcohols, which behave as nucleophiles (lone pairs on the oxygen atom). In the case under consideration, boron is an *electrophile*, and so it can react with the *acetoxyl anion:*

The group $\overline{B}(OH)_2OCOMe$ now has a powerful +I effect and so is *o/p*-directing. The observation that the *o*-derivative predominates is not easy to explain (see *Text* for a discussion of this problem).

6. (i) This is an example of the von Richter reaction; it does *not* take place by the benzyne mechanism (see *Text*, Ch. 21).

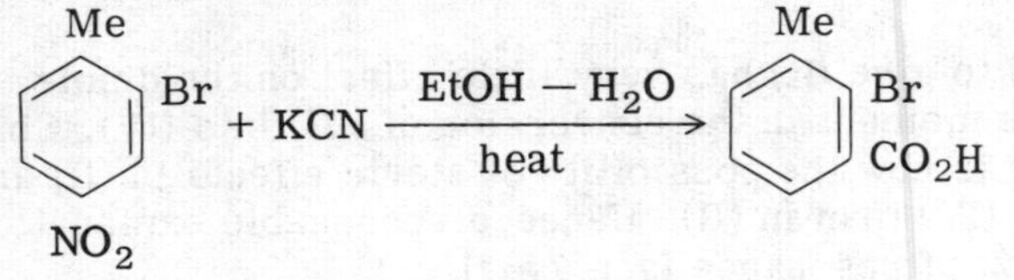

(ii)

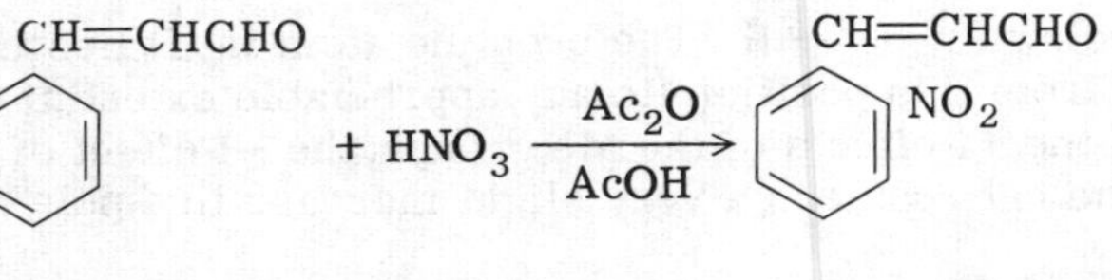

Although the side-chain is electron-withdrawing, substitution occurs in the *o*- and *p*-positions. In this case, the correct orientation is obtained from a consideration of the stabilities of the various carbonium ions: *o*, *m* and *p* (see *Text*, Ch. 20).

The major product is the *o*-compound. This is often the case when acetyl nitrate (HNO_3—Ac_2O) is the nitrating agent. The explanation is uncertain (see *Text*).

(iii)

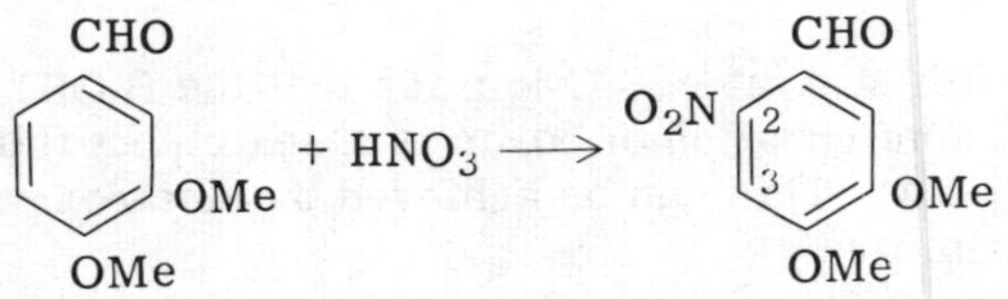

The aldehyde group, although sensitive to oxidiation, remains unchanged (see *Text*, Ch. 27). The OMe groups are *o/p*-directing and the CHO group is *m*-directing; the former take precedence. The product is the 2-nitro-derivative (*p* to one OMe group). However, since the 3-position is 'assisted' by both the CHO group and one OMe group (*o* to 3), it might have been expected that the 3-nitro-derivative would be the major product. One possible explanation could be steric effects at the 3-position arising from the orientation of the Me groups due to resonance (the two Me groups will be as far apart as possible).

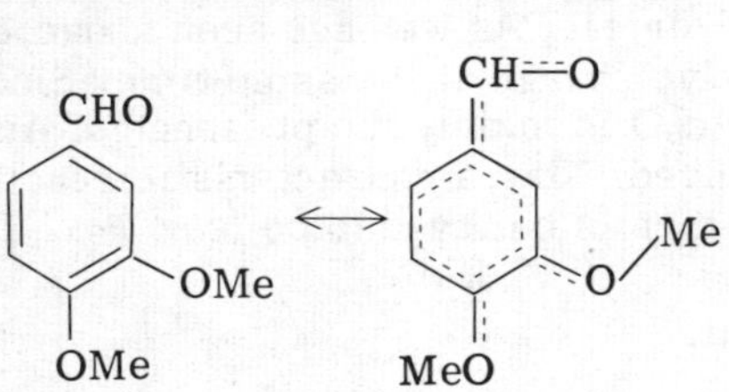

(iv)

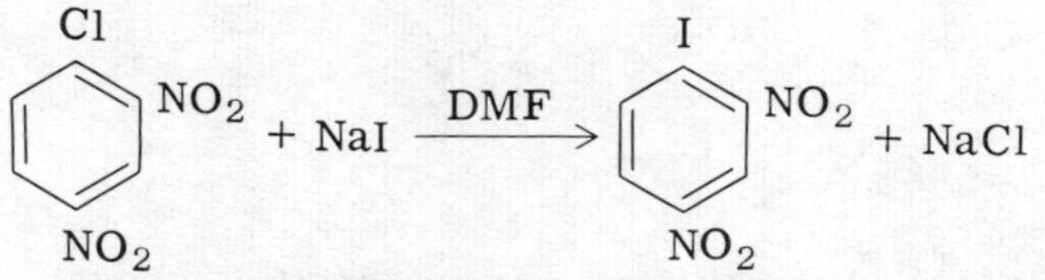

This is an example of activated nucleophilic substitution (also compare with the Finkelstein reaction; see *Text*, Ch. 3).

(v)

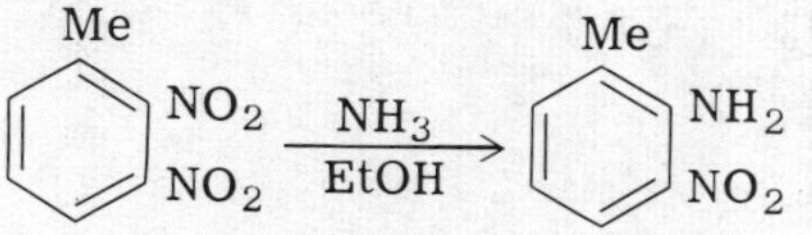

This is an example of activated nucleophilic substitution. Each nitro-group is activated by the other *o*-nitro-group; hence there are two possible complexes (see *Text*, Ch. 21):

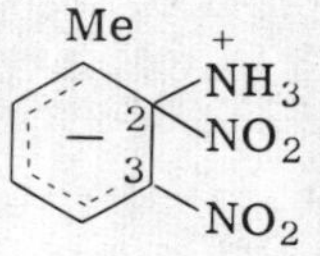

(I)

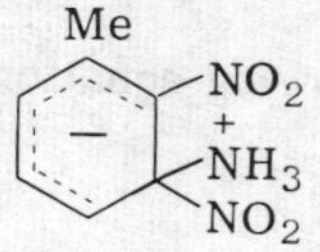

(II)

In (I), C-2 is sp^3 hybridised and the 3-NO_2 group can become coplanar with the ring and so the complex is stabilised by resonance. In (II), C-3 is sp^3 hybridised, but the 2-NO_2 group cannot become coplanar with the ring because of the steric effect due to the 1-Me group. Hence, (II) is not stabilised by resonance (steric inhibition of resonance), and so reaction proceeds through the more stable complex (I).

(vi)

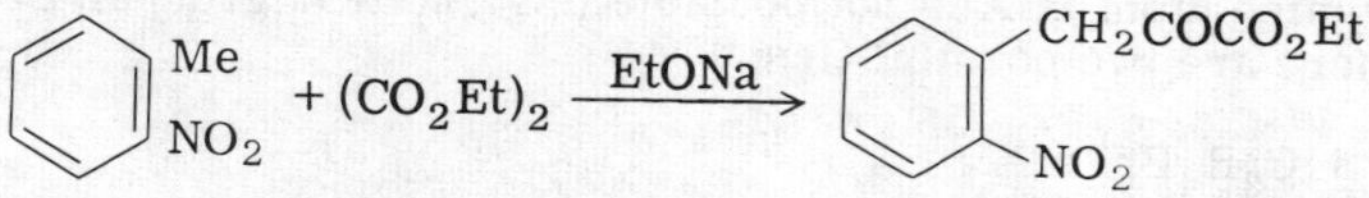

This is a Claisen type of condensation and takes place because the Me group is sufficiently activated by the *o*-NO_2 group for a proton to be removed. This activation arises from the fact that the carbanion is stabilised by conjugation with the *o*-nitro-group in one of the resonating structures (now have extended conjugation).

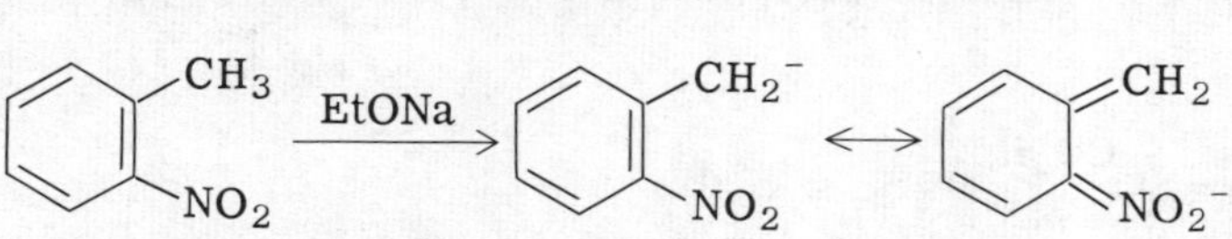

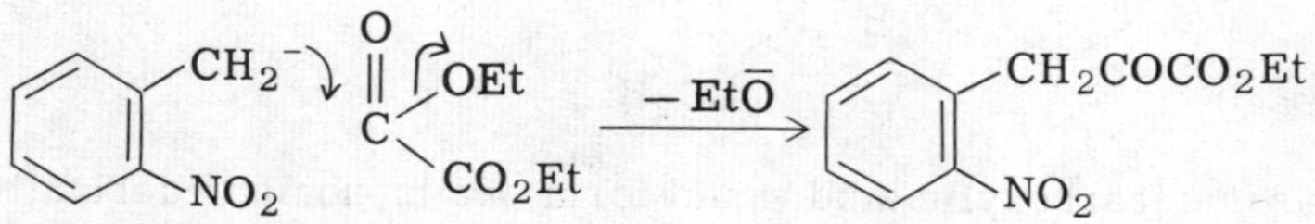

(vii)

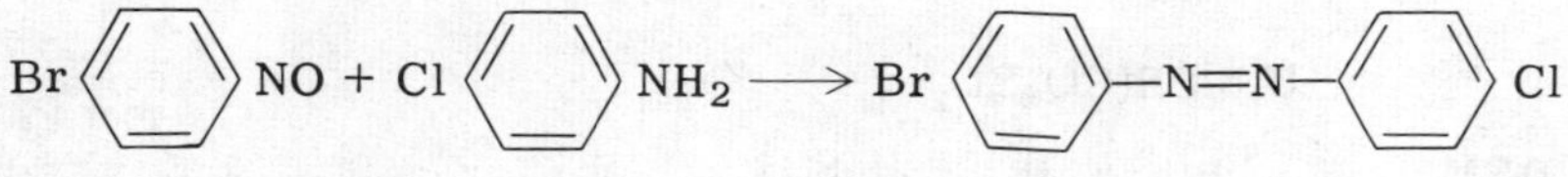

Aryl nitroso-compounds behave like carbonyl compounds in that they undergo similar condensation reactions (see *Text*; see also (viii), below).

(viii)

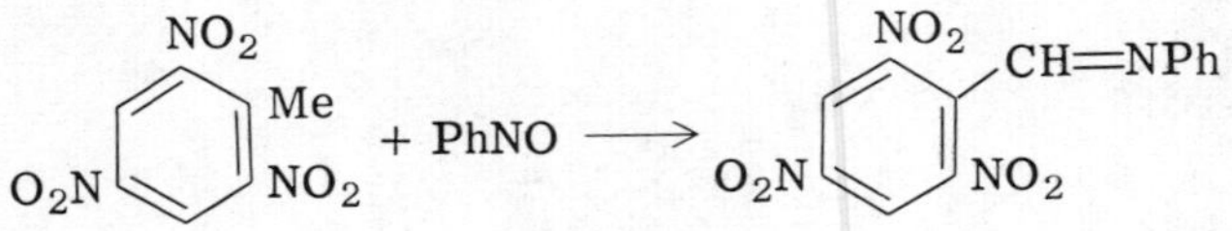

The methyl group is strongly activated by the *o,p*-nitro-groups (see (vi), above).

The mechanism of reactions of this type (including (vii), above) is not certain. A possibility is:

$$\text{Z–CH}_3 + \text{Ph–}\ddot{\text{N}}\text{=O} \longrightarrow \text{Ph–}\overset{+}{\text{N}}(\text{H})\text{=O} + \text{Z–CH}_2^-$$

$$\longrightarrow \text{Ph–N}^+(\text{H})(\text{O}^-)\text{–CH}_2\text{–Z} \longrightarrow \text{Ph–}\ddot{\text{N}}(\text{OH})\text{–CH}_2\text{–Z} \xrightarrow{-\text{H}_2\text{O}} \text{Ph–N=CH–Z}$$

7. From the given data, it is possible to *suggest* the following equations:

$$\underset{\textbf{A}}{\text{C}_8\text{H}_9\text{Br}} \xrightarrow[\text{EtONa}]{\text{CH}_2(\text{CO}_2\text{Et})_2} \text{C}_8\text{H}_9\text{CH}(\text{CO}_2\text{Et})_2 \xrightarrow{\text{H}_2\text{SO}_4}$$

$$\underset{\textbf{B}\ (\text{C}_{10}\text{H}_{12}\text{O}_2)}{\text{C}_8\text{H}_9\text{CH}_2\text{CO}_2\text{H}} \xrightarrow{[\text{O}]} \underset{\textbf{C}\ (\text{C}_8\text{H}_6\text{O}_4)}{\text{C}_6\text{H}_4(\text{CO}_2\text{H})_2} \xrightarrow{\text{HNO}_3}$$

$$\underset{\textbf{D}}{\text{C}_6\text{H}_3(\text{CO}_2\text{H})_2(\text{NO}_2)}$$

Since activation of the bromine atom in **A** is not possible (e.g., no NO_2 groups), Br must be in a side-chain. Hence, there are two possibilities:

(*a*) $C_8H_9Br \equiv C_6H_5 + C_2H_4Br$

(*b*) $C_8H_9Br \equiv C_6H_4 + CH_3 + CH_2Br$

Oxidation of **B** to a dicarboxylic acid is possible only if **A** contains *two* side-chains. Therefore, (*a*) is eliminated and so leaves (*b*). Hence, there are three possible isomers for **C** (remember that long side-chains are finally oxidised to CO_2H directly attached to the ring; see *Text*, Ch. 20).

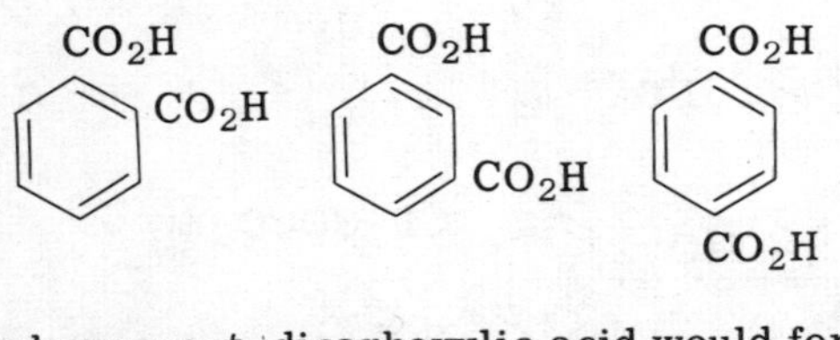

Only benzene-*p*-dicarboxylic acid would form *one* mononitro-derivative. Hence, **A** must be the *p*-disubstituted derivative. We can now formulate the reactions as follows:

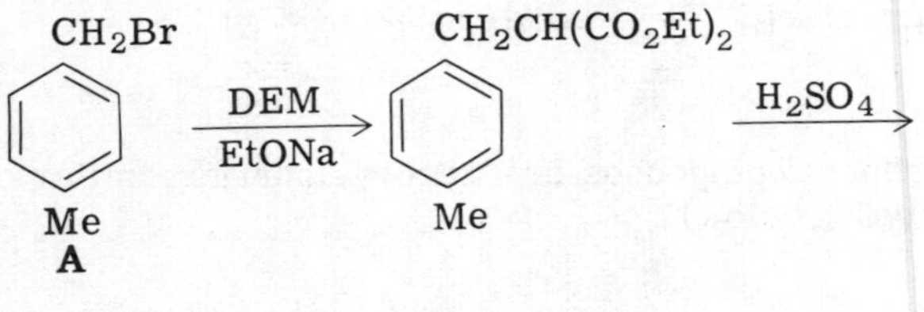

A

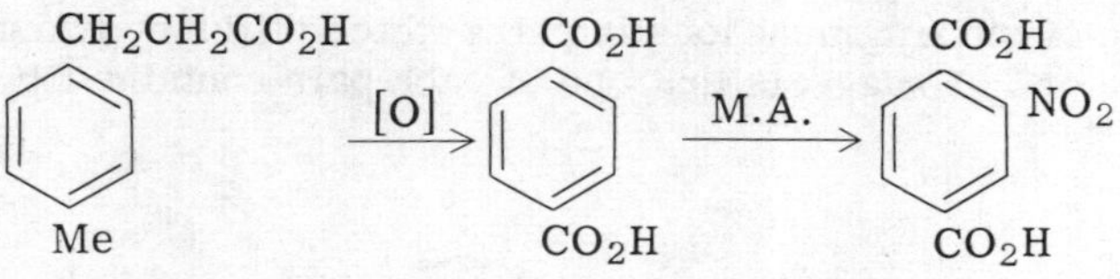

B **C** **D**

8. The acetamido group (NHCOMe) is *o/p*-directing and so nitration with M.A. can be expected to give a mixture of *o*- and *p*-nitroacetanilides. The predominance of the *p*-derivative has been explained in several ways, e.g., the steric effect at the *o*-position decreases attack here. Other explanations have also been proposed (see *Text*).

A common feature of nitration with HNO_3—Ac_2O is the large amount of *o*-derivative when the substituent group contains a hetero-atom. The mechanism is not fully understood, and one of the difficulties is that the nature of the nitrating species is not certain. According to one proposal, the nitrating species is dinitrogen pentoxide, which reacts as follows:

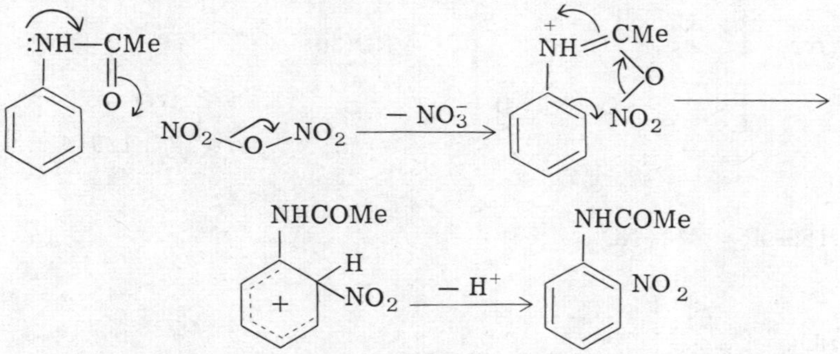

Attack by N_2O_5 at the *p*-position produces the *p*-derivative, but because this involves the polarisation of a 'double bond' of the benzene ring, which occurs with greater difficulty than polarisation of the C=O double bond, the *o*-product predominates.

9. Since the reaction is zero-order with respect to toluene, this is not involved in the rate-determining step. Hence, the r/d step precedes the step involving toluene (see *Text*, Ch. 2). Now,

$$k = A \exp(-E/RT)$$

A and *R* are constants, and if *k* varies with *T*, the reaction rate must be controlled by an activation energy factor. Therefore, nitric acid must undergo some change that requires an activation energy. Let us first consider the possibility that the step is protonation, in which case the nitrating species is the nitracidium ion.

$$HNO_3 + HNO_3 \rightleftharpoons H_2NO_3^+ + NO_3^-$$

Bonds are formed and broken, and so an activation energy is required. It is well-known, however, that acid-base proton exchanges are extremely rapid. Hence, the activation energy of this reaction is very low and can be expected to be very little affected by temperature. Therefore, we require some other reaction involving only nitric acid to be the r/d step. Since bond fission requires energy to bring it about, we can postulate the following scheme:

$$2HNO_3 \underset{}{\overset{\text{fast}}{\rightleftharpoons}} NO_3^- + H_2\overset{+}{O}\text{—}NO_2 \underset{}{\overset{\text{r/d}}{\rightleftharpoons}} H_2O + NO_2^+$$

The rate of the second step can therefore be expected to be temperature-dependent. If this were the mechanism, the nitrating species is the nitronium ion. Further work is necessary to confirm (or refute) this.

10. The presence of the fragment ions in pairs which differ by two units and appear roughly in the ratio of 3 : 1 suggests that one of each pair contains Cl(35) and the other Cl(37).

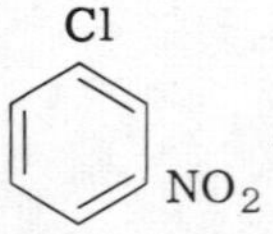

$= C_6H_4ClNO_2$; *M.W.* = 157, 159.

If we now use the characteristic fragmentation patterns of aromatic nitro-compounds (and, if necessary, Table 1. 6), we can account for the data as follows.

$M^+ = 157, 159.$
$m/e\ 127 = 157–30 = M^+–NO$
$m/e\ 129 = 159–30 = M^+–NO$

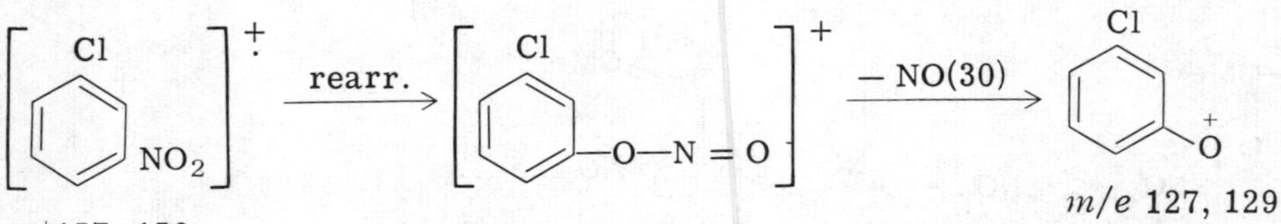

$M^{+\cdot}$ 157, 159 — *m/e* 127, 129

$m/e\ 111, 113 = 157, 159–46 = M^+–NO_2$

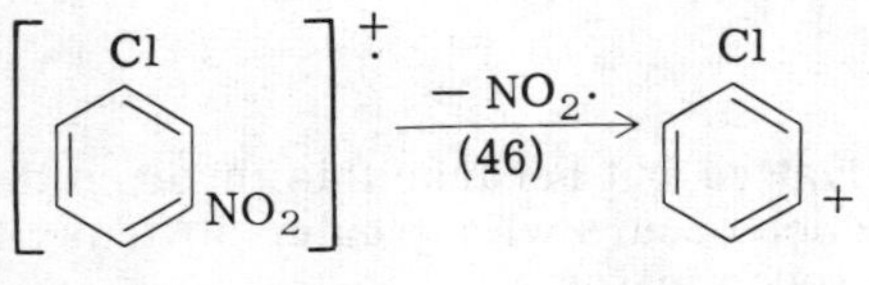

$M^{+\cdot}$ 157, 159 — *m/e* 111, 113

Now, a characteristic fragmentation ion formed from benzene is *m/e* 51, and is produced by loss of acetylene (26) from the benzene carbonium ion (see *Text*, Ch. 20). Since 85(87) = 111(113)–26, this suggests:

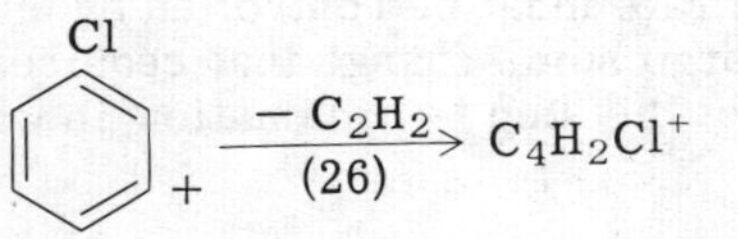

m/e 111, 113 — *m/e* 85, 87

This now leaves one ion, *m/e* 76, to be accounted for. The fact that a pair is not given in the data suggests that there is no chlorine in this ion. Since 76 + 35(37) = 111(113), the fragmentation path is:

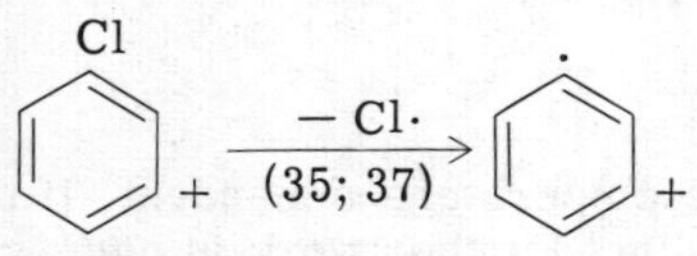

m/e 111, 113 — *m/e* 76

In view of the fact that aryl halides can eliminate halogen (see *Text,* Ch. 21) and that one of the fragment ions described above does this, one might have expected the ion m/e 122, formed as follows:

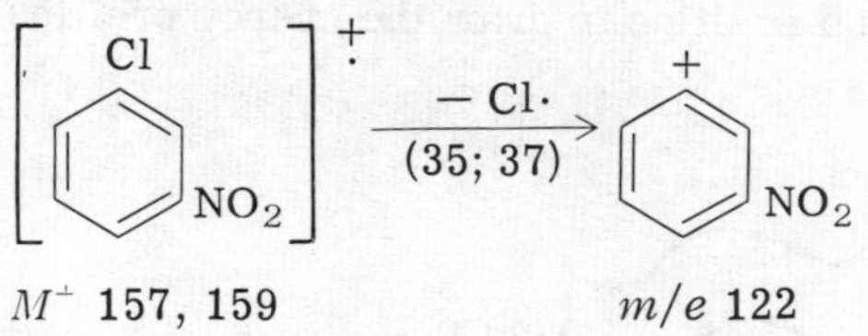

23 Aromatic amino-compounds

(Problems on page 42)

1. The overall equation may be written as:

However, Z^+ must be generated from some precursor and H^+ is not liberated as such (see Ch. 22, Prob. 2). We therefore cannot estimate the heat of reaction from this equation and so cannot estimate the relative energy levels of the reactant and product systems. In view of the fact that energy profiles are not always what they are meant to represent (see Ch. 8, Prob. 3), we can tackle the problem as follows. We can start with substrate and Z^+ and draw the relevant parts of the energy profiles to show: (i) that the step involving the substrate and Z^+ to form the σ-complex is rate-determining (in accordance with the accepted mechanisms); (ii) that the relative activation energies for *o*-, *m*- and *p*-substitution depend on the nature of the substituent group (Y).

For convenience, we can draw both energy profiles in one diagram, using the same energy level for both reactant systems. In this way, we can readily show how the various activation energies are related to one another. Both substrates preferentially undergo *o*/*p*-substitution, but since we do not know their relative amounts, we shall represent their energy profiles by a single curve. Also, because *m*-substitution is far more difficult, its activation energy is higher than that of *o/p*-substitution. For aniline, the ring is activated (hence relatively low activation energies), whereas for chlorobenzene, the ring is deactivated (hence relatively high activation energies).

With these arguments as our basis, we are now in a position to draw the energy profiles.

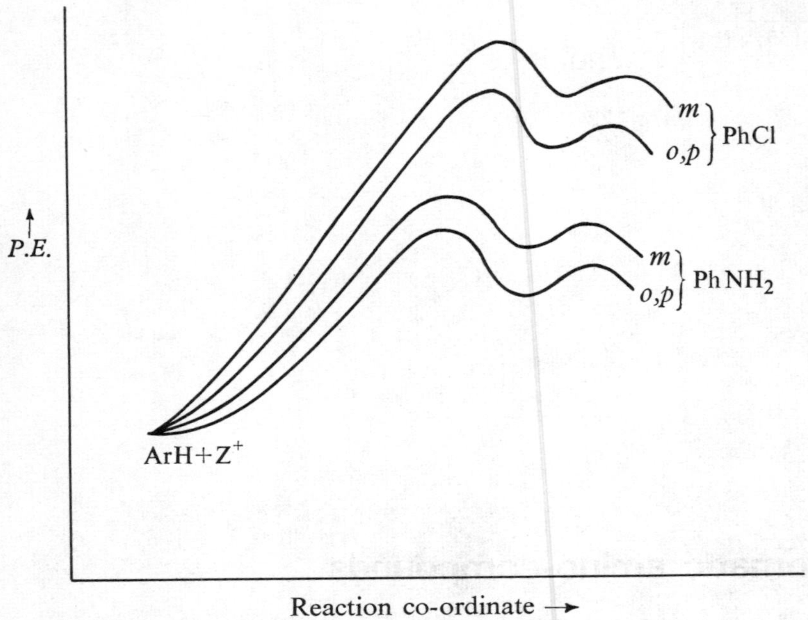

2. Because of the positive charge on the nitrogen and the phosphorus atom, these groups exert a powerful —I effect and so are *m*-directing. However, N and P differ in that the latter has an empty *d*-orbital and can expand its valency shell to a decet (see *Text*, Ch. 14). Hence:

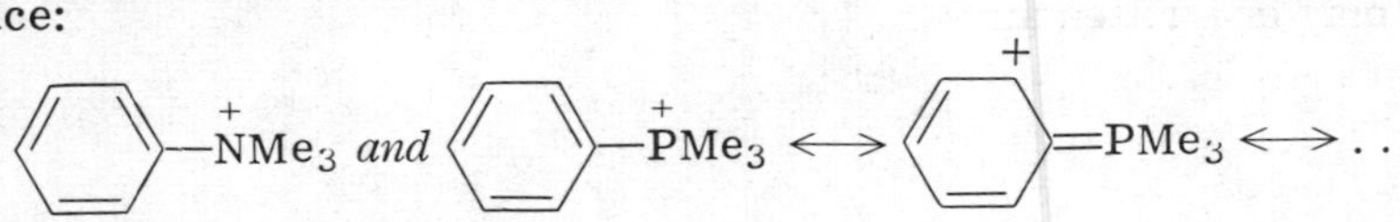

Thus, in the phosphorus compound, the ring is deactivated by *both* the —I and —R effects. It can therefore be expected that the rate of *m*-substitution in the phosphorus compound will be slower than that in the nitrogen compound.

3. (i) $PhCH{=}CH_2 + Et_2NH \longrightarrow$ no reaction

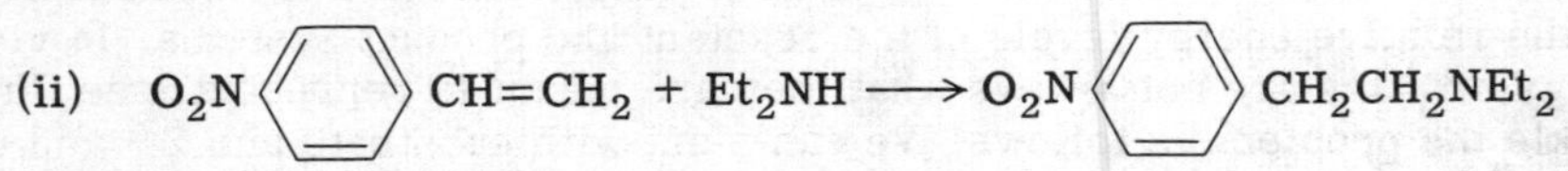

Since the structures of both substrates are closely related and yet one undergoes reaction and the other does not, it must mean that the introduction of the *p*-nitro-group has a powerful effect on the behaviour of the vinyl group. Now, the characteristic property of a nitro-group is its —I and/or —R effect. Let us therefore consider the nitro-compound from this point of view. Thus, the molecule can be represented as a resonance hybrid:

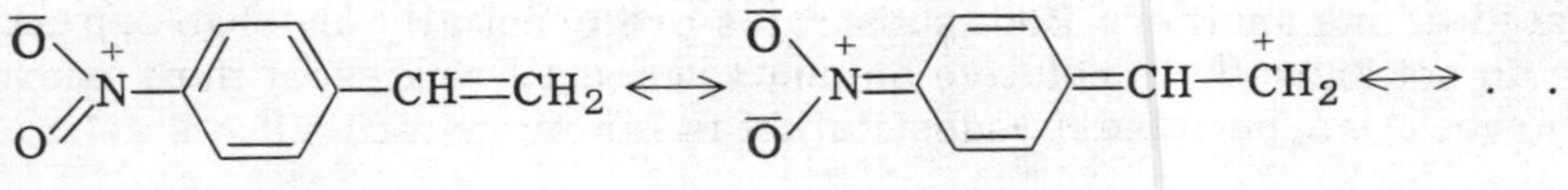

The first thing to note is that the terminal methylene group carries a positive charge and therefore can behave as an electrophilic centre. However, to behave as an electrophile, a nucleophile (or nucleophilic centre) must also be present. This now leads us to examine diethylamine, and we see that this is a nucleophile (lone-pair of electrons on N). We can now formulate the reaction as follows:

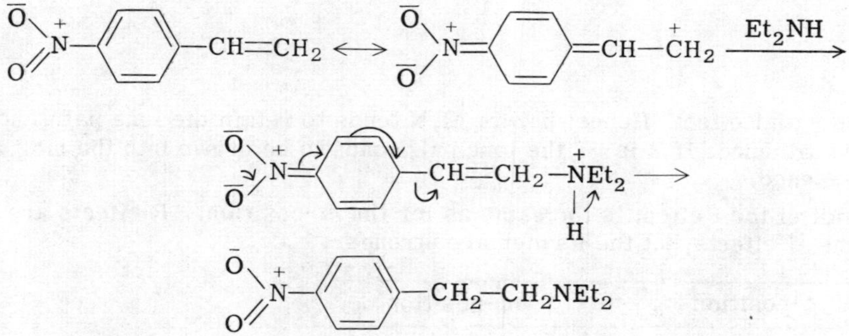

What we have done so far is to offer an explanation why reaction (ii) *does* proceed. To complete our answer, we must offer an explanation why reaction (i) does *not* proceed. Styrene can be represented as a resonance hybrid:

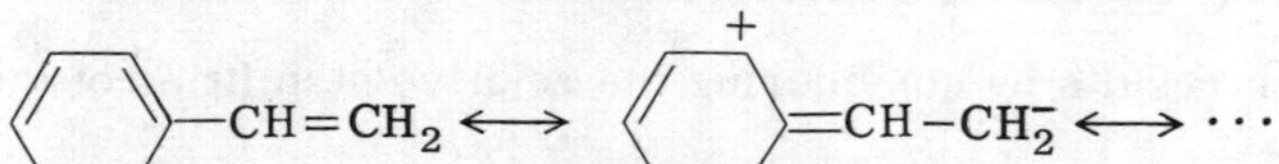

In this case, the terminal methylene group carries a negative charge and can therefore behave as a *nucleophile*. Since Et_2NH can behave *only* as a nucleophile, no reaction will occur.

Attack at the benzene ring does not occur because the benzene R.E. will be lost.

4. The usual reaction of nitrous acid is the formation of diazonium salts with aromatic primary amines and *N*-nitroso-compounds with secondary amines. If, however, neither of these groups is present, the *o*- and/or *p*-positions, if sufficiently activated, react to form *C*-nitroso-compounds.

(i)

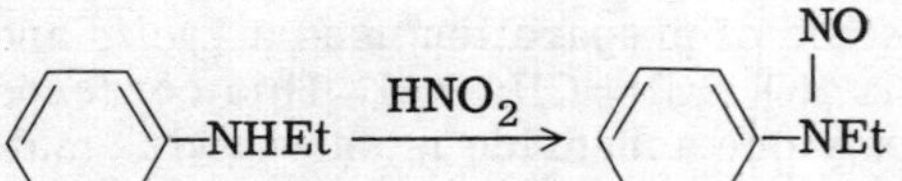

(ii)

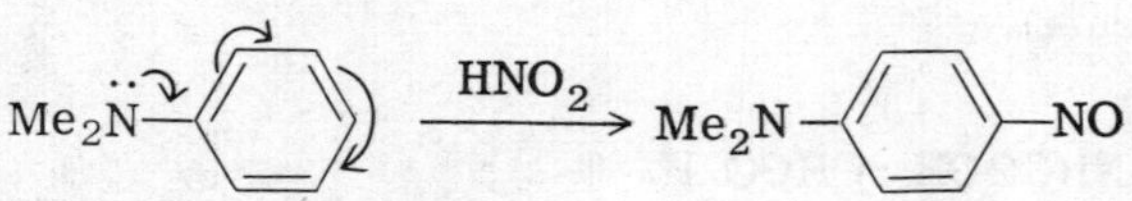

(iii)

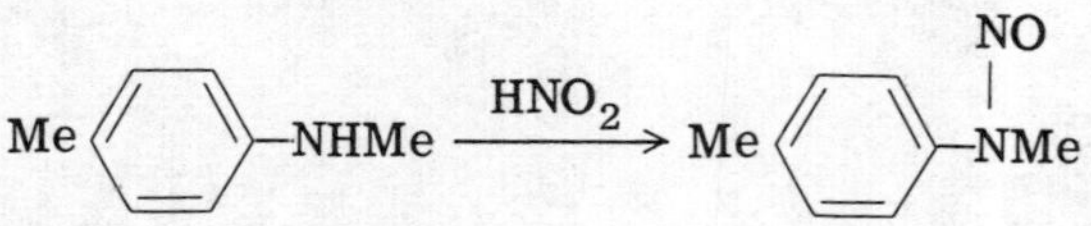

(iv)

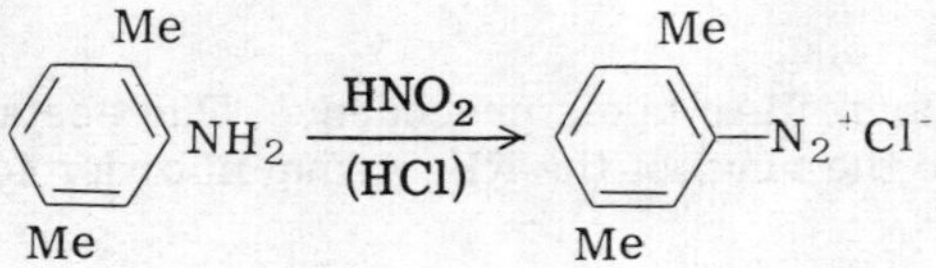

5. Since the basicity of the anilines is dependent on the availability of the lone pair of electrons on the N atom, we have:

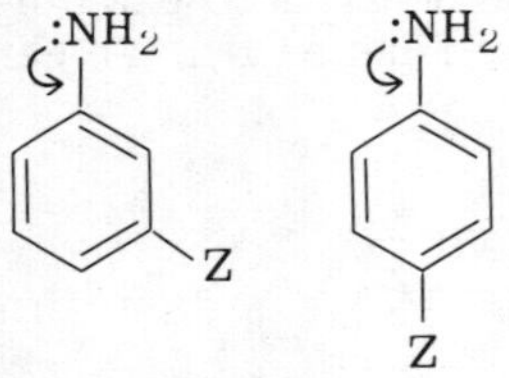

***m*-position.** Z has only an I effect. Hence: if Z is +I, N tends to retain the lone pair, and so the basicity is strengthened. If Z is —I, the lone pair tends to be drawn into the ring, and so the basicity is weakened.

***p*-position.** The result of the I effect is the same as for the *m*-position. ±R effects are similar to those of the ±I effects, but the former are stronger.

	Group	*p*-position	*m*-position
(i)	NO_2	—R; weakens	—I; weakens
(ii)	OR	+R; strengthens	—I; weakens
(iii)	CO_2R	—R; weakens	—I; weakens
(iv)	NR_2	+R; strengthens	—I; weakens

The student should also verify these results by considering the relative stabilities of the aniline and its conjugated acid.

6. (i) Isocyanic acid is generated and reacts with the NH_2 group with rearrangement to give a substituted urea (see *Text*, Ch. 16).

$$\text{NaNCO} \xrightarrow{\text{AcOH}} \text{HNCO}$$

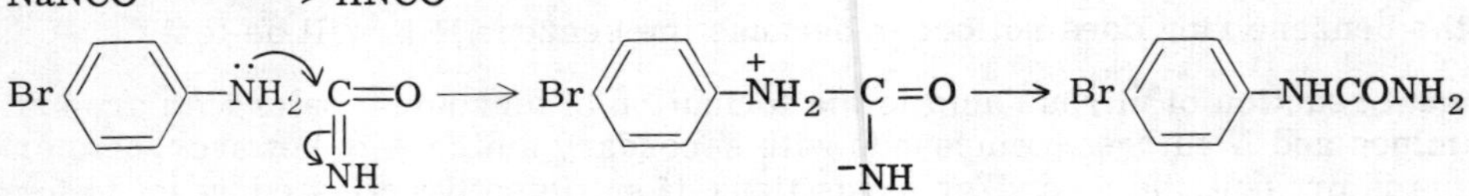

(ii) PhNH—COCH=$CHCO_2H$

This is a 'half' acid amide, and a common method of preparation uses a *cyclic* anhydride. (see *Text*, Ch. 16). The structure of the *acid* is HO_2CCH=$CHCO_2H$. This corresponds to maleic acid (*cis*) and fumaric acid (*trans*). Only one anhydride is known, viz., maleic anhydride (see *Text*, Ch. 17). We must therefore use this (which we may do, since no mention is made of the geometry of the product).

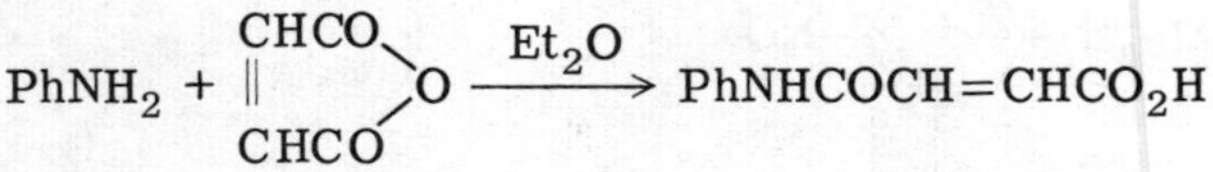

(iii)

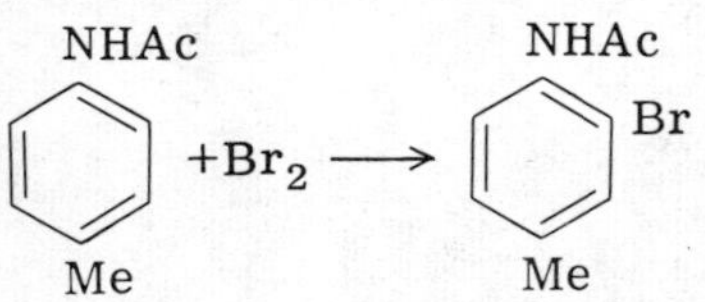

Both substituent groups are *o*/*p*-directing, and NHAc takes precedence. The acetamido-group is used to decrease the activating effect (on the ring) of the NH_2 group in order to introduce *one* bromine atom.

(iv)

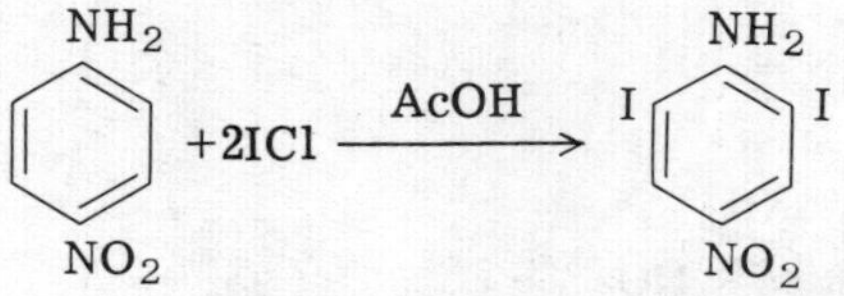

ICl ($\overset{\delta+}{I}-\overset{\delta-}{Cl}$) is a very good iodinating agent for activated rings.

(v)

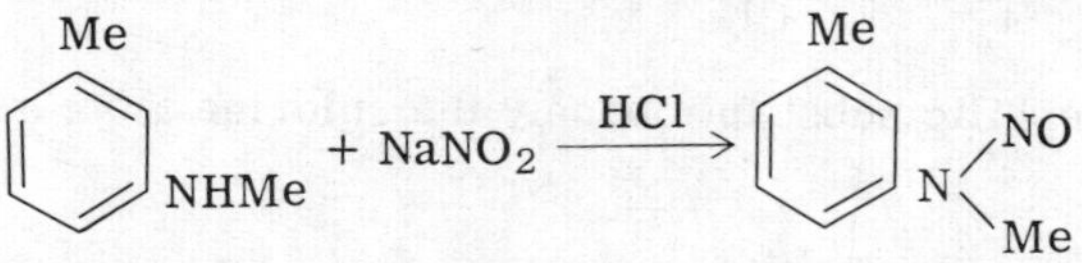

The MeNH group does not activate the *p*-position sufficiently to get a *C*-nitroso-derivative (see also Prob. 4, above).

(vi) This reaction requires the introduction of the carbethoxyl group on nitrogen. This is usually done with ethyl chloroformate.

$PhCH_2NH_2 + ClCO_2Et \longrightarrow PhCH_2NHCO_2Et$

(vii) The carbonyl chloride group can be introduced on nitrogen by means of carbonyl chloride.

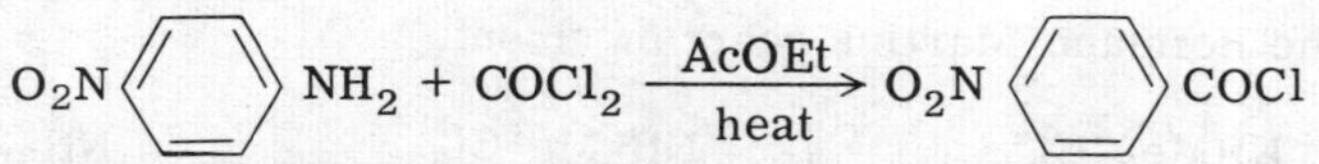

(viii) Under the conditions of the experiment, urea is decomposed into isocyanic acid, which then reacts as shown (see (i), above for the mechanism).

$NH_2CONH_2 \longrightarrow NH_3 + HNCO$

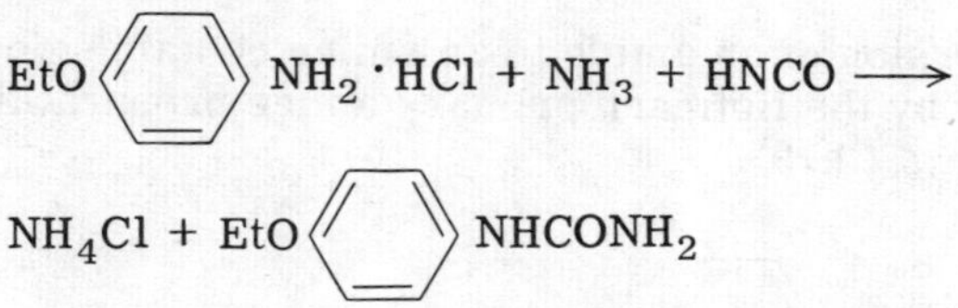

(ix)

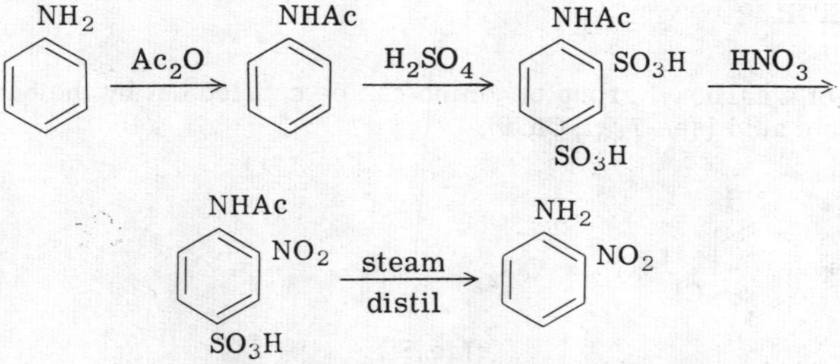

The acetamido group is almost exclusively *p*-directing. Hence, to obtain the *o*-derivative, the *p*-position must be protected. A good way of doing this is to introduce a sulphonic acid group, since this is readily removed by heating with dilute acid or in aqueous solution (see *Text*, Ch. 25). Because of the conditions used to sulphonate acetanilide, *two* sulphonic acid groups are introduced. Also note the replacement of the *o*-SO_3H group by NO_2. Since the removal of the SO_3H group results in the formation of sulphuric acid, hydrolysis of the amide is readily effected (in acid solution).

(x) $$Ph_2NH \xrightarrow{Li} Ph_2NLi \xrightarrow[Cu\,(cat.)]{PhI} Ph_3N + LiI$$

(xi)

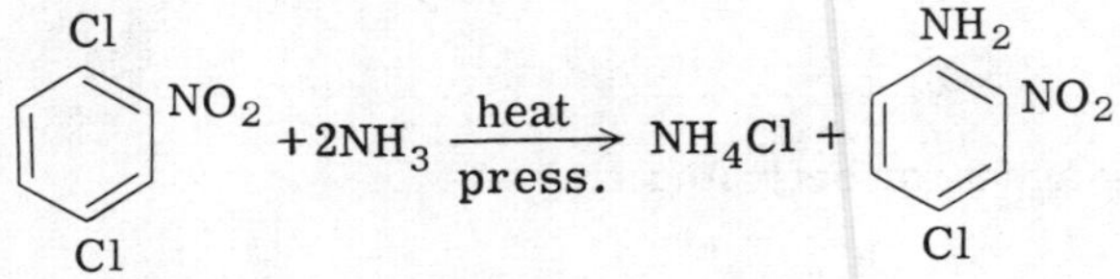

This is an example of activated nucleophilic substitution; only the chlorine atom *o*- to the nitro-group is activated.

(xii) $Ph_3N + MeI \longrightarrow$ no reaction (see *Text*).

(xiii)

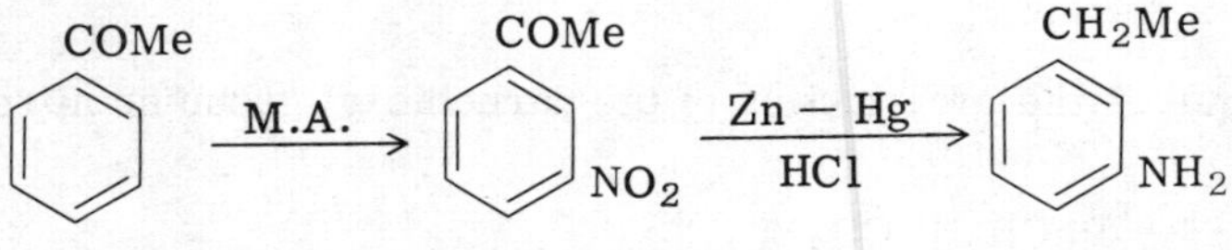

The Clemmensen reduction converts CO into CH_2, but at the same time the reagent (metal + acid) reduces the nitro-group to the amino-group.

(xiv) This is an example of the Hofmann-Martius rearrangement.

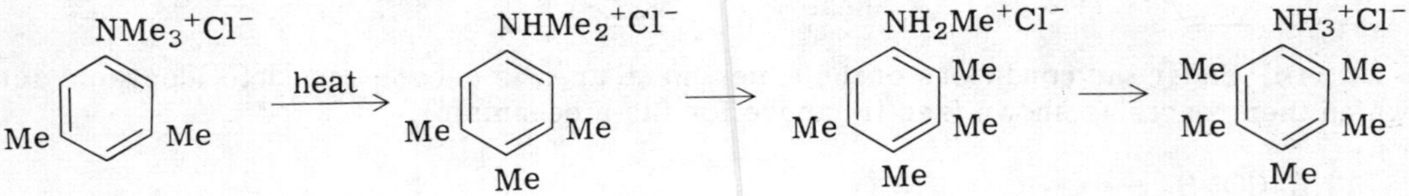

(xv) This is the conversion of an amide to an amine containing one carbon atom less. Hence, it can be carried out by the Hofmann primary amine preparation; Br_2/KOH or NaOCl is the usual reagent (see *Text*, Ch. 9).

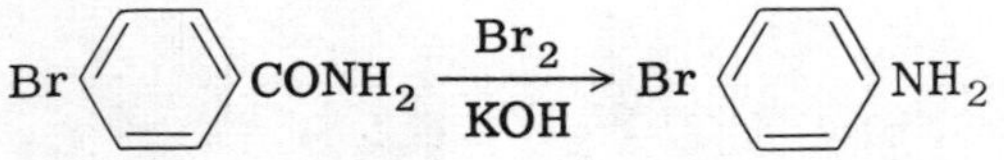

(xvi) Replacement of a carboxyl group by amino can be carried out by the Schmidt reaction; this uses hydrazoic acid (see *Text*, Ch. 9).

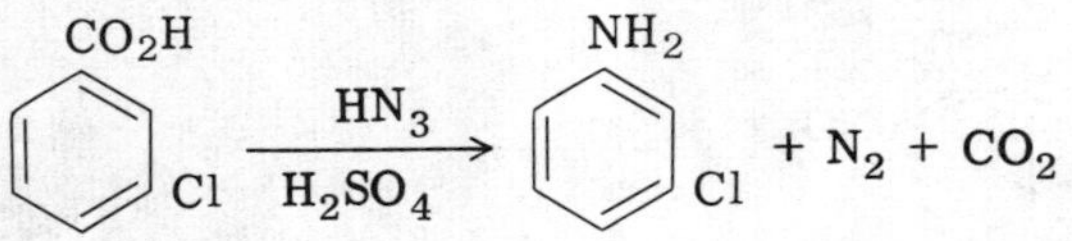

(xvii) $$PhNH_2 + PhSO_2Cl \longrightarrow PhNHSO_2Ph \xrightarrow[NaOH]{Me_2SO_4}$$

$$PhNMeSO_2Ph \xrightarrow{H_3O^+} PhNHMe + PhSO_3H$$

It is difficult to prepare *N*-methylaniline by direct methylation; the product is a mixture of monomethyl- and dimethyl-derivatives. The preparation via the sulphonamide gives pure monomethylaniline.

(xviii)

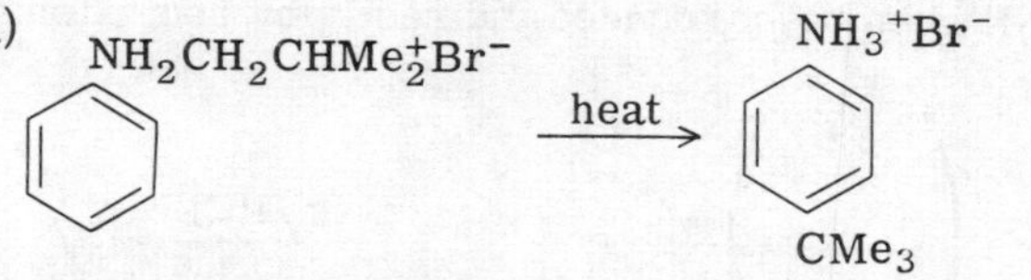

This is an example of the Hofmann-Martius rearrangement (see *Text*). The mechanism involves the formation of an alkyl carbonium ion, and if this is capable of rearranging, it does so to form a more stable ion. Thus:

$$Ph\overset{+}{N}H_2CH_2CHMe_2Br^- \underset{}{\overset{\text{heat}}{\rightleftharpoons}} PhNH_2 + Me_2CHCH_2Br \rightleftharpoons Br^- + Me_2CHCH_2^+$$

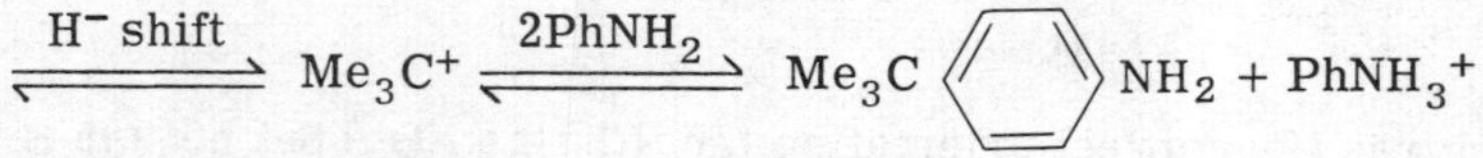

$$\text{(xix)}\quad PhN = NPh \xleftarrow{\text{LAH}} PhNO_2 \xrightarrow[\text{Pd} - \text{C}]{\text{NaBH}_4} PhNH_2$$

7. The product formed in each case is the *p*-nitroso derivative. The reactive species is the nitrosonium ion (electrophile) and, in the absence of steric effects, the most likely site of attack will be that with the highest electron density. Thus, the reactions may be formulated:

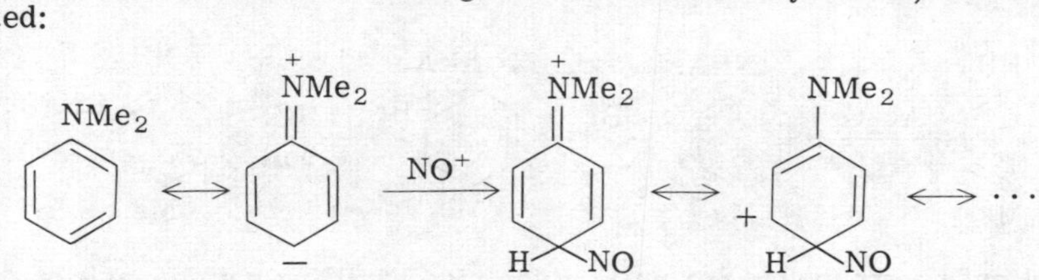

(I) (II) (III) (IV)

The *p*-position has a high electron density and the σ-complex is stabilised by resonance.

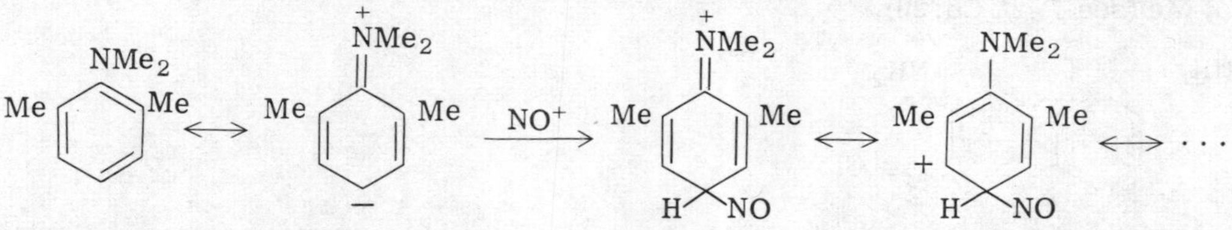

(V) (VI) (VII) (VIII)

In the 2, 6-dimethyl derivative, because of steric inhibition of resonance, the major resonating structure is (V). Hence, the electron density at the *p*-position in (V) is much lower than that at the *p*-position in (I). Also, the contributions of (VII) and VIII) will be much less than those of (III) and (IV) to the corresponding complexes. Hence, attack by the nitrosonium ion is facilitated in (I) and retarded in (V), and the stability of the σ-complex from (I) is far greater than that from (V). The result is that the activation energy of the reaction with dimethylaniline is lower than that with the 2, 6-dimethyl derivative. Therefore, the rate of reaction of the former is faster than that of the latter.

8. Revise (if necessary) Ch. 22, Prob. 3.

(i)

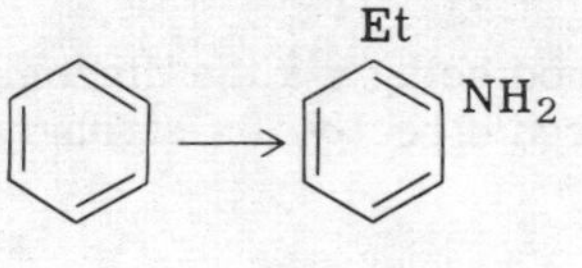

Et (o/p) can be introduced directly; NH_2 can be introduced indirectly by reduction of NO_2 (m). Hence:

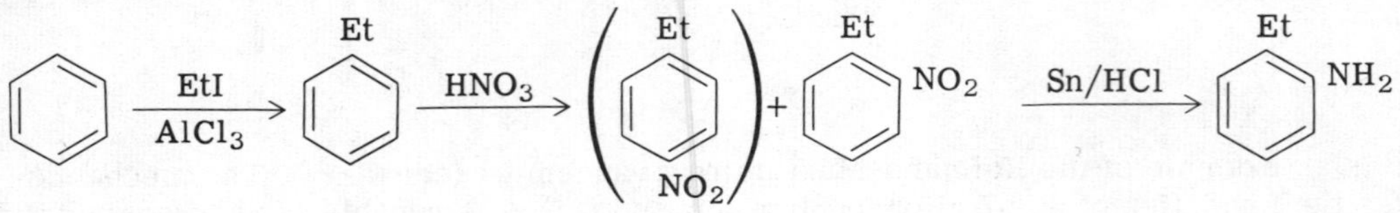

(ii)

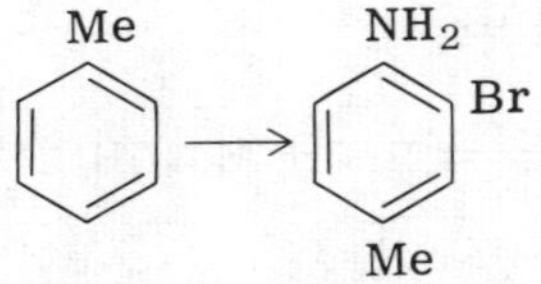

Me (o/p) gives the correct orientation for NH_2 (as NO_2) but not for Br. NH_2 gives the correct orientation for Br. Hence:

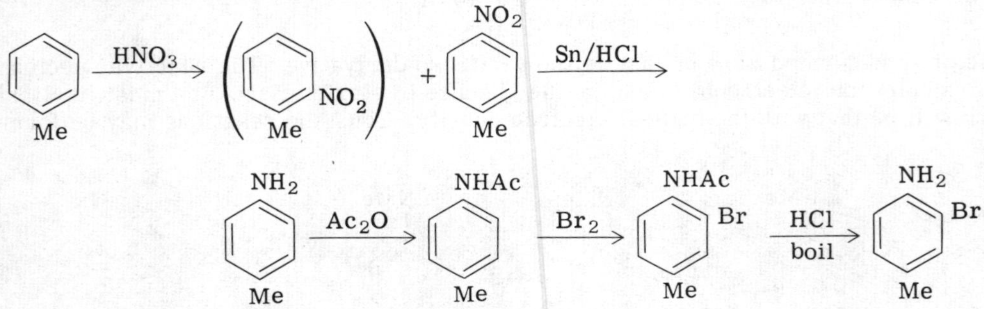

NH_2 is a very strong activating group, and normally gives rise to filling both o-positions and the p-position. Hence, to introduce *one* group, the NH_2 group is acetylated. The acetamido-group is almost exclusively p-directing, but if this position is blocked, the entering group takes up the o-position exclusively. Also, the directive power of NHAc is much greater than that of Me (see *Text*, Ch. 20).

(iii)

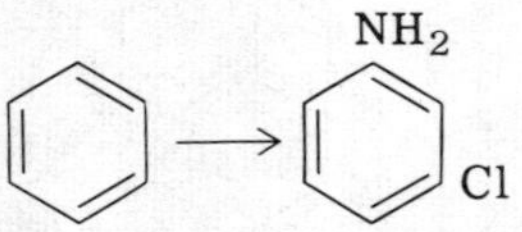

$NH_2 \equiv NO_2(m)$; hence NO_2 gives the correct orientation; Cl (o/p) does not.

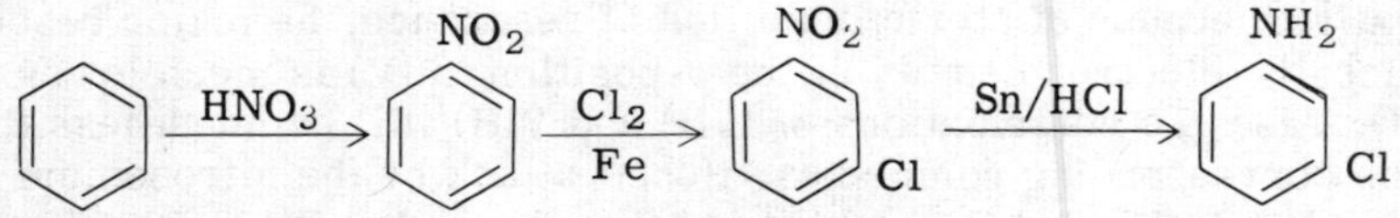

(iv)

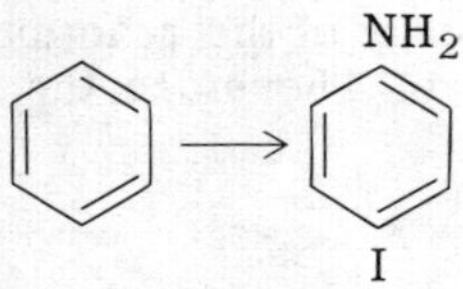

Iodine is not usually introduced directly, the common method being via the diazonium salt (see *Text*, Ch. 24). However, iodine can be readily introduced directly if a strongly activating substituent is present, e.g., NH_2. Hence (see also *Text*):

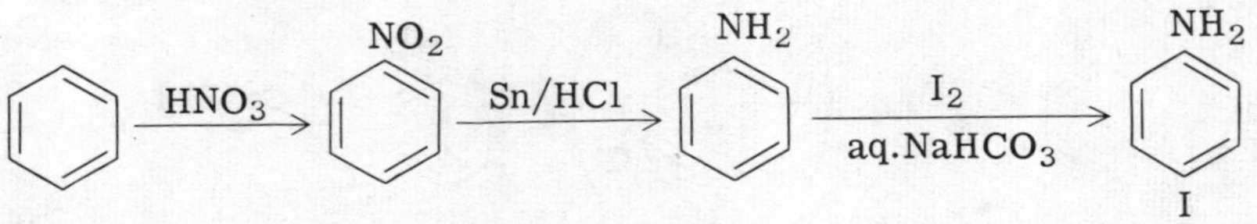

Because iodine is far less reactive than chlorine or bromine, it is possible to isolate the mono-iodo compound (*p*-).

(v)

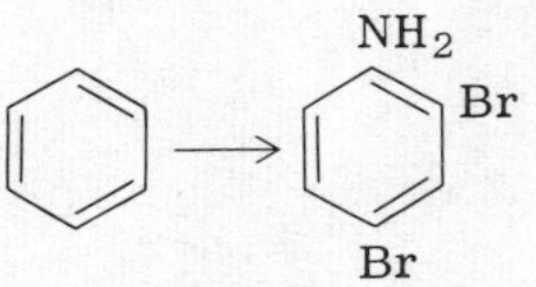

NH_2 (from NO_2) is *o*/*p*-directing, but since *three* bromine atoms will be introduced (*o*, *o*, *p*), it seems reasonable to try the use of the acetamido-group, the first Br going to the *p*-position and then, because this position is now blocked, introducing the second Br into the *o*-position (see (ii), above).

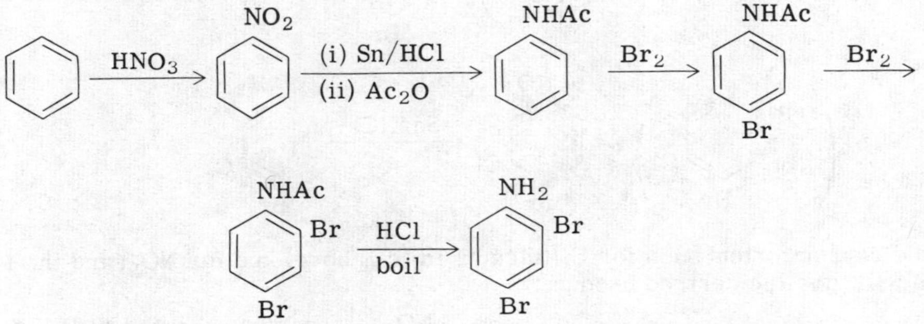

We can also use an alternative method. If we introduce one Br first, this will give the correct orientation for NO_2. We then proceed as follows:

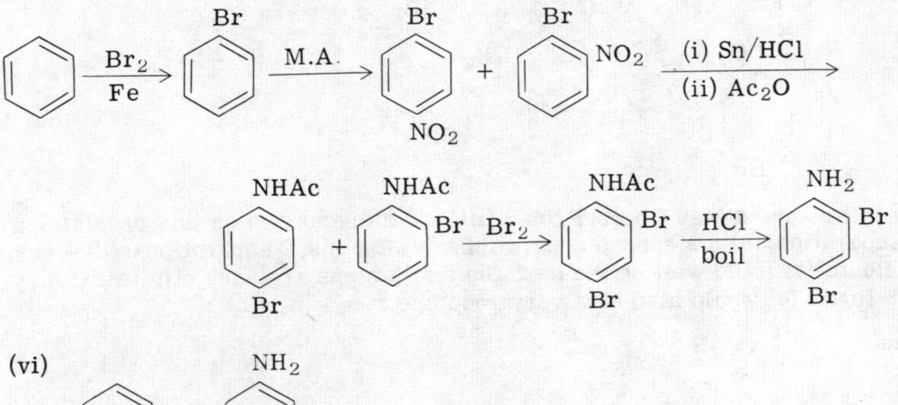

(vi)

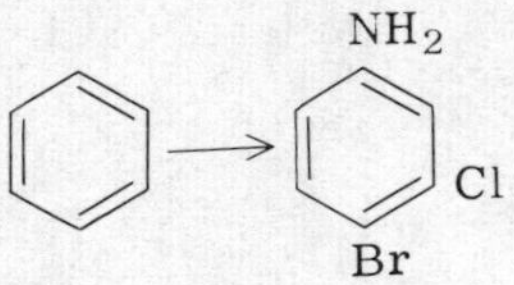

(*a*) NO_2 gives the correct orientation for Cl, and the Cl, NO_2-derivative can give the required product.

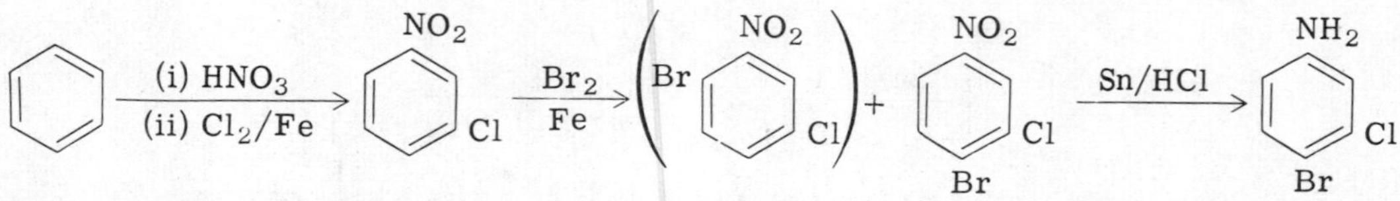

(*b*) The above synthesis may be modified as follows:

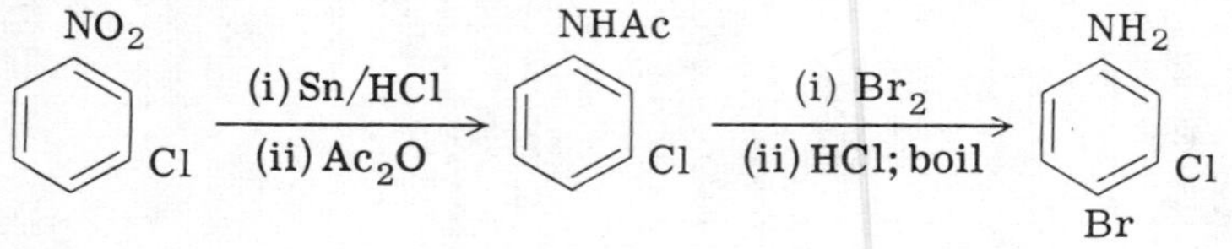

(*c*) Cl gives the correct orientation for Br, and the Br, Cl-derivative gives the correct orientation for NO_2.

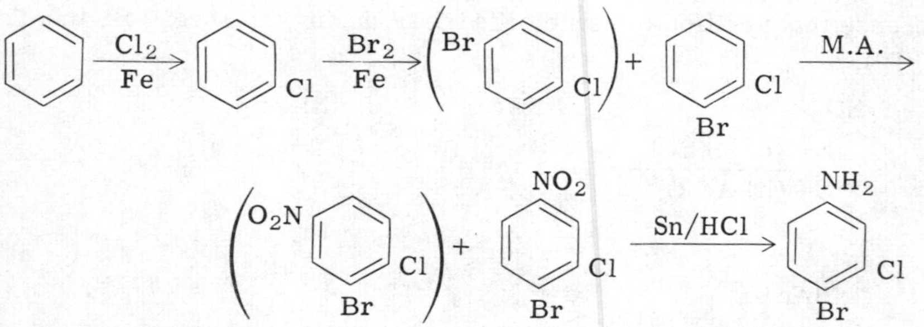

(*d*) Br gives the correct orientation for Cl (then as in (*c*), above) and for NO_2, and the Br, NO_2-derivative will give the desired product.

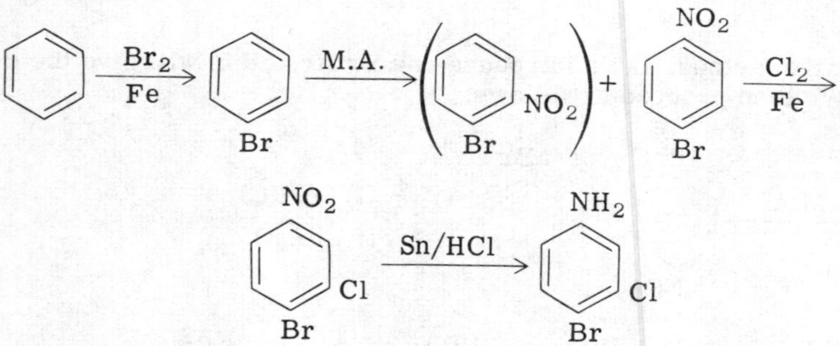

If we compare these routes (see Ch.22, Prob.3 (iii) for a discussion on this problem), we have: (*a*) 4 steps, 1 separation; (*b*) 6 steps, no separations; (*c*) 4 steps, 2 separations; (*d*) 4 steps, 1 separation. Route (*b*) could well be the best; the last 3 steps are very efficient (i.e., yields are excellent). Route (*d*) would also be a very good one.

(vii)

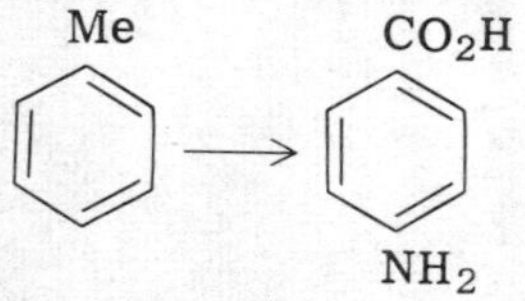

Me (*o*/*p*) can be oxidised to carboxyl *after* the NO_2 is introduced, and then the product reduced. Reduction cannot be carried out *before* oxidation, unless the NH_2 group is protected (see *Text*, Ch. 20).

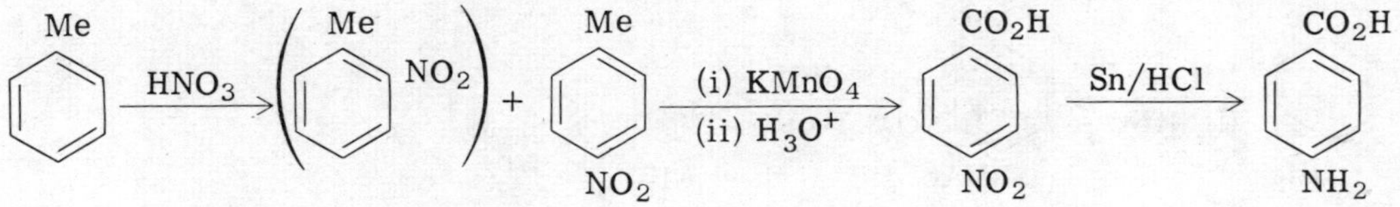

(viii)

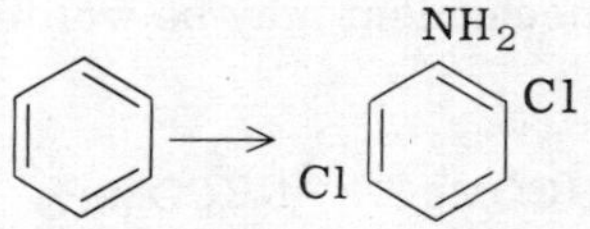

NH_2 (from NO_2) gives the correct orientation for 2-Cl, but not for 5-Cl. NO_2 gives the correct orientation for 5-Cl, and the Cl, NO_2-derivative would give the desired product.

(*a*)

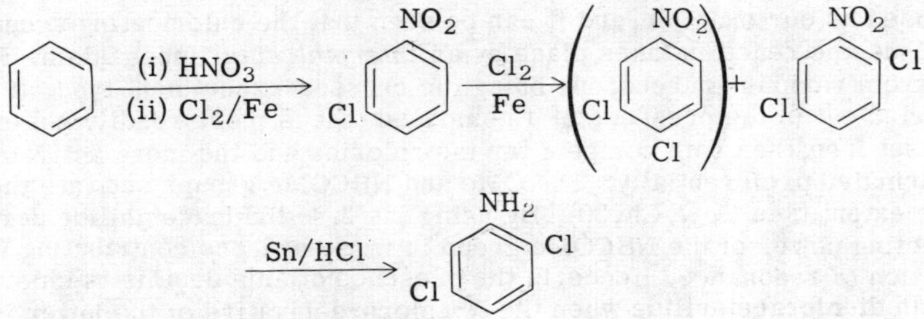

(*b*) 2-Cl gives the correct orientation for NO_2 ($\equiv NH_2$) and for 5-Cl.

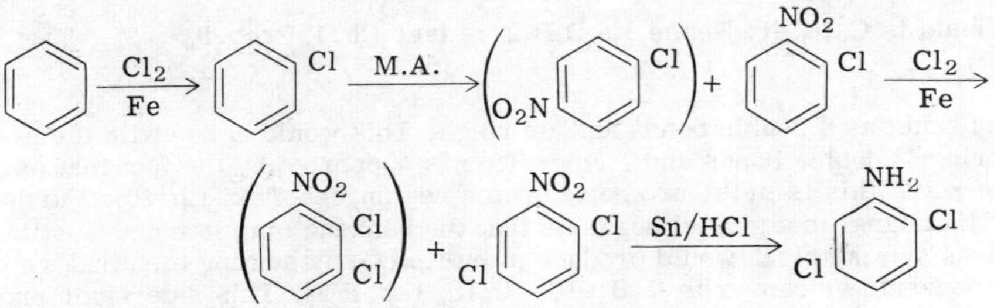

(*c*) An alternative to (*b*) is:

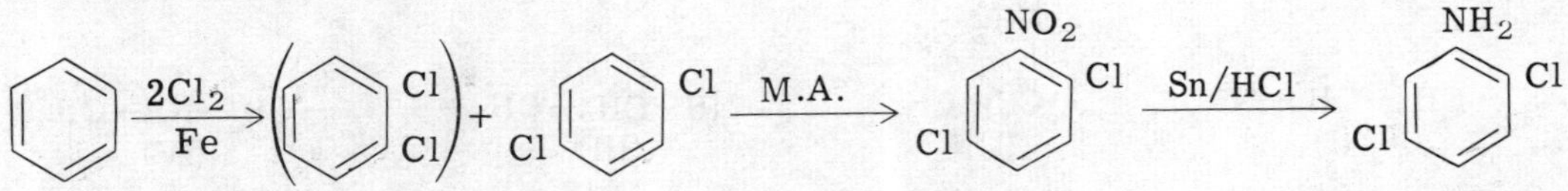

Probably (*c*) is the best route (3 steps, 1 separation).

(ix)

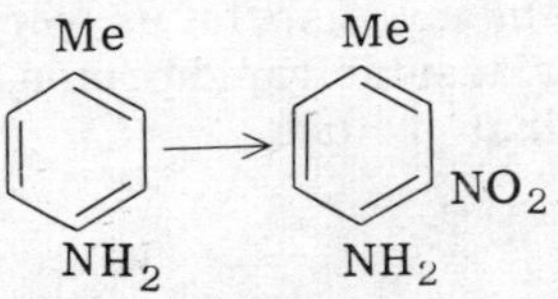

NH_2 gives the correct orientation, but must be deactivated to introduce only *one* NO_2 group. At the same time, this protects the ring from oxidation. Also remember that NHAc directs *o*/*p* more strongly than does Me.

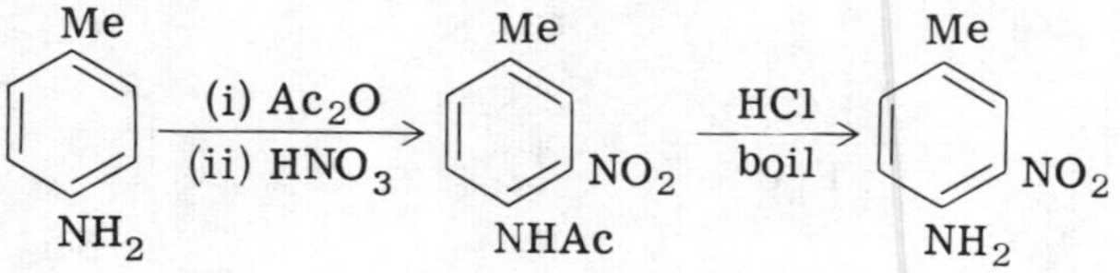

9. This is an example of the Orton rearrangement; the mechanism may be written in simplified form as follows (see also *Text*):

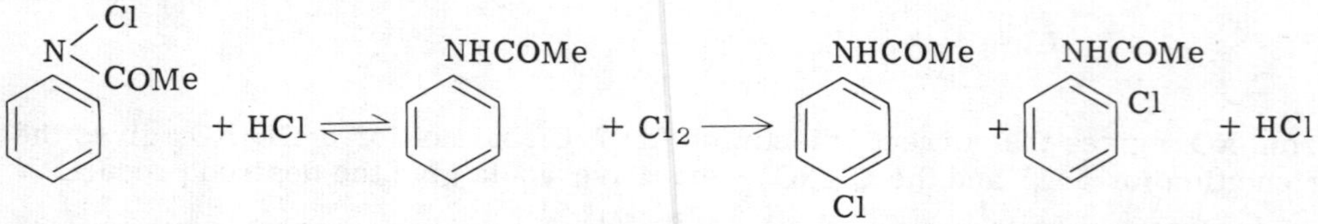

Acetanilide has been used as our example, and it can be seen that the chlorinating agent is molecular chlorine. Thus, the reaction takes place by an *intermolecular* mechanism. Evidence for this comes from various sources, one being the crossed-experiment type. If the rearrangement is carried out in the presence of a compound that is more readily chlorinated than the anilide set free, then both compete for the chlorine and the more activated compound will be chlorinated preferentially. The OMe and NHCOMe groups activate the ring to about the same extent (see *Text*, Ch. 20). By using the 2,4-dichloroacetanilide derivative, however, the directing power of the NHCOMe group is weakened, one contributing factor being steric inhibition of resonance. Hence, in the presence of anisole, this is chlorinated more rapidly than dichloracetanilide when the *N*-chloro-derivative of the latter is used.

10. The four signals are: τ 8·6 (d), 8·4 (s), 5·9 (q), and 2·7 (s). Hence, there are four different types of protons.

The molecular formula is $C_8H_{11}N$. Hence, its D.B.E. is (see Ch. 1, Prob. 9):

$$8 + 1 - (11 - 1)/2 = 4$$

Thus, the compound contains 4 double bonds and/or rings. This would agree with the presence of a benzene ring: 3 double bonds and 1 ring. This is supported by the fact that one signal (singlet) has τ 2·7 (this is in the aromatic proton region; see *Text*, Ch. 20). Furthermore, the fact that the signal is a singlet suggests that the benzene ring is monosubstituted (disubstitution, unless symmetrical, would produce a multiplet). Assuming then that we have a monosubstituted benzene, we can write $C_8H_{11}N = C_6H_5 + C_2H_6N$. This side-chain produces three signals, and so must contain three kinds of protons. There are four possibilities for C_2H_6N:

(I) $-CH_2-CH_2-NH_2$

(II) $-CH(NH_2)-CH_3$

(III) $-NH-CH_2-CH_3$

(IV) $-CH_2-NH-CH_3$

(III) is almost certainly ruled out, since the lone pair on the nitrogen atom could enter into resonance with the ring, thereby changing the electron densities at the *o*-, *m*-, and *p*-positions, in which case the ring signal will be a multiplet. In any case, let us work out the multiplicities expected for these side-chains. Since we may assume rapid proton exchange for the NH or NH_2 group, there will be no coupling with vicinal protons.

(I) $-CH_2-CH_2-NH_2$: t, t, s

(II) $-CH(NH_2)-CH_3$: CH q, NH_2 s, CH_3 d

(III) $-NH-CH_2-CH_3$: s, q, t

(IV) $-CH_2-NH-CH_3$: s, s, s

Only (II) agrees with the observed signals. Hence, the compound is (V), and it is now possible to make the following assignments: Ph, τ 2·7 (s); CH, τ 5·9 (q); CH_3, τ 8·6 (d); NH_2, τ 8·4 (s).

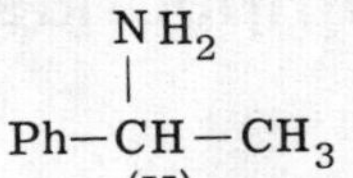

(V)

The large downfield shift for the CH proton may be attributed to deshielding by the benzene ring current and by the —I effect of nitrogen atom (see also Table 3. 2).

11.

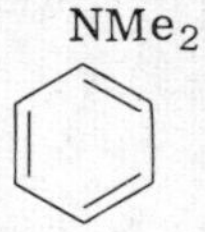

$= C_8H_{11}N$; M.W. = 121 = M^+

We now make use of the characteristic fragmentation patterns (and Table 1. 6 where necessary).

m/e 120 = 121 − 1 = M^+ − H

This can be formed as follows (see also aliphatic amines; *Text*, Ch. 13).

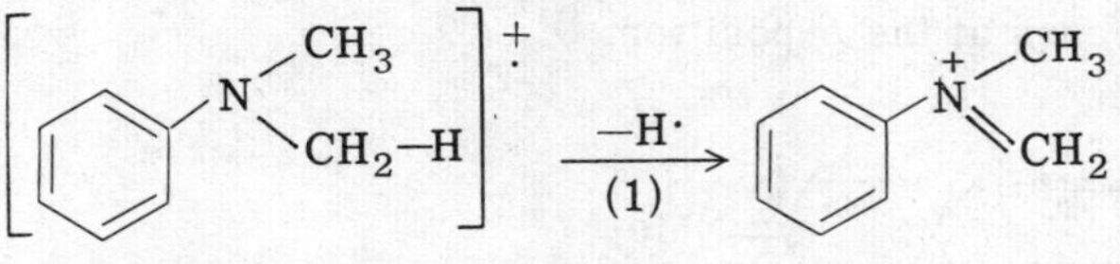

M^+ 121 m/e 120

m/e 105 = 120 − 15 = 120 − Me. Hence:

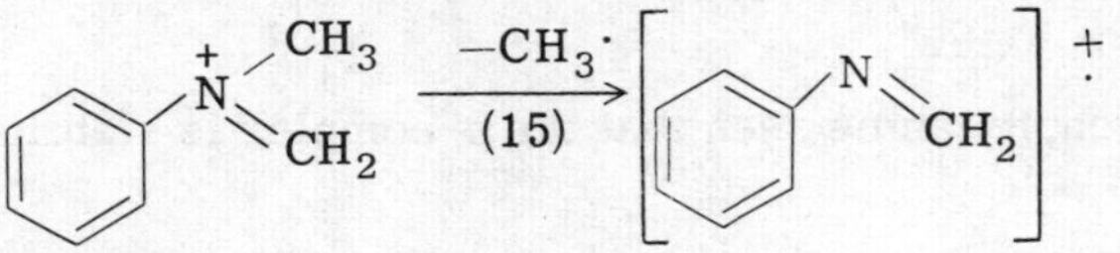

m/e 120 m/e 105

m/e 77 = 105 − 28 = 105 − H_2CN

We are led to suggest this path because m/e 77 ($C_6H_5^+$) is a typical ion of the fragmentation of benzene derivatives, and this is confirmed by the formation of the ion m/e 51 (see *Text*, Ch. 20).

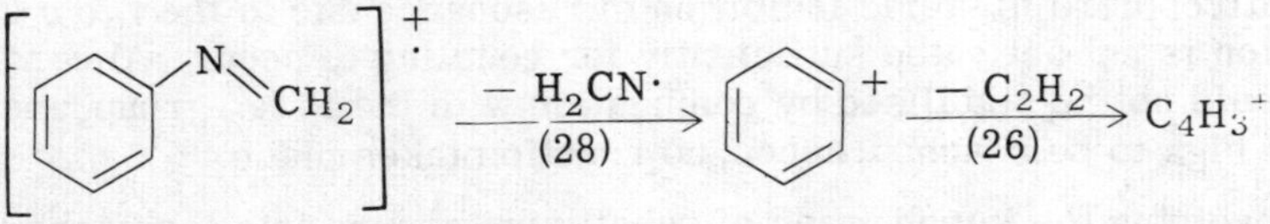

m/e 105 m/e 77 m/e 51

24 Diazonium salts and their related compounds

(Problems on page 44)

1. The diazonium cation is an electrophile and hence, from the charge-distribution approach, can be expected to couple most readily at the site of highest electron density. $PhNMe_2$ may be represented as a resonance hybrid:

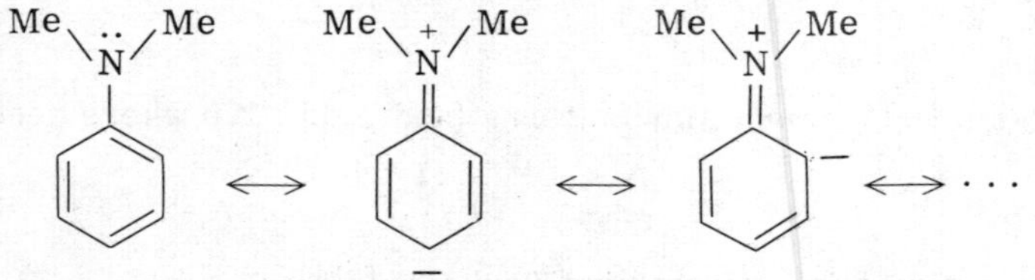

Thus, the *o*- and *p*-positions are activated (towards electrophiles), and because of steric hindrance at the *o*-positions, coupling occurs at the *p*-position.

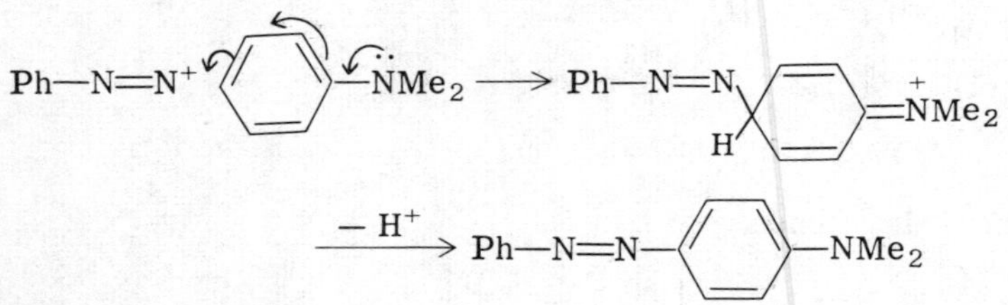

From the carbonium-ion stability approach, it can be seen that the σ-complex is stabilised by conjugation with the NMe_2 group.

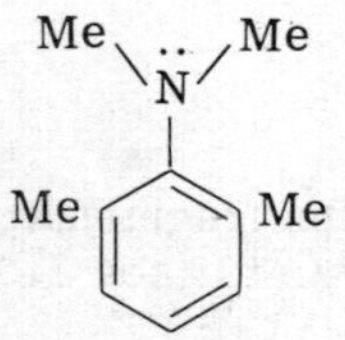

In the 2, 6-dimethyl-derivative, there is steric inhibition of resonance due to the two *o*-Me groups. Hence the *p*-position is not activated sufficiently for coupling to occur. Alternatively, the *p*- σ-complex would not be stabilised by conjugation with the NMe_2 group, and so its activation energy is too high to be reached; hence, no reaction takes place.

2. Whenever syntheses involving the introduction of substituent groups into the benzene ring are required, the direct method should always be tried first (see Ch. 22, Prob. 3).

(i)

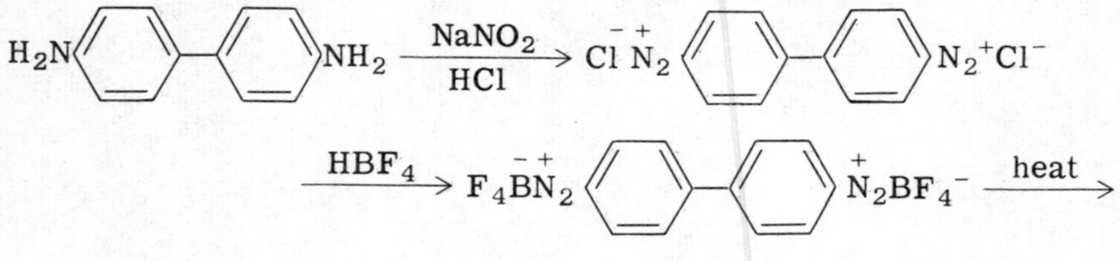

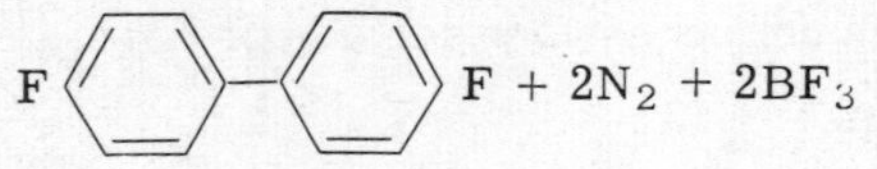

Alternative methods are available (see *Text*).

(ii)

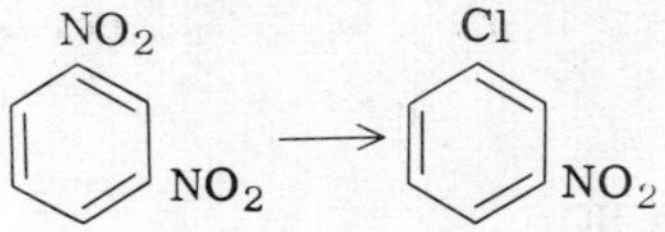

Since only *one* NO_2 group is replaced by Cl (via the diazonium salt), only *one* NO_2 group must be reduced. One method is (via the Sandmeyer reaction):

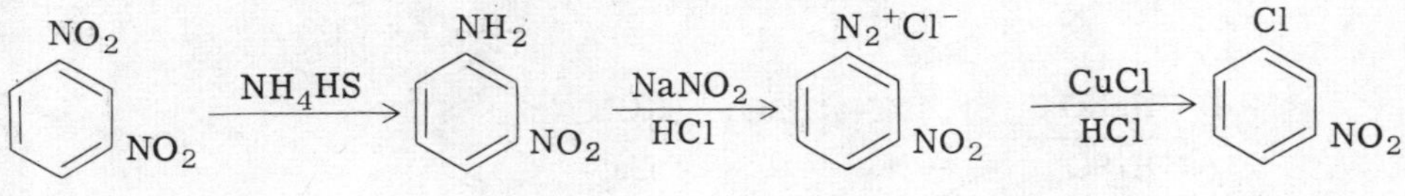

(iii)

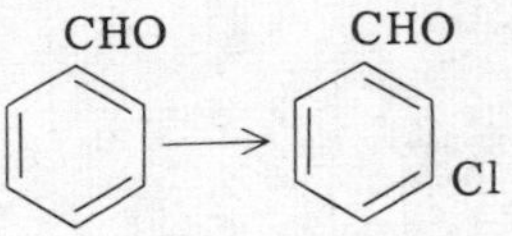

The formyl group (CHO) is *m*-directing, but direct chlorination results in the formation of benzoyl chloride (PhCOCl; see *Text*, Ch. 28). Hence, an indirect method is used. It is worth noting that in this case, nitration gives a fair amount of *o*-derivative (see *Text*, Ch. 27).

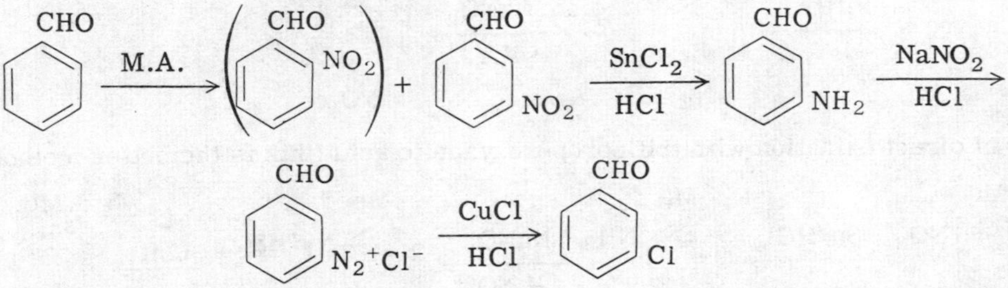

It can be seen that the method requires selective reduction of the NO_2 group. Alkaline ferrous sulphate may also be used for this reduction.

(iv)

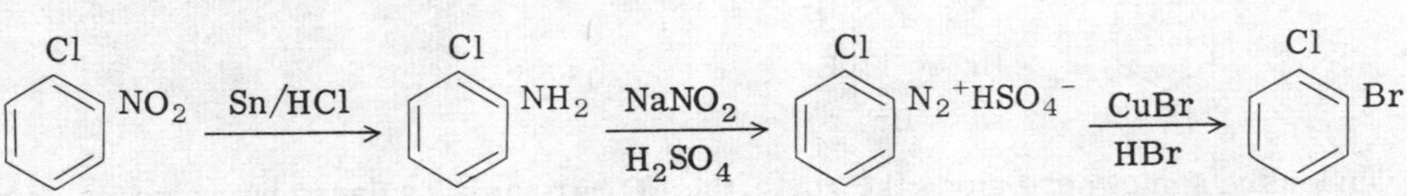

$NaNO_2/HBr$ can be used instead of $NaNO_2/H_2SO_4$, but there is very little improvement in the yield.

(v)

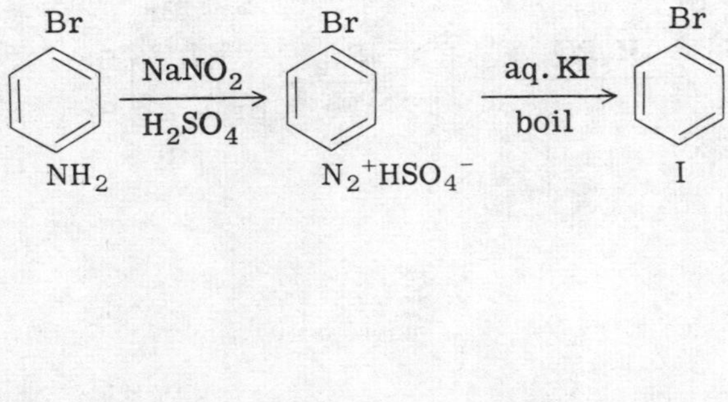

Note the use of KI alone (i.e., CuI is not necessary; see *Text*).

(vi)

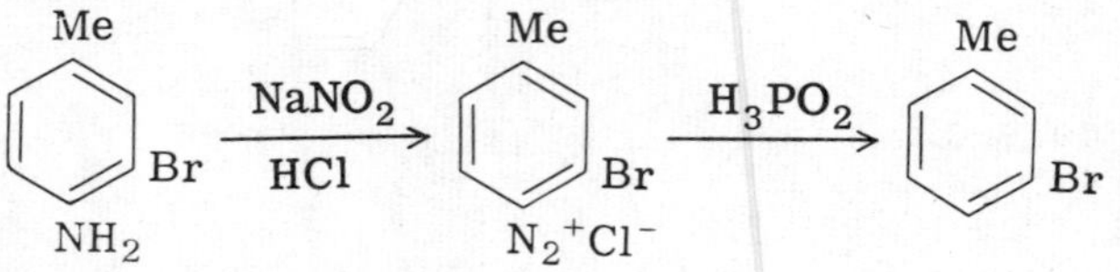

Other methods are available (see *Text*).

(vii)

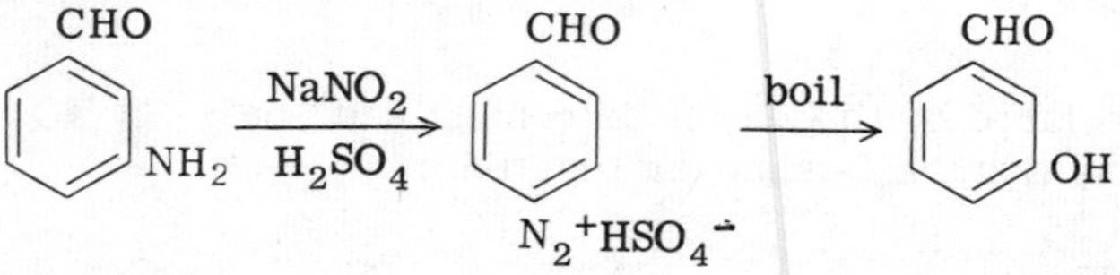

(viii)

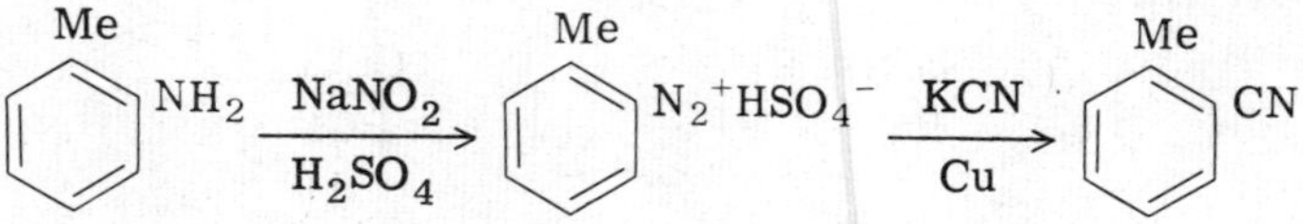

Other methods are available (see *Text*).

(ix)

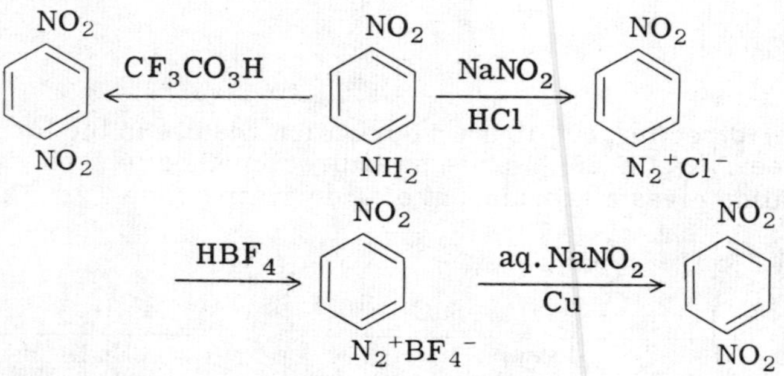

Note the use of direct oxidation with trifluoroperoxyacetic acid (this is the better method).

(x)

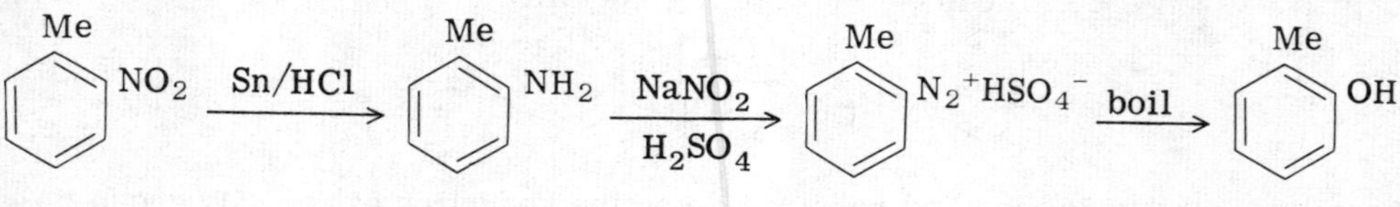

(xi)

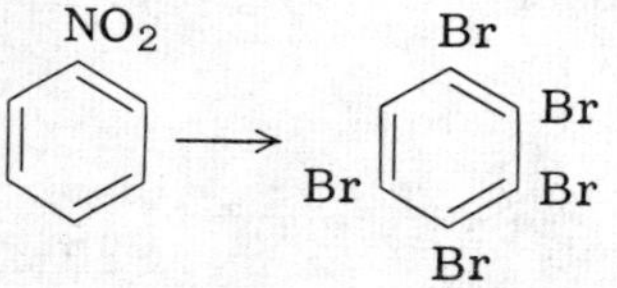

Obviously, somewhere along the route, the NO_2 group is replaced by hydrogen or by bromine. The continued bromination of benzene can lead to pentabromobenzene (iron catalyst). Hence, one possibility is:

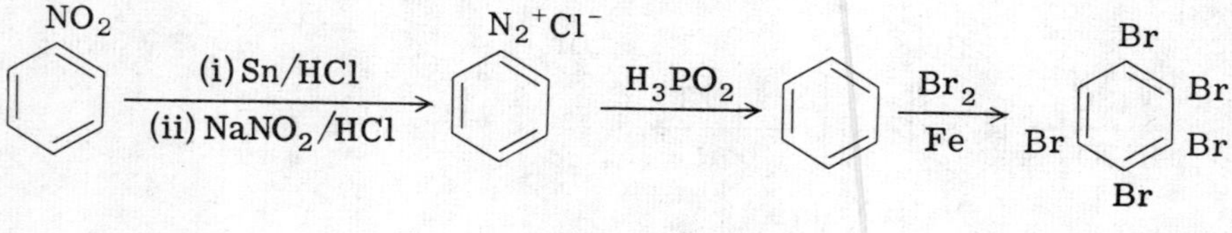

Although the product can be obtained, the process is difficult because the introduction of each successive bromine atom (1 → 2 → 3 → 4) deactivates the ring more and more. Because of this problem of deactivation, a synthesis involving the *complete* bromination of nitrobenzene and then removal of the nitro-group (via the diazonium salt) is more difficult than the previous one. All this leads to the idea that if we could *activate* the ring somewhere along the route, we might perhaps achieve a practicable synthesis. This immediately suggests the amino-group, whose presence will facilitate the introduction of three bromine atoms (*o*, *o*, *p*). Thus:

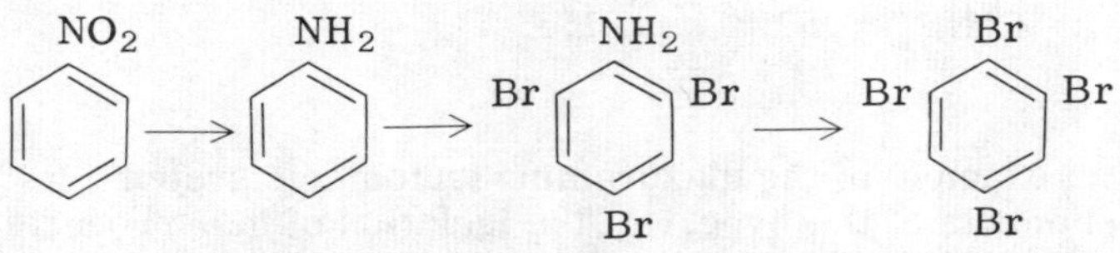

If we now replace NH_2 by Br, we obtain 1, 2, 3, 5-tetrabromobenzene. We would now have to introduced the fifth bromine atom directly. Since this brings us back to the original difficulty, we must try again. Inspection of the tetrabromo-compound shows that the 'missing' (fifth) Br is *meta* to the original NO_2 group. This suggests that we might try introducing *this* Br into the *nitro-compound* and then proceeding as before, i.e.,

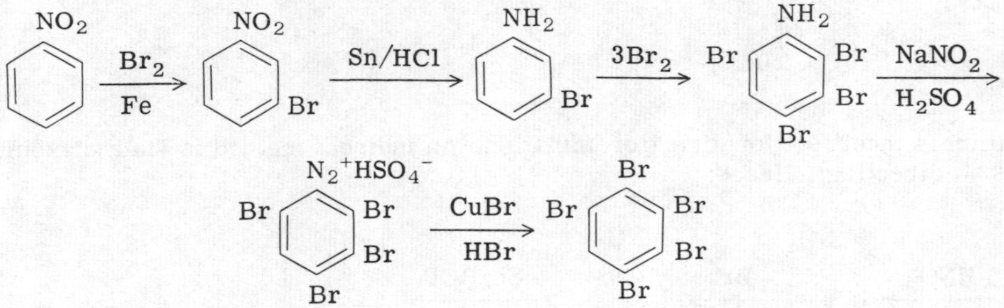

3. Since diazonium salts can couple with aromatic primary amines to form diazoamino-compounds, we may write the equation as follows:

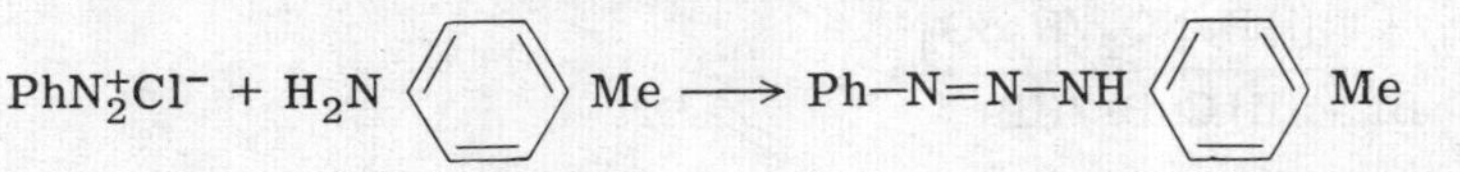

The product, on boiling with sulphuric acid, would be expected to give two products, phenol and *p*-toluidine (via reversal of the formation of the diazoamino-compound):

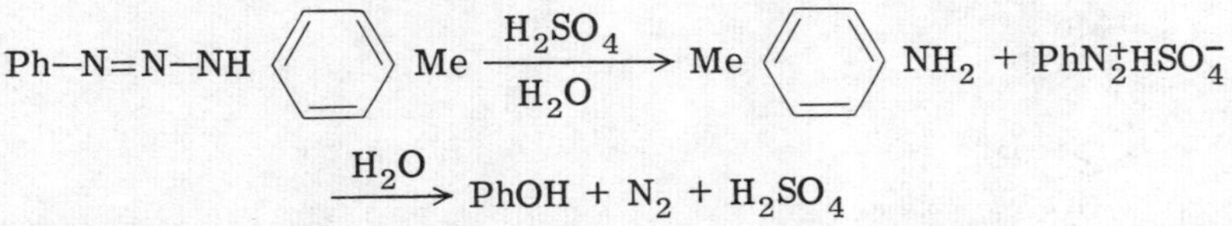

Since two other products are also obtained, it is reasonable to suppose that the breakdown of the diazoamino-compound can occur by another route (simultaneously with the first) or that the diazoamino-compound can undergo some structural change and then breaks down by the first route to give these two other products. Strictly speaking, we ought to know what these other products are in order to decide what is happening. However, we could argue as follows (but remember, we could be wrong!). Whenever a compound behaves in two (or more) different ways, there is always the possibility of a rearrangement being involved. Since a common type of rearrangement is that shown by a tautomeric system, let us try this. Inspection of the structure of the diazoamino-compound shows the most likely possibility is:

—N=N—NH— ⇌ —NH—N=N—

On this assumption, we can then explain the formation of *four* products.

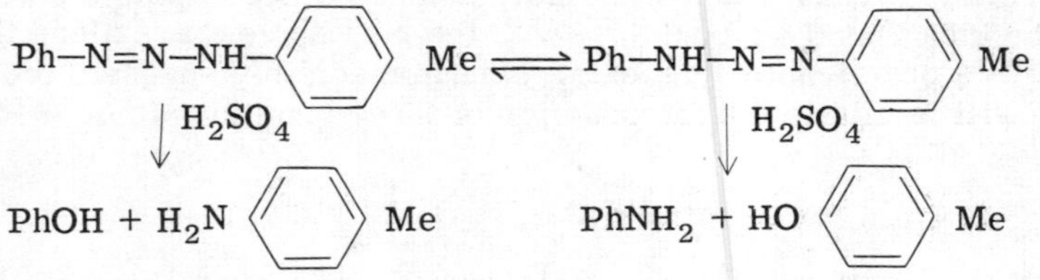

What we have been discussing is an example of the diazo-amino tautomeric system. Its *existence* was established by experiments of this type, i.e., the isolation of *four* products and their *identification*. Therefore, in discussing the question, the student was expected to know how diazoamino-compounds break down and that they are tautomeric compounds. If, however, the student had forgotten the second point, the sort of approach given above may help him to recall it (or even discover it?).

4. (i)

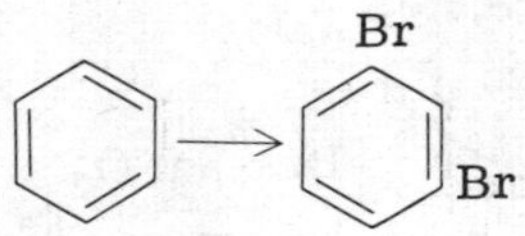

The orientation is incorrect for direct bromination. An indirect method is via a diazonium salt (NO_2 is *m*-directing). Hence:

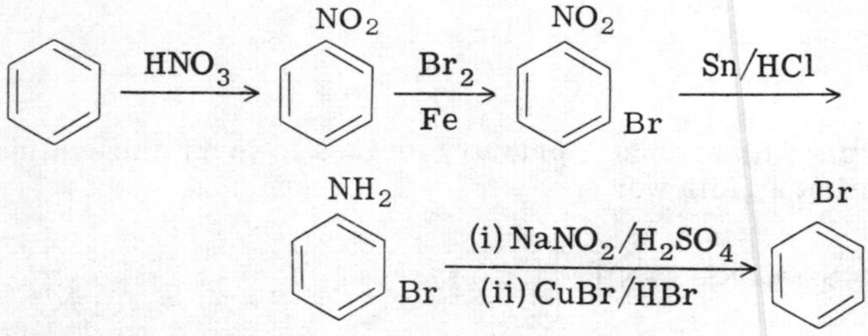

There is no point in preparing the *m*-dinitro-compound, etc. (see also (iii), below).

(ii)

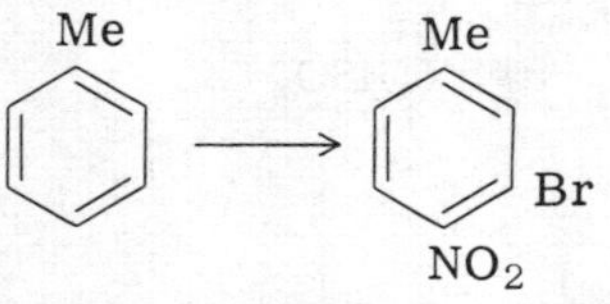

Direct introduction of the groups cannot be done. Let us try NH_2 instead of NO_2; this will give the correct orientation for Br (see also Ch. 23, Prob. 8 ii).

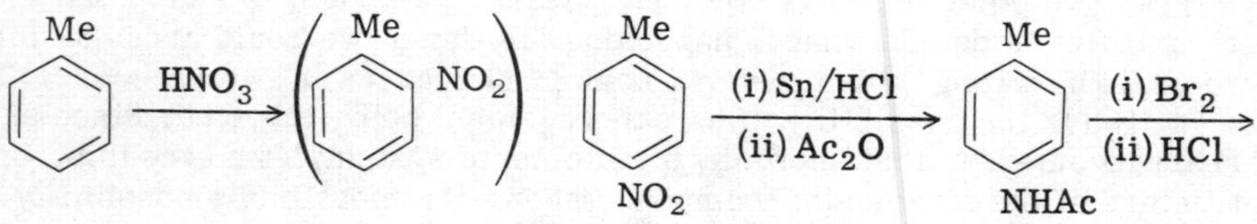

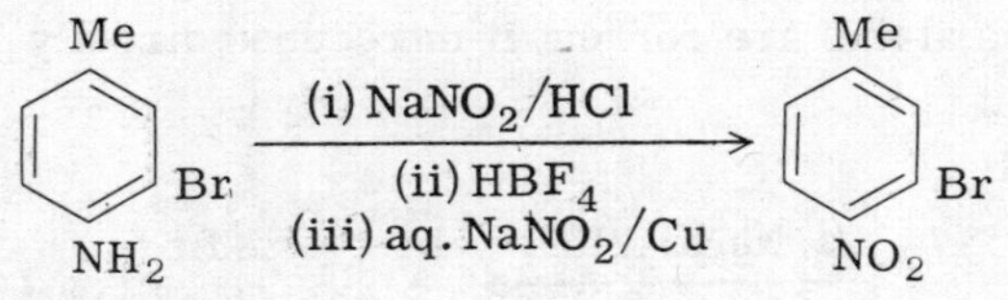

(iii)

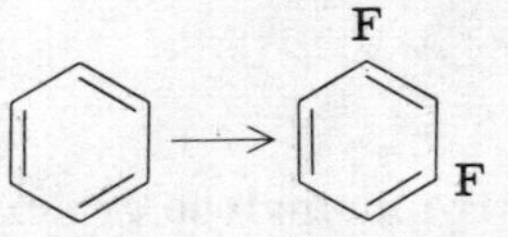

The direct introduction of fluorine is not possible. Hence, we require the *m*-dinitro-compound. However, it is best to reduce these one at a time. If both are reduced to give *m*-diaminobenzene, diazotisation of this usually results in the formation of azo-compounds (see *Text*, Ch. 23).

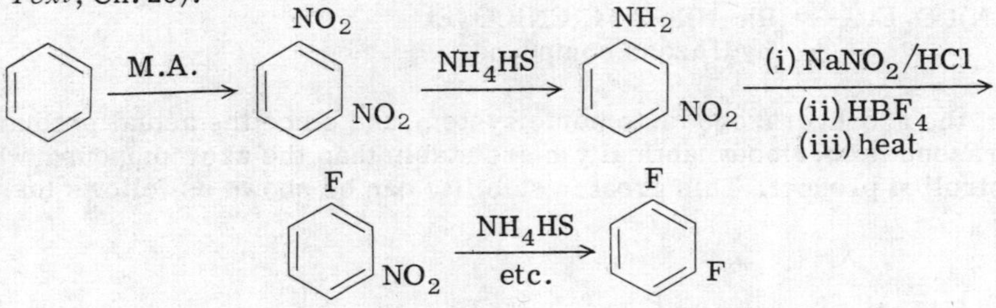

(iv)

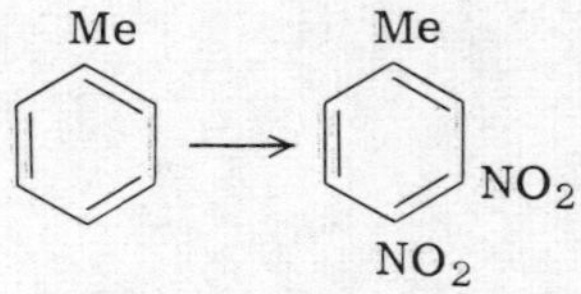

The direct introduction of the two groups in the correct orientation is not possible. Since the *p*-nitro-group can be introduced directly, a possible route is:

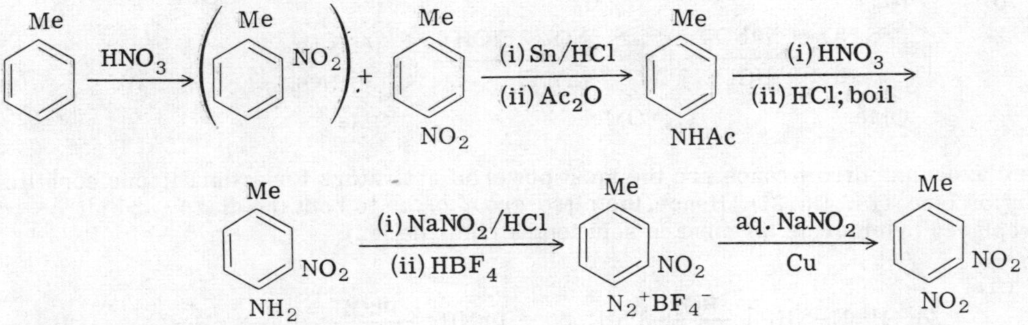

Acetylation deactivates the ring (which is very strongly activated by the amino-group) so that only *one* nitro-group is introduced. At the same time, because aromatic primary amines are very sensitive to oxidising agents, it is best to protect the amino-group.

(v)

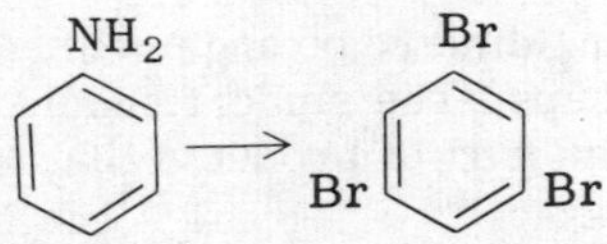

The positions of the bromine atoms are correct if introduced directly into aniline. Hence:

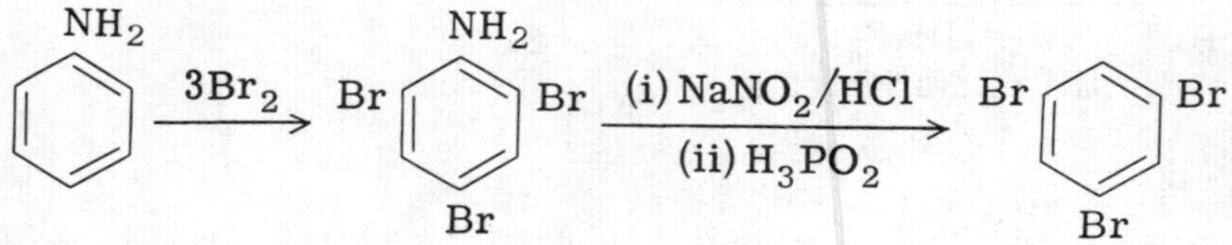

5. Since we have an electrophile and a compound with an active methylene group, we can formulate the reaction as follows:

$$CH_2(CN)CO_2Et \xrightarrow{AcO^-} AcOH + \bar{C}H(CN)CO_2Et \xrightarrow{PhN_2^+}$$

$$\underset{\text{azo-compound}}{Ph{-}N{=}N{-}CH(CN)CO_2Et} \longrightarrow \underset{\text{hydrazone compound}}{Ph{-}NH{-}N{=}C(CN)CO_2Et}$$

This is an example of the azo-hydrazone tautomeric system, and since the actual product is the latter, the hydrazone is thermodynamically more stable than the azo-compound, which is the kinetically controlled product. This greater stability can be shown as follows (using Table 2. 2).

$$-N{=}N{-}\overset{|}{\underset{|}{C}}{-}H \rightleftharpoons -\overset{H}{\overset{|}{N}}{-}N{=}\overset{|}{\underset{|}{C}}$$

+418·4 (N=N)	−389·1 (N—H)
+284·5 (N—C)	−163·2 (N—N)
+414·2 (C—H)	−615·1 (N=C)

$\Delta H^{\ominus} = -50{\cdot}3$ kJ, and equating this with $\Delta G^{\ominus}$, we can see that the conversion of the azo-compound into the hydrazone is thermodynamically favourable (see also Ch. 20, Prob. 18).

6. (i)

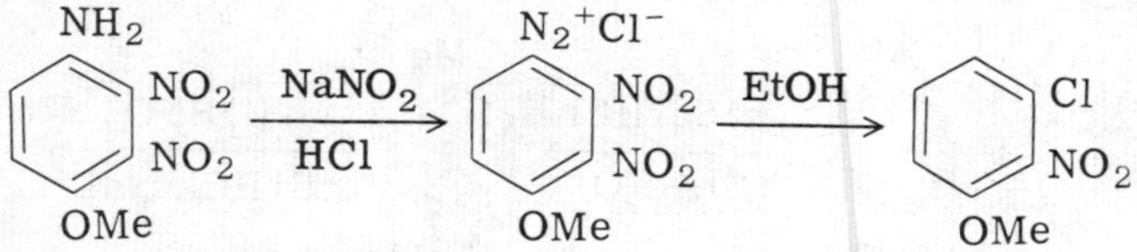

The diazo- and nitro-groups are the most powerful activators for aromatic nucleophilic substitution (see *Text*, Ch. 21). Hence, the nitro-group *ortho* to both the diazo- and nitro-groups is replaced by chlorine (present in solution as chloride ion).

(ii)

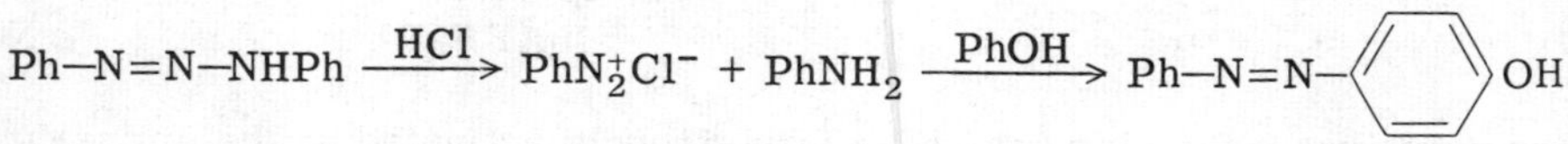

Phenol, in strongly acid solution, is more reactive towards diazonium salts than is aniline (which is in the form of $PhNH_3^+$, and so the ring is deactivated).

(iii) Questions (iii)-(viii) are all examples of the benzidine rearrangement. The nature of the product depends on the nature of the substituent groups. The student should not try to learn them by heart; the object of the exercise is to see the sort of products that are obtained (however, (viii) is important).

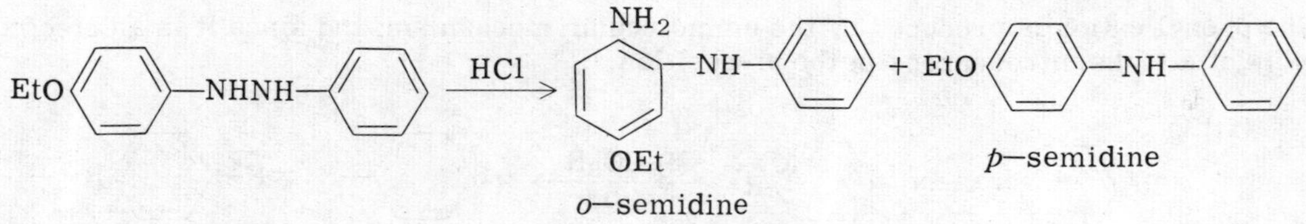

(iv)

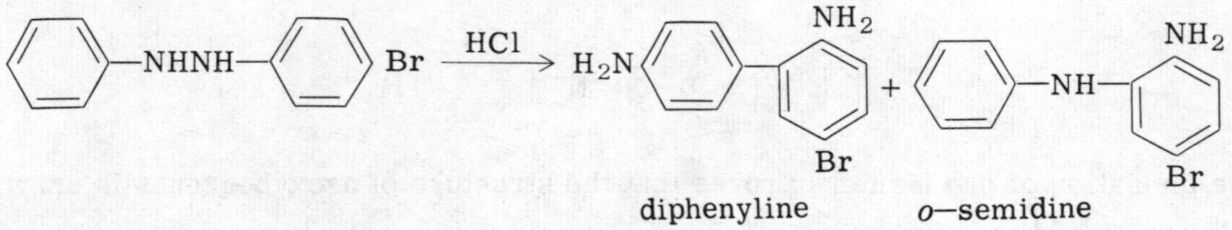

(v)

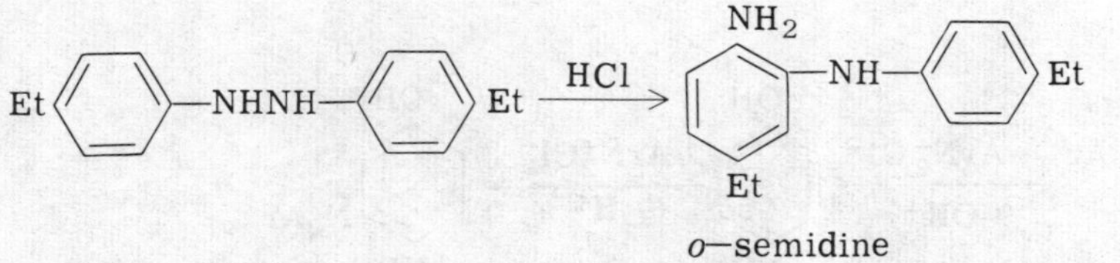

(vi)

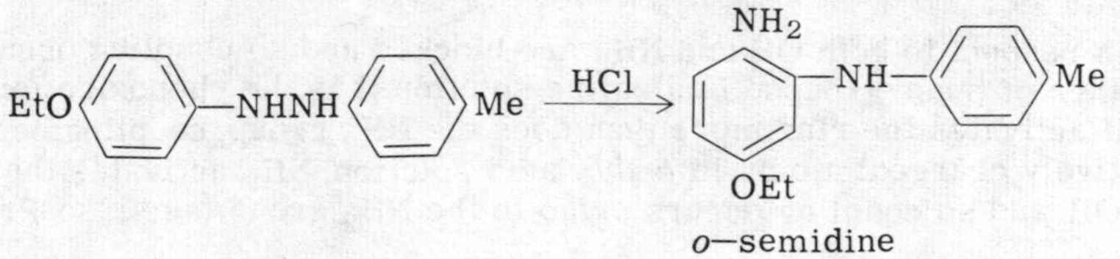

(vii)

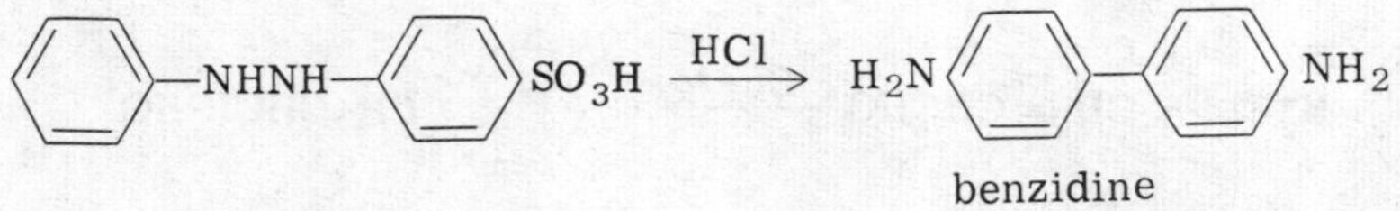

A sulphonic acid- or carboxyl-group in the 4-position is eliminated in the rearrangement.

(viii)

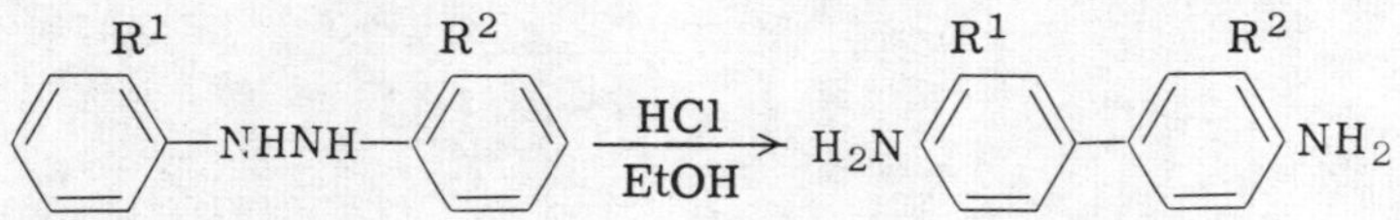

(I) + (II)
(I): R^1=R^2=OMe
(II): R^1=R^2=OEt

(III) + (IV)
(III): R^1=R^2-OMe
(IV): R^1=R^2=OEt

Each hydrazo-compound rearranges separately, and since no crossed products are obtained, the rearrangement is intramolecular.

(ix)

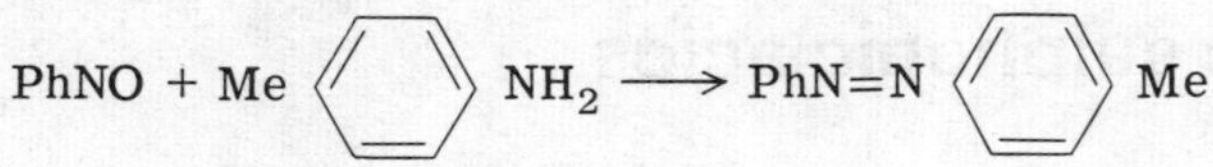

(x)

$$PhN_2^+BF_4^- \xrightarrow{\text{heat}} Ph^+ + N_2 + BF_4^-$$

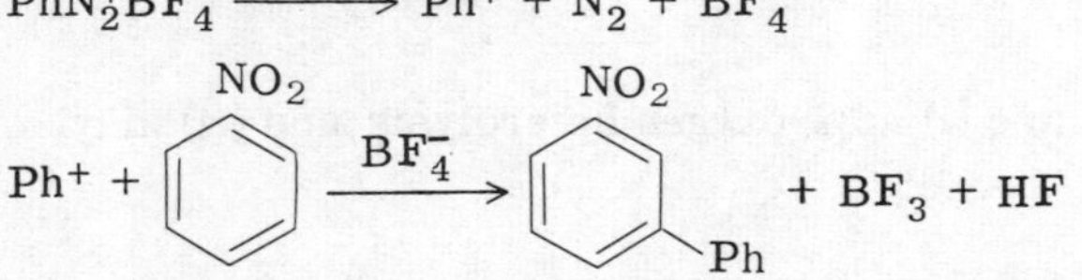

The phenyl cation is produced by the unimolecular mechanism, and since it is an *electrophile*, it attacks nitrobenzene in the *m*-position.

(xi)

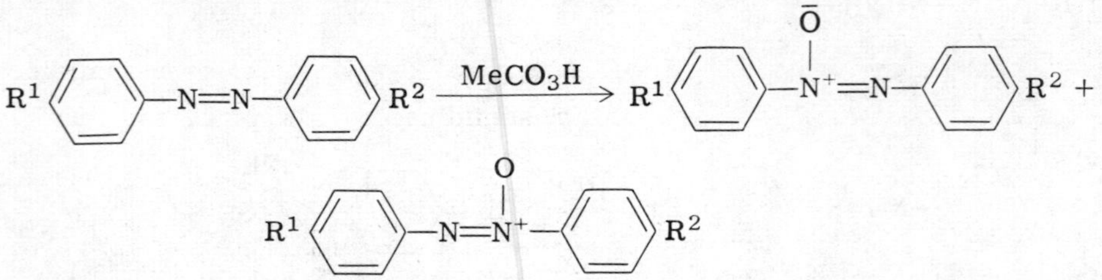

The formation of *two* isomers proves that the structure of azoxybenzenes is unsymmetrical (hence *not* —N—N— with the O bridging both N atoms).

(xii)

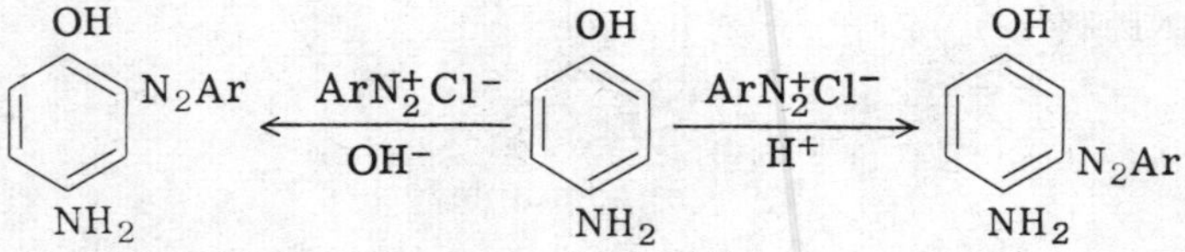

The *p*-positions with respect to both OH and NH_2 are blocked and so coupling occurs in the position *ortho* to either of these groups. In alkaline solution it is the phenoxide ion that is present, and since it activates the ring more than does the NH_2 group, coupling occurs *ortho* to the oxygen (negatively charged) atom. In *feebly* acid solution, NH_2 activates the ring more strongly than does OH, and so coupling occurs *ortho* to the NH_2 group (see also Prob. (ii), above).

(xiii)

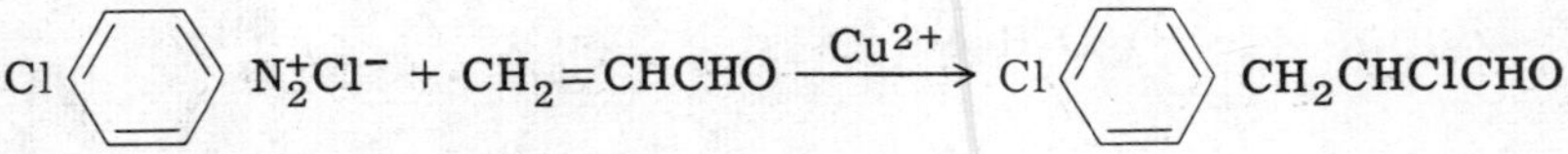

This is an example of the Meerwein reaction (see *Text*).

25 Aromatic sulphonic acids

(Problems on page 46)

1. The two possible mechanisms are (i) acyl-oxygen heterolysis, and (ii) alkyl-oxygen heterolysis (see Table 9.2).

(i) $B_{AC}2$

$$ArSO_2{-}OR + \overline{O}H \longrightarrow ArSO_2OH + RO^- \longrightarrow ArSO_2O^- + ROH$$

The $B_{AC}1$ mechanism has not been observed.

(ii) $\boldsymbol{B_{AL}2}$ $\quad ArSO_2O{-}R\ \overline{O}H \longrightarrow ArSO_2\overline{O} + ROH$

$\boldsymbol{B_{AL}1}$ $\quad ArSO_2O{-}R \rightleftharpoons ArSO_2\overline{O} + R^+ \xrightarrow[-H^+]{H_2O} ROH$

To ascertain whether mechanism (i) or (ii) is operating, we can use H_2O labelled with ^{18}O. If acyl-oxygen heterolysis, (i), is the mechanism, the alcohol will *not* contain ^{18}O. On the other hand, if alkyl-oxygen heterolysis, (ii), is the mechanism, the alcohol *will* contain ^{18}O. This method, it should be noted, does not distinguish between $B_{AL}2$ and $B_{AL}1$ (kinetic studies can be used for this purpose).

2. (i)

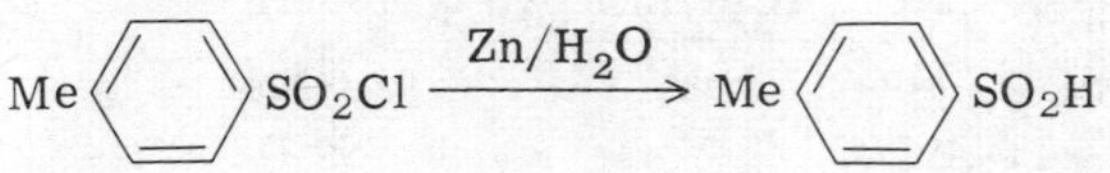

Other reducing agents may be used (see *Text*).

(ii)

$$PhSO_2Cl \xrightarrow[or\ LAH]{Zn/H_2SO_4} PhSH$$

(iii)

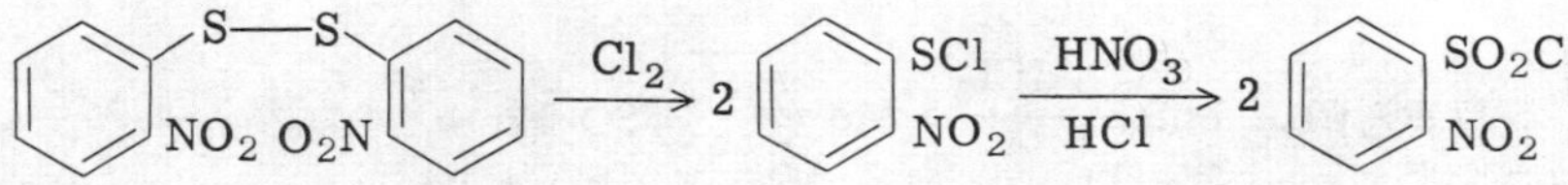

(iv)

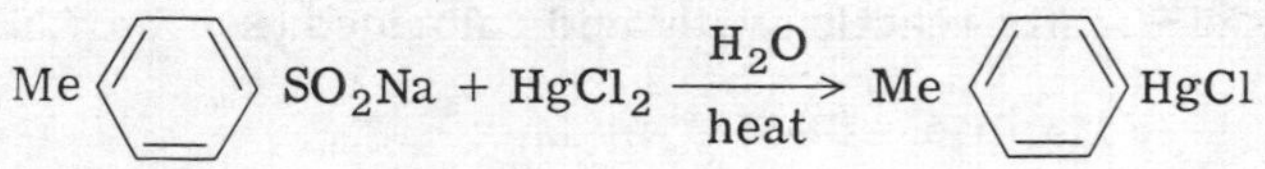

(v)

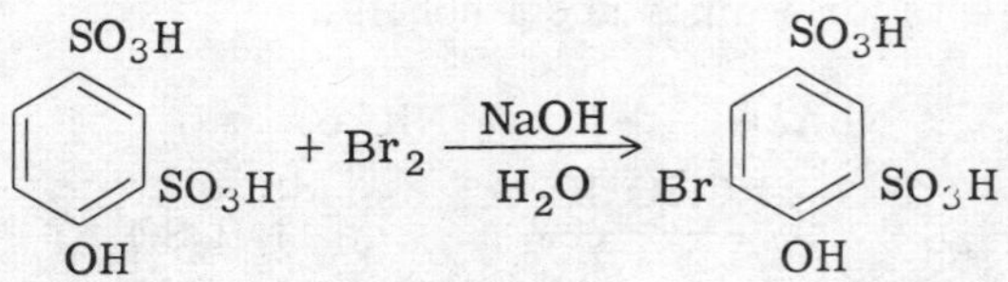

(vi)

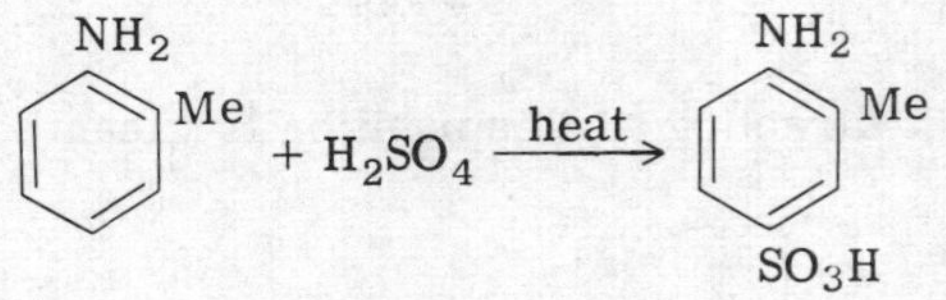

It should be noted that since the medium is strongly acid, the *m*-acid (3-amino-4-methylbenzenesulphonic acid) might have been expected (NH_3^+ is *m*-directing; Me is *o*/*p*-directing). See *Text* for a possible mechanism.

(vii)

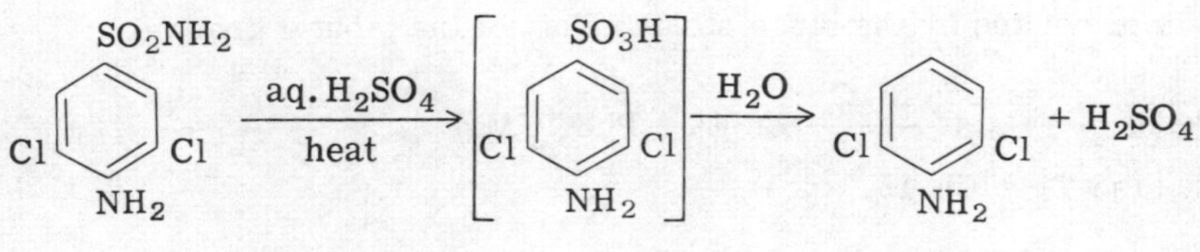

(viii)

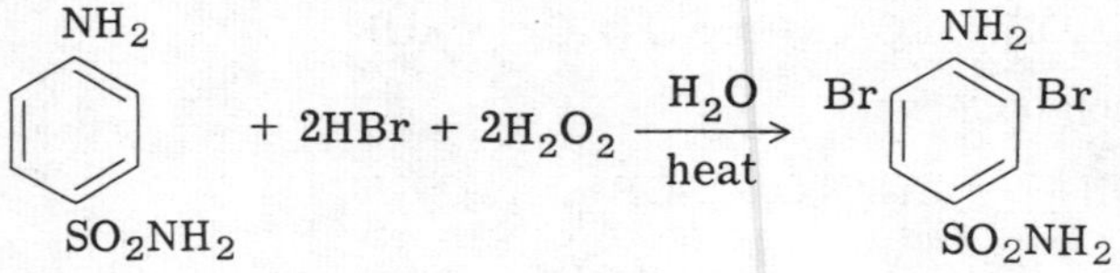

A mixture of hydrobromic acid and hydrogen peroxide is a convenient source of bromine:

$$2HBr + H_2O_2 \longrightarrow Br_2 + 2H_2O$$

When substitution occurs, HBr is also generated, and this is again oxidised by hydrogen peroxide.

(ix)

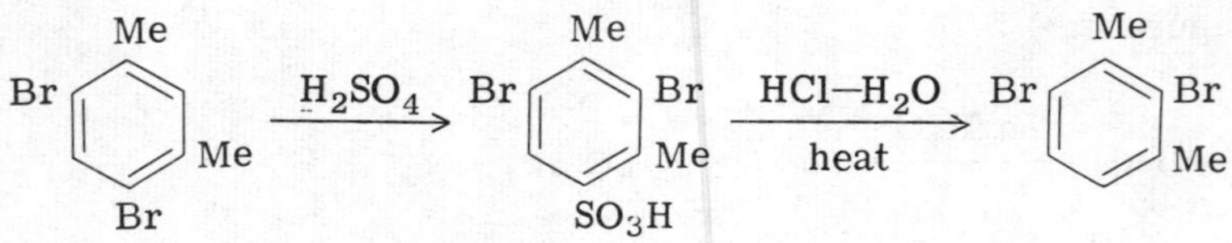

This is an example of the Jacobsen rearrangement (see *Text*).

(x) $PhSO_3^-Na^+ + PCl_5 \longrightarrow PhSO_2Cl + NaCl + POCl_3$

This should be compared with the analogous reaction involving the salt of a carboxylic acid (see *Text*, Ch. 9).

(xi)

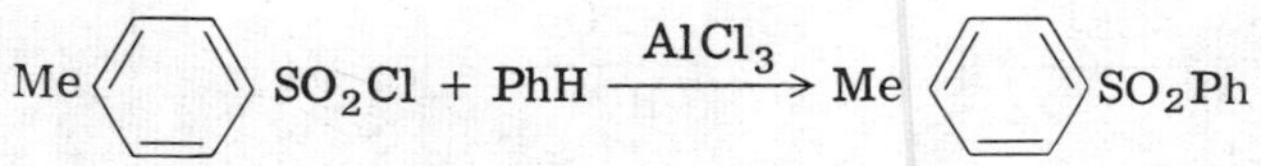

This is an example of the Friedel–Crafts reaction with acid chlorides (see *Text*, Ch. 20).

(xii) $PhSO_3Et + 2MeNH_2 \longrightarrow MeNHEt + PhSO_3^-\overset{+}{N}H_3Me$

Sulphonic esters are good alkylating agents for amines and alcohols.

(xiii)

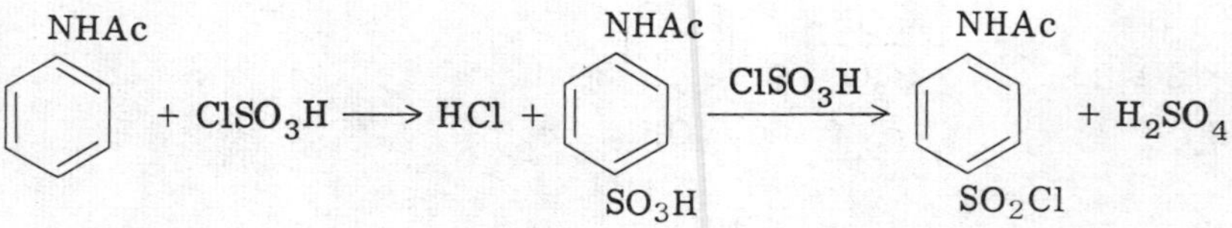

The acetamido-group is almost exclusively *p*-directing (if this position is vacant).

(xiv)

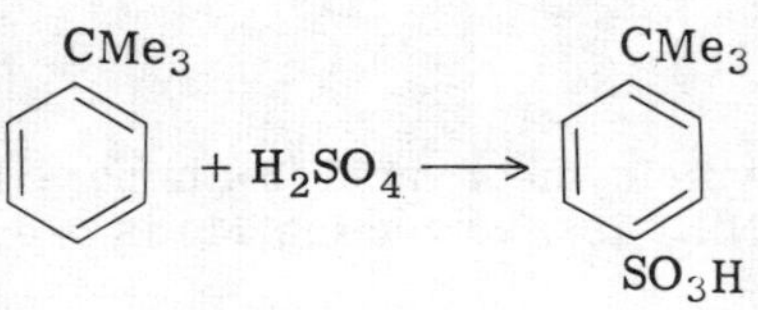

o-Substitution is prevented by the large steric effect of the t-butyl group.

(xv) $PhSO_3Et + EtMgI \xrightarrow{Et_2O} Et\text{-}Et + PhSO_3^-\overset{+}{M}gI$

See (xii), above; also *Text*, Ch. 15.

3.

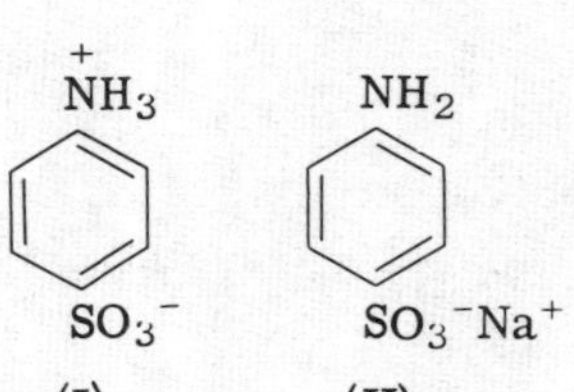

(I) (II)

The free acid exists as the dipolar ion (I). Because there is not an available lone pair of electrons on the nitrogen atom (to donate to the carbon atom of a CO group in Ac_2O), it is not possible to acetylate (I). On the other hand, in the sodium salt (II), the lone pair is available and so (II) is readily acetylated.

4. By analogy with the Friedel-Crafts reaction involving carbonyl chlorides (RCOCl), we may propose the following mechanism (see *Text*, Ch. 20):

(i) $$Cl{-}SO_2{-}Cl + AlCl_3 \rightleftharpoons Cl{-}SO_2{-}\overset{+}{Cl}{-}\overset{-}{Al}Cl_3 \rightleftharpoons \overset{+}{S}O_2Cl + AlCl_4^-$$

(ii) $$ArH + \overset{+}{S}O_2Cl \rightleftharpoons Ar^{+}\langle^{H}_{SO_2Cl} \xrightarrow{AlCl_4^-} ArSO_2Cl + HCl + AlCl_3$$

5. Bis (4-aminophenyl) sulphone is

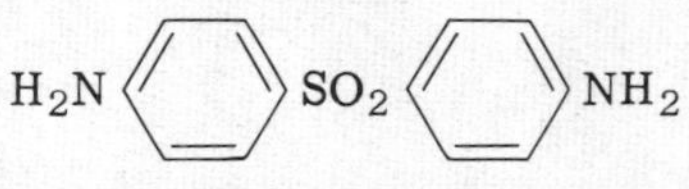

Sulphones may be prepared by the oxidation of diaryl sulphides, and since chlorine in *p*-chloronitrobenzene is activated, this suggests that the first step in the synthesis could be:

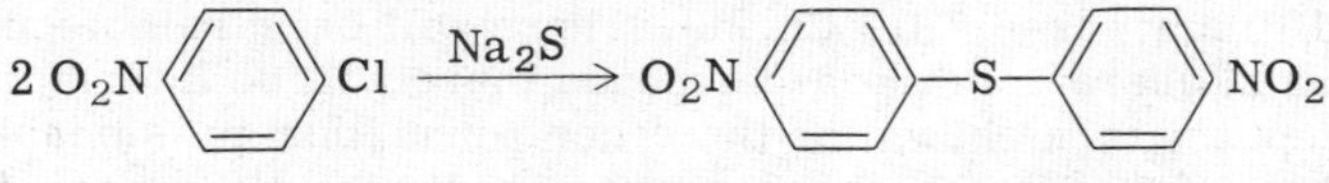

The question now is: Do we oxidise to the sulphone first (with hydrogen peroxide) and then reduce the nitro-group to amino, or vice versa? Sulphones are resistant to the action of reducing agents (see *Text*, Ch. 14), but there is the possibility that *o*—H, because it is activated by the nitro-group, may be displaced by oxidation to form a phenol (see *Text*, Ch. 22). Hence, it would appear better to reduce first and then oxidise. If, however, we do this we must protect the amino-groups, otherwise the rings will be oxidised. Protection is readily achieved by acetylation and this is satisfactory because the hydrogen peroxide is used in acetic acid (no hydrolysis will occur). Hence:

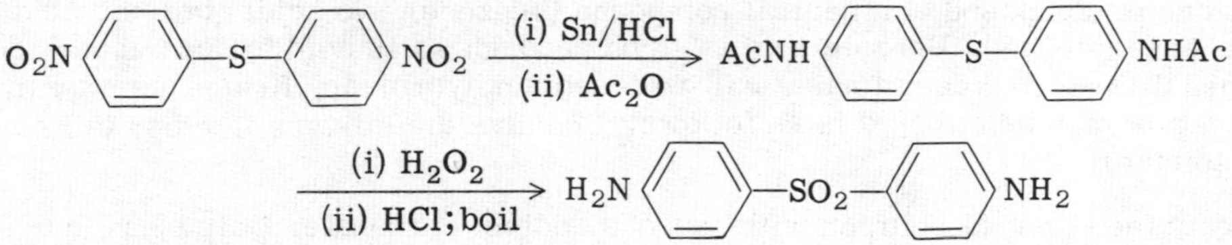

26 Phenols and quinones

(Problems on page 47)

1.

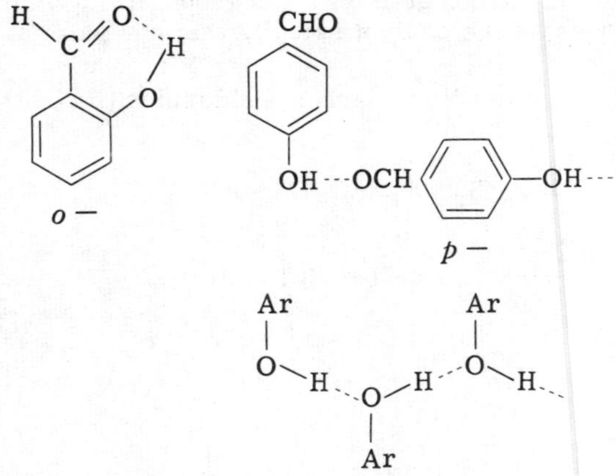

o — *p* — *p* —

The benzene disubstitution patterns are different, one showing the *o*-pattern (770-735 cm^{-1}) and the other the *p*-pattern (840-810 cm^{-1}; see *Text*, Ch. 20). There will be intramolecular hydrogen bonding in the *o*-compound, whereas in the *p*-compound there will be intermolecular hydrogen bonding. If we accept the type shown for the *o*-compound, then presumably this is also possible for intermolecular bonding in the *p*-compound. On the other hand, the more usual type of hydrogen bonding between hydroxyl groups is possible for the *p*-compound, and this is probably the one that predominates (or may be exclusive). Whichever type of hydrogen-bonding operates in the *p*-compound, its i.r. range will be different from that of the intramolecularly hydrogen bonded *o*-compound. Also, in solution, the position of the bond in the *o*-compound is little affected on dilution, whereas that in the *p*-compound moves progressively to the higher frequencies due to decreasing association (see also Prob. 5, below).

2. The mixture is composed of a neutral compound (benzene), two acids (benzoic acid and phenol), and a base (aniline). Benzoic acid is strong enough as an acid to decompose sodium carbonate; phenol is not, but does form a salt with sodium hydroxide. Hence, one way in which a separation may be effected is as follows. Dissolve the mixture in ether (all components are soluble).

(i) Extract the ethereal solution with aqueous sodium carbonate. Only benzoic acid enters the aqueous layer, and may be recovered on acidification.

(ii) Extract the ethereal solution from (i) with aqueous sodium hydroxide. Only phenol enters the aqueous layer, and may be recovered on acidification.

(iii) Extract the ethereal solution from (ii) with hydrochloric acid. Only aniline enters the aqueous layer, and may be recovered on basifying the solution.

(iv) Evaporate off the ether from the ethereal solution from (iii). This leaves benzene.

3. The strength of an acid depends on the ease with which the compound can transfer a proton to a base (the proton acceptor). The strength may be measured by determining the position of equilibrium when the acid is dissolved in water (the base), and the more the equilibrium lies to the right (in the following equations), the stronger is the acid (see *Text*, Ch. 2).

(i) $ROH + H_2O \rightleftharpoons RO^- + H_3O^+$

(ii) $ArOH + H_2O \rightleftharpoons Ar\bar{O} + H_3O^+$

In alcohols, only the +I effect of the alkyl group can operate, and it is this which decides the strength of the alcohol as an acid.

$R \rightarrow O \rightarrow H$

The stronger the +I effect, the more difficult it is for the hydrogen atom to be transferred as a proton, i.e., the weaker is the alcohol as an acid (see *Text*, Ch. 6). Furthermore, since *any* alkyl group has a +I effect with respect to hydrogen, alcohols are weaker acids than water. Thus, equilibrium (i) lies completely to the left.

When phenols ionise, the phenoxide ion produced is stabilised by resonance (only a negative charge is spread) more than is the phenol itself (both a positive and a negative charge are now present).

$Ar{-}OH \longleftrightarrow \overline{Ar}{=}\overset{+}{O}H; \quad Ar{-}\bar{O} \longleftrightarrow \overline{Ar}{=}O$

Thus, since the phenoxide ion is more stable than the un-ionised phenol, this is the driving force that shifts the equilibrium towards the right in (ii). Hence, phenols are stronger acids than alcohols. (It might be noted that resonance stabilisation is not possible for either the alcohol or its alkoxide ion.)

Alcohols and phenols behave as nucleophiles by virtue of the fact that both contain a hydroxyl group, the oxygen atom of which is the donor atom (of a pair of electrons). The more this lone pair is available for donation, the greater is the nucleophilic power. Because alcohols are not resonance hybrids, a lone pair on the oxygen atom is readily available. On the other hand, because phenols are resonance hybrids, the lone pair on the oxygen atom is now far less available, and the oxygen atom carries a positive charge (the latter discourages attack at electrophilic centres). Hence, alcohols are stronger nucleophiles than phenols.

4. (i)

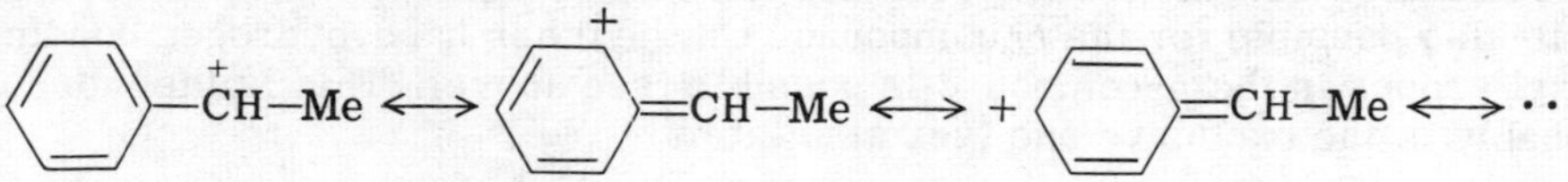

(I)

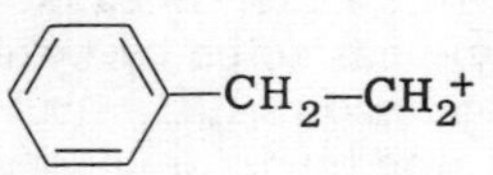

(II)

(I) can be represented as a resonance hybrid, because conjugation with the ring is possible. On the other hand, (II) cannot be represented as a resonance hybrid. Hence, (I) is more stable than (II). This is in keeping with the general rule that secondary carbonium ions are more stable than primary.

(ii)

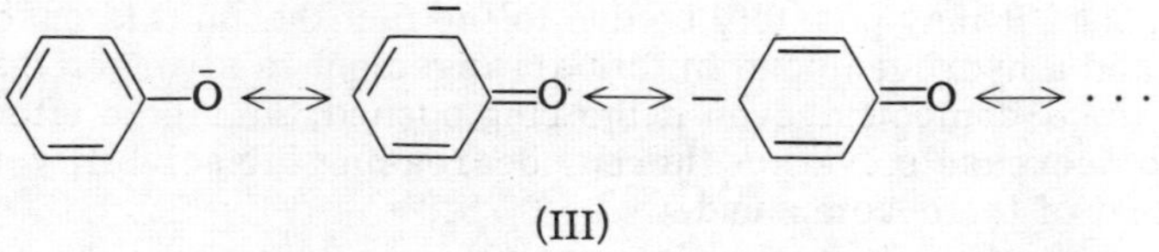

(III)

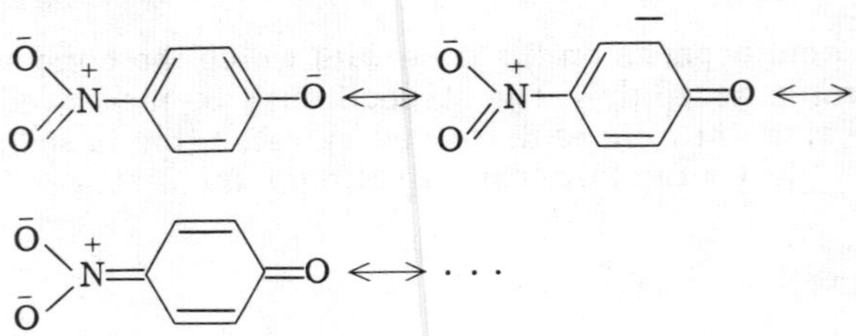

(IV)

(III) can be represented as a resonance hybrid; some of the resonating structures have carbon atoms carrying a negative charge. (IV) is also a resonance hybrid, but now there is extended conjugation and there are also contributing structures in which the negative charge resides on an oxygen atom (of the nitro-group). Since oxygen is much more electronegative than carbon, this also makes these structures more stable than those for (III). Hence, (IV) is much more stable than (III).

5.

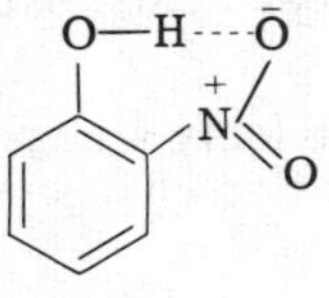

o

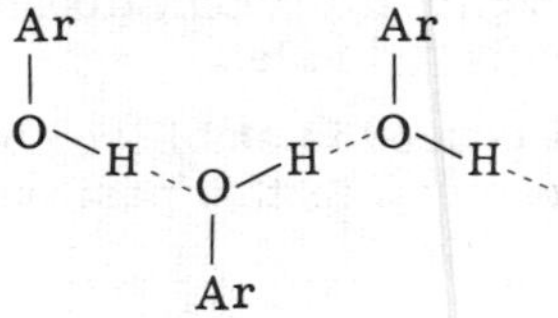

p

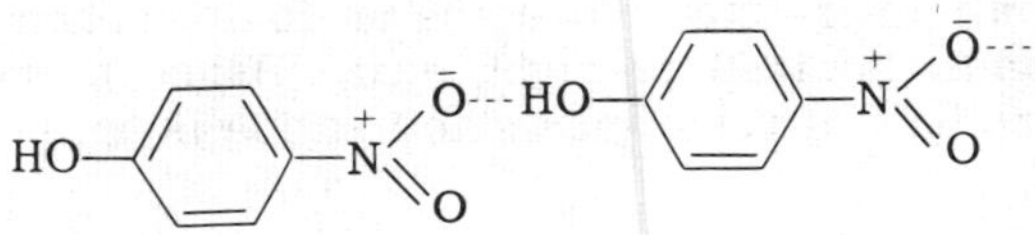

p

In the *o*-compound, the hydrogen bonding is intramolecular, and so the molecule behaves as a 'single unit'. Hydrogen bonding in the *p*-compound is intermolecular. Thus, we are dealing with associated molecules, which behave as a large molecule with high molecular weight, thereby giving rise to a high boiling point (much higher than that of the monomer). Since we have assumed that the *o*-compound has the type of hydrogen bonding shown, this presumably is also possible for the *p*-compound. On the other hand, hydrogen bonding between hydroxyl groups in the *p*-compound is possible (see above). This is the more usual type and probably is the exclusive one (see also later).

Solubility depends on a number of factors, one being the ability of the solute to form hydrogen bonds with the solvent, which is water in this case. Because the intramolecular hydrogen bonding in the *o*-compound is (apparently) unaffected by water, little hydrogen bonding occurs between the two, and so the *o*-compound shows little tendency to dissolve. On the other hand, the *p*-compound in contact with water tends to undergo disassociation into the monomeric molecules and these can now form hydrogen bonds with the water, thereby achieving a degree of solubility.

Nitrobenzene is almost insoluble in water. Hence, on the basis of the argument given, it would appear that the nitro-group does not form hydrogen bonds with water. This would mean that the intermolecular hydrogen bonding for the *p*-compound must be between hydroxyl groups. It also means that if the hydrogen bonding drawn for the *o*-compound is correct, it must be very strong. It may possibly be due to the fact that in this case a six-membered ring is formed, and a hydrogen bond in this arrangement is stronger than one in an open chain. Also, for the *o*-compound, even if the ring opened, the steric effect of the *o*-nitro-group would tend to decrease solvation, thereby decreasing the solubility of the *o*-compound with respect to that of the *p*-compound.

6.

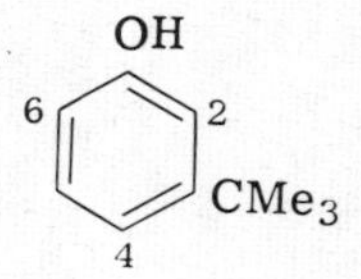

Positions 2,4 and 6 are strongly activated by the hydroxyl group (and reinforced by the *o*/*p*-directing effect of the t-butyl group). Now, phenol itself readily forms the 2,4,6-trihalogeno-compound with chlorine or bromine; iodination requires the presence of, e.g., mercuric oxide. Since the introduction of a *m*-t-butyl group changes the course of the reaction (with respect to phenol), this suggests a steric effect is operating (the t-butyl group has a very large bulk). If the steric effect were due to the t-butyl group alone, then the behaviour of the halogens would be expected to follow a similar pattern. Since this pattern is not the same, there must be some property of the halogens that also plays a part. It is unlikely to be their relative reactivities (chlorine and bromine behave in the same way with phenol). A reasonable explanation is that a steric effect involving the halogens themselves is also operating. This is suggested by the fact that the number of halogen atoms introduced is in the order: I < Br < Cl, and this is the *reverse* of the order of the size of the atom, i.e.: I > Br > Cl. Hence, on the basis of two steric effects (one due to the t-butyl group and the other to the size of the halogen atom), the halogeno-compounds are:

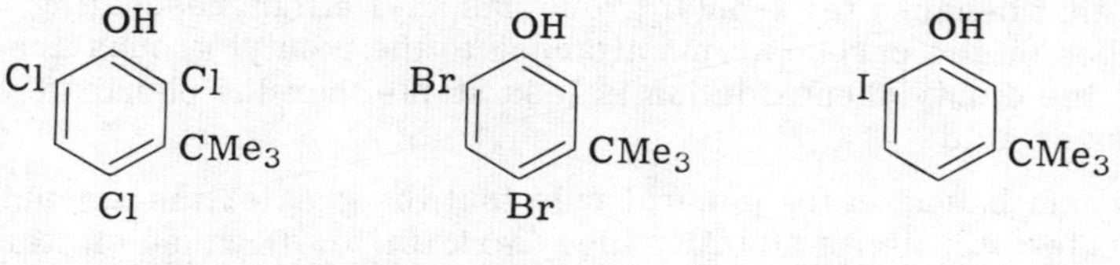

7. *Charge distribution approach.*

(*a*)

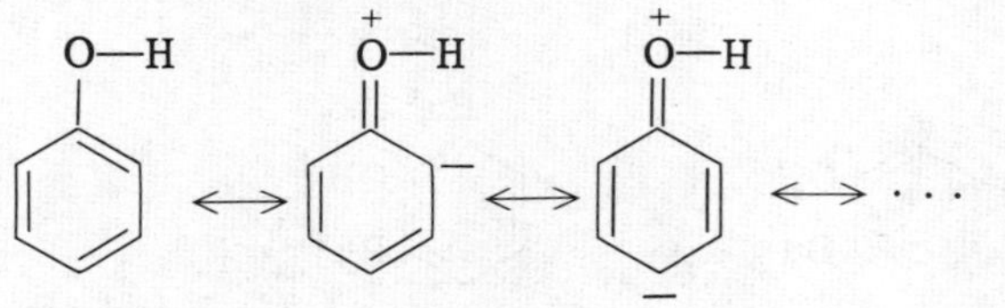

Stability of ions

(*b*)

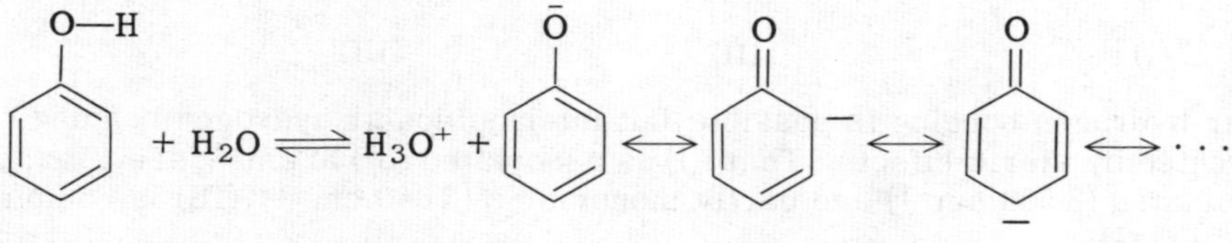

Because of the positive charge on the oxygen atom in (*a*), or because of the increased stability of the phenoxide ion with respect to the un-ionised phenol molecule, phenol has acidic property (see also Prob. 3, above).

(*c*)

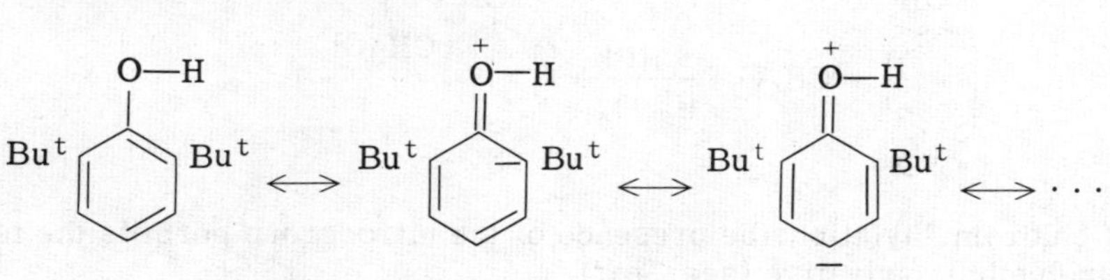

(*d*)

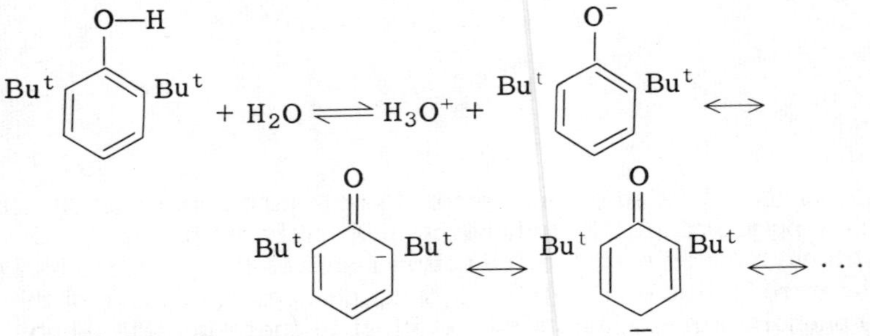

Let us first consider (*c*). The t-butyl groups exert a very large steric effect and this will tend to prevent the hydroxyl group becoming coplanar with the ring, i.e., there will be steric inhibition of resonance. Consequently, the positive charge on the oxygen atom is much smaller than that for phenol in (*a*), and so phenol is the stronger acid.

In (*d*), in going from the phenol to the corresponding phenoxide ion, because of the loss of the proton there is far less steric inhibition of resonance, and so there is now a driving force to ionise. Unless we argued that there is still some steric inhibition of resonance in this ion, we cannot compare the strength of this dibutylphenol with that of phenol. Since the dibutylphenol is the weaker acid, we should try to find some other argument to justify this fact. One possibility could be that because the t-butyl group has a +I effect, resonating structures with negative charges on the carbon atoms attached to the t-butyl groups are destabilised, and consequently the dibutylphenoxide ion is less stable than the phenoxide ion. Hence, phenol is the stronger acid.

We can also use another argument based on the general rule that the greater the amount of the solvation of an ion, the greater will be its stability (see *Text*, Ch. 5). Because of steric effects in the dibutylphenoxide ion (particularly at the oxygen atom), this ion will be far less solvated than the phenoxide ion. Consequently, the former is less stable than the latter and so the former is the weaker acid.

8.

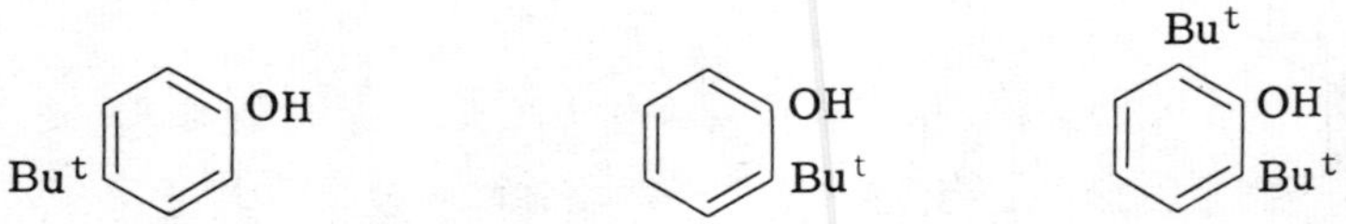

	(I)	(II)	(III)
OH(cm^{-1})	3500	3500 3640	3640

No intramolecular hydrogen bonding is possible, but intermolecular hydrogen bonding can occur if not prevented by steric effects. Thus, (I) is associated (3 500 cm^{-1}; see *Text*, Ch. 6); (II) is partly associated (3 500 cm^{-1}) and partly monomeric (3 640 cm^{-1}); (III) is completely monomeric (3 640 cm^{-1}).

9. (i) $CH_2(OEt)_2 \xrightarrow[H_2O]{HCl} CH_2O + 2EtOH$

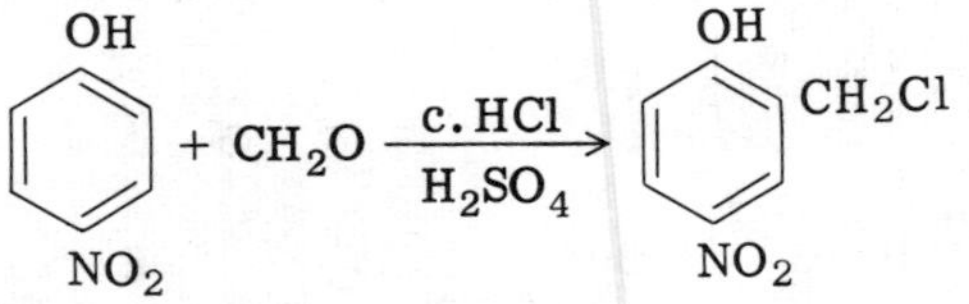

This is an example of chloromethylation; the presence of the nitro-group permits the isolation of the monochloromethyl derivative (see *Text*).

(ii)

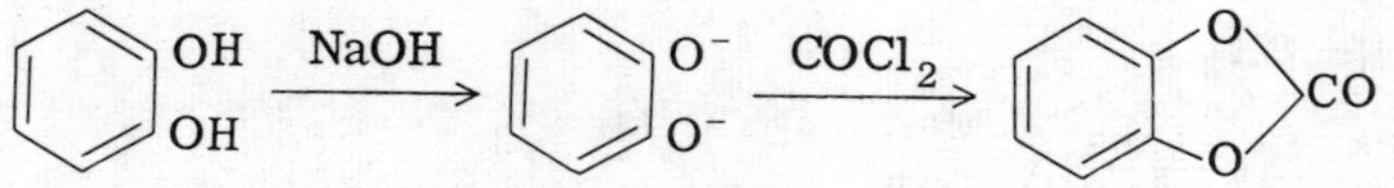

(iii)

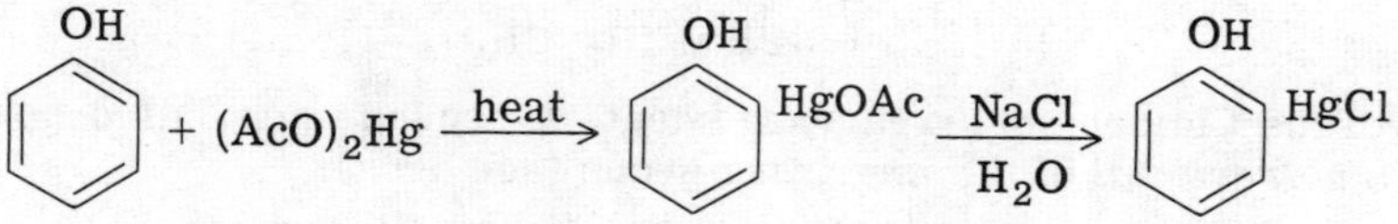

(iv)

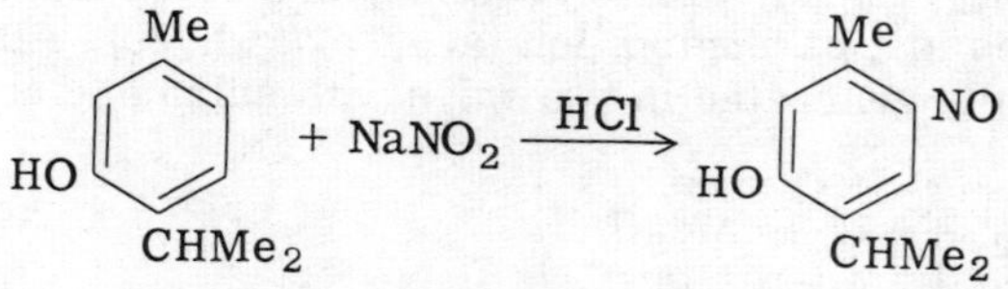

Nitrosation of phenols usually occurs preferentially in the *p*-position. In any case, for this substrate, the *p*-position is less hindered than the *o*.

(v) $PhOH \xrightarrow{NaOH} PhO^-Na^+ \xrightarrow{ClCH_2CO_2^-Na^+} PhOCH_2CO_2^-Na^+$

(vi)

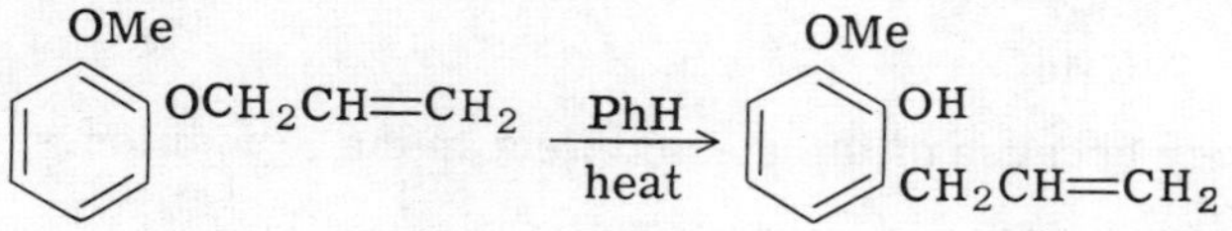

This is an example of the Claisen *ortho*-rearrangement.

(vii)

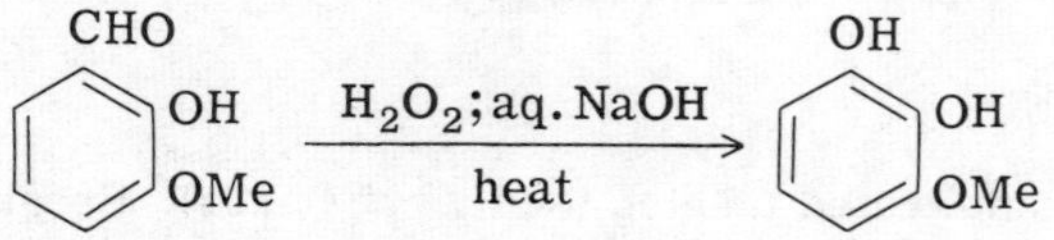

This is an example of the Dakin reaction.

(viii)

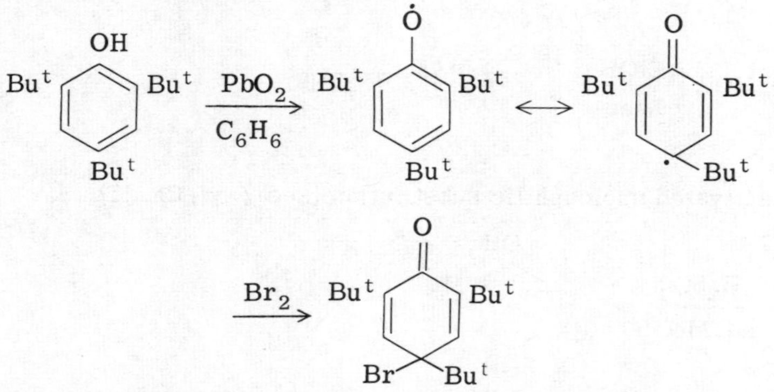

This is an example of one-electron transfer oxidation (see *Text*).

(ix) $PhOH + CH_2{=}CHCH_2Br \xrightarrow{K_2CO_3} PhOCH_2CH{=}CH_2$

Under the conditions of the experiment (usually in acetone solution), the mechanism is S_N2. For the preparation of alkyl phenyl ethers, a strong base, e.g., NaOH, is used. In this case, the phenoxide ion, a much stronger nucleophile than phenol, is necessary. With allyl bromide, however, because of its high reactivity, the weaker nucleophile, phenol, is capable of undergoing reaction (see also *Text*, Ch. 12).

(x)

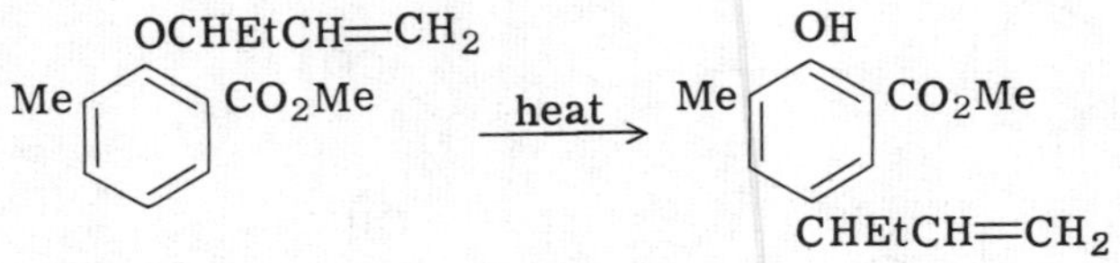

This is an example of the Claisen *para*-rearrangement. It occurs when both *o*-positions are blocked, and there is *no reversal* of the migrating group (see *Text*).

(xi) This is an alkylation reaction (Friedel–Crafts reaction), and since sulphuric acid is the catalyst, the reactive species is a carbonium ion (see *Text*, Ch. 20). Also, because the ion is t-butyl, this can be conveniently generated in two ways with sulphuric acid.

(*a*) $$Me_3COH \xrightarrow{H_2SO_4} Me_3C\text{—}OH_2^+ \xrightarrow{-H_2O} Me_3C^+$$

(*b*) $$Me_2C{=}CH_2 \xrightarrow{H_2SO_4} Me_3C^+$$

Then:

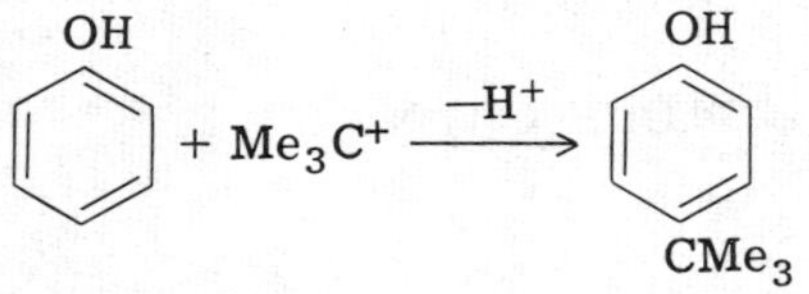

p-Substitution occurs exclusively because of the steric effect in the *o*-position.

(xii)

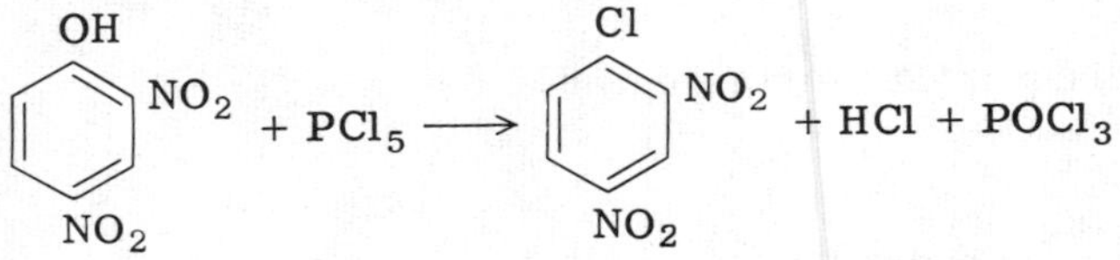

Phenol itself reacts in a much more complicated fashion (see *Text*, Ch. 21). Because of the strong activating effect of the *o*- and *p*-nitro-groups, the hydroxyl group readily undergoes nucleophilic substitution by chloride ion.

(xiii)

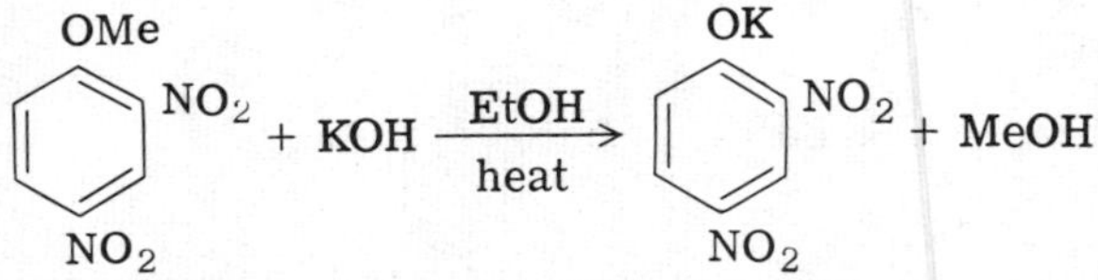

This is an example of activated nucleophilic substitution (see *Text*, Ch. 21).

(xiv)

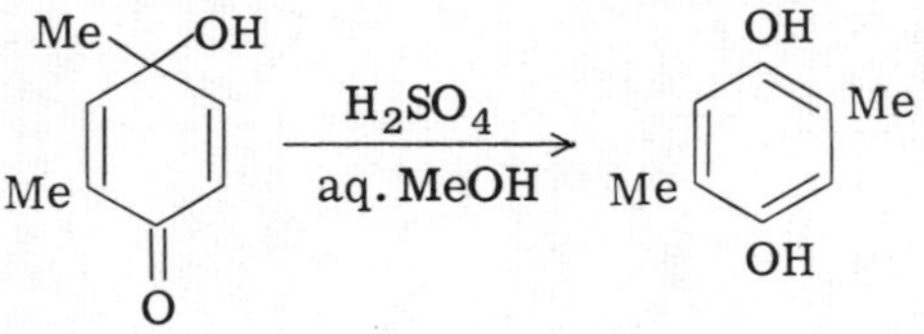

This is an example of the dienone-phenol rearrangement (see *Text*). The methyl group migrates to the *o*-position, in this case to the one *para* to the nuclear methyl group because of the steric effect.

(xv) This is an example of the formation of a 2,5-disubstituted product by 1,4-addition; the methylene group in the cyano-ester is highly activated. The redox potential of the *p*-benzoquinone system is higher than that of the product system and so the latter is oxidised to the quinone derivative, which then undergoes a second addition.

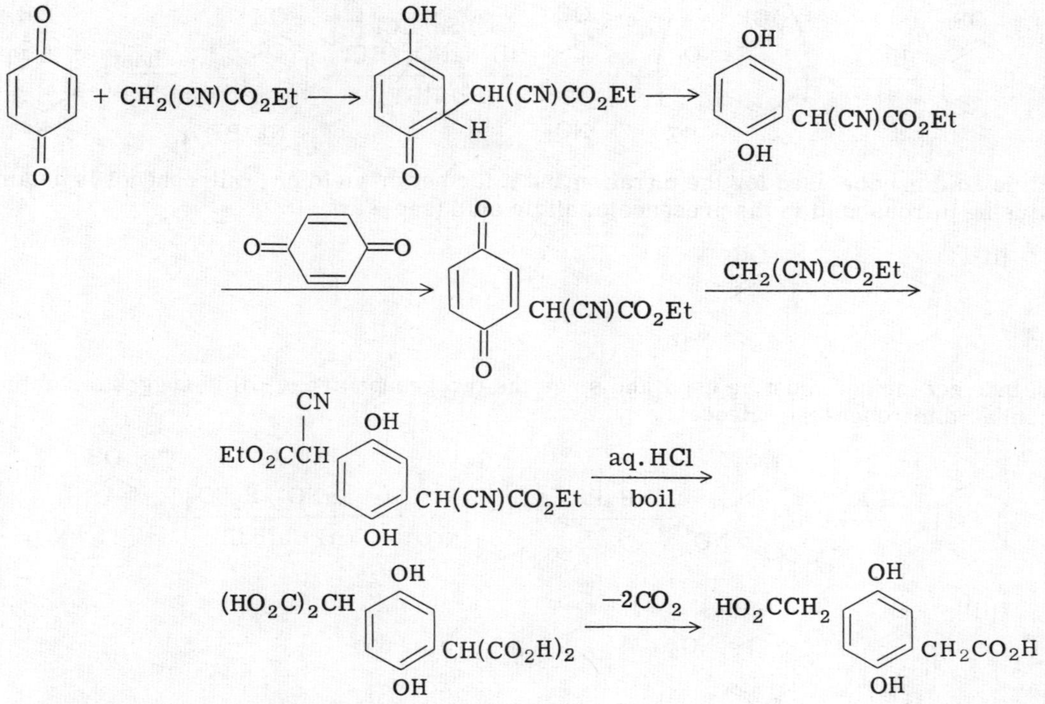

10. The ether is readily split by acid as follows:

$$Ph{-}O{-}CPh_3 \underset{}{\overset{H^+}{\rightleftharpoons}} Ph{-}\overset{+}{O}(H){-}CPh_3 \rightleftharpoons PhOH + Ph_3\overset{+}{C}$$

The two possible ions are Ph^+ and Ph_3C^+, but since the latter is far more stable than the former (by resonance stabilisation), fission occurs as shown (see *Text*, Ch. 7). Phenol has its *o*- and *p*-positions activated and is attacked preferentially at the *p*-position by the triphenylmethyl carbonium ion (which has an extremely large bulk) because of the steric effect exerted by the hydroxyl group at the *o*-positions.

11. $Ph{-}O{-}CH_2{-}CH_2{-}O{-}Ph$

Since the molecule is symmetrical (about the centre), the two phenyl groups give one signal only, and the two methylene groups also give one signal only. Hence, the compound shows two NMR signals.

12. (i)

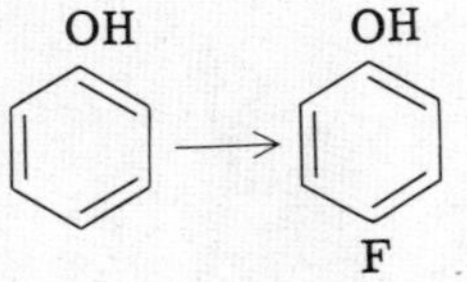

Fluorine must be introduced indirectly. Hence, a possible route is:

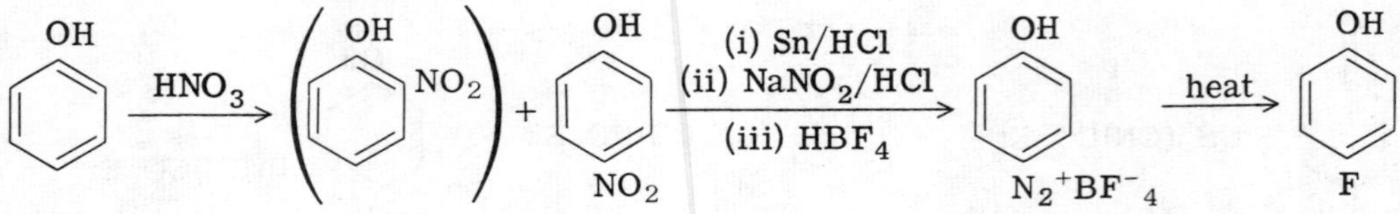

Nitric acid may be used for the nitration, but a far better yield of *p*-nitrophenol is obtained by using nitrous acid in the presence of nitric acid (see *Text*).

(ii)

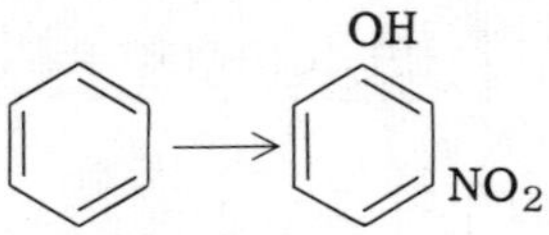

An indirect method must be used, and since the two groups are *m*-, this suggests starting from *m*-dinitrobenzene. Hence:

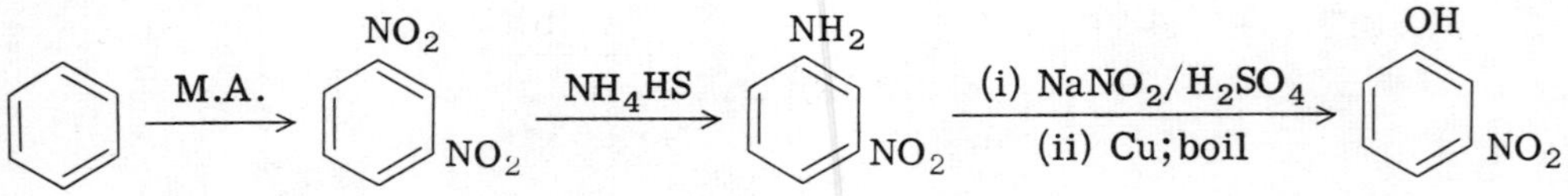

(iii)

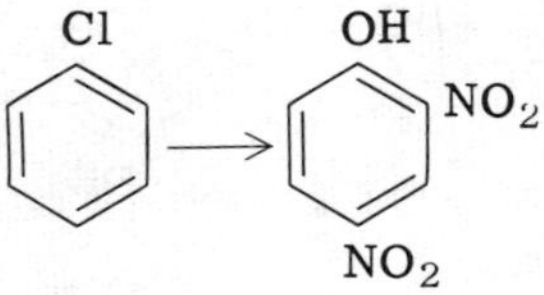

Chlorine is *o*/*p*-directing, and since the *o*/*p*-nitro-groups activate the chlorine atom, this can be replaced by the hydroxyl group (activated nucleophilic substitution).

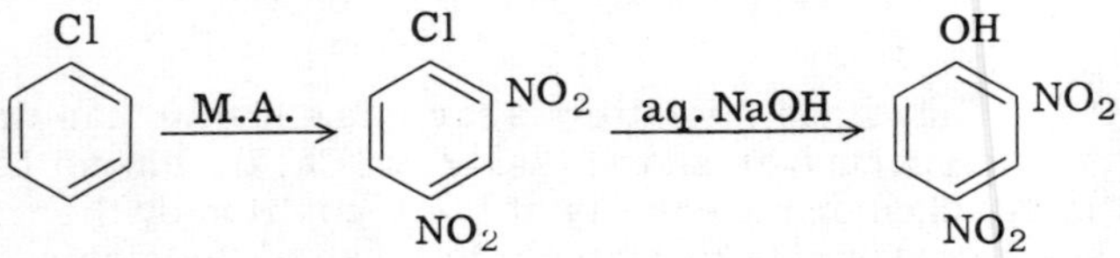

(iv)

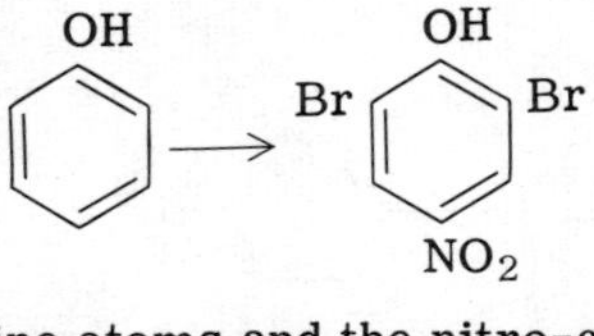

The bromine atoms and the nitro-group can be introduced directly with the correct orientation. However, nitration is carried out first, since bromination gives 2, 4, 6-tribromophenol. Hence (see also (i), above):

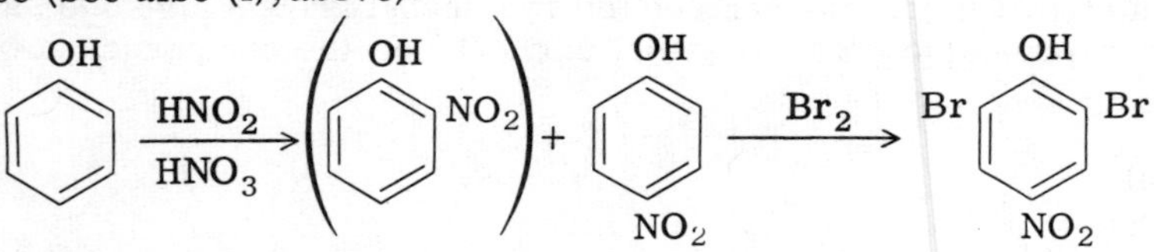

(v)

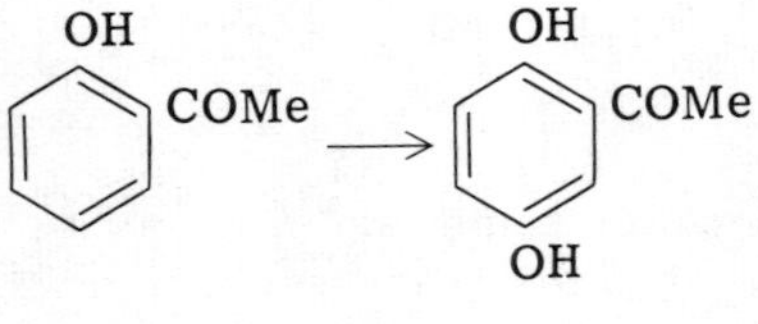

The Elbs persulphate oxidation is a method of directly introducing a hydroxyl group *para* to one already present (see *Text*).

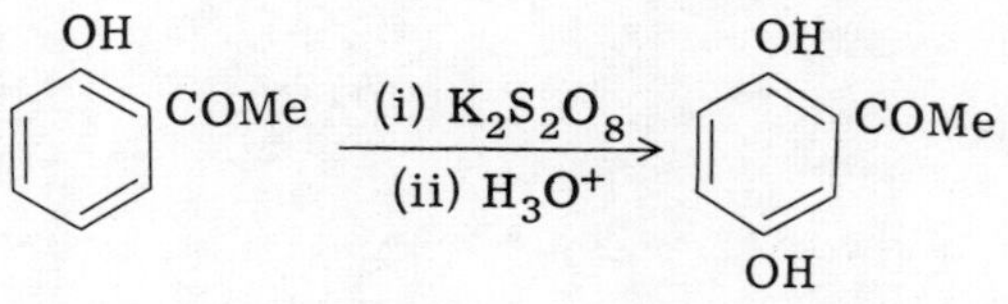

If the student were unaware of this reaction (or had forgotten it), he could try some of the more standard methods, e.g.,

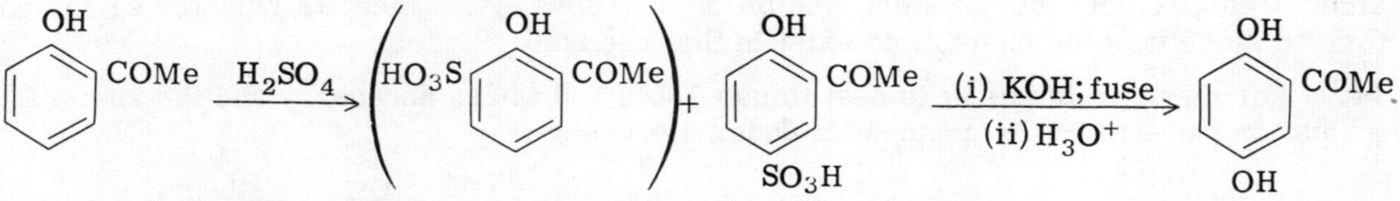

It is difficult to say what sort of yield would be obtained. Alkaline fusion is a somewhat drastic reaction and may split the ketone group. However, controlled conditions may give some of the desired product.

(vi)

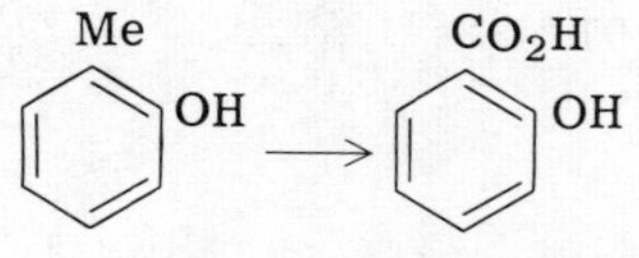

The direct oxidation of methyl to carboxyl is possible, but the hydroxyl group must be protected in order to prevent the breakdown of the ring (see *Text*, Ch. 20). Two reasonable methods are (Ts = p-$MeC_6H_4SO_2^-$):

(*a*)

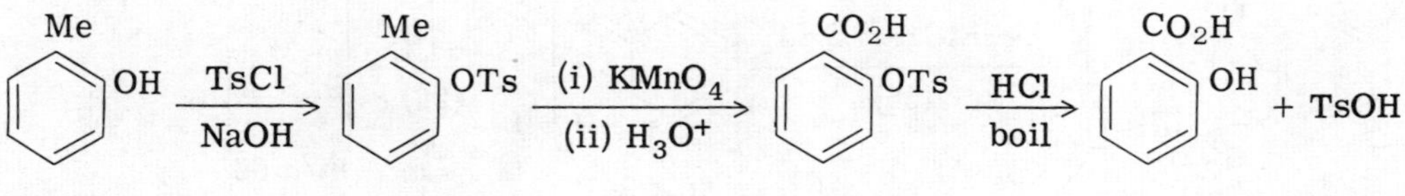

(*b*)

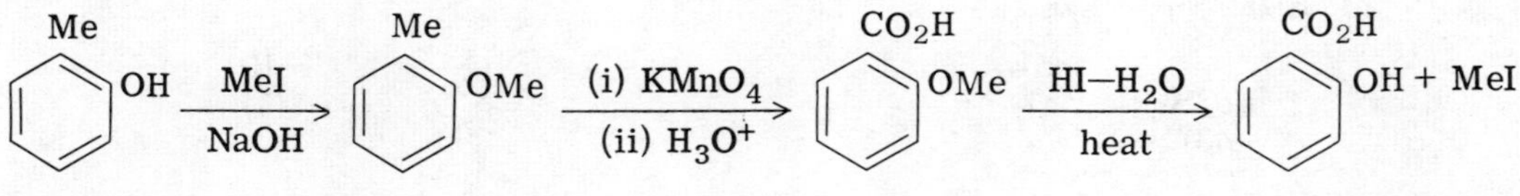

13.

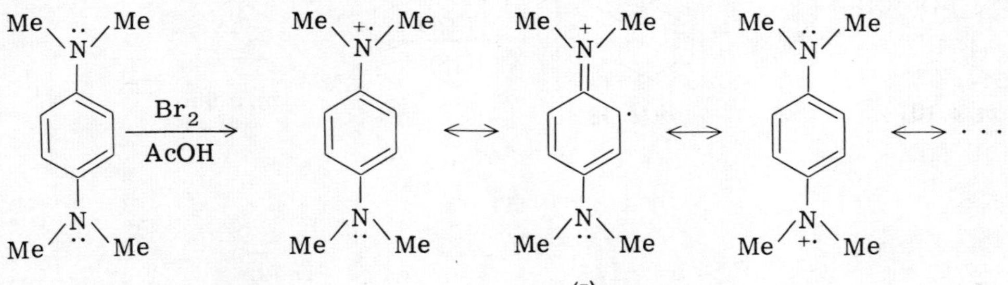

(I)

This Wurster salt (cationic free radical) is stabilised by resonance.

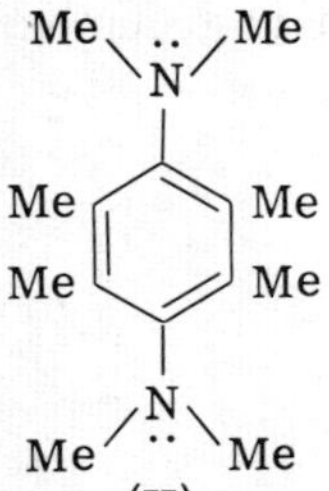

(II)

The corresponding 2,3,5,6-tetramethyl derivative (II) cannot form such contributing structures as (I) because of steric inhibition of resonance. Hence, its Wurster salt is not formed since it is too unstable to exist in the free state.

14. For these problems, it is best to use Table 1.6 where necessary and the known fragmentation patterns of the compounds under discussion.

(i)

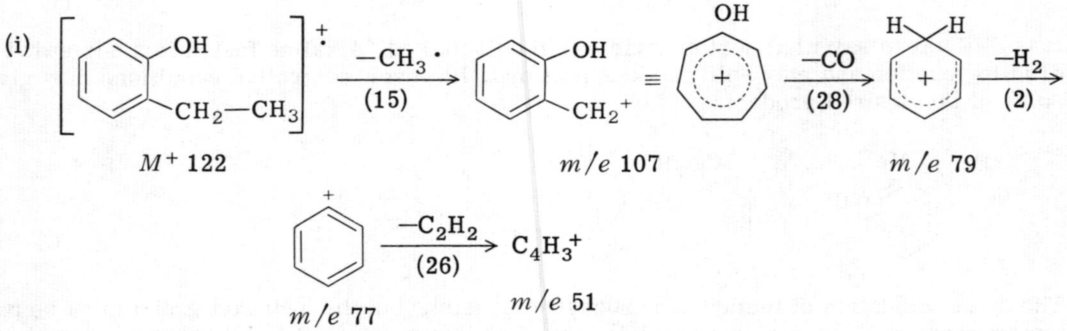

(ii) Note the McLafferty rearrangement.

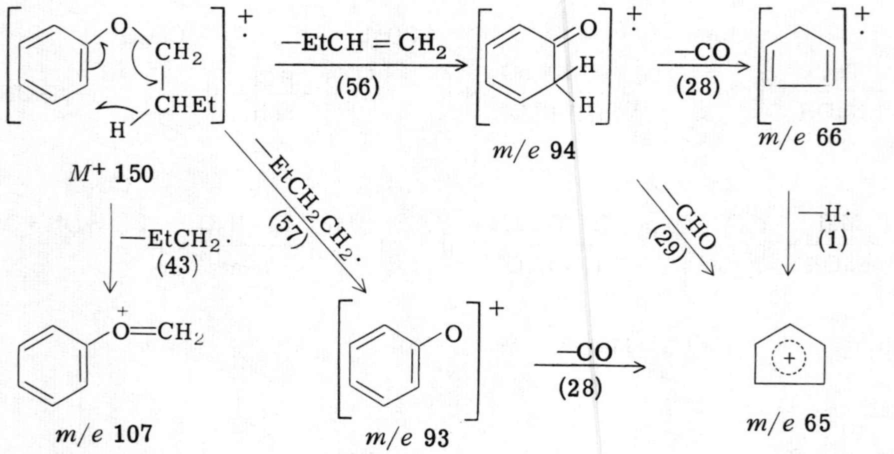

27 Aromatic alcohols, aldehydes and ketones

(Problems on page 49)

1.

PhCOEt
(I)

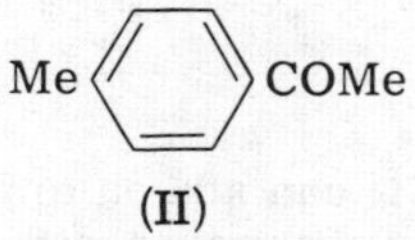

(II)

(i) (I) does not undergo the haloform reaction, whereas (II) does.

(ii) The i.r. spectrum of (I) will show the monosubstitution pattern of a benzene derivative, whereas (II) will show the *p*-disubstitution pattern. Alternatively, the NMR spectrum of (I) will show a quartet (CH_2) and a triplet (CH_3) for the ethyl group. This pair of signals is not shown by (II); each methyl will appear as a singlet.

2. (i)

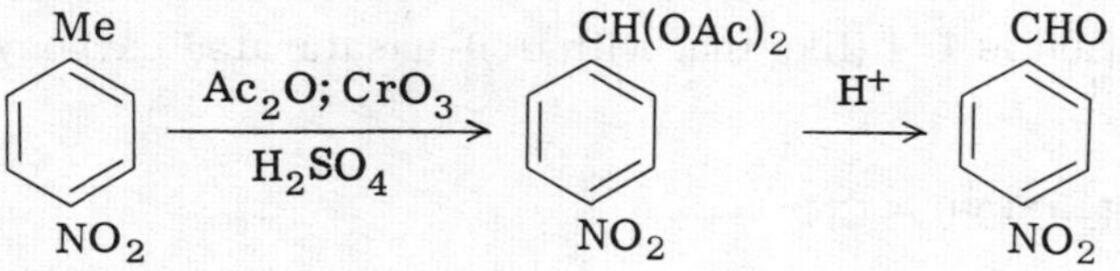

(ii) $$Ph_2CO \xrightarrow[\text{NaOH}]{\text{Zn}} Ph_2CHOH$$

Zinc and *acid* gives *bimolecular* reduction (see *Text*).

(iii)

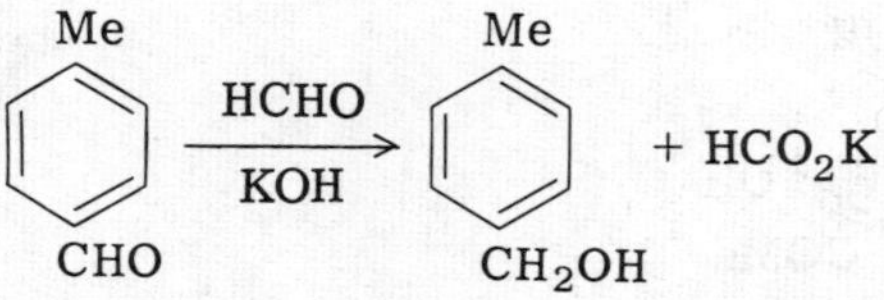

This is an example of the crossed Cannizzaro reaction (see *Text*, Ch. 8).

(iv)

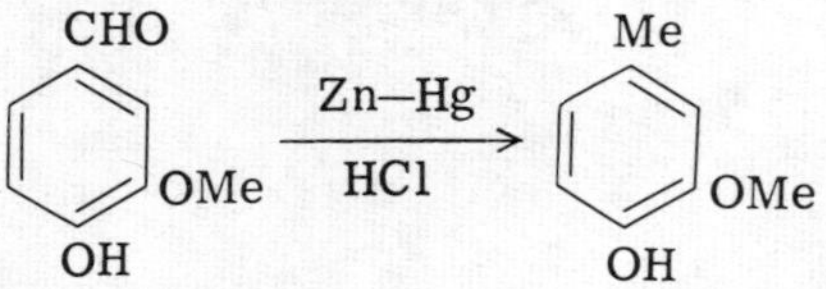

This is an example of the Clemmensen reduction of aldehydes.

(v)

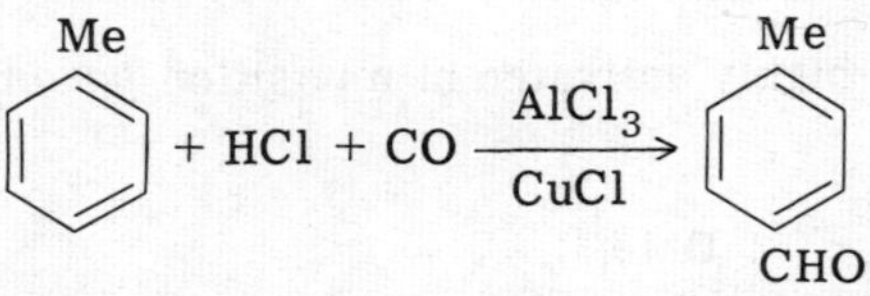

This is an example of the Gattermann-Koch aldehyde synthesis. A mixture of hydrogen cyanide and hydrogen chloride may also be used (Gattermann aldehyde synthesis).

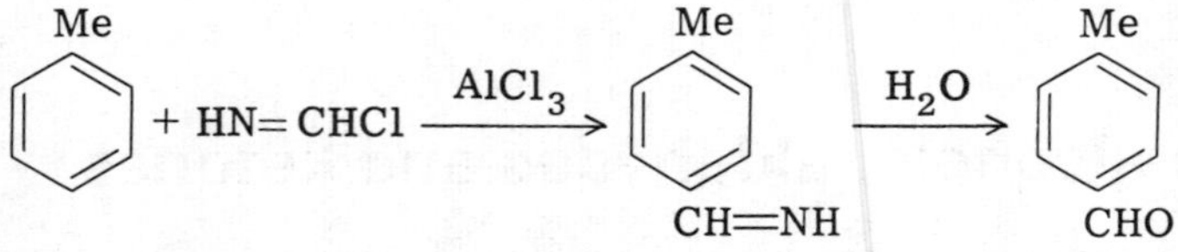

(vi) The product requires another 'PhCHCN' group. This suggests the possibility that benzaldehyde condenses with this group, which contains an active methylene group. Hence, we could try:

$$\mathrm{PhCHO + PhCH_2CN \xrightarrow[H_2O]{MeOH} PhCH{=}C(CN)Ph}$$

If this double bond is activated, then a molecule of HCN can add on to give the desired product (see *Text*, Ch. 12). Since the cyano-group is strongly electron-attracting (like a carbonyl group), we may formulate the reaction as:

$$\mathrm{NaCN + H_2O \rightleftharpoons NaOH + HCN}$$

$$\mathrm{PhCH{=}C(CN)Ph + HCN \rightleftharpoons Ph\overset{+}{C}H{-}CH(CN)Ph + CN^-}$$

$$\mathrm{\longrightarrow PhCH(CN)CH(CN)Ph}$$

It may be possible that the addition is 1, 4 (like that with α,β-unsaturated carbonyl compounds), i.e.,

$$\mathrm{PhCH{=}C(C{\equiv}N){-}Ph \xrightarrow{HCN} Ph\overset{+}{C}H{-}C({=}C{=}NH){-}Ph + CN^- \longrightarrow}$$

$$\mathrm{PhCH(CN){-}C({=}C{=}NH){-}Ph \longrightarrow PhCH(CN){-}CH(CN)Ph}$$

(vii)

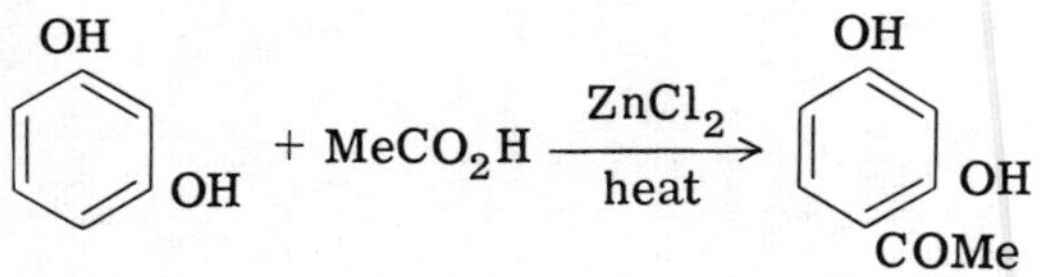

The mechanism of this reaction is uncertain. Since zinc chloride is a Lewis acid, a possibility is (*cf. Text,* Ch. 5):

$$\mathrm{Me{-}\overset{O}{\overset{\|}{C}}{-}OH + ZnCl_2 \rightleftharpoons Me{-}\overset{O}{\overset{\|}{C}}{-}\overset{+}{O}H(\bar{Z}nCl_2)}$$

$$\mathrm{\longrightarrow Me\overset{+}{C}{=}O + [HO{-}ZnCl_2]^-}$$

The acyl cation then attacks the ring, which is highly activated at a position *para* to one hydroxyl group and *ortho* to the other.

(viii)

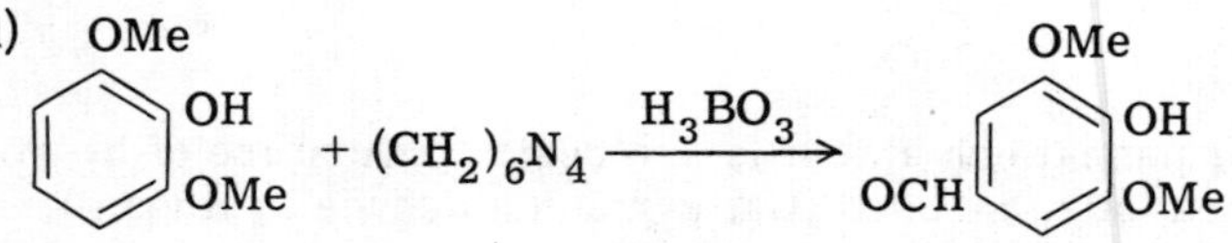

This is an example of the Duff reaction. The formyl group is normally introduced *ortho* to the hydroxyl group, but since these positions are blocked (by activating groups), the formyl group enters the *p*-position.

(ix)

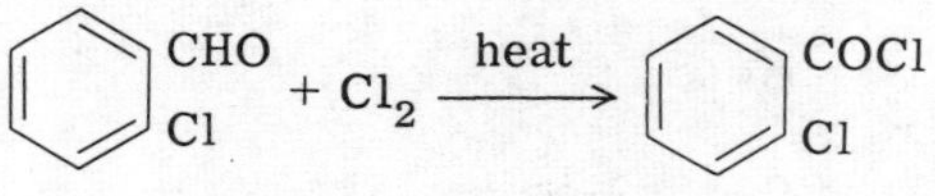

This reaction occurs in the *absence* of a halogen carrier (no α-hydrogen atom is present).

(x) Compare this reaction with (vi), above.

$$KCN + AcOH \longrightarrow HCN + AcOK$$

$$PhCH{=}CH{-}\overset{O}{\overset{\|}{C}}Ph \xrightarrow{HCN} Ph\overset{+}{C}H{-}CH{=}\overset{OH}{\overset{|}{C}}Ph \xrightarrow{CN^-}$$

$$PhCH(CN)CH{=}\overset{OH}{\overset{|}{C}}Ph \longrightarrow PhCH(CN)CH_2COPh$$

(xi) $PhCH_2OH + COCl_2 \longrightarrow PhCH_2OCOCl + HCl$

(xii) $PhCHBrCHBrCHO \xrightarrow[heat]{K_2CO_3} PhCH{=}CBrCHO$

Potassium carbonate is a *mild* base and so can remove only *one* molecule of HBr. Because of the strong—I effect of the aldehyde group, the proton on the α-carbon atom is more easily removed than that on the β-carbon atom. The mechanism is E2. A strong base (e.g., EtO^-) is required to remove the β-proton (E2 mechanism) to form a triple bond (see *Text*, Ch. 4).

(xiii) $PhCH(CN)COMe \xrightarrow[heat]{aq.\ H_2SO_4} PhCH(CO_2H)COMe$

$$\xrightarrow{-CO_2} PhCH_2COMe$$

β-Ketoacids readily eliminate carbon dioxide.

(xiv) $PhCOMe \xrightarrow{SeO_2} PhCOCHO$

Selenium dioxide oxidises an active methylene group to a carbonyl group (see *Text*, Ch. 8).

(xv)

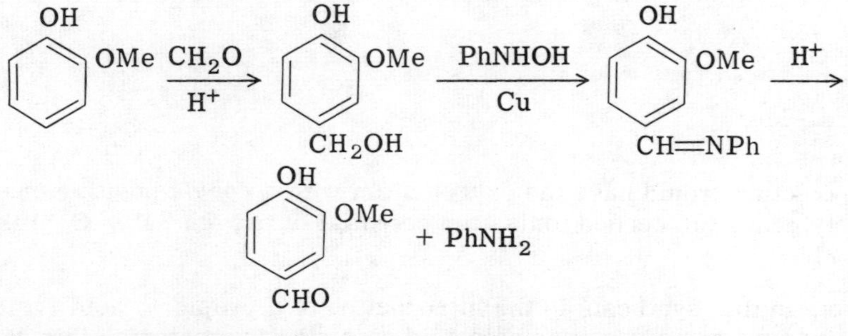

The first step is an example of the Lederer-Manasse reaction (see *Text*, Ch. 26).

(xvi)

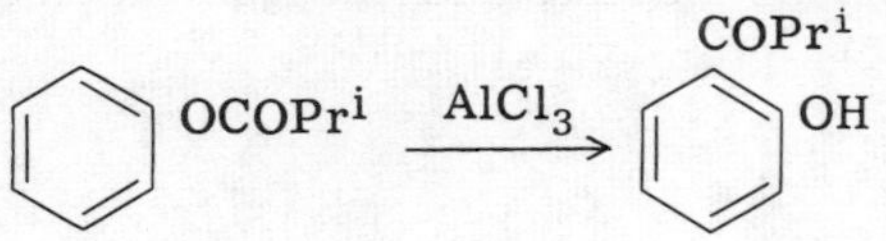

This is an example of the Fries rearrangement (see *Text*, Ch. 26).

(xvii) $PhCO_2H \xrightarrow{SOCl_2} PhCOCl \xrightarrow{Me_2NH} PhCONMe_2 \xrightarrow{Li(OEt)_2AlH_2} PhCHO$

See *Text*, Ch. 8.

3.

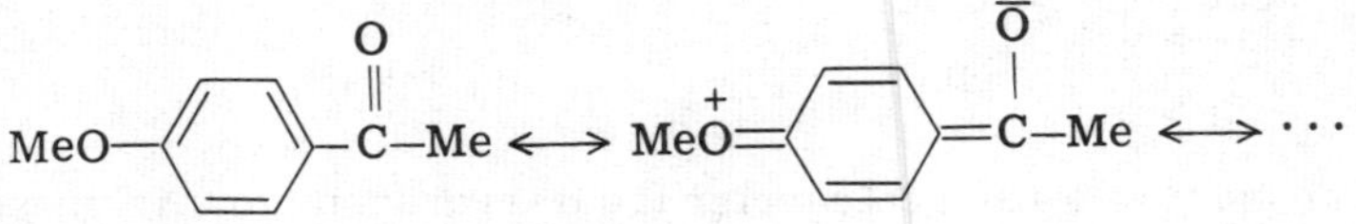

(I); MeO is +R

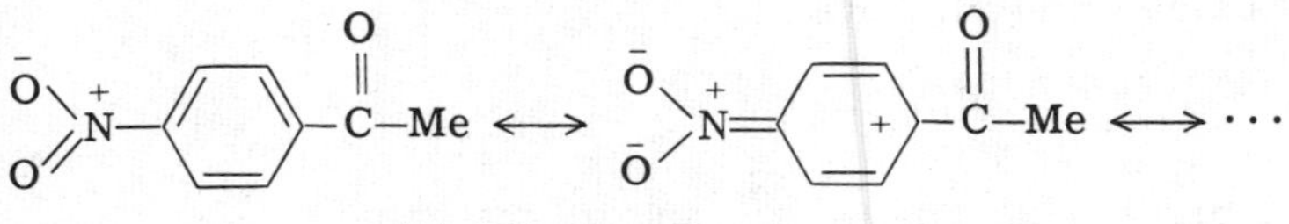

(II); NO_2 is −R

It is reasonable to assume (but not necessarily true) that the greater the negative charge on the oxygen atom of the carbonyl group, the more readily it is protonated in acid solution. Hence, (I) will be more readily protonated than (II).

Alternatively, protonated (I) is more stabilised by resonance than is protonated (II).

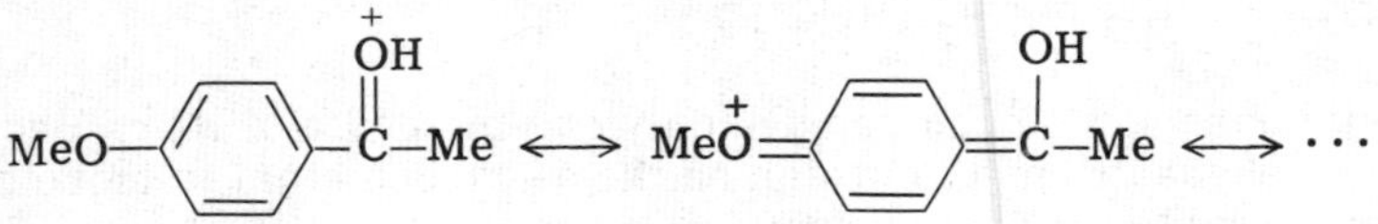

(I*a*)

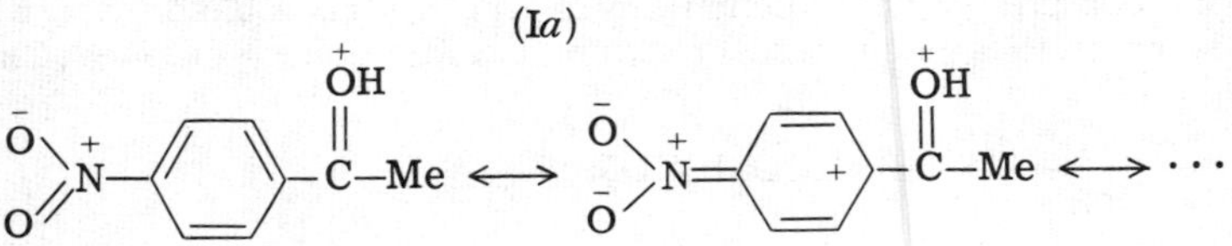

(II*a*)

In (I*a*), there is extended conjugation and only one charge (positive) is involved. In (II*a*), there is not this extended conjugation and the relative close proximity of two positive charges is a destablising factor. Hence, (I*a*) is more stable than (II*a*) and so the former is produced more readily than the latter.

The student might be inclined to argue that extended conjugation is possible in (II*a*) as follows:

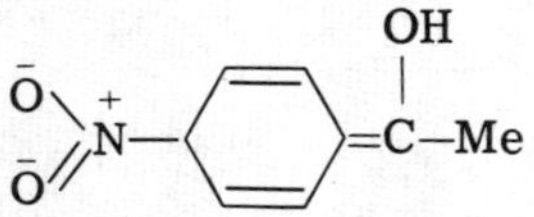

Such a resonating structure would have the oxygen atom with a *double* positive charge. This is extremely unlikely, since the second ionisation potential of oxygen ($O^+ \rightarrow O^{++}$) is extremely high (see *Text*, Ch. 2).

4. A necessary step in this synthesis is the introduction of a propionic acid residue into the benzene ring. Since we are given propionic acid as a starting material, this can be introduced as a ketone group to give PhCOEt. This then suggests the application of the Willgerodt reaction. Hence, a possible synthesis is:

$$MeCH_2CO_2H \xrightarrow{SOCl_2} MeCH_2COCl$$

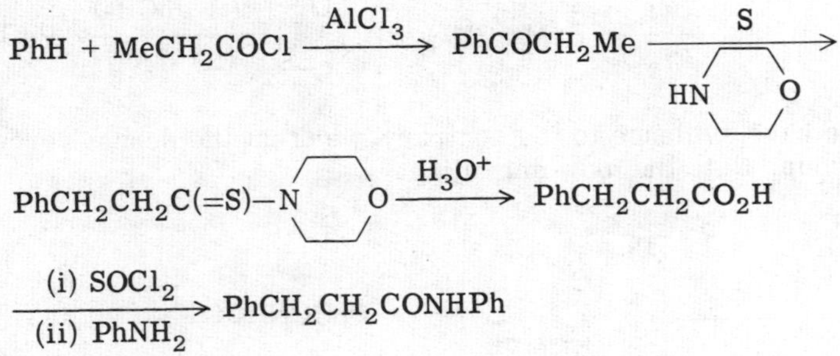

5. $PhCHO \longrightarrow PhCH{=}CHCOPh$

An obvious synthesis is:

$$PhCHO + CH_3COPh \xrightarrow{OH^-} PhCH{=}CHCOPh$$

Hence, we need to convert benzaldehyde into acetophenone. The latter is a ketone and can be prepared by the oxidation of a secondary alcohol. Therefore, one route is:

$$PhCHO \xrightarrow[\text{(ii) } H_3O^+]{\text{(i) MeMgI}} PhCHOHCH_3 \xrightarrow[\text{dichr.}]{\text{acid}} PhCOCH_3$$

A second synthesis is not so easy to devise as the first one. However, we start with a carbonyl compound and we end up with an alkene. The conditions are in keeping with the Wittig synthesis (see *Text*, Ch. 14). What we require is:

$$PhCHO + Ph_3P{=}Z \longrightarrow PhCH{=}Z + Ph_3PO$$

where Z = CHCOPh. Since the preparation of the phosphorane involves removal of HX from the quaternary phosphonium salt, the salt we require is $[Ph_3PCH_2COPh]^+X^-$.

Thus, we can start with ω-bromoacetophenone. Hence, the synthesis is:

$$PhCOCH_3 \xrightarrow[AlCl_3]{Br_2} PhCOCH_2Br$$

$$Ph_3P + PhCOCH_2Br \longrightarrow Ph_3\overset{+}{P}CH_2COPh \xrightarrow{Na_2CO_3}$$

$$Ph_3P{=}CHCOPh \xrightarrow{PhCHO} PhCH{=}CHCOPh + Ph_3PO$$

Removal of a molecule of HX from the phosphonium salt is usually done with a strong base, e.g., EtO^-. In our case, we have used the weak base, sodium carbonate, because the proton to be removed is activated by the adjacent carbonyl group (the ethoxide ion could have been used).

6. The starting material is a bromo-compound and the product is a secondary alcohol. This suggests the use of a Grignard reagent and an aldehyde, i.e.

$$PhCHO + BrMgZ \longrightarrow PhCHOHZ$$

The difficulty here is that Z contains an aldehyde group; this would react with the Grignard reagent. This can be prevented by acetal formation, and since dioxalans are more stable than acetals of monohydric alcohols (and easier to prepare), we can try the following synthesis (see also *Text*, Ch. 11):

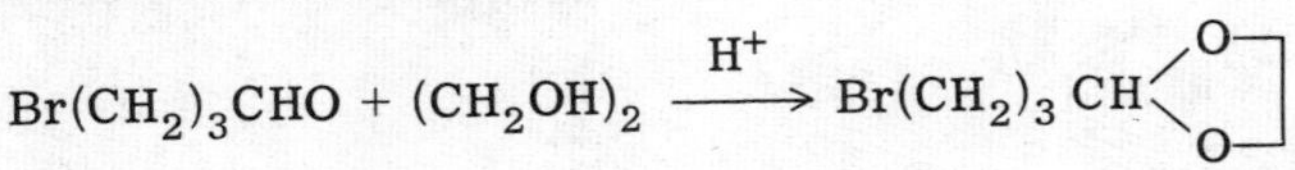

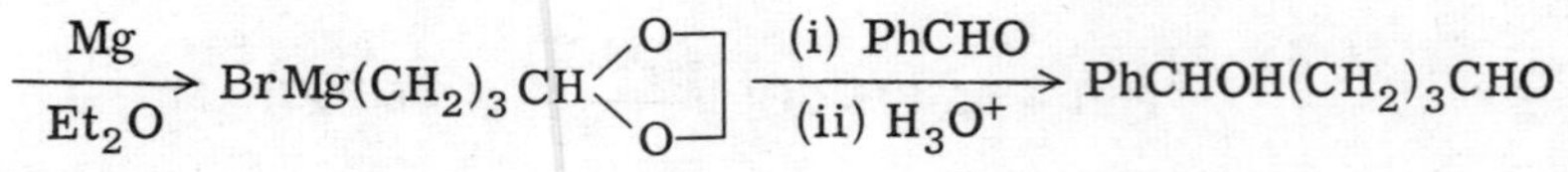

7. Unless we have evidence to the contrary, we treat the benzoyl group in the same way as a phenyl group. Note the *anti*-migration.

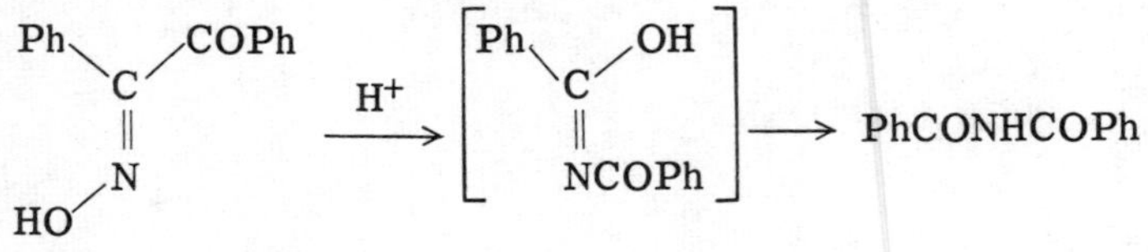

syn-phenyl

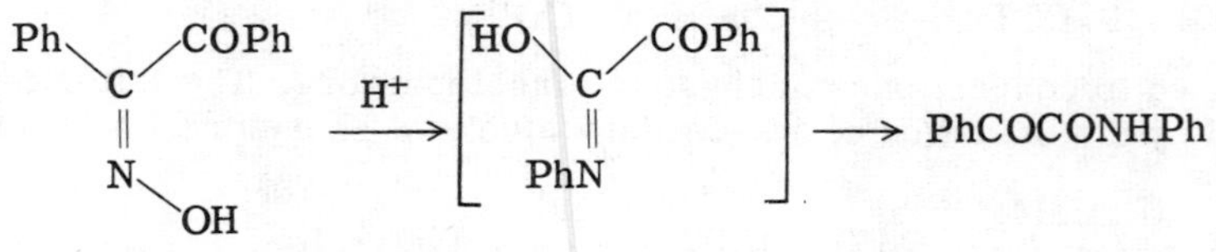

anti-phenyl

8. $Ph_2CHCHO \xrightarrow{H^+} PhCH_2COPh$

Both compounds have the same molecular formula, and (i) a phenyl group has migrated to an adjacent carbon atom; (ii) a hydrogen atom has done the same, but in the opposite direction. These are the characteristics of the 1, 2-shift. Moreover, since the rearrangement is acid-catalysed, the function of the proton is to create an electrophilic centre. To achieve this, the only reasonable point of attack for the proton is the oxygen of the carbonyl group. Hence, a possible mechanism is the formation of a phenonium ion, followed by a hydride ion shift.

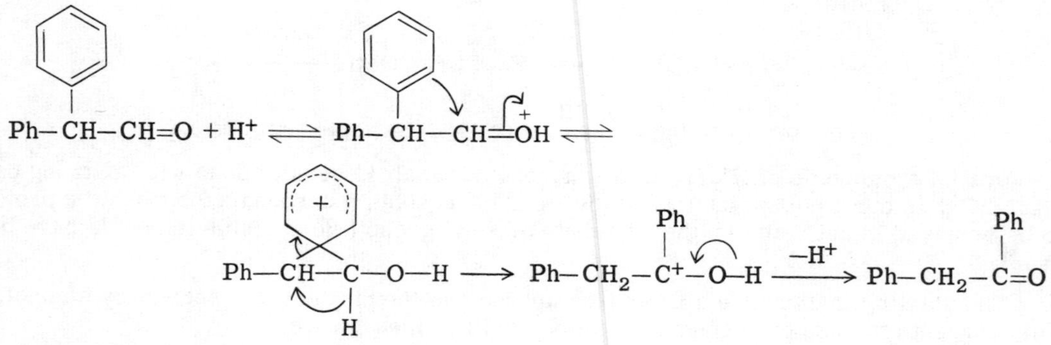

9. (i)

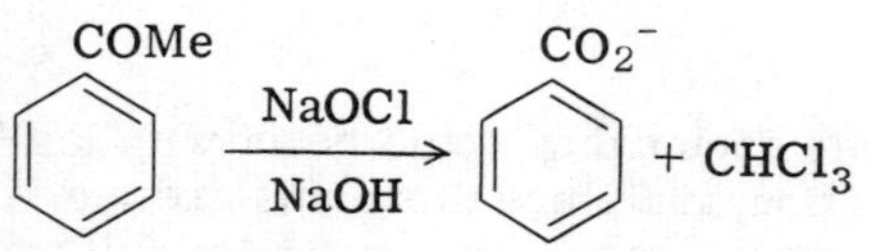

(ii)

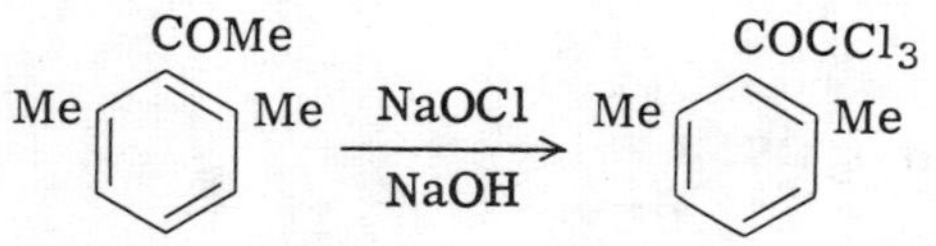

Inspection of the two structures suggests that there is the possibility that a steric effect could operate in (ii), and that this is therefore the cause of the difference in behaviour. The student should realise that this approach has been adopted because of the presence of the two methyl groups *ortho* to the functional group. We may be on the wrong track, but we can only find this out by pursuing the argument. Now, the accepted mechanism of the haloform reaction is (see *Text*, Ch. 8):

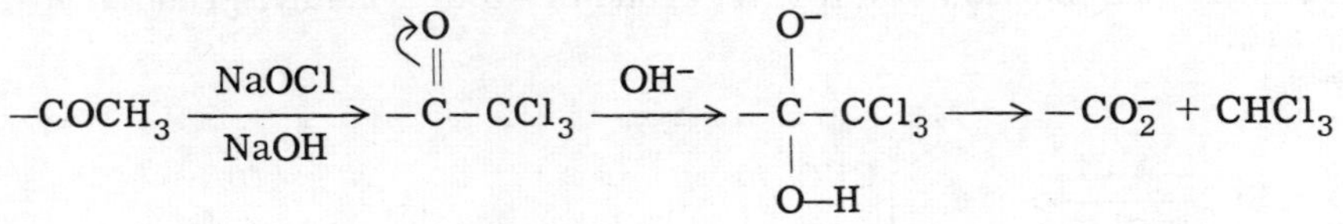

(I)

Both (i) and (ii) proceed to the trichloromethyl compound, but whereas (i) proceeds further to give (I) and then the final products, (ii) cannot form (I). In the formation of (I), the carbon atom of the carbonyl group changes from sp^2 to sp^3 hybridisation, the latter leading to greater crowding round the central carbon atom. Hence, on the basis of the steric effects of the *o*-methyl groups, (I) cannot be formed in reaction (ii), and so the trichloromethyl compound is isolated.

10.

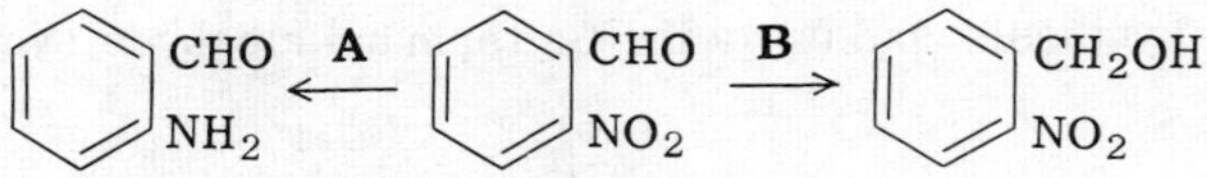

A: $FeSO_4/NH_4OH$ *or* $SnCl_2/HCl$.

B: $NaBH_4$ *or* LAH + $AlCl_3$ *or* $Al(OPr^i)_3$.

The student should also refer to Table 6.1 (and the discussion in that chapter).

11. Since the amine contains *two* phenyl groups attached to *one* carbon atom, oxidation of this type of structure would be expected to give diphenyl ketone (benzophenone). In order to get only benzoic acid, one phenyl group must have migrated; this suggests a 1,2-shift. Hence, a possible mechanism is (see also Prob. 8, above):

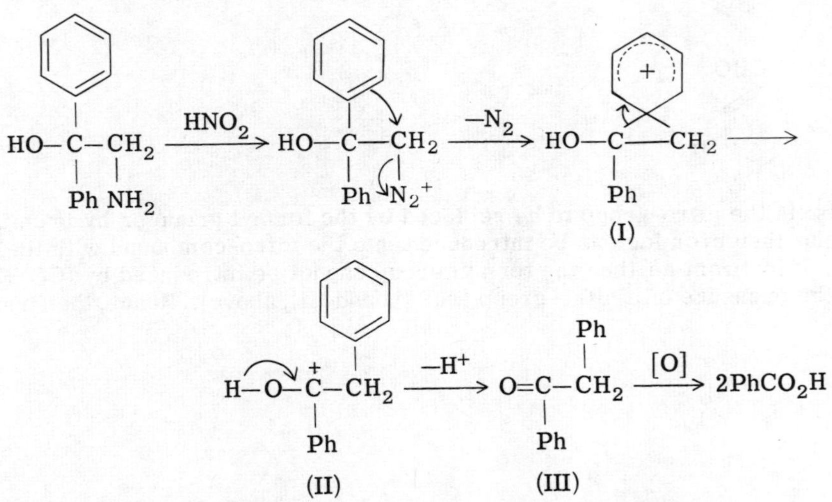

The conversion of (I) into (III) may possibly occur in one step, i.e., (I) loses a proton at the same time as the phenonium ion completes the migration.

12. (i)

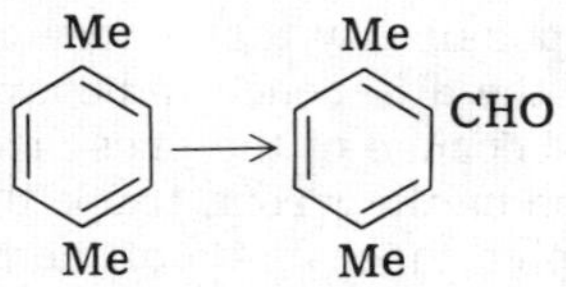

All that is required is the introduction of a formyl group *ortho* to a methyl group (*o*/*p*-directing). Two methods are:

(*a*)

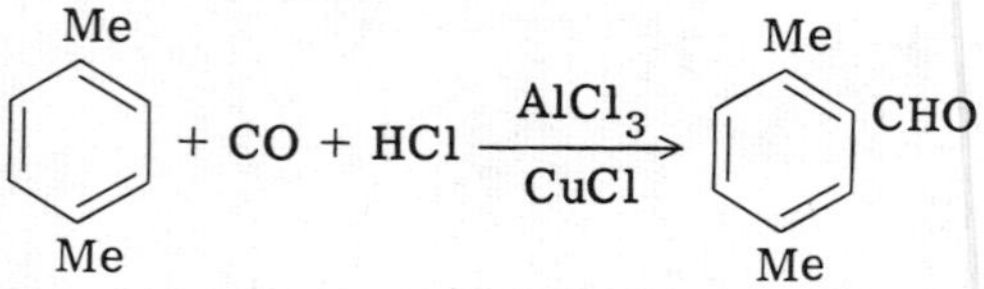

(*b*)

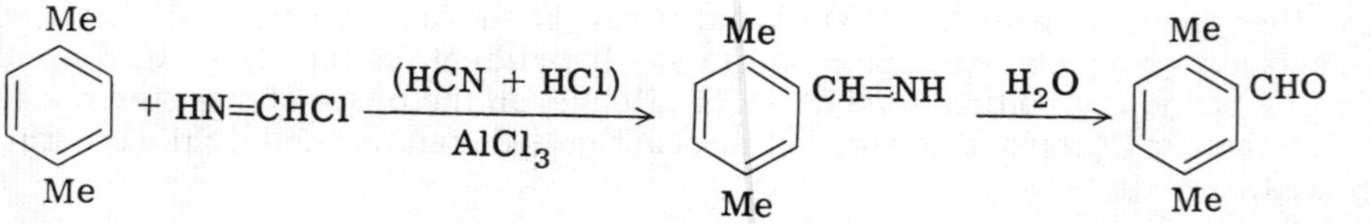

(*a*) is the Gattermann-Koch aldehyde synthesis, and (*b*) is the Gattermann aldehyde synthesis.

(ii)

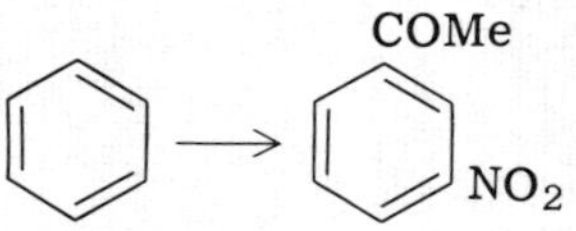

The Friedel-Crafts acylation is the first step; the acetyl group is *m*-directing. The synthesis cannot be carried out the other way round, since the nitro-group, although orienting correctly (*m*), inhibits the Friedel-Crafts reaction (see *Text*, Ch. 20).

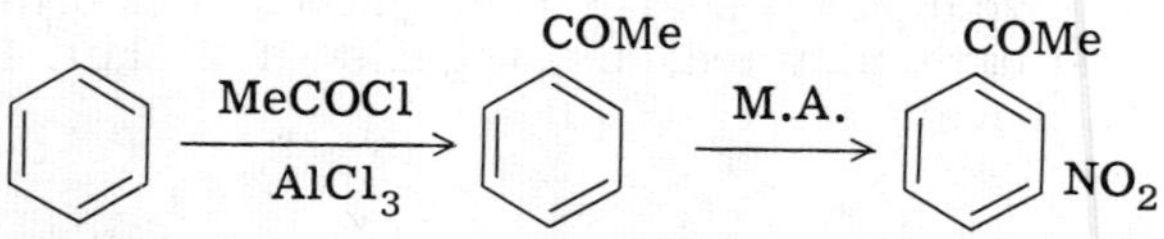

(iii)

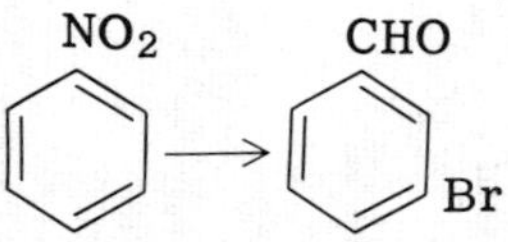

The first question is: Is the nitro-group to be replaced by the formyl group or by bromine? If by the formyl group, then bromine can be introduced into the nitro-compound with the correct orientation. If by bromine, then the formyl group cannot be introduced by (CO +HCl) or (HCN + HCl) in the presence of a nitro-group (see (i) and (ii), above). Hence, the first step could be:

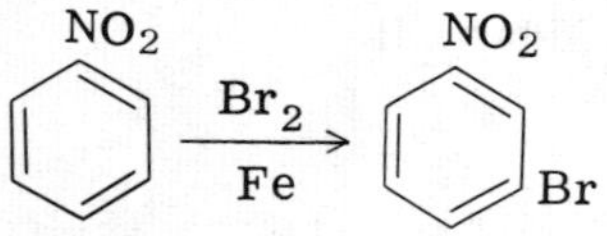

Since the formyl group can be prepared from the cyano-group, we can now proceed as follows (Stephen's method):

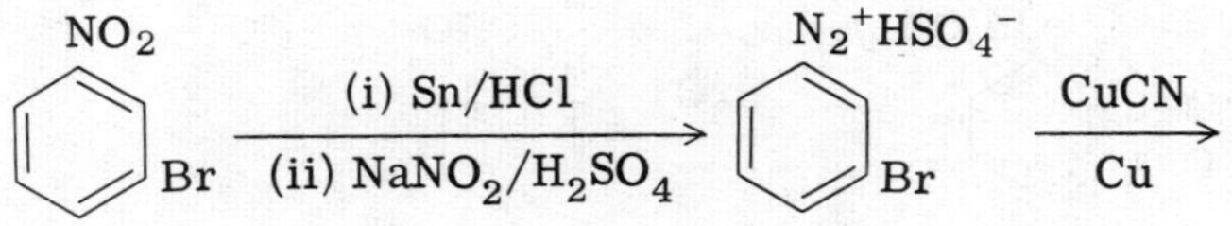

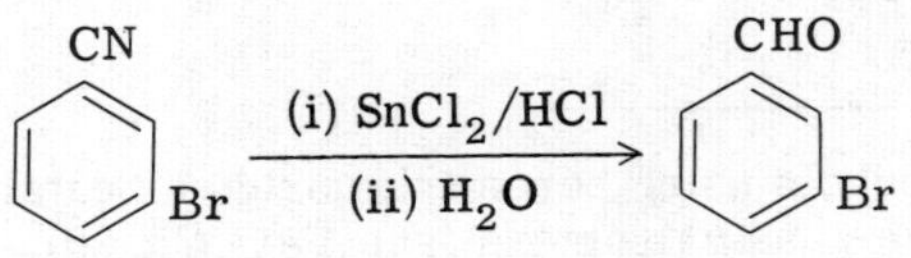

An alternative route involving a diazonium salt may also be used; this is by direct reaction with formaldoxime (see *Text*, Ch. 24).

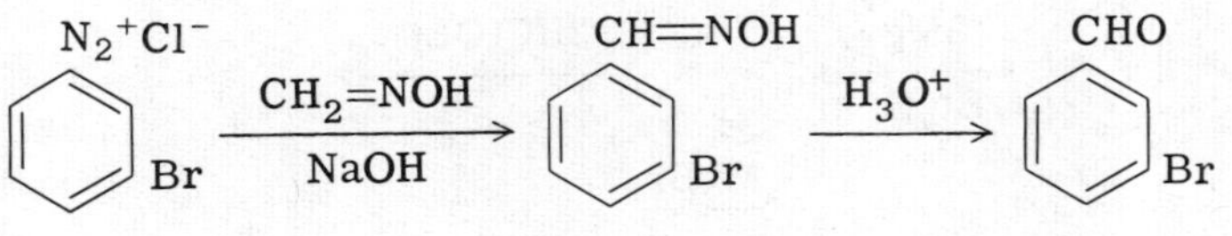

(iv)

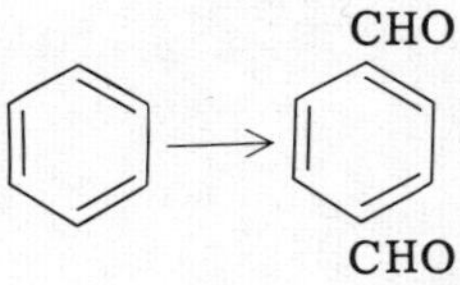

In view of the answer given in (iii), above, a possible synthesis would require the preparation of *p*-dicyanobenzene. This can be carried out via *p*-diaminobenzene, and since the two nitro-groups cannot be directly introduced with the required orientation, we may proceed as follows.

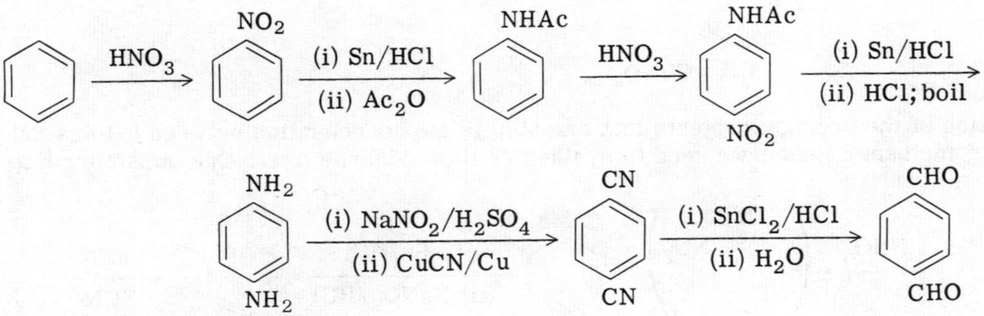

The method involving formaldoxime may also be used (see (iii), above).

Still another method is the following. Because of the difficulty of preparing *p*-dialkylbenzenes by the Friedel-Crafts reaction, an alternative route is by means of the Wurtz-Fittig reaction.

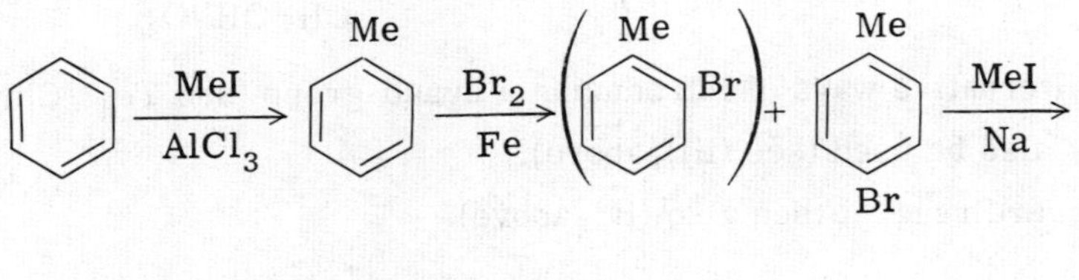

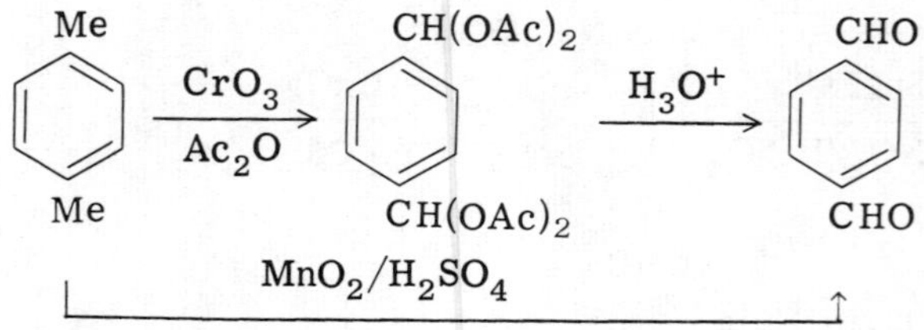

The student might also note that alkylation with alcohols using boron trifluoride as catalyst at low temperatures gives mainly a mixture of the *o*- and *p*-isomers (see *Text*, Ch. 20).

It is also worth describing one other method which involves the use of a Grignard reagent (see also *Text*, Ch. 15).

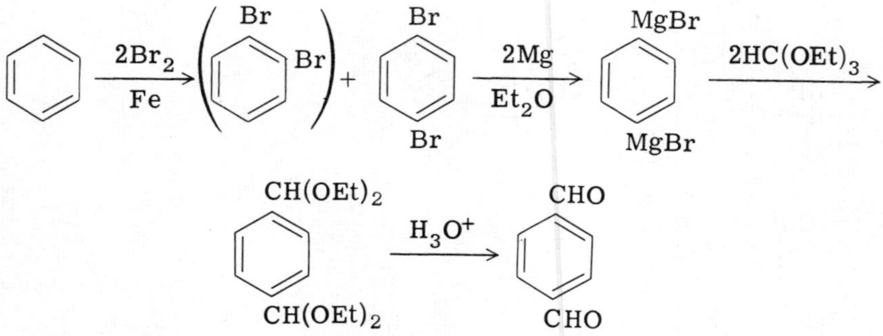

All of these syntheses are practical ones. It is not easy to say which is the best. It could well be the first; no separations are involved and every step is efficient (i.e., gives a high yield).

(v)

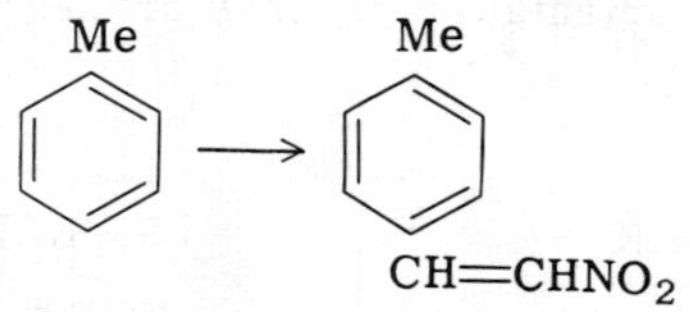

Inspection of the product suggests that one step is the condensation between *p*-tolualdehyde and nitromethane. Hence, we need to synthesise this aldehyde first. One possible route is:

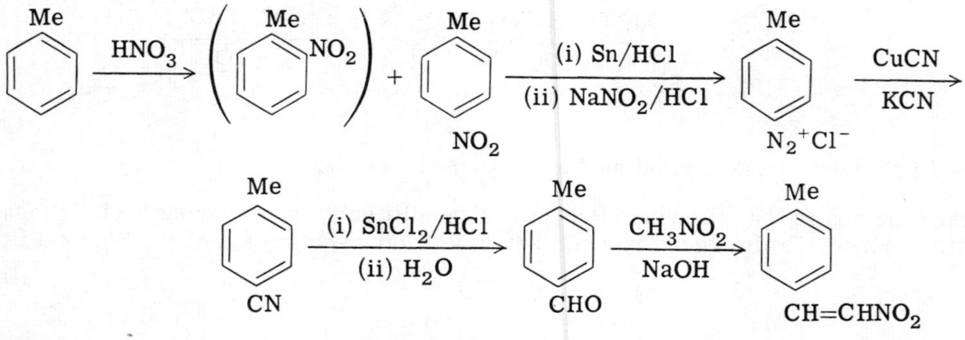

The student should note the alternative ways of introducing a cyano-group (see *Text*, Ch. 24).

The formaldoxime route may also be used (see (iii), above).

Another method is via a Grignard reagent (see also (iv), above).

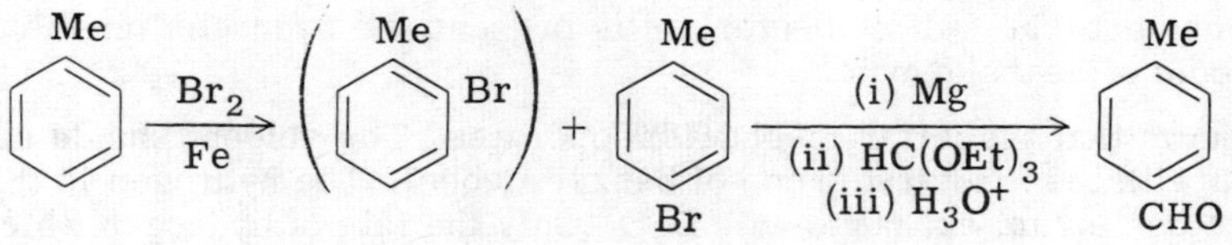

This method could well be the best one.

13. Since the final product is benzoic acid, we can assume that **A** contains one side-chain only. Now:

$$\mathbf{A}\ C_8H_6O_2 = C_6H_5 + C_2HO_2$$

Also, the D.B.E. of **A** is (see Ch. 1, Prob. 9): $8 + 1 - 6/2 = 6$

Three double bonds and one ring account for benzene. Hence, the side-chain contains two double bonds. A structure that fits these facts is:

$$C_8H_6O_2 = C_6H_5{-}\overset{\overset{\displaystyle O}{\|}}{C}{-}CHO$$

A, on treatment with NaOH and then acid, gives **B**, $C_8H_8O_3$. The side-chain is now $C_2H_3O_3$. Since acidification was necessary, this suggests that a sodium salt was formed first, i.e., the side-chain now contains a carboxyl group. Also, the D.B.E. of **B** is:

$$8 + 1 - 8/2 = 5$$

Hence, the side-chain now contains *one* double bond (4 accounted for by benzene). A structure of **B** which fits the facts is:

$$C_8H_8O_3 = C_6H_5{-}CHOH{-}CO_2H$$

Therefore, under the influence of alkali, we have:

$$C_6H_5{-}CO{-}CHO \xrightarrow{OH^-} C_6H_5{-}CHOH{-}CO_2H$$

Since the keto-group has been reduced and the aldehyde group oxidised, this then is a case of the *internal* Cannizzaro reaction. We can now formulate the reactions as follows:

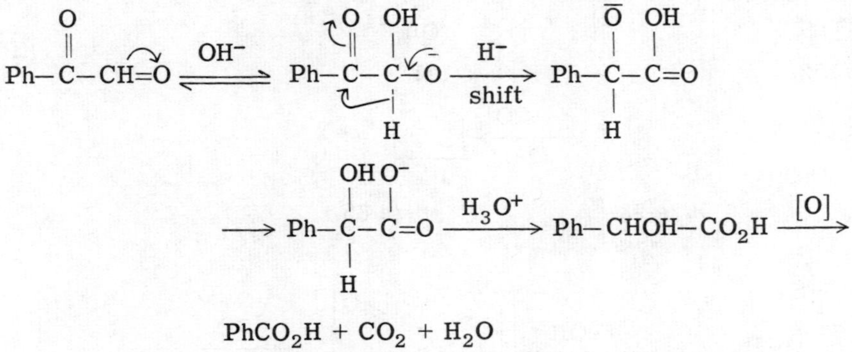

14. The mechanism of the Cannizzaro reaction, *whatever* it may be, results in the loss of hydrogen from one benzaldehyde molecule and addition of hydrogen to another benzaldehyde molecule. One possibility is that hydrogen is transferred directly from one molecule to the other. Another possibility is that one molecule transfers hydrogen to the solvent and the solvent transfers hydrogen to the other molecule of benzaldehyde. These two possibilities may be distinguished by carrying out the reaction in a solution of sodium hydroxide in water containing deuterium oxide (D_2O). If the solvent supplies the hydrogen, the benzyl alcohol produced will contain some deuterium. If no deuterium is present, the hydrogen must have come from another molecule of benzaldehyde.

In practice, it has been shown that the latter mechanism operates. The student should note that we are *not* concerned with the *hydroxyl* group of benzyl alcohol. The hydrogen in this group will *always* undergo deuterium exchange with D_2O. Only the fate of hydrogen which does not undergo this type of exchange can be investigated by this method.

15. For this type of problem, it is best to use Table 1.6 in conjunction with the known fragmentation patterns of *o*-substituted benzyl alcohols. The important point to remember is the effect of the *o*-substituent on the course of the fragmentations.

(i)

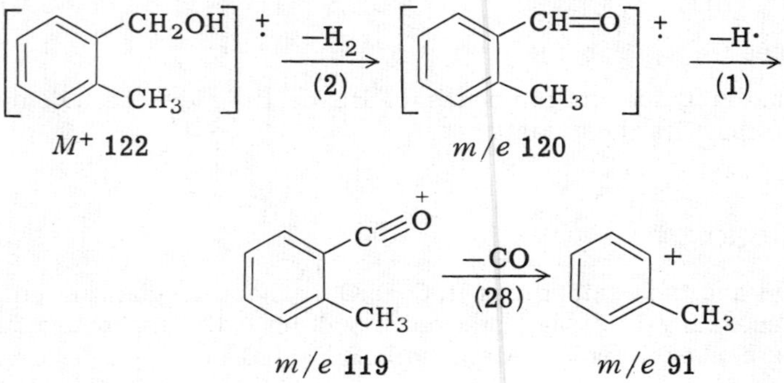

(ii)

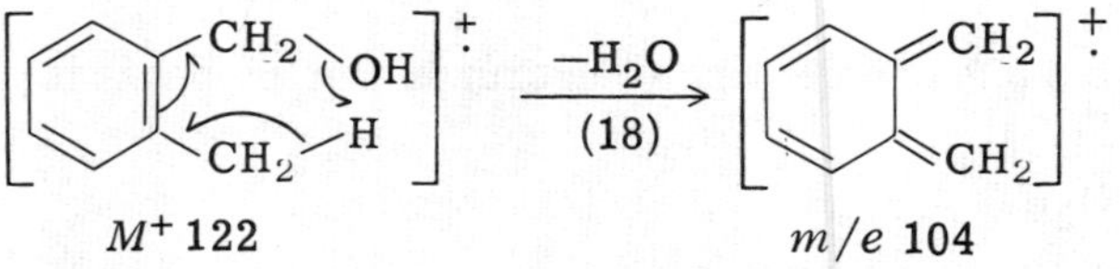

16. See the remarks made in Prob. 15, above.

(i)

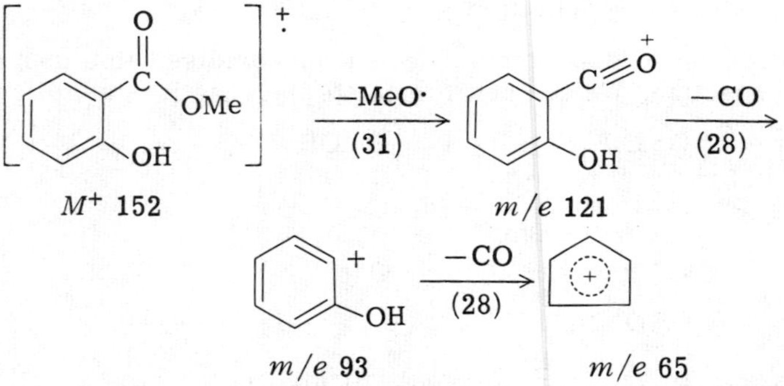

(ii)

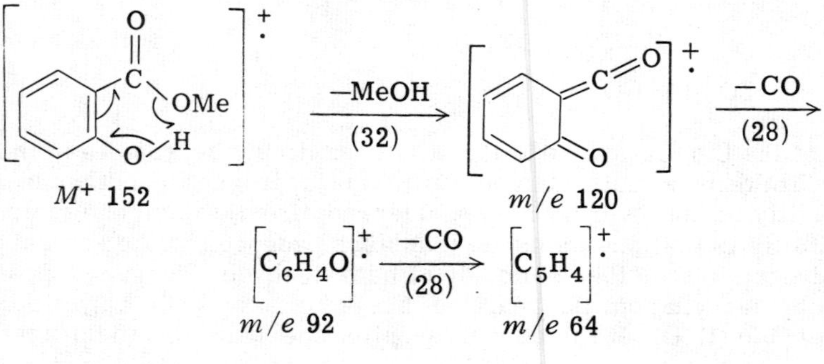

28 Aromatic acids

(Problems on page 51)

1. If we assume that the mechanism of hydrolysis involves addition of a molecule of water as the rate-determining step (see *Text*, Ch. 9), we have:

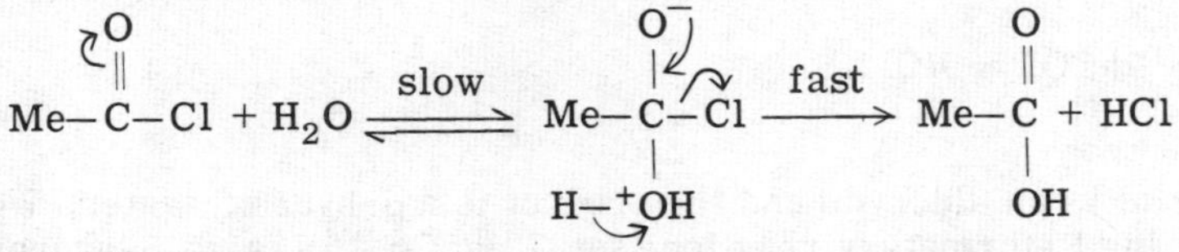

On the other hand, benzoyl chloride may be written as a resonance hybrid:

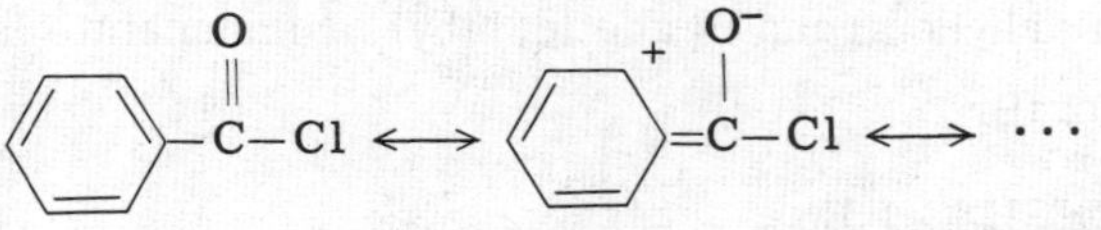

Hence, the electrophilic character of the carbon atom of the carbonyl group is much less than that in MeCOCl and so reaction with water (the nucleophile) is very much slower in benzoyl chloride than in acetyl chloride.

2. (i) $$PhCO(CH_2)_2CO_2H \xrightarrow[HCl]{Zn{-}Hg} Ph(CH_2)_3CO_2H$$

This is an example of the Clemmensen reduction.

(ii) $$Ph_2C(OH)CO_2H \xrightarrow[heat]{I_2/P} [Ph_2CICO_2H + HI] \longrightarrow Ph_2CHCO_2H + I_2$$

Hydrogen iodide is a very powerful reducing agent (see *Text*, Ch. 3).

(iii)

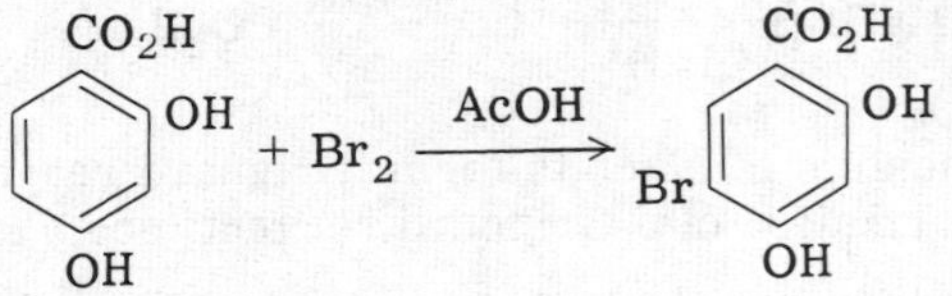

If the reaction is carried out in aqueous or in aqueous alkaline solution, the carboxyl group is also replaced by bromine (see (xv), below).

(iv)

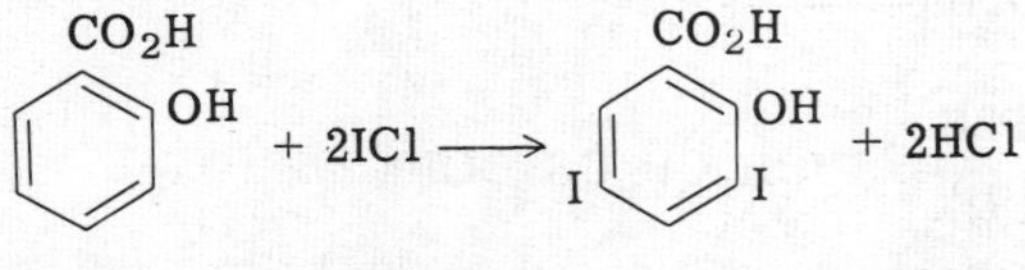

Iodine monochloride is a good iodinating reagent ($I \rightarrow Cl$).

(v) $$KNCO + AcOH \xrightarrow{H_2O} HNCO + AcOK$$

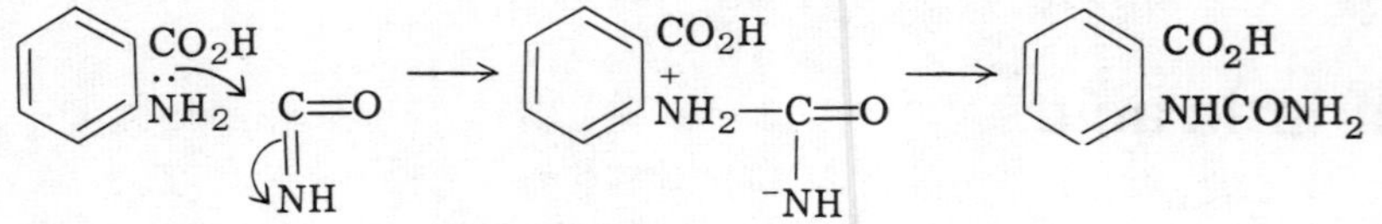

This type of reaction is commonly used for preparing substituted ureas (see *Text*, Ch. 16).

(vi)

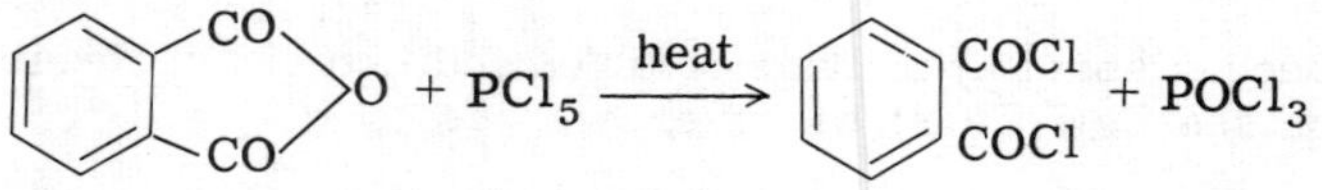

(vii) $PhCOCl + HF \longrightarrow PhCOF + HCl$

See *Text*, Ch. 9.

(viii) The starting material is an aldehyde and the product is a β-hydroxy-ester. We therefore have an example of the Reformatsky reaction (see *Text*, Ch. 15). Also, β-hydroxy-esters are dehydrated on heating (see *Text*, Ch. 17).

Since the ester residue to be added to the aldehyde group is acetic, ethyl bromoacetate is used. Hence:

$$PhCHO + BrCH_2CO_2Et + Zn \longrightarrow Ph\overset{\displaystyle OZnBr}{\overset{|}{C}}HCH_2CO_2Et$$

$$\xrightarrow{H_3O^+} PhCHOHCH_2CO_2Et \xrightarrow[-H_2O]{heat} PhCH{=}CHCO_2Et$$

(ix)

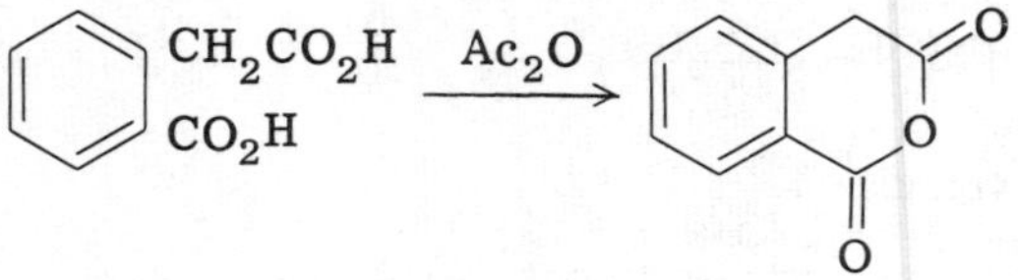

1, 4- and 1, 5-Dicarboxylic acids are readily dehydrated by acetic anhydride to give the corresponding cyclic anhydride (see *Text*, Ch. 16).

(x)

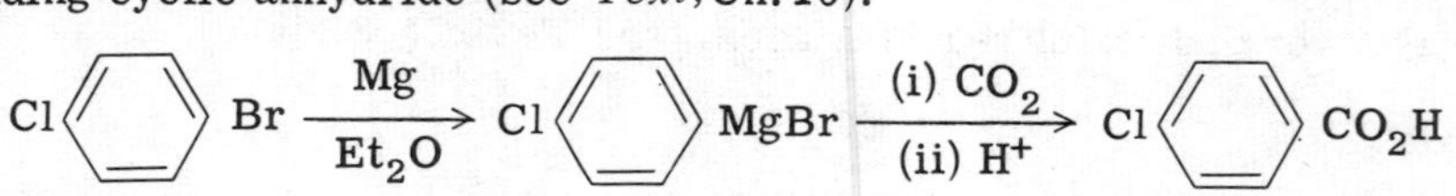

Aryl bromides and iodides react with magnesium in *ether* to form Grignard reagents. Aryl chlorides require THF as solvent for the formation of a Grignard reagent (see *Text*, Ch. 21).

(xi) γ- and δ-Lactones react with concentrated hydrochloric acid to form the corresponding chloroacid (see *Text*, Ch. 17).

$$\underbrace{CH_2CH_2CH_2CO}_{O} \overset{HCl}{\rightleftharpoons} Cl(CH_2)_3CO_2H \xrightarrow{SOCl_2}$$

$$Cl(CH_2)_3COCl \xrightarrow[AlCl_3]{PhH} ClCH_2CH_2CH_2COPh$$

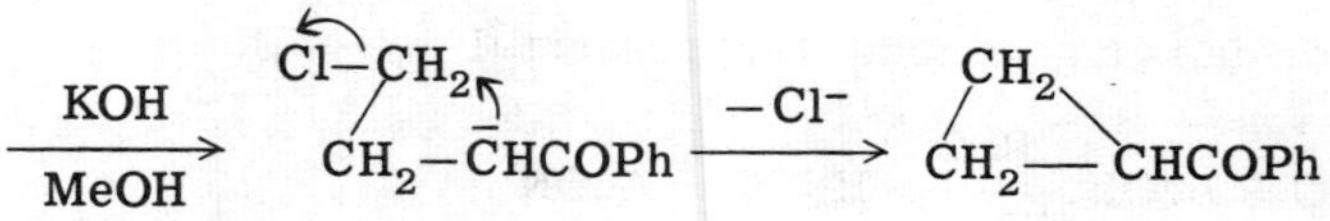

The chlorine atom of the carbonyl group is far more reactive than the other chlorine atom (of the alkyl chloride type), and so the Friedel-Crafts reaction leads to a ketone. The methylene group adjacent to the carbonyl group is activated. Hence, a carbanion is formed (at this carbon atom), and this is followed by ring closure to give a cyclopropane derivative.

(xii) The added residue in the product is $OCCH{=}CHCO_2H$. This suggests the possibility of using $HO_2CCH{=}CHCO_2H$ (maleic and/or fumaric acid). Since aluminium chloride is given in the equation, a Friedel-Crafts reaction is almost certainly involved. This reaction can be carried out with cyclic anhydrides to give keto-acids. However, because only maleic acid forms an anhydride, this is the one to be used (see *Text*, Ch. 17).

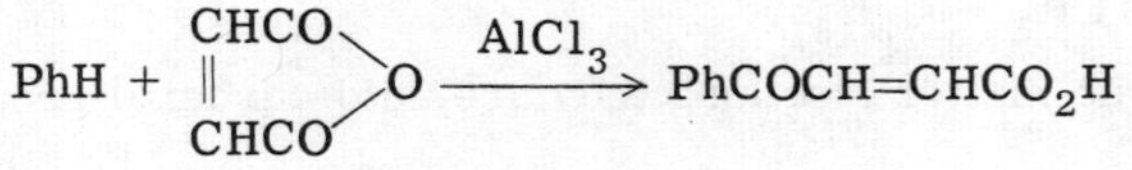

(xiii)

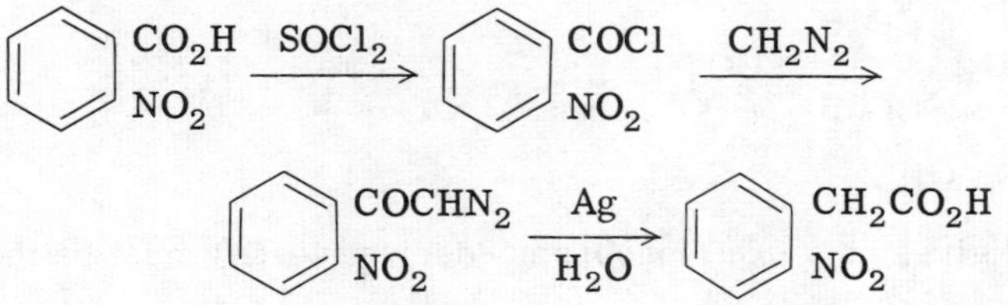

Diazoketones undergo the Wolff rearrangement to form acids (see *Text*, Ch. 13).

(xiv) $$PhCOCH_3 \xrightarrow[HCl]{NaNO_2} [PhCOCH_2NO] \longrightarrow PhCOCH{=}NOH \xrightarrow{Ac_2O} PhCOCN \xrightarrow{H^+} PhCOCO_2H$$

An activated methylene group readily reacts with nitrous acid to form a nitroso-compound, which rearranges to the oxime (this is characteristic of the CH—NO group; see *Text*, Ch. 8). Aldoximes are readily dehydrated to cyanides.

(xv)

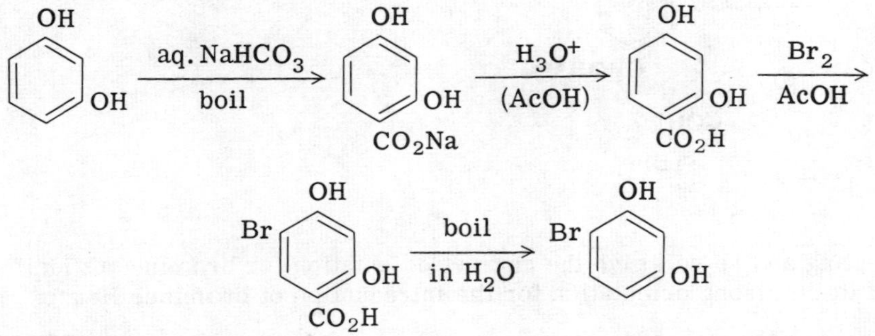

The first step is an example of the Kolbe-Schmitt reaction. When the ring contains two *m*-hydroxyl groups, the position *ortho* to one and *para* to the other is now sufficiently activated to be carboxylated under mild conditions (compared to the presence of only one hydroxyl group). Aqueous sodium carbonate, on boiling, generates the carbon dioxide for the reaction. The presence of hydroxyl groups *ortho* and/or *para* to a carboxyl group causes replacement of the latter by bromine when treated with bromine-water (or bromine in alkaline solution). In the absence of water, the usual product is the brominated acid which, on boiling with water, eliminates carbon dioxide (see *Text*, and also *Text*, Ch. 26).

3. The high unsaturation and the use of Br_2/Fe suggest that **(A)** is aromatic. Also, the D.B.E. of **(A)** is (see Ch. 1, Prob. 9):

$9 + 1 - 12/2 = 4 \equiv$ benzene

Furthermore, if **(A)** were acyclic, oxidation with permanganate would give a number of fragments, i.e., there would be a loss of more than one carbon atom.

(B) $C_8H_6O_4$ D.B.E. $= 8 + 1 - 6/2 = 6$

Since the benzene ring accounts for the D.B.E. of 4, **(B)** contains two more double bonds than **(A)**. Also, since **(B)** contains four oxygen atoms, this suggests the possibility of two carboxyl groups (= two double bonds), i.e., **(B)** may be written as $C_6H_4(CO_2H)_2 = C_8H_6O_4$. Hence, **(A)** must have two side-chains.

(A) $C_9H_{12} = C_6H_4 + C_3H_8 = C_6H_4 + CH_3 + C_2H_5$

Vigorous oxidation converts side-chains, no matter how long they are, into carboxyl groups directly attached to the ring. Thus:

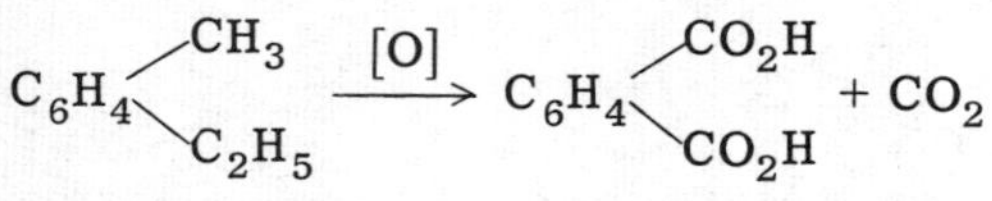

(A) **(B)**

Since the bromination of **(B)** gives only *one* monobromo-derivative, **(B)** must be the *p*-isomer. Hence:

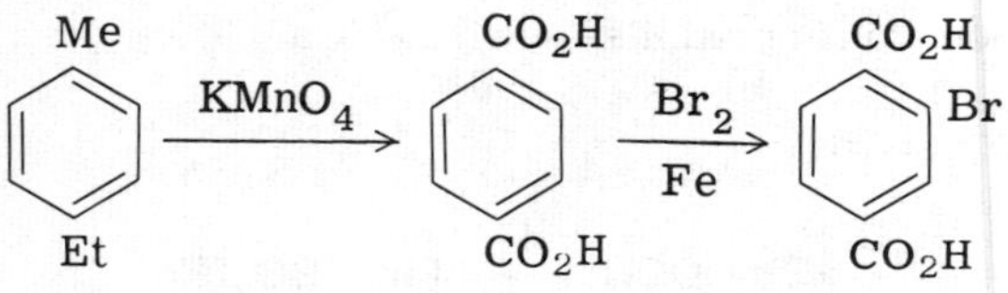

(A) **(B)** **(C)**

4. (i)

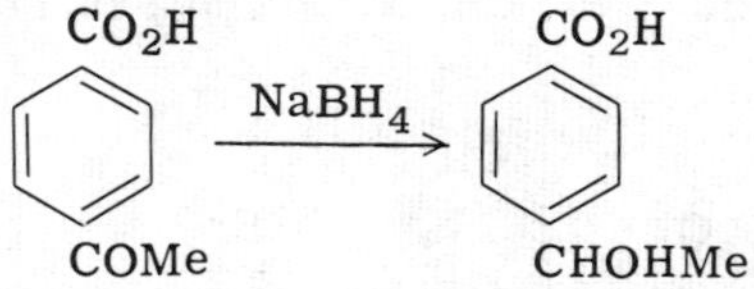

(ii)

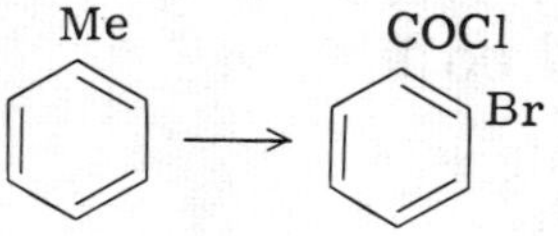

Me is *o*/*p*-directing and so will give the correct orientation for bromine. Oxidation of Me to CO_2H will lead to the wrong orientation for the introduction of bromine. Hence:

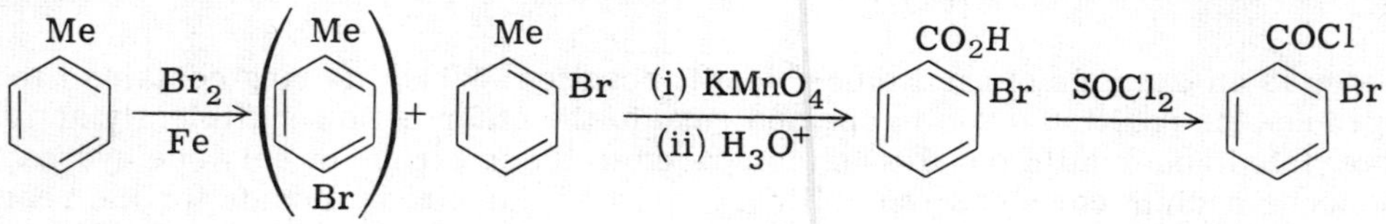

(iii)

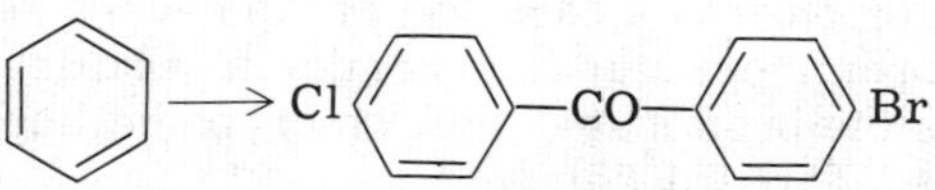

Since the product is a ketone, this suggests a Friedel-Crafts reaction with an acid chloride. There are two possible combinations:

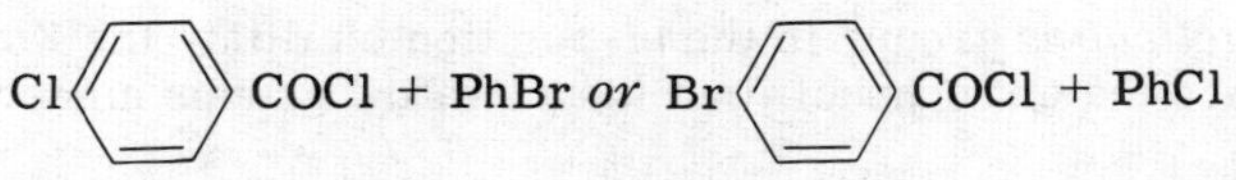

There appears to be little to choose between the two; we shall describe the first one.

(*a*)

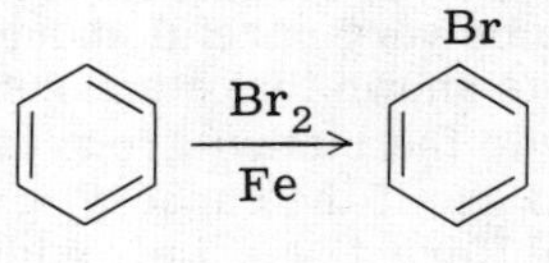

(*b*)

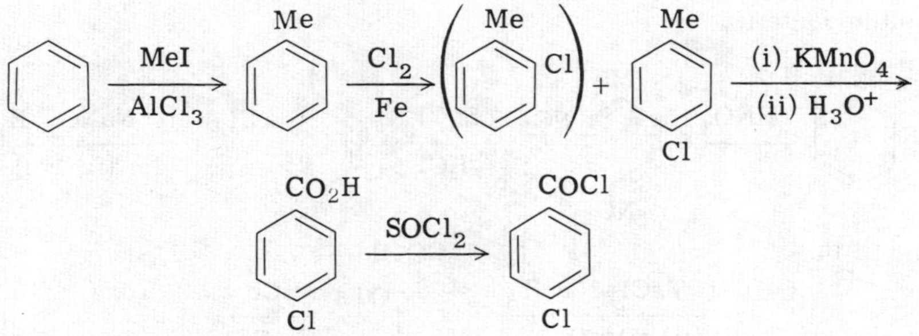

An alternative route is:

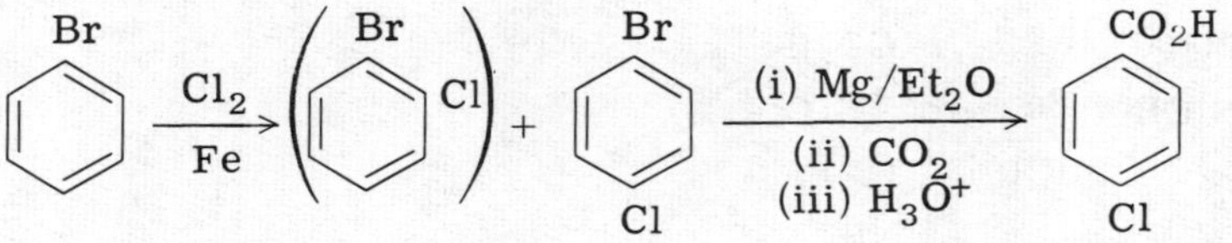

(*c*)

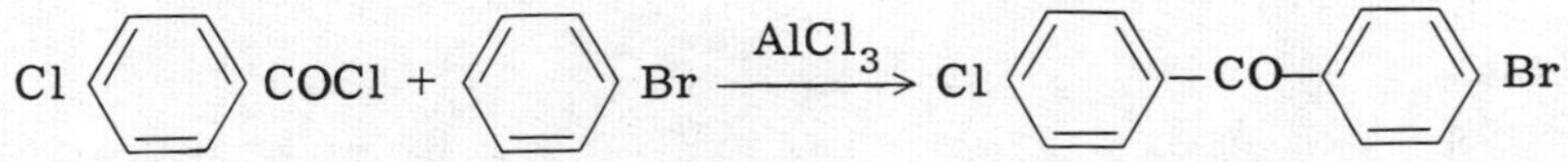

Very little *o*-product is formed (largely a steric effect).

(iv)

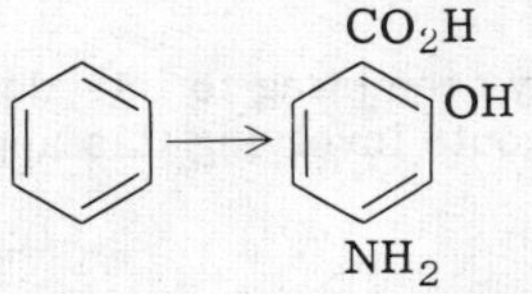

The presence of a carboxyl group *ortho* to a hydroxyl group suggests the possibility of the Kolbe-Schmitt reaction. The *m*-orientation of the OH and NH_2 groups suggests the introduction of the OH via a diazonium salt. Thus, we require *m*-dinitrobenzene. Hence, a possible route is:

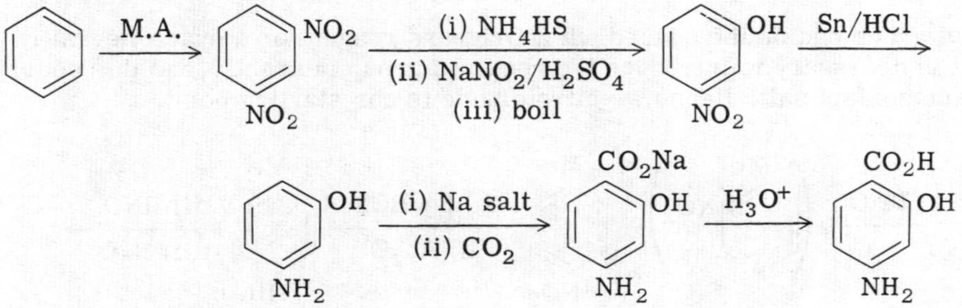

As we have seen in Prob. 2(xv), two *activating* groups in the *m*-position facilitate the Kolbe-Schmidt reaction. We therefore reduce the nitrophenol first (deactivating groups hinder the reaction).

Whenever the benzene ring contains a carboxyl group, it is always tempting to try the oxidation of a methyl side-chain to obtain it. If we introduce a methyl group, then we get the correct orientation for the two nitro-groups. We would now have to be very careful about selective reduction and also about the stage of the oxidation. It is known that stannous chloride-hydrochloric acid selectively reduces the nitro-group *ortho* to the methyl group (see *Text*, Ch. 23), but, unfortunately, it is not certain which reagent to use if the carboxyl group is present instead of the methyl. It is therefore better to try a different way, i.e., oxidise at a later stage. A possible route is:

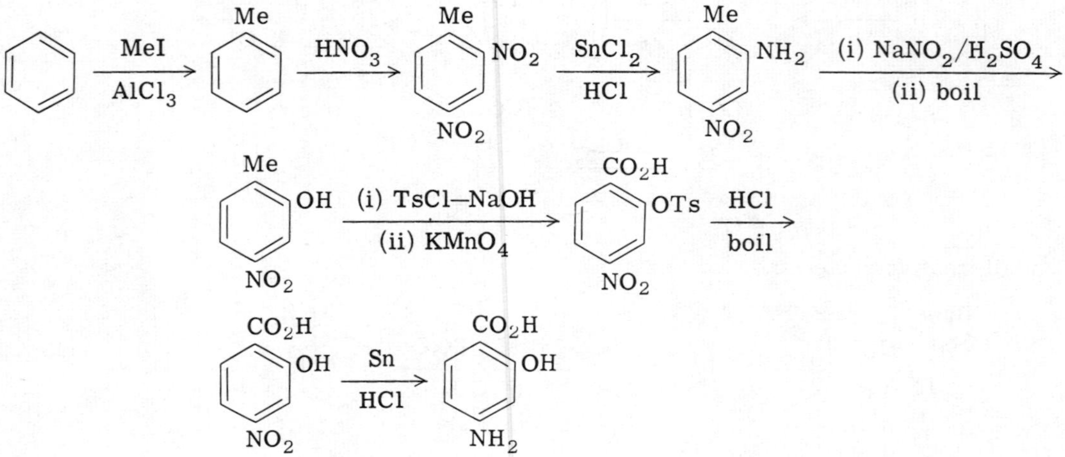

5.

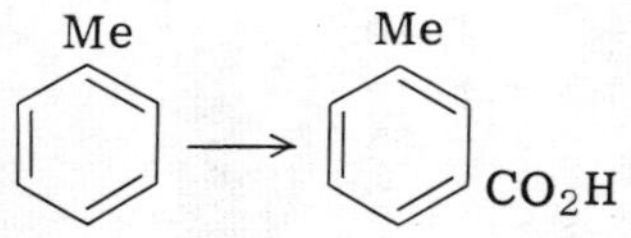

(*a*) It was pointed out in Prob. 4 (iv), above, that it is always tempting to introduce a carboxyl group by oxidation of a methyl group. A possible route involving this approach is (see *Text*, Ch. 20):

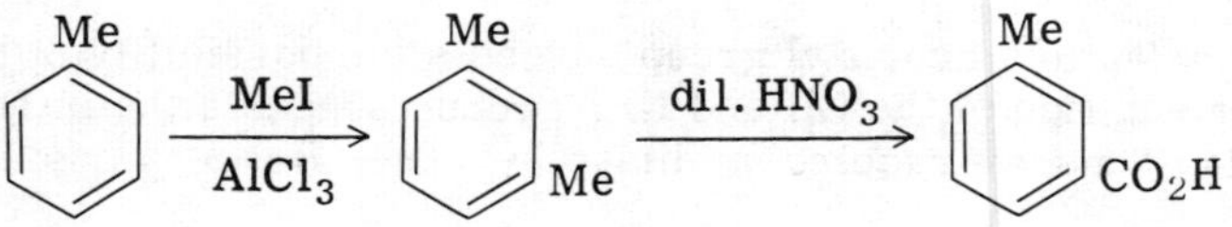

(*b*) A carboxyl group can be introduced via a Grignard reagent or a cyano-derivative. In both cases, it is necessary to introduce the required group indirectly, and this could be done via the *m*-diazonium salt. Hence, *m*-nitrotoluene is our starting point.

(i)

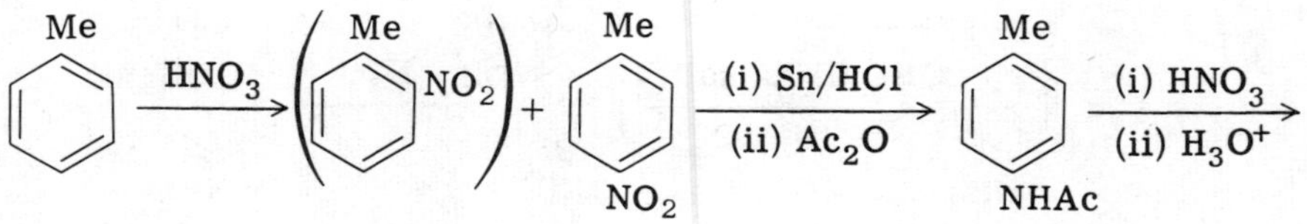

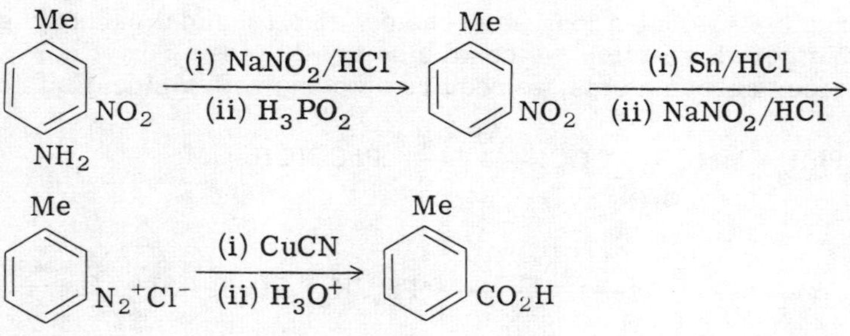

(ii)

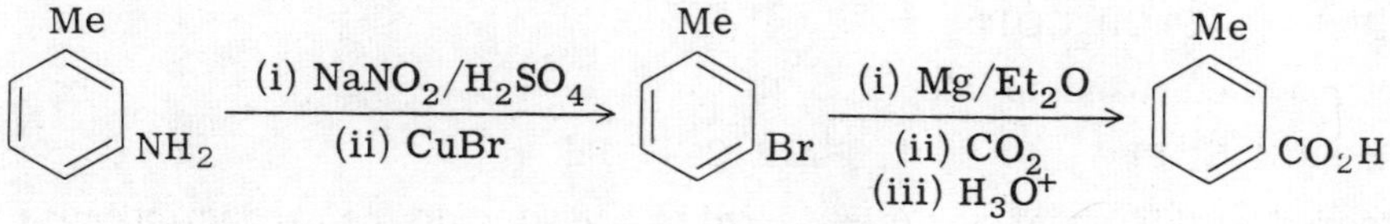

6.

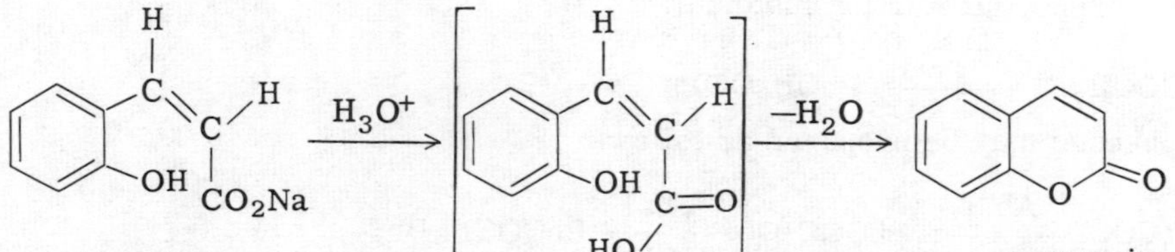

coumarinic acid (salt)

coumarin

Because of the proximity of the hydroxyl and carboxyl groups, formation of the six-membered lactone occurs extremely readily.

Since the corresponding 3-nitro-derivative gives the free acid under the same conditions, the nitro-group must prevent lactone formation.

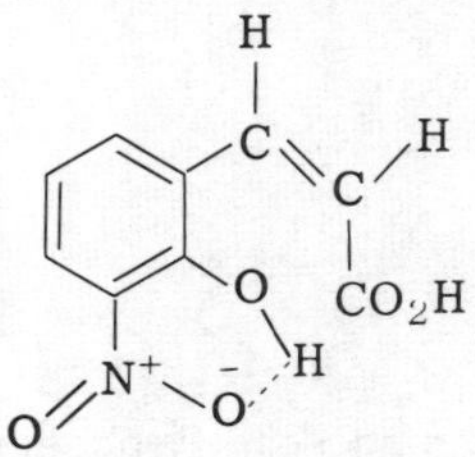

One possibility could be the steric effect. Another possibility is that the nitro-group forms a hydrogen bond with the *o*-hydroxyl group and that this is strong enough to stabilise the acid, thereby preventing lactone formation. Further information is required to decide which (if either) is the correct explanation, e.g., i.r. spectroscopy would show the presence of an intramolecular hydrogen bond. According to the literature, hydrogen bonding is the correct explanation.

7. (i) $PhH \longrightarrow Ph(CH_2)_3CO_2H$

One method is the Friedel-Crafts reaction with a cyclic anhydride, followed by a Clemmensen reduction. Since the side-chain contains four carbon atoms, we require succinic anhydride.

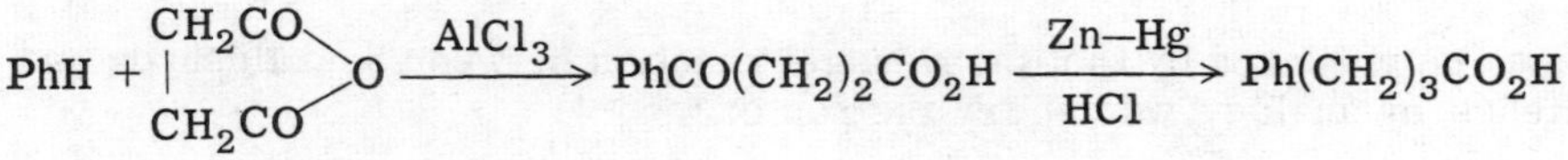

Alternatively, since a long side-chain with a terminal carboxyl group suggests the use of the Willgerodt reaction, we could also try this (see *Text*, Ch. 27). Since the side-chain contains four carbon atoms, we require a ketone with a side-chain of four carbon atoms.

$$PhH + Me(CH_2)_2COCl \xrightarrow{AlCl_3} PhCO(CH_2)_2Me$$

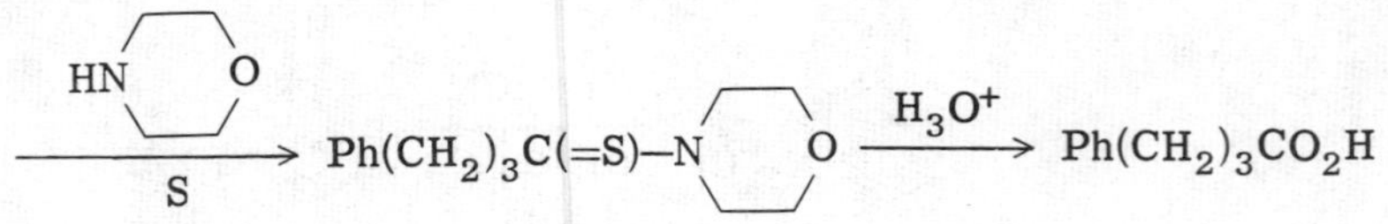

(ii) $PhH \longrightarrow PhCH_2COPh$

The product may be broken up into:

(*a*) $Ph + CH_2COPh$ *or* (*b*) $PhCH_2CO + Ph$

We have ignored, at this stage, any other ways of breaking up the product because both the ways given use benzene, the starting material.

(*a*) $PhH + PhCOCH_2Br \xrightarrow{AlCl_3} PhCH_2COPh$

ω-Bromoacetophenone may be prepared as follows:

$$PhH + CH_3COCl \xrightarrow{AlCl_3} PhCOCH_3 \xrightarrow[AlCl_3]{Br_2} PhCOCH_2Br$$

(*b*) $PhH + PhCH_2COCl \xrightarrow{AlCl_3} PhCH_2COPh$

Phenylacetyl chloride may be prepared:

(i) $PhCOCH_3 \xrightarrow{\text{Willgerodt}} PhCH_2CO_2H \xrightarrow{SOCl_2} PhCH_2COCl$

(ii)

$$PhH \xrightarrow[\text{(ii) } Mg/Et_2O]{\text{(i) } Br_2/Fe} PhMgBr \xrightarrow[\text{(ii) } H_3O^+]{\text{(i) } CH_2-CH_2 \text{ (O bridged)}}$$

$$PhCH_2CH_2OH \xrightarrow[\text{dichr.}]{\text{acid}} PhCH_2CO_2H$$

If we now break up the product as $PhCH_2$ + COPh, we see that there is no readily observable method of joining them. In this case, since we have offered reasonable syntheses, there is little point in pursuing the matter.

The student who is interested in some other methods can look up stilbene, diphenylacetylene, and benzoin in *Text*, Ch. 29, and see whether he can devise any syntheses from these compounds.

(iii)

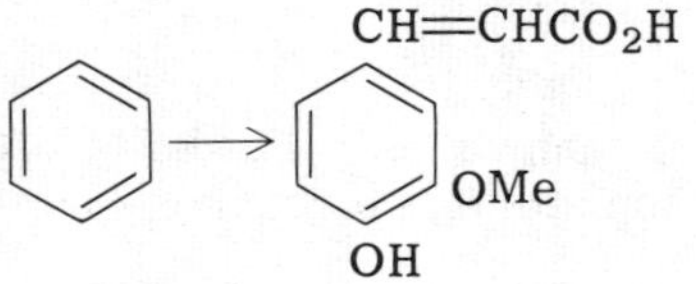

Cinnamic acid may be prepared by the Knoevenagel reaction between benzaldehyde and malonic acid. Hence, by analogy, we can try the route:

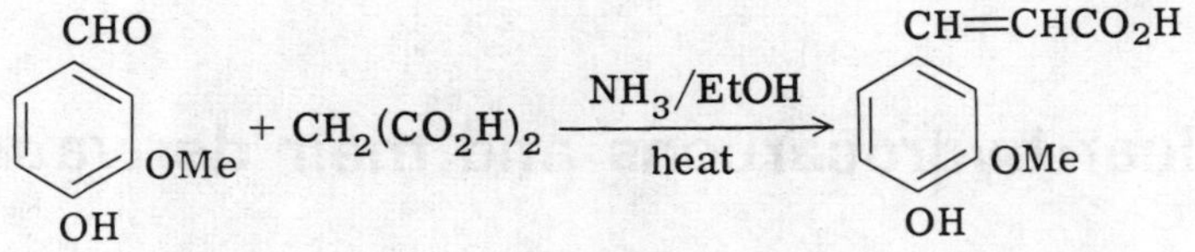

What we now have to do is to synthesise 4-hydroxy-3-methoxybenzaldehyde from benzene. This may be done as follows, the key compound being catechol (*o*-dihydroxybenzene).

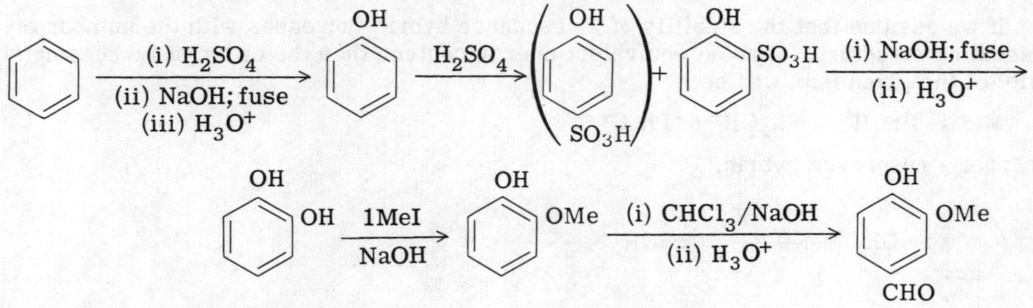

The last step is the Reimer-Tiemann reaction (see *Text*, Ch. 27). This can also be used, together with the Dakin reaction to prepare catechol (see *Text*, Ch. 26).

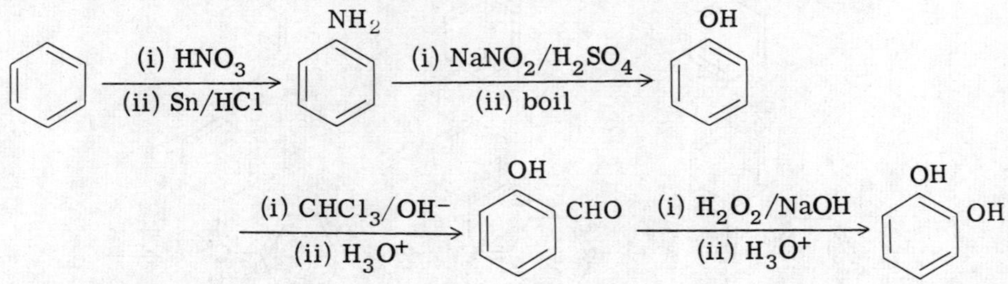

Note the alternative way of preparing phenol from benzene.

The student should also refer to Ch. 27, Prob. 2 (xv) for another method of preparing 4-hydroxy-3-methoxybenzaldehyde.

29 Polynuclear hydrocarbons and their derivatives

(Problems on page 53)

1. If we assume that the stability of a resonance hybrid increases with the number of resonating structures of about equivalent energy content, then the order of increasing stability of the given ions will be:

$Me^+ < PhCH_2^+ < Ph_2CH^+ < Ph_3C^+$

Me^+; not a resonance hybrid.

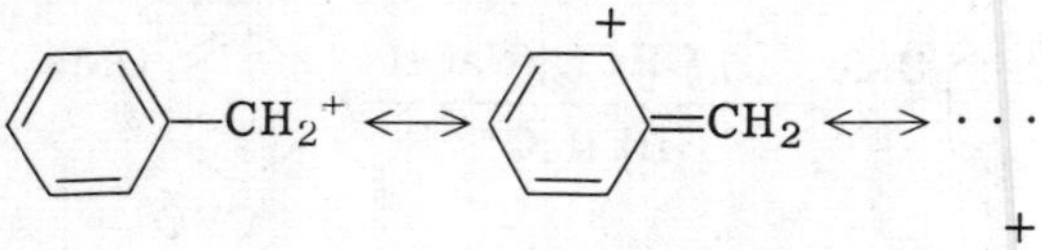

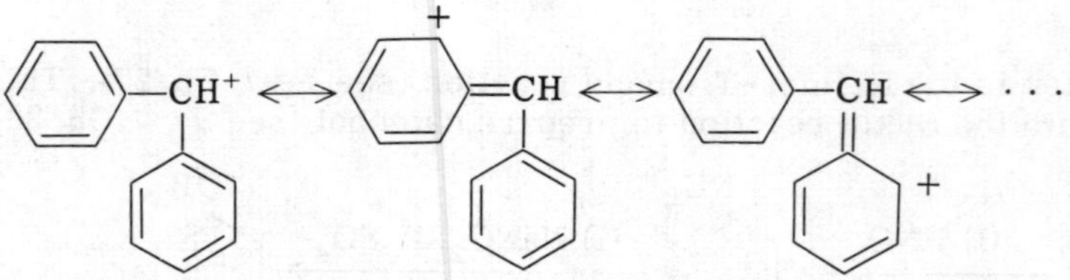

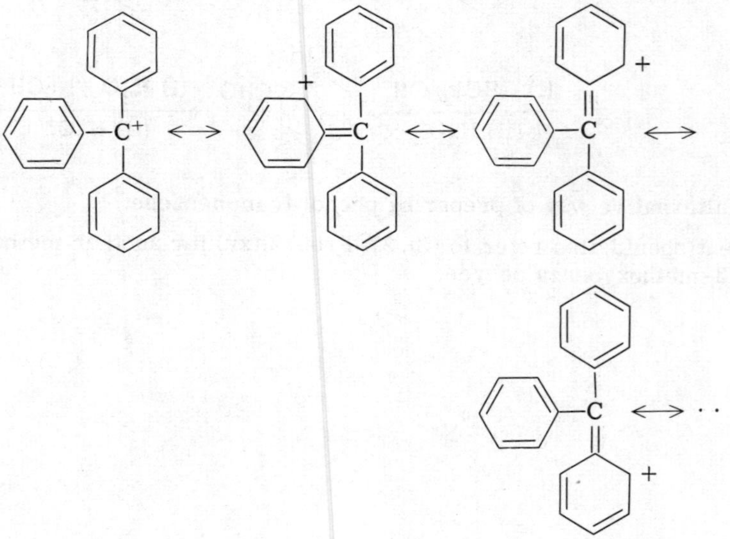

2. Triphenylmethanol is a tertiary alcohol; this type of alcohol may be prepared from a Grignard reagent and:

(i) a ketone; (ii) an ester other than formic (also the acid may be used); (iii) diethyl carbonate (see *Text*, Ch. 15).

(i) **Benzophenone** (1 mole of Grignard)

$$Ph_2CO + PhMgBr \longrightarrow Ph_3COMgBr \xrightarrow{H_3O^+} Ph_3COH$$

(ii) **Ethyl benzoate** (2 moles of Grignard)

$$PhCO_2Et + PhMgBr \longrightarrow Ph_2C\begin{matrix}OMgBr\\OEt\end{matrix} \xrightarrow{PhMgBr} Ph_3COMgBr \xrightarrow{H_3O^+} Ph_3COH$$

(iii) **Diethyl carbonate** (3 moles of Grignard)

$$O{=}C(OEt)_2 + PhMgBr \longrightarrow PhC\begin{matrix}OMgBr\\OEt\\OEt\end{matrix} \xrightarrow{PhMgBr} Ph_2C\begin{matrix}OMgBr\\OEt\end{matrix} \xrightarrow{PhMgBr}$$

$$Ph_3COMgBr \xrightarrow{H_3O^+} Ph_3COH$$

3. The two likely mechanisms that are possible for this reaction are SN1 and SN2 (see *Text*, Ch. 5).

$$\text{S}_\text{N}1\ Ph_2CHCl \underset{}{\overset{\text{slow}}{\rightleftharpoons}} Cl^- + Ph_2CH^+ \xrightarrow[\text{fast}]{F^-} Ph_2CHF$$

$$\text{S}_\text{N}2\ Ph_2CHCl + F^- \longrightarrow Ph_2CHF + Cl^-$$

In the SN2 mechanism, since there is only one reaction step which involves the two reactant molecules, the addition of chloride ions would not be expected to affect the rate. On the other hand, if the mechanism is SN1, the rate is determined by the heterolytic step. In this case, the addition of chloride ions results in a common ion effect. The equilibrium constant, K, for the ionisation step, is given by:

$$K = \frac{[Ph_2CH^+][Cl^-]}{[Ph_2CHCl]}$$

Since K is a *constant* at a fixed temperature, increasing the concentration of the chloride ion (by the addition of sodium chloride), increases the value of the numerator and hence, to keep K constant, the denominator must increase, i.e., the equilibrium is shifted to the left. Thus, the concentration of the benzhydryl carbonium ion is decreased, and consequently its rate of reaction with the fluoride ion is decreased. This means, in effect, that in these circumstances, both the chloride and fluoride ions are competing for the benzhydryl carbonium ion, and as the chloride ion concentration increases, the rate of the combination of the carbonium ion with fluoride ion decreases. Hence, a point may be reached where the step involving the fluoride ion becomes rate-determining, i.e., the mechanism is SN2(C^+) [see *Text*, Ch. 5].

4. If the rearrangement is intermolecular, the reaction may be written as:

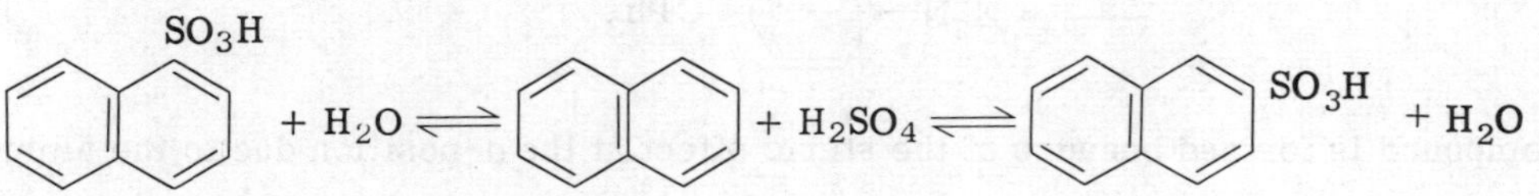

Whatever are the details of the removal of the sulphonic acid group, it is removed and then re-introduced. If, then, the 1-sulphonic acid is labelled with ^{35}S, and this is heated in concentrated sulphuric acid, some ^{32}S will be incorporated into the 2-sulphonic acid if the reaction is intermolecular, but none will be incorporated if the reaction is intramolecular. Alternatively, if the 1-sulphonic acid containing ^{32}S is heated in concentrated sulphuric acid labelled with ^{35}S, this will be incorporated into the 2-acid if the reaction is intermolecular, but will not be incorporated if the reaction is intramolecular.

Both of these experiments have actually been carried out, and both showed that the reaction is intermolecular. Scholl *et al.* (1955) used the second technique and Vorozhtsov *et al.* (1961) used the first.

^{35}S is radioactive, and so may be readily identified.

5. (i)

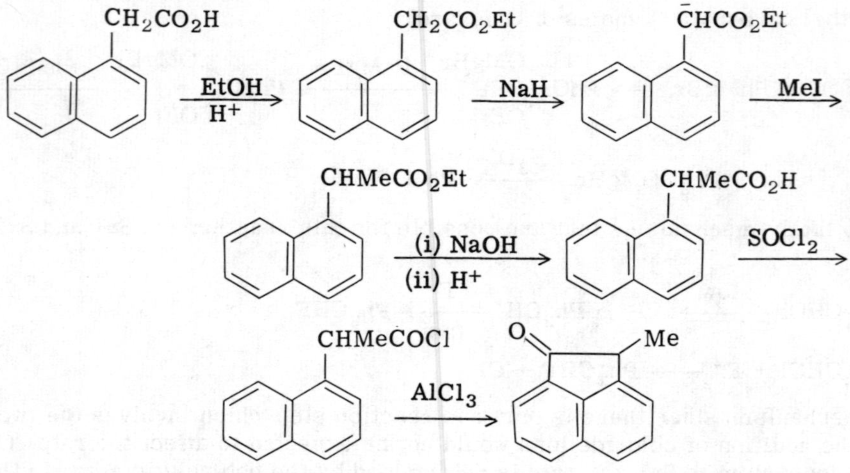

Ring-closure at the 8-position to give a *five-membered* ring is preferred to ring-closure at the 2-position to give a *four-membered* ring (see *Text*, Ch. 19).

(ii) (*a*) $CHBr_3 + t\text{–}BuOK \longrightarrow CBr_2 + t\text{–}BuOH + KBr$

(*b*)

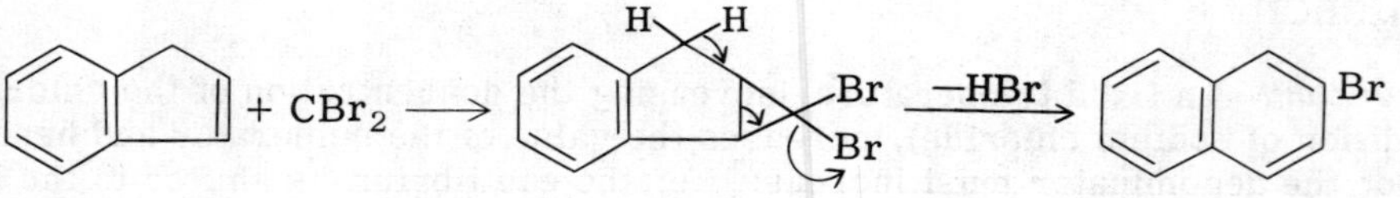

(iii)

$$PhNH_2{\cdot}HCl \xrightleftharpoons{AcOH} PhNH_2 + HCl$$

$$Ph_3COH + HCl \xrightleftharpoons{-H_2O} Ph_3C^+ + Cl^-$$

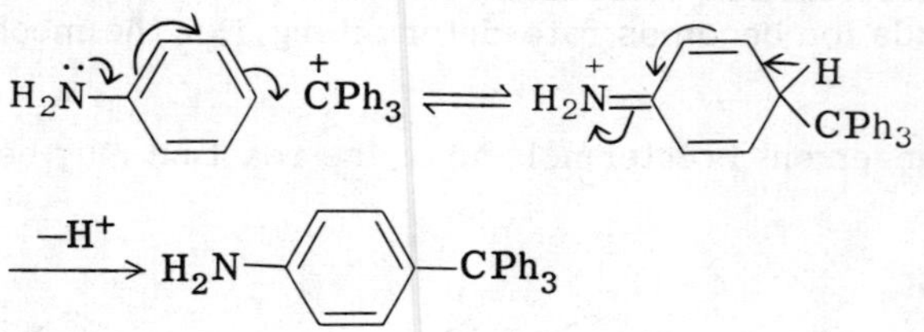

The *p*-compound is formed because of the steric effect at the *o*-position due to the amino-group.

(iv)

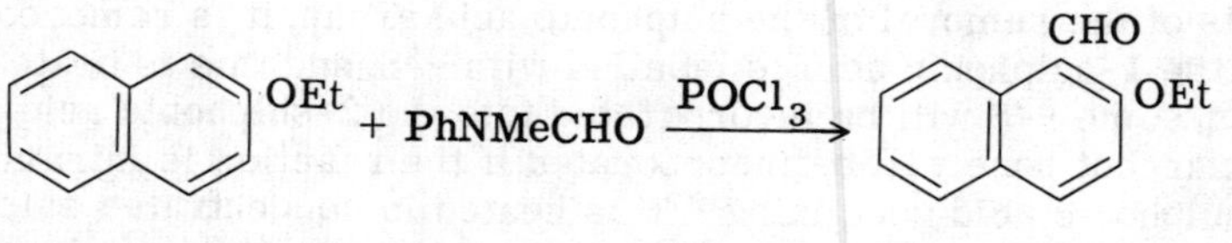

(See also Prob. 15, below). This is an example of the Vilsmeier-Haak formylation.

(v)

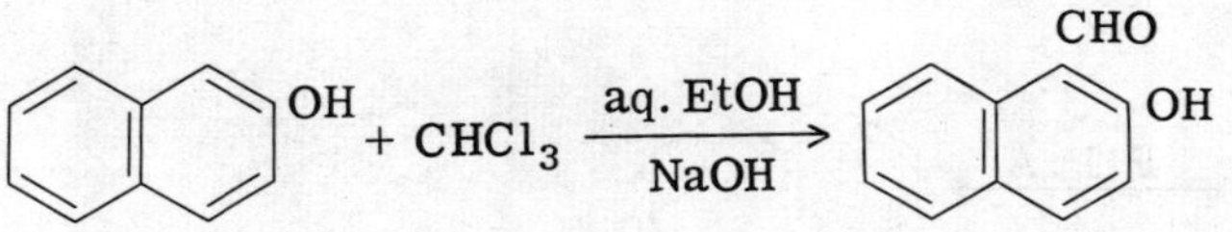

(See also Prob. 15, below). This is an example of the Reimer-Tiemann reaction.

(vi)

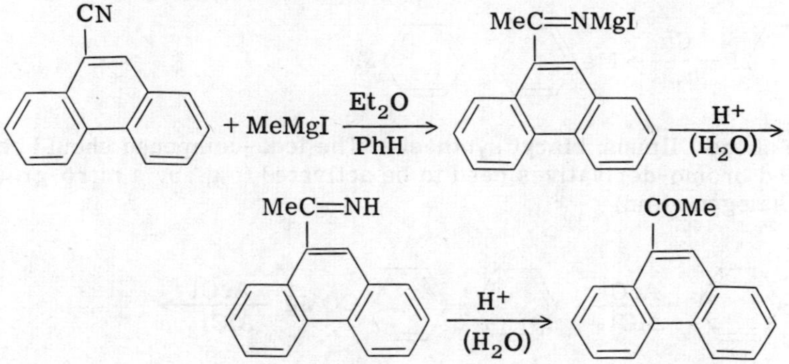

(vii)

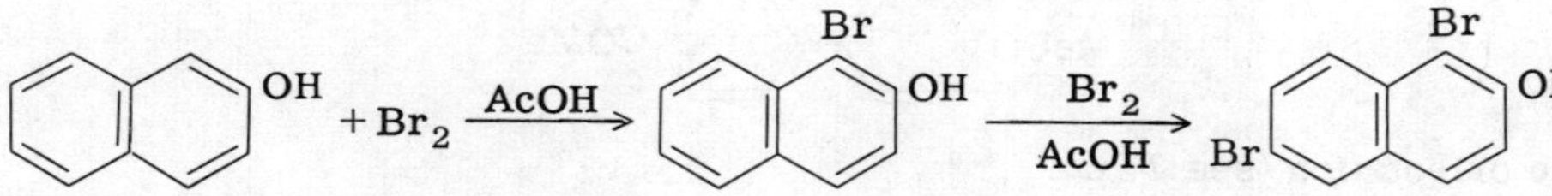

(See also Prob. 15, below).

(viii) $PhCHOHCOPh \xrightarrow[\text{aq. NaOH}]{NaBrO_3} PhCOCOPh \xrightarrow[\text{heat}]{\text{aq. NaOH}} Ph_2C(OH)CO_2^-Na^+$

The last step is the benzilic acid rearrangement.

(ix)

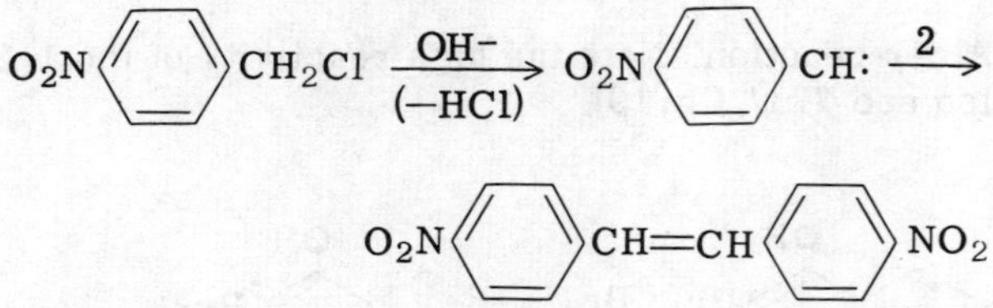

The arylmethylene is readily formed, since it is stabilised by resonance (extended conjugation with the *p*-nitro-group).

(x)

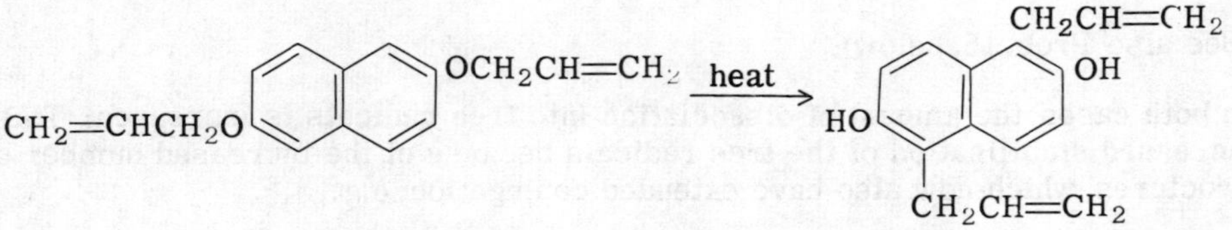

This is an example of the Claisen *ortho*-rearrangement (see Text, Ch. 26), and migration occurs to give the 1, 5-diallyl derivative and not the 3, 7- (see also Prob. 15, below).

(xi)

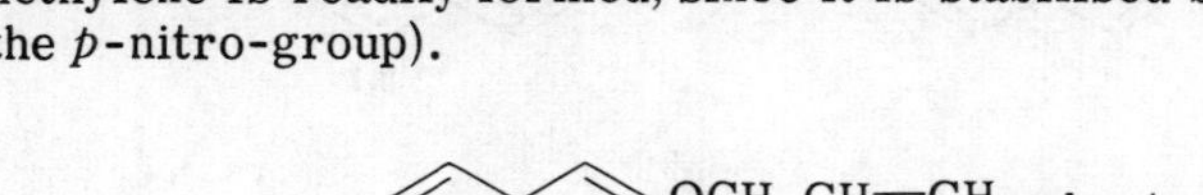

(See also Prob. 15, below).

(xii)

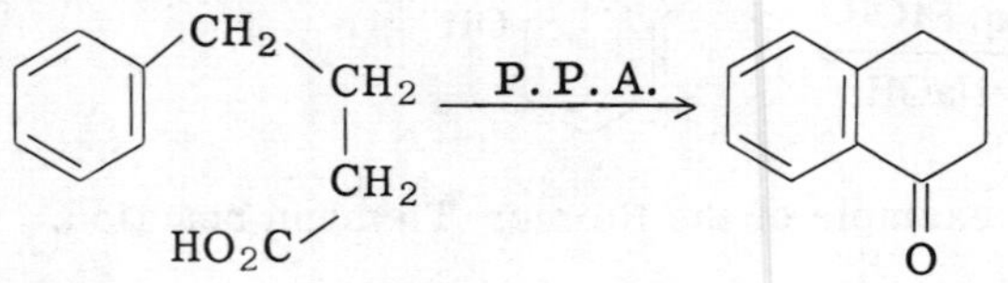

(xiii)

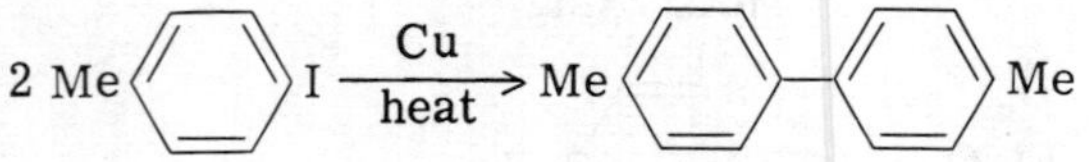

This is an example of the Ullmann biaryl synthesis. The iodo-compound should be used, since the chloro- and bromo-derivatives need to be activated (e.g., by a nitro-group *ortho* and/or *para* to the halogen atom).

(xiv)

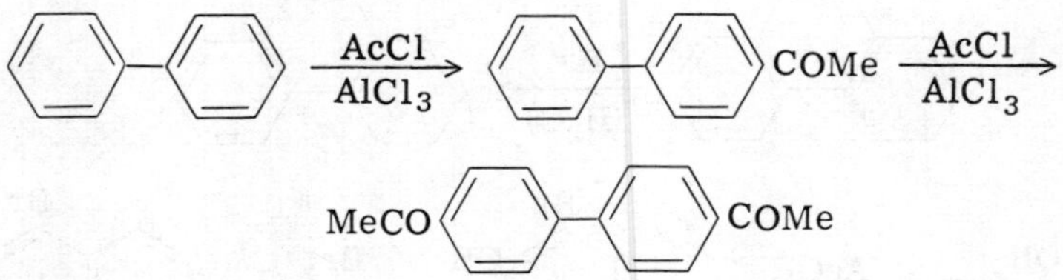

Note the orientation (see *Text*).

(xv)

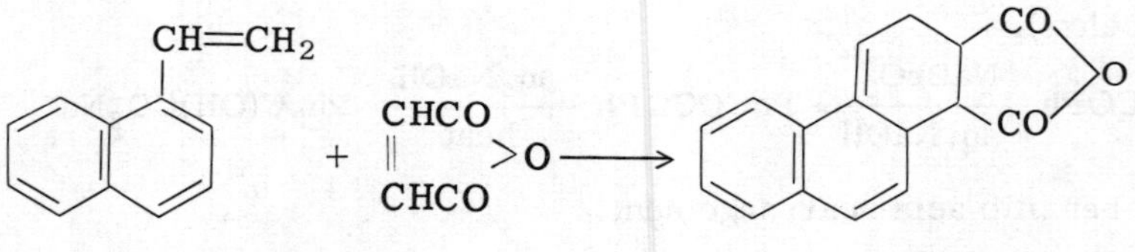

This is an example of the Diels-Alder reaction. Note the high reactivity of the 1, 2-'double bond' in naphthalene (see *Text*; also see *Text*, Ch. 19).

(xvi)

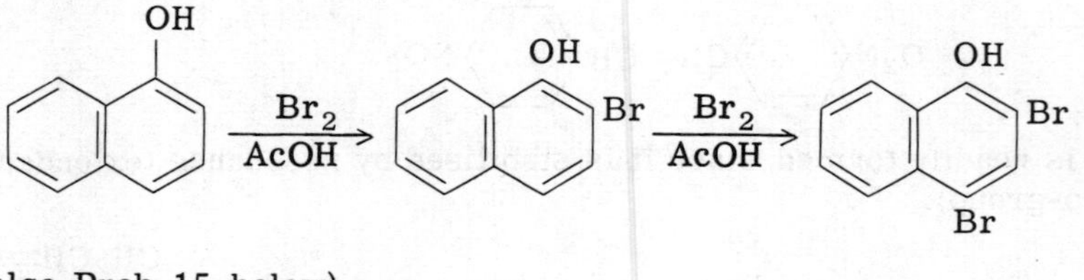

(See also Prob. 15, below).

6. In both cases, the amount of dissociation into free radicals is increased. This is due to the increased stabilisation of the free radicals because of the increased number of resonating structures, which now also have extended conjugation, e.g.,

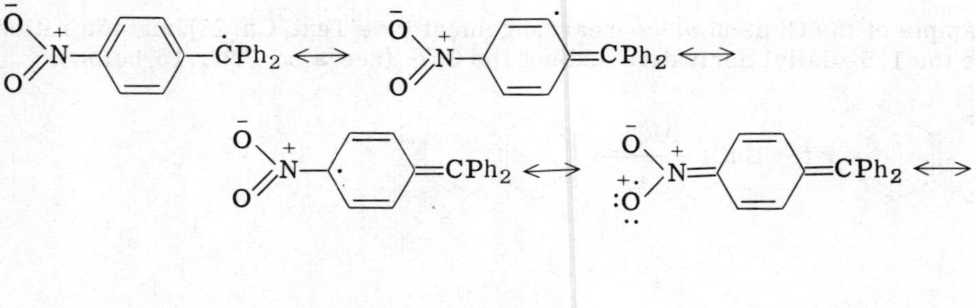

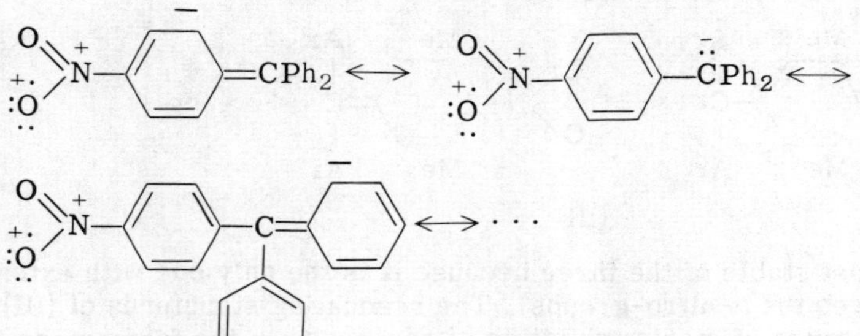

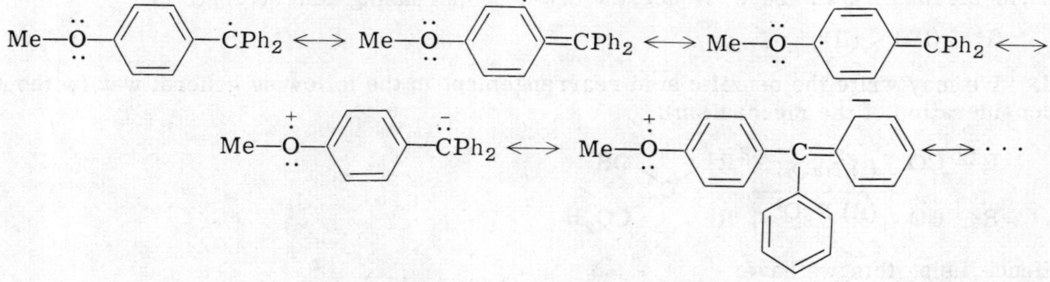

7.

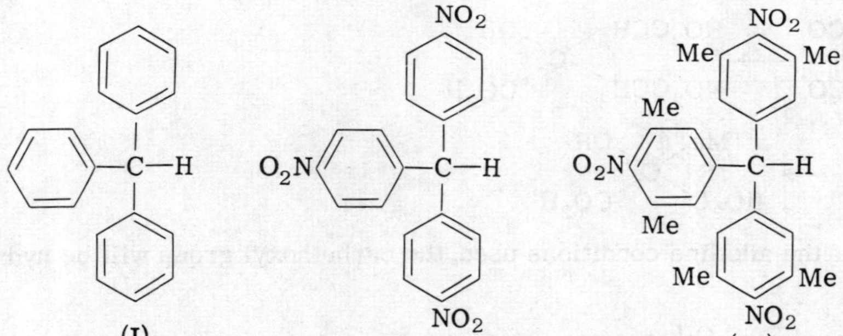

(I) (II) (III)

Any factor that facilitates release of the proton of the methine group (CH) and/or stabilises the carbanion (conjugate base) with respect to its acid, will increase the strength of the acid.

(i) The nitro-group exerts a —I effect and so will facilitate release of the proton in (II) and (III), and hence these are stronger acids than (I).

(ii) The cB can be represented as a resonance hybrid in each case.

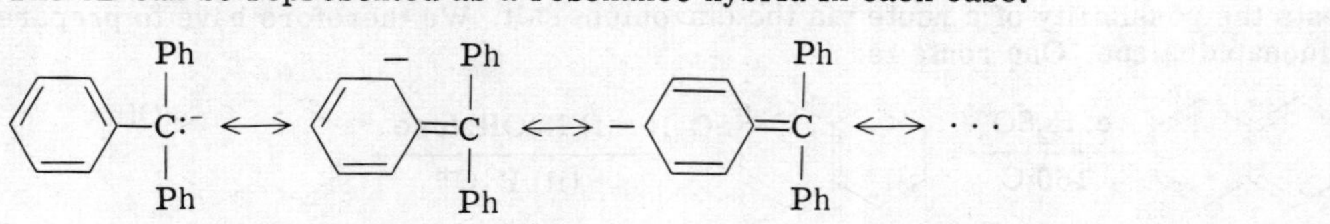

(I)

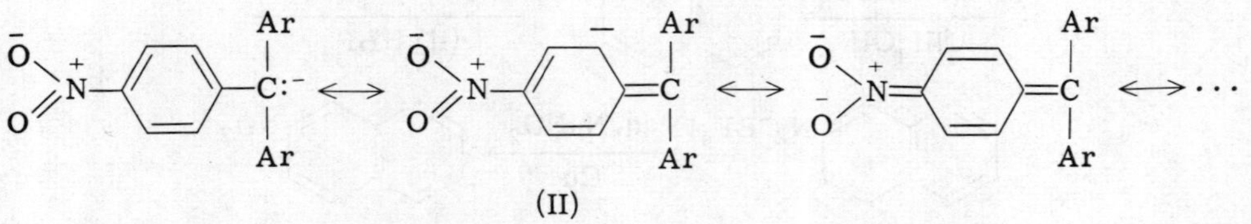

(II)

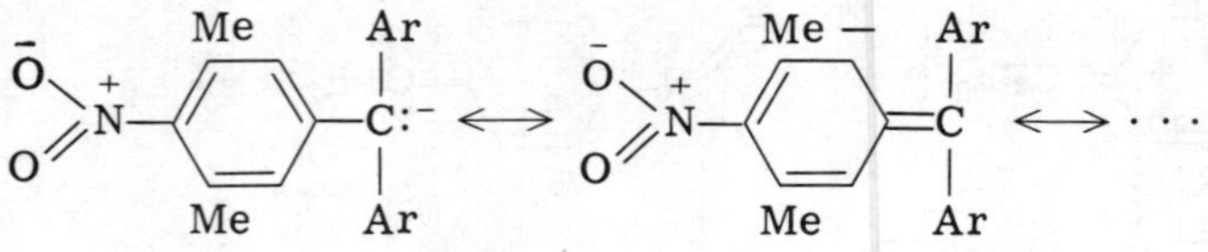

(III)

(II) is the most stable of the three because it is the only one with extended conjugation (which involves the *p*-nitro-groups). The resonating structures of (III) are similar to those of (II), but because of steric inhibition of resonance in the former, the latter is the stronger acid. However, the *p*-nitro-groups still exert a —I effect, and so (III) is a stronger acid than (I), as already shown above. Hence, the order of increasing acid strength is:

(I) < (III) < (II)

8. We may write the benzilic acid rearrangement in the following general way (without consideration of the mechanism):

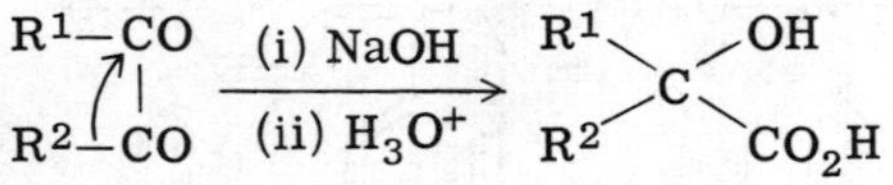

Hence, using this, we have:

(i)

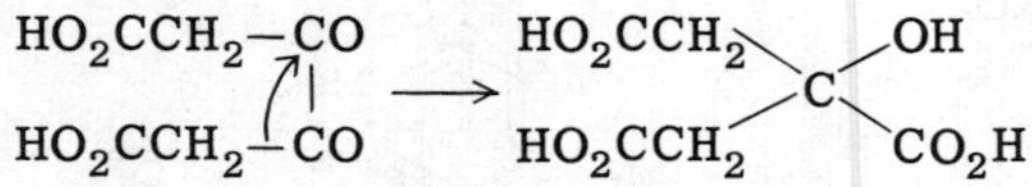

(ii)

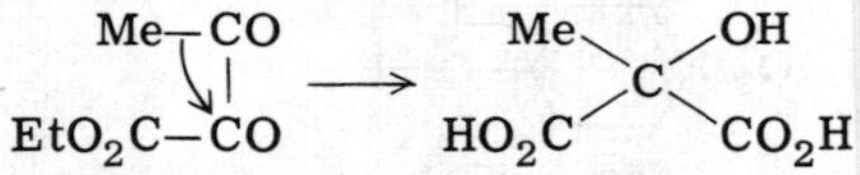

In this case, because of the alkaline conditions used, the carbethoxyl group will be hydrolysed (Me migrates).

(iii)

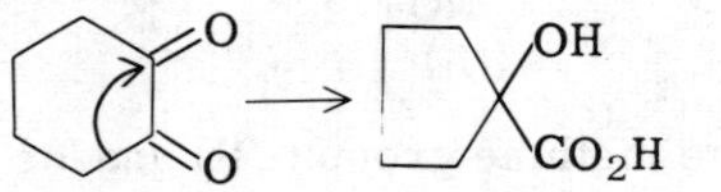

9.

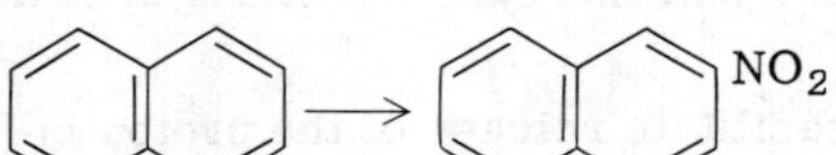

Since direct nitration gives the 1-nitro-derivative, an indirect method must be used. This suggests the possibility of a route via the diazonium salt. We therefore have to prepare 2-aminonaphthalene. One route is:

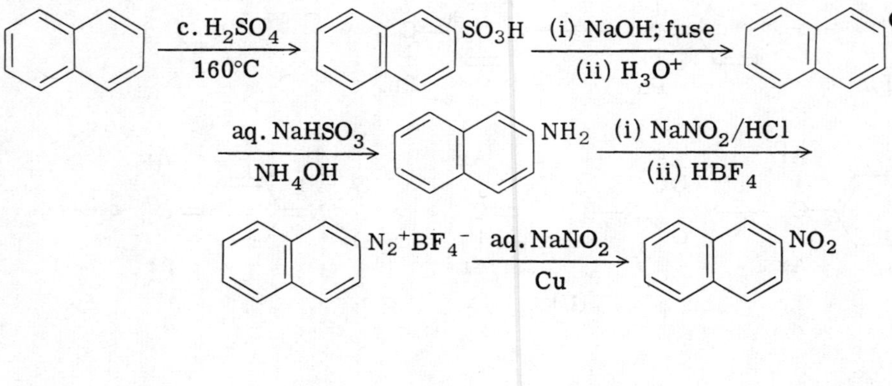

An alternative indirect route is also a common one in which a group is introduced which directs the required group into the desired position, and then this group is removed. This approach also generally involves the use of a diazonium salt. Hence:

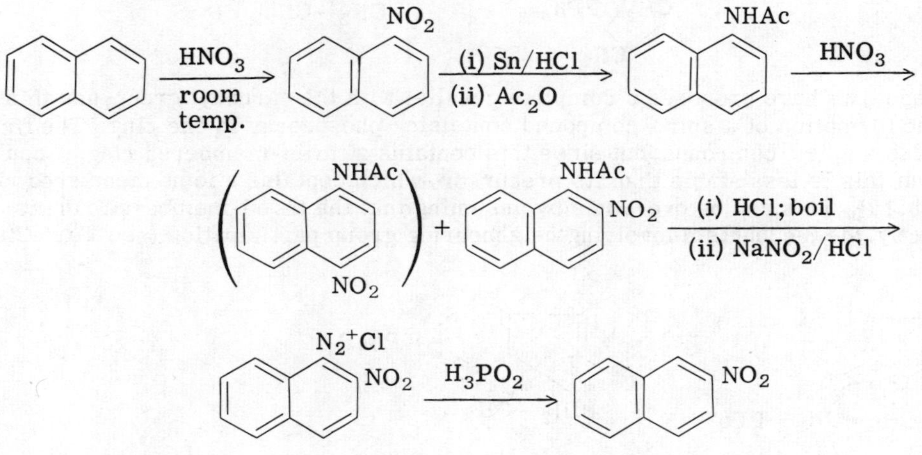

10.

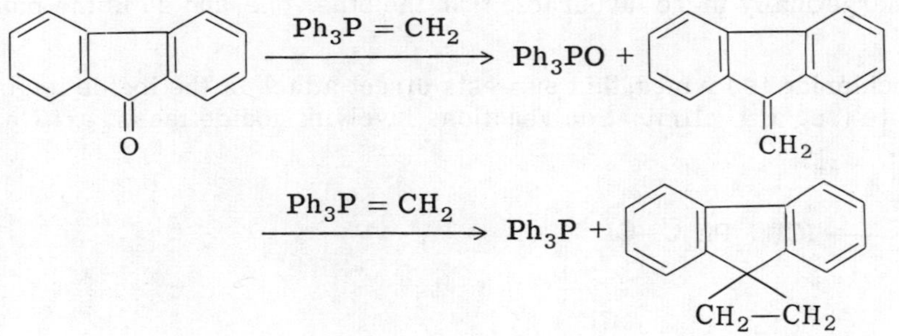

Since there is no carbonyl group in the first product, the double bond of the exocyclic methylene group has reacted with methylenephosphorane. We may therefore propose a mechanism based on the assumption that the exocyclic double bond behaves like the double bond in a carbonyl group. The latter reaction is (see *Text*, Ch. 14):

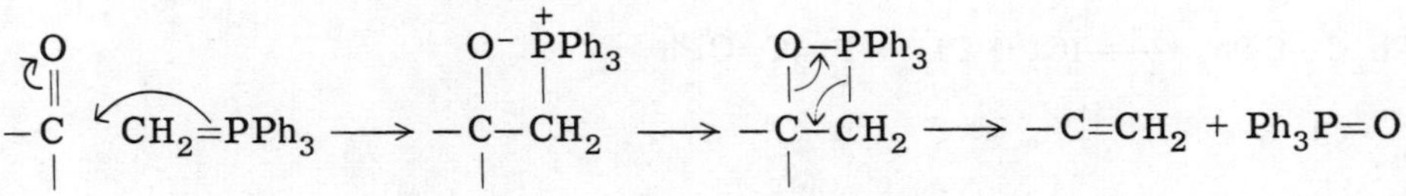

However, in the case under consideration, there is no oxygen atom, phosphine and not phosphine oxide is eliminated, and a single and not a double bond is formed with the methylene group. This can be accommodated by replacing O^- by $>C^-$ (of the five-membered ring), and then as follows:

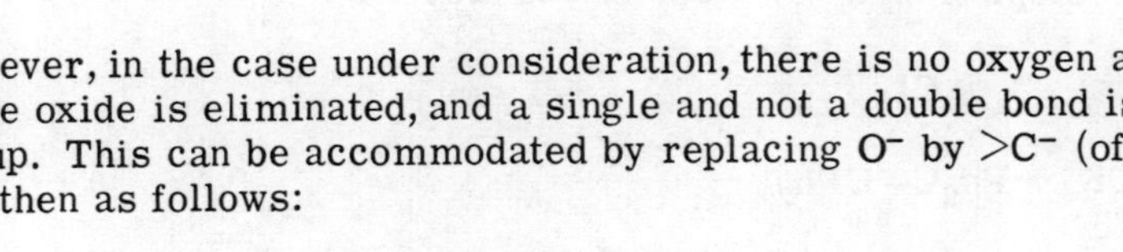

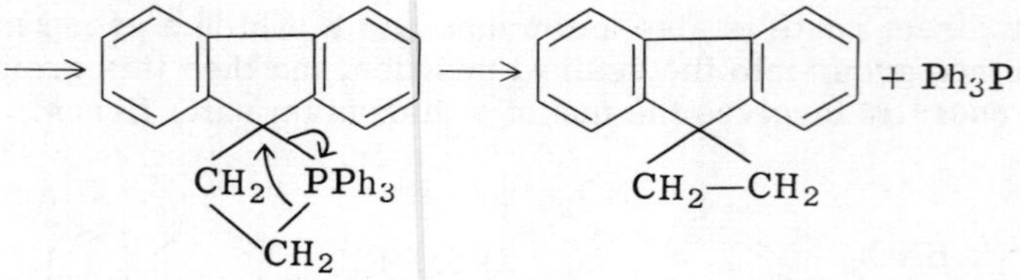

In this sequence, we have proposed a complete parallel with the carbonyl group, and this results in the formation of a spiro-compound containing phosphorus in one ring. The final product is also a spiro-compound, but since this contains a three-membered ring, it could be argued that this is less stable than its precursor—which contains a four-membered ring (see *Text*, Ch. 19). This can be overcome by assuming that the three-membered ring is formed directly, the mechanism involving neighbouring group participation (see *Text*, Ch. 17), i.e.,

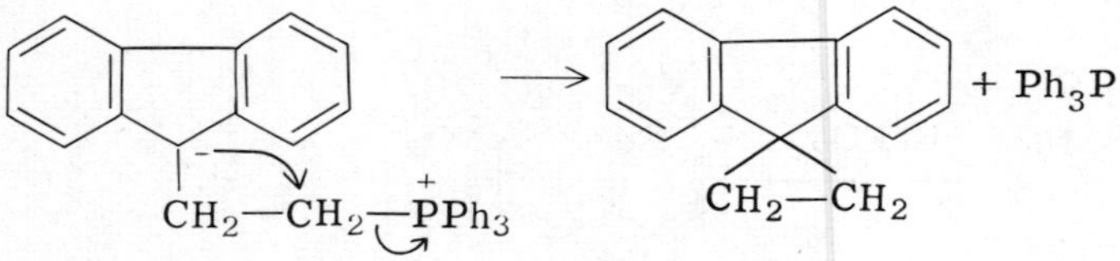

This route would be energetically more favourable than the other one, and so is the more likely one.

11. Since iodine monochloride is formed, this suggests direct attack of the iodide ion on chlorine in the substrate (see also elimination reactions involving iodide ions, *Text*, Ch. 4). Hence, a possibility is:

(i)

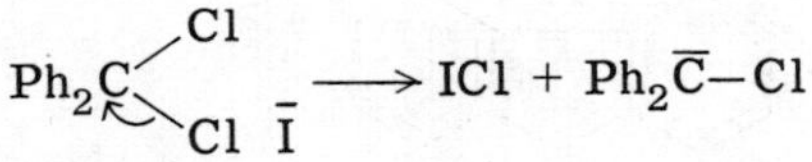

(ii)

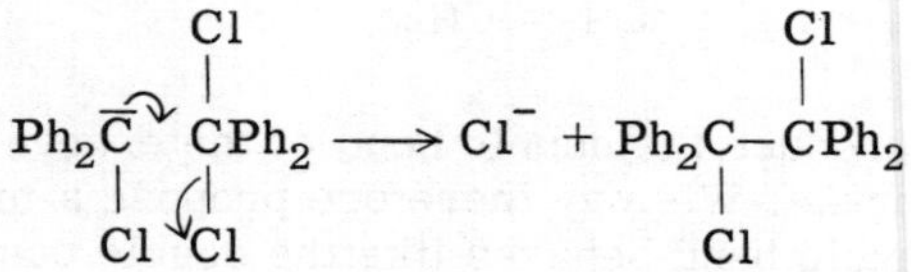

(iii)

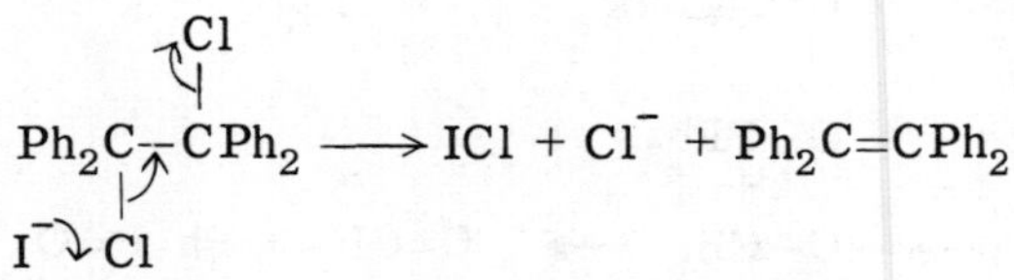

Alternatively, it is possible that steps (i) and (ii) occur in *one* step, i.e.,

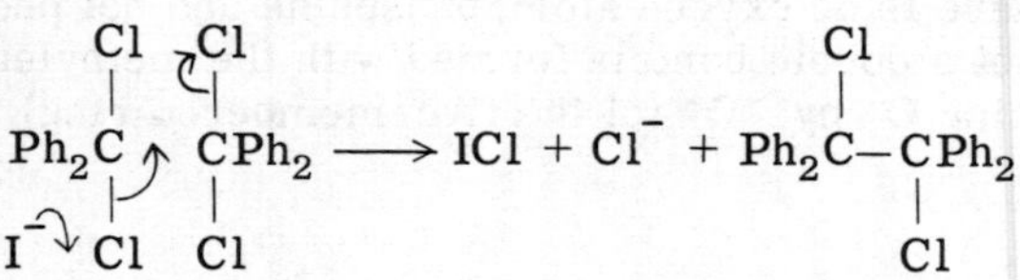

In this case, we have a third-order reaction, and also the geometry of the three molecules to form the T.S. must be correct. These points are made on the assumption that this step is rate-determining. It would therefore appear that the second-order reaction is the more likely one (kinetic studies would settle this problem).

12. Since we require a ring-closure, this suggests the presence of a side-chain with a terminal carboxyl group. The butyric acid residue can be readily introduced as a ketone, and this suggests the Willgerodt reaction (see *Text*, Ch. 27). Hence, a possible synthesis is:

(i) $MeCH_2CH_2CO_2H \xrightarrow{SOCl_2} MeCH_2CH_2COCl$

(ii)

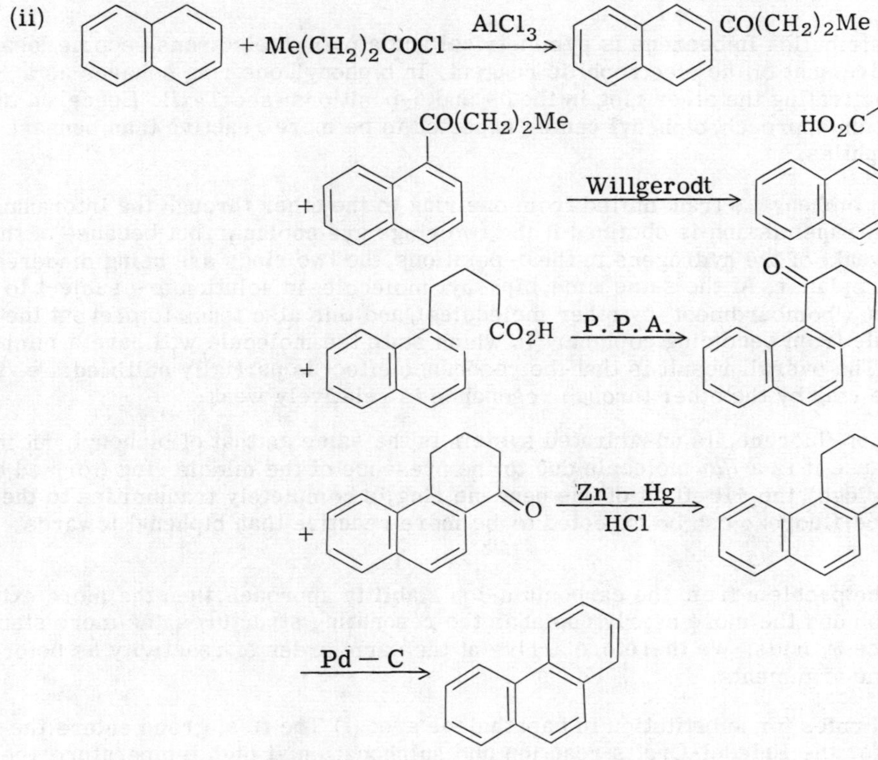

13.

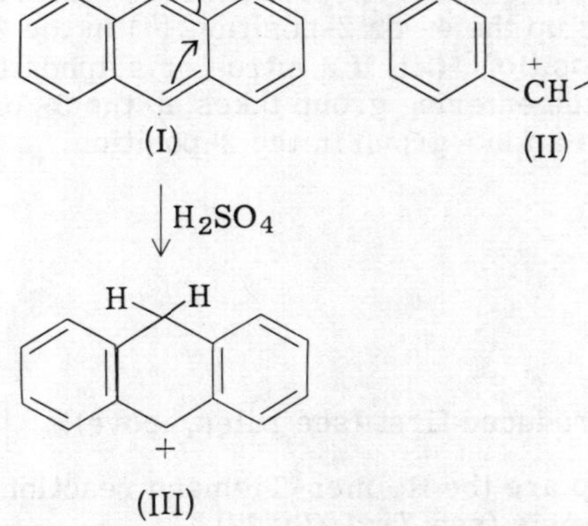

Since the u.v. spectra are similar, this suggests that both solutions contain ions that closely resemble each other in structure. Also, because the NMR spectrum of anthracene in TFA (trifluoroacetic acid) shows the presence of a CH_2 group, a proton has added on at some position. Hence, it is this protonated species which has the closely similar structure to Ph_2CH^+. Now, it is known that the 9- and 10-positions in anthracene are particularly reactive and so, if we add a proton to the 9-position to give (III), the *unsaturated* framework of (III) is the same as (II). In this case, (II) and (III) can be expected to give similar u.v. spectra (see *Text*, Ch. 31).

14.

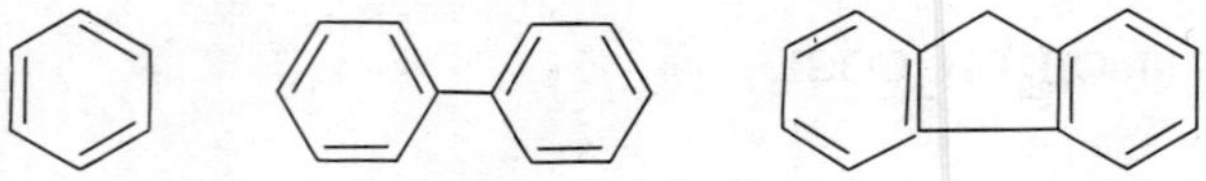

The electron distribution in benzene is symmetrical and a pair of electrons become localised at the requirement of the electrophilic reagent. In biphenyl, one ring behaves as a +R group, thereby activating the other ring in the *o*- and *p*-positions (see *Text*). Hence, on the charge distribution approach, biphenyl can be expected to be more reactive than benzene towards electrophiles.

The +R effect in biphenyl is transmitted from one ring to the other through the interannular bond. Maximum transmission is obtained if the two rings are coplanar, but because of the steric effects (weak) of the hydrogens in the *o*-positions, the two rings are being hindered from becoming coplanar. At the same time, biphenyl molecules in solution are subject to thermal agitation ('bombardment' by other molecules), and this also tends to prevent the biphenyl molecule from becoming coplanar (in which state the molecule will have a minimum entropy). The overall result is that the resonance effect is partially nullified, i.e., the activation of one ring by the other through resonance is relatively weak.

If we now consider fluorene, its unsaturated system is the same as that of biphenyl, but in the former, because it is a *flat* molecule due to the presence of the middle ring (formed by the methylene bridge), the +R effect of one benzene ring is completely transmitted to the other ring. Hence, fluorene can be expected to be more reactive than biphenyl towards electrophiles.

If we consider the problem from the carbonium-ion stability approach, then the more extended the conjugation and the more nearly coplanar the resonating structures, the more stable are the resonance hybrids. We therefore arrive at the same order of reactivity as before by using the same arguments.

15. The general rules for substitution in naphthalene are: (i) The first group enters the 1-position except for the Friedel-Crafts reaction and sulphonation at high temperature; the predominant product is now the 2-derivative. (ii) If the substituent present is *o*/*p*-directing and is: (*a*) in the 1-position, the entering group takes up the 4- or 2-position; (*b*) in the 2-position, the entering group usually takes up the 1-position. (iii) If a nitro- or sulphonic acid group (*m*-directing) is in the 1- or 2-position, the entering group takes up the 5- or 8-position. This is also the case for halogen and the amino-group in the 2-position.

(i)

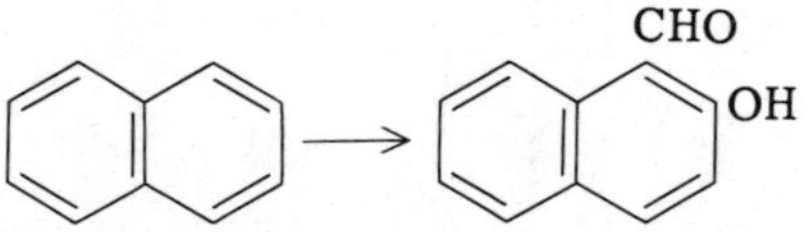

The hydroxyl group is *o*/*p*-directing, and so it is introduced first (see rules, above).

Two convenient ways of introducing the formyl group are the Reimer-Tiemann reaction (see *Text*, Ch. 27) and the Gattermann aldehyde synthesis (see *Text*, Ch. 27).

(*a*)

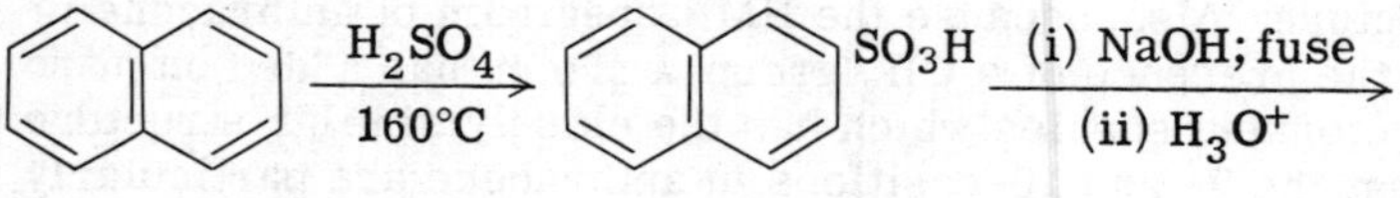

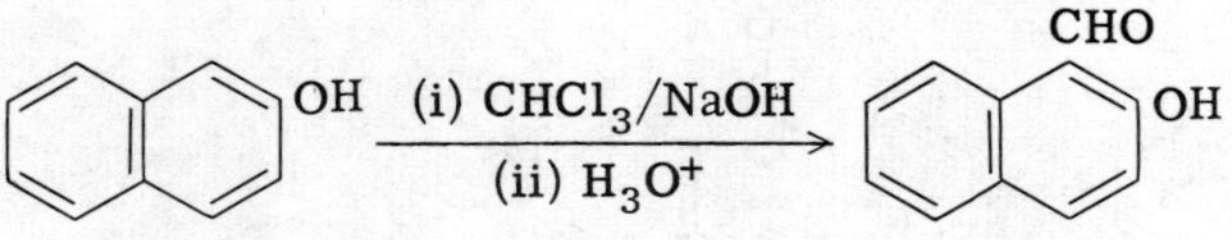

(*b*)

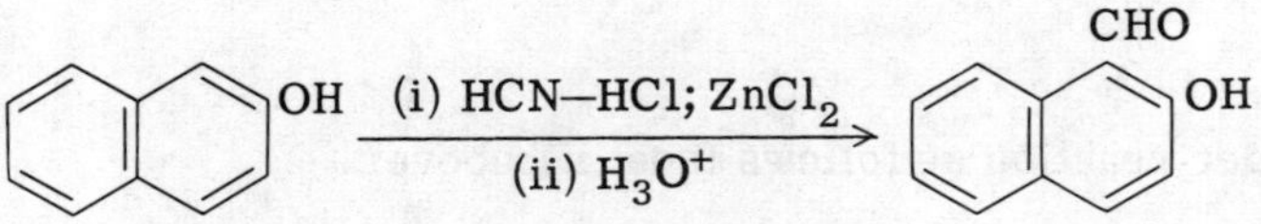

(ii)

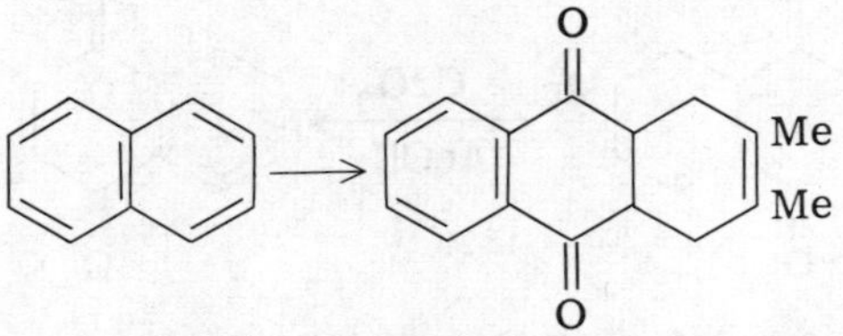

The product is a partially hydrogenated anthraquinone. Since one double bond in the 1, 4-naphthaquinone residue is now a single bond, and the rest of the molecule is a six-membered ring with one double bond, this leads to the possibility of a Diels-Alder reaction (see *Text,* Ch. 19). Hence, a possible synthesis is (remember, quinones can behave as dienophiles):

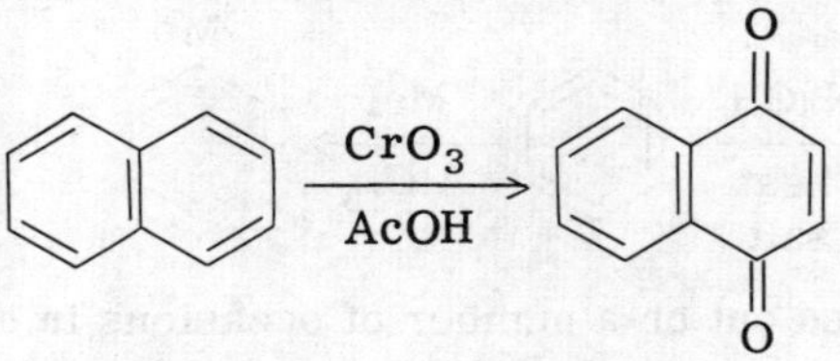

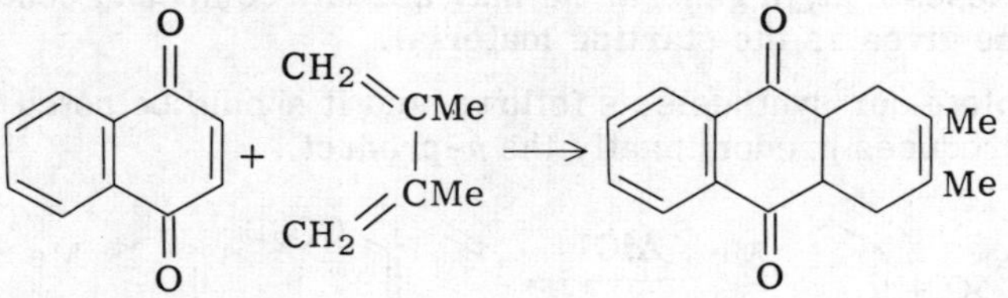

(iii)

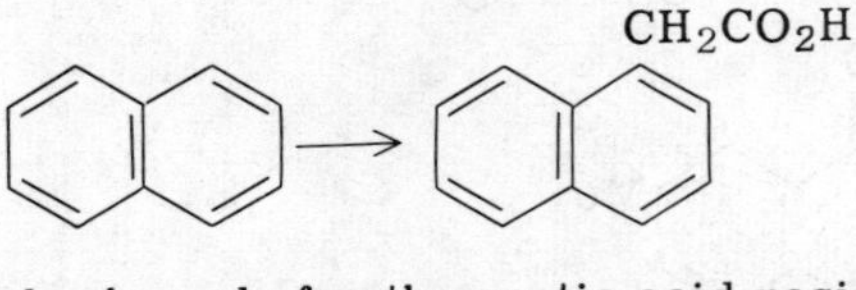

If we work backwards for the acetic acid residue, we have:

$$-CH_2CO_2H \longleftarrow -CH_2CN \longleftarrow -\mathbf{CH_2X} \xrightarrow{Mg/CO_2} -CH_2CO_2H$$

Hence, two possible routes are:

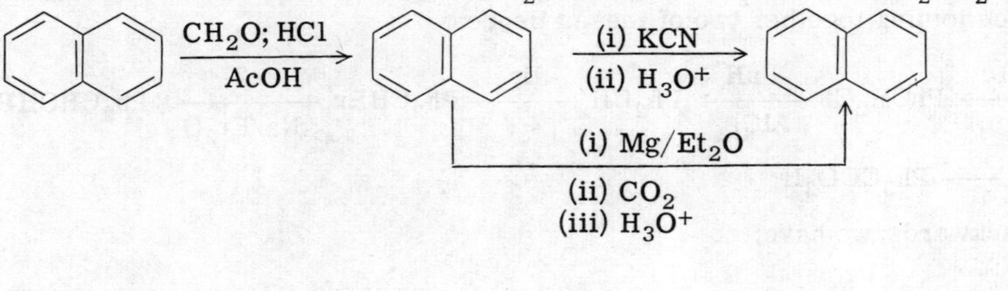

(iv)

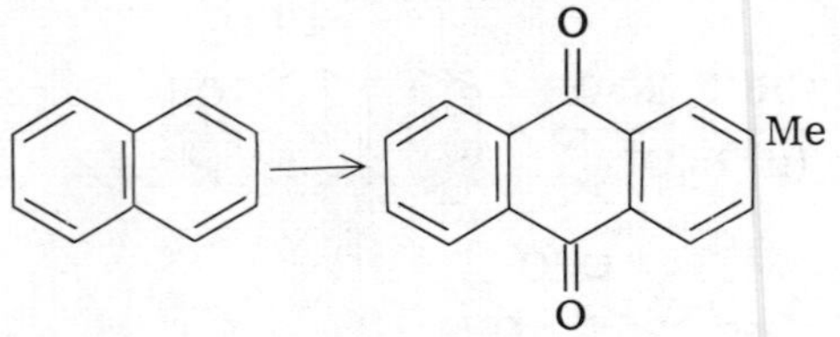

One method is via the Diels-Alder reaction as follows (see (ii), above).

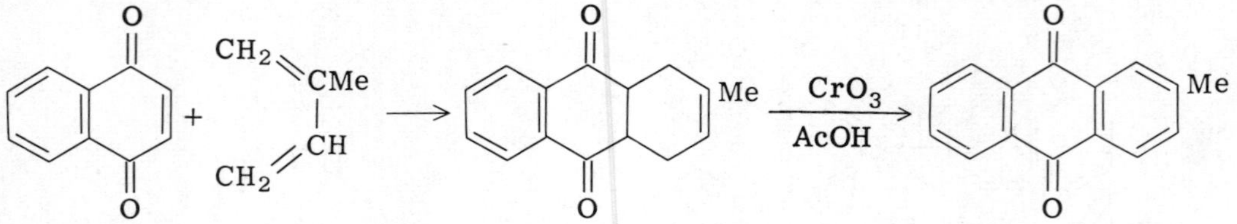

Since the product is a methylanthraquinone, it is always worth trying a method analogous to that used for anthraquinone itself, i.e., the one involving phthalic anhydride. The other compound we need is toluene, and this may be used as such, i.e., it need not be prepared from naphthalene. If the student feels that it should be prepared from naphthalene, one possible route is via phthalic anhydride:

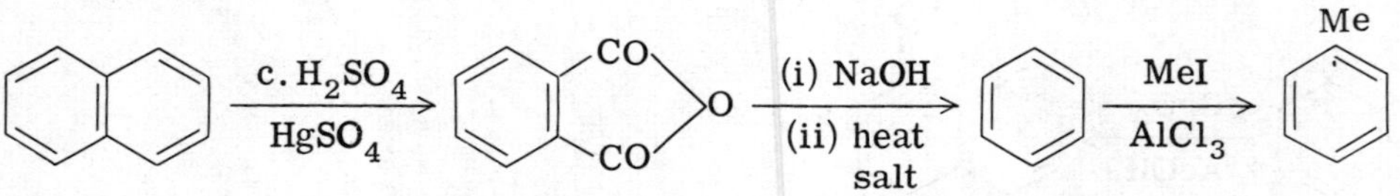

The student should remember that it has been pointed out on a number of occasions in connection with syntheses that, in general, he may use any compound containing *fewer* carbon atoms than the one given as the starting material.

We can now complete our synthesis as follows, and it should be noted that the Friedel-Crafts reaction produces predominantly the *p*-product.

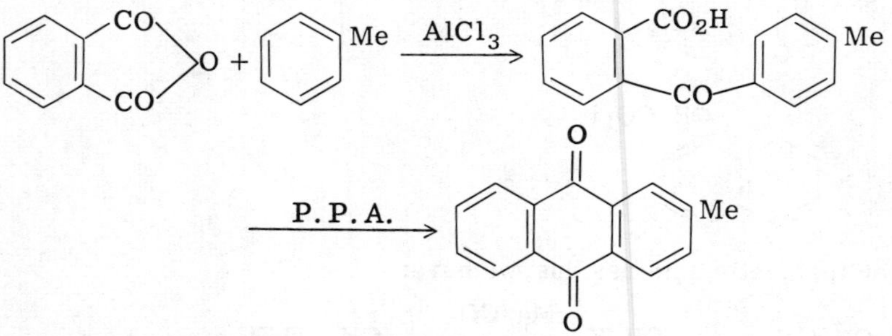

(v) $PhMe \longrightarrow Ph_2CHCHPh_2$

Since the product contains two Ph_2CH fragments, this suggests the possibility of $PhMe \longrightarrow Ph_2CH_2$, and then joining together two of these. Hence:

$$PhCH_3 \xrightarrow[\text{b.p.}]{Cl_2} PhCH_2Cl \xrightarrow[AlCl_3]{PhH} Ph_2CH_2 \xrightarrow{Br_2} Ph_2CHBr \xrightarrow[Na/Et_2O]{2} Ph_2CHCHPh_2$$

(vi) $PhH \longrightarrow Ph_3CCO_2H$

Working backwards, we have:

$$-CO_2H \longleftarrow -CN \longleftarrow -\mathbf{X} \xrightarrow[CO_2]{Mg} -CO_2H$$

The key compound is therefore Ph_3CX. Hence, one route is (see also (iii), above):

$$3PhH + CCl_4 \xrightarrow{AlCl_3} Ph_3CCl \xrightarrow[\text{(ii) } CO_2 \text{ (iii) } H_3O^+]{\text{(i) } Mg/Et_2O} Ph_3CCO_2H$$

(vii)

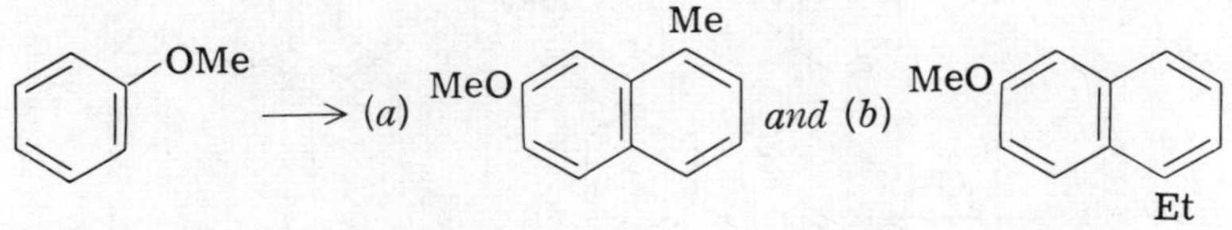

The method required is the Haworth synthesis; the methoxyl group is predominantly *p*-directing.

(*a*)

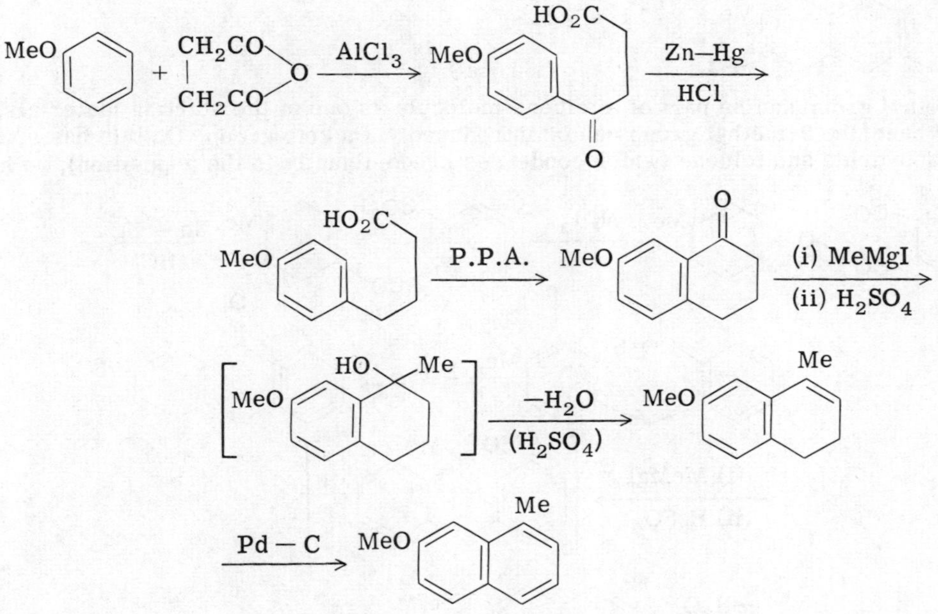

(*b*)

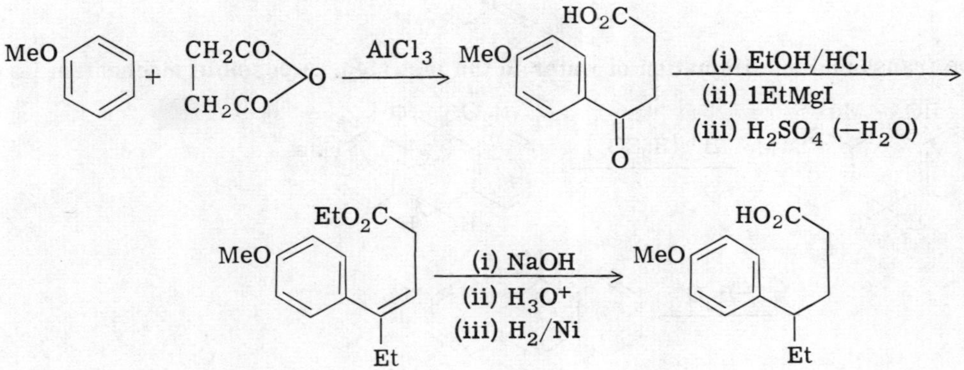

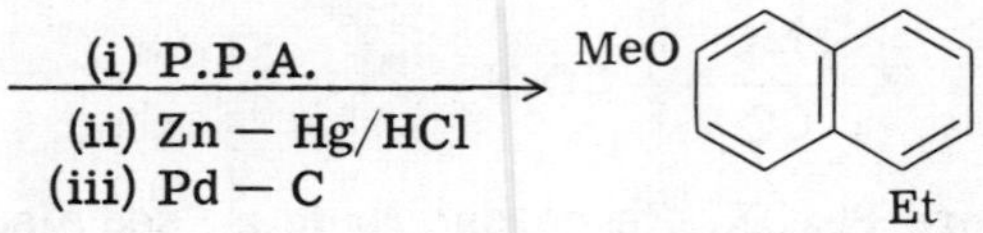

It should be noted that *one* mole of Grignard reagent is used (the keto group reacts preferentially).

(viii)

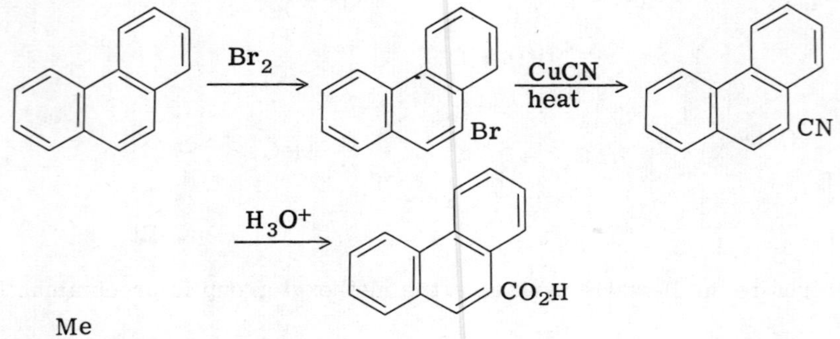

16.

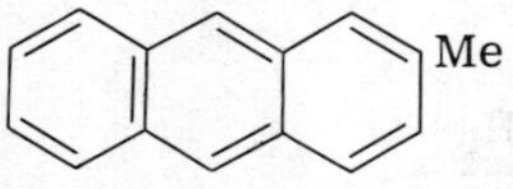

The 2-methyl group can be part of a toluene molecule as one of the starting materials. On the other hand, the 9-methyl group can be introduced via a keto group. On this basis, using phthalic anhydride and toluene (which condenses predominantly in the *p*-position), we have:

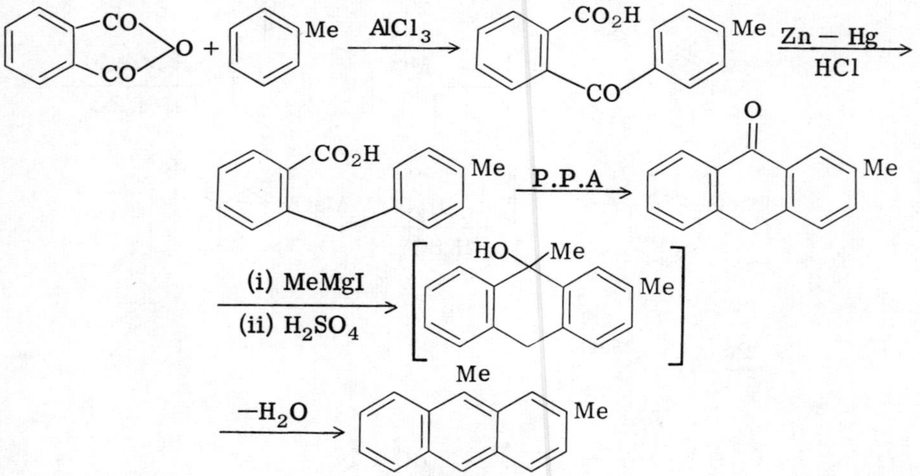

Note the transannular elimination of water in the last step. A possible mechanism is:

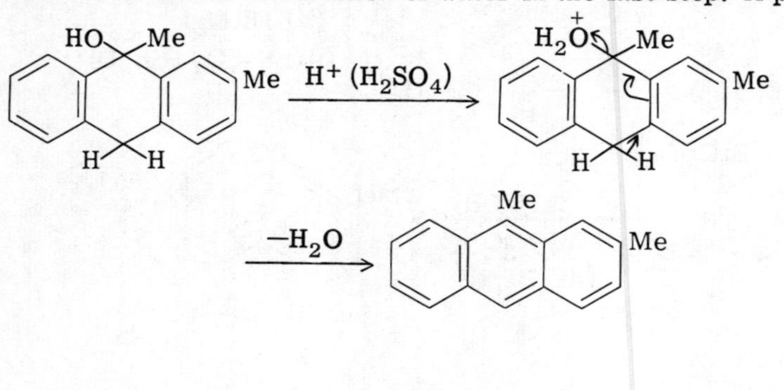

30 Heterocyclic compounds

(Problems on page 55)

1. In this context, by 'more reactive' we mean that pyrrole is more readily substituted by electrophiles than is furan.

(I) (II) (III) (IV)

Hence, from the charge distribution approach, since 2- (or 5-) substitution occurs preferentially in both compounds, this means that the electron density is higher in this position in pyrrole than in furan. Oxygen is more electronegative than nitrogen and so electron-withdrawal from all positions in furan can be expected to be greater than in pyrrole (see *Text*, NMR spectra). Also, for the same reason (of different electronegativities), resonating structures of furan, (II), in which the oxygen carries a positive charge, can be expected to make smaller contributions than the corresponding resonating structures of pyrrole, (I). Hence, electron-withdrawal is greater, and electron-donation is smaller in furan than in pyrrole. Therefore, pyrrole is more reactive than furan.

Let us now consider the problem from the carbonium-ion stability approach. In the two resonating structures, (III) and (IV), of the carbonium ions, both the nitrogen and oxygen carry positive charges, but since the former can accommodate this charge more readily than the latter (the former is less electronegative than the latter), the pyrrole carbonium ion is more stable than that of furan. Hence, the former is produced more readily than the latter, i.e., pyrrole is more reactive than furan.

2. (i)

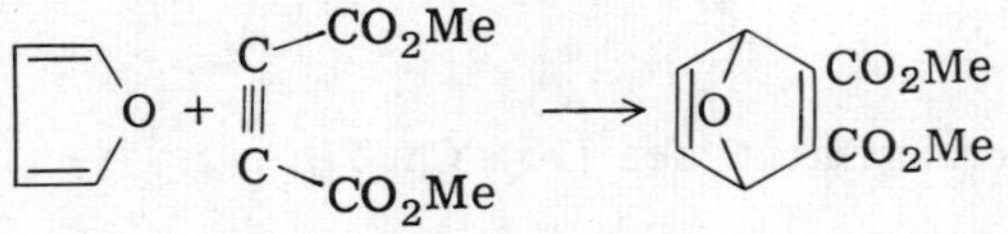

Furan is less aromatic than pyrrole or thiophen, and is the only one of these that behaves as a diene in the Diels-Alder reaction (*cf.* with Prob. 1, above).

(ii)

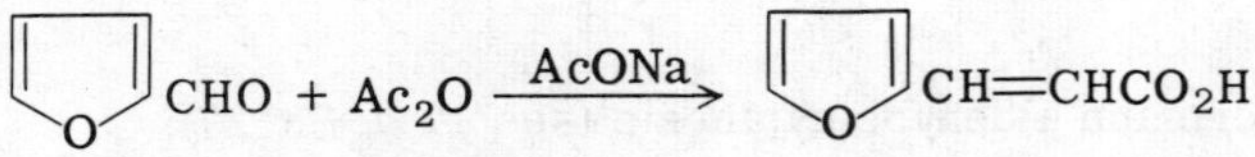

This is an example of the Perkin reaction (see *Text*, Ch. 27).

(iii)

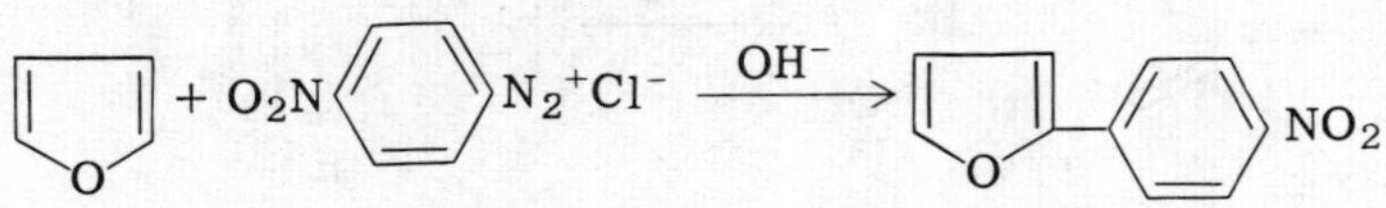

This is an example of the Gomberg reaction (this takes place by a free-radical mechanism; see *Text*, Ch. 24). This reaction with furan differs from that with pyrrole (see *Text*). The former is apparently not reactive enough to combine with a diazonium cation, but readily combines with free radicals (which are extremely reactive).

(iv)

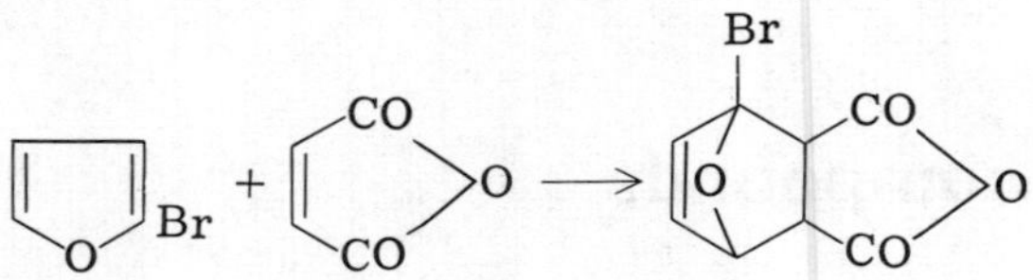

Cf. with (i), above.

(v) The group required in the 2-position is methyl, but since this must be activated in order to undergo the condensation, this is the purpose of introducing the nitro-group into the 5-position (*'para'* to the methyl group).

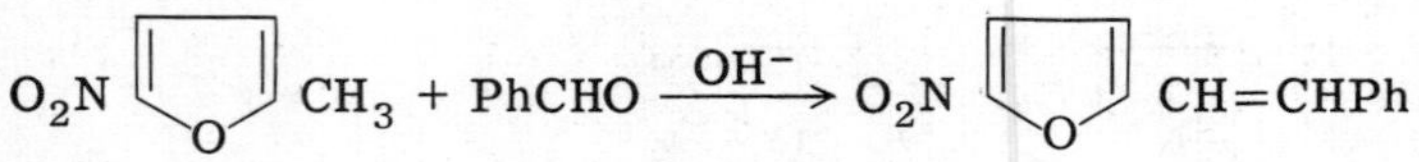

(vi)

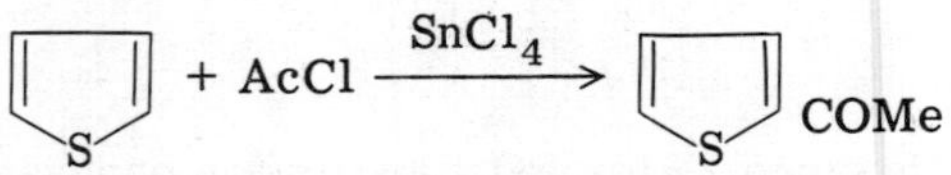

Aluminium chloride attacks thiophen; hence the use of the milder catalyst, stannic chloride, in the Friedel-Crafts reaction.

(vii)

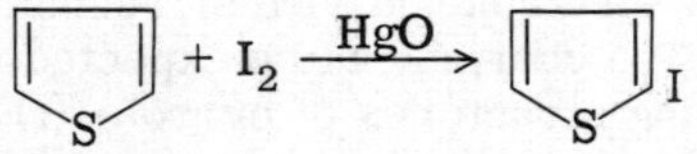

(viii)

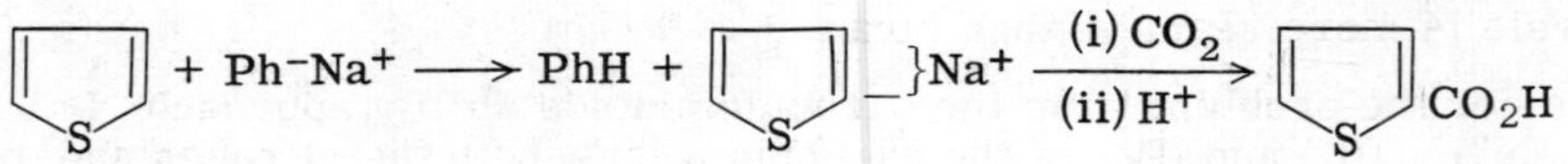

(ix)

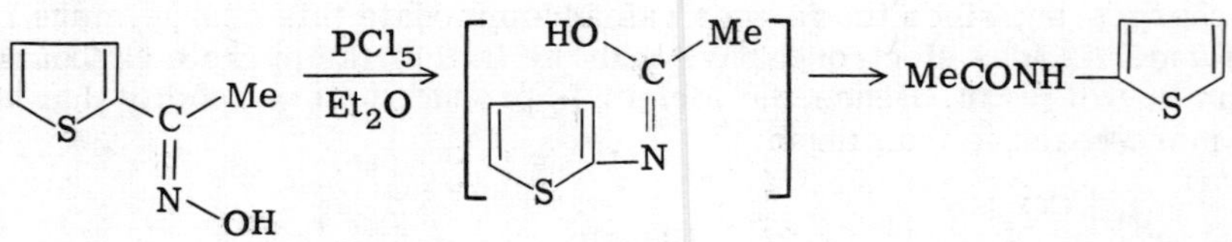

This is an example of the Beckmann rearrangement (see *Text*, Ch. 27).

(x)

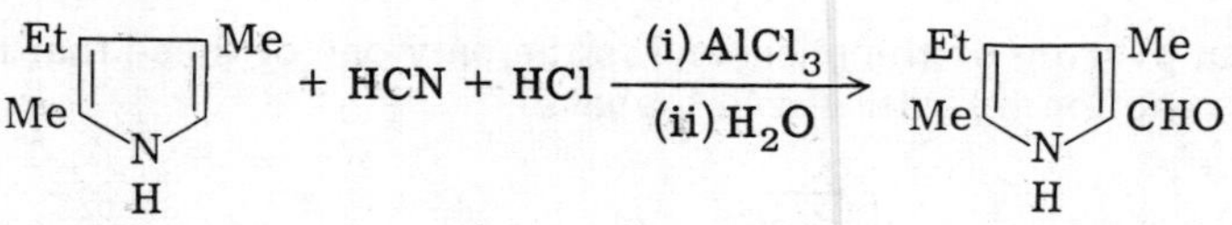

This is an example of the Gattermann aldehyde synthesis (see *Text*, Ch. 27).

(xi)

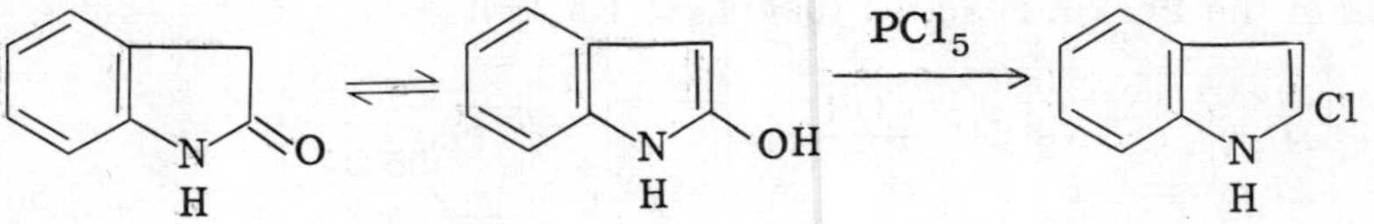

(xii)

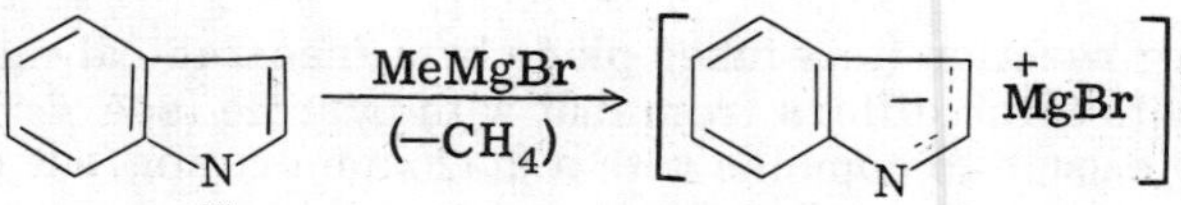

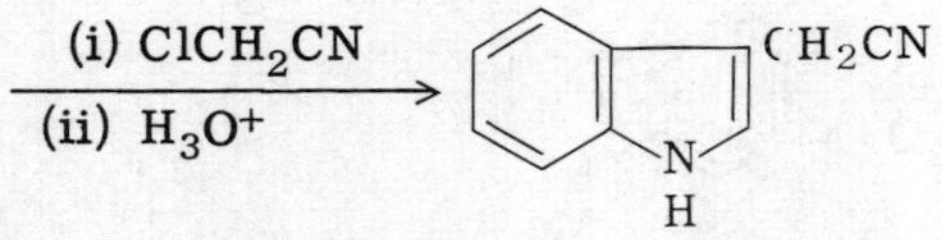

(xiii) $MeLi + CH_2Cl_2 \longrightarrow MeH + LiCl + :CHCl$

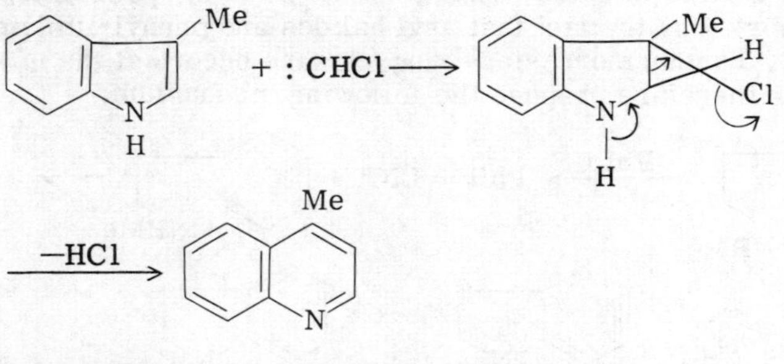

(xiv) This is an example of the Fischer indole synthesis.

$$(MeO)_2CHCH_2CH_2CO_2Et \xrightarrow[EtOH]{H_2SO_4} OCHCH_2CH_2CO_2Et$$

If we now put $R{=}CH_2CO_2Et$, we have:

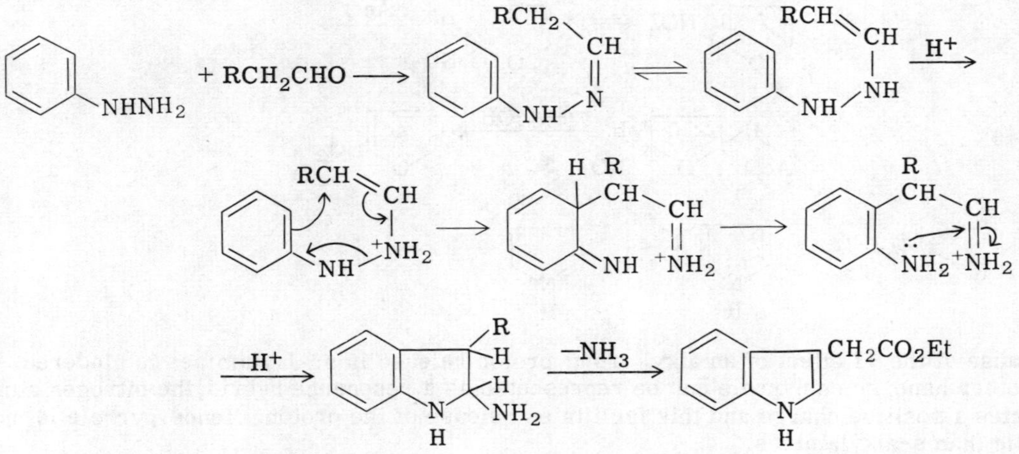

3.

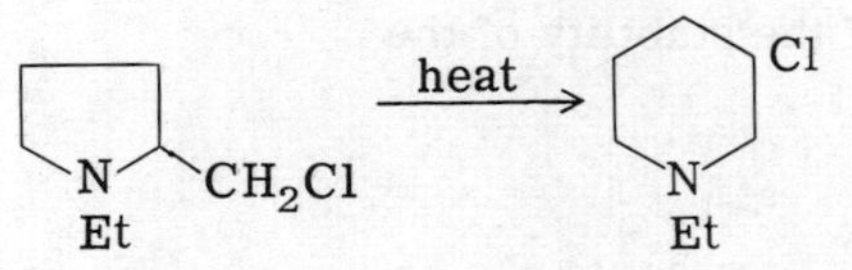

Since the ring has expanded, the nitrogen atom in the pyrrolidine must have linked up with the exocyclic methylene group. A possible mechanism is therefore (note the neighbouring group participation):

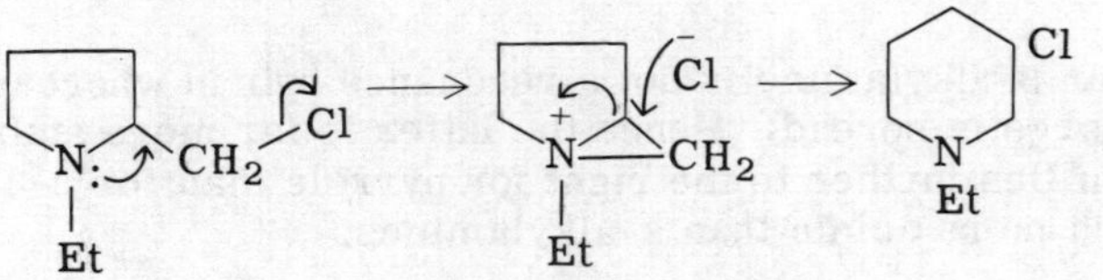

4. 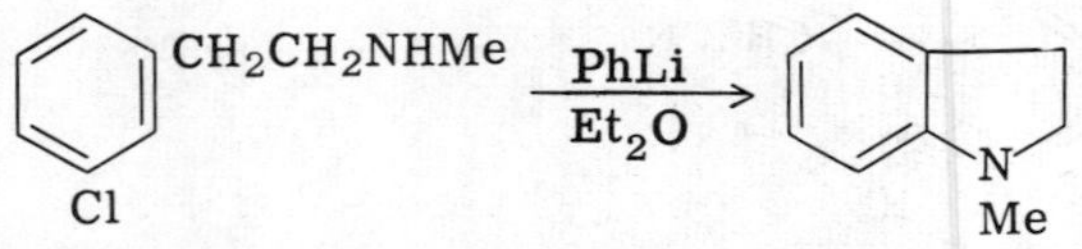

Ring-closure has occurred in the position *ortho* to the side-chain and chlorine, *in the* m-*position*, has been eliminated. This reminds us of the possibility of the benzyne mechanism. This is supported by the fact that aryl halides and phenyl-lithium can give benzynes (see *Text,* Ch. 21). Furthermore, since ring-closure occurs at the *nitrogen atom* of the side-chain, we can therefore propose the following mechanism:

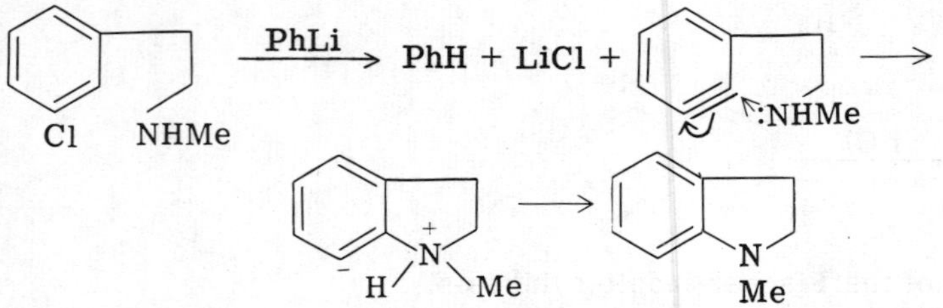

5. We shall assume that the nitrating species is acetyl nitrate and that it breaks up as shown.

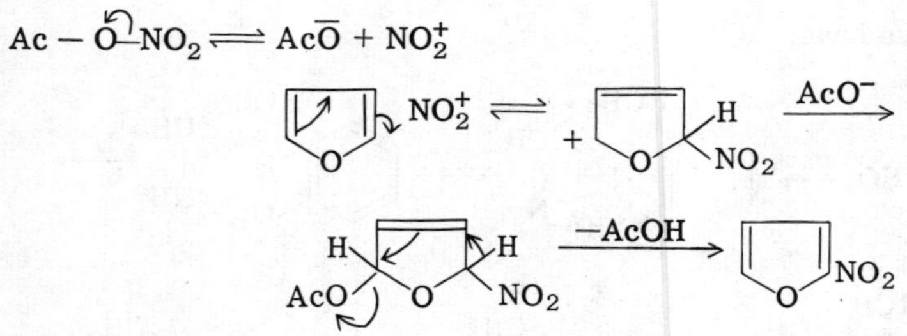

6. 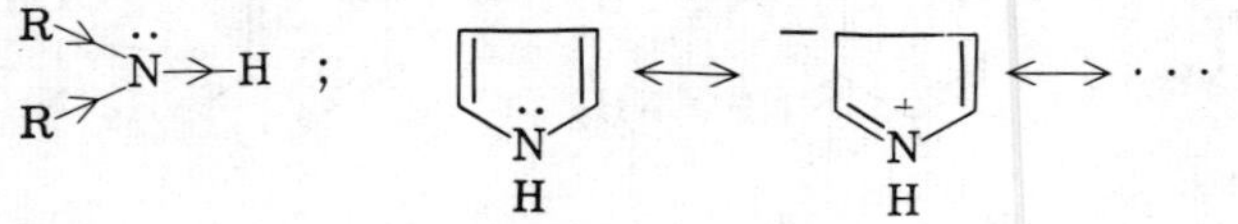

Because of the +I effect of an alkyl group, proton release in s-alkylamines is hindered. On the other hand, since pyrrole can be represented as a resonance hybrid, the nitrogen atom carries a positive charge and this facilitates release of the proton. Hence, pyrrole is more acidic than s-alkylamines.

If we consider the problem from the point of view of the stability of the

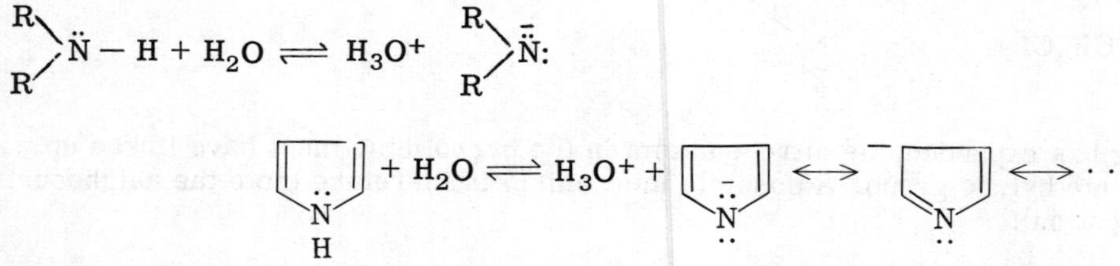

conjugate base, we see that the cB of an s-alkylamine is not a resonance hybrid whereas that of pyrrole is (and the negative charge is spread). Hence the latter is far more stable than the former and so the equilibrium lies further to the right for pyrrole than for s-alkylamines. Consequently, pyrrole is much more acidic than s-alkylamines.

7. The condensation between formaldehyde, a secondary amine and a compound containing an *active hydrogen atom*, is an example of the Mannich reaction (see *Text*, Ch. 13). Since this reaction fails with furan but proceeds with 2-methylfuran, it therefore means that hydrogen is not sufficiently active in the former but is in the latter. It also follows that it is the 2-methyl group (+I) which has brought about this activation. We also know that pyrrole is more reactive than furan (see Prob. 1, above), and since reaction proceeds with pyrrole, its hydrogen is sufficiently activated. The student should note that although one compound is more reactive than another, the more reactive compound does not necessarily undergo a reaction which the less reactive one does not. The electrophilic power of the reagent also plays a part. What actually happens is decided by experiment, and then we attempt to explain the result. We can predict that one compound is more likely (or even far more likely) to undergo a given reaction, but there is no certainty that it must do so.

The final point to consider is the site of attack. Since furan and pyrrole undergo electrophilic substitution in the 2- (or 5-) position, and since the Mannich reaction proceeds by an electrophilic mechanism (according to the accepted one), we may formulate the reactions as follows:

(A)

$$Me_2NH + CH_2O \overset{H^+}{\rightleftharpoons} Me_2\overset{+}{N}{=}CH_2 \longleftrightarrow Me_2N{-}\overset{+}{C}H_2$$

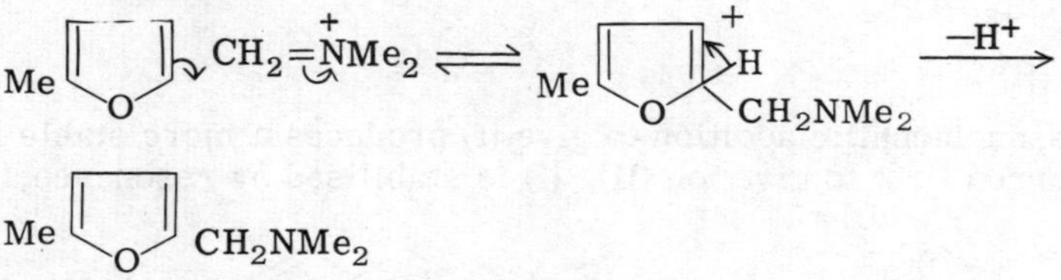

(B)

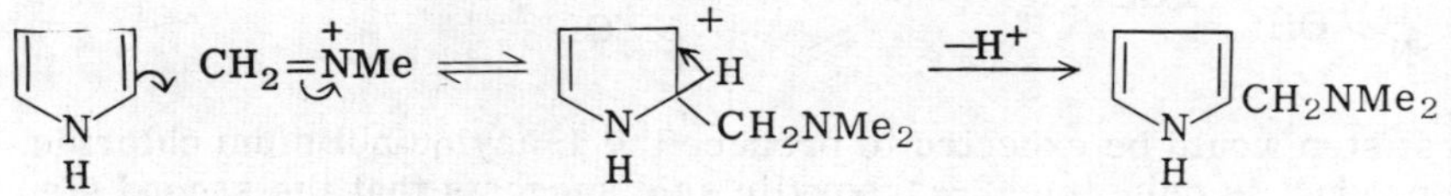

8. (i)

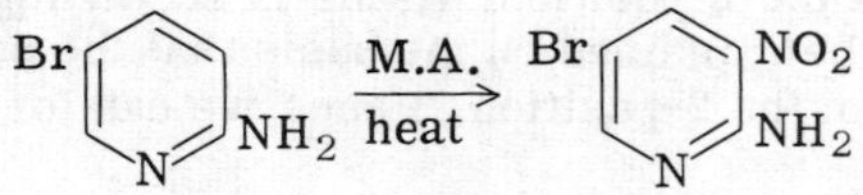

These conditions of nitration are typical for aromatics. Pyridine normally requires vigorous conditions, but the presence of the 2-amino-group activates the positions *ortho* (3) and *para* (5) to it.

(ii)

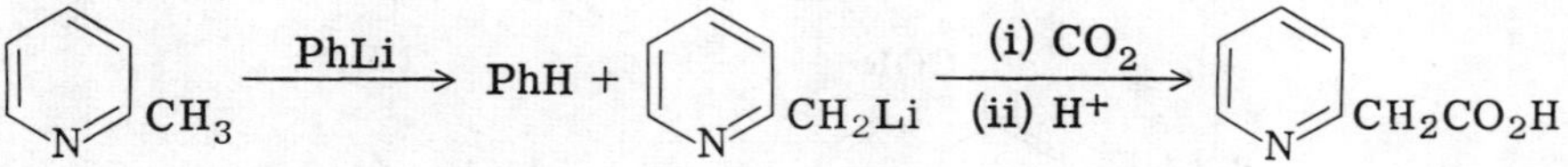

The methyl group in the 2- or 4-position in pyridine is activated.

(iii)

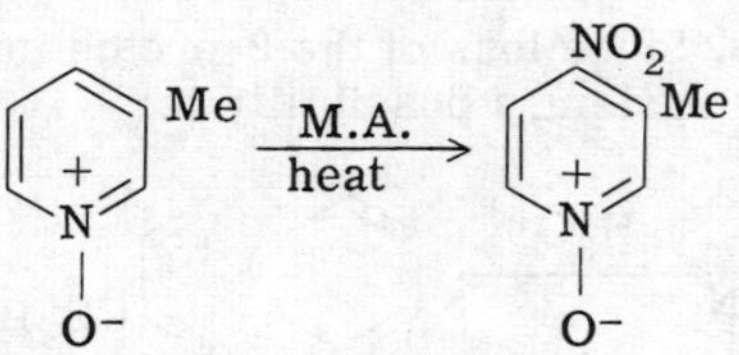

Pyridine-I-oxide is activated in the 4-position (and also in the 2-) towards electrophiles.

(iv)

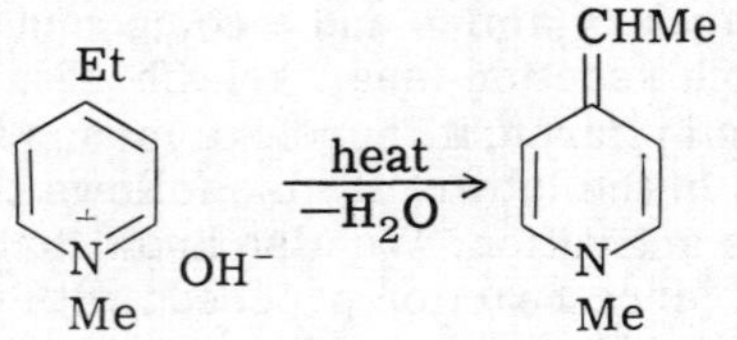

Anhydro-bases are formed when a 2- or 4-alkyl pyridinium hydroxide is heated.

(v)

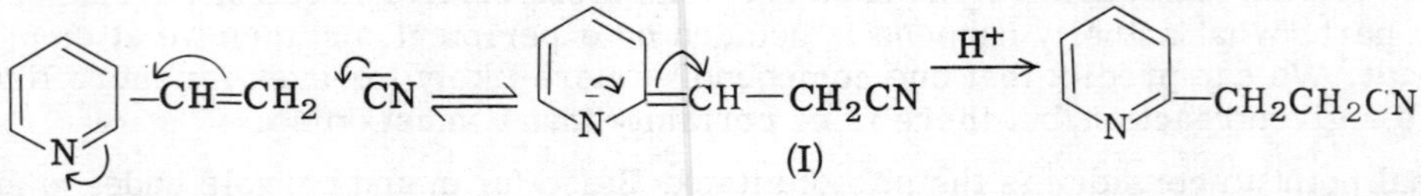

Polarisation of the double bond in the side-chain is *towards* the ring due to the —I effect of the nitrogen atom. This is 'assisted' by attack of the cyanide ion at the terminal carbon atom, if this addition is the first step.

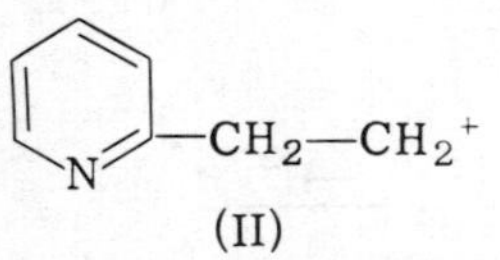

Furthermore, nucleophilic addition to give (I) produces a more stable ion than had proton addition occurred first to give ion (II). (I) is stabilised by resonance, (II) is not.

(vi)

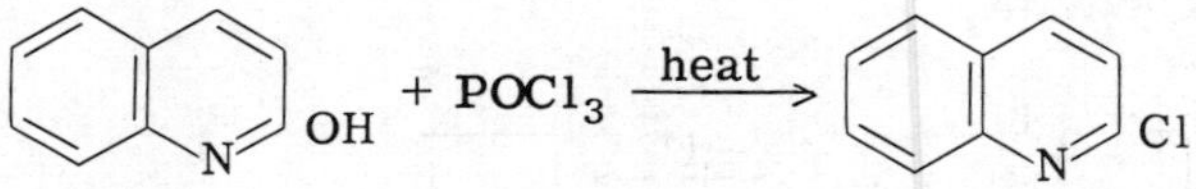

(vii) The first step would be expected to produce the 1-acylquinolinium chloride. The fact that the final product is quinoline-2-carboxylic acid suggests that the second step is the addition of the cyanide ion (a nucleophile) at the 2-position. If this is so, then position 2 behaves as an electrophilic centre. This can be explained on the basis that the positively charged nitrogen atom withdraws electrons from the 2-position. Hence, we can formulate these steps as follows:

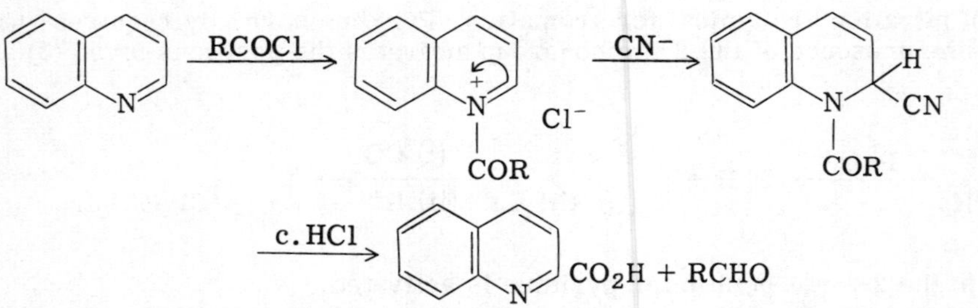

The final problem is the mechanism of the last step. Since loss of the 2-proton in the *Reissert compound* restores aromaticity to the ring system, a possibility is:

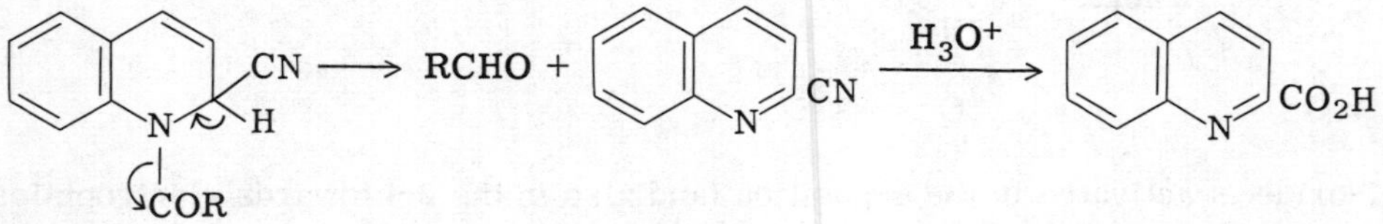

Another possibility is that hydrolysis to the acid occurs first.

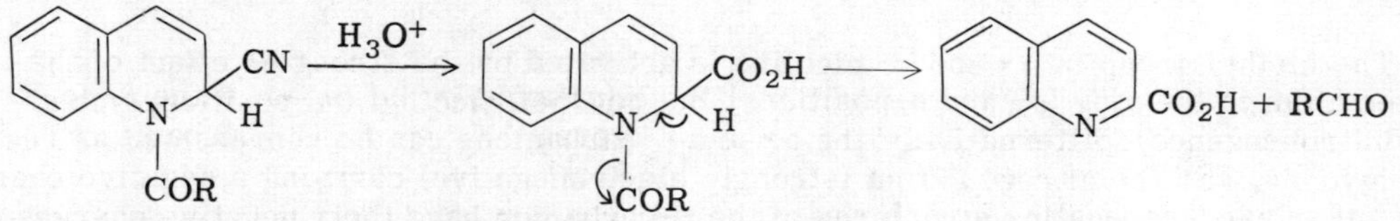

With the data at our disposal, this is about as far as we can go. We can, of course, vary the sequence for the formation of the aldehyde, in the sense that the 2-proton is removed by Cl^- (H_2O?) and $R\overline{C}O$ acquires a proton from H_3O^+.

In actual fact, there is evidence to show that the mechanism of the hydrolysis is far more complicated than the 'reasonable' ones given above. The following is one that has been proposed, and bears a similarity to that given for the von Richter reaction (see *Text*, Ch. 21).

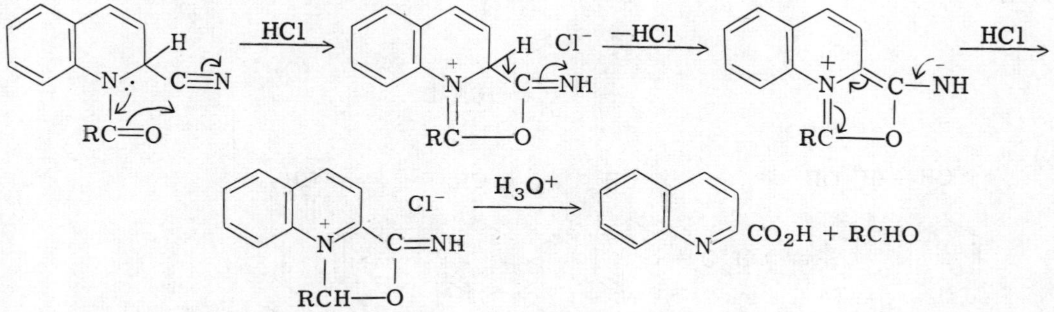

One other point that may be mentioned is the conversion of an acyl chloride into an aldehyde via a Reissert compound.

(viii)

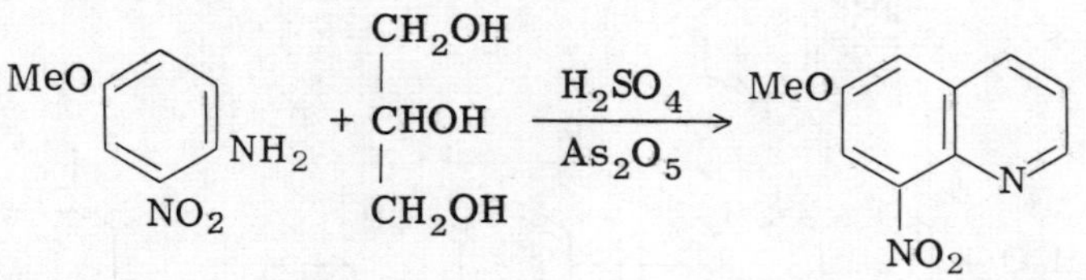

This is an example of the Skraup synthesis (see also below).

(ix)

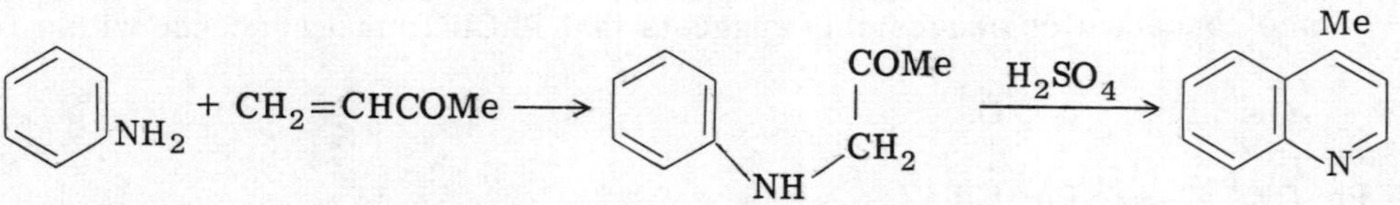

When glycerol is used in the Skraup synthesis, acraldehyde is produced and this then undergoes 1, 4-addition with aniline. Hence the analogous 1, 4-addition given above.

(x)

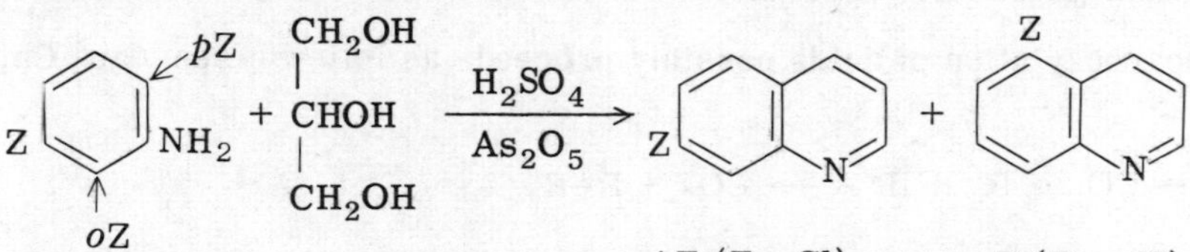

(Z=Cl, NO_2, OMe)

*p*Z (Z = Cl)
*p*Z (Z = OMe)
*p*Z (Z = NO_2)

*o*Z (Z = Cl)
*o*Z (Z = NO_2; major product)

The direction of ring-closure is very much affected by the directing influence of Z (see *Text*).

9. The methyl group in α- and γ- picoline is activated by the strong $-I$ effect of the nitrogen atom in the ring (*o*- and *p*-positions), but not the β-methyl (*m*-position; *cf.* the methylnitrobenzenes). Alternatively, the α- and γ-carbanions can be represented as resonance hybrids, with the nitrogen atom (strongly electronegative) carrying a negative charge. On the other hand, resonating structures of the β-carbanion have their negative charges on carbon atoms only and are fewer in number, and consequently are less stable than the α- and γ-carbanions (see *Text*, Ch. 2). The result is that only the α- and γ-carbanions are formed in the presence of hydroxide ions, and so only these two can incorporate deuterium from deuterium oxide.

(α)

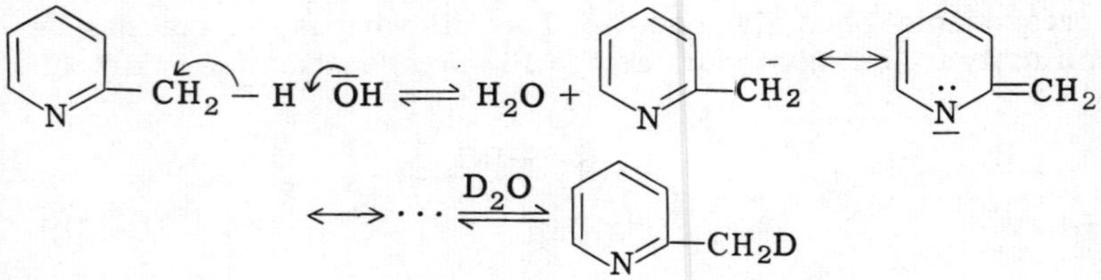

(γ)

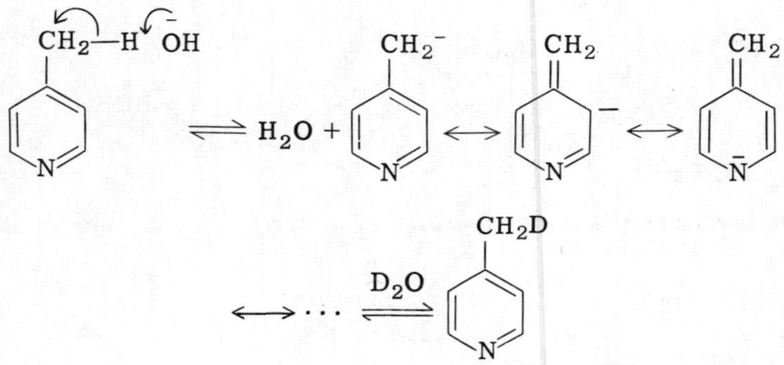

(β)

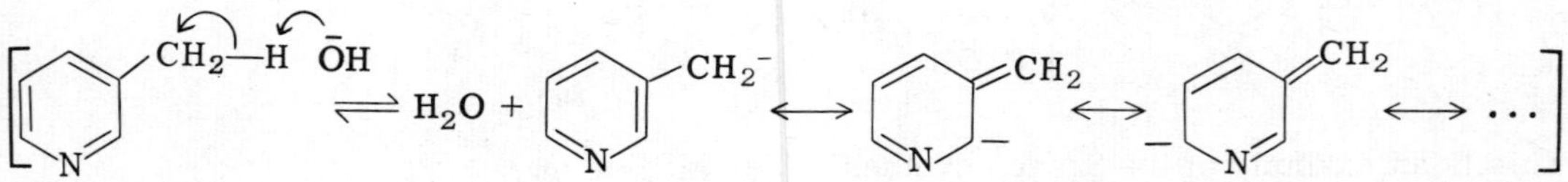

10. Since condensation occurs, this suggests that PhCHO must condense with a carbanion, i.e.,

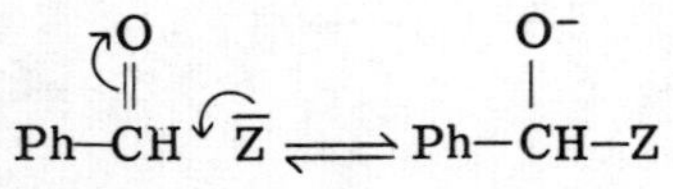

This leads to the possibility that quinoline-2-carboxylic acid forms a carbanion (with the negative charge at the 2-position).

Now, the thermal decarboxylation of acids possibly proceeds as follows (see *Text*, Ch. 9):

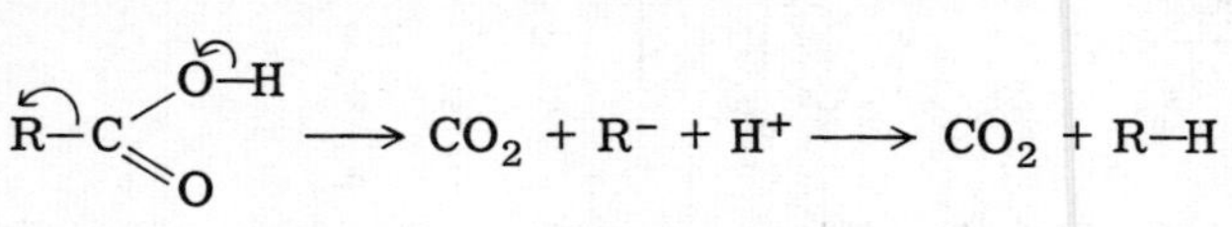

On the other hand, *salts* of the carboxylic acids also undergo decarboxylation on heating, but the conditions are milder (lower temperature) than for the acids.

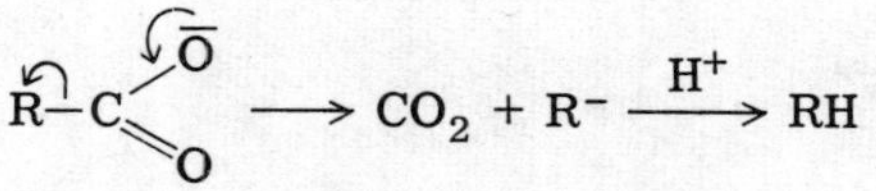

In either case, the carbanion is formed as an intermediate. Hence, by analogy, we can argue that either mechanism (or possibly both) operates with quinoline-2-carboxylic acid. The possibility of the carboxylate ion mechanism is reasonable because the basic properties of quinoline would give rise to a dipolar ion (however small its amount would be). If we accept that decarboxylation of the carboxylate ion is energetically more favourable, we can propose the following mechanism:

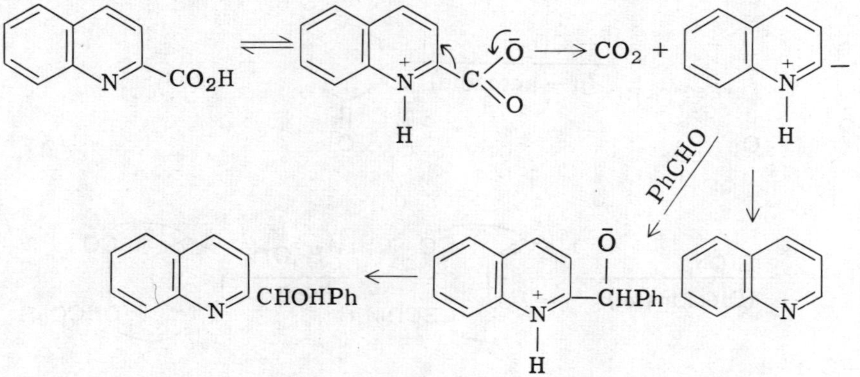

11. (i)

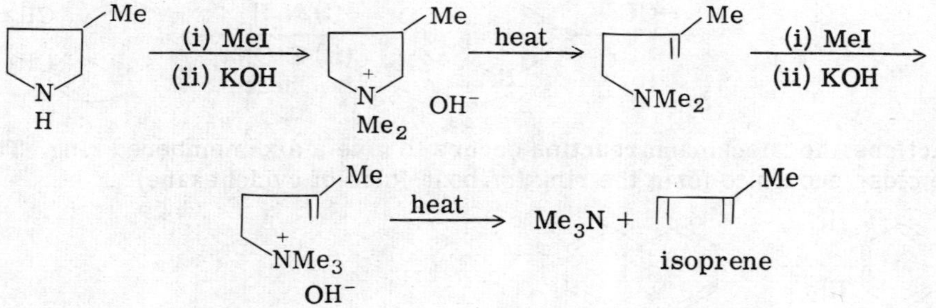

(ii)

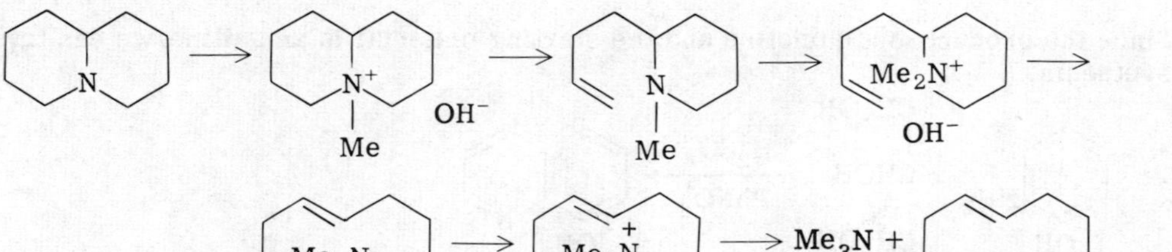

(iii)

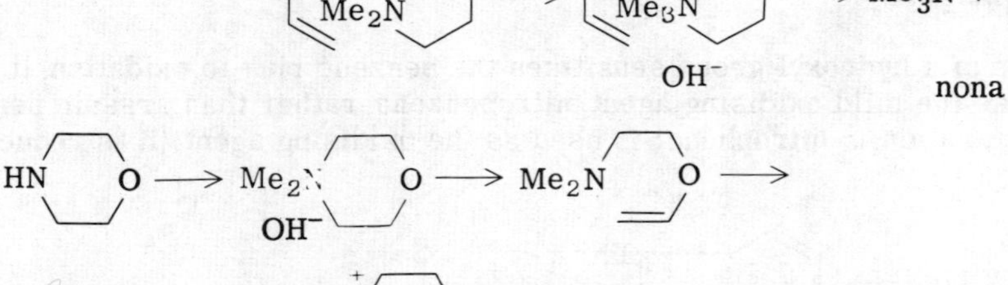

12. (i) $2CH_2{=}CHCO_2Et + MeNH_2 \xrightarrow{\text{1, 4-addition}}$

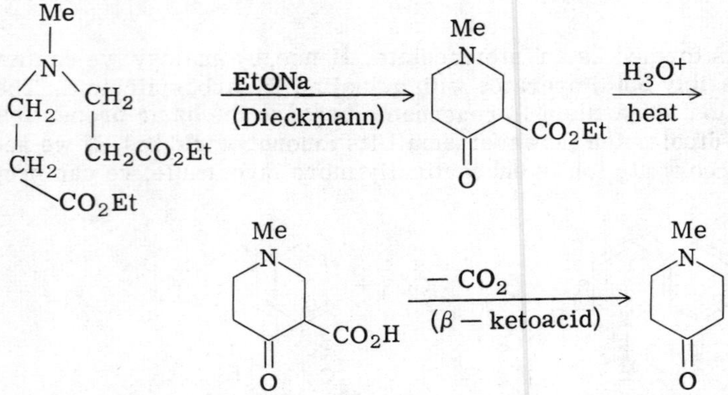

(ii)

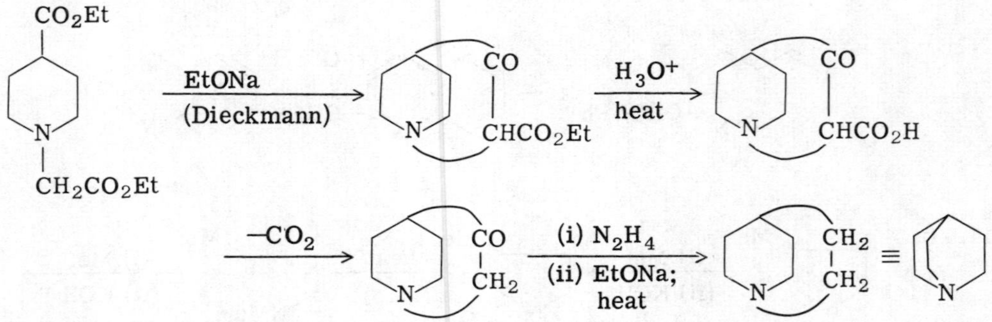

In both reactions, the Dieckmann reaction occurs to give a six-membered ring. The reacting groups are close enough to form the ring (*cf.* boat-form of cyclohexane).

13. (i)

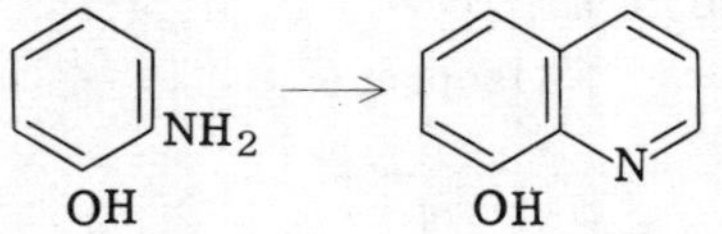

Since the product is a quinoline and the starting material is an aniline, we can try the Skraup synthesis.

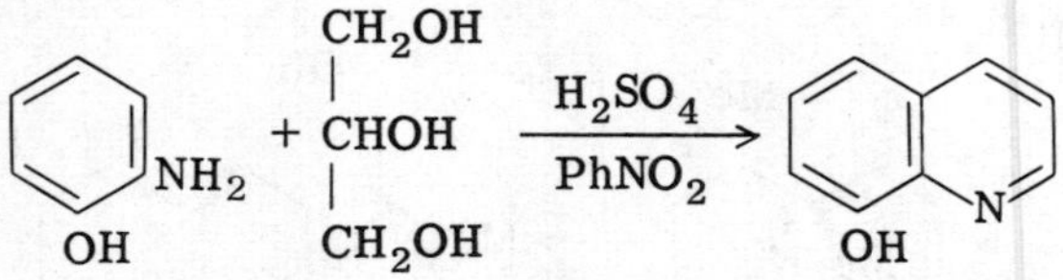

Since the presence of a hydroxyl group sensitises the benzene ring to oxidation, it is better to play safe and use the mild oxidising agent, nitrobenzene, rather than arsenic pentoxide. In the industrial preparation, *o*-nitrophenol is used as the oxidising agent (it is reduced to the amine).

(ii)

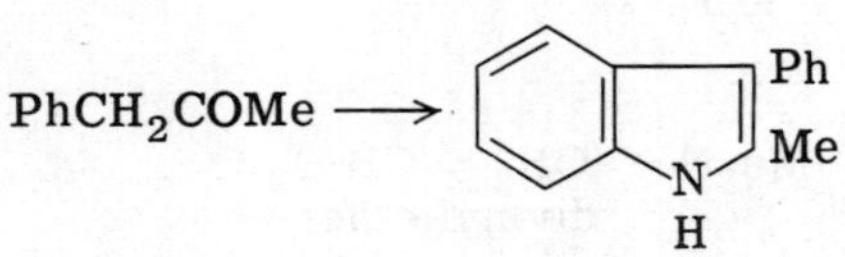

The starting material is a ketone and the product is an indole. This suggests that we can try the Fischer indole synthesis (see Prob. 2 (xiv), above, for the mechanism).

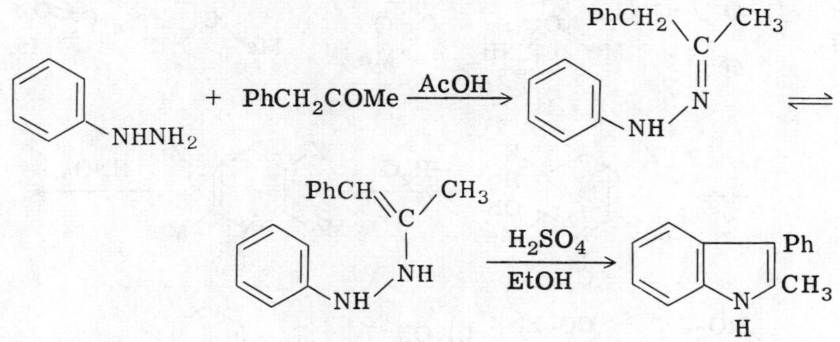

It should be noted that tautomerisation occurs with the benzyl group and not with the methyl group; hydrogen in the methylene group of benzyl is strongly activated.

(iii)

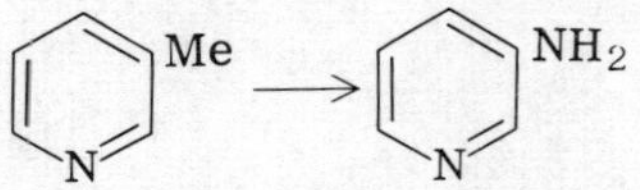

It would appear that the synthesis requires the conversion of the methyl group into amino. As we have seen on a number of occasions, it is always worth trying the oxidation of methyl to carboxyl and then considering what possibilities may be tried. Now we have to convert carboxyl to amino; this suggests the use of the Hofmann primary amine preparation (or the Schmidt reaction). Only the former is given here (the latter is somewhat hazardous).

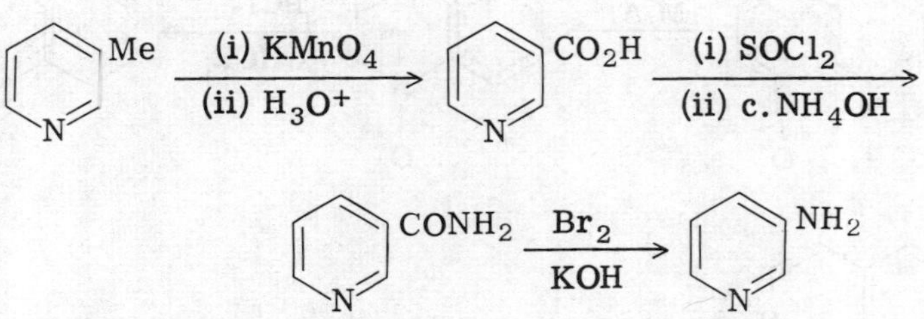

(iv)

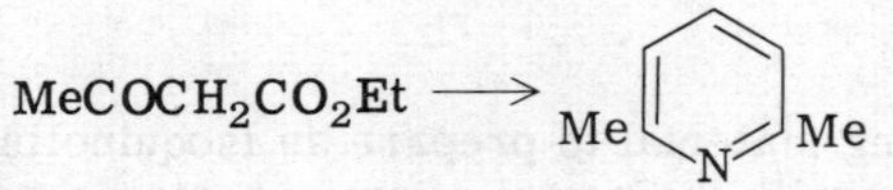

The starting material suggests the use of the Hantzsch pyridine synthesis. Since the two methyl groups in positions 2 and 6 come from E.A.A., we require two molecules of E.A.A. and an aldehyde with no alkyl group (which would have ended up at position 4; see *Text*). Hence, the required aldehyde is formaldehyde. To save ourselves from writing unnecessarily complicated equations, we shall use Z for CO_2Et, and change back to CO_2Et where necessary.

(*a*) $MeCOCH_2Z \xrightarrow{NH_3} MeC(NH_2){=}CHZ$

(*b*) $CH_2O + MeCOCH_2Z \xrightarrow{NH_3} CH_2{=}CZCOMe$

(*c*)

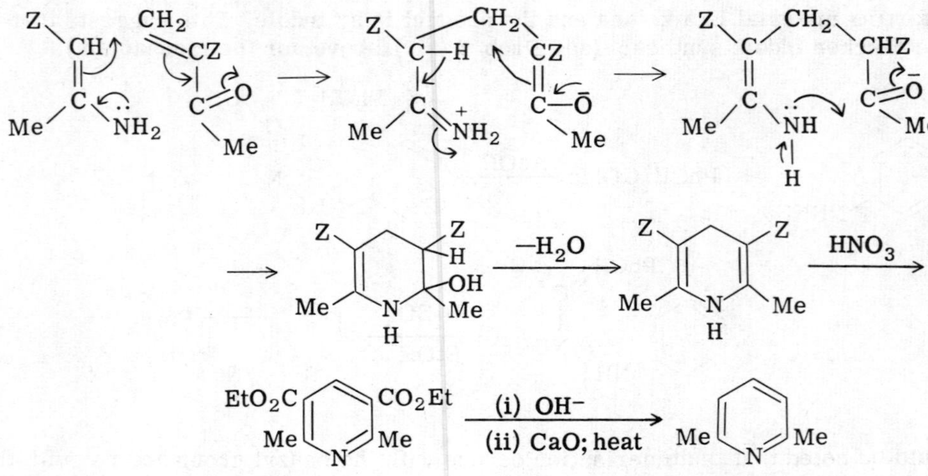

(v)

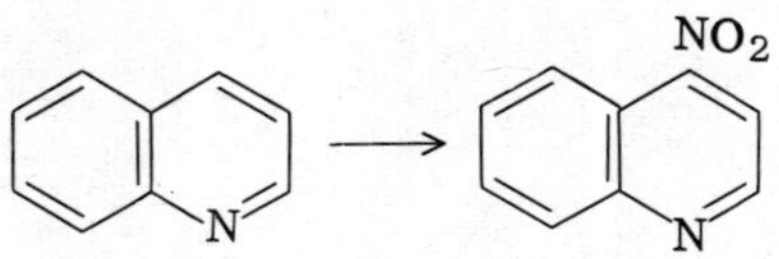

Since quinoline is attacked preferentially at the 8-position by electrophiles, we must use an indirect method. However, pyridine-1-oxide is readily nitrated in the 4-position and so, by analogy, we can try this route.

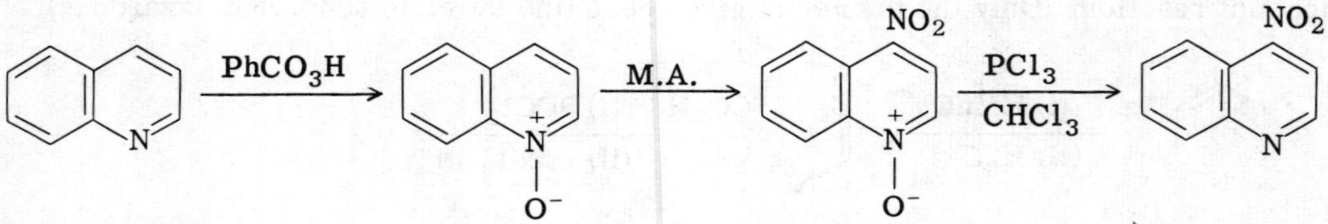

(vi)

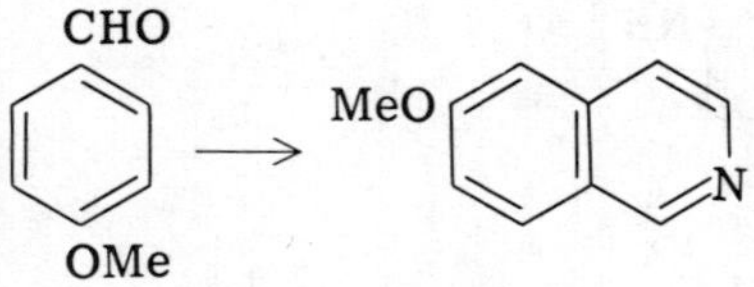

Since we have an aromatic aldehyde as starting material to prepare an isoquinoline, this suggests using the Pomeranz-Fritsch reaction with the acetal of aminoacetaldehyde.

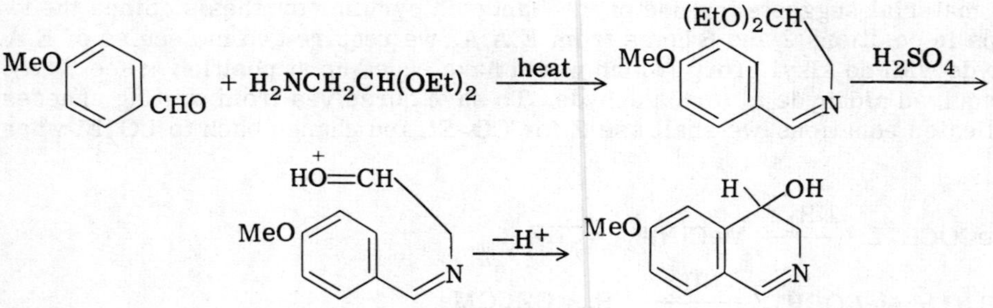

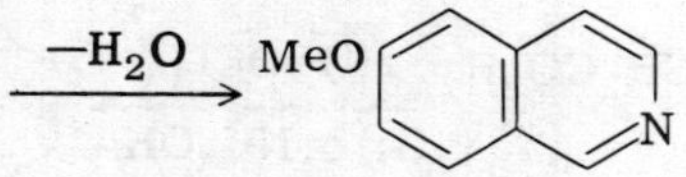

(vii)

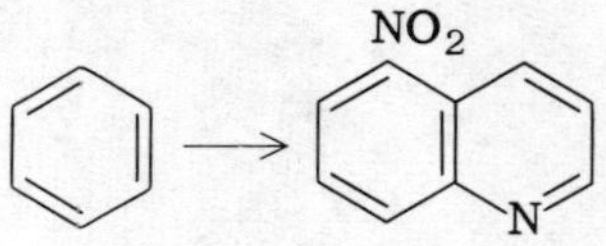

The question is: Do we synthesise quinoline first and then nitrate, or vice versa? Since the first approach gives the 8-nitro-derivative (see (v), above), let us consider the second route. Substituted anilines can be used in the Skraup synthesis. The nitro-group, which is *m*-directing, gives ring-closure predominantly in the *o*-position (see also Prob. 10(x), above). Hence:

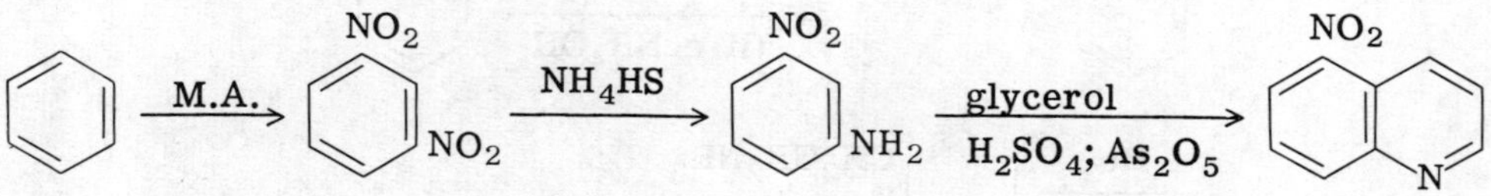

(viii)

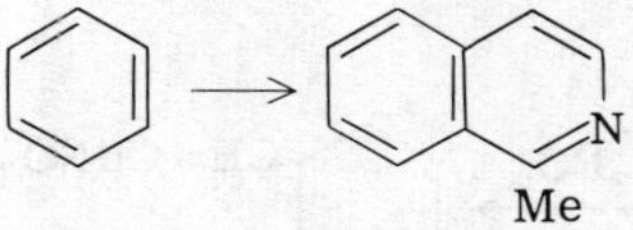

One method we can try is the Pomeranz-Fritsch reaction, using acetophenone (*cf.* with (vi), above).

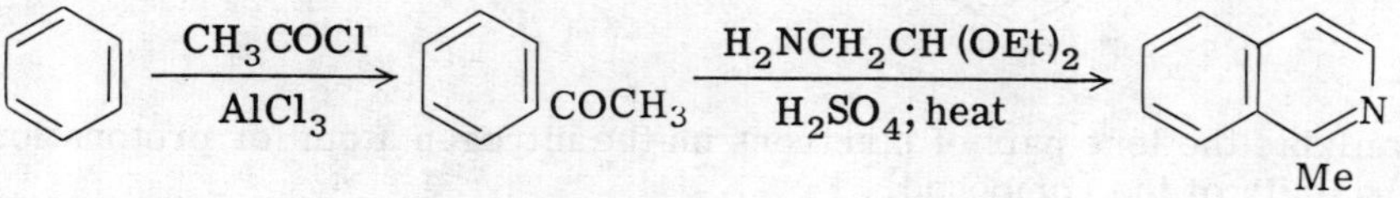

It is also instructive to consider the application of the Bischler-Napieralski reaction. This would be:

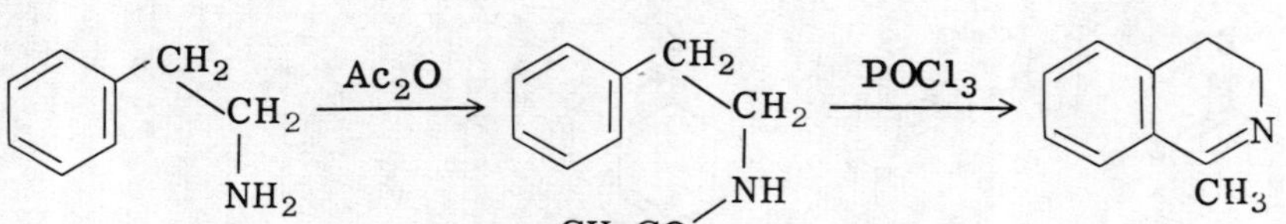

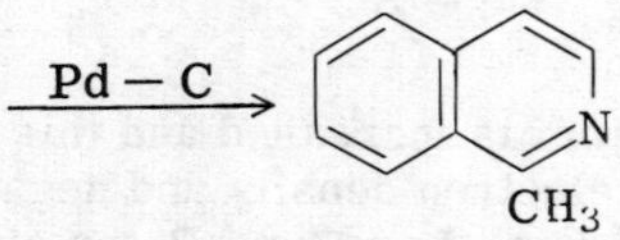

Hence, we now have to synthesise β-phenylethylamine from benzene. The following are some of the possible routes.

(*a*)

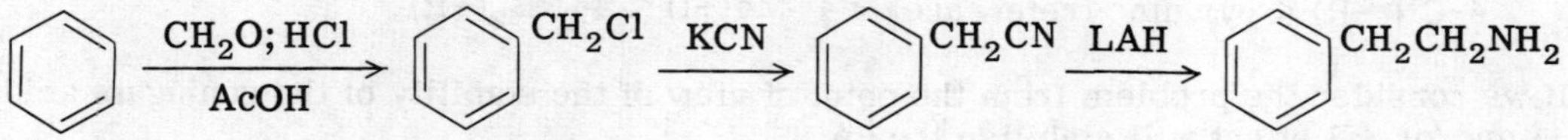

(*b*)

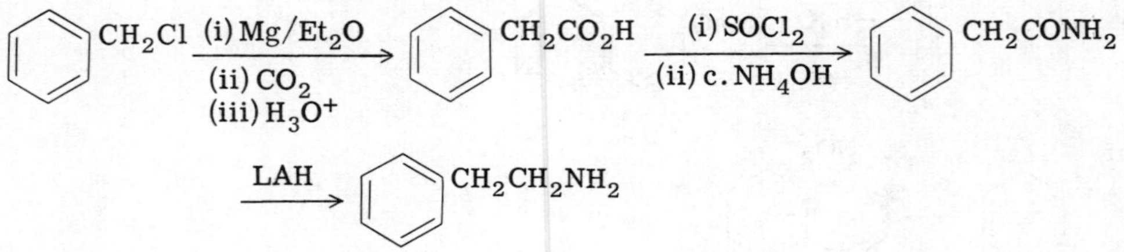

(*c*)

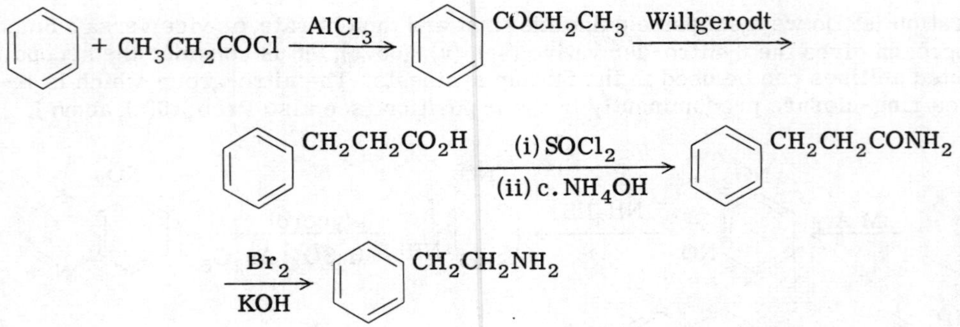

(*d*)

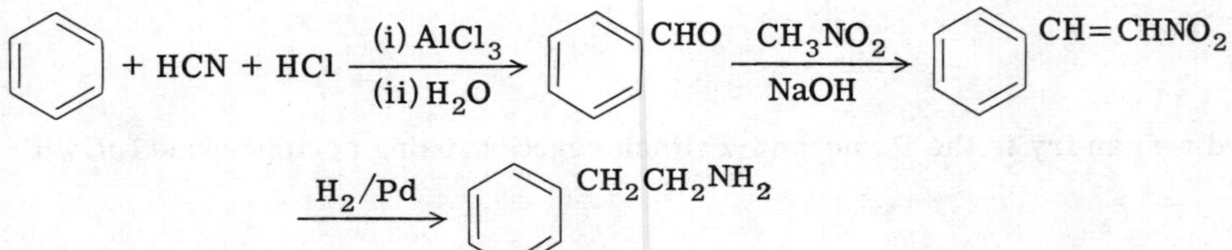

14. The more available the lone pair of electrons on the nitrogen atom for protonation, the greater is the basicity of the compound.

A substituent in the 4-position has an I and/or R effect. For a +I

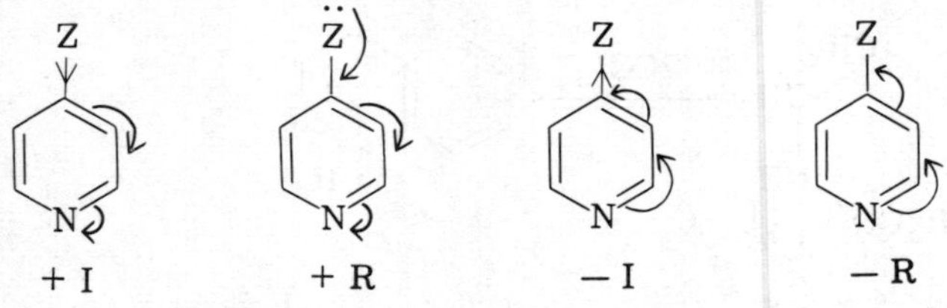

effect, the electron density on the nitrogen atom is increased and this atom becomes negatively charged. If Z has a +R effect, then the electron density and negative charge are larger than for the +I effect. If the effect of Z is —I and/or —R, the situation is reversed; the electron density at the nitrogen atom is decreased and this atom becomes positively charged (this will repel a proton). Hence, the order of increasing basicity is:

4-CN(—R) < pyridine (reference) < 4—Me(+I) < 4—NH_2(+R)

If we consider the problem from the point of view of the stability of the conjugate acid, a +I and/or +R effect will stabilise the cA:

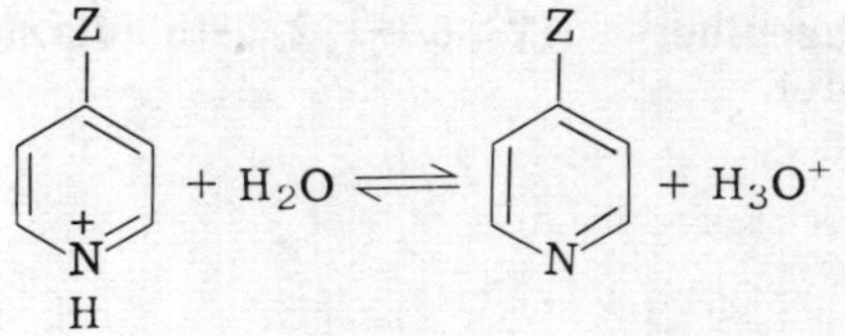

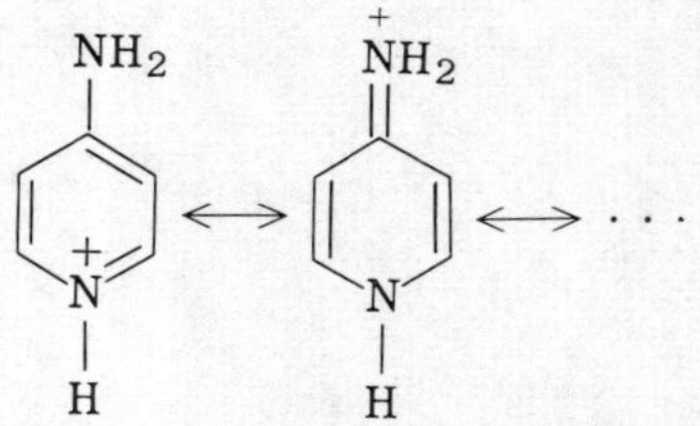

Thus, there is a number of resonating structures and the single positive charge is spread. If we now consider the +I and/or +R effect, we have, e.g.,

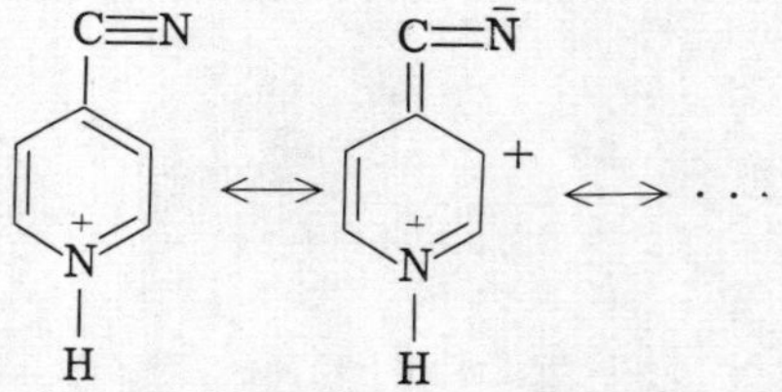

In this case, the number of resonating structures is fewer and there are unlike charges. Hence, this cA is far less stable than that of the 4-aminopyridine. This leads to the same order of stability as before.

15. Many of the common heterocyclic systems are given in the *Text*. Their names, structures, and rules of numbering should be learned by heart. The rules for naming complicated fused-ring systems should not be learned by heart, but it is worthwhile to practise their application.

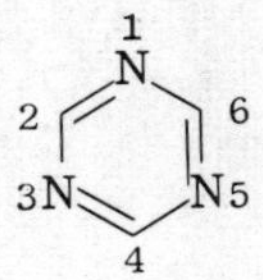

(I)

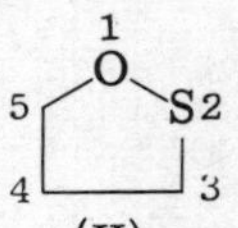

(II)

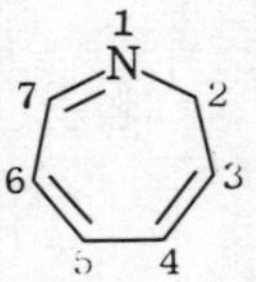

(III)

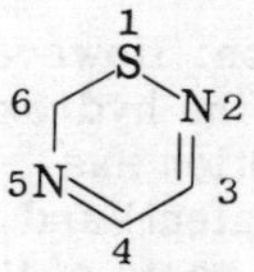

(IV)

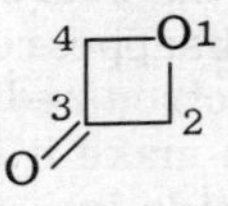

(V)

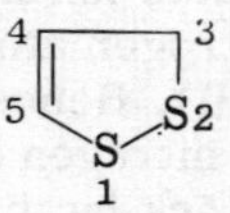

(VI)

(I), 1, 3, 5-triazine; (II) 1, 2-oxathiolan; (III), 2*H*-azepine; (IV), 6*H*-1, 2, 5-thiadiazine; (V), oxetan-3-one; (VI), 1, 2-dithiole (3*H*-1, 2-dithiole).

16. See the remarks made in Prob. 15, above.

(I), 4*H*-imidazo[4, 5-*d*]thiazole;

(II), 7*H*-pyrazino[2, 3-*c*]carbazole;

(III), 4*H*-1, 3-oxathiolo[5, 4-*b*]pyrrole;

(IV), thiazolo[5, 4-*b*]pyridine.

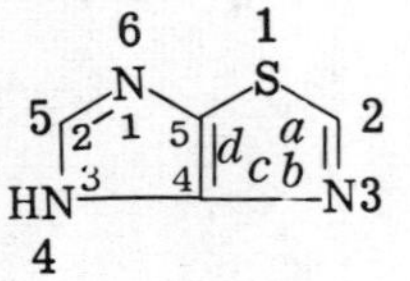

(I)

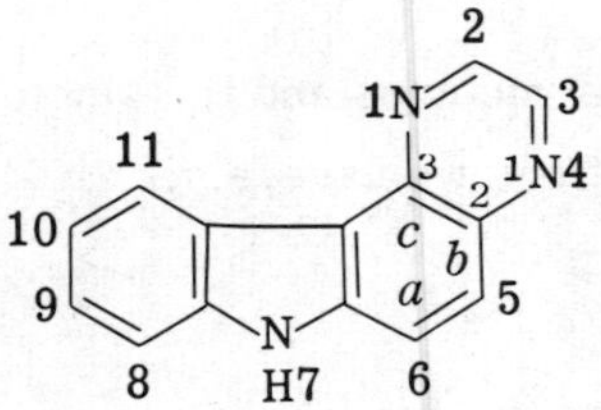

(II)

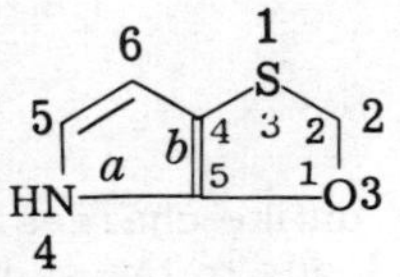

(III)

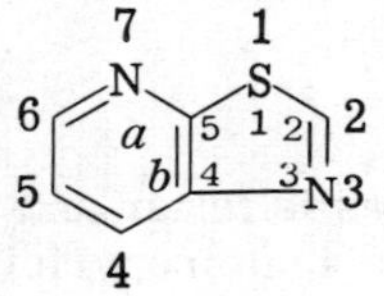

(IV)

17. The structure of pyridine methiodide is (I) and its molecular formula is (II).

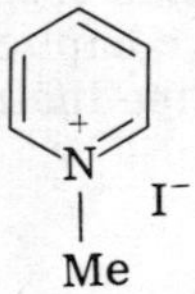

$= C_6H_8N^+ I^-$
(II)

(I)

Hence, we may write the equations as:

$$\underset{\text{(II)}}{C_6H_8N^+I^-} \xrightarrow{OH^-} \underset{\textbf{(A)}}{C_6H_9NO} \xrightarrow{[O]} \underset{\textbf{(B)}}{C_6H_7NO}$$

A has been formed by replacement of the iodide ion in (II) by the hydroxide ion. However, A is not a salt; the positive charge on the nitrogen atom has disappeared and the hydroxide ion must have formed a covalent bond with (II). Hence, an electron redistribution has occurred, two electrons now residing on the nitrogen atom (to make it tercovalent) and an electrophilic centre has been created for attack by the hydroxide ion. On the basis of these arguments, we can formulate the reaction steps as follows:

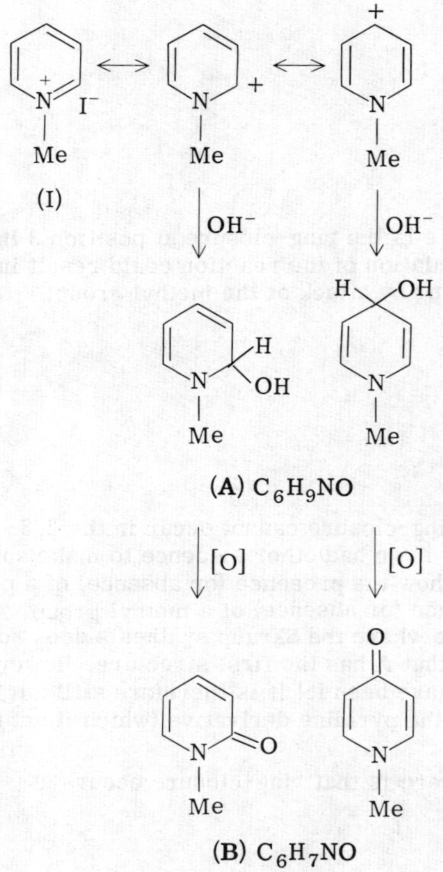

According to this scheme, **A** is a secondary alcohol, and on oxidation would be expected to form a ketone by loss of two hydrogen atoms. Thus, **B** would be a ketone, i.e., a 1-methylpyridone.

For the data given, **A** and **B** could be one or other of the corresponding pairs (or *each* could be a mixture of the two isomers). In actual fact, other evidence has shown that **B** is the 2-pyridone (and this fixes **A** as the 2-hydroxy-compound).

18.

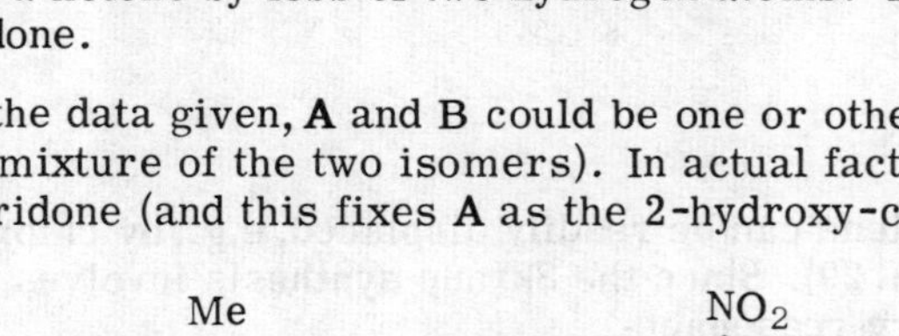

The difference between the molecular formulæ of **A** and **B** is CH_2. Also, **B** contains only one nitrogen atom, although its precursor contains two. All of this suggests that the methyl group is retained in **A** and that there is no nitro-group in **B**. Since a Skraup synthesis has been carried out in each case, let us see how far we can get on the assumption that each reaction has gone 'according to plan' to give a quinoline derivative.

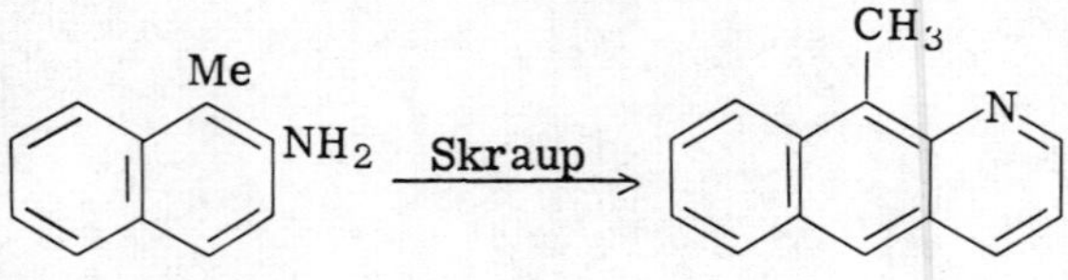

(A) $C_{14}H_{11}N$

The formula is correct for A, but the unusual feature is the ring-closure at position 3 in naphthalene (see *Text*, Ch. 29). An alternative formulation of the reaction could result in the formation of a seven-membered heterocyclic ring by attack at the methyl group.

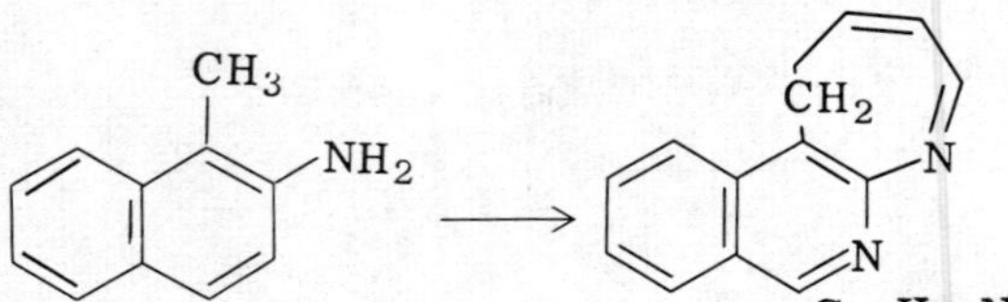

$C_{14}H_{11}N$

If we argued that this is **A**, the implication is that ring-closure cannot occur in the 2, 3-position in naphthalene. Obviously, then, it is better if we had other evidence to make sure which structure is **A**, e.g., i.r. spectroscopy would show the presence (or absence) of a pyridine ring; NMR spectroscopy would show the presence (or absence) of a methyl group. As a matter of interest, there appears to be no example where the Skraup synthesis does not give the pyridine derivative. Hence, we can accept that **A** has the first structure. However, since an exception may turn up one day, this could have been it! It is therefore still advisable to obtain other information to justify that **A** is the pyridine derivative (which it actually is).

Let us now consider B. A reasonable explanation here is that ring-closure occurs at position 1 with ejection of the nitro-group.

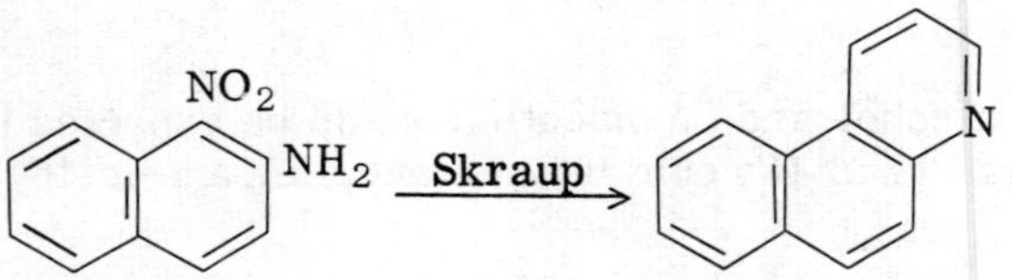

(B) $C_{13}H_9N$

It is known that the 1-nitro-group in naphthalene can be readily displaced, e.g., by chlorine using phosphorus pentachloride (see *Text*, Ch. 29). Since the Skraup synthesis involves the use of vigorous conditions, the analogy is very reasonable.

31 Dyes and photochemistry

(Problems on page 58)

1. The 'expected' product is adipic acid.

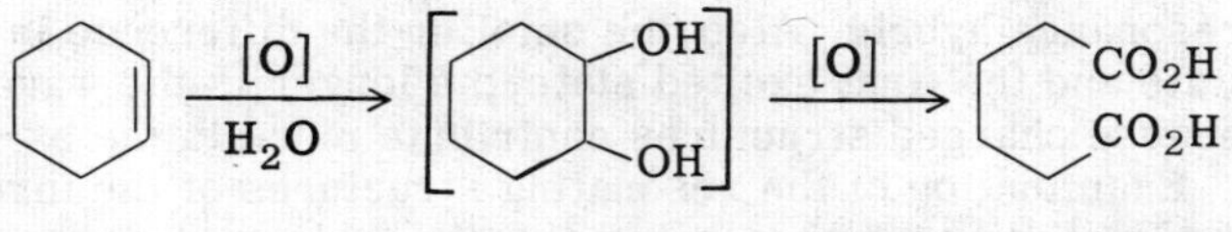

The oxidation of cyclohexene, C_6H_{10}, to give the compound C_6H_8O has resulted in the loss of two hydrogen atoms and the addition of one oxygen atom. Since opening of the ring at the double bond would be expected to incorporate more than one oxygen atom, a reasonable interpretation of the results is that a methylene group in cyclohexene has been oxidised to a carbonyl group. Now, the u.v. absorption band for an isolated double bond is at ~175 nm and that for an isolated keto group is at ~190 nm. Since the actual band is at 225 nm, this suggests that the double bond and keto group are conjugated (see Table 31.2). Hence, the oxidation may be formulated as follows:

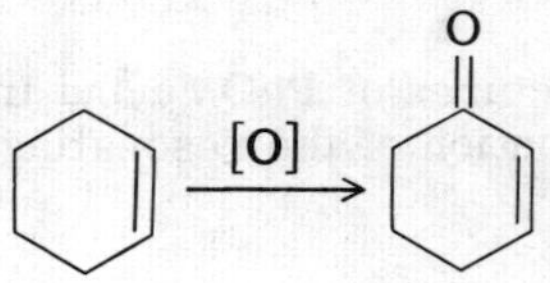

(C_6H_8O)

2.

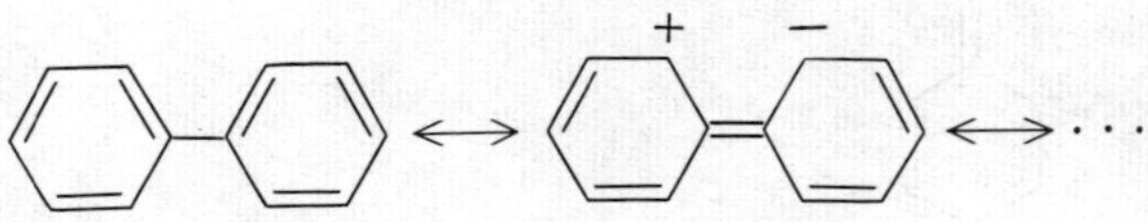

(I) λ_{max} 251·5 nm (ϵ 18, 300)

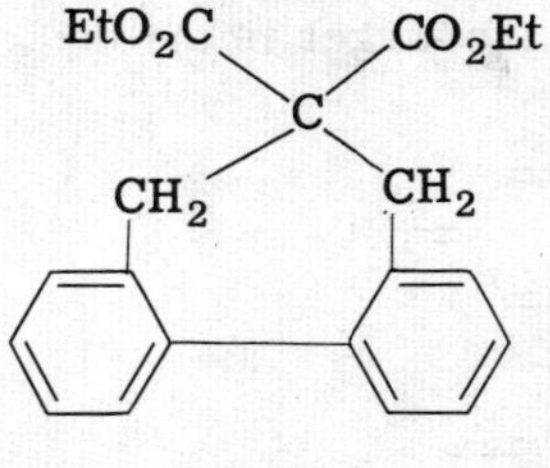

(II) λ_{max} 249 nm (ϵ 17,000)

Since (II) has been resolved, the molecule cannot be coplanar. The absorption band of CO in CO_2Et is at 200-205 nm (see *Text*). Therefore, the chromophore that absorbs at 249 nm must be the two benzene rings in the biphenyl moiety. Now, this band and its molar absorptivity are very close to those of (I). Biphenyl, (I), is a resonance hybrid, but because of steric effects due to the *o*-hydrogen atoms, the molecule is not completely coplanar (see also Ch. 29, Prob. 14). Hence, the two rings in (II), which must be twisted with respect to each other, are out of coplanarity by an angle which is not much greater than that in biphenyl.

3. If the u.v. spectra of PhMe, Ph_2CH_2, and Ph_3CH are similar, then the chromophores in all three will be very closely similar. All three contain benzene rings and these must be the chromophores in each compound. Also, because the rings are isolated (i.e., cannot conjugate as in biphenyl), all three can be expected to show absorption maxima which will be at about the same wavelength in each. However, because the number of rings is 1, 2, and 3, the molar absorptivities will also be in this ratio of 1 : 2 : 3.

4. (i)

PhOH	PhOMe	PhO^-
210 nm	217 nm	235 nm

The larger the amount of conjugation and the larger the number of *charged* resonating structures (with fairly close energy content), the greater is the resonance energy and so the lower is the energy content of the resonance hybrid. Also, the smaller the difference in energy levels between the ground state and the first excited state, the longer is the wavelength of the absorption. This is because charged structures contribute more to the excited state than to the ground state. Examination of the resonating structures of the three compounds shows that the contributions of charged resonating structures to each resonance hybrid increases in the order: PhOH < PhOMe < PhO^-. Hence, λ_{max} increases in this order.

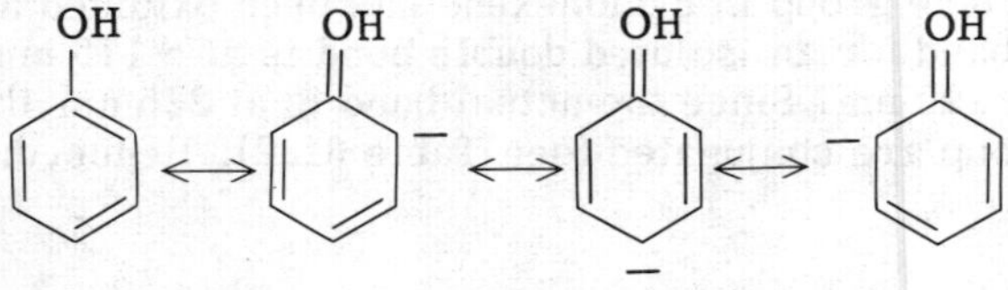

Replacement of H by Me gives the corresponding resonating structures of PhOMe, but in this case, because of the +I effect of the methyl group, the contribution of the resonating structures is increased (above that in PhOH).

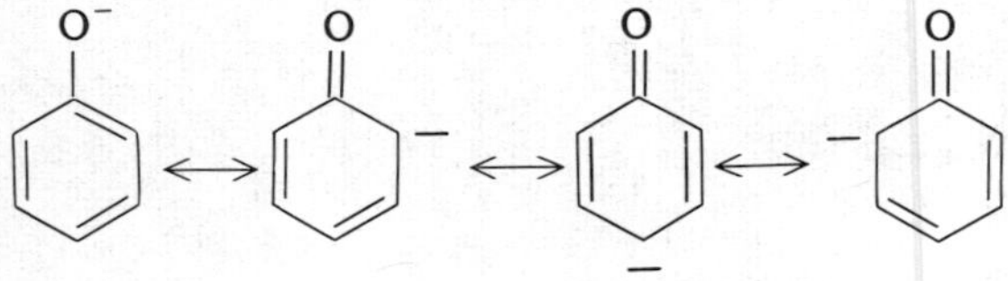

In the phenoxide ion, *all* of the contributing resonating structures are charged, and since only a negative (i.e., one kind of) charge is delocalised, this increases the stability of the resonance hybrid still further (than those of the other two). Hence, the larger is the increase in the wavelength ($PhO^- \gg$ PhOMe > PhOH).

(ii)

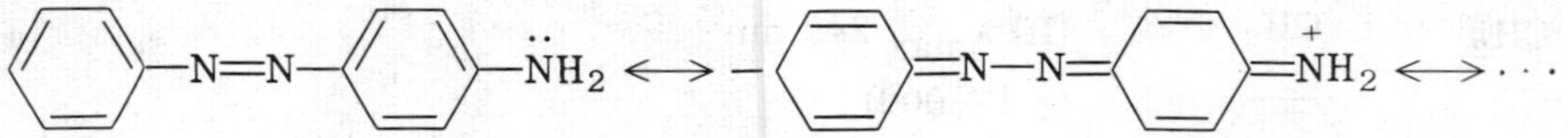

(I) yellow

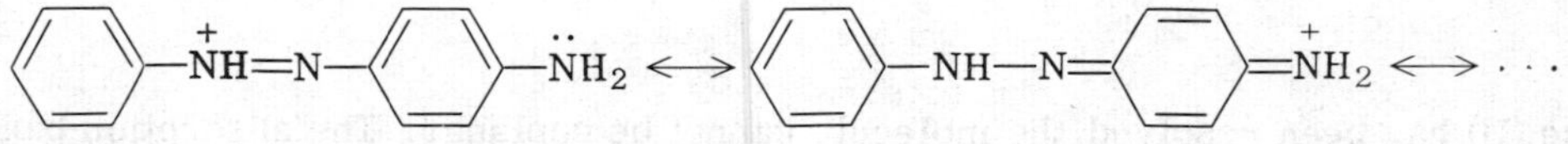

(II) violet

Since the colour *observed* is the complementary colour of the band *absorbed*, λ_{max} of the yellow compound, (I), is 435-480 nm, and that of the violet compound, (II), is 560-580 nm (see Table 31.1). Hence, (II) absorbs in the longer wavelength region, and this can be explained on the basis that *all* the resonating structures are charged, and some carry only one charge (positive).

5.

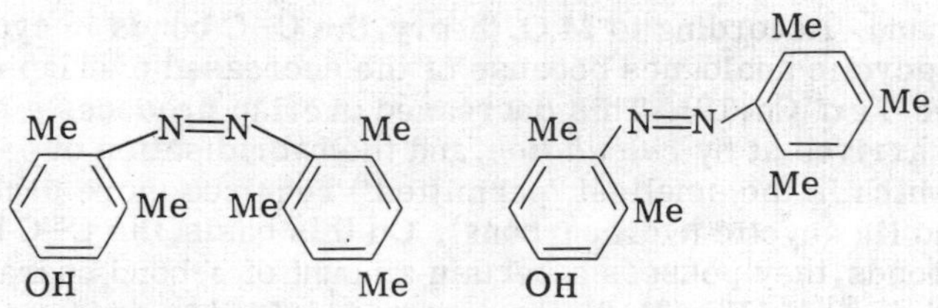

cis; colourless *trans*; coloured

Because of the steric effects of the four *o*-Me groups in the *cis*-isomer, the two benzene rings cannot become coplanar (steric inhibition of resonance). The result is that the chromophores, the two benzene rings and azo-group, cannot enter into extended conjugation; hence, they absorb separately. Since both chromophores absorb in the u.v. region (below 400 nm), the *cis*-compound is colourless.

The situation with the *trans*-isomer is entirely different. There is no steric inhibition of resonance, and because the molecule can become coplanar, this isomer can exist as a resonance hybrid of extended conjugated charged resonating structures. In these circumstances, the difference between the energy levels of the ground state and first excited state is low enough for absorption to occur in the visible region (400-750 nm) and hence is coloured (complementary colour).

6. For a solvent to have an effect on the position of the absorption maximum of the solute, there must be some interaction between solute and solvent to cause a change in the charge distribution in the solute. This interaction usually involves a lone pair of electrons in the chromophore of the solute. Since alkenes do not contain lone pairs, the polarity of the solvent will have no effect on the position of the absorption maximum.

7. CH_4 C_2H_6 

125 nm 135 nm 190 nm

Since only single bonds are present in these compounds, absorption must be due to $\sigma \longrightarrow \sigma^*$ transitions. In methane, the only transition possible is that of the C—H bond. In ethane, two transitions are possible, one for the C—H bond and the other for the C—C bond. The latter transition requires less energy to bring it about and so this occurs at a longer wavelength than that for a C—H bond. On this basis, we may accept that the absorption band at 135 nm is due to the transition involving the C—C bond in ethane.

Since cyclopropane can apparently undergo only $\sigma \longrightarrow \sigma^*$ transitions (C—C and C—H), the very large increase in the wavelength of the absorption band (compared to the other two hydrocarbons) is unexpected. Thus, some factor operates in cyclopropane to bring the energy levels of the ground state and the first excited state (of C—C or C—H) closer together. As we have seen, according to classical theory, cyclopropane is under large strain (see *Text*, Ch. 19). Hence, it is this strain factor that has caused the large shift to the longer wavelength. An interesting point about this is that we have offered an explanation in terms of another explanation which has subsequently been shown to be untenable. Obviously, then, we must try to find an explanation which is consistent with current theories of light absorption. One way of doing this is to accept that 'strain' is the cause, but now we must try to re-interpret the meaning of this term and then see if we can come up with an acceptable explanation. This does not necessarily mean that everyone will accept what is proposed. All it means is that the explanation offered is reasonable. Those who do not accept it usually offer an alternative explanation which is also reasonable. Cases like this, involving *theoretical* discussions, are quite common, e.g., mechanisms of reactions. However, as new facts are discovered, it may be possible to eliminate one (or more) of the alternatives. This does not mean that later on the generally accepted explanation will not be changed.

We now return to the problem in hand. According to M.O. theory, the C—C bonds in cyclopropane are weaker than those in acyclic analogues because of the decreased overlap of the orbitals forming the bonds (see *Text*, Ch. 19). This decreased overlap produces a bond angle of about 104°. This angle is arrived at by *calculation*, and the hybridisation of *s*- and *p*-orbitals to achieve this angle (which is the smallest 'permitted') requires more *p*-character than that in a normal C—C bond (in acyclic hydrocarbons). On this basis, the C—C bonds in cyclopropane are not 'pure' σ-bonds; they possess a certain amount of π-bond character (π-bonds are *p—p* bonds). As we have seen (Fig. 31.2), the energy required to promote a $\sigma \longrightarrow \sigma^*$ transition is *greater* than that of any other type of transition. Hence, because of the partial π-bond character of the C—C bonds in cyclopropane, the energy levels of the ground state and first excited state are closer together than those of a 'pure' σ-bond in ethane. This produces a shift in the absorption maximum of cyclopropane towards the longer wavelength region.

8.

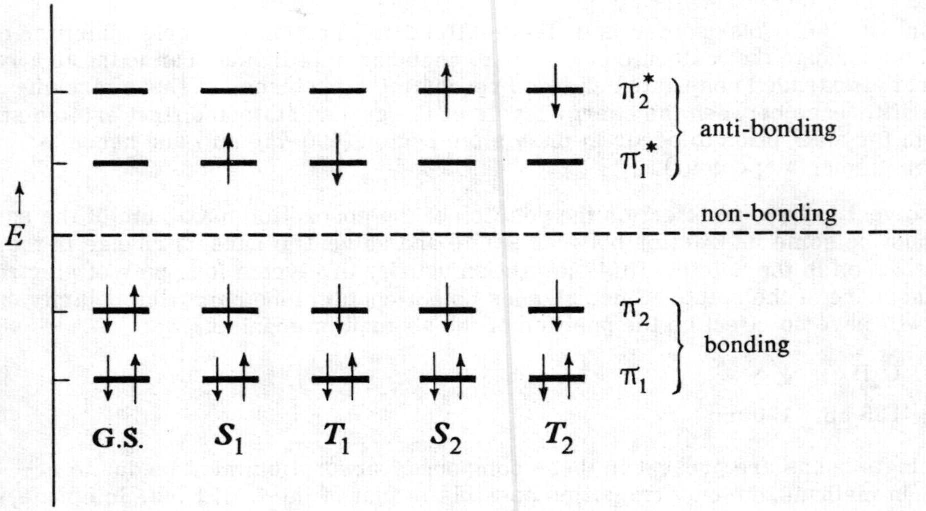

9. (i)

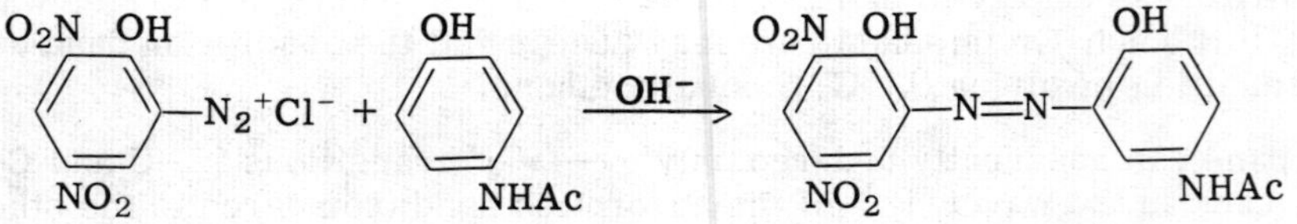

Coupling occurs preferentially in the position *para* to the hydroxyl group, but when this is occupied, coupling occurs in the *o*-position. The reaction is carried out in faintly alkaline solution (see *Text*, Ch. 24).

(ii)

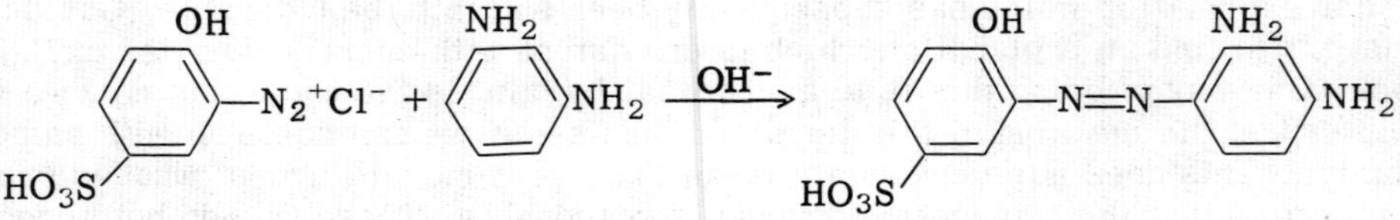

(iii)

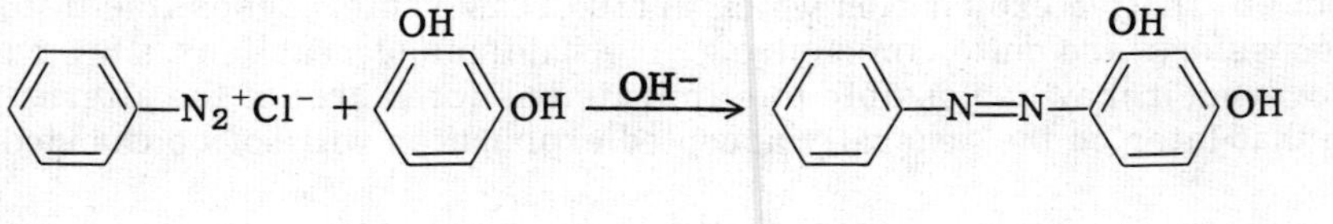

(iv)

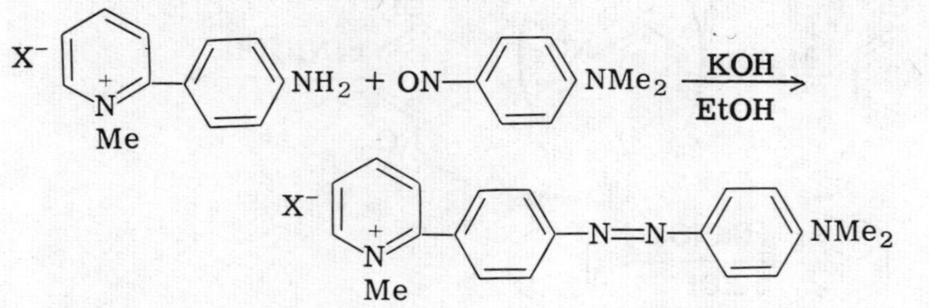

See *Text,* Ch. 22.

10. These problems are basically a question of preparing the necessary components. Many of the principles required have been dealt with in problems given earlier (see appropriate chapters). Diazonium salts normally couple with arenes containing OH and/or NH_2.

(i)

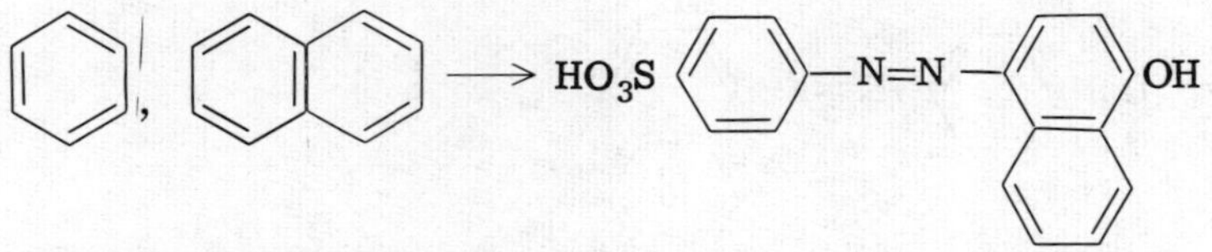

The required coupling components are:

diazotised sulphanilic acid

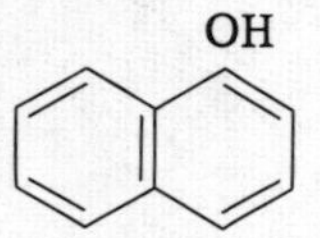

1-naphthol

(*a*)

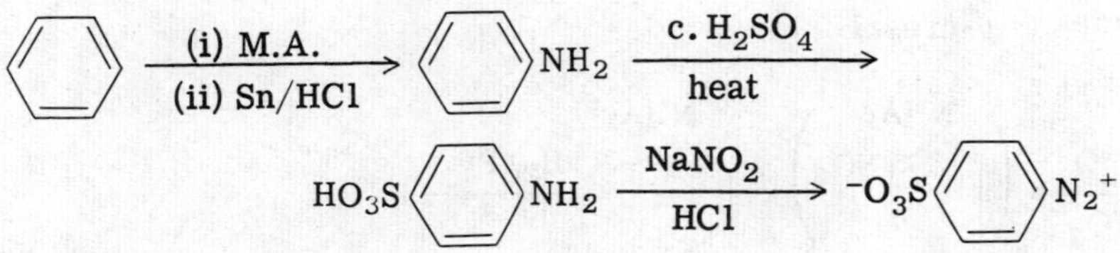

(*b*)

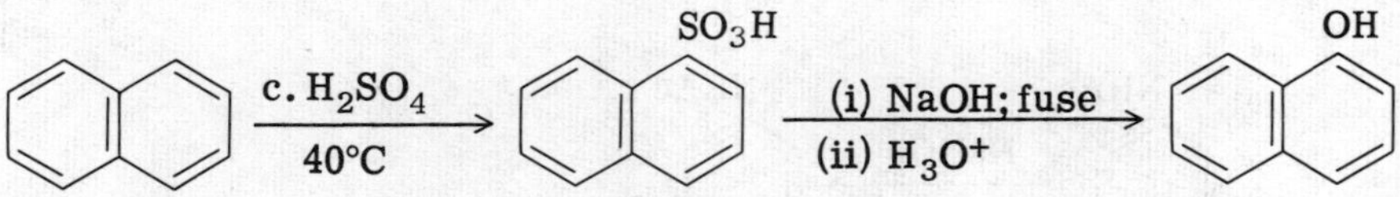

(*c*)

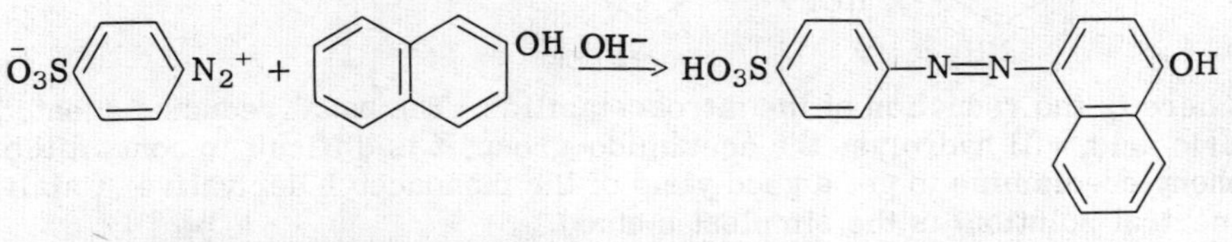

(ii) The required coupling components are:

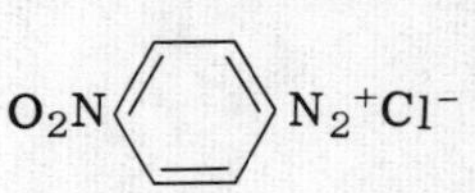

diazotised *p*-nitroaniline

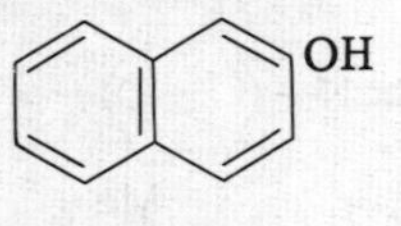

2-naphthol

(*a*)

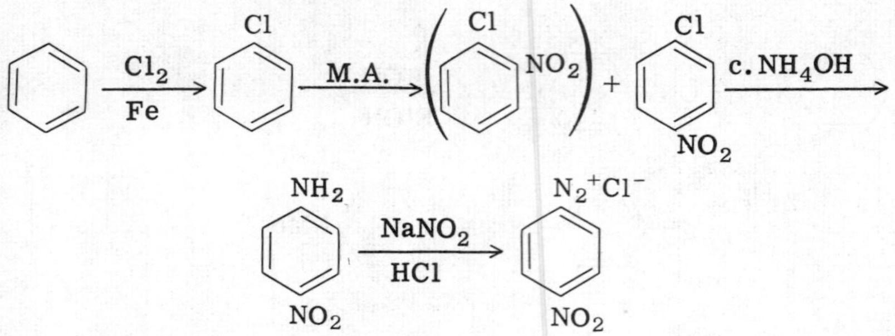

(*b*)

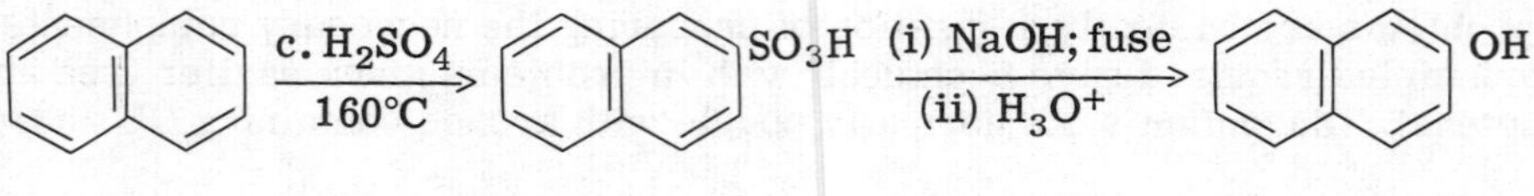

(*c*)

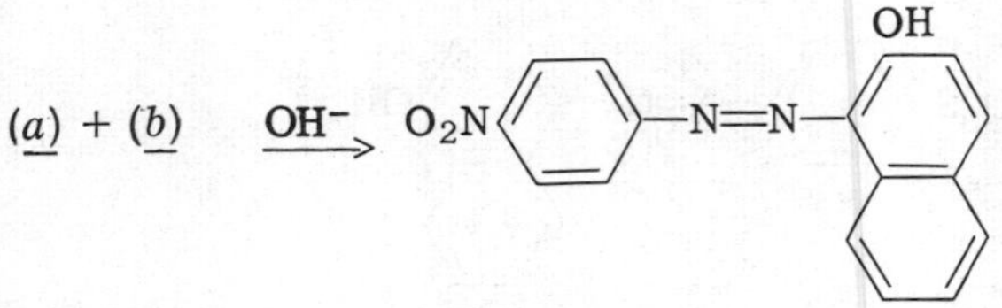

(iii) The required coupling components are:

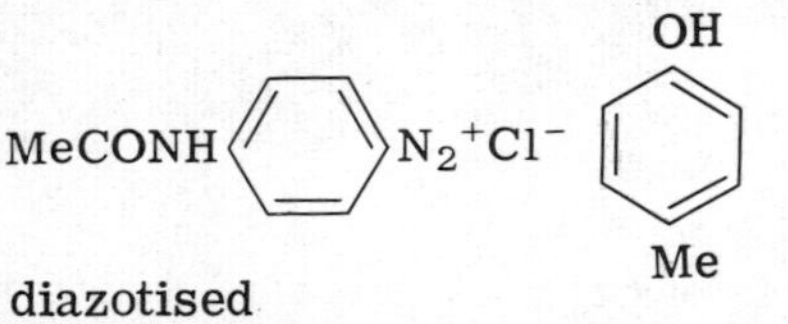

diazotised *p*-aminoacetanilide — *p*-cresol

(*a*)

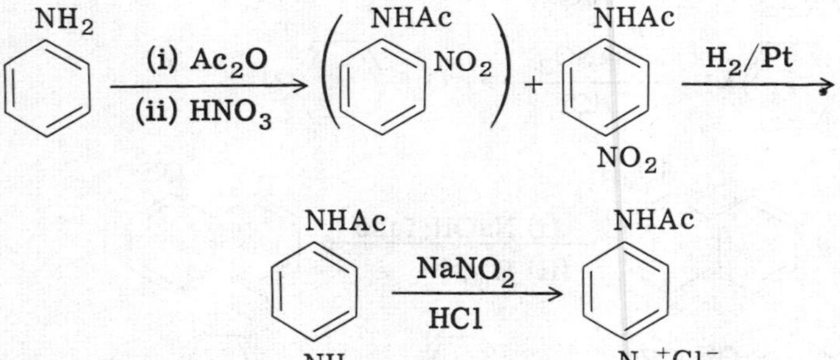

The difficulty here is the reduction of the nitroacetanilide. The usual reducing agent, tin and hydrochloric acid, will hydrolyse the acetamido-group; it is difficult to control the acetylation of *p*-phenylenediamine to get a good yield of the monoacetyl-derivative. Catalytic reduction (in neutral solution) is the simplest method.

(*b*)

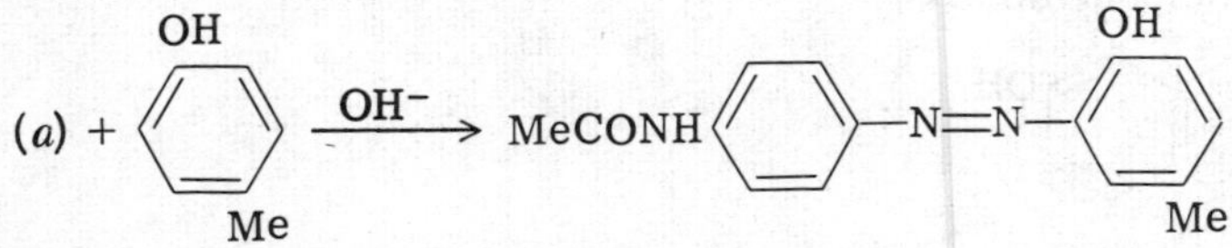

(iv) The required coupling components are:

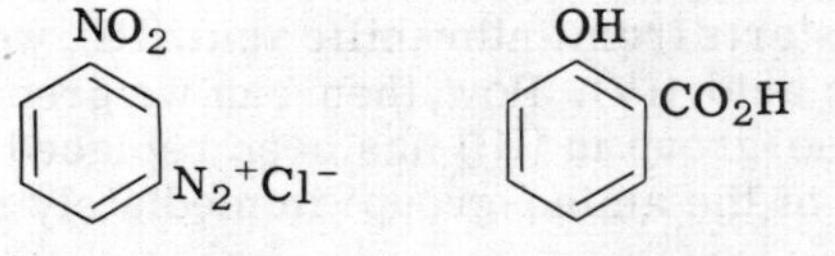

diazotised *m*-nitroaniline   salicylic acid

(*a*)

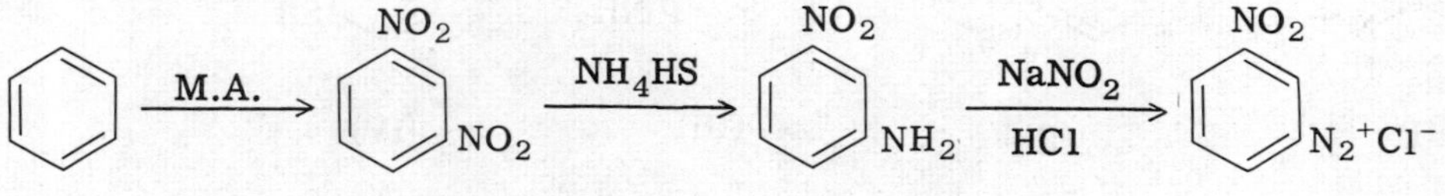

(*b*)

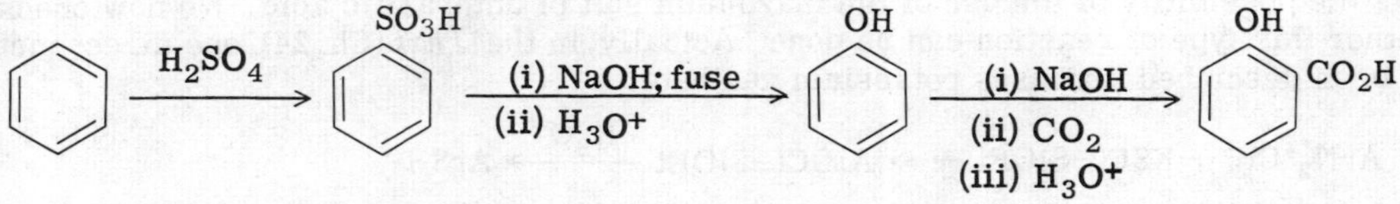

Phenol may also be prepared via benzenediazonium chloride.

(*c*)

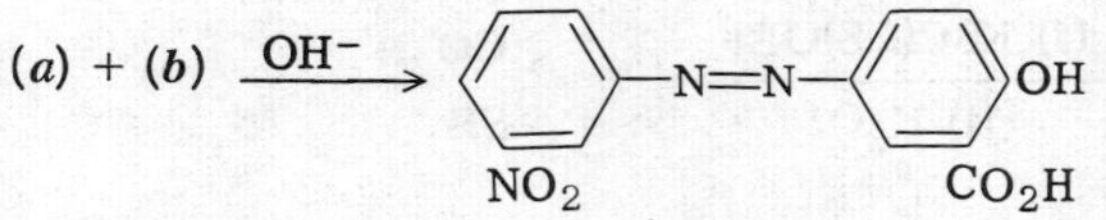

(v)

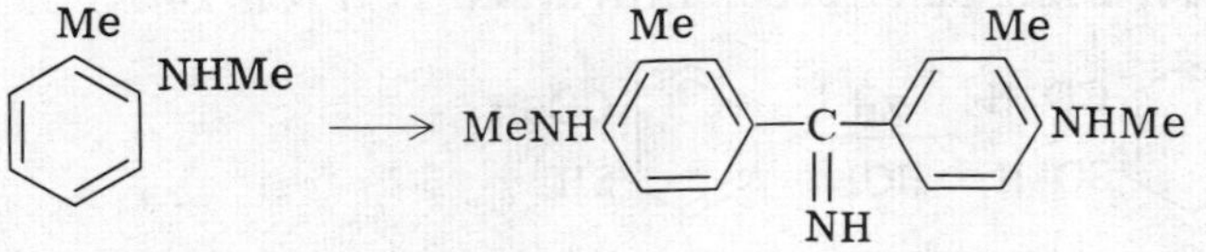

This type of imino-derivative may be prepared by heating a diarylmethane with sulphur in a current of ammonia (see *Text*). Hence:

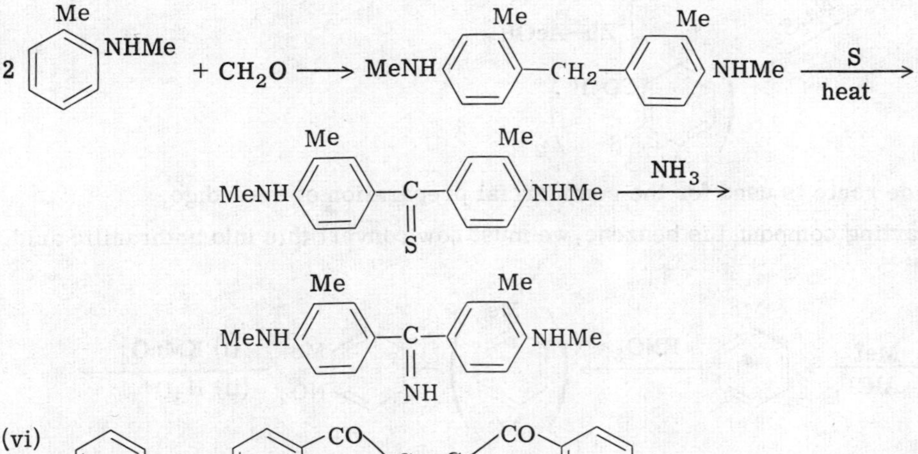

(vi)

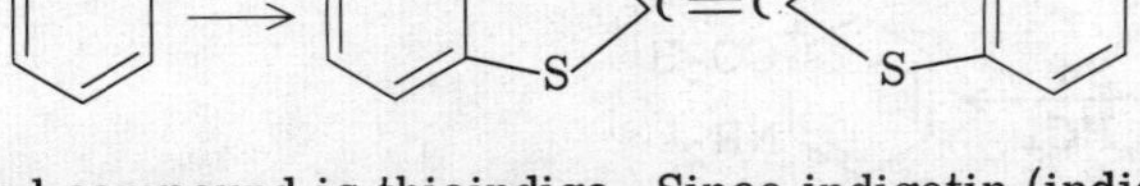

The desired compound is thioindigo. Since indigotin (indigo) is prepared by the oxidation of indoxyl, (I), we can try to prepare thioindigo by the oxidation of the analogous sulphur compound, thioindoxyl, (II). Also, since one route to indoxyl starts from anthranilic acid, (III), we need to prepare the analogous sulphur compound, thiosalicylic acid, (IV). How, then, can we prepare (IV)? Comparison of (III) with (IV) shows that the amino-group in (III) has been replaced by the thiol group to give (IV). The phrase 'replacement of the amino-group' immediately sug-

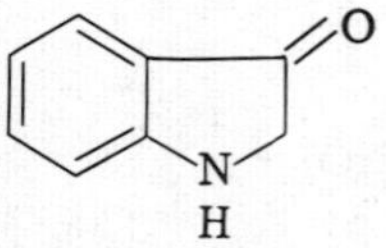

(I)

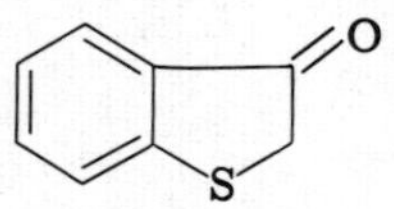

(II)

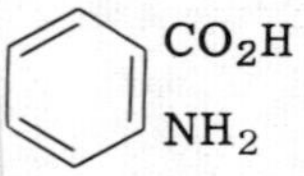

(III)

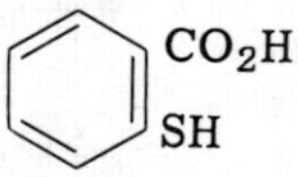

(IV)

gests the possibility of the use of the diazonium salt of anthranilic acid. We now check up whether this type of reaction can be done. Actually, in the *Text* (Ch. 24), one direct method has been described that uses potassium xanthate.

$$ArN_2^+Cl^- + KSC(=S)OEt \longrightarrow ArSC(=S)OEt \xrightarrow{H_3O^+} ArSH$$

Hence, one route is:

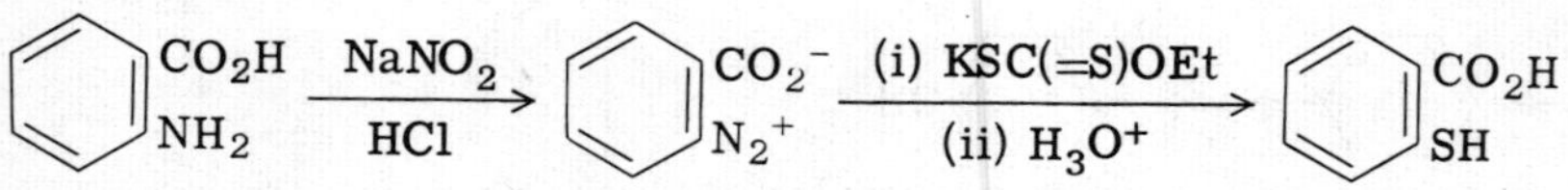

It is also instructive to consider some other possibilities; see *Text* (Ch. 24).

(*a*)

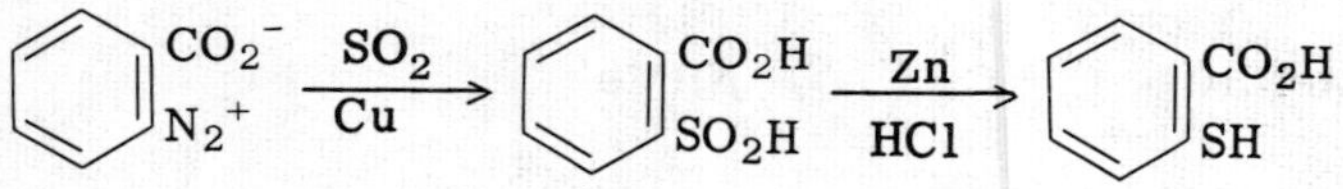

(*b*)

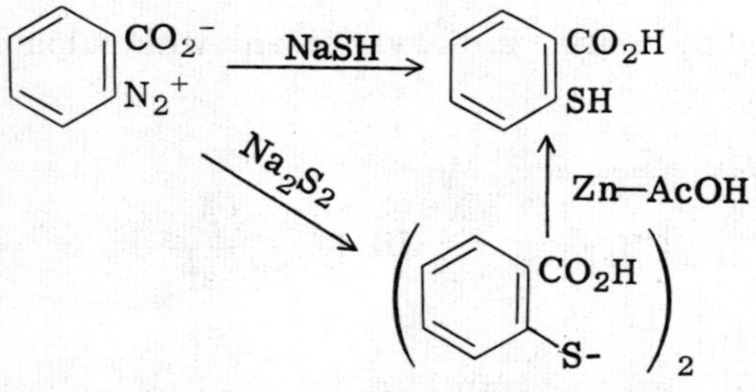

The disulphide route is used for the commercial preparation of thioindigo.

Since our starting compound is benzene, we must now convert this into anthranilic acid. One route is:

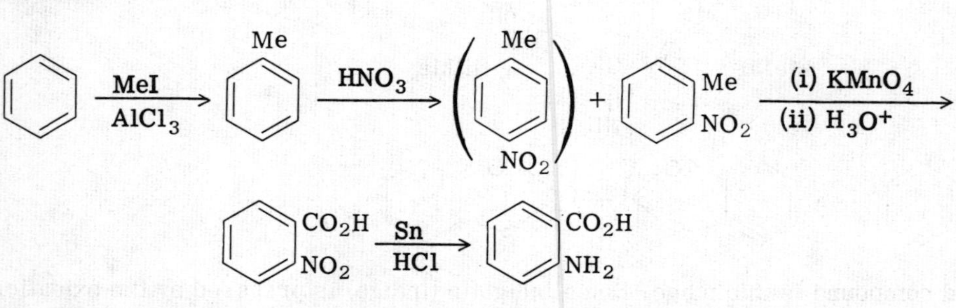

The final steps in the thioindigo synthesis are (by analogy with indigotin):

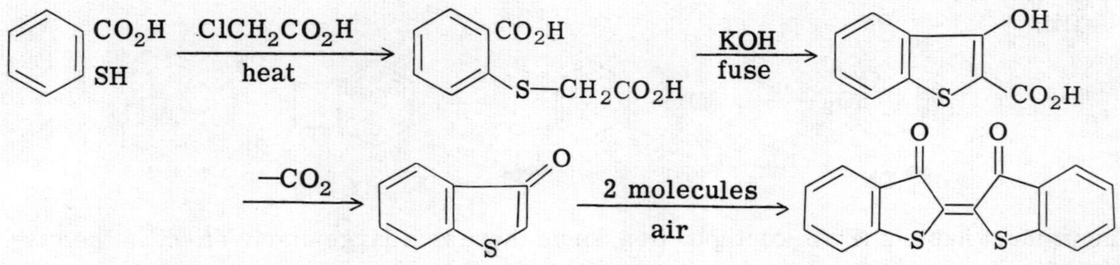

11. Since the substrate is an aliphatic ketone, photochemical decomposition can cause the elimination of carbon monoxide (this is the Norrish type I process; see *Text*). If, however, a six-membered ring transition state involving the oxygen atom (of the CO group) and a γ-hydrogen atom is possible, the main products are a ketone and an alkene. This is the Norrish type II process; it is analogous to the McLafferty rearrangement in mass spectrometry (see *Text*, Ch. 8)

The predominant products obtained from methyl neopentyl ketone, since a γ-hydrogen atom is present, are isobutene and acetone, which are formed as follows:

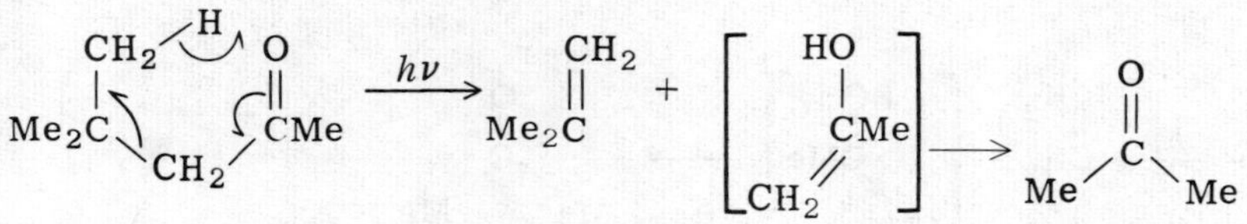

12. (i)

(*a*)

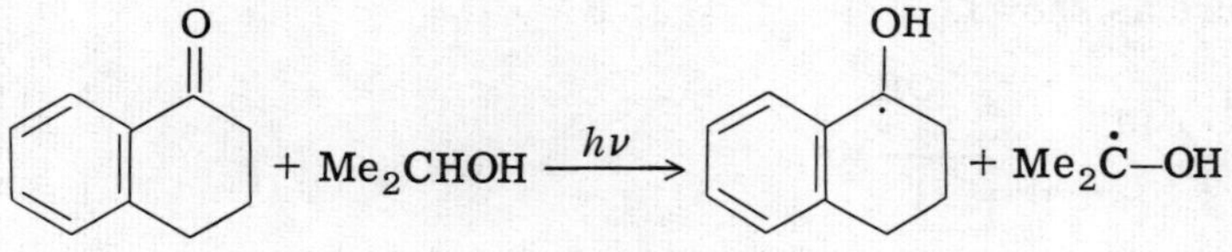

(*b*)

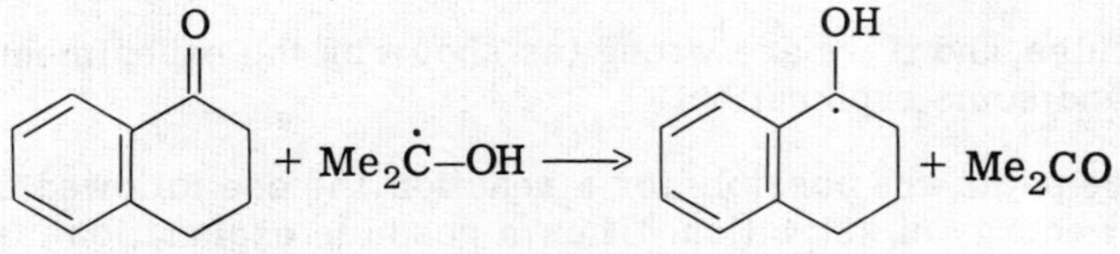

(*c*)

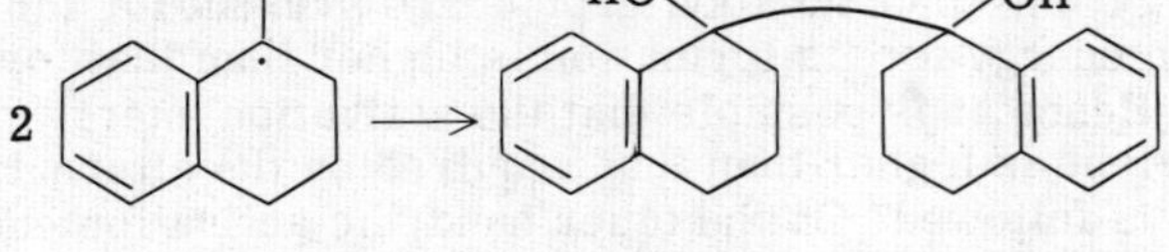

(ii) This substrate does not possess a γ-hydrogen atom; hence, the route followed is:

(*a*) $Me_2CHO{-}NO \xrightarrow{h\nu} Me_2CHO\cdot + NO$

(*b*)

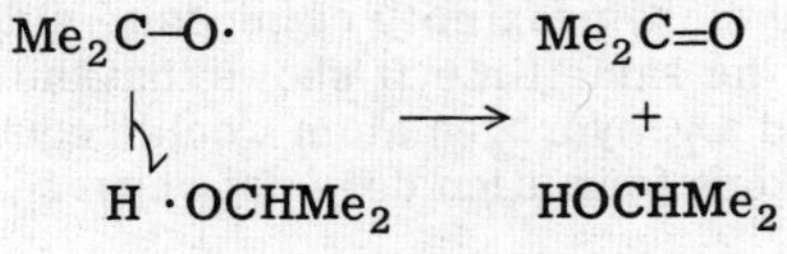

(iii)

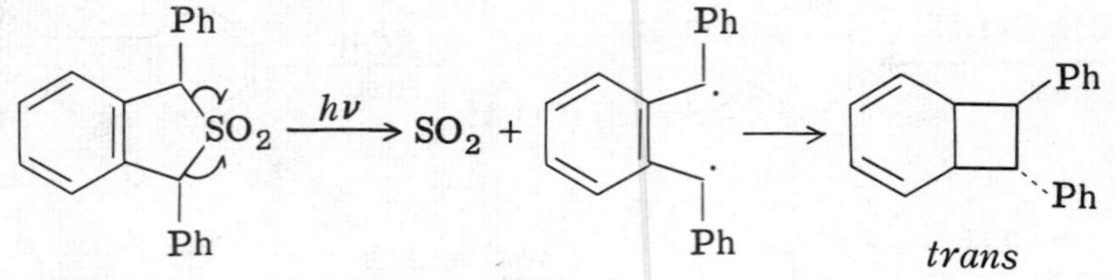

The product has the *trans*-configuration, since there is a large steric effect in the *cis*-isomer.

(iv) This substrate possesses allylic hydrogen atoms. Hence, the peroxide is formed.

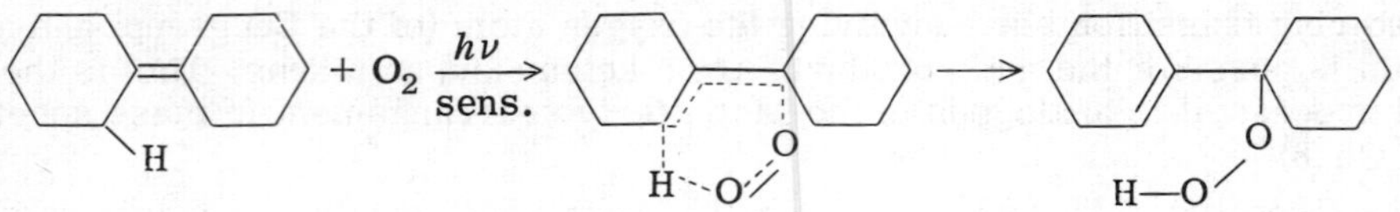

(v) The substrate contains a γ-hydrogen atom, and so the Norrish type II process is expected to predominate (*cf.* Prob. 11, above). Hence, if we try this, we have:

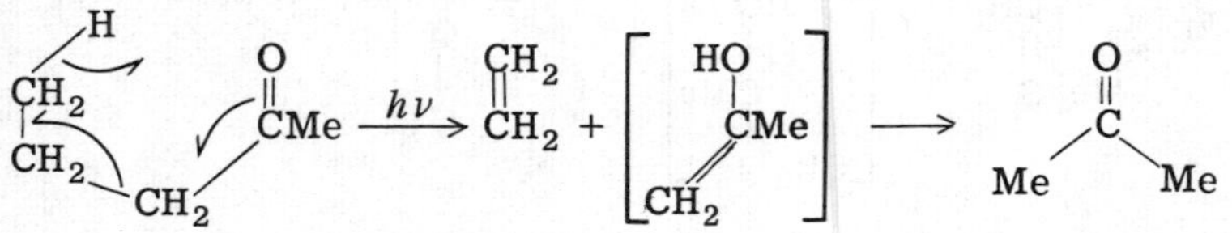

However, in actual fact, this is not the *predominant* path; the main product is a cyclobutane, and is formed by ring-closure as shown.

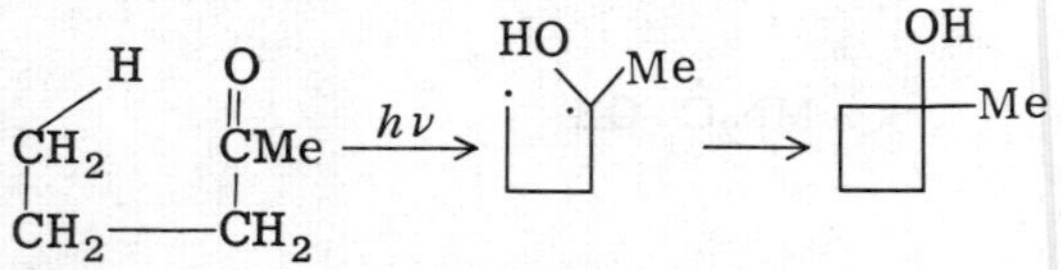

We can only follow the general rules, and if we are wrong (as shown by the experimental results), then we have to try to find some explanation.

As we have seen, when alternative paths are possible for a reaction, the one followed is usually that which has the lowest energy of activation. Thus, a possible explanation may lie in the fact the 'expected' alkene is ethylene. In the formation of substituted ethylenes, alkyl groups enter into hyperconjugation with the partly formed double bond in the transition state. This lowers the activation energy, but since no alkyl groups are present for the formation of ethylene itself, the activation energy for its formation is higher than that for any substituted ethylene (see *Text*, Ch. 5). Hence, it is possible that the activation energy required to cause the elimination of ethylene is higher than that required by the alternative path of ring-closure, and so the latter is followed. On the other hand, the activation energy required for the elimination of substituted ethylenes is lower than that required for ring-closure, and so the former path is followed.

(vi) The substrate is cyclodecanone, and its conformation is derived from two chair forms of cyclohexane (see *Text*, Ch. 19). For our purpose, it is convenient to draw the molecule as shown in the equations, since this indicates the close proximity of opposite sides of the ring that gives rise to transannular reactions. At the same time, it shows that the formation of a six-membered ring T.S. with O (of CO) and a γ-hydrogen atom would require a very large amount of distortion of the ring. Hence, alkene formation does not occur.

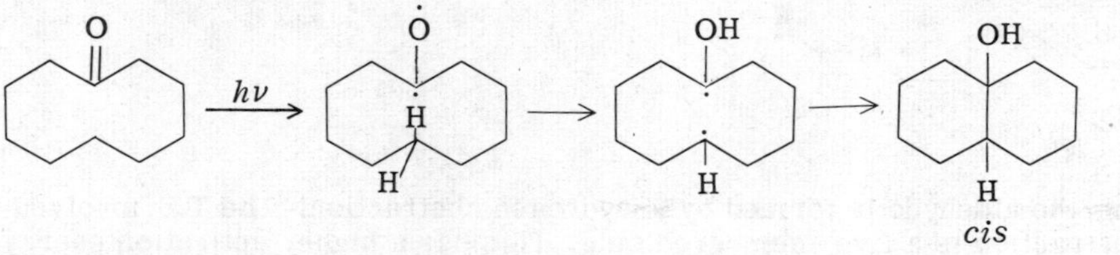

(vii)

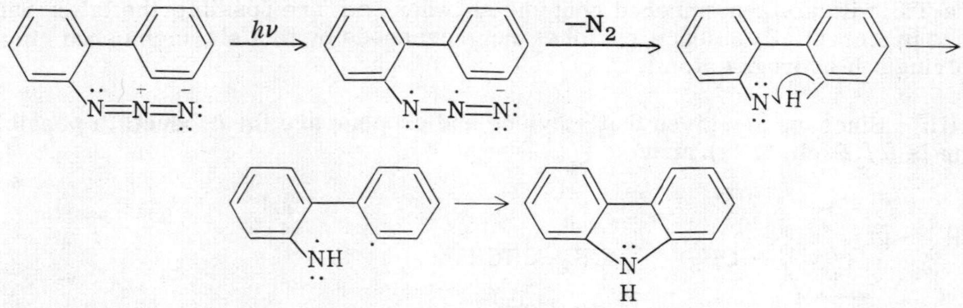

The azide is converted into the triplet state, and this eliminates nitrogen to form a nitrene (see *Text*, Ch. 13).

(viii) Compare with (vii), above.

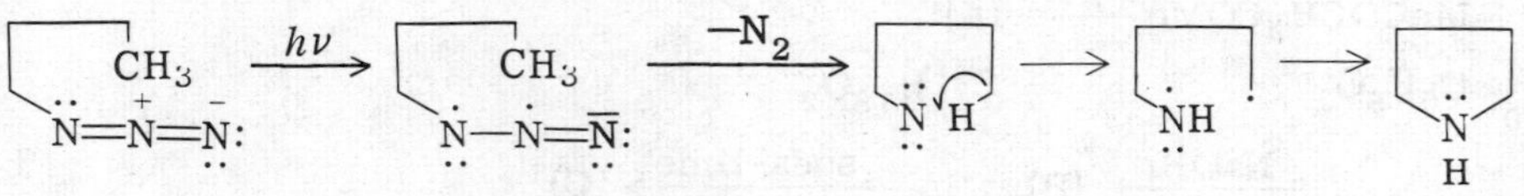

13. The isomerisation of 5-methylhex-3-en-2-one, (I), may be represented as follows:

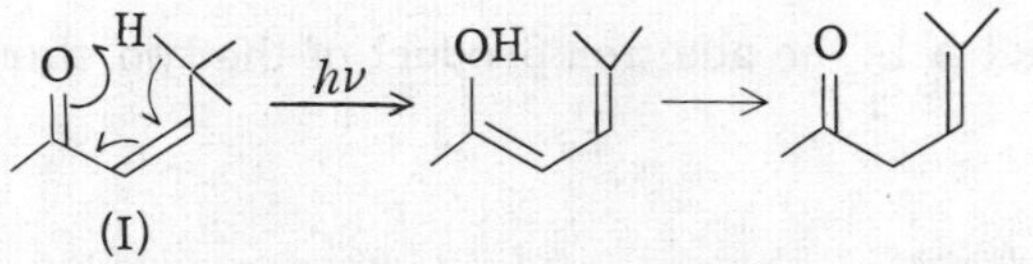

(I)

The enol, which is formed first, tautomerises to the more stable keto form.
If isomerisation occurred with pent-3-en-2-one, (II), the reaction would be written as:

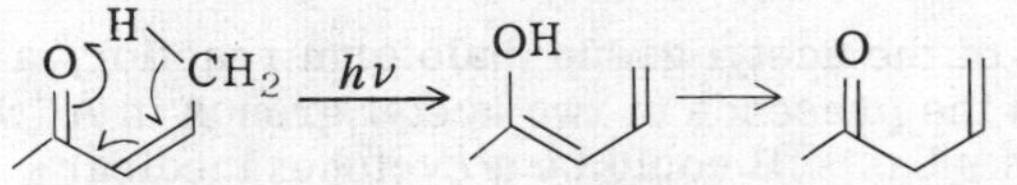

(II)

Since isomerisation occurs when (I) is irradiated with light of *low* energy and does *not* with (II), the implication is that the γ-hydrogen atom in (I) can be transferred to oxygen but not that in (II). What, then, is the difference between these two γ-hydrogens? In (I), the γ-hydrogen is tertiary, whereas that in (II) is primary. If we refer to Table 2.2, we see that the the *dissociation* energy of a tertiary C—H bond is 374·5 kJ mol^{-1} and that of a primary C—H bond is 405·8 kJ mol^{-1}. Thus, irradiation of (I) with light of just sufficient energy to effect its isomerisation will not isomerise (II).

14. In a case like this, we follow the mechanism described for cyclopentanone (see *Text*) and use reasonable modifications, where necessary, to account for the products obtained.

(i)

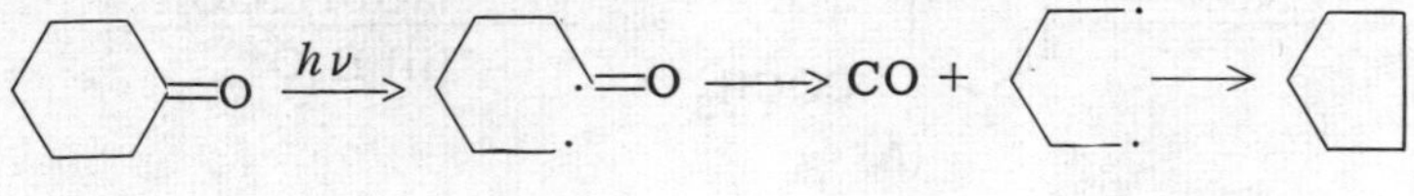

(ii) 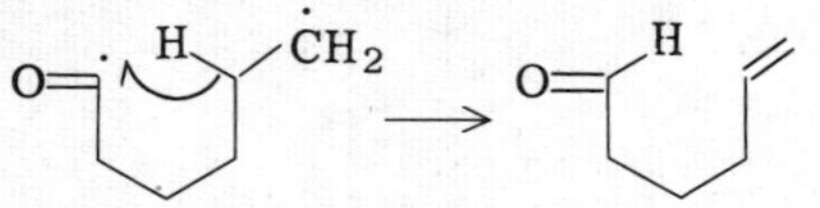

In cyclopentanone, the aldehyde is formed by γ-hydrogen abstraction. The T.S. involved in this hydrogen abstraction is a five-membered ring. This has a higher activation energy than a T.S. with a six-membered ring and so, when both are possible, the latter will be formed preferentially. Hence, cyclohexanone proceeds by this six-membered ring T.S. (involving a δ-hydrogen atom).

(iii) Since we are given that ethylene and propene are the products, a possible mechanism is (*cf.* Prob. 11 (v), above):

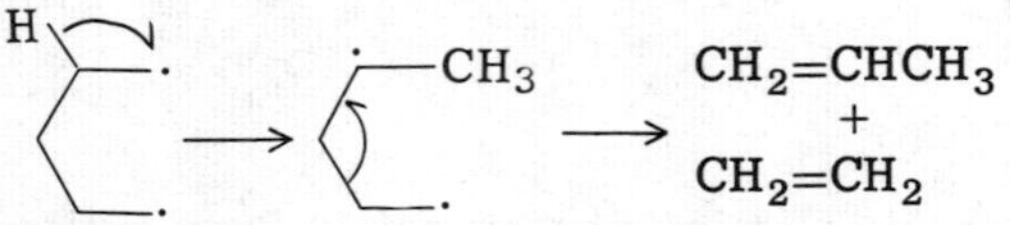

15. 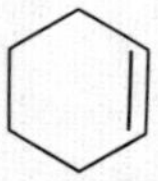 + MeCOCH$_2$COMe $\xrightarrow{h\nu}$ **(A)**

C_6H_{10} $\quad C_5H_8O_2$ $\qquad C_{11}H_{18}O_2$

$\xrightarrow[\text{NaOH}]{\text{NaOBr}}$ **(B)** $\xrightarrow[\text{heat}]{\text{soda-lime}}$ **(C)**

$C_9H_{14}O_4$ $\qquad C_7H_{14}$

It can be seen from the molecular formulæ that **A** is the addition product of the two starting materials. Since the D.B.E. of **A** is:

$11 + 1 - 18/2 = 3,$

A contains three double bonds and/or rings (see Ch. 1, Prob. 9). Hence, it is reasonable to assume (as a working hypothesis) that **A** contains the cyclohexane ring and two carbonyl groups.

Oxidation with alkaline hypobromite is typical of the acetyl group (haloform reaction), and since two carbon atoms are lost, this suggests the presence of two acetyl groups in **A** (which is supported by the D.B.E. value). Hence, on this basis, B could be a cyclohexanedicarboxylic acid. This is in agreement with the presence of four oxygen atoms and a D.B.E. = $9 + 1 - 14/2 = 3$ (one ring and two carbonyl groups). Decarboxylation of **B** to give **C**, C_7H_{14}, then leads to the conclusion that **C** is methylcyclohexane.

All of the foregoing account can be formulated on the basis that the dione undergoes fission at the CO—CH$_2$ bond to give two free radicals which add to the double bond in cyclohexene.

$CH_3COCH_2COCH_3 \xrightarrow{h\nu} CH_3\dot{C}O + CH_3COCH_2\cdot$

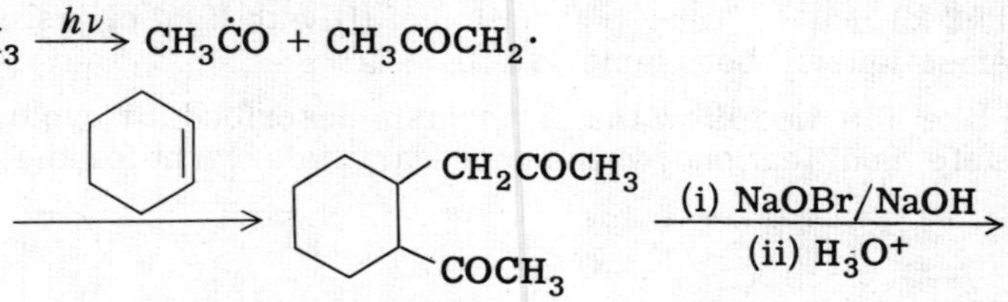

(A)

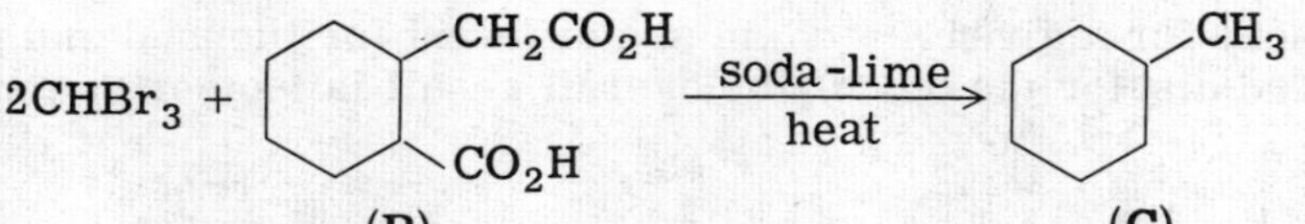

(B) (C)

16. The rules for intramolecular electrocyclic reactions are given in Table 31.3; these decide the geometry of the product.

Whichever is the starting material, the acyclic polyene or its cyclic isomer, the motions (disrotatory or conrotatory) in either direction (acyclic ⇌ cyclic) are the same for a given set of conditions. This follows from the principle of microscopic reversibility.

(i)

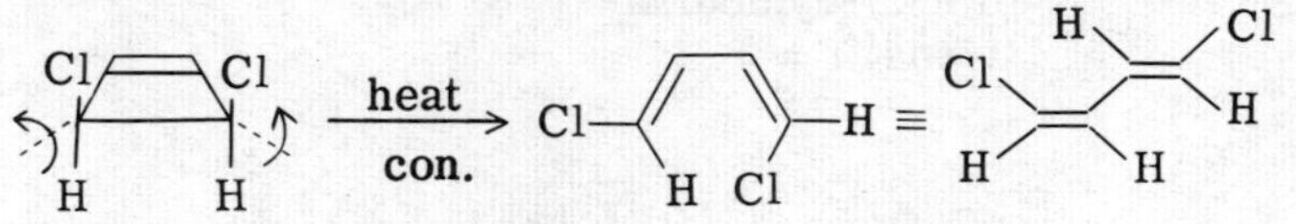

cis — *cis, trans*

$(4n-2)$-cycloalkene; $n = 1$ — $4n$-acyclic polyene; $n = 1$

(ii)

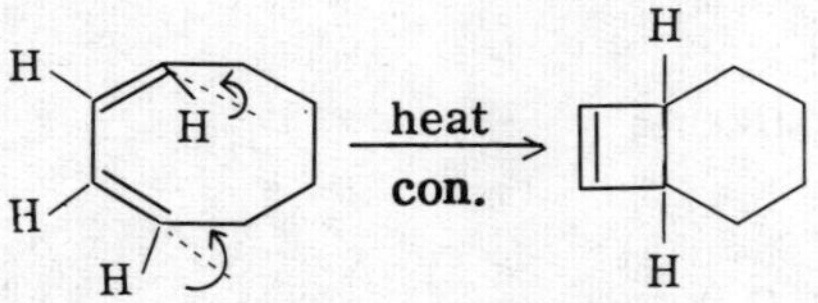

cis, trans — *cis*

$4n$-acyclic polyene; $n = 1$ — $(4n-2)$-cycloalkene; $n = 1$

(iii)

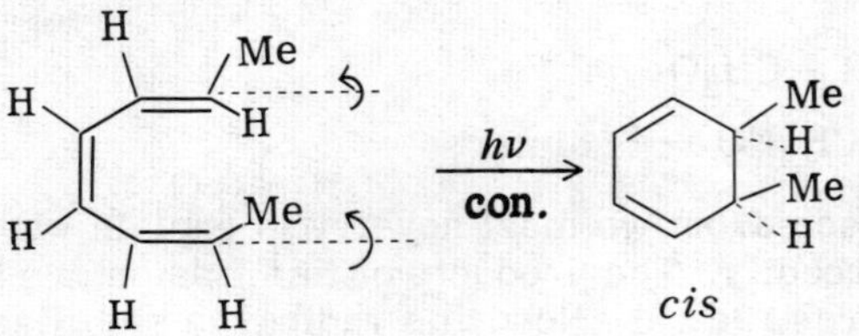

cis, cis, trans — *cis*

$(4n+2)$-acyclic polyene; $n = 1$ — $4n$-cycloalkadiene; $n = 1$

(iv)

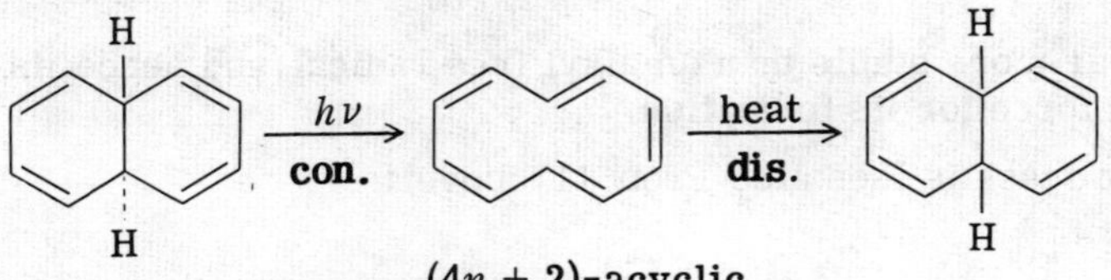

trans — $(4n+2)$-acyclic polyene; $n = 1$ — *cis*

$4n$-cycloalkadiene; $n = 1$ — $4n$-cycloalkadiene; $n = 1$

The polyene system involved in ring-opening and ring-closure is limited to only part of the system (see *Text*).

Since the motions for a given n-system are different for thermal and photochemical control (con. for one and dis. for the other), the overall result is isomerisation of the *trans*-isomer into the *cis*-.

(v)

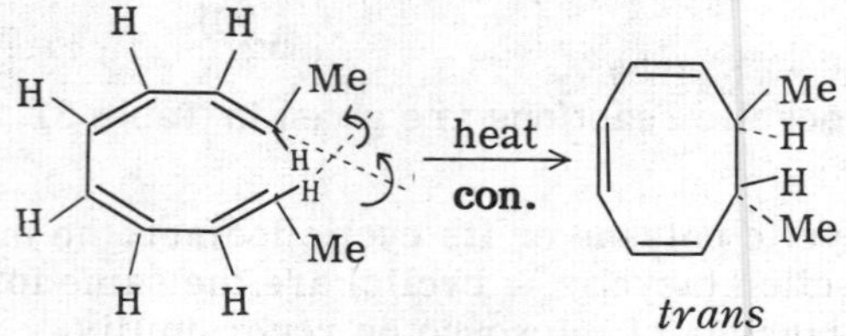

trans

trans, cis, cis, trans

4 n-acyclic polyene; $n = 2$

(4 n-2)-cycloalkatriene; $n = 2$

(vi)

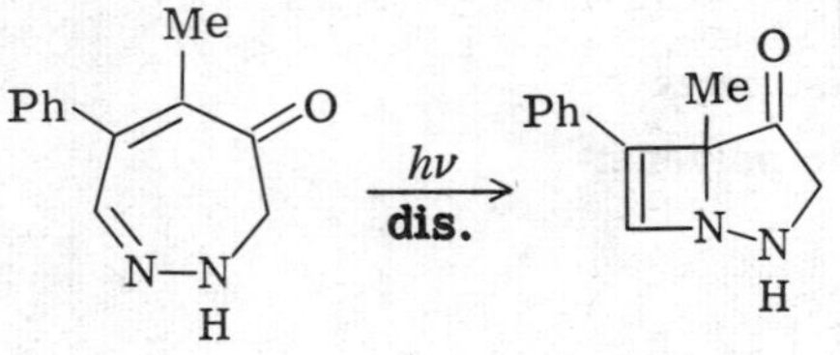

4n-acyclic polyene; $n = 1$

(4n-2)-cycloalkene; $n = 1$

17. $2PhCH_2CH_2ONO \xrightarrow{h\nu} (PhCH_2NO)_2 + 2CH_2O$

Although the substrate does not contain a γ-hydrogen atom, the 'expected' products, aldehyde and hydroxy-compound, formed by the elimination of nitric oxide, are *not* obtained (see also Prob. 12 (ii), above). Since formaldehyde is eliminated, fission must have occurred at the CH_2—CH_2 bond. We can, therefore, propose the following mechanism:

$$PhCH_2{-}CH_2O{-}NO \xrightarrow{h\nu} NO + PhCH_2{-}CH_2O\cdot$$

$$\longrightarrow CH_2O + NO + PhCH_2\cdot \longrightarrow PhCH_2NO \longrightarrow \text{dimer}$$

The first step is what we might have expected, but we must now try to explain why the fission in the second step has occurred so readily. The products of this fission are formaldehyde (a stable molecule) and the benzyl free radical. Now, this radical is stabilised by resonance (*cf.* the benzyl carbonium ion and cation):

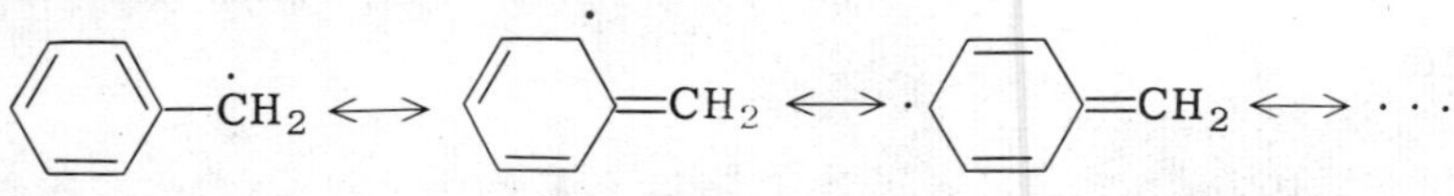

Thus, this free radical is far more stable than an alkyl free radical, and hence its resonance stabilisation is the driving force for its formation.

18. The Norrish type I process is (see also Prob. 11, above):

$$R^1{-}CO{-}R^2 \xrightarrow{h\nu} R^1\cdot + R^2{-}\dot{C}O \longrightarrow R^2\cdot + CO$$

If diaryl ketones followed this route, we would have:

$$Ar^1{-}CO{-}Ar^2 \xrightarrow{h\nu} Ar^1\cdot + Ar^2{-}\dot{C}O \longrightarrow Ar^2\cdot + CO$$

Now, aryl free radicals are extremely unstable; they are *not* stabilised by resonance. Since we have argued in Prob. 17, above, that the stability of a product can act as a driving force for the reaction to proceed along that path then, if we accept this, it is also reasonable to argue that the reverse is true, i.e., the more unstable the product is, the less likely that path will be followed. Hence, the Norrish type I process is energetically unfavourable for diaryl ketones (see also the *Text* for a discussion of the use of benzophenone as a sensitiser).

A point that is added here is that the term *unstable* is possibly not the appropriate one; the term *highly reactive* is preferred by some authors. Since aryl free radicals are not stabilised by resonance, they are certainly very unstable when compared to the benzyl free radical. On the other hand, we talk about carbonium ions being very reactive rather than being very unstable. The student will remember that the energy profiles of reactions involving carbonium-ion intermediates always have the energy levels of these ions much higher than their precursors (see, e.g., Ch. 4, Prob. 14). Since the *stability* of a molecule is measured by its enthalpy of formation, the less negative this value is, the less stable is the molecule (with respect ot its constituent elements). Hence, from this point of view, carbonium ions are less stable than the molecules from which they are derived.